PhET SIMULATIONS

Available in the Pearson eText and in the Study Area of MasteringPhysics ®

Extended Edition includes Chapters 1–44. Standard Edition includes Chapters 1–37.
Three-volume edition: Volume 1 includes Chapters 1–20, Volume 2 includes Chapters 21–37,
and Volume 3 includes Chapters 37–44.

Section		Page
1.6	Estimation	10
1.7	Vector Addition	13
2.4	*Forces in 1 Dimension	47
2.4	*The Moving Man	49
2.5	Lunar Lander	52
3.2	Maze Game	76
3.3	*Projectile Motion	79
3.4	Ladybug Revolution, Motion in 2D	87
5.2	Lunar Lander	146
5.3	Forces in 1 Dimension, Friction, *The Ramp	149
6.2	*The Ramp	181
6.3	Molecular Motors, Stretching DNA	188
7.3	*The Ramp	222
7.5	*Energy Skate Park	229
9.3	Ladybug Revolution	286
10.6	Torque	326
12.3	Balloons & Buoyancy	380
13.2	Lunar Lander	406
13.4	My Solar System	412
14.2	Motion in 2D	443
14.3	*Masses & Springs	446
14.5	*Pendulum Lab	453
15.8	Fourier: Making Waves, Waves on a String	495
16.6	Sound, Wave Interference	529
17.6	States of Matter	566
17.7	The Greenhouse Effect	570
18.3	Balloons & Buoyancy, Friction, Gas Properties	599
18.6	States of Matter	612
21.2	Balloons and Static Electricity, John Travoltage	691
21.6	*Charges and Fields, Electric Field of Dreams, Electric Field Hockey	708
21.7	Microwaves	711
23.2	*Charges & Fields	761
24.5	Molecular Motors, Optical Tweezers and Applications, Stretching DNA	806
25.3	Resistance in a Wire	825
25.4	Battery Voltage, Signal Circuit	829
25.5	Battery-Resistor Circuit, *Circuit Construction Kit (AC+DC), *Circuit Construction Kit (DC Only), Ohm's Law	834

Section		Page
25.6	Conductivity	838
26.4	*Circuit Construction Kit (AC+DC), *Circuit Construction Kit (DC Only)	866
27.3	Magnet and Compass, Magnets and Electromagnets	891
28.5	Faraday's Electromagnetic Lab, Magnets and Electromagnets	933
29.2	Faraday's Electromagnetic Lab, Faraday's Law, Generator	962
31.3	*Circuit Construction Kit (AC+DC), Faraday's Electromagnetic Lab	1031
32.3	Radio Waves & Electromagnetic Fields	1061
32.5	Microwaves	1070
34.4	*Geometric Optics	1131
34.6	Color Vision	1142
35.2	*Wave Interference	1168
36.2	*Wave Interference	1192
38.1	Photoelectric Effect	1262
38.4	Fourier: Making Waves, Quantum Wave Interference	1274
39.2	Davisson-Germer: Electron Diffraction	1287
39.2	Rutherford Scattering	1294
39.3	Models of the Hydrogen Atom	1297
39.3	Neon Lights and Other Discharge Lamps	1304
39.4	Lasers	1307
39.5	Blackbody Spectrum, The Greenhouse Effect	1310
40.1	Fourier: Making Waves	1328
40.1	Quantum Tunneling and Wave Packets	1337
40.3	Double Wells & Covalent Bonds, Quantum Bound States	1343
40.4	Quantum Tunneling and Wave Packets	1347
41.5	Stern-Gerlach Experiment	1383
42.1	Double Wells and Covalent Bonds	1406
42.2	The Greenhouse Effect	1409
42.4	Band Structure, Conductivity	1417
42.6	Semiconductors, Conductivity	1422
43.1	Simplified MRI	1444
43.3	Alpha Decay	1450
43.7	Nuclear Fission	1464

*Indicates an associated tutorial available in the MasteringPhysics Item Library.

ActivPhysics OnLine™
ACTIVITIES

1.1 Analyzing Motion Using Diagrams
1.2 Analyzing Motion Using Graphs
1.3 Predicting Motion from Graphs
1.4 Predicting Motion from Equations
1.5 Problem-Solving Strategies for Kinematics
1.6 Skier Races Downhill
1.7 Balloonist Drops Lemonade
1.8 Seat Belts Save Lives
1.9 Screeching to a Halt
1.10 Pole-Vaulter Lands
1.11 Car Starts, Then Stops
1.12 Solving Two-Vehicle Problems
1.13 Car Catches Truck
1.14 Avoiding a Rear-End Collision
2.1.1 Force Magnitudes
2.1.2 Skydiver
2.1.3 Tension Change
2.1.4 Sliding on an Incline
2.1.5 Car Race
2.2 Lifting a Crate
2.3 Lowering a Crate
2.4 Rocket Blasts Off
2.5 Truck Pulls Crate
2.6 Pushing a Crate Up a Wall
2.7 Skier Goes Down a Slope
2.8 Skier and Rope Tow
2.9 Pole-Vaulter Vaults
2.10 Truck Pulls Two Crates
2.11 Modified Atwood Machine
3.1 Solving Projectile Motion Problems
3.2 Two Balls Falling
3.3 Changing the x-Velocity
3.4 Projectile x- and y-Accelerations
3.5 Initial Velocity Components
3.6 Target Practice I
3.7 Target Practice II
4.1 Magnitude of Centripetal Acceleration
4.2 Circular Motion Problem Solving
4.3 Cart Goes Over Circular Path
4.4 Ball Swings on a String
4.5 Car Circles a Track
4.6 Satellites Orbit
5.1 Work Calculations
5.2 Upward-Moving Elevator Stops
5.3 Stopping a Downward-Moving Elevator
5.4 Inverse Bungee Jumper
5.5 Spring-Launched Bowler
5.6 Skier Speed
5.7 Modified Atwood Machine
6.1 Momentum and Energy Change
6.2 Collisions and Elasticity
6.3 Momentum Conservation and Collisions
6.4 Collision Problems
6.5 Car Collision: Two Dimensions
6.6 Saving an Astronaut
6.7 Explosion Problems
6.8 Skier and Cart
6.9 Pendulum Bashes Box
6.10 Pendulum Person-Projectile Bowling
7.1 Calculating Torques
7.2 A Tilted Beam: Torques and Equilibrium
7.3 Arm Levers
7.4 Two Painters on a Beam
7.5 Lecturing from a Beam

7.6 Rotational Inertia
7.7 Rotational Kinematics
7.8 Rotoride–Dynamics Approach
7.9 Falling Ladder
7.10 Woman and Flywheel Elevator–Dynamics Approach
7.11 Race Between a Block and a Disk
7.12 Woman and Flywheel Elevator–Energy Approach
7.13 Rotoride–Energy Approach
7.14 Ball Hits Bat
8.1 Characteristics of a Gas
8.2 Maxwell-Boltzmann Distribution–Conceptual Analysis
8.3 Maxwell-Boltzmann Distribution–Quantitative Analysis
8.4 State Variables and Ideal Gas Law
8.5 Work Done By a Gas
8.6 Heat, Internal Energy, and First Law of Thermodynamics
8.7 Heat Capacity
8.8 Isochoric Process
8.9 Isobaric Process
8.10 Isothermal Process
8.11 Adiabatic Process
8.12 Cyclic Process–Strategies
8.13 Cyclic Process–Problems
8.14 Carnot Cycle
9.1 Position Graphs and Equations
9.2 Describing Vibrational Motion
9.3 Vibrational Energy
9.4 Two Ways to Weigh Young Tarzan
9.5 Ape Drops Tarzan
9.6 Releasing a Vibrating Skier I
9.7 Releasing a Vibrating Skier II
9.8 One-and Two-Spring Vibrating Systems
9.9 Vibro-Ride
9.10 Pendulum Frequency
9.11 Risky Pendulum Walk
9.12 Physical Pendulum
10.1 Properties of Mechanical Waves
10.2 Speed of Waves on a String
10.3 Speed of Sound in a Gas
10.4 Standing Waves on Strings
10.5 Tuning a Stringed Instrument: Standing Waves
10.6 String Mass and Standing Waves
10.7 Beats and Beat Frequency
10.8 Doppler Effect: Conceptual Introduction
10.9 Doppler Effect: Problems
10.10 Complex Waves: Fourier Analysis
11.1 Electric Force: Coulomb's Law
11.2 Electric Force: Superposition Principle
11.3 Electric Force: Superposition Principle (Quantitative)
11.4 Electric Field: Point Charge
11.5 Electric Field Due to a Dipole
11.6 Electric Field: Problems
11.7 Electric Flux
11.8 Gauss's Law
11.9 Motion of a Charge in an Electric Field: Introduction
11.10 Motion in an Electric Field: Problems
11.11 Electric Potential: Qualitative Introduction

11.12 Electric Potential, Field, and Force
11.13 Electrical Potential Energy and Potential
12.1 DC Series Circuits (Qualitative)
12.2 DC Parallel Circuits
12.3 DC Circuit Puzzles
12.4 Using Ammeters and Voltmeters
12.5 Using Kirchhoff's Laws
12.6 Capacitance
12.7 Series and Parallel Capacitors
12.8 RC Circuit Time Constants
13.1 Magnetic Field of a Wire
13.2 Magnetic Field of a Loop
13.3 Magnetic Field of a Solenoid
13.4 Magnetic Force on a Particle
13.5 Magnetic Force on a Wire
13.6 Magnetic Torque on a Loop
13.7 Mass Spectrometer
13.8 Velocity Selector
13.9 Electromagnetic Induction
13.10 Motional emf
14.1 The RL Circuit
14.2 The RLC Oscillator
14.3 The Driven Oscillator
15.1 Reflection and Refraction
15.2 Total Internal Reflection
15.3 Refraction Applications
15.4 Plane Mirrors
15.5 Spherical Mirrors: Ray Diagrams
15.6 Spherical Mirror: The Mirror Equation
15.7 Spherical Mirror: Linear Magnification
15.8 Spherical Mirror: Problems
15.9 Thin-Lens Ray Diagrams
15.10 Converging Lens Problems
15.11 Diverging Lens Problems
15.12 Two-Lens Optical Systems
16.1 Two-Source Interference: Introduction
16.2 Two-Source Interference: Qualitative Questions
16.3 Two-Source Interference: Problems
16.4 The Grating: Introduction and Qualitative Questions
16.5 The Grating: Problems
16.6 Single-Slit Diffraction
16.7 Circular Hole Diffraction
16.8 Resolving Power
16.9 Polarization
17.1 Relativity of Time
17.2 Relativity of Length
17.3 Photoelectric Effect
17.4 Compton Scattering
17.5 Electron Interference
17.6 Uncertainty Principle
17.7 Wave Packets
18.1 The Bohr Model
18.2 Spectroscopy
18.3 The Laser
19.1 Particle Scattering
19.2 Nuclear Binding Energy
19.3 Fusion
19.4 Radioactivity
19.5 Particle Physics
20.1 Potential Energy Diagrams
20.2 Particle in a Box
20.3 Potential Wells
20.4 Potential Barriers

SEARS AND ZEMANSKY'S

UNIVERSITY PHYSICS

VOLUME 3
13TH EDITION

HUGH D. YOUNG
CARNEGIE MELLON
UNIVERSITY

ROGER A. FREEDMAN
UNIVERSITY
OF CALIFORNIA,
SANTA BARBARA

CONTRIBUTING AUTHOR
A. LEWIS FORD
TEXAS A&M UNIVERSITY

Addison-Wesley

Boston   Columbus   Indianapolis
New York   San Francisco   Upper Saddle River
Amsterdam   Cape Town   Dubai   London
Madrid   Milan   Munich   Paris   Montréal   Toronto
Delhi   Mexico City   São Paulo   Sydney
Hong Kong   Seoul   Singapore   Taipei   Tokyo

Publisher: Jim Smith
Executive Editor: Nancy Whilton
Project Editor: Chandrika Madhavan
Director of Development: Michael Gillespie
Editorial Manager: Laura Kenney
Senior Development Editor: Margot Otway
Editorial Assistant: Steven Le
Associate Media Producer: Kelly Reed
Managing Editor: Corinne Benson
Production Project Manager: Beth Collins

Production Management and Composition: Nesbitt Graphics
Copyeditor: Carol Reitz
Interior Designer: Elm Street Publishing Services
Cover Designer: Derek Bacchus
Illustrators: Rolin Graphics
Senior Art Editor: Donna Kalal
Photo Researcher: Eric Shrader
Manufacturing Buyer: Jeff Sargent
Senior Marketing Manager: Kerry Chapman

Cover Photo Credits: Getty Images/Mirko Cassanelli; Mirko Cassanelli

Credits and acknowledgments borrowed from other sources and reproduced, with permission, in this textbook appear on the appropriate page within the text or on p. C-1.

Library of Congress Cataloging-in-Publication Data

Young, Hugh D.
  Sears and Zemansky's university physics : with modern physics. -- 13th ed.
/ Hugh D. Young, Roger A. Freedman ; contributing author, A. Lewis Ford.
    p. cm.
  Includes bibliographical references and index.
  ISBN-13: 978-0-321-69686-1 (student ed. : alk. paper)
  ISBN-10: 0-321-69686-7 (student ed. : alk. paper)
  ISBN-13: 978-0-321-69685-4 (exam copy)
  ISBN-10: 0-321-69685-9 (exam copy)
  1. Physics--Textbooks. I. Freedman, Roger A. II. Ford, A. Lewis (Albert
Lewis) III. Sears, Francis Weston, 1898-1975. University physics. IV. Title.
V. Title: University physics.
  QC21.3.Y68 2012
  530--dc22
                                                2010044896

ISBN 13: 978-0-321-75120-1; ISBN 10: 0-321-75120-5 (Student edition)
ISBN 13: 978-0-321-69685-4; ISBN 10: 0-321-69685-9 (Exam copy)

Addison-Wesley
is an imprint of

www.pearsonhighered.com   1 2 3 4 5 6 7 8 9 10—VHC—14 13 12 11 10

BRIEF CONTENTS

MECHANICS

1 Units, Physical Quantities, and Vectors   1
2 Motion Along a Straight Line   35
3 Motion in Two or Three Dimensions   69
4 Newton's Laws of Motion   104
5 Applying Newton's Laws   134
6 Work and Kinetic Energy   176
7 Potential Energy and Energy Conservation   207
8 Momentum, Impulse, and Collisions   241
9 Rotation of Rigid Bodies   278
10 Dynamics of Rotational Motion   308
11 Equilibrium and Elasticity   344
12 Fluid Mechanics   373
13 Gravitation   402
14 Periodic Motion   437

WAVES/ACOUSTICS

15 Mechanical Waves   472
16 Sound and Hearing   509

THERMODYNAMICS

17 Temperature and Heat   551
18 Thermal Properties of Matter   590
19 The First Law of Thermodynamics   624
20 The Second Law of Thermodynamics   652

ELECTROMAGNETISM

21 Electric Charge and Electric Field   687
22 Gauss's Law   725
23 Electric Potential   754
24 Capacitance and Dielectrics   788
25 Current, Resistance, and Electromotive Force   818
26 Direct-Current Circuits   850

27 Magnetic Field and Magnetic Forces   883
28 Sources of Magnetic Field   923
29 Electromagnetic Induction   957
30 Inductance   991
31 Alternating Current   1021
32 Electromagnetic Waves   1051

OPTICS

33 The Nature and Propagation of Light   1080
34 Geometric Optics   1114
35 Interference   1163
36 Diffraction   1190

MODERN PHYSICS

37 Relativity   1223
38 Photons: Light Waves Behaving as Particles   1261
39 Particles Behaving as Waves   1286
40 Quantum Mechanics   1328
41 Atomic Structure   1364
42 Molecules and Condensed Matter   1405
43 Nuclear Physics   1439
44 Particle Physics and Cosmology   1480

APPENDICES

A The International System of Units   A-1
B Useful Mathematical Relations   A-3
C The Greek Alphabet   A-4
D Periodic Table of Elements   A-5
E Unit Conversion Factors   A-6
F Numerical Constants   A-7

Answers to Odd-Numbered Problems   A-9

ABOUT THE AUTHORS

Hugh D. Young is Emeritus Professor of Physics at Carnegie Mellon University. He earned both his undergraduate and graduate degrees from that university. He earned his Ph.D. in fundamental particle theory under the direction of the late Richard Cutkosky. He joined the faculty of Carnegie Mellon in 1956 and retired in 2004. He also had two visiting professorships at the University of California, Berkeley.

Dr. Young's career has centered entirely on undergraduate education. He has written several undergraduate-level textbooks, and in 1973 he became a coauthor with Francis Sears and Mark Zemansky for their well-known introductory texts. In addition to his role on Sears and Zemansky's *University Physics,* he is also author of Sears and Zemansky's *College Physics*.

Dr. Young earned a bachelor's degree in organ performance from Carnegie Mellon in 1972 and spent several years as Associate Organist at St. Paul's Cathedral in Pittsburgh. He has played numerous organ recitals in the Pittsburgh area. Dr. Young and his wife, Alice, usually travel extensively in the summer, especially overseas and in the desert canyon country of southern Utah.

Roger A. Freedman is a Lecturer in Physics at the University of California, Santa Barbara. Dr. Freedman was an undergraduate at the University of California campuses in San Diego and Los Angeles, and did his doctoral research in nuclear theory at Stanford University under the direction of Professor J. Dirk Walecka. He came to UCSB in 1981 after three years teaching and doing research at the University of Washington.

At UCSB, Dr. Freedman has taught in both the Department of Physics and the College of Creative Studies, a branch of the university intended for highly gifted and motivated undergraduates. He has published research in nuclear physics, elementary particle physics, and laser physics. In recent years, he has worked to make physics lectures a more interactive experience through the use of classroom response systems.

In the 1970s Dr. Freedman worked as a comic book letterer and helped organize the San Diego Comic-Con (now the world's largest popular culture convention) during its first few years. Today, when not in the classroom or slaving over a computer, Dr. Freedman can be found either flying (he holds a commercial pilot's license) or driving with his wife, Caroline, in their 1960 Nash Metropolitan convertible.

A. Lewis Ford is Professor of Physics at Texas A&M University. He received a B.A. from Rice University in 1968 and a Ph.D. in chemical physics from the University of Texas at Austin in 1972. After a one-year postdoc at Harvard University, he joined the Texas A&M physics faculty in 1973 and has been there ever since. Professor Ford's research area is theoretical atomic physics, with a specialization in atomic collisions. At Texas A&M he has taught a variety of undergraduate and graduate courses, but primarily introductory physics.

HOW TO SUCCEED IN PHYSICS BY REALLY TRYING

Mark Hollabaugh *Normandale Community College*

Physics encompasses the large and the small, the old and the new. From the atom to galaxies, from electrical circuitry to aerodynamics, physics is very much a part of the world around us. You probably are taking this introductory course in calculus-based physics because it is required for subsequent courses you plan to take in preparation for a career in science or engineering. Your professor wants you to learn physics and to enjoy the experience. He or she is very interested in helping you learn this fascinating subject. That is part of the reason your professor chose this textbook for your course. That is also the reason Drs. Young and Freedman asked me to write this introductory section. We want you to succeed!

The purpose of this section of *University Physics* is to give you some ideas that will assist your learning. Specific suggestions on how to use the textbook will follow a brief discussion of general study habits and strategies.

Preparation for This Course

If you had high school physics, you will probably learn concepts faster than those who have not because you will be familiar with the language of physics. If English is a second language for you, keep a glossary of new terms that you encounter and make sure you understand how they are used in physics. Likewise, if you are farther along in your mathematics courses, you will pick up the mathematical aspects of physics faster. Even if your mathematics is adequate, you may find a book such as Arnold D. Pickar's *Preparing for General Physics: Math Skill Drills and Other Useful Help (Calculus Version)* to be useful. Your professor may actually assign sections of this math review to assist your learning.

Learning to Learn

Each of us has a different learning style and a preferred means of learning. Understanding your own learning style will help you to focus on aspects of physics that may give you difficulty and to use those components of your course that will help you overcome the difficulty. Obviously you will want to spend more time on those aspects that give you the most trouble. If you learn by hearing, lectures will be very important. If you learn by explaining, then working with other students will be useful to you. If solving problems is difficult for you, spend more time learning how to solve problems. Also, it is important to understand and develop good study habits. Perhaps the most important thing you can do for yourself is to set aside adequate, regularly scheduled study time in a distraction-free environment.

Answer the following questions for yourself:

- Am I able to use fundamental mathematical concepts from algebra, geometry and trigonometry? (If not, plan a program of review with help from your professor.)
- In similar courses, what activity has given me the most trouble? (Spend more time on this.) What has been the easiest for me? (Do this first; it will help to build your confidence.)

- Do I understand the material better if I read the book before or after the lecture? (You may learn best by skimming the material, going to lecture, and then undertaking an in-depth reading.)
- Do I spend adequate time in studying physics? (A rule of thumb for a class like this is to devote, on the average, 2.5 hours out of class for each hour in class. For a course meeting 5 hours each week, that means you should spend about 10 to 15 hours per week studying physics.)
- Do I study physics every day? (Spread that 10 to 15 hours out over an entire week!) At what time of the day am I at my best for studying physics? (Pick a specific time of the day and stick to it.)
- Do I work in a quiet place where I can maintain my focus? (Distractions will break your routine and cause you to miss important points.)

Working with Others

Scientists or engineers seldom work in isolation from one another but rather work cooperatively. You will learn more physics and have more fun doing it if you work with other students. Some professors may formalize the use of cooperative learning or facilitate the formation of study groups. You may wish to form your own informal study group with members of your class who live in your neighborhood or dorm. If you have access to e-mail, use it to keep in touch with one another. Your study group is an excellent resource when reviewing for exams.

Lectures and Taking Notes

An important component of any college course is the lecture. In physics this is especially important because your professor will frequently do demonstrations of physical principles, run computer simulations, or show video clips. All of these are learning activities that will help you to understand the basic principles of physics. Don't miss lectures, and if for some reason you do, ask a friend or member of your study group to provide you with notes and let you know what happened.

Take your class notes in outline form, and fill in the details later. It can be very difficult to take word for word notes, so just write down key ideas. Your professor may use a diagram from the textbook. Leave a space in your notes and just add the diagram later. After class, edit your notes, filling in any gaps or omissions and noting things you need to study further. Make references to the textbook by page, equation number, or section number.

Make sure you ask questions in class, or see your professor during office hours. Remember the only "dumb" question is the one that is not asked. Your college may also have teaching assistants or peer tutors who are available to help you with difficulties you may have.

Examinations

Taking an examination is stressful. But if you feel adequately prepared and are well-rested, your stress will be lessened. Preparing for an exam is a continual process; it begins the moment the last exam is over. You should immediately go over the exam and understand any mistakes you made. If you worked a problem and made substantial errors, try this: Take a piece of paper and divide it down the middle with a line from top to bottom. In one column, write the proper solution to the problem. In the other column, write what you did and why, if you know, and why your solution was incorrect. If you are uncertain why you made your mistake, or how to avoid making it again, talk with your professor. Physics continually builds on fundamental ideas and it is important to correct any misunderstandings immediately. *Warning:* While cramming at the last minute may get you through the present exam, you will not adequately retain the concepts for use on the next exam.

PREFACE

This book is the product of more than six decades of leadership and innovation in physics education. When the first edition of *University Physics* by Francis W. Sears and Mark W. Zemansky was published in 1949, it was revolutionary among calculus-based physics textbooks in its emphasis on the fundamental principles of physics and how to apply them. The success of *University Physics* with generations of several million students and educators around the world is a testament to the merits of this approach, and to the many innovations it has introduced subsequently.

In preparing this new Thirteenth Edition, we have further enhanced and developed *University Physics* to assimilate the best ideas from education research with enhanced problem-solving instruction, pioneering visual and conceptual pedagogy, the first systematically enhanced problems, and the most pedagogically proven and widely used online homework and tutorial system in the world.

New to This Edition

- Included in each chapter, **Bridging Problems** provide a transition between the single-concept Examples and the more challenging end-of-chapter problems. Each Bridging Problem poses a difficult, multiconcept problem, which often incorporates physics from earlier chapters. In place of a full solution, it provides a skeleton **Solution Guide** consisting of questions and hints, which helps train students to approach and solve challenging problems with confidence.

- **All Examples, Conceptual Examples, and Problem-Solving Strategies are revised** to enhance conciseness and clarity for today's students.

- The **core modern physics chapters** (Chapters 38–41) are revised extensively to provide a more idea-centered, less historical approach to the material. Chapters 42–44 are also revised significantly.

- **The fluid mechanics chapter now precedes the chapters on gravitation and periodic motion,** so that the latter immediately precedes the chapter on mechanical waves.

- **Additional bioscience applications** appear throughout the text, mostly in the form of marginal photos with explanatory captions, to help students see how physics is connected to many breakthroughs and discoveries in the biosciences.

- The **text has been streamlined** for tighter and more focused language.

- Using data from MasteringPhysics, changes to the end-of-chapter content include the following:
  - **15%–20% of problems are new.**
  - The number and level of **calculus-requiring problems** has been increased.
  - Most chapters include **five to seven biosciences-related problems.**
  - The number of **cumulative problems** (those incorporating physics from earlier chapters) has been increased.

- **Over 70 PhET simulations** are linked to the Pearson eText and provided in the Study Area of the MasteringPhysics website (with icons in the print text). These powerful simulations allow students to interact productively with the physics concepts they are learning. PhET clicker questions are also included on the Instructor Resource DVD.

- **Video Tutors bring key content to life throughout the text:**
  - **Dozens of Video Tutors feature "pause-and-predict" demonstrations of key physics concepts** and incorporate assessment as the student progresses to actively engage the student in understanding the key conceptual ideas underlying the physics principles.

Standard, Extended, and Three-Volume Editions

With MasteringPhysics:

- **Standard Edition:** Chapters 1–37 (ISBN 978-0-321-69688-5)
- **Extended Edition:** Chapters 1–44 (ISBN 978-0-321-67546-0)

Without MasteringPhysics:

- **Standard Edition:** Chapters 1–37 (ISBN 978-0-321-69689-2)
- **Extended Edition:** Chapters 1–44 (ISBN 978-0-321-69686-1)
- **Volume 1:** Chapters 1–20 (ISBN 978-0-321-73338-2)
- **Volume 2:** Chapters 21–37 (ISBN 978-0-321-75121-8)
- **Volume 3:** Chapters 37–44 (ISBN 978-0-321-75120-1)

- **Every Worked Example in the book is accompanied by a Video Tutor Solution** that walks students through the problem-solving process, providing a virtual teaching assistant on a round-the-clock basis.
- **All of these Video Tutors play directly through links within the Pearson eText.** Many also appear in the Study Area within MasteringPhysics.

Key Features of *University Physics*

- Deep and extensive **problem sets** cover a wide range of difficulty and exercise both physical understanding and problem-solving expertise. Many problems are based on complex real-life situations.
- This text offers a larger number of **Examples** and **Conceptual Examples** than any other leading calculus-based text, allowing it to explore problem-solving challenges not addressed in other texts.
- A research-based **problem-solving approach (Identify, Set Up, Execute, Evaluate)** is used not just in every Example but also in the Problem-Solving Strategies and throughout the Student and Instructor Solutions Manuals and the Study Guide. This consistent approach teaches students to tackle problems thoughtfully rather than cutting straight to the math.
- **Problem-Solving Strategies** coach students in how to approach specific types of problems.
- The **Figures** use a simplified graphical style to focus on the physics of a situation, and they incorporate **explanatory annotation.** Both techniques have been demonstrated to have a strong positive effect on learning.
- Figures that illustrate Example solutions often take the form of black-and-white **pencil sketches,** which directly represent what a student should draw in solving such a problem.
- The popular **Caution paragraphs** focus on typical misconceptions and student problem areas.
- End-of-section **Test Your Understanding** questions let students check their grasp of the material and use a multiple-choice or ranking-task format to probe for common misconceptions.
- **Visual Summaries** at the end of each chapter present the key ideas in words, equations, and thumbnail pictures, helping students to review more effectively.

Instructor Supplements

Note: For convenience, all of the following instructor supplements (except for the Instructor Resource DVD) can be downloaded from the Instructor Area, accessed via the left-hand navigation bar of MasteringPhysics (www.masteringphysics.com).

Instructor Solutions, prepared by A. Lewis Ford (Texas A&M University) and Wayne Anderson, contain complete and detailed solutions to all end-of-chapter problems. All solutions follow consistently the same Identify/Set Up/ Execute/Evaluate problem-solving framework used in the textbook. Download only from the MasteringPhysics Instructor Area or from the Instructor Resource Center (www.pearsonhighered.com/irc).

The cross-platform **Instructor Resource DVD** (ISBN 978-0-321-69661-8) provides a comprehensive library of more than 420 applets from ActivPhysics OnLine as well as all line figures from the textbook in JPEG format. In addition, all the key equations, problem-solving strategies, tables, and chapter summaries are provided in editable Word format. In-class weekly multiple-choice questions for use with various Classroom Response Systems (CRS) are also provided, based on the Test Your Understanding questions in the text. Lecture outlines in PowerPoint are also included along with over 70 PhET simulations.

MasteringPhysics® (www.masteringphysics.com) is the most advanced, educationally effective, and widely used physics homework and tutorial system in the world. Eight years in development, it provides instructors with a library of extensively pre-tested end-of-chapter problems and rich, multipart, multistep tutorials that incorporate a wide variety of answer types, wrong answer feedback, individualized help (comprising hints or simpler sub-problems upon request), all driven by the largest metadatabase of student problem-solving in the world. NSF-sponsored published research (and subsequent studies) show that Mastering-Physics has dramatic educational results. MasteringPhysics allows instructors to build wide-ranging homework assignments of just the right difficulty and length and provides them with efficient tools to analyze both class trends, and the work of any student in unprecedented detail.

MasteringPhysics routinely provides instant and individualized feedback and guidance to more than 100,000 students every day. A wide range of tools and support make MasteringPhysics fast and easy for instructors and students to learn to use. Extensive class tests show that by the end of their course, an unprecedented eight of nine students recommend MasteringPhysics as their preferred way to study physics and do homework.

MasteringPhysics enables instructors to:
- Quickly build homework assignments that combine regular end-of-chapter problems and tutoring (through additional multi-step tutorial problems that offer wrong-answer feedback and simpler problems upon request).
- Expand homework to include the widest range of automatically graded activities available—from numerical problems with randomized values, through algebraic answers, to free-hand drawing.
- Choose from a wide range of nationally pre-tested problems that provide accurate estimates of time to complete and difficulty.
- After an assignment is completed, quickly identify not only the problems that were the trickiest for students but the individual problem types where students had trouble.
- Compare class results against the system's worldwide average for each problem assigned, to identify issues to be addressed with just-in-time teaching.
- Check the work of an individual student in detail, including time spent on each problem, what wrong answers they submitted at each step, how much help they asked for, and how many practice problems they worked.

ActivPhysics OnLine™ (which is accessed through the Study Area within www.masteringphysics.com) provides a comprehensive library of more than 420 tried and tested ActivPhysics applets updated for web delivery using the latest online technologies. In addition, it provides a suite of highly regarded applet-based tutorials developed by education pioneers Alan Van Heuvelen and Paul D'Alessandris. Margin icons throughout the text direct students to specific exercises that complement the textbook discussion.

The online exercises are designed to encourage students to confront misconceptions, reason qualitatively about physical processes, experiment quantitatively, and learn to think critically. The highly acclaimed ActivPhysics OnLine companion workbooks help students work through complex concepts and understand them more clearly. More than 420 applets from the ActivPhysics OnLine library are also available on the Instructor Resource DVD for this text.

The **Test Bank** contains more than 2,000 high-quality problems, with a range of multiple-choice, true/false, short-answer, and regular homework-type questions. Test files are provided both in TestGen (an easy-to-use, fully networkable program for creating and editing quizzes and exams) and Word format. Download only from the MasteringPhysics Instructor Area or from the Instructor Resource Center (www.pearsonhighered.com/irc).

Five Easy Lessons: Strategies for Successful Physics Teaching (ISBN 978-0-805-38702-5) by Randall D. Knight (California Polytechnic State University, San Luis Obispo) is packed with creative ideas on how to enhance any physics course. It is an invaluable companion for both novice and veteran physics instructors.

Student Supplements

The **Study Guide** by Laird Kramer reinforces the text's emphasis on problem-solving strategies and student misconceptions. The *Study Guide for Volume 1* (ISBN 978-0-321-69665-6) covers Chapters 1–20, and the *Study Guide for Volumes 2 and 3* (ISBN 978-0-321-69669-4) covers Chapters 21–44.

The **Student Solutions Manual** by Lewis Ford (Texas A&M University) and Wayne Anderson contains detailed, step-by-step solutions to more than half of the odd-numbered end-of-chapter problems from the textbook. All solutions follow consistently the same Identify/Set Up/Execute/Evaluate problem-solving framework used in the textbook. The *Student Solutions Manual for Volume 1* (ISBN 978-0-321-69668-7) covers Chapters 1–20, and the *Student Solutions Manual for Volumes 2 and 3* (ISBN 978-0-321-69667-0) covers Chapters 21–44.

 MasteringPhysics® (www.masteringphysics.com) is a homework, tutorial, and assessment system based on years of research into how students work physics problems and precisely where they need help. Studies show that students who use MasteringPhysics significantly increase their scores compared to hand-written homework. MasteringPhysics achieves this improvement by providing students with instantaneous feedback specific to their wrong answers, simpler sub-problems upon request when they get stuck, and partial credit for their method(s). This individualized, 24/7 Socratic tutoring is recommended by nine out of ten students to their peers as the most effective and time-efficient way to study.

Pearson eText is available through MasteringPhysics, either automatically when MasteringPhysics is packaged with new books, or available as a purchased upgrade online. Allowing students access to the text wherever they have access to the Internet, Pearson eText comprises the full text, including figures that can be enlarged for better viewing. With eText, students are also able to pop up definitions and terms to help with vocabulary and the reading of the material. Students can also take notes in eText using the annotation feature at the top of each page.

Pearson Tutor Services (www.pearsontutorservices.com). Each student's subscription to MasteringPhysics also contains complimentary access to Pearson Tutor Services, powered by Smarthinking, Inc. By logging in with their MasteringPhysics ID and password, students will be connected to highly qualified e-instructors who provide additional interactive online tutoring on the major concepts of physics. Some restrictions apply; offer subject to change.

 ActivPhysics OnLine™ (which is accessed through the Study Area within www.masteringphysics.com) provides students with a suite of highly regarded applet-based tutorials (see above). The following workbooks help students work through complex concepts and understand them more clearly.

ActivPhysics OnLine Workbook, Volume 1: Mechanics * Thermal Physics * Oscillations & Waves (978-0-805-39060-5)

ActivPhysics OnLine Workbook, Volume 2: Electricity & Magnetism * Optics * Modern Physics (978-0-805-39061-2)

Acknowledgments

We would like to thank the hundreds of reviewers and colleagues who have offered valuable comments and suggestions over the life of this textbook. The continuing success of *University Physics* is due in large measure to their contributions.

Edward Adelson (Ohio State University), Ralph Alexander (University of Missouri at Rolla), J. G. Anderson, R. S. Anderson, Wayne Anderson (Sacramento City College), Alex Azima (Lansing Community College), Dilip Balamore (Nassau Community College), Harold Bale (University of North Dakota), Arun Bansil (Northeastern University), John Barach (Vanderbilt University), J. D. Barnett, H. H. Barschall, Albert Bartlett (University of Colorado), Marshall Bartlett (Hollins University), Paul Baum (CUNY, Queens College), Frederick Becchetti (University of Michigan), B. Bederson, David Bennum (University of Nevada, Reno), Lev I. Berger (San Diego State University), Robert Boeke (William Rainey Harper College), S. Borowitz, A. C. Braden, James Brooks (Boston University), Nicholas E. Brown (California Polytechnic State University, San Luis Obispo), Tony Buffa (California Polytechnic State University, San Luis Obispo), A. Capecelatro, Michael Cardamone (Pennsylvania State University), Duane Carmony (Purdue University), Troy Carter (UCLA), P. Catranides, John Cerne (SUNY at Buffalo), Tim Chupp (University of Michigan), Shinil Cho (La Roche College), Roger Clapp (University of South Florida), William M. Cloud (Eastern Illinois University), Leonard Cohen (Drexel University), W. R. Coker (University of Texas, Austin), Malcolm D. Cole (University of Missouri at Rolla), H. Conrad, David Cook (Lawrence University), Gayl Cook (University of Colorado), Hans Courant (University of Minnesota), Bruce A. Craver (University of Dayton), Larry Curtis (University of Toledo), Jai Dahiya (Southeast Missouri State University), Steve Detweiler (University of Florida), George Dixon (Oklahoma State University), Donald S. Duncan, Boyd Edwards (West Virginia University), Robert Eisenstein (Carnegie Mellon University), Amy Emerson Missourn (Virginia Institute of Technology), William Faissler (Northeastern University), William Fasnacht (U.S. Naval Academy), Paul Feldker (St. Louis Community College), Carlos Figueroa (Cabrillo College), L. H. Fisher, Neil Fletcher (Florida State University), Robert Folk, Peter Fong (Emory University), A. Lewis Ford (Texas A&M University), D. Frantszog, James R. Gaines (Ohio State University), Solomon Gartenhaus (Purdue University), Ron Gautreau (New Jersey Institute of Technology), J. David Gavenda (University of Texas, Austin), Dennis Gay (University of North Florida), James Gerhart (University of Washington), N. S. Gingrich, J. L. Glathart, S. Goodwin, Rich Gottfried (Frederick Community College), Walter S. Gray (University of Michigan), Paul Gresser (University of Maryland), Benjamin Grinstein (UC San Diego), Howard Grotch (Pennsylvania State University), John Gruber (San Jose State University), Graham D. Gutsche (U.S. Naval Academy), Michael J. Harrison (Michigan State University), Harold Hart (Western Illinois University), Howard Hayden (University of Connecticut), Carl Helrich (Goshen College), Laurent Hodges (Iowa State University), C. D. Hodgman, Michael Hones (Villanova University), Keith Honey (West Virginia Institute of Technology), Gregory Hood (Tidewater Community College), John Hubisz (North Carolina State University), M. Iona, John Jaszczak (Michigan Technical University), Alvin Jenkins (North Carolina State University), Robert P. Johnson (UC Santa Cruz), Lorella Jones (University of Illinois), John Karchek (GMI Engineering & Management Institute), Thomas Keil (Worcester Polytechnic Institute), Robert Kraemer (Carnegie Mellon University), Jean P. Krisch (University of Michigan), Robert A. Kromhout, Andrew Kunz (Marquette University), Charles Lane (Berry College), Thomas N. Lawrence (Texas State University), Robert J. Lee, Alfred Leitner (Rensselaer Polytechnic University), Gerald P. Lietz (De Paul University), Gordon Lind (Utah State University), S. Livingston, Elihu Lubkin (University of Wisconsin, Milwaukee), Robert Luke (Boise State University), David Lynch (Iowa State University), Michael Lysak (San Bernardino Valley College), Jeffrey Mallow (Loyola University), Robert Mania (Kentucky State University), Robert Marchina (University of Memphis), David Markowitz (University of Connecticut), R. J. Maurer, Oren Maxwell (Florida International University), Joseph L. McCauley (University of Houston), T. K. McCubbin, Jr. (Pennsylvania State University), Charles McFarland (University of Missouri at Rolla), James Mcguire (Tulane University), Lawrence McIntyre (University of Arizona), Fredric Messing (Carnegie-Mellon University), Thomas Meyer (Texas A&M University), Andre Mirabelli (St. Peter's College, New Jersey), Herbert Muether (S.U.N.Y., Stony Brook), Jack Munsee (California State University, Long Beach), Lorenzo Narducci (Drexel University), Van E. Neie (Purdue University), David A. Nordling (U. S. Naval Academy), Benedict Oh (Pennsylvania State University), L. O. Olsen, Jim Pannell (DeVry Institute of Technology), W. F. Parks (University of Missouri), Robert Paulson (California State University, Chico), Jerry Peacher (University of Missouri at Rolla), Arnold Perlmutter (University of Miami), Lennart Peterson (University of Florida), R. J. Peterson (University of Colorado, Boulder), R. Pinkston, Ronald Poling (University of Minnesota), J. G. Potter, C. W. Price (Millersville University), Francis Prosser (University of Kansas), Shelden H. Radin, Roberto Ramos (Drexel University), Michael Rapport (Anne Arundel Community College), R. Resnick, James A. Richards, Jr., John S. Risley (North Carolina State University), Francesc Roig (University of California, Santa Barbara), T. L. Rokoske, Richard Roth (Eastern Michigan University), Carl Rotter (University of West Virginia), S. Clark Rowland (Andrews University), Rajarshi Roy (Georgia Institute of Technology), Russell A. Roy (Santa Fe Community College), Dhiraj Sardar (University of Texas, San Antonio), Bruce Schumm (UC Santa Cruz), Melvin Schwartz (St. John's University), F. A. Scott, L. W. Seagondollar, Paul Shand (University of Northern Iowa), Stan Shepherd (Pennsylvania State University), Douglas Sherman (San Jose State), Bruce Sherwood (Carnegie Mellon University), Hugh Siefkin (Greenville College), Tomasz Skwarnicki (Syracuse University), C. P. Slichter, Charles W. Smith (University of Maine, Orono), Malcolm Smith (University of Lowell), Ross Spencer (Brigham Young University), Julien Sprott (University of Wisconsin), Victor Stanionis (Iona College), James Stith (American Institute of Physics), Chuck Stone (North Carolina A&T State

University), Edward Strother (Florida Institute of Technology), Conley Stutz (Bradley University), Albert Stwertka (U.S. Merchant Marine Academy), Kenneth Szpara-DeNisco (Harrisburg Area Community College), Martin Tiersten (CUNY, City College), David Toot (Alfred University), Somdev Tyagi (Drexel University), F. Verbrugge, Helmut Vogel (Carnegie Mellon University), Robert Webb (Texas A & M), Thomas Weber (Iowa State University), M. Russell Wehr, (Pennsylvania State University), Robert Weidman (Michigan Technical University), Dan Whalen (UC San Diego), Lester V. Whitney, Thomas Wiggins (Pennsylvania State University), David Willey (University of Pittsburgh, Johnstown), George Williams (University of Utah), John Williams (Auburn University), Stanley Williams (Iowa State University), Jack Willis, Suzanne Willis (Northern Illinois University), Robert Wilson (San Bernardino Valley College), L. Wolfenstein, James Wood (Palm Beach Junior College), Lowell Wood (University of Houston), R. E. Worley, D. H. Ziebell (Manatee Community College), George O. Zimmerman (Boston University)

In addition, we both have individual acknowledgments we would like to make.

I want to extend my heartfelt thanks to my colleagues at Carnegie Mellon, especially Professors Robert Kraemer, Bruce Sherwood, Ruth Chabay, Helmut Vogel, and Brian Quinn, for many stimulating discussions about physics pedagogy and for their support and encouragement during the writing of several successive editions of this book. I am equally indebted to the many generations of Carnegie Mellon students who have helped me learn what good teaching and good writing are, by showing me what works and what doesn't. It is always a joy and a privilege to express my gratitude to my wife Alice and our children Gretchen and Rebecca for their love, support, and emotional sustenance during the writing of several successive editions of this book. May all men and women be blessed with love such as theirs. — H. D. Y.

I would like to thank my past and present colleagues at UCSB, including Rob Geller, Carl Gwinn, Al Nash, Elisabeth Nicol, and Francesc Roig, for their whole-hearted support and for many helpful discussions. I owe a special debt of gratitude to my early teachers Willa Ramsay, Peter Zimmerman, William Little, Alan Schwettman, and Dirk Walecka for showing me what clear and engaging physics teaching is all about, and to Stuart Johnson for inviting me to become a co-author of *University Physics* beginning with the 9th edition. I want to express special thanks to the editorial staff at Addison-Wesley and their partners: to Nancy Whilton for her editorial vision; to Margot Otway for her superb graphic sense and careful development of this edition; to Peter Murphy for his contributions to the worked examples; to Jason J. B. Harlow for his careful reading of the page proofs; and to Chandrika Madhavan, Steven Le, and Cindy Johnson for keeping the editorial and production pipeline flowing. Most of all, I want to express my gratitude and love to my wife Caroline, to whom I dedicate my contribution to this book. Hey, Caroline, the new edition's done at last — let's go flying! — R. A. F.

Please Tell Us What You Think!

We welcome communications from students and professors, especially concerning errors or deficiencies that you find in this edition. We have devoted a lot of time and effort to writing the best book we know how to write, and we hope it will help you to teach and learn physics. In turn, you can help us by letting us know what still needs to be improved! Please feel free to contact us either electronically or by ordinary mail. Your comments will be greatly appreciated.

December 2010

Hugh D. Young
Department of Physics
Carnegie Mellon University
Pittsburgh, PA 15213
hdy@andrew.cmu.edu

Roger A. Freedman
Department of Physics
University of California, Santa Barbara
Santa Barbara, CA 93106-9530
airboy@physics.ucsb.edu
http://www.physics.ucsb.edu/~airboy/

DETAILED CONTENTS

MODERN PHYSICS

37 RELATIVITY 1223

37.1 Invariance of Physical Laws 1223
37.2 Relativity of Simultaneity 1227
37.3 Relativity of Time Intervals 1228
37.4 Relativity of Length 1233
37.5 The Lorentz Transformations 1237
37.6 The Doppler Effect for
      Electromagnetic Waves 1241
37.7 Relativistic Momentum 1243
37.8 Relativistic Work and Energy 1246
37.9 Newtonian Mechanics and
      Relativity 1249
      Summary 1252
      Questions/Exercises/Problems 1253

**38 PHOTONS: LIGHT WAVES
      BEHAVING AS PARTICLES** 1261

38.1 Light Absorbed as Photons:
      The Photoelectric Effect 1261
38.2 Light Emitted as Photons:
      X-Ray Production 1266
38.3 Light Scattered as Photons: Compton
      Scattering and Pair Production 1269
38.4 Wave–Particle Duality, Probability,
      and Uncertainty 1273
      Summary 1280
      Questions/Exercises/Problems 1281

**39 PARTICLES BEHAVING
      AS WAVES** 1286

39.1 Electron Waves 1286
39.2 The Nuclear Atom and Atomic
      Spectra 1292
39.3 Energy Levels and the Bohr Model
      of the Atom 1297
39.4 The Laser 1307
39.5 Continuous Spectra 1310
39.6 The Uncertainty Principle Revisited 1314
      Summary 1318
      Questions/Exercises/Problems 1319

40 QUANTUM MECHANICS 1328

40.1 Wave Functions and the One-Dimensional
      Schrödinger Equation 1328
40.2 Particle in a Box 1338
40.3 Potential Wells 1343
40.4 Potential Barriers and Tunneling 1347
40.5 The Harmonic Oscillator 1350
      Summary 1355
      Questions/Exercises/Problems 1356

41 ATOMIC STRUCTURE 1364

41.1 The Schrödinger Equation in
      Three Dimensions 1365
41.2 Particle in a Three-Dimensional Box 1366
41.3 The Hydrogen Atom 1372
41.4 The Zeeman Effect 1379
41.5 Electron Spin 1383
41.6 Many-Electron Atoms
      and the Exclusion Principle 1387
41.7 X-Ray Spectra 1393
      Summary 1397
      Questions/Exercises/Problems 1399

**42 MOLECULES AND
      CONDENSED MATTER** 1405

42.1 Types of Molecular Bonds 1405
42.2 Molecular Spectra 1408
42.3 Structure of Solids 1412
42.4 Energy Bands 1416
42.5 Free-Electron Model of Metals 1418
42.6 Semiconductors 1422
42.7 Semiconductor Devices 1425
42.8 Superconductivity 1430
      Summary 1431
      Questions/Exercises/Problems 1432

43 NUCLEAR PHYSICS 1439

43.1 Properties of Nuclei 1439
43.2 Nuclear Binding and Nuclear
      Structure 1444

43.3   Nuclear Stability and Radioactivity   1449
43.4   Activities and Half-Lives   1456
43.5   Biological Effects of Radiation   1459
43.6   Nuclear Reactions   1462
43.7   Nuclear Fission   1464
43.8   Nuclear Fusion   1469
       Summary   1472
       Questions/Exercises/Problems   1473

**44   PARTICLE PHYSICS
        AND COSMOLOGY   1480**

44.1   Fundamental Particles—A History   1480
44.2   Particle Accelerators and Detectors   1485
44.3   Particles and Interactions   1490
44.4   Quarks and the Eightfold Way   1496
44.5   The Standard Model
       and Beyond   1499

44.6   The Expanding Universe   1501
44.7   The Beginning of Time   1508
       Summary   1517
       Questions/Exercises/Problems   1518

APPENDICES

A   The International System of Units   A-1
B   Useful Mathematical Relations   A-3
C   The Greek Alphabet   A-4
D   Periodic Table of Elements   A-5
E   Unit Conversion Factors   A-6
F   Numerical Constants   A-7

    Answers to Odd-Numbered Problems   A-9
    Photo Credits   C-1
    Index   I-1

37

RELATIVITY

? At Brookhaven National Laboratory in New York, atomic nuclei are accelerated to 99.995% of the ultimate speed limit of the universe—the speed of light. Is there also an upper limit on the *kinetic energy* of a particle?

LEARNING GOALS

By studying this chapter, you will learn:

- The two postulates of Einstein's special theory of relativity, and what motivates these postulates.

- Why different observers can dis- agree about whether two events are simultaneous.

- How relativity predicts that moving clocks run slow, and that experi- mental evidence confirms this.

- How the length of an object changes due to the object's motion.

- How the velocity of an object depends on the frame of reference from which it is observed.

- How the theory of relativity modifies the relationship between velocity and momentum.

- How to solve problems involving work and kinetic energy for particles moving at relativistic speeds.

- Some of the key concepts of Einstein's general theory of relativity.

When the year 1905 began, Albert Einstein was an unknown 25-year-old clerk in the Swiss patent office. By the end of that amazing year he had published three papers of extraordinary importance. One was an analysis of Brownian motion; a second (for which he was awarded the Nobel Prize) was on the photoelectric effect. In the third, Einstein introduced his **special theory of relativity,** proposing drastic revisions in the Newtonian concepts of space and time.

The special theory of relativity has made wide-ranging changes in our under-standing of nature, but Einstein based it on just two simple postulates. One states that the laws of physics are the same in all inertial frames of reference; the other states that the speed of light in vacuum is the same in all inertial frames. These innocent-sounding propositions have far-reaching implications. Here are three: (1) Events that are simultaneous for one observer may not be simultaneous for another. (2) When two observers moving relative to each other measure a time interval or a length, they may not get the same results. (3) For the conservation principles for momentum and energy to be valid in all inertial systems, Newton's second law and the equations for momentum and kinetic energy have to be revised.

Relativity has important consequences in *all* areas of physics, including elec-tromagnetism, atomic and nuclear physics, and high-energy physics. Although many of the results derived in this chapter may run counter to your intuition, the theory is in solid agreement with experimental observations.

37.1 Invariance of Physical Laws

Let's take a look at the two postulates that make up the special theory of relativ-ity. Both postulates describe what is seen by an observer in an *inertial frame of reference,* which we introduced in Section 4.2. The theory is "special" in the sense that it applies to observers in such special reference frames.

Einstein's First Postulate

Einstein's first postulate, called the **principle of relativity,** states: **The laws of physics are the same in every inertial frame of reference.** If the laws differed, that difference could distinguish one inertial frame from the others or make one frame somehow more "correct" than another. Here are two examples. Suppose you watch two children playing catch with a ball while the three of you are aboard a train moving with constant velocity. Your observations of the motion *of the ball,* no matter how carefully done, can't tell you how fast (or whether) the train is moving. This is because Newton's laws of motion are the same in every inertial frame.

Another example is the electromotive force (emf) induced in a coil of wire by a nearby moving permanent magnet. In the frame of reference in which the *coil* is stationary (Fig. 37.1a), the moving magnet causes a change of magnetic flux through the coil, and this induces an emf. In a different frame of reference in which the *magnet* is stationary (Fig. 37.1b), the motion of the coil through a magnetic field induces the emf. According to the principle of relativity, both of these frames of reference are equally valid. Hence the same emf must be induced in both situations shown in Fig. 37.1. As we saw in Chapter 29, this is indeed the case, so Faraday's law is consistent with the principle of relativity. Indeed, *all* of the laws of electromagnetism are the same in every inertial frame of reference.

Equally significant is the prediction of the speed of electromagnetic radiation, derived from Maxwell's equations (see Section 32.2). According to this analysis, light and all other electromagnetic waves travel in vacuum with a constant speed, now defined to equal exactly 299,792,458 m/s. (We often use the approximate value $c = 3.00 \times 10^8$ m/s, which is within one part in 1000 of the exact value.) As we will see, the speed of light in vacuum plays a central role in the theory of relativity.

37.1 The same emf is induced in the coil whether (a) the magnet moves relative to the coil or (b) the coil moves relative to the magnet.

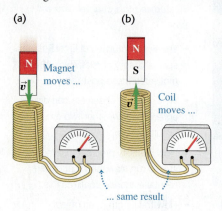

(a)   (b)

N   Magnet
    moves ...

$\vec{v}$

N
S

Coil
$\vec{v}$ moves ...

... same result

Einstein's Second Postulate

During the 19th century, most physicists believed that light traveled through a hypothetical medium called the *ether,* just as sound waves travel through air. If so, the speed of light measured by observers would depend on their motion relative to the ether and would therefore be different in different directions. The Michelson-Morley experiment, described in Section 35.5, was an effort to detect motion of the earth relative to the ether. Einstein's conceptual leap was to recognize that if Maxwell's equations are valid in all inertial frames, then the speed of light in vacuum should also be the same in all frames and in all directions. In fact, Michelson and Morley detected *no* ether motion across the earth, and the ether concept has been discarded. Although Einstein may not have known about this negative result, it supported his bold hypothesis of the constancy of the speed of light in vacuum.

Einstein's second postulate states: **The speed of light in vacuum is the same in all inertial frames of reference and is independent of the motion of the source.**

Let's think about what this means. Suppose two observers measure the speed of light in vacuum. One is at rest with respect to the light source, and the other is moving away from it. Both are in inertial frames of reference. According to the principle of relativity, the two observers must obtain the same result, despite the fact that one is moving with respect to the other.

If this seems too easy, consider the following situation. A spacecraft moving past the earth at 1000 m/s fires a missile straight ahead with a speed of 2000 m/s (relative to the spacecraft) (Fig. 37.2). What is the missile's speed relative to the earth? Simple, you say; this is an elementary problem in relative velocity (see Section 3.5). The correct answer, according to Newtonian mechanics, is 3000 m/s.

37.2 (a) Newtonian mechanics makes correct predictions about relatively slow-moving objects; (b) it makes incorrect predictions about the behavior of light.

(a)

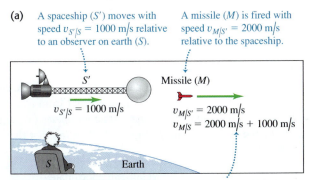

A spaceship (S') moves with speed $v_{S'/S} = 1000$ m/s relative to an observer on earth (S). A missile (M) is fired with speed $v_{M/S'} = 2000$ m/s relative to the spaceship.

S'

Missile (M)

$v_{S'/S} = 1000$ m/s

$v_{M/S'} = 2000$ m/s
$v_{M/S} = 2000$ m/s + 1000 m/s

S Earth

NEWTONIAN MECHANICS HOLDS: Newtonian mechanics tells us correctly that the missile moves with speed $v_{M/S} = 3000$ m/s relative to the observer on earth.

(b)

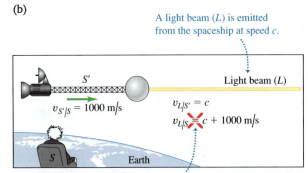

A light beam (L) is emitted from the spaceship at speed c.

S' Light beam (L)

$v_{S'/S} = 1000$ m/s

$v_{L/S'} = c$
$v_{L/S} \neq c + 1000$ m/s

S Earth

NEWTONIAN MECHANICS FAILS: Newtonian mechanics tells us *incorrectly* that the light moves at a speed greater than c relative to the observer on earth ... which would contradict Einstein's second postulate.

But now suppose the spacecraft turns on a searchlight, pointing in the same direction in which the missile was fired. An observer on the spacecraft measures the speed of light emitted by the searchlight and obtains the value c. According to Einstein's second postulate, the motion of the light after it has left the source cannot depend on the motion of the source. So the observer on earth who measures the speed of this same light must also obtain the value c, *not* $c + 1000$ m/s. This result contradicts our elementary notion of relative velocities, and it may not appear to agree with common sense. But "common sense" is intuition based on everyday experience, and this does not usually include measurements of the speed of light.

The Ultimate Speed Limit

Einstein's second postulate immediately implies the following result:

It is impossible for an inertial observer to travel at c, the speed of light in vacuum.

We can prove this by showing that travel at c implies a logical contradiction. Suppose that the spacecraft S' in Fig. 37.2b is moving at the speed of light relative to an observer on the earth, so that $v_{S'/S} = c$. If the spacecraft turns on a headlight, the second postulate now asserts that the earth observer S measures the headlight beam to be also moving at c. Thus this observer measures that the headlight beam and the spacecraft move together and are always at the same point in space. But Einstein's second postulate also asserts that the headlight beam moves at a speed c relative to the spacecraft, so they *cannot* be at the same point in space. This contradictory result can be avoided only if it is impossible for an inertial observer, such as a passenger on the spacecraft, to move at c. As we go through our discussion of relativity, you may find yourself asking the question Einstein asked himself as a 16-year-old student, "What would I see if I were traveling at the speed of light?" Einstein realized only years later that his question's basic flaw was that he could *not* travel at c.

The Galilean Coordinate Transformation

Let's restate this argument symbolically, using two inertial frames of reference, labeled S for the observer on earth and S' for the moving spacecraft, as shown in Fig. 37.3. To keep things as simple as possible, we have omitted the z-axes. The x-axes of the two frames lie along the same line, but the origin O' of frame S' moves relative to the origin O of frame S with constant velocity u along the common x-x'-axis. We on earth set our clocks so that the two origins coincide at time $t = 0$, so their separation at a later time t is ut.

37.3 The position of particle P can be described by the coordinates x and y in frame of reference S or by x' and y' in frame S'.

Frame S' moves relative to frame S with constant velocity u along the common x-x'-axis.

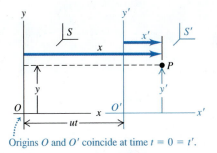

Origins O and O' coincide at time $t = 0 = t'$.

CAUTION **Choose your inertial frame coordinates wisely** Many of the equations derived in this chapter are true *only* if you define your inertial reference frames as stated in the preceding paragraph. For instance, the positive *x*-direction must be the direction in which the origin O' moves relative to the origin O. In Fig. 37.3 this direction is to the right; if instead O' moves to the left relative to O, you must define the positive *x*-direction to be to the left. ▌

Now think about how we describe the motion of a particle P. This might be an exploratory vehicle launched from the spacecraft or a pulse of light from a laser. We can describe the *position* of this particle by using the earth coordinates (x, y, z) in S or the spacecraft coordinates (x', y', z') in S'. Figure 37.3 shows that these are simply related by

$$x = x' + ut \qquad y = y' \qquad z = z' \qquad \begin{matrix} \text{(Galilean coordinate} \\ \text{transformation)} \end{matrix} \qquad (37.1)$$

These equations, based on the familiar Newtonian notions of space and time, are called the **Galilean coordinate transformation.**

If particle P moves in the *x*-direction, its instantaneous velocity v_x as measured by an observer stationary in S is $v_x = dx/dt$. Its velocity v'_x as measured by an observer stationary in S' is $v'_x = dx'/dt$. We can derive a relationship between v_x and v'_x by taking the derivative with respect to t of the first of Eqs. (37.1):

$$\frac{dx}{dt} = \frac{dx'}{dt} + u$$

Now dx/dt is the velocity v_x measured in S, and dx'/dt is the velocity v'_x measured in S', so we get the *Galilean velocity transformation* for one-dimensional motion:

$$v_x = v'_x + u \qquad \text{(Galilean velocity transformation)} \qquad (37.2)$$

Although the notation differs, this result agrees with our discussion of relative velocities in Section 3.5.

Now here's the fundamental problem. Applied to the speed of light in vacuum, Eq. (37.2) says that $c = c' + u$. Einstein's second postulate, supported subsequently by a wealth of experimental evidence, says that $c = c'$. This is a genuine inconsistency, not an illusion, and it demands resolution. If we accept this postulate, we are forced to conclude that Eqs. (37.1) and (37.2) *cannot* be precisely correct, despite our convincing derivation. These equations have to be modified to bring them into harmony with this principle.

The resolution involves some very fundamental modifications in our kinematic concepts. The first idea to be changed is the seemingly obvious assumption that the observers in frames S and S' use the same *time scale,* formally stated as $t = t'$. Alas, we are about to show that this everyday assumption cannot be correct; the two observers *must* have different time scales. We must define the velocity v' in frame S' as $v' = dx'/dt'$, not as dx'/dt; the two quantities are not the same. The difficulty lies in the concept of *simultaneity,* which is our next topic. A careful analysis of simultaneity will help us develop the appropriate modifications of our notions about space and time.

Test Your Understanding of Section 37.1 As a high-speed spaceship flies past you, it fires a strobe light that sends out a pulse of light in all directions. An observer aboard the spaceship measures a spherical wave front that spreads away from the spaceship with the same speed c in all directions. (a) What is the shape of the wave front that *you* measure? (i) spherical; (ii) ellipsoidal, with the longest axis of the ellipsoid along the direction of the spaceship's motion; (iii) ellipsoidal, with the shortest axis of the ellipsoid along the direction of the spaceship's motion; (iv) not enough information is given to decide. (b) Is the wave front centered on the spaceship? ▌

37.2 Relativity of Simultaneity

Measuring times and time intervals involves the concept of **simultaneity.** In a given frame of reference, an **event** is an occurrence that has a definite position and time (Fig. 37.4). When you say that you awoke at seven o'clock, you mean that two events (your awakening and your clock showing 7:00) occurred *simultaneously.* The fundamental problem in measuring time intervals is this: In general, two events that are simultaneous in one frame of reference are *not* simultaneous in a second frame that is moving relative to the first, even if both are inertial frames.

A Thought Experiment in Simultaneity

This may seem to be contrary to common sense. To illustrate the point, here is a version of one of Einstein's *thought experiments*—mental experiments that follow concepts to their logical conclusions. Imagine a train moving with a speed comparable to *c*, with uniform velocity (Fig. 37.5). Two lightning bolts strike a passenger car, one near each end. Each bolt leaves a mark on the car and one on the ground at the instant the bolt hits. The points on the ground are labeled *A* and *B* in the figure, and the corresponding points on the car are *A'* and *B'*. Stanley is stationary on the ground at *O*, midway between *A* and *B*. Mavis is moving with the train at *O'* in the middle of the passenger car, midway between *A'* and *B'*. Both Stanley and Mavis see both light flashes emitted from the points where the lightning strikes.

Suppose the two wave fronts from the lightning strikes reach Stanley at *O* simultaneously. He knows that he is the same distance from *B* and *A*, so Stanley concludes that the two bolts struck *B* and *A* simultaneously. Mavis agrees that the two wave fronts reached Stanley at the same time, but she disagrees that the flashes were emitted simultaneously.

Stanley and Mavis agree that the two wave fronts do not reach Mavis at the same time. Mavis at *O'* is moving to the right with the train, so she runs into the wave front from *B'* *before* the wave front from *A'* catches up to her. However, because she is in the middle of the passenger car equidistant from *A'* and *B'*, her observation is that both wave fronts took the same time to reach her because both moved the same distance at the same speed *c*. (Recall that the speed of each wave front with respect to *either* observer is *c*.) Thus she concludes that the lightning bolt at *B'* struck *before* the one at *A'*. Stanley at *O* measures the two events to be simultaneous, but Mavis at *O'* does not! *Whether or not two events at different x-axis locations are simultaneous depends on the state of motion of the observer.*

You may want to argue that in this example the lightning bolts really *are* simultaneous and that if Mavis at *O'* could communicate with the distant points without the time delay caused by the finite speed of light, she would realize this. But that would be erroneous; the finite speed of information transmission is not the real issue. If *O'* is midway between *A'* and *B'*, then in her frame of reference the time for a signal to travel from *A'* to *O'* is the same as that from *B'* to *O'*. Two signals arrive simultaneously at *O'* only if they were emitted simultaneously at *A'* and *B'*. In this example they do *not* arrive simultaneously at *O'*, and so Mavis must conclude that the events at *A'* and *B'* were *not* simultaneous.

Furthermore, there is no basis for saying that Stanley is right and Mavis is wrong, or vice versa. According to the principle of relativity, no inertial frame of reference is more correct than any other in the formulation of physical laws. Each observer is correct *in his or her own frame of reference.* In other words, simultaneity is not an absolute concept. Whether two events are simultaneous depends on the frame of reference. As we mentioned at the beginning of this section, simultaneity plays an essential role in measuring time intervals. It follows that *the time interval between two events may be different in different frames of reference.* So our next task is to learn how to compare time intervals in different frames of reference.

37.4 An event has a definite position and time—for instance, on the pavement directly below the center of the Eiffel Tower at midnight on New Year's Eve.

37.5 A thought experiment in simultaneity.

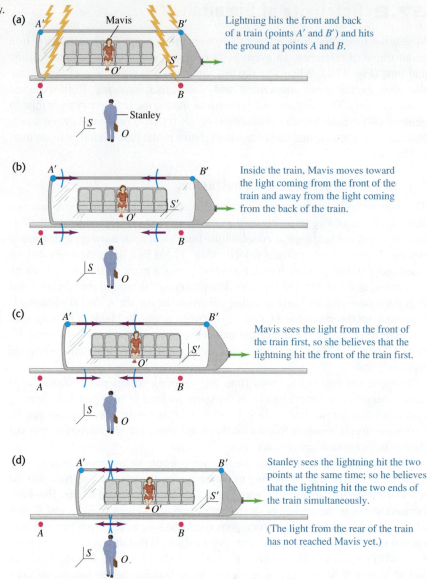

(a) Lightning hits the front and back of a train (points A' and B') and hits the ground at points A and B.

(b) Inside the train, Mavis moves toward the light coming from the front of the train and away from the light coming from the back of the train.

(c) Mavis sees the light from the front of the train first, so she believes that the lightning hit the front of the train first.

(d) Stanley sees the lightning hit the two points at the same time; so he believes that the lightning hit the two ends of the train simultaneously.

(The light from the rear of the train has not reached Mavis yet.)

Test Your Understanding of Section 37.2   Stanley, who works for the rail system shown in Fig. 37.5, has carefully synchronized the clocks at all of the rail stations. At the moment that Stanley measures all of the clocks striking noon, Mavis is on a high-speed passenger car traveling from Ogdenville toward North Haverbrook. According to Mavis, when the Ogdenville clock strikes noon, what time is it in North Haverbrook? (i) noon; (ii) before noon; (iii) after noon.

MasteringPHYSICS

ActivPhysics 17.1: Relativity of Time

37.3 Relativity of Time Intervals

We can derive a quantitative relationship between time intervals in different coordinate systems. To do this, let's consider another thought experiment. As before, a frame of reference S' moves along the common x-x'-axis with constant speed u relative to a frame S. As discussed in Section 37.1, u must be less than the speed of light c. Mavis, who is riding along with frame S', measures the time interval between two events that occur at the *same* point in space. Event 1 is when a flash of light from a light source leaves O'. Event 2 is when the flash returns to O', having been reflected from a mirror a distance d away, as shown in Fig. 37.6a. We label the time interval Δt_0, using the subscript zero as a reminder that the apparatus is at rest, with zero velocity, in frame S'. The flash of light moves a total distance $2d$, so the time interval is

37.6 (a) Mavis, in frame of reference S', observes a light pulse emitted from a source at O' and reflected back along the same line. (b) How Stanley (in frame of reference S) and Mavis observe the same light pulse. The positions of O' at the times of departure and return of the pulse are shown.

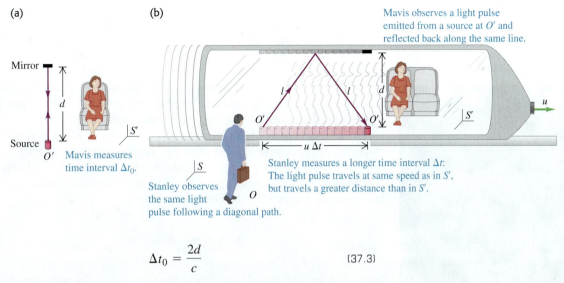

(a)

Mirror

d

Source

O' Mavis measures time interval Δt_0.

(b)

S

Stanley observes the same light pulse following a diagonal path.

O

Mavis observes a light pulse emitted from a source at O' and reflected back along the same line.

l l

d

O' O'

$u \Delta t$

u

Stanley measures a longer time interval Δt: The light pulse travels at same speed as in S', but travels a greater distance than in S'.

$$\Delta t_0 = \frac{2d}{c} \tag{37.3}$$

The round-trip time measured by Stanley in frame S is a different interval Δt; in his frame of reference the two events occur at *different* points in space. During the time Δt, the source moves relative to S a distance $u\,\Delta t$ (Fig. 37.6b). In S' the round-trip distance is $2d$ perpendicular to the relative velocity, but the round-trip distance in S is the longer distance $2l$, where

$$l = \sqrt{d^2 + \left(\frac{u\,\Delta t}{2}\right)^2}$$

In writing this expression, we have assumed that both observers measure the same distance d. We will justify this assumption in the next section. The speed of light is the same for both observers, so the round-trip time measured in S is

$$\Delta t = \frac{2l}{c} = \frac{2}{c}\sqrt{d^2 + \left(\frac{u\,\Delta t}{2}\right)^2} \tag{37.4}$$

We would like to have a relationship between Δt and Δt_0 that is independent of d. To get this, we solve Eq. (37.3) for d and substitute the result into Eq. (37.4), obtaining

$$\Delta t = \frac{2}{c}\sqrt{\left(\frac{c\,\Delta t_0}{2}\right)^2 + \left(\frac{u\,\Delta t}{2}\right)^2} \tag{37.5}$$

Now we square this and solve for Δt; the result is

$$\Delta t = \frac{\Delta t_0}{\sqrt{1 - u^2/c^2}}$$

Since the quantity $\sqrt{1 - u^2/c^2}$ is less than 1, Δt is greater than Δt_0: Thus Stanley measures a *longer* round-trip time for the light pulse than does Mavis.

Time Dilation

We may generalize this important result. In a particular frame of reference, suppose that two events occur at the same point in space. The time interval between these events, as measured by an observer at rest in this same frame (which we call the *rest frame* of this observer), is Δt_0. Then an observer in a second frame moving with constant speed u relative to the rest frame will measure the time interval to be Δt, where

37.7 This image shows an exploding star, called a *supernova,* within a distant galaxy. The brightness of a typical supernova decays at a certain rate. But supernovae that are moving away from us at a substantial fraction of the speed of light decay more slowly, in accordance with Eq. (37.6). The decaying supernova is a moving "clock" that runs slow.

37.8 The quantity $\gamma = 1/\sqrt{1 - u^2/c^2}$ as a function of the relative speed u of two frames of reference.

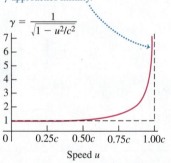

$$\Delta t = \frac{\Delta t_0}{\sqrt{1 - u^2/c^2}} \quad \text{(time dilation)} \qquad (37.6)$$

We recall that no inertial observer can travel at $u = c$ and we note that $\sqrt{1 - u^2/c^2}$ is imaginary for $u > c$. Thus Eq. (37.6) gives sensible results only when $u < c$. The denominator of Eq. (37.6) is always smaller than 1, so Δt is always *larger* than Δt_0. Thus we call this effect **time dilation.**

Think of an old-fashioned pendulum clock that has one second between ticks, as measured by Mavis in the clock's rest frame; this is Δt_0. If the clock's rest frame is moving relative to Stanley, he measures a time between ticks Δt that is longer than one second. In brief, *observers measure any clock to run slow if it moves relative to them* (Fig. 37.7). Note that this conclusion is a direct result of the fact that the speed of light in vacuum is the same in both frames of reference.

The quantity $1/\sqrt{1 - u^2/c^2}$ in Eq. (37.6) appears so often in relativity that it is given its own symbol γ (the Greek letter gamma):

$$\gamma = \frac{1}{\sqrt{1 - u^2/c^2}} \qquad (37.7)$$

In terms of this symbol, we can express the time dilation formula, Eq. (37.6), as

$$\Delta t = \gamma \, \Delta t_0 \quad \text{(time dilation)} \qquad (37.8)$$

As a further simplification, u/c is sometimes given the symbol β (the Greek letter beta); then $\gamma = 1/\sqrt{1 - \beta^2}$.

Figure 37.8 shows a graph of γ as a function of the relative speed u of two frames of reference. When u is very small compared to c, u^2/c^2 is much smaller than 1 and γ is very nearly *equal* to 1. In that limit, Eqs. (37.6) and (37.8) approach the Newtonian relationship $\Delta t = \Delta t_0$, corresponding to the same time interval in all frames of reference.

If the relative speed u is great enough that γ is appreciably greater than 1, the speed is said to be *relativistic;* if the difference between γ and 1 is negligibly small, the speed u is called *nonrelativistic.* Thus $u = 6.00 \times 10^7$ m/s $= 0.200c$ (for which $\gamma = 1.02$) is a relativistic speed, but $u = 6.00 \times 10^4$ m/s $= 0.000200c$ (for which $\gamma = 1.00000002$) is a nonrelativistic speed.

Proper Time

There is only one frame of reference in which a clock is at rest, and there are infinitely many in which it is moving. Therefore the time interval measured between two events (such as two ticks of the clock) that occur at the same point in a particular frame is a more fundamental quantity than the interval between events at different points. We use the term **proper time** to describe the time interval Δt_0 between two events that occur *at the same point.*

CAUTION **Measuring time intervals** It is important to note that the time interval Δt in Eq. (37.6) involves events that occur *at different space points* in the frame of reference S. Note also that any differences between Δt and the proper time Δt_0 are *not* caused by differences in the times required for light to travel from those space points to an observer at rest in S. We assume that our observer is able to correct for differences in light transit times, just as an astronomer who's observing the sun understands that an event seen now on earth actually occurred 500 s ago on the sun's surface. Alternatively, we can use *two* observers, one stationary at the location of the first event and the other at the second, each with his or her own clock. We can synchronize these two clocks without difficulty, as long as they are at rest in the same frame of reference. For example, we could send a light pulse simultaneously to the two clocks from a point midway between them. When the pulses arrive, the observers set their clocks to a prearranged time. (But note that clocks that are synchronized in one frame of reference *are not* in general synchronized in any other frame.) ∎

In thought experiments, it's often helpful to imagine many observers with synchronized clocks at rest at various points in a particular frame of reference. We can picture a frame of reference as a coordinate grid with lots of synchronized clocks distributed around it, as suggested by Fig. 37.9. Only when a clock is moving relative to a given frame of reference do we have to watch for ambiguities of synchronization or simultaneity.

Throughout this chapter we will frequently use phrases like "Stanley *observes* that Mavis passes the point $x = 5.00$ m, $y = 0$, $z = 0$ at time 2.00 s." This means that Stanley is using a grid of clocks in his frame of reference, like the grid shown in Fig. 37.9, to record the time of an event. We could restate the phrase as "When Mavis passes the point at $x = 5.00$ m, $y = 0$, $z = 0$, the clock at that location in Stanley's frame of reference reads 2.00 s." We will avoid using phrases like "Stanley *sees* that Mavis is a certain point at a certain time," because there is a time delay for light to travel to Stanley's eye from the position of an event.

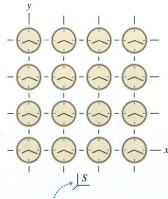

37.9 A frame of reference pictured as a coordinate system with a grid of synchronized clocks.

The grid is three dimensional; identical planes of clocks lie in front of and behind the page, connected by grid lines perpendicular to the page.

Problem-Solving Strategy 37.1 | Time Dilation

IDENTIFY *the relevant concepts:* The concept of time dilation is used whenever we compare the time intervals between events as measured by observers in different inertial frames of reference.

SET UP *the problem* using the following steps:
1. First decide what two events define the beginning and the end of the time interval. Then identify the two frames of reference in which the time interval is measured.
2. Identify the target variable.

EXECUTE *the solution* as follows:
1. In many problems, the time interval as measured in one frame of reference is the *proper* time Δt_0. This is the time interval

between two events in a frame of reference in which the two events occur at the same point in space. In a second frame of reference that has a speed u relative to that first frame, there is a longer time interval Δt between the same two events. In this second frame the two events occur at different points. You will need to decide in which frame the time interval is Δt_0 and in which frame it is Δt.
2. Use Eq. (37.6) or (37.8) to relate Δt_0 and Δt, and then solve for the target variable.

EVALUATE *your answer:* Note that Δt is never smaller than Δt_0, and u is never greater than c. If your results suggest otherwise, you need to rethink your calculation.

Example 37.1 | Time dilation at 0.990c

High-energy subatomic particles coming from space interact with atoms in the earth's upper atmosphere, in some cases producing unstable particles called *muons*. A muon decays into other particles with a mean lifetime of 2.20 μs $= 2.20 \times 10^{-6}$ s as measured in a reference frame in which it is at rest. If a muon is moving at $0.990c$ relative to the earth, what will an observer on earth measure its mean lifetime to be?

SOLUTION

IDENTIFY and SET UP: The muon's lifetime is the time interval between two events: the production of the muon and its subsequent decay. Our target variable is the lifetime in your frame of reference on earth, which we call frame S. We are given the lifetime in a frame S' in which the muon is at rest; this is its *proper* lifetime, $\Delta t_0 = 2.20$ μs. The relative speed of these two frames is

$u = 0.990c$. We use Eq. (37.6) to relate the lifetimes in the two frames.

EXECUTE: The muon moves relative to the earth between the two events, so the two events occur at different positions as measured in S and the time interval in that frame is Δt (the target variable). From Eq. (37.6),

$$\Delta t = \frac{\Delta t_0}{\sqrt{1 - u^2/c^2}} = \frac{2.20 \ \mu\text{s}}{\sqrt{1 - (0.990)^2}} = 15.6 \ \mu\text{s}$$

EVALUATE: Our result predicts that the mean lifetime of the muon in the earth frame (Δt) is about seven times longer than in the muon's frame (Δt_0). This prediction has been verified experimentally; indeed, this was the first experimental confirmation of the time dilation formula, Eq. (37.6).

Example 37.2 Time dilation at airliner speeds

An airplane flies from San Francisco to New York (about 4800 km, or 4.80×10^6 m) at a steady speed of 300 m/s (about 670 mi/h). How much time does the trip take, as measured by an observer on the ground? By an observer in the plane?

SOLUTION

IDENTIFY and SET UP: Here we're interested in the time interval between the airplane departing from San Francisco and landing in New York. The target variables are the time intervals as measured in the frame of reference of the ground S and in the frame of reference of the airplane S'.

EXECUTE: As measured in S the two events occur at different positions (San Francisco and New York), so the time interval measured by ground observers corresponds to Δt in Eq. (37.6). To find it, we simply divide the distance by the speed $u = 300$ m/s:

$$\Delta t = \frac{4.80 \times 10^6 \text{ m}}{300 \text{ m/s}} = 1.60 \times 10^4 \text{ s} \quad (\text{about } 4\tfrac{1}{2} \text{ hours})$$

In the airplane's frame S', San Francisco and New York passing under the plane occur at the same point (the position of the plane). Hence the time interval in the airplane is a proper time, corresponding to Δt_0 in Eq. (37.6). We have

$$\frac{u^2}{c^2} = \frac{(300 \text{ m/s})^2}{(3.00 \times 10^8 \text{ m/s})^2} = 1.00 \times 10^{-12}$$

From Eq. (37.6),

$$\Delta t_0 = (1.60 \times 10^4 \text{ s})\sqrt{1 - 1.00 \times 10^{-12}}$$

The square root can't be evaluated with adequate precision with an ordinary calculator. But we can approximate it using the binomial theorem (see Appendix B):

$$(1 - 1.00 \times 10^{-12})^{1/2} = 1 - \left(\tfrac{1}{2}\right)(1.00 \times 10^{-12}) + \cdots$$

The remaining terms are of the order of 10^{-24} or smaller and can be discarded. The approximate result for Δt_0 is

$$\Delta t_0 = (1.60 \times 10^4 \text{ s})(1 - 0.50 \times 10^{-12})$$

The proper time Δt_0, measured in the airplane, is very slightly less (by less than one part in 10^{12}) than the time measured on the ground.

EVALUATE: We don't notice such effects in everyday life. But present-day atomic clocks (see Section 1.3) can attain a precision of about one part in 10^{13}. A cesium clock traveling a long distance in an airliner has been used to measure this effect and thereby verify Eq. (37.6) even at speeds much less than c.

Example 37.3 Just when is it proper?

Mavis boards a spaceship and then zips past Stanley on earth at a relative speed of $0.600c$. At the instant she passes him, they both start timers. (a) A short time later Stanley measures that Mavis has traveled 9.00×10^7 m beyond him and is passing a space station. What does Stanley's timer read as she passes the space station? What does Mavis's timer read? (b) Stanley starts to blink just as Mavis flies past him, and Mavis measures that the blink takes 0.400 s from beginning to end. According to Stanley, what is the duration of his blink?

SOLUTION

IDENTIFY and SET UP: This problem involves time dilation for two *different* sets of events measured in Stanley's frame of reference (which we call S) and in Mavis's frame of reference (which we call S'). The two events of interest in part (a) are when Mavis passes Stanley and when Mavis passes the space station; the target variables are the time intervals between these two events as measured in S and in S'. The two events in part (b) are the start and finish of Stanley's blink; the target variable is the time interval between these two events as measured in S.

EXECUTE: (a) The two events, Mavis passing the earth and Mavis passing the space station, occur at different positions in Stanley's frame but at the same position in Mavis's frame. Hence Stanley measures time interval Δt, while Mavis measures the *proper* time Δt_0. As measured by Stanley, Mavis moves at $0.600c = 0.600(3.00 \times 10^8 \text{ m/s}) = 1.80 \times 10^8$ m/s and travels 9.00×10^7 m in time $\Delta t = (9.00 \times 10^7 \text{ m})/(1.80 \times 10^8 \text{ m/s}) = 0.500$ s. From Eq. (37.6), Mavis's timer reads an elapsed time of

$$\Delta t_0 = \Delta t \sqrt{1 - u^2/c^2} = 0.500 \text{ s} \sqrt{1 - (0.600)^2} = 0.400 \text{ s}$$

(b) It is tempting to answer that Stanley's blink lasts 0.500 s in his frame. But this is wrong, because we are now considering a *different* pair of events than in part (a). The start and finish of Stanley's blink occur at the same point in his frame S but at different positions in Mavis's frame S', so the time interval of 0.400 s that she measures between these events is equal to Δt. The duration of the blink measured on Stanley's timer is the proper time Δt_0:

$$\Delta t_0 = \Delta t \sqrt{1 - u^2/c^2} = 0.400 \text{ s} \sqrt{1 - (0.600)^2} = 0.320 \text{ s}$$

EVALUATE: This example illustrates the relativity of simultaneity. In Mavis's frame she passes the space station at the same instant that Stanley finishes his blink, 0.400 s after she passed Stanley. Hence these two events are simultaneous to Mavis in frame S'. But these two events are *not* simultaneous to Stanley in his frame S: According to his timer, he finishes his blink after 0.320 s and Mavis passes the space station after 0.500 s.

The Twin Paradox

Equations (37.6) and (37.8) for time dilation suggest an apparent paradox called the **twin paradox.** Consider identical twin astronauts named Eartha and Astrid.

Eartha remains on earth while her twin Astrid takes off on a high-speed trip through the galaxy. Because of time dilation, Eartha observes Astrid's heartbeat and all other life processes proceeding more slowly than her own. Thus to Eartha, Astrid ages more slowly; when Astrid returns to earth she is younger (has aged less) than Eartha.

Now here is the paradox: All inertial frames are equivalent. Can't Astrid make exactly the same arguments to conclude that Eartha is in fact the younger? Then each twin measures the other to be younger when they're back together, and that's a paradox.

To resolve the paradox, we recognize that the twins are *not* identical in all respects. While Eartha remains in an approximately inertial frame at all times, Astrid must *accelerate* with respect to that inertial frame during parts of her trip in order to leave, turn around, and return to earth. Eartha's reference frame is always approximately inertial; Astrid's is often far from inertial. Thus there is a real physical difference between the circumstances of the two twins. Careful analysis shows that Eartha is correct; when Astrid returns, she *is* younger than Eartha.

Test Your Understanding of Section 37.3 Samir (who is standing on the ground) starts his stopwatch at the instant that Maria flies past him in her spaceship at a speed of 0.600*c*. At the same instant, Maria starts her stopwatch. (a) As measured in Samir's frame of reference, what is the reading on Maria's stopwatch at the instant that Samir's stopwatch reads 10.0 s? (i) 10.0 s; (ii) less than 10.0 s; (iii) more than 10.0 s. (b) As measured in Maria's frame of reference, what is the reading on Samir's stopwatch at the instant that Maria's stopwatch reads 10.0 s? (i) 10.0 s; (ii) less than 10.0 s; (iii) more than 10.0 s.

Application Who's the Grandmother?

The answer to this question may seem obvious, but it could depend on which person had traveled to a distant planet at relativistic speeds. Imagine that a 20-year-old woman had given birth to a child and then immediately left on a 100-light-year trip (50 light-years out and 50 light-years back) at 99.5% the speed of light. Because of time dilation for the traveler, only 10 years would pass, and she would be 30 years old when she returned, even though 100 years had passed by for people on earth. Meanwhile, the child she left behind at home could have had a baby 20 years after her departure, and this grandchild would now be 80 years old!

37.4 Relativity of Length

Not only does the time interval between two events depend on the observer's frame of reference, but the *distance* between two points may also depend on the observer's frame of reference. The concept of simultaneity is involved. Suppose you want to measure the length of a moving car. One way is to have two assistants make marks on the pavement at the positions of the front and rear bumpers. Then you measure the distance between the marks. But your assistants have to make their marks *at the same time*. If one marks the position of the front bumper at one time and the other marks the position of the rear bumper half a second later, you won't get the car's true length. Since we've learned that simultaneity isn't an absolute concept, we have to proceed with caution.

Mastering PHYSICS

ActivPhysics 17.2: Relativity of Length

Lengths Parallel to the Relative Motion

To develop a relationship between lengths that are measured parallel to the direction of motion in various coordinate systems, we consider another thought experiment. We attach a light source to one end of a ruler and a mirror to the other end. The ruler is at rest in reference frame S', and its length in this frame is l_0 (Fig. 37.10a). Then the time Δt_0 required for a light pulse to make the round trip from source to mirror and back is

$$\Delta t_0 = \frac{2l_0}{c} \tag{37.9}$$

This is a proper time interval because departure and return occur at the same point in S'.

In reference frame S the ruler is moving to the right with speed u during this travel of the light pulse (Fig. 37.10b). The length of the ruler in S is l, and the time of travel from source to mirror, as measured in S, is Δt_1. During this interval

37.10 (a) A ruler is at rest in Mavis's frame S'. A light pulse is emitted from a source at one end of the ruler, reflected by a mirror at the other end, and returned to the source position. (b) Motion of the light pulse as measured in Stanley's frame S.

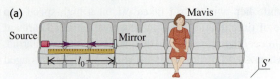

(a)

Source   Mirror   Mavis

l_0

S'

The ruler is stationary in Mavis's frame of reference S'. The light pulse travels a distance l_0 from the light source to the mirror.

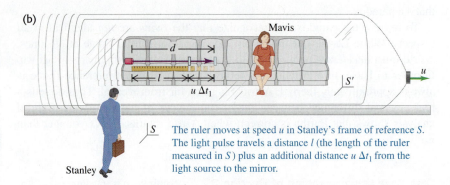

(b)

Mavis

d

l

$u \, \Delta t_1$

S'

u

S The ruler moves at speed u in Stanley's frame of reference S. The light pulse travels a distance l (the length of the ruler measured in S) plus an additional distance $u \, \Delta t_1$ from the light source to the mirror.

Stanley

the ruler, with source and mirror attached, moves a distance $u \, \Delta t_1$. The total length of path d from source to mirror is not l, but rather

$$d = l + u \, \Delta t_1 \tag{37.10}$$

The light pulse travels with speed c, so it is also true that

$$d = c \, \Delta t_1 \tag{37.11}$$

Combining Eqs. (37.10) and (37.11) to eliminate d, we find

$$c \, \Delta t_1 = l + u \, \Delta t_1 \quad \text{or}$$

$$\Delta t_1 = \frac{l}{c - u} \tag{37.12}$$

(Dividing the distance l by $c - u$ does *not* mean that light travels with speed $c - u$, but rather that the distance the pulse travels in S is greater than l.)

In the same way we can show that the time Δt_2 for the return trip from mirror to source is

$$\Delta t_2 = \frac{l}{c + u} \tag{37.13}$$

The *total* time $\Delta t = \Delta t_1 + \Delta t_2$ for the round trip, as measured in S, is

$$\Delta t = \frac{l}{c - u} + \frac{l}{c + u} = \frac{2l}{c(1 - u^2/c^2)} \tag{37.14}$$

We also know that Δt and Δt_0 are related by Eq. (37.6) because Δt_0 is a proper time in S'. Thus Eq. (37.9) for the round-trip time in the rest frame S' of the ruler becomes

$$\Delta t \sqrt{1 - \frac{u^2}{c^2}} = \frac{2l_0}{c} \tag{37.15}$$

Finally, combining Eqs. (37.14) and (37.15) to eliminate Δt and simplifying, we obtain

$$l = l_0 \sqrt{1 - \frac{u^2}{c^2}} = \frac{l_0}{\gamma} \quad \text{(length contraction)} \tag{37.16}$$

[We have used the quantity $\gamma = 1/\sqrt{1 - u^2/c^2}$ defined in Eq. (37.7).] Thus the length l measured in S, in which the ruler is moving, is *shorter* than the length l_0 measured in its rest frame S'.

CAUTION **Length contraction is real** This is *not* an optical illusion! The ruler really is shorter in reference frame S than it is in S'. ▌

A length measured in the frame in which the body is at rest (the rest frame of the body) is called a **proper length**; thus l_0 is a proper length in S', and the length measured in any other frame moving relative to S' is *less than* l_0. This effect is called **length contraction.**

When u is very small in comparison to c, γ approaches 1. Thus in the limit of small speeds we approach the Newtonian relationship $l = l_0$. This and the corresponding result for time dilation show that Eqs. (37.1), the Galilean coordinate transformation, are usually sufficiently accurate for relative speeds much smaller than c. If u is a reasonable fraction of c, however, the quantity $\sqrt{1 - u^2/c^2}$ can be appreciably less than 1. Then l can be substantially smaller than l_0, and the effects of length contraction can be substantial (Fig. 37.11).

Lengths Perpendicular to the Relative Motion

We have derived Eq. (37.16) for lengths measured in the direction *parallel* to the relative motion of the two frames of reference. Lengths that are measured *perpendicular* to the direction of motion are *not* contracted. To prove this, consider two identical meter sticks. One stick is at rest in frame S and lies along the positive y-axis with one end at O, the origin of S. The other is at rest in frame S' and lies along the positive y'-axis with one end at O', the origin of S'. Frame S' moves in the positive x-direction relative to frame S. Observers Stanley and Mavis, at rest in S and S' respectively, station themselves at the 50-cm mark of their sticks. At the instant the two origins coincide, the two sticks lie along the same line. At this instant, Mavis makes a mark on Stanley's stick at the point that coincides with her own 50-cm mark, and Stanley does the same to Mavis's stick.

Suppose for the sake of argument that Stanley observes Mavis's stick as longer than his own. Then the mark Stanley makes on her stick is *below* its center. In that case, Mavis will think Stanley's stick has become shorter, since half of its length coincides with *less* than half her stick's length. So Mavis observes moving sticks getting shorter and Stanley observes them getting longer. But this implies an asymmetry between the two frames that contradicts the basic postulate of relativity that tells us all inertial frames are equivalent. We conclude that consistency with the postulates of relativity requires that both observers measure the rulers as having the *same* length, even though to each observer one of them is stationary and the other is moving (Fig. 37.12). So *there is no length contraction perpendicular to the direction of relative motion of the coordinate systems.* We used this result in our derivation of Eq. (37.6) in assuming that the distance d is the same in both frames of reference.

37.11 The speed at which electrons traverse the 3-km beam line of the SLAC National Accelerator Laboratory is slower than c by less than 1 cm/s. As measured in the reference frame of such an electron, the beam line (which extends from the top to the bottom of this photograph) is only about 15 cm long!

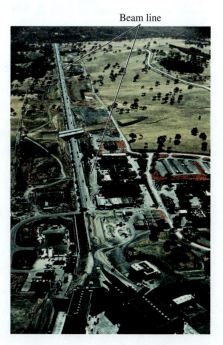

Beam line

37.12 The meter sticks are perpendicular to the relative velocity. For any value of u, both Stanley and Mavis measure either meter stick to have a length of 1 meter.

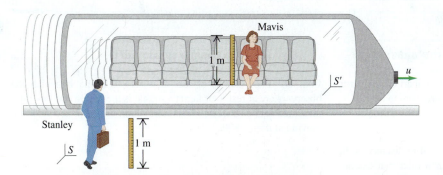

For example, suppose a moving rod of length l_0 makes an angle θ_0 with the direction of relative motion (the x-axis) as measured in its rest frame. Its length component in that frame parallel to the motion, $l_0 \cos\theta_0$, is contracted to $(l_0 \cos\theta_0)/\gamma$. However, its length component perpendicular to the motion, $l_0 \sin\theta_0$, remains the same.

Problem-Solving Strategy 37.2   Length Contraction

IDENTIFY *the relevant concepts:* The concept of length contraction is used whenever we compare the length of an object as measured by observers in different inertial frames of reference.

SET UP *the problem* using the following steps:
1. Decide what defines the length in question. If the problem describes an object such as a ruler, it is just the distance between the ends of the object. If the problem is about a distance between two points in space, it helps to envision an object like a ruler that extends from one point to the other.
2. Identify the target variable.

EXECUTE *the solution* as follows:
1. Determine the reference frame in which the object in question is at rest. In this frame, the length of the object is its proper length l_0. In a second reference frame moving at speed u relative to the first frame, the object has contracted length l.
2. Keep in mind that length contraction occurs only for lengths parallel to the direction of relative motion of the two frames. Any length that is perpendicular to the relative motion is the same in both frames.
3. Use Eq. (37.16) to relate l and l_0, and then solve for the target variable.

EVALUATE *your answer:* Check that your answers make sense: l is never larger than l_0, and u is never greater than c.

Example 37.4   How long is the spaceship?

A spaceship flies past earth at a speed of $0.990c$. A crew member on board the spaceship measures its length, obtaining the value 400 m. What length do observers measure on earth?

SOLUTION

IDENTIFY and SET UP: This problem is about the nose-to-tail length of the spaceship as measured on the spaceship and on earth. This length is along the direction of relative motion (Fig. 37.13), so there will be length contraction. The spaceship's 400-m length is the *proper* length l_0 because it is measured in the frame in which the spaceship is at rest. Our target variable is the length l measured in the earth frame, relative to which the spaceship is moving at $u = 0.990c$.

EXECUTE: From Eq. (37.16), the length in the earth frame is

$$l = l_0 \sqrt{1 - \frac{u^2}{c^2}} = (400 \text{ m}) \sqrt{1 - (0.990)^2} = 56.4 \text{ m}$$

EVALUATE: The spaceship is shorter in a frame in which it is in motion than in a frame in which it is at rest. To measure the length l, two earth observers with synchronized clocks could measure the

37.13 Measuring the length of a moving spaceship.

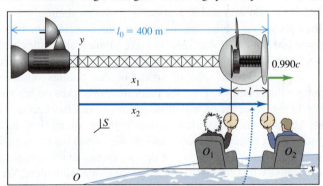

The two observers on earth (S) must measure x_2 and x_1 simultaneously to obtain the correct length $l = x_2 - x_1$ in their frame of reference.

positions of the two ends of the spaceship simultaneously in the earth's reference frame, as shown in Fig. 37.13. (These two measurements will *not* appear simultaneous to an observer in the spaceship.)

Example 37.5   How far apart are the observers?

Observers O_1 and O_2 in Fig. 37.13 are 56.4 m apart on the earth. How far apart does the spaceship crew measure them to be?

SOLUTION

IDENTIFY and SET UP: In this example the 56.4-m distance is the *proper* length l_0. It represents the length of a ruler that extends from O_1 to O_2 and is at rest in the earth frame in which the observers are at rest. Our target variable is the length l of this ruler measured in the spaceship frame, in which the earth and ruler are moving at $u = 0.990c$.

EXECUTE: As in Example 37.4, but with $l_0 = 56.4$ m,

$$l = l_0\sqrt{1 - \frac{u^2}{c^2}} = (56.4 \text{ m})\sqrt{1 - (0.990)^2} = 7.96 \text{ m}$$

EVALUATE: This answer does *not* say that the crew measures their spaceship to be both 400 m long and 7.96 m long. As measured on earth, the tail of the spacecraft is at the position of O_1 at the same instant that the nose of the spacecraft is at the position of O_2. Hence the length of the spaceship measured on earth equals the 56.4-m distance between O_1 and O_2. But in the spaceship frame O_1 and O_2 are only 7.96 m apart, and the nose (which is 400 m in front of the tail) passes O_2 before the tail passes O_1.

How an Object Moving Near c Would Appear

Let's think a little about the visual appearance of a moving three-dimensional body. If we could see the positions of all points of the body simultaneously, it would appear to shrink only in the direction of motion. But we *don't* see all the points simultaneously; light from points farther from us takes longer to reach us than does light from points near to us, so we see the farther points at the positions they had at earlier times.

Suppose we have a rectangular rod with its faces parallel to the coordinate planes. When we look end-on at the center of the closest face of such a rod at rest, we see only that face. (See the center rod in computer-generated Fig. 37.14a.) But when that rod is moving past us toward the right at an appreciable fraction of the speed of light, we may also see its left side because of the earlier-time effect just described. That is, we can see some points that we couldn't see when the rod was at rest because the rod moves out of the way of the light rays from those points to us. Conversely, some light that can get to us when the rod is at rest is blocked by the moving rod. Because of all this, the rods in Figs. 37.14b and 37.14c appear rotated and distorted.

Test Your Understanding of Section 37.4 A miniature spaceship is flying past you, moving horizontally at a substantial fraction of the speed of light. At a certain instant, you observe that the nose and tail of the spaceship align exactly with the two ends of a meter stick that you hold in your hands. Rank the following distances in order from longest to shortest: (i) the proper length of the meter stick; (ii) the proper length of the spaceship; (iii) the length of the spaceship measured in your frame of reference; (iv) the length of the meter stick measured in the spaceship's frame of reference.

37.14 Computer simulation of the appearance of an array of 25 rods with square cross section. The center rod is viewed end-on. The simulation ignores color changes in the array caused by the Doppler effect (see Section 37.6).

(a) Array at rest

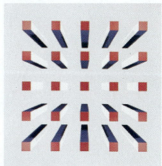

(b) Array moving to the right at 0.2c

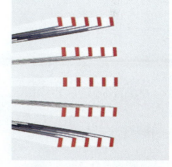

(c) Array moving to the right at 0.9c

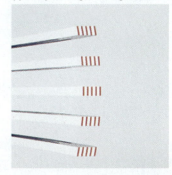

37.5 The Lorentz Transformations

In Section 37.1 we discussed the Galilean coordinate transformation equations, Eqs. (37.1). They relate the coordinates (x, y, z) of a point in frame of reference S to the coordinates (x', y', z') of the point in a second frame S'. The second frame moves with constant speed u relative to S in the positive direction along the common x-x'-axis. This transformation also assumes that the time scale is the same in the two frames of reference, as expressed by the additional relationship $t = t'$. This Galilean transformation, as we have seen, is valid only in the limit when u approaches zero. We are now ready to derive more general transformations that are consistent with the principle of relativity. The more general relationships are called the **Lorentz transformations.**

The Lorentz Coordinate Transformation

Our first question is this: When an event occurs at point (x, y, z) at time t, as observed in a frame of reference S, what are the coordinates (x', y', z') and time t' of the event as observed in a second frame S' moving relative to S with constant speed u in the $+x$-direction?

To derive the coordinate transformation, we refer to Fig. 37.15 (next page), which is the same as Fig. 37.3. As before, we assume that the origins coincide at the initial time $t = 0 = t'$. Then in S the distance from O to O' at time t is

37.15 As measured in frame of reference S, x' is contracted to x'/γ, so $x = ut + x'/\gamma$ and $x' = \gamma(x - ut)$.

Frame S' moves relative to frame S with constant velocity u along the common x-x'-axis.

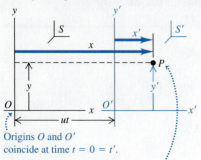

Origins O and O' coincide at time $t = 0 = t'$.

The Lorentz coordinate transformation relates the spacetime coordinates of an event as measured in the two frames: (x, y, z, t) in frame S and (x', y', z', t') in frame S'.

still ut. The coordinate x' is a *proper length* in S', so in S it is contracted by the factor $1/\gamma = \sqrt{1 - u^2/c^2}$, as in Eq. (37.16). Thus the distance x from O to P, as seen in S, is not simply $x = ut + x'$, as in the Galilean coordinate transformation, but

$$x = ut + x'\sqrt{1 - \frac{u^2}{c^2}} \tag{37.17}$$

Solving this equation for x', we obtain

$$x' = \frac{x - ut}{\sqrt{1 - u^2/c^2}} \tag{37.18}$$

Equation (37.18) is part of the Lorentz coordinate transformation; another part is the equation giving t' in terms of x and t. To obtain this, we note that the principle of relativity requires that the *form* of the transformation from S to S' be identical to that from S' to S. The only difference is a change in the sign of the relative velocity component u. Thus from Eq. (37.17) it must be true that

$$x' = -ut' + x\sqrt{1 - \frac{u^2}{c^2}} \tag{37.19}$$

We now equate Eqs. (37.18) and (37.19) to eliminate x'. This gives us an equation for t' in terms of x and t. We leave the algebraic details for you to work out; the result is

$$t' = \frac{t - ux/c^2}{\sqrt{1 - u^2/c^2}} \tag{37.20}$$

As we discussed previously, lengths perpendicular to the direction of relative motion are not affected by the motion, so $y' = y$ and $z' = z$.

Collecting all these transformation equations, we have

$$x' = \frac{x - ut}{\sqrt{1 - u^2/c^2}} = \gamma(x - ut)$$

$$y' = y$$

$$z' = z \qquad \text{(Lorentz coordinate transformation)} \tag{37.21}$$

$$t' = \frac{t - ux/c^2}{\sqrt{1 - u^2/c^2}} = \gamma(t - ux/c^2)$$

These equations are the *Lorentz coordinate transformation,* the relativistic generalization of the Galilean coordinate transformation, Eqs. (37.1) and $t = t'$. For values of u that approach zero, the radicals in the denominators and γ approach 1, and the ux/c^2 term approaches zero. In this limit, Eqs. (37.21) become identical to Eqs. (37.1) along with $t = t'$. In general, though, both the coordinates and time of an event in one frame depend on its coordinates and time in another frame. *Space and time have become intertwined; we can no longer say that length and time have absolute meanings independent of the frame of reference.* For this reason, we refer to time and the three dimensions of space collectively as a four-dimensional entity called **spacetime,** and we call (x, y, z, t) together the **spacetime coordinates** of an event.

The Lorentz Velocity Transformation

We can use Eqs. (37.21) to derive the relativistic generalization of the Galilean velocity transformation, Eq. (37.2). We consider only one-dimensional motion along the x-axis and use the term "velocity" as being short for the "x-component of the velocity." Suppose that in a time dt a particle moves a distance dx, as measured in frame S. We obtain the corresponding distance dx' and time dt' in S' by taking differentials of Eqs. (37.21):

$$dx' = \gamma(dx - u\,dt)$$

$$dt' = \gamma(dt - u\,dx/c^2)$$

We divide the first equation by the second and then divide the numerator and denominator of the result by dt to obtain

$$\frac{dx'}{dt'} = \frac{\dfrac{dx}{dt} - u}{1 - \dfrac{u}{c^2}\dfrac{dx}{dt}}$$

Now dx/dt is the velocity v_x in S, and dx'/dt' is the velocity v'_x in S', so we finally obtain the relativistic generalization

$$v'_x = \frac{v_x - u}{1 - uv_x/c^2} \qquad \text{(Lorentz velocity transformation)} \qquad (37.22)$$

When u and v_x are much smaller than c, the denominator in Eq. (37.22) approaches 1, and we approach the nonrelativistic result $v'_x = v_x - u$. The opposite extreme is the case $v_x = c$; then we find

$$v'_x = \frac{c - u}{1 - uc/c^2} = \frac{c(1 - u/c)}{1 - u/c} = c$$

This says that anything moving with velocity $v_x = c$ measured in S also has velocity $v'_x = c$ measured in S', despite the relative motion of the two frames. So Eq. (37.22) is consistent with Einstein's postulate that the speed of light in vacuum is the same in all inertial frames of reference.

The principle of relativity tells us there is no fundamental distinction between the two frames S and S'. Thus the expression for v_x in terms of v'_x must have the same form as Eq. (37.22), with v_x changed to v'_x, and vice versa, and the sign of u reversed. Carrying out these operations with Eq. (37.22), we find

$$v_x = \frac{v'_x + u}{1 + uv'_x/c^2} \qquad \text{(Lorentz velocity transformation)} \qquad (37.23)$$

This can also be obtained algebraically by solving Eq. (37.22) for v_x. Both Eqs. (37.22) and (37.23) are *Lorentz velocity transformations* for one-dimensional motion.

CAUTION **Use the correct reference frame coordinates** Keep in mind that the Lorentz transformation equations given by Eqs. (37.21), (37.22), and (37.23) assume that frame S' is moving in the positive x-direction with velocity u relative to frame S. You should always set up your coordinate system to follow this convention. ▮

When u is less than c, the Lorentz velocity transformations show us that a body moving with a speed less than c in one frame of reference always has a speed less than c in *every other* frame of reference. This is one reason for concluding that no material body may travel with a speed equal to or greater than that of light in vacuum, relative to *any* inertial frame of reference. The relativistic generalizations of energy and momentum, which we will explore later, give further support to this hypothesis.

Problem-Solving Strategy 37.3 | Lorentz Transformations

IDENTIFY *the relevant concepts:* The Lorentz *coordinate* transformation equations relate the spacetime coordinates of an event in one inertial reference frame to the coordinates of the same event in a second inertial frame. The Lorentz *velocity* transformation equations relate the velocity of an object in one inertial reference frame to its velocity in a second inertial frame.

SET UP *the problem* using the following steps:
1. Identify the target variable.
2. Define the two inertial frames S and S'. Remember that S' moves relative to S at a constant velocity u in the $+x$-direction.
3. If the coordinate transformation equations are needed, make a list of spacetime coordinates in the two frames, such as x_1, x_1', t_1, t_1', and so on. Label carefully which of these you know and which you don't.
4. In velocity-transformation problems, clearly identify u (the relative velocity of the two frames of reference), v_x (the velocity of the object relative to S), and v_x' (the velocity of the object relative to S').

EXECUTE *the solution* as follows:
1. In a coordinate-transformation problem, use Eqs. (37.21) to solve for the spacetime coordinates of the event as measured in S' in terms of the corresponding values in S. (If you need to solve for the spacetime coordinates in S in terms of the corresponding values in S', you can easily convert the expressions in Eqs. (37.21): Replace all of the primed quantities with unprimed ones, and vice versa, and replace u with $-u$.)
2. In a velocity-transformation problem, use either Eq. (37.22) or Eq. (37.23), as appropriate, to solve for the target variable.

EVALUATE *your answer:* Don't be discouraged if some of your results don't seem to make sense or if they disagree with "common sense." It takes time to develop intuition about relativity; you'll gain it with experience.

Example 37.6 | Was it received before it was sent?

Winning an interstellar race, Mavis pilots her spaceship across a finish line in space at a speed of $0.600c$ relative to that line. A "hooray" message is sent from the back of her ship (event 2) at the instant (in her frame of reference) that the front of her ship crosses the line (event 1). She measures the length of her ship to be 300 m. Stanley is at the finish line and is at rest relative to it. When and where does he measure events 1 and 2 to occur?

SOLUTION

IDENTIFY and SET UP: This example involves the Lorentz coordinate transformation. Our derivation of this transformation assumes that the origins of frames S and S' coincide at $t = 0 = t'$. Thus for simplicity we fix the origin of S at the finish line and the origin of S' at the front of the spaceship so that Stanley and Mavis measure event 1 to be at $x = 0 = x'$ and $t = 0 = t'$.

Mavis in S' measures her spaceship to be 300 m long, so she has the "hooray" sent from 300 m behind her spaceship's front at the instant she measures the front to cross the finish line. That is, she measures event 2 at $x' = -300$ m and $t' = 0$.

Our target variables are the coordinate x and time t of event 2 that Stanley measures in S.

EXECUTE: To solve for the target variables, we modify the first and last of Eqs. (37.21) to give x and t as functions of x' and t'. We do so in the same way that we obtained Eq. (37.23) from Eq. (37.22). We remove the primes from x' and t', add primes to x and t, and replace each u with $-u$. The results are

$$x = \gamma(x' + ut') \quad \text{and} \quad t = \gamma(t' + ux'/c^2)$$

From Eq. (37.7), $\gamma = 1.25$ for $u = 0.600c = 1.80 \times 10^8$ m/s. We also substitute $x' = -300$ m, $t' = 0$, $c = 3.00 \times 10^8$ m/s, and $u = 1.80 \times 10^8$ m/s in the equations for x and t to find $x = -375$ m at $t = -7.50 \times 10^{-7}$ s $= -0.750$ μs for event 2.

EVALUATE: Mavis says that the events are simultaneous, but Stanley says that the "hooray" was sent *before* Mavis crossed the finish line. This does not mean that the effect preceded the cause. The fastest that Mavis can send a signal the length of her ship is 300 m/$(3.00 \times 10^8$ m/s$) = 1.00$ μs. She cannot send a signal from the front at the instant it crosses the finish line that would cause a "hooray" to be broadcast from the back at the same instant. She would have to send that signal from the front at least 1.00 μs before then, so she had to slightly anticipate her success.

Example 37.7 | Relative velocities

(a) A spaceship moving away from the earth at $0.900c$ fires a robot space probe in the same direction as its motion at $0.700c$ relative to the spaceship. What is the probe's velocity relative to the earth? (b) A scoutship is sent to catch up with the spaceship by traveling at $0.950c$ relative to the earth. What is the velocity of the scoutship relative to the spaceship?

SOLUTION

IDENTIFY and SET UP: This example uses the Lorentz velocity transformation. Let the earth and spaceship reference frames be S and S', respectively (Fig. 37.16); their relative velocity is $u = 0.900c$. In part (a) we are given the probe velocity $v_x' = 0.700c$ with respect to S', and the target variable is the velocity v_x of the

37.16 The spaceship, robot space probe, and scoutship.

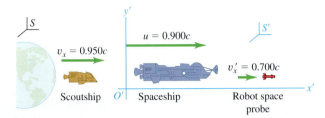

probe relative to S. In part (b) we are given the velocity $v_x = 0.950c$ of the scoutship relative to S, and the target variable is its velocity v'_x relative to S'.

EXECUTE: (a) We use Eq. (37.23) to find the probe velocity relative to the earth:

$$v_x = \frac{v'_x + u}{1 + uv'_x/c^2} = \frac{0.700c + 0.900c}{1 + (0.900c)(0.700c)/c^2} = 0.982c$$

(b) We use Eq. (37.22) to find the scoutship velocity relative to the spaceship:

$$v'_x = \frac{v_x - u}{1 - uv_x/c^2} = \frac{0.950c - 0.900c}{1 - (0.900c)(0.950c)/c^2} = 0.345c$$

EVALUATE: What would the Galilean velocity transformation formula, Eq. (37.2), say? In part (a) we would have found the probe's velocity relative to the earth to be $v_x = v'_x + u = 0.700c + 0.900c = 1.600c$, which is greater than c and hence impossible. In part (b), we would have found the scoutship's velocity relative to the spaceship to be $v'_x = v_x - u = 0.950c - 0.900c = 0.050c$; the relativistically correct value, $v'_x = 0.345c$, is almost seven times greater than the incorrect Galilean value.

Test Your Understanding of Section 37.5 (a) In frame S events P_1 and P_2 occur at the same x-, y-, and z-coordinates, but event P_1 occurs before event P_2. In frame S', which event occurs first? (b) In frame S events P_3 and P_4 occur at the same time t and the same y- and z-coordinates, but event P_3 occurs at a less positive x-coordinate than event P_4. In frame S', which event occurs first?   ❙

37.6 The Doppler Effect for Electromagnetic Waves

An additional important consequence of relativistic kinematics is the Doppler effect for electromagnetic waves. In our previous discussion of the Doppler effect (see Section 16.8) we quoted without proof the formula, Eq. (16.30), for the frequency shift that results from motion of a source of electromagnetic waves relative to an observer. We can now derive that result.

Here's a statement of the problem. A source of light is moving with constant speed u toward Stanley, who is stationary in an inertial frame (Fig. 37.17). As measured in its rest frame, the source emits light waves with frequency f_0 and period $T_0 = 1/f_0$. What is the frequency f of these waves as received by Stanley?

Let T be the time interval between *emission* of successive wave crests as observed in Stanley's reference frame. Note that this is *not* the interval between the *arrival* of successive crests at his position, because the crests are emitted at different points in Stanley's frame. In measuring only the frequency f he receives, he does not take into account the difference in transit times for successive crests. Therefore the frequency he receives is *not* $1/T$. What is the equation for f?

During a time T the crests ahead of the source move a distance cT, and the source moves a shorter distance uT in the same direction. The distance λ between

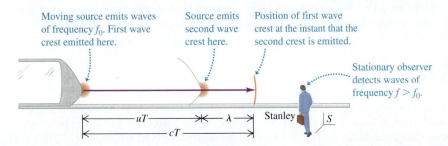

Moving source emits waves of frequency f_0. First wave crest emitted here.

Source emits second wave crest here.

Position of first wave crest at the instant that the second crest is emitted.

Stationary observer detects waves of frequency $f > f_0$.

Stanley

37.17 The Doppler effect for light. A light source moving at speed u relative to Stanley emits a wave crest, then travels a distance uT toward an observer and emits the next crest. In Stanley's reference frame S, the second crest is a distance λ behind the first crest.

successive crests—that is, the wavelength—is thus $\lambda = (c - u)T$, as measured in Stanley's frame. The frequency that he measures is c/λ. Therefore

$$f = \frac{c}{(c - u)T} \tag{37.24}$$

So far we have followed a pattern similar to that for the Doppler effect for sound from a moving source (see Section 16.8). In that discussion our next step was to equate T to the time T_0 between emissions of successive wave crests by the source. However, due to time dilation it is *not* relativistically correct to equate T to T_0. The time T_0 is measured in the rest frame of the source, so it is a proper time. From Eq. (37.6), T_0 and T are related by

$$T = \frac{T_0}{\sqrt{1 - u^2/c^2}} = \frac{cT_0}{\sqrt{c^2 - u^2}}$$

or, since $T_0 = 1/f_0$,

$$\frac{1}{T} = \frac{\sqrt{c^2 - u^2}}{cT_0} = \frac{\sqrt{c^2 - u^2}}{c}f_0$$

Remember, $1/T$ is not equal to f. We must substitute this expression for $1/T$ into Eq. 37.24 to find f:

$$f = \frac{c}{c - u}\frac{\sqrt{c^2 - u^2}}{c}f_0$$

Using $c^2 - u^2 = (c - u)(c + u)$ gives

$$f = \sqrt{\frac{c + u}{c - u}}f_0 \qquad \text{(Doppler effect, electromagnetic waves, source approaching observer)} \tag{37.25}$$

This shows that when the source moves *toward* the observer, the observed frequency f is *greater* than the emitted frequency f_0. The difference $f - f_0 = \Delta f$ is called the Doppler frequency shift. When u/c is much smaller than 1, the fractional shift $\Delta f/f$ is also small and is approximately equal to u/c:

$$\frac{\Delta f}{f} = \frac{u}{c}$$

When the source moves *away from* the observer, we change the sign of u in Eq. (37.25) to get

$$f = \sqrt{\frac{c - u}{c + u}}f_0 \qquad \text{(Doppler effect, electromagnetic waves, source moving away from observer)} \tag{37.26}$$

This agrees with Eq. (16.30), which we quoted previously, with minor notation changes.

With light, unlike sound, there is no distinction between motion of source and motion of observer; only the *relative* velocity of the two is significant. The last four paragraphs of Section 16.8 discuss several practical applications of the Doppler effect with light and other electromagnetic radiation; we suggest you review those paragraphs now. Figure 37.18 shows one common application.

37.18 This handheld radar gun emits a radio beam of frequency f_0, which in the frame of reference of an approaching car has a higher frequency f given by Eq. (37.25). The reflected beam also has frequency f in the car's frame, but has an even higher frequency f' in the police officer's frame. The radar gun calculates the car's speed by comparing the frequencies of the emitted beam and the doubly Doppler-shifted reflected beam. (Compare Example 16.18 in Section 16.8.)

Example 37.8 A jet from a black hole

Many galaxies have supermassive black holes at their centers (see Section 13.8). As material swirls around such a black hole, it is heated, becomes ionized, and generates strong magnetic fields. The resulting magnetic forces steer some of the material into high-speed jets that blast out of the galaxy and into intergalactic space (Fig. 37.19). The light we observe from the jet in Fig. 37.19 has a

37.19 This image shows a fast-moving jet 5000 light-years in length emanating from the center of the galaxy M87. The light from the jet is emitted by fast-moving electrons spiraling around magnetic field lines (see Fig. 27.18).

frequency of 6.66×10^{14} Hz (in the far ultraviolet region of the electromagnetic spectrum; see Fig. 32.4), but in the reference frame of the jet material the light has a frequency of 5.55×10^{13} Hz (in the infrared). What is the speed of the jet material with respect to us?

SOLUTION

IDENTIFY and SET UP: This problem involves the Doppler effect for electromagnetic waves. The frequency we observe is $f = 6.66 \times 10^{14}$ Hz, and the frequency in the frame of the source is $f_0 = 5.55 \times 10^{13}$ Hz. Since $f > f_0$, the jet is approaching us and we use Eq. (37.25) to find the target variable u.

EXECUTE: We need to solve Eq. (37.25) for u. We'll leave it as an exercise for you to show that the result is

$$u = \frac{(f/f_0)^2 - 1}{(f/f_0)^2 + 1} c$$

We have $f/f_0 = (6.66 \times 10^{14} \text{ Hz})/(5.55 \times 10^{13} \text{ Hz}) = 12.0$, so

$$u = \frac{(12.0)^2 - 1}{(12.0)^2 + 1} c = 0.986c$$

EVALUATE: Because the frequency shift is quite substantial, it would have been erroneous to use the approximate expression $\Delta f/f = u/c$. Had you done so, you would have found $u = c(\Delta f/f_0) = c(6.66 \times 10^{14} \text{ Hz} - 5.55 \times 10^{13} \text{ Hz})/(5.55 \times 10^{13} \text{ Hz}) = 11.0c$. This result cannot be correct because the jet material cannot travel faster than light.

37.7 Relativistic Momentum

Newton's laws of motion have the same form in all inertial frames of reference. When we use transformations to change from one inertial frame to another, the laws should be *invariant* (unchanging). But we have just learned that the principle of relativity forces us to replace the Galilean transformations with the more general Lorentz transformations. As we will see, this requires corresponding generalizations in the laws of motion and the definitions of momentum and energy.

The principle of conservation of momentum states that *when two bodies interact, the total momentum is constant,* provided that the net external force acting on the bodies in an inertial reference frame is zero (for example, if they form an isolated system, interacting only with each other). If conservation of momentum is a valid physical law, it must be valid in *all* inertial frames of reference. Now, here's the problem: Suppose we look at a collision in one inertial coordinate system S and find that momentum is conserved. Then we use the Lorentz transformation to obtain the velocities in a second inertial system S'. We find that if we use the Newtonian definition of momentum ($\vec{p} = m\vec{v}$), momentum is *not* conserved in the second system! If we are convinced that the principle of relativity and the Lorentz transformation are correct, the only way to save momentum conservation is to generalize the *definition* of momentum.

We won't derive the correct relativistic generalization of momentum, but here is the result. Suppose we measure the mass of a particle to be m when it is at rest relative to us: We often call m the **rest mass.** We will use the term *material particle* for a particle that has a nonzero rest mass. When such a particle has a velocity $\vec{v}$, its **relativistic momentum** $\vec{p}$ is

$$\vec{p} = \frac{m\vec{v}}{\sqrt{1 - v^2/c^2}} \qquad \text{(relativistic momentum)} \qquad (37.27)$$

37.20 Graph of the magnitude of the momentum of a particle of rest mass m as a function of speed v. Also shown is the Newtonian prediction, which gives correct results only at speeds much less than c.

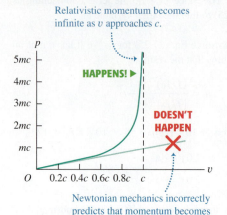

Relativistic momentum becomes infinite as v approaches c.

HAPPENS! ▶

DOESN'T HAPPEN

Newtonian mechanics incorrectly predicts that momentum becomes infinite only if v becomes infinite.

When the particle's speed v is much less than c, this is approximately equal to the Newtonian expression $\vec{p} = m\vec{v}$, but in general the momentum is greater in magnitude than mv (Fig. 37.20). In fact, as v approaches c, the momentum approaches infinity.

Relativity, Newton's Second Law, and Relativistic Mass

What about the relativistic generalization of Newton's second law? In Newtonian mechanics the most general form of the second law is

$$\vec{F} = \frac{d\vec{p}}{dt} \tag{37.28}$$

That is, the net force $\vec{F}$ on a particle equals the time rate of change of its momentum. Experiments show that this result is still valid in relativistic mechanics, provided that we use the relativistic momentum given by Eq. 37.27. That is, the relativistically correct generalization of Newton's second law is

$$\vec{F} = \frac{d}{dt} \frac{m\vec{v}}{\sqrt{1 - v^2/c^2}} \tag{37.29}$$

Because momentum is no longer directly proportional to velocity, the rate of change of momentum is no longer directly proportional to the acceleration. As a result, *constant force does not cause constant acceleration*. For example, when the net force and the velocity are both along the x-axis, Eq. 37.29 gives

$$F = \frac{m}{(1 - v^2/c^2)^{3/2}} a \qquad (\vec{F} \text{ and } \vec{v} \text{ along the same line}) \tag{37.30}$$

where a is the acceleration, also along the x-axis. Solving Eq. (37.30) for the acceleration a gives

$$a = \frac{F}{m} \left(1 - \frac{v^2}{c^2} \right)^{3/2}$$

We see that as a particle's speed increases, the acceleration caused by a given force continuously *decreases*. As the speed approaches c, the acceleration approaches zero, no matter how great a force is applied. Thus it is impossible to accelerate a particle with nonzero rest mass to a speed equal to or greater than c. We again see that the speed of light in vacuum represents an ultimate speed limit.

Equation (37.27) for relativistic momentum is sometimes interpreted to mean that a rapidly moving particle undergoes an increase in mass. If the mass at zero velocity (the rest mass) is denoted by m, then the "relativistic mass" m_{rel} is given by

$$m_{rel} = \frac{m}{\sqrt{1 - v^2/c^2}}$$

Indeed, when we consider the motion of a system of particles (such as rapidly moving ideal-gas molecules in a stationary container), the total rest mass of the system is the sum of the relativistic masses of the particles, not the sum of their rest masses.

However, if blindly applied, the concept of relativistic mass has its pitfalls. As Eq. (37.29) shows, the relativistic generalization of Newton's second law is *not* $\vec{F} = m_{rel}\vec{a}$, and we will show in Section 37.8 that the relativistic kinetic energy of a particle is *not* $K = \frac{1}{2}m_{rel}v^2$. The use of relativistic mass has its supporters and detractors, some quite strong in their opinions. We will mostly deal with individual particles, so we will sidestep the controversy and use Eq. (37.27) as the generalized definition of momentum with m as a constant for each particle, independent of its state of motion.

We will use the abbreviation

$$\gamma = \frac{1}{\sqrt{1 - v^2/c^2}}$$

We used this abbreviation in Section 37.3 with v replaced by u, the relative speed of two coordinate systems. Here v is the speed of a particle in a particular coordinate system—that is, the speed of the particle's *rest frame* with respect to that system. In terms of γ, Eqs. (37.27) and (37.30) become

$$\vec{p} = \gamma m \vec{v} \qquad \text{(relativistic momentum)} \qquad (37.31)$$

$$F = \gamma^3 m a \qquad (\vec{F} \text{ and } \vec{v} \text{ along the same line}) \qquad (37.32)$$

In linear accelerators (used in medicine as well as nuclear and elementary-particle physics; see Fig. 37.11) the net force $\vec{F}$ and the velocity $\vec{v}$ of the accelerated particle are along the same straight line. But for much of the path in most *circular* accelerators the particle moves in uniform circular motion at constant speed v. Then the net force and velocity are perpendicular, so the force can do no work on the particle and the kinetic energy and speed remain constant. Thus the denominator in Eq. (37.29) is constant, and we obtain

$$F = \frac{m}{(1 - v^2/c^2)^{1/2}} a = \gamma m a \qquad (\vec{F} \text{ and } \vec{v} \text{ perpendicular}) \qquad (37.33)$$

Recall from Section 3.4 that if the particle moves in a circle, the net force and acceleration are directed inward along the radius r, and $a = v^2/r$.

What about the general case in which $\vec{F}$ and $\vec{v}$ are neither along the same line nor perpendicular? Then we can resolve the net force $\vec{F}$ at any instant into components parallel to and perpendicular to $\vec{v}$. The resulting acceleration will have corresponding components obtained from Eqs. (37.32) and (37.33). Because of the different γ^3 and γ factors, the acceleration components will not be proportional to the net force components. That is, *unless the net force on a relativistic particle is either along the same line as the particle's velocity or perpendicular to it, the net force and acceleration vectors are not parallel.*

Example 37.9   Relativistic dynamics of an electron

An electron (rest mass 9.11×10^{-31} kg, charge -1.60×10^{-19} C) is moving opposite to an electric field of magnitude $E = 5.00 \times 10^5$ N/C. All other forces are negligible in comparison to the electric-field force. (a) Find the magnitudes of momentum and of acceleration at the instants when $v = 0.010c$, $0.90c$, and $0.99c$. (b) Find the corresponding accelerations if a net force of the same magnitude is perpendicular to the velocity.

SOLUTION

IDENTIFY and SET UP: In addition to the expressions from this section for relativistic momentum and acceleration, we need the relationship between electric force and electric field from Chapter 21. In part (a) we use Eq. (37.31) to determine the magnitude of momentum; the force acts along the same line as the velocity, so we use Eq. (37.32) to determine the magnitude of acceleration. In part (b) the force is perpendicular to the velocity, so we use Eq. (37.33) rather than Eq. (37.32).

EXECUTE: (a) For $v = 0.010c$, $0.90c$, and $0.99c$ we have $\gamma = \sqrt{1 - v^2/c^2} = 1.00$, 2.29, and 7.09, respectively. The values of the momentum magnitude $p = \gamma m v$ are

$$p_1 = (1.00)(9.11 \times 10^{-31} \text{ kg})(0.010)(3.00 \times 10^8 \text{m/s})$$
$$= 2.7 \times 10^{-24} \text{ kg} \cdot \text{m/s at } v_1 = 0.010c$$

$$p_2 = (2.29)(9.11 \times 10^{-31} \text{ kg})(0.90)(3.00 \times 10^8 \text{ m/s})$$
$$= 5.6 \times 10^{-22} \text{ kg} \cdot \text{m/s at } v_2 = 0.90c$$

$$p_3 = (7.09)(9.11 \times 10^{-31} \text{ kg})(0.99)(3.00 \times 10^8 \text{ m/s})$$
$$= 1.9 \times 10^{-21} \text{ kg} \cdot \text{m/s at } v_3 = 0.99c$$

From Eq. (21.4), the magnitude of the force on the electron is

$$F = |q|E = (1.60 \times 10^{-19} \text{ C})(5.00 \times 10^5 \text{ N/C})$$
$$= 8.00 \times 10^{-14} \text{ N}$$

Continued

From Eq. (37.32), $a = F/\gamma^3 m$. For $v = 0.010c$ and $\gamma = 1.00$,

$$a_1 = \frac{8.00 \times 10^{-14}\,\text{N}}{(1.00)^3(9.11 \times 10^{-31}\,\text{kg})} = 8.8 \times 10^{16}\,\text{m/s}^2$$

The accelerations at the two higher speeds are smaller than the non-relativistic value by factors of $\gamma^3 = 12.0$ and 356, respectively:

$$a_2 = 7.3 \times 10^{15}\,\text{m/s}^2 \qquad a_3 = 2.5 \times 10^{14}\,\text{m/s}^2$$

(b) From Eq. (37.33), $a = F/\gamma m$ if $\vec{F}$ and $\vec{v}$ are perpendicular. When $v = 0.010c$ and $\gamma = 1.00$,

$$a_1 = \frac{8.00 \times 10^{-14}\,\text{N}}{(1.00)(9.11 \times 10^{-31}\,\text{kg})} = 8.8 \times 10^{16}\,\text{m/s}^2$$

Now the accelerations at the two higher speeds are smaller by factors of $\gamma = 2.29$ and 7.09, respectively:

$$a_2 = 3.8 \times 10^{16}\,\text{m/s}^2 \qquad a_3 = 1.2 \times 10^{16}\,\text{m/s}^2$$

These accelerations are larger than the corresponding ones in part (a) by factors of γ^2.

EVALUATE: Our results in part (a) show that at higher speeds, the relativistic values of momentum differ more and more from the nonrelativistic values calculated from $p = mv$. The momentum at $0.99c$ is more than three times as great as at $0.90c$ because of the increase in the factor γ. Our results also show that the acceleration drops off very quickly as v approaches c.

Test Your Understanding of Section 37.7   According to relativistic mechanics, when you double the speed of a particle, the magnitude of its momentum increases by (i) a factor of 2; (ii) a factor greater than 2; (iii) a factor between 1 and 2 that depends on the mass of the particle. ❙

37.8 Relativistic Work and Energy

When we developed the relationship between work and kinetic energy in Chapter 6, we used Newton's laws of motion. When we generalize these laws according to the principle of relativity, we need a corresponding generalization of the equation for kinetic energy.

Relativistic Kinetic Energy

We use the work–energy theorem, beginning with the definition of work. When the net force and displacement are in the same direction, the work done by that force is $W = \int F\,dx$. We substitute the expression for F from Eq. (37.30), the applicable relativistic version of Newton's second law. In moving a particle of rest mass m from point x_1 to point x_2,

$$W = \int_{x_1}^{x_2} F\,dx = \int_{x_1}^{x_2} \frac{ma\,dx}{(1 - v^2/c^2)^{3/2}} \tag{37.34}$$

To derive the generalized expression for kinetic energy K as a function of speed v, we would like to convert this to an integral on v. To do this, first remember that the kinetic energy of a particle equals the net work done on it in moving it from rest to the speed v: $K = W$. Thus we let the speeds be zero at point x_1 and v at point x_2. So as not to confuse the variable of integration with the final speed, we change v to v_x in Eq. 37.34. That is, v_x is the varying x-component of the velocity of the particle as the net force accelerates it from rest to a speed v. We also realize that dx and dv_x are the infinitesimal changes in x and v_x, respectively, in the time interval dt. Because $v_x = dx/dt$ and $a = dv_x/dt$, we can rewrite $a\,dx$ in Eq. (37.34) as

$$a\,dx = \frac{dv_x}{dt}\,dx = dx\frac{dv_x}{dt} = \frac{dx}{dt}\,dv_x = v_x dv_x$$

Making these substitutions gives us

$$K = W = \int_0^v \frac{mv_x dv_x}{(1 - v_x^2/c^2)^{3/2}} \tag{37.35}$$

We can evaluate this integral by a simple change of variable; the final result is

$$K = \frac{mc^2}{\sqrt{1 - v^2/c^2}} - mc^2 = (\gamma - 1)mc^2 \qquad \text{(relativistic kinetic energy)} \tag{37.36}$$

As v approaches c, the kinetic energy approaches infinity. If Eq. (37.36) is correct, it must also approach the Newtonian expression $K = \frac{1}{2}mv^2$ when v is much smaller than c (Fig. 37.21). To verify this, we expand the radical, using the binomial theorem in the form

$$(1 + x)^n = 1 + nx + n(n - 1)x^2/2 + \cdots$$

In our case, $n = -\frac{1}{2}$ and $x = -v^2/c^2$, and we get

$$\gamma = \left(1 - \frac{v^2}{c^2}\right)^{-1/2} = 1 + \frac{1}{2}\frac{v^2}{c^2} + \frac{3}{8}\frac{v^4}{c^4} + \cdots$$

Combining this with $K = (\gamma - 1)mc^2$, we find

$$K = \left(1 + \frac{1}{2}\frac{v^2}{c^2} + \frac{3}{8}\frac{v^4}{c^4} + \cdots - 1\right)mc^2$$

$$= \frac{1}{2}mv^2 + \frac{3}{8}\frac{mv^4}{c^2} + \cdots \qquad (37.37)$$

When v is much smaller than c, all the terms in the series in Eq. (37.37) except the first are negligibly small, and we obtain the Newtonian expression $\frac{1}{2}mv^2$.

Rest Energy and $E = mc^2$

Equation (37.36) for the kinetic energy of a moving particle includes a term $mc^2/\sqrt{1 - v^2/c^2}$ that depends on the motion and a second energy term mc^2 that is independent of the motion. It seems that the kinetic energy of a particle is the difference between some **total energy** E and an energy mc^2 that it has even when it is at rest. Thus we can rewrite Eq. (37.36) as

$$E = K + mc^2 = \frac{mc^2}{\sqrt{1 - v^2/c^2}} = \gamma mc^2 \qquad \begin{array}{l}\text{(total energy of} \\ \text{a particle)}\end{array} \qquad (37.38)$$

For a particle at rest $(K = 0)$, we see that $E = mc^2$. The energy mc^2 associated with rest mass m rather than motion is called the **rest energy** of the particle.

There is in fact direct experimental evidence that rest energy really does exist. The simplest example is the decay of a neutral *pion*. This is an unstable subatomic particle of rest mass m_π; when it decays, it disappears and electromagnetic radiation appears. If a neutral pion has no kinetic energy before its decay, the total energy of the radiation after its decay is found to equal exactly $m_\pi c^2$. In many other fundamental particle transformations the sum of the rest masses of the particles changes. In every case there is a corresponding energy change, consistent with the assumption of a rest energy mc^2 associated with a rest mass m.

Historically, the principles of conservation of mass and of energy developed quite independently. The theory of relativity shows that they are actually two special cases of a single broader conservation principle, the *principle of conservation of mass and energy*. In some physical phenomena, neither the sum of the rest masses of the particles nor the total energy other than rest energy is separately conserved, but there is a more general conservation principle: In an isolated system, when the sum of the rest masses changes, there is always a change in $1/c^2$ times the total energy other than the rest energy. This change is equal in magnitude but opposite in sign to the change in the sum of the rest masses.

This more general mass-energy conservation law is the fundamental principle involved in the generation of power through nuclear reactions. When a uranium nucleus undergoes fission in a nuclear reactor, the sum of the rest masses of the resulting fragments is *less than* the rest mass of the parent nucleus. An amount of energy is released that equals the mass decrease multiplied by c^2. Most of this energy can be used to produce steam to operate turbines for electric power generators.

37.21 Graph of the kinetic energy of a particle of rest mass m as a function of speed v. Also shown is the Newtonian prediction, which gives correct results only at speeds much less than c.

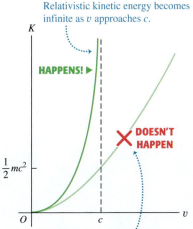

Relativistic kinetic energy becomes infinite as v approaches c.

HAPPENS! ▶

✗ **DOESN'T HAPPEN**

Newtonian mechanics incorrectly predicts that kinetic energy becomes infinite only if v becomes infinite.

Application Monitoring Mass–Energy Conversion
Although the control room of a nuclear power plant is very complex, the physical principle on which such a plant operates is a simple one: Part of the rest energy of atomic nuclei is converted to thermal energy, which in turn is used to produce steam to drive electric generators.

We can also relate the total energy E of a particle (kinetic energy plus rest energy) directly to its momentum by combining Eq. (37.27) for relativistic momentum and Eq. (37.38) for total energy to eliminate the particle's velocity. The simplest procedure is to rewrite these equations in the following forms:

$$\left(\frac{E}{mc^2}\right)^2 = \frac{1}{1 - v^2/c^2} \quad \text{and} \quad \left(\frac{p}{mc}\right)^2 = \frac{v^2/c^2}{1 - v^2/c^2}$$

Subtracting the second of these from the first and rearranging, we find

$$E^2 = (mc^2)^2 + (pc)^2 \qquad \text{(total energy, rest energy, and momentum)} \qquad (37.39)$$

Again we see that for a particle at rest ($p = 0$), $E = mc^2$.

Equation (37.39) also suggests that a particle may have energy and momentum even when it has no rest mass. In such a case, $m = 0$ and

$$E = pc \qquad \text{(zero rest mass)} \qquad (37.40)$$

In fact, zero rest mass particles do exist. Such particles always travel at the speed of light in vacuum. One example is the *photon*, the quantum of electromagnetic radiation (to be discussed in Chapter 38). Photons are emitted and absorbed during changes of state of an atomic or nuclear system when the energy and momentum of the system change.

Example 37.10   Energetic electrons

(a) Find the rest energy of an electron ($m = 9.109 \times 10^{-31}$ kg, $q = -e = -1.602 \times 10^{-19}$ C) in joules and in electron volts. (b) Find the speed of an electron that has been accelerated by an electric field, from rest, through a potential increase of 20.0 kV or of 5.00 MV (typical of a high-voltage x-ray machine).

SOLUTION

IDENTIFY and SET UP: This problem uses the ideas of rest energy, relativistic kinetic energy, and (from Chapter 23) electric potential energy. We use $E = mc^2$ to find the rest energy and Eqs. (37.7) and (37.38) to find the speed that gives the stated total energy.

EXECUTE: (a) The rest energy is

$$mc^2 = (9.109 \times 10^{-31}\text{ kg})(2.998 \times 10^8\text{ m/s})^2$$
$$= 8.187 \times 10^{-14}\text{ J}$$

From the definition of the electron volt in Section 23.2, 1 eV = 1.602×10^{-19} J. Using this, we find

$$mc^2 = (8.187 \times 10^{-14}\text{ J})\frac{1\text{ eV}}{1.602 \times 10^{-19}\text{ J}}$$
$$= 5.11 \times 10^5\text{ eV} = 0.511\text{ MeV}$$

(b) In calculations such as this, it is often convenient to work with the quantity $\gamma = 1/\sqrt{1 - v^2/c^2}$ from Eq. (37.38). Solving this for v, we find

$$v = c\sqrt{1 - (1/\gamma)^2}$$

The total energy E of the accelerated electron is the sum of its rest energy mc^2 and the kinetic energy eV_{ba} that it gains from the work done on it by the electric field in moving from point a to point b:

$$E = \gamma mc^2 = mc^2 + eV_{ba} \quad \text{or}$$
$$\gamma = 1 + \frac{eV_{ba}}{mc^2}$$

An electron accelerated through a potential increase of $V_{ba} = 20.0$ kV gains 20.0 keV of energy, so for this electron

$$\gamma = 1 + \frac{20.0 \times 10^3\text{ eV}}{0.511 \times 10^6\text{ eV}} = 1.039$$

and

$$v = c\sqrt{1 - (1/1.039)^2} = 0.272c = 8.15 \times 10^7\text{ m/s}$$

Repeating the calculation for $V_{ba} = 5.00$ MV, we find $eV_{ba}/mc^2 = 9.78$, $\gamma = 10.78$, and $v = 0.996c$.

EVALUATE: With $V_{ba} = 20.0$ kV, the added kinetic energy of 20.0 keV is less than 4% of the rest energy of 0.511 MeV, and the final speed is about one-fourth the speed of light. With $V_{ba} = 5.00$ MV, the added kinetic energy of 5.00 MeV is much greater than the rest energy and the speed is close to c.

CAUTION **Three electron energies** All electrons have *rest* energy 0.511 MeV. An electron accelerated from rest through a 5.00-MeV potential increase has *kinetic* energy 5.00 MeV (we call it a "5.00-MeV electron") and *total* energy 5.51 MeV. Be careful to distinguish these energies from one another. ▌

Example 37.11   A relativistic collision

Two protons (each with mass $M_p = 1.67 \times 10^{-27}$ kg) are initially moving with equal speeds in opposite directions. They continue to exist after a head-on collision that also produces a neutral pion of mass $M_\pi = 2.40 \times 10^{-28}$ kg (Fig. 37.22). If all three particles are at rest after the collision, find the initial speed of the protons. Energy is conserved in the collision.

37.22 In this collision the kinetic energy of two protons is transformed into the rest energy of a new particle, a pion.

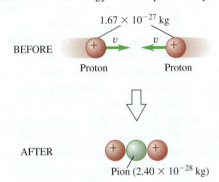

1.67 × 10⁻²⁷ kg

BEFORE

Proton   Proton

AFTER

Pion (2.40 × 10⁻²⁸ kg)

SOLUTION

IDENTIFY and SET UP: Relativistic total energy is conserved in the collision, so we can equate the (unknown) total energy of the two protons before the collision to the combined rest energies of the two protons and the pion after the collision. We then use Eq. (37.38) to find the speed of each proton.

EXECUTE: The total energy of each proton before the collision is γMc^2. By conservation of energy,

$$2(\gamma M_p c^2) = 2(M_p c^2) + M_\pi c^2$$

$$\gamma = 1 + \frac{M_\pi}{2M_p} = 1 + \frac{2.40 \times 10^{-28}\ \text{kg}}{2(1.67 \times 10^{-27}\ \text{kg})} = 1.072$$

From Eq. (37.38), the initial proton speed is

$$v = c\sqrt{1 - (1/\gamma)^2} = 0.360c$$

EVALUATE: The proton rest energy is 938 MeV, so the initial kinetic energy of each proton is $(\gamma - 1)Mc^2 = 0.072Mc^2 = (0.072)(938\ \text{MeV}) = 67.5$ MeV. You can verify that the rest energy $M_\pi c^2$ of the pion is twice this, or 135 MeV. All the kinetic energy "lost" in this completely inelastic collision is transformed into the rest energy of the pion.

Test Your Understanding of Section 37.8 A proton is accelerated from rest by a constant force that always points in the direction of the particle's motion. Compared to the amount of kinetic energy that the proton gains during the first meter of its travel, how much kinetic energy does the proton gain during one meter of travel while it is moving at 99% of the speed of light? (i) the same amount; (ii) a greater amount; (iii) a smaller amount.

37.9 Newtonian Mechanics and Relativity

The sweeping changes required by the principle of relativity go to the very roots of Newtonian mechanics, including the concepts of length and time, the equations of motion, and the conservation principles. Thus it may appear that we have destroyed the foundations on which Newtonian mechanics is built. In one sense this is true, yet the Newtonian formulation is still accurate whenever speeds are small in comparison with the speed of light in vacuum. In such cases, time dilation, length contraction, and the modifications of the laws of motion are so small that they are unobservable. In fact, every one of the principles of Newtonian mechanics survives as a special case of the more general relativistic formulation.

The laws of Newtonian mechanics are not *wrong;* they are *incomplete.* They are a limiting case of relativistic mechanics. They are *approximately* correct when all speeds are small in comparison to *c,* and they become exactly correct in the limit when all speeds approach zero. Thus relativity does not completely destroy the laws of Newtonian mechanics but *generalizes* them. This is a common pattern in the development of physical theory. Whenever a new theory is in partial conflict with an older, established theory, the new must yield the same predictions as the old in areas in which the old theory is supported by experimental evidence. Every new physical theory must pass this test, called the **correspondence principle.**

The General Theory of Relativity

At this point we may ask whether the special theory of relativity gives the final word on mechanics or whether *further* generalizations are possible or necessary.

37.23 Without information from outside the spaceship, the astronaut cannot distinguish situation (b) from situation (c).

(a) An astronaut is about to drop her watch in a spaceship.

(b) In gravity-free space, the floor accelerates upward at $a = g$ and hits the watch.

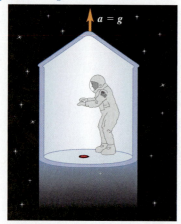

$a = g$

(c) On the earth's surface, the watch accelerates downward at $a = g$ and hits the floor.

$a = 0$

Spaceship

g

For example, inertial frames have occupied a privileged position in our discussion. Can the principle of relativity be extended to noninertial frames as well?

Here's an example that illustrates some implications of this question. A student decides to go over Niagara Falls while enclosed in a large wooden box. During her free fall she doesn't fall to the floor of the box because both she and the box are in free fall with a downward acceleration of 9.8 m/s². But an alternative interpretation, from her point of view, is that she doesn't fall to the floor because her gravitational interaction with the earth has suddenly been turned off. As long as she remains in the box and it remains in free fall, she cannot tell whether she is indeed in free fall or whether the gravitational interaction has vanished.

A similar problem occurs in a space station in orbit around the earth. Objects in the space station *seem* to be weightless, but without looking outside the station there is no way to determine whether gravity has been turned off or whether the station and all its contents are accelerating toward the center of the earth. Figure 37.23 makes a similar point for a spaceship that is not in free fall but may be accelerating relative to an inertial frame or be at rest on the earth's surface.

These considerations form the basis of Einstein's **general theory of relativity.** If we cannot distinguish experimentally between a uniform gravitational field at a particular location and a uniformly accelerated reference frame, then there cannot be any real distinction between the two. Pursuing this concept, we may try to represent *any* gravitational field in terms of special characteristics of the coordinate system. This turns out to require even more sweeping revisions of our space-time concepts than did the special theory of relativity. In the general theory of relativity the geometric properties of space are affected by the presence of matter (Fig. 37.24).

The general theory of relativity has passed several experimental tests, including three proposed by Einstein. One test has to do with understanding the rotation of the axes of the planet Mercury's elliptical orbit, called the *precession of the perihelion.* (The perihelion is the point of closest approach to the sun.) A second test concerns the apparent bending of light rays from distant stars when they pass near the sun. The third test is the *gravitational red shift,* the increase in wavelength of light proceeding outward from a massive source. Some details of the general theory are more difficult to test, but this theory has played a central role in investigations of the formation and evolution of stars, black holes, and studies of the evolution of the universe.

The general theory of relativity may seem to be an exotic bit of knowledge with little practical application. In fact, this theory plays an essential role in the

37.24 A two-dimensional representation of curved space. We imagine the space (a plane) as being distorted as shown by a massive object (the sun). Light from a distant star (solid line) follows the distorted surface on its way to the earth. The dashed line shows the direction from which the light *appears* to be coming. The effect is greatly exaggerated; for the sun, the maximum deviation is only 0.00048°.

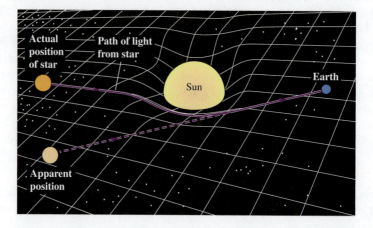

Actual position of star

Path of light from star

Sun

Earth

Apparent position

global positioning system (GPS), which makes it possible to determine your position on the earth's surface to within a few meters using a handheld receiver (Fig. 37.25). The heart of the GPS system is a collection of more than two dozen satellites in very precise orbits. Each satellite emits carefully timed radio signals, and a GPS receiver simultaneously detects the signals from several satellites. The receiver then calculates the time delay between when each signal was emitted and when it was received, and uses this information to calculate the receiver's position. To ensure the proper timing of the signals, it's necessary to include corrections due to the special theory of relativity (because the satellites are moving relative to the receiver on earth) as well as the general theory (because the satellites are higher in the earth's gravitational field than the receiver). The corrections due to relativity are small—less than one part in 10^9—but are crucial to the superb precision of the GPS system.

37.25 A GPS receiver uses radio signals from the orbiting GPS satellites to determine its position. To account for the effects of relativity, the receiver must be tuned to a slightly higher frequency (10.23 MHz) than the frequency emitted by the satellites (10.22999999543 MHz).

Invariance of physical laws, simultaneity: All of the fundamental laws of physics have the same form in all inertial frames of reference. The speed of light in vacuum is the same in all inertial frames and is independent of the motion of the source. Simultaneity is not an absolute concept; events that are simultaneous in one frame are not necessarily simultaneous in a second frame moving relative to the first.

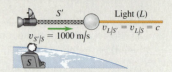

Time dilation: If two events occur at the same space point in a particular frame of reference, the time interval Δt_0 between the events as measured in that frame is called a proper time interval. If this frame moves with constant velocity u relative to a second frame, the time interval Δt between the events as observed in the second frame is longer than Δt_0. (See Examples 37.1–37.3.)

$$\Delta t = \frac{\Delta t_0}{\sqrt{1 - u^2/c^2}} = \gamma\,\Delta t_0 \qquad (37.6),\ (37.8)$$

$$\gamma = \frac{1}{\sqrt{1 - u^2/c^2}} \qquad (37.7)$$

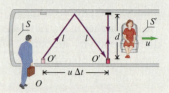

Length contraction: If two points are at rest in a particular frame of reference, the distance l_0 between the points as measured in that frame is called a proper length. If this frame moves with constant velocity u relative to a second frame and the distances are measured parallel to the motion, the distance l between the points as measured in the second frame is shorter than l_0. (See Examples 37.4 and 37.5.)

$$l = l_0\sqrt{1 - u^2/c^2} = \frac{l_0}{\gamma} \qquad (37.16)$$

The Lorentz transformations: The Lorentz coordinate transformations relate the coordinates and time of an event in an inertial frame S to the coordinates and time of the same event as observed in a second inertial frame S' moving at velocity u relative to the first. For one-dimensional motion, a particle's velocities v_x in S and v_x' in S' are related by the Lorentz velocity transformation. (See Examples 37.6 and 37.7.)

$$x' = \frac{x - ut}{\sqrt{1 - u^2/c^2}} = \gamma(x - ut)$$

$$y' = y \qquad z' = z \qquad (37.21)$$

$$t' = \frac{t - ux/c^2}{\sqrt{1 - u^2/c^2}} = \gamma(t - ux/c^2)$$

$$v_x' = \frac{v_x - u}{1 - uv_x/c^2} \qquad (37.22)$$

$$v_x = \frac{v_x' + u}{1 + uv_x'/c^2} \qquad (37.23)$$

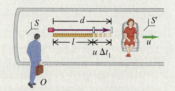

The Doppler effect for electromagnetic waves: The Doppler effect is the frequency shift in light from a source due to the relative motion of source and observer. For a source moving toward the observer with speed u, Eq. (37.25) gives the received frequency f in terms of the emitted frequency f_0. (See Example 37.8.)

$$f = \sqrt{\frac{c + u}{c - u}}\,f_0 \qquad (37.25)$$

Moving source emits light of frequency f_0. Stationary observer detects light of frequency $f > f_0$.

Relativistic momentum and energy: For a particle of rest mass m moving with velocity $\vec{v}$, the relativistic momentum $\vec{p}$ is given by Eq. (37.27) or (37.31) and the relativistic kinetic energy K is given by Eq. (37.36). The total energy E is the sum of the kinetic energy and the rest energy mc^2. The total energy can also be expressed in terms of the magnitude of momentum p and rest mass m. (See Examples 37.9–37.11.)

$$\vec{p} = \frac{m\vec{v}}{\sqrt{1 - v^2/c^2}} = \gamma m\vec{v} \qquad (37.27),\ (37.31)$$

$$K = \frac{mc^2}{\sqrt{1 - v^2/c^2}} - mc^2 = (\gamma - 1)mc^2 \qquad (37.36)$$

$$E = K + mc^2 = \frac{mc^2}{\sqrt{1 - v^2/c^2}} = \gamma mc^2 \qquad (37.38)$$

$$E^2 = (mc^2)^2 + (pc)^2 \qquad (37.39)$$

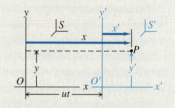

BRIDGING PROBLEM | Colliding Protons

In an experiment, two protons are shot directly toward each other. Their speeds are such that in the frame of reference of each proton, the other proton is moving at $0.500c$. (a) What does an observer in the laboratory measure for the speed of each proton? (b) What is the kinetic energy of each proton as measured by an observer in the laboratory? (c) What is the kinetic energy of each proton as measured by the other proton?

SOLUTION GUIDE

See MasteringPhysics® study area for a Video Tutor solution.

IDENTIFY and SET UP

1. This problem uses the Lorentz velocity transformation, which allows us to relate the velocity v_x of a proton in one frame to its velocity v_x' in a different frame. It also uses the idea of relativistic kinetic energy.
2. Take the x-axis to be the line of motion of the protons, and take the $+x$-direction to be to the right. In the frame in which the left-hand proton is at rest, the right-hand proton has velocity $-0.500c$. In the laboratory frame the two protons have velocities

$-\alpha c$ and $+\alpha c$, where α (each proton's laboratory speed as a fraction of c) is our first target variable. Given this we can find the laboratory kinetic energy of each proton.

EXECUTE

3. Write a Lorentz velocity-transformation equation that relates the velocity of the right-hand proton in the laboratory frame to its velocity in the frame of the left-hand proton. Solve this equation for α. (*Hint:* Remember that α cannot be greater than 1. Why?)
4. Use your result from step 3 to find the laboratory kinetic energy of each proton.
5. Find the kinetic energy of the right-hand proton as measured in the frame of the left-hand proton.

EVALUATE

6. How much total kinetic energy must be imparted to the protons by a scientist in the laboratory? If the experiment were to be repeated with one proton stationary, what kinetic energy would have to be given to the other proton for the collision to be equivalent?

Problems

For instructor-assigned homework, go to www.masteringphysics.com

•, ••, •••: Problems of increasing difficulty. **CP**: Cumulative problems incorporating material from earlier chapters. **CALC**: Problems requiring calculus. **BIO**: Biosciences problems.

DISCUSSION QUESTIONS

Q37.1 You are standing on a train platform watching a high-speed train pass by. A light inside one of the train cars is turned on and then a little later it is turned off. (a) Who can measure the proper time interval for the duration of the light: you or a passenger on the train? (b) Who can measure the proper length of the train car: you or a passenger on the train? (c) Who can measure the proper length of a sign attached to a post on the train platform: you or a passenger on the train? In each case explain your answer.

Q37.2 If simultaneity is not an absolute concept, does that mean that we must discard the concept of causality? If event A is to *cause* event B, A must occur first. Is it possible that in some frames A appears to be the cause of B, and in others B appears to be the cause of A? Explain.

Q37.3 A rocket is moving to the right at $\frac{1}{2}$ the speed of light relative to the earth. A light bulb in the center of a room inside the rocket suddenly turns on. Call the light hitting the front end of the room event A and the light hitting the back of the room event B (Fig. Q37.3). Which event occurs first, A or B, or are they simultaneous, as viewed by (a) an astronaut riding in the rocket and (b) a person at rest on the earth?

Figure **Q37.3**

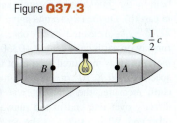

Q37.4 What do you think would be different in everyday life if the speed of light were 10 m/s instead of 3.00×10^8 m/s?

Q37.5 The average life span in the United States is about 70 years. Does this mean that it is impossible for an average person to travel a distance greater than 70 light-years away from the earth? (A light-year is the distance light travels in a year.) Explain.

Q37.6 You are holding an elliptical serving platter. How would you need to travel for the serving platter to appear round to another observer?

Q37.7 Two events occur at the same space point in a particular inertial frame of reference and are simultaneous in that frame. Is it possible that they may not be simultaneous in a different inertial frame? Explain.

Q37.8 A high-speed train passes a train platform. Larry is a passenger on the train, Adam is standing on the train platform, and David is riding a bicycle toward the platform in the same direction as the train is traveling. Compare the length of a train car as measured by Larry, Adam, and David.

Q37.9 The theory of relativity sets an upper limit on the speed that a particle can have. Are there also limits on the energy and momentum of a particle? Explain.

Q37.10 A student asserts that a material particle must always have a speed slower than that of light, and a massless particle must always move at exactly the speed of light. Is she correct? If so, how do massless particles such as photons and neutrinos acquire this speed? Can't they start from rest and accelerate? Explain.

Q37.11 The speed of light relative to still water is 2.25×10^8 m/s. If the water is moving past us, the speed of light we measure depends on the speed of the water. Do these facts violate Einstein's second postulate? Explain.

Q37.12 When a monochromatic light source moves toward an observer, its wavelength appears to be shorter than the value measured when the source is at rest. Does this contradict the hypothesis that the speed of light is the same for all observers? Explain.

Q37.13 In principle, does a hot gas have more mass than the same gas when it is cold? Explain. In practice, would this be a measurable effect? Explain.

Q37.14 Why do you think the development of Newtonian mechanics preceded the more refined relativistic mechanics by so many years?

EXERCISES

Section 37.2 Relativity of Simultaneity

37.1 • Suppose the two lightning bolts shown in Fig. 37.5a are simultaneous to an observer on the train. Show that they are *not* simultaneous to an observer on the ground. Which lightning strike does the ground observer measure to come first?

Section 37.3 Relativity of Time Intervals

37.2 • The positive muon (μ^+), an unstable particle, lives on average 2.20×10^{-6} s (measured in its own frame of reference) before decaying. (a) If such a particle is moving, with respect to the laboratory, with a speed of $0.900c$, what average lifetime is measured in the laboratory? (b) What average distance, measured in the laboratory, does the particle move before decaying?

37.3 • How fast must a rocket travel relative to the earth so that time in the rocket "slows down" to half its rate as measured by earth-based observers? Do present-day jet planes approach such speeds?

37.4 • A spaceship flies past Mars with a speed of $0.985c$ relative to the surface of the planet. When the spaceship is directly overhead, a signal light on the Martian surface blinks on and then off. An observer on Mars measures that the signal light was on for 75.0 μs. (a) Does the observer on Mars or the pilot on the spaceship measure the proper time? (b) What is the duration of the light pulse measured by the pilot of the spaceship?

37.5 • The negative pion (π^-) is an unstable particle with an average lifetime of 2.60×10^{-8} s (measured in the rest frame of the pion). (a) If the pion is made to travel at very high speed relative to a laboratory, its average lifetime is measured in the laboratory to be 4.20×10^{-7} s. Calculate the speed of the pion expressed as a fraction of c. (b) What distance, measured in the laboratory, does the pion travel during its average lifetime?

37.6 •• As you pilot your space utility vehicle at a constant speed toward the moon, a race pilot flies past you in her spaceracer at a constant speed of $0.800c$ relative to you. At the instant the spaceracer passes you, both of you start timers at zero. (a) At the instant when you measure that the spaceracer has traveled 1.20×10^8 m past you, what does the race pilot read on her timer? (b) When the race pilot reads the value calculated in part (a) on her timer, what does she measure to be your distance from her? (c) At the instant when the race pilot reads the value calculated in part (a) on her timer, what do you read on yours?

37.7 •• A spacecraft flies away from the earth with a speed of 4.80×10^6 m/s relative to the earth and then returns at the same speed. The spacecraft carries an atomic clock that has been carefully synchronized with an identical clock that remains at rest on earth. The spacecraft returns to its starting point 365 days (1 year) later, as measured by the clock that remained on earth. What is the difference in the elapsed times on the two clocks, measured in hours? Which clock, the one in the spacecraft or the one on earth, shows the shorter elapsed time?

37.8 • An alien spacecraft is flying overhead at a great distance as you stand in your backyard. You see its searchlight blink on for 0.190 s. The first officer on the spacecraft measures that the searchlight is on for 12.0 ms. (a) Which of these two measured times is the proper time? (b) What is the speed of the spacecraft relative to the earth expressed as a fraction of the speed of light c?

Section 37.4 Relativity of Length

37.9 • A spacecraft of the Trade Federation flies past the planet Coruscant at a speed of $0.600c$. A scientist on Coruscant measures the length of the moving spacecraft to be 74.0 m. The spacecraft later lands on Coruscant, and the same scientist measures the length of the now stationary spacecraft. What value does she get?

37.10 • A meter stick moves past you at great speed. Its motion relative to you is parallel to its long axis. If you measure the length of the moving meter stick to be 1.00 ft (1 ft = 0.3048 m)—for example, by comparing it to a 1-foot ruler that is at rest relative to you—at what speed is the meter stick moving relative to you?

37.11 •• **Why Are We Bombarded by Muons?** Muons are unstable subatomic particles that decay to electrons with a mean lifetime of 2.2 μs. They are produced when cosmic rays bombard the upper atmosphere about 10 km above the earth's surface, and they travel very close to the speed of light. The problem we want to address is why we see any of them at the earth's surface. (a) What is the greatest distance a muon could travel during its 2.2-μs lifetime? (b) According to your answer in part (a), it would seem that muons could never make it to the ground. But the 2.2-μs lifetime is measured in the frame of the muon, and muons are moving very fast. At a speed of $0.999c$, what is the mean lifetime of a muon as measured by an observer at rest on the earth? How far would the muon travel in this time? Does this result explain why we find muons in cosmic rays? (c) From the point of view of the muon, it still lives for only 2.2 μs, so how does it make it to the ground? What is the thickness of the 10 km of atmosphere through which the muon must travel, as measured by the muon? Is it now clear how the muon is able to reach the ground?

37.12 • An unstable particle is created in the upper atmosphere from a cosmic ray and travels straight down toward the surface of the earth with a speed of $0.99540c$ relative to the earth. A scientist at rest on the earth's surface measures that the particle is created at an altitude of 45.0 km. (a) As measured by the scientist, how much time does it take the particle to travel the 45.0 km to the surface of the earth? (b) Use the length-contraction formula to calculate the distance from where the particle is created to the surface of the earth as measured in the particle's frame. (c) In the particle's frame, how much time does it take the particle to travel from where it is created to the surface of the earth? Calculate this time both by the time dilation formula and from the distance calculated in part (b). Do the two results agree?

37.13 • As measured by an observer on the earth, a spacecraft runway on earth has a length of 3600 m. (a) What is the length of the runway as measured by a pilot of a spacecraft flying past at a speed of 4.00×10^7 m/s relative to the earth? (b) An observer on earth measures the time interval from when the spacecraft is directly over one end of the runway until it is directly over the other end. What result does she get? (c) The pilot of the spacecraft measures the time it takes him to travel from one end of the runway to the other end. What value does he get?

37.14 • A rocket ship flies past the earth at 85.0% of the speed of light. Inside, an astronaut who is undergoing a physical examination is having his height measured while he is lying down parallel to the direction the rocket ship is moving. (a) If his height is measured to be 2.00 m by his doctor inside the ship, what height would a person watching this from earth measure for his height? (b) If the earth-based person had measured 2.00 m, what would the doctor in the spaceship have measured for the astronaut's height? Is this a reasonable height? (c) Suppose the astronaut in part (a) gets up after the examination and stands with his body perpendicular to the direction of motion. What would the doctor in the rocket and the observer on earth measure for his height now?

Section 37.5 The Lorentz Transformations

37.15 • An observer in frame S' is moving to the right ($+x$-direction) at speed $u = 0.600c$ away from a stationary observer in frame S. The observer in S' measures the speed v' of a particle moving to the right away from her. What speed v does the observer in S measure for the particle if (a) $v' = 0.400c$; (b) $v' = 0.900c$; (c) $v' = 0.990c$?

37.16 • Space pilot Mavis zips past Stanley at a constant speed relative to him of $0.800c$. Mavis and Stanley start timers at zero when the front of Mavis's ship is directly above Stanley. When Mavis reads 5.00 s on her timer, she turns on a bright light under the front of her spaceship. (a) Use the Lorentz coordinate transformation derived in Example 37.6 to calculate x and t as measured by Stanley for the event of turning on the light. (b) Use the time dilation formula, Eq. (37.6), to calculate the time interval between the two events (the front of the spaceship passing overhead and turning on the light) as measured by Stanley. Compare to the value of t you calculated in part (a). (c) Multiply the time interval by Mavis's speed, both as measured by Stanley, to calculate the distance she has traveled as measured by him when the light turns on. Compare to the value of x you calculated in part (a).

37.17 •• A pursuit spacecraft from the planet Tatooine is attempting to catch up with a Trade Federation cruiser. As measured by an observer on Tatooine, the cruiser is traveling away from the planet with a speed of $0.600c$. The pursuit ship is traveling at a speed of $0.800c$ relative to Tatooine, in the same direction as the cruiser. (a) For the pursuit ship to catch the cruiser, should the velocity of the cruiser relative to the pursuit ship be directed toward or away from the pursuit ship? (b) What is the speed of the cruiser relative to the pursuit ship?

37.18 • An extraterrestrial spaceship is moving away from the earth after an unpleasant encounter with its inhabitants. As it departs, the spaceship fires a missile toward the earth. An observer on earth measures that the spaceship is moving away with a speed of $0.600c$. An observer in the spaceship measures that the missile is moving away from him at a speed of $0.800c$. As measured by an observer on earth, how fast is the missile approaching the earth?

37.19 •• Two particles are created in a high-energy accelerator and move off in opposite directions. The speed of one particle, as measured in the laboratory, is $0.650c$, and the speed of each particle relative to the other is $0.950c$. What is the speed of the second particle, as measured in the laboratory?

37.20 •• Two particles in a high-energy accelerator experiment are approaching each other head-on, each with a speed of $0.9520c$ as measured in the laboratory. What is the magnitude of the velocity of one particle relative to the other?

37.21 •• Two particles in a high-energy accelerator experiment approach each other head-on with a relative speed of $0.890c$. Both particles travel at the same speed as measured in the laboratory. What is the speed of each particle, as measured in the laboratory?

37.22 •• An enemy spaceship is moving toward your starfighter with a speed, as measured in your frame, of $0.400c$. The enemy ship fires a missile toward you at a speed of $0.700c$ relative to the

enemy ship (Fig. E37.22). (a) What is the speed of the missile relative to you? Express your answer in terms of the speed of light. (b) If you measure that the enemy ship is 8.00×10^6 km away from you when the missile is fired, how much time, measured in your frame, will it take the missile to reach you?

Figure **E37.22**

Enemy   Starfighter

37.23 • An imperial spaceship, moving at high speed relative to the planet Arrakis, fires a rocket toward the planet with a speed of $0.920c$ relative to the spaceship. An observer on Arrakis measures that the rocket is approaching with a speed of $0.360c$. What is the speed of the spaceship relative to Arrakis? Is the spaceship moving toward or away from Arrakis?

Section 37.6 The Doppler Effect for Electromagnetic Waves

37.24 • Electromagnetic radiation from a star is observed with an earth-based telescope. The star is moving away from the earth with a speed of $0.600c$. If the radiation has a frequency of 8.64×10^{14} Hz in the rest frame of the star, what is the frequency measured by an observer on earth?

37.25 • **Tell It to the Judge.** (a) How fast must you be approaching a red traffic light ($\lambda = 675$ nm) for it to appear yellow ($\lambda = 575$ nm)? Express your answer in terms of the speed of light. (b) If you used this as a reason not to get a ticket for running a red light, how much of a fine would you get for speeding? Assume that the fine is $1.00 for each kilometer per hour that your speed exceeds the posted limit of 90 km/h.

37.26 • A source of electromagnetic radiation is moving in a radial direction relative to you. The frequency you measure is 1.25 times the frequency measured in the rest frame of the source. What is the speed of the source relative to you? Is the source moving toward you or away from you?

Section 37.7 Relativistic Momentum

37.27 • A proton has momentum with magnitude p_0 when its speed is $0.400c$. In terms of p_0, what is the magnitude of the proton's momentum when its speed is doubled to $0.800c$?

37.28 • **When Should You Use Relativity?** As you have seen, relativistic calculations usually involve the quantity γ. When γ is appreciably greater than 1, we must use relativistic formulas instead of Newtonian ones. For what speed v (in terms of c) is the value of γ (a) 1.0% greater than 1; (b) 10% greater than 1; (c) 100% greater than 1?

37.29 • (a) At what speed is the momentum of a particle twice as great as the result obtained from the nonrelativistic expression mv? Express your answer in terms of the speed of light. (b) A force is applied to a particle along its direction of motion. At what speed is the magnitude of force required to produce a given acceleration twice as great as the force required to produce the same acceleration when the particle is at rest? Express your answer in terms of the speed of light.

37.30 • As measured in an earth-based frame, a proton is moving in the $+x$-direction at a speed of 2.30×10^8 m/s. (a) What force (magnitude and direction) is required to produce an acceleration in the $-x$-direction that has magnitude 2.30×10^8 m/s^2? (b) What magnitude of acceleration does the force calculated in part (a) give to a proton that is initially at rest?

37.31 • An electron is acted upon by a force of 5.00×10^{-15} N due to an electric field. Find the acceleration this force produces in each case: (a) The electron's speed is 1.00 km/s. (b) The electron's speed is 2.50×10^8 m/s and the force is parallel to the velocity.

37.32 • **Relativistic Baseball.** Calculate the magnitude of the force required to give a 0.145-kg baseball an acceleration $a = 1.00$ m/s^2 in the direction of the baseball's initial velocity when this velocity has a magnitude of (a) 10.0 m/s; (b) 0.900c; (c) 0.990c. (d) Repeat parts (a), (b), and (c) if the force and acceleration are perpendicular to the velocity.

Section 37.8 Relativistic Work and Energy

37.33 •• What is the speed of a particle whose kinetic energy is equal to (a) its rest energy and (b) five times its rest energy?

37.34 • If a muon is traveling at 0.999c, what are its momentum and kinetic energy? (The mass of such a muon at rest in the laboratory is 207 times the electron mass.)

37.35 • A proton (rest mass 1.67×10^{-27} kg) has total energy that is 4.00 times its rest energy. What are (a) the kinetic energy of the proton; (b) the magnitude of the momentum of the proton; (c) the speed of the proton?

37.36 •• (a) How much work must be done on a particle with mass m to accelerate it (a) from rest to a speed of 0.090c and (b) from a speed of 0.900c to a speed of 0.990c? (Express the answers in terms of mc^2.) (c) How do your answers in parts (a) and (b) compare?

37.37 • **CP** (a) By what percentage does your rest mass increase when you climb 30 m to the top of a ten-story building? Are you aware of this increase? Explain. (b) By how many grams does the mass of a 12.0-g spring with force constant 200 N/cm change when you compress it by 6.0 cm? Does the mass increase or decrease? Would you notice the change in mass if you were holding the spring? Explain.

37.38 • A 60.0-kg person is standing at rest on level ground. How fast would she have to run to (a) double her total energy and (b) increase her total energy by a factor of 10?

37.39 • **An Antimatter Reactor.** When a particle meets its antiparticle, they annihilate each other and their mass is converted to light energy. The United States uses approximately 1.0×10^{20} J of energy per year. (a) If all this energy came from a futuristic antimatter reactor, how much mass of matter and antimatter fuel would be consumed yearly? (b) If this fuel had the density of iron (7.86 g/cm^3) and were stacked in bricks to form a cubical pile, how high would it be? (Before you get your hopes up, antimatter reactors are a *long* way in the future—if they ever will be feasible.)

37.40 •• Electrons are accelerated through a potential difference of 750 kV, so that their kinetic energy is 7.50×10^5 eV. (a) What is the ratio of the speed v of an electron having this energy to the speed of light, c? (b) What would the speed be if it were computed from the principles of classical mechanics?

37.41 • A particle has rest mass 6.64×10^{-27} kg and momentum 2.10×10^{-18} kg·m/s. (a) What is the total energy (kinetic plus rest energy) of the particle? (b) What is the kinetic energy of the particle? (c) What is the ratio of the kinetic energy to the rest energy of the particle?

37.42 •• A 0.100-μg speck of dust is accelerated from rest to a speed of 0.900c by a constant 1.00×10^6 N force. (a) If the nonrelativistic mechanics is used, how far does the object travel to reach its final speed? (b) Using the correct relativistic treatment of Section 37.8, how far does the object travel to reach its final speed? (c) Which distance is greater? Why?

37.43 • Compute the kinetic energy of a proton (mass 1.67×10^{-27} kg) using both the nonrelativistic and relativistic expressions, and compute the ratio of the two results (relativistic divided by nonrelativistic) for speeds of (a) 8.00×10^7 m/s and (b) 2.85×10^8 m/s.

37.44 • What is the kinetic energy of a proton moving at (a) 0.100c; (b) 0.500c; (c) 0.900c? How much work must be done to (d) increase the proton's speed from 0.100c to 0.500c and (e) increase the proton's speed from 0.500c to 0.900c? (f) How do the last two results compare to results obtained in the nonrelativistic limit?

37.45 • (a) Through what potential difference does an electron have to be accelerated, starting from rest, to achieve a speed of 0.980c? (b) What is the kinetic energy of the electron at this speed? Express your answer in joules and in electron volts.

37.46 • **Creating a Particle.** Two protons (each with rest mass $M = 1.67 \times 10^{-27}$ kg) are initially moving with equal speeds in opposite directions. The protons continue to exist after a collision that also produces an η^0 particle (see Chapter 44). The rest mass of the η^0 is $m = 9.75 \times 10^{-28}$ kg. (a) If the two protons and the η^0 are all at rest after the collision, find the initial speed of the protons, expressed as a fraction of the speed of light. (b) What is the kinetic energy of each proton? Express your answer in MeV. (c) What is the rest energy of the η^0, expressed in MeV? (d) Discuss the relationship between the answers to parts (b) and (c).

37.47 • The sun produces energy by nuclear fusion reactions, in which matter is converted into energy. By measuring the amount of energy we receive from the sun, we know that it is producing energy at a rate of 3.8×10^{26} W. (a) How many kilograms of matter does the sun lose each second? Approximately how many tons of matter is this (1 ton = 2000 lbs)? (b) At this rate, how long would it take the sun to use up all its mass?

PROBLEMS

37.48 • Inside a spaceship flying past the earth at three-fourths the speed of light, a pendulum is swinging. (a) If each swing takes 1.50 s as measured by an astronaut performing an experiment inside the spaceship, how long will the swing take as measured by a person at mission control on earth who is watching the experiment? (b) If each swing takes 1.50 s as measured by a person at mission control on earth, how long will it take as measured by the astronaut in the spaceship?

37.49 • After being produced in a collision between elementary particles, a positive pion (π^+) must travel down a 1.90-km-long tube to reach an experimental area. A π^+ particle has an average lifetime (measured in its rest frame) of 2.60×10^{-8} s; the π^+ we are considering has this lifetime. (a) How fast must the π^+ travel if it is not to decay before it reaches the end of the tube? (Since u will be very close to c, write $u = (1 - \Delta)c$ and give your answer in terms of Δ rather than u.) (b) The π^+ has a rest energy of 139.6 MeV. What is the total energy of the π^+ at the speed calculated in part (a)?

37.50 •• A cube of metal with sides of length a sits at rest in a frame S with one edge parallel to the x-axis. Therefore, in S the cube has volume a^3. Frame S' moves along the x-axis with a speed u. As measured by an observer in frame S', what is the volume of the metal cube?

37.51 ••• The starships of the Solar Federation are marked with the symbol of the federation, a circle, while starships of the Denebian Empire are marked with the empire's symbol, an ellipse whose major axis is 1.40 times longer than its minor axis ($a = 1.40b$ in Fig. P37.51). How fast, relative to an observer, does an empire ship have to travel for its marking to be confused with the marking of a federation ship?

Figure **P37.51**

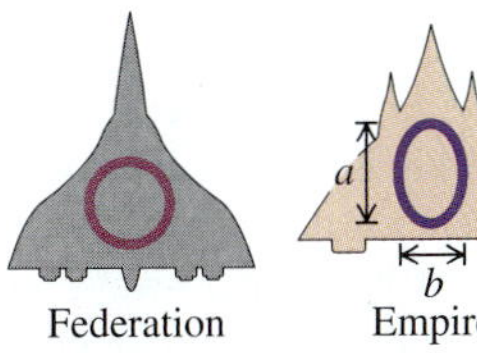

Federation   Empire

37.52 •• A space probe is sent to the vicinity of the star Capella, which is 42.2 light-years from the earth. (A light-year is the distance light travels in a year.) The probe travels with a speed of $0.9930c$. An astronaut recruit on board is 19 years old when the probe leaves the earth. What is her biological age when the probe reaches Capella?

37.53 • A particle is said to be *extremely relativistic* when its kinetic energy is much greater than its rest energy. (a) What is the speed of a particle (expressed as a fraction of c) such that the total energy is ten times the rest energy? (b) What is the percentage difference between the left and right sides of Eq. (37.39) if $(mc^2)^2$ is neglected for a particle with the speed calculated in part (a)?

37.54 •• **Everyday Time Dilation.** Two atomic clocks are carefully synchronized. One remains in New York, and the other is loaded on an airliner that travels at an average speed of 250 m/s and then returns to New York. When the plane returns, the elapsed time on the clock that stayed behind is 4.00 h. By how much will the readings of the two clocks differ, and which clock will show the shorter elapsed time? (*Hint:* Since $u \ll c$, you can simplify $\sqrt{1 - u^2/c^2}$ by a binomial expansion.)

37.55 • **The Large Hadron Collider (LHC).** Physicists and engineers from around the world have come together to build the largest accelerator in the world, the Large Hadron Collider (LHC) at the CERN Laboratory in Geneva, Switzerland. The machine will accelerate protons to kinetic energies of 7 TeV in an underground ring 27 km in circumference. (For the latest news and more information on the LHC, visit www.cern.ch.) (a) What speed v will protons reach in the LHC? (Since v is very close to c, write $v = (1 - \Delta)c$ and give your answer in terms of Δ.) (b) Find the relativistic mass, m_{rel}, of the accelerated protons in terms of their rest mass.

37.56 • **CP** A nuclear bomb containing 12.0 kg of plutonium explodes. The sum of the rest masses of the products of the explosion is less than the original rest mass by one part in 10^4. (a) How much energy is released in the explosion? (b) If the explosion takes place in 4.00 μs, what is the average power developed by the bomb? (c) What mass of water could the released energy lift to a height of 1.00 km?

37.57 • **CP Čerenkov Radiation.** The Russian physicist P. A. Čerenkov discovered that a charged particle traveling in a solid with a speed exceeding the speed of light in that material radiates electromagnetic radiation. (This is analogous to the sonic boom produced by an aircraft moving faster than the speed of sound in air; see Section 16.9. Čerenkov shared the 1958 Nobel Prize for this discovery.) What is the minimum kinetic energy (in electron volts) that an electron must have while traveling inside a slab of crown glass ($n = 1.52$) in order to create this Čerenkov radiation?

37.58 •• A photon with energy E is emitted by an atom with mass m, which recoils in the opposite direction. (a) Assuming that the motion of the atom can be treated nonrelativistically, compute the recoil speed of the atom. (b) From the result of part (a), show that the recoil speed is much less than c whenever E is much less than the rest energy mc^2 of the atom.

37.59 •• In an experiment, two protons are shot directly toward each other, each moving at half the speed of light relative to the laboratory. (a) What speed does one proton measure for the other

proton? (b) What would be the answer to part (a) if we used only nonrelativistic Newtonian mechanics? (c) What is the kinetic energy of each proton as measured by (i) an observer at rest in the laboratory and (ii) an observer riding along with one of the protons? (d) What would be the answers to part (c) if we used only nonrelativistic Newtonian mechanics?

37.60 •• Two protons are moving away from each other. In the frame of each proton, the other proton has a speed of $0.600c$. What does an observer in the rest frame of the earth measure for the speed of each proton?

37.61 •• Frame S' has an x-component of velocity u relative to frame S, and at $t = t' = 0$ the two frames coincide (see Fig. 37.3). A light pulse with a spherical wave front is emitted at the origin of S' at time $t' = 0$. Its distance x' from the origin after a time t' is given by $x'^2 = c^2 t'^2$. Use the Lorentz coordinate transformation to transform this equation to an equation in x and t, and show that the result is $x^2 = c^2 t^2$; that is, the motion appears exactly the same in frame of reference S as it does in S'; the wave front is observed to be spherical in both frames.

37.62 • In certain radioactive beta decay processes, the beta particle (an electron) leaves the atomic nucleus with a speed of 99.95% the speed of light relative to the decaying nucleus. If this nucleus is moving at 75.00% the speed of light in the laboratory reference frame, find the speed of the emitted electron relative to the laboratory reference frame if the electron is emitted (a) in the same direction that the nucleus is moving and (b) in the opposite direction from the nucleus's velocity. (c) In each case in parts (a) and (b), find the kinetic energy of the electron as measured in (i) the laboratory frame and (ii) the reference frame of the decaying nucleus.

37.63 •• **CALC** A particle with mass m accelerated from rest by a constant force F will, according to Newtonian mechanics, continue to accelerate without bound; that is, as $t \to \infty$, $v \to \infty$. Show that according to relativistic mechanics, the particle's speed approaches c as $t \to \infty$. [*Note:* A useful integral is $\int (1 - x^2)^{-3/2} dx = x/\sqrt{1 - x^2}$.]

37.64 •• Two events are observed in a frame of reference S to occur at the same space point, the second occurring 1.80 s after the first. In a second frame S' moving relative to S, the second event is observed to occur 2.35 s after the first. What is the difference between the positions of the two events as measured in S'?

37.65 ••• Two events observed in a frame of reference S have positions and times given by (x_1, t_1) and (x_2, t_2), respectively. (a) Frame S' moves along the x-axis just fast enough that the two events occur at the same position in S'. Show that in S', the time interval $\Delta t'$ between the two events is given by

$$\Delta t' = \sqrt{(\Delta t)^2 - \left(\frac{\Delta x}{c}\right)^2}$$

where $\Delta x = x_2 - x_1$ and $\Delta t = t_2 - t_1$. Hence show that if $\Delta x > c \, \Delta t$, there is *no* frame S' in which the two events occur at the same point. The interval $\Delta t'$ is sometimes called the *proper time interval* for the events. Is this term appropriate? (b) Show that if $\Delta x > c \, \Delta t$, there is a different frame of reference S' in which the two events occur *simultaneously*. Find the distance between the two events in S'; express your answer in terms of Δx, Δt, and c. This distance is sometimes called a *proper length*. Is this term appropriate? (c) Two events are observed in a frame of reference S' to occur simultaneously at points separated by a distance of 2.50 m. In a second frame S moving relative to S' along the line joining the two points in S', the two events appear to be separated by 5.00 m. What is the time interval between the events as measured in S? [*Hint:* Apply the result obtained in part (b).]

37.66 •• **Albert in Wonderland.** Einstein and Lorentz, being avid tennis players, play a fast-paced game on a court where they stand 20.0 m from each other. Being very skilled players, they play without a net. The tennis ball has mass 0.0580 kg. You can ignore gravity and assume that the ball travels parallel to the ground as it travels between the two players. Unless otherwise specified, all measurements are made by the two men. (a) Lorentz serves the ball at 80.0 m/s. What is the ball's kinetic energy? (b) Einstein slams a return at 1.80×10^8 m/s. What is the ball's kinetic energy? (c) During Einstein's return of the ball in part (a), a white rabbit runs beside the court in the direction from Einstein to Lorentz. The rabbit has a speed of 2.20×10^8 m/s relative to the two men. What is the speed of the rabbit relative to the ball? (d) What does the rabbit measure as the distance from Einstein to Lorentz? (e) How much time does it take for the rabbit to run 20.0 m, according to the players? (f) The white rabbit carries a pocket watch. He uses this watch to measure the time (as he sees it) for the distance from Einstein to Lorentz to pass by under him. What time does he measure?

37.67 • One of the wavelengths of light emitted by hydrogen atoms under normal laboratory conditions is $\lambda = 656.3$ nm, in the red portion of the electromagnetic spectrum. In the light emitted from a distant galaxy this same spectral line is observed to be Doppler-shifted to $\lambda = 953.4$ nm, in the infrared portion of the spectrum. How fast are the emitting atoms moving relative to the earth? Are they approaching the earth or receding from it?

37.68 •• **Measuring Speed by Radar.** A baseball coach uses a radar device to measure the speed of an approaching pitched baseball. This device sends out electromagnetic waves with frequency f_0 and then measures the shift in frequency Δf of the waves reflected from the moving baseball. If the fractional frequency shift produced by a baseball is $\Delta f/f_0 = 2.86 \times 10^{-7}$, what is the baseball's speed in km/h? (*Hint:* Are the waves Doppler-shifted a second time when reflected off the ball?)

37.69 • **Space Travel?** Travel to the stars requires hundreds or thousands of years, even at the speed of light. Some people have suggested that we can get around this difficulty by accelerating the rocket (and its astronauts) to very high speeds so that they will age less due to time dilation. The fly in this ointment is that it takes a great deal of energy to do this. Suppose you want to go to the immense red giant Betelgeuse, which is about 500 light-years away. (A light-year is the distance that light travels in a year.) You plan to travel at constant speed in a 1000-kg rocket ship (a little over a ton), which, in reality, is far too small for this purpose. In each case that follows, calculate the time for the trip, as measured by people on earth and by astronauts in the rocket ship, the energy needed in joules, and the energy needed as a percentage of U.S. yearly use (which is 1.0×10^{20} J). For comparison, arrange your results in a table showing v_{rocket}, t_{earth}, t_{rocket}, E (in J), and E (as % of U.S. use). The rocket ship's speed is (a) 0.50c; (b) 0.99c; (c) 0.9999c. On the basis of your results, does it seem likely that any government will invest in such high-speed space travel any time soon?

37.70 •• A spaceship moving at constant speed u relative to us broadcasts a radio signal at constant frequency f_0. As the spaceship approaches us, we receive a higher frequency f; after it has passed, we receive a lower frequency. (a) As the spaceship passes by, so it is instantaneously moving neither toward nor away from us, show that the frequency we receive is not f_0, and derive an expression for the frequency we do receive. Is the frequency we receive higher or lower than f_0? (*Hint:* In this case, successive wave crests move the same distance to the observer and so they

have the same transit time. Thus f equals $1/T$. Use the time dilation formula to relate the periods in the stationary and moving frames.) (b) A spaceship emits electromagnetic waves of frequency $f_0 = 345$ MHz as measured in a frame moving with the ship. The spaceship is moving at a constant speed $0.758c$ relative to us. What frequency f do we receive when the spaceship is approaching us? When it is moving away? In each case what is the shift in frequency, $f - f_0$? (c) Use the result of part (a) to calculate the frequency f and the frequency shift $(f - f_0)$ we receive at the instant that the ship passes by us. How does the shift in frequency calculated here compare to the shifts calculated in part (b)?

37.71 • **CP** In a particle accelerator a proton moves with constant speed $0.750c$ in a circle of radius 628 m. What is the net force on the proton?

37.72 •• **CP** The French physicist Armand Fizeau was the first to measure the speed of light accurately. He also found experimentally that the speed, relative to the lab frame, of light traveling in a tank of water that is itself moving at a speed V relative to the lab frame is

$$v = \frac{c}{n} + kV$$

where $n = 1.333$ is the index of refraction of water. Fizeau called k the dragging coefficient and obtained an experimental value of $k = 0.44$. What value of k do you calculate from relativistic transformations?

CHALLENGE PROBLEMS

37.73 ••• **CALC Lorentz Transformation for Acceleration.** Using a method analogous to the one in the text to find the Lorentz transformation formula for velocity, we can find the Lorentz transformation for *acceleration*. Let frame S' have a constant x-component of velocity u relative to frame S. An object moves relative to frame S along the x-axis with instantaneous velocity v_x and instantaneous acceleration a_x. (a) Show that its instantaneous acceleration in frame S' is

$$a'_x = a_x \left(1 - \frac{u^2}{c^2}\right)^{3/2}\left(1 - \frac{uv_x}{c^2}\right)^{-3}$$

[*Hint:* Express the acceleration in S' as $a'_x = dv'_x/dt'$. Then use Eq. (37.21) to express dt' in terms of dt and dx, and use Eq. (37.22) to express dv'_x in terms of u and dv_x. The velocity of the object in S is $v_x = dx/dt$.] (b) Show that the acceleration in frame S can be expressed as

$$a_x = a'_x \left(1 - \frac{u^2}{c^2}\right)^{3/2}\left(1 + \frac{uv'_x}{c^2}\right)^{-3}$$

where $v'_x = dx'/dt'$ is the velocity of the object in frame S'.

37.74 ••• **CALC A Realistic Version of the Twin Paradox.** A rocket ship leaves the earth on January 1, 2100. Stella, one of a pair of twins born in the year 2075, pilots the rocket (reference frame S'); the other twin, Terra, stays on the earth (reference frame S). The rocket ship has an acceleration of constant magnitude g in its own reference frame (this makes the pilot feel at home, since it simulates the earth's gravity). The path of the rocket ship is a straight line in the $+x$-direction in frame S. (a) Using the results of Challenge Problem 37.73, show that in Terra's earth frame S, the rocket's acceleration is

$$\frac{du}{dt} = g\left(1 - \frac{u^2}{c^2}\right)^{3/2}$$

where u is the rocket's instantaneous velocity in frame S. (b) Write the result of part (a) in the form $dt = f(u) \, du$, where $f(u)$ is a function of u, and integrate both sides. (*Hint:* Use the integral given in Problem 37.63.) Show that in Terra's frame, the time when Stella attains a velocity v_{1x} is

$$t_1 = \frac{v_{1x}}{g\sqrt{1 - v_{1x}^2/c^2}}$$

(c) Use the time dilation formula to relate dt and dt' (infinitesimal time intervals measured in frames S and S', respectively). Combine this result with the result of part (a) and integrate as in part (b) to show the following: When Stella attains a velocity v_{1x} relative to Terra, the time t_1' that has elapsed in frame S' is

$$t_1' = \frac{c}{g}\operatorname{arctanh}\left(\frac{v_{1x}}{c}\right)$$

Here arctanh is the inverse hyperbolic tangent. (*Hint:* Use the integral given in Challenge Problem 5.124.) (d) Combine the results of parts (b) and (c) to find t_1 in terms of t_1', g, and c alone. (e) Stella accelerates in a straight-line path for five years (by her clock), slows down at the same rate for five years, turns around, accelerates for five years, slows down for five years, and lands back on the earth. According to Stella's clock, the date is January 1, 2120. What is the date according to Terra's clock?

37.75 ••• CP Determining the Masses of Stars. Many of the stars in the sky are actually *binary stars,* in which two stars orbit about their common center of mass. If the orbital speeds of the stars are high enough, the motion of the stars can be detected by the Doppler shifts of the light they emit. Stars for which this is the case are called *spectroscopic binary stars.* Figure P37.75 shows the simplest case of a spectroscopic binary star: two identical stars, each with mass m, orbiting their center of mass in a circle of radius R. The plane of the stars' orbits is edge-on to the line of sight of an observer on the earth. (a) The light produced by heated hydrogen gas in a laboratory on the earth has a frequency of 4.568110×10^{14} Hz. In the light received from the stars by a telescope on the earth, hydrogen light is observed to vary in frequency between 4.567710×10^{14} Hz and 4.568910×10^{14} Hz. Determine whether the binary star system as a whole is moving toward or away from the earth, the speed of this motion, and the orbital speeds of the stars. (*Hint:* The speeds involved are much less than c, so you may use the approximate result $\Delta f/f = u/c$ given in Section 37.6.) (b) The light from each star in the binary system varies from its maximum frequency to its minimum frequency and back again in 11.0 days. Determine the orbital radius R and the mass m of each star. Give your answer for m in kilograms and as a multiple of the mass of the sun, 1.99×10^{30} kg. Compare the value of R to the distance from the earth to the sun, 1.50×10^{11} m. (This technique is actually used in astronomy to determine the masses of stars. In practice, the problem is more complicated because the two stars in

Figure **P37.75**

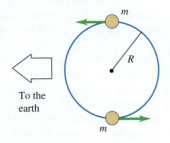

a binary system are usually not identical, the orbits are usually not circular, and the plane of the orbits is usually tilted with respect to the line of sight from the earth.)

37.76 ••• CP CALC Relativity and the Wave Equation. (a) Consider the Galilean transformation along the x-direction: $x' = x - vt$ and $t' = t$. In frame S the wave equation for electromagnetic waves in a vacuum is

$$\frac{\partial^2 E(x,t)}{\partial x^2} - \frac{1}{c^2}\frac{\partial^2 E(x,t)}{\partial t^2} = 0$$

where E represents the electric field in the wave. Show that by using the Galilean transformation the wave equation in frame S' is found to be

$$\left(1 - \frac{v^2}{c^2}\right)\frac{\partial^2 E(x',t')}{\partial x'^2} + \frac{2v}{c^2}\frac{\partial^2 E(x',t')}{\partial x' \partial t'} - \frac{1}{c^2}\frac{\partial^2 E(x',t')}{\partial t'^2} = 0$$

This has a different form than the wave equation in S. Hence the Galilean transformation *violates* the first relativity postulate that all physical laws have the same form in all inertial reference frames. (*Hint:* Express the derivatives $\partial/\partial x$ and $\partial/\partial t$ in terms of $\partial/\partial x'$ and $\partial/\partial t'$ by use of the chain rule.) (b) Repeat the analysis of part (a), but use the Lorentz coordinate transformations, Eqs. (37.21), and show that in frame S' the wave equation has the same form as in frame S:

$$\frac{\partial^2 E(x',t')}{\partial x^2} - \frac{1}{c^2}\frac{\partial^2 E(x',t')}{\partial t'^2} = 0$$

Explain why this shows that the speed of light in vacuum is c in both frames S and S'.

37.77 ••• CP Kaon Production. In high-energy physics, new particles can be created by collisions of fast-moving projectile particles with stationary particles. Some of the kinetic energy of the incident particle is used to create the mass of the new particle. A proton–proton collision can result in the creation of a negative kaon (K^-) and a positive kaon (K^+)

$$p + p \rightarrow p + p + K^- + K^+$$

(a) Calculate the minimum kinetic energy of the incident proton that will allow this reaction to occur if the second (target) proton is initially at rest. The rest energy of each kaon is 493.7 MeV, and the rest energy of each proton is 938.3 MeV. (*Hint:* It is useful here to work in the frame in which the total momentum is zero. But note that the Lorentz transformation must be used to relate the velocities in the laboratory frame to those in the zero-total-momentum frame.) (b) How does this calculated minimum kinetic energy compare with the total rest mass energy of the created kaons? (c) Suppose that instead the two protons are both in motion with velocities of equal magnitude and opposite direction. Find the minimum combined kinetic energy of the two protons that will allow the reaction to occur. How does this calculated minimum kinetic energy compare with the total rest mass energy of the created kaons? (This example shows that when colliding beams of particles are used instead of a stationary target, the energy requirements for producing new particles are reduced substantially.)

Answers

Chapter Opening Question ?

No. While the speed of light c is the ultimate "speed limit" for any particle, there is *no* upper limit on a particle's kinetic energy (see Fig. 37.21). As the speed approaches c, a small increase in speed corresponds to a large increase in kinetic energy.

Test Your Understanding Questions

37.1 Answers: (a) (i), (b) no You, too, will measure a spherical wave front that expands at the same speed c in all directions. This is a consequence of Einstein's second postulate. The wave front that you measure is *not* centered on the current position of the spaceship; rather, it is centered on the point P where the spaceship was located at the instant that it emitted the light pulse. For example, suppose the spaceship is moving at speed $c/2$. When your watch shows that a time t has elapsed since the pulse of light was emitted, your measurements will show that the wave front is a sphere of radius ct centered on P and that the spaceship is a distance $ct/2$ from P.

37.2 Answer: (iii) In Mavis's frame of reference, the two events (the Ogdenville clock striking noon and the North Haverbrook clock striking noon) are not simultaneous. Figure 37.5 shows that the event toward the front of the rail car occurs first. Since the rail car is moving toward North Haverbrook, that clock struck noon before the one on Ogdenville. So, according to Mavis, it is after noon in North Haverbrook.

37.3 Answers: (a) (ii), (b) (ii) The statement that moving clocks run slow refers to any clock that is moving relative to an observer. Maria and her stopwatch are moving relative to Samir, so Samir measures Maria's stopwatch to be running slow and to have ticked off fewer seconds than his own stopwatch. Samir and his stopwatch are moving relative to Maria, so she likewise measures Samir's stopwatch to be running slow. Each observer's measurement is correct for his or her own frame of reference. *Both* observers conclude that a moving stopwatch runs slow. This is consistent with the principle of relativity (see Section 37.1), which states that the laws of physics are the same in all inertial frames of reference.

37.4 Answer: (ii), (i) and (iii) (tie), (iv) You measure the rest length of the stationary meter stick and the contracted length of the moving spaceship to both be 1 meter. The rest length of the spaceship is greater than the contracted length that you measure, and so must be greater than 1 meter. A miniature observer on board the spaceship would measure a contracted length for the meter stick of less than 1 meter. Note that in your frame of reference the nose and tail of the spaceship can simultaneously align with the two ends of the meter stick, since in your frame of reference they have the same length of 1 meter. In the spaceship's frame these two alignments cannot happen simultaneously because the meter stick is shorter than the spaceship. Section 37.2 tells us that this shouldn't be a surprise; two events that are simultaneous to one observer may not be simultaneous to a second observer moving relative to the first one.

37.5 Answers: (a) P_1, (b) P_4 (a) The last of Eqs. (37.21) tells us the times of the two events in S': $t_1' = \gamma(t_1 - ux_1/c^2)$ and $t_2' = \gamma(t_2 - ux_2/c^2)$. In frame S the two events occur at the same x-coordinate, so $x_1 = x_2$, and event P_1 occurs before event P_2, so $t_1 < t_2$. Hence you can see that $t_1' < t_2'$ and event P_1 happens before P_2 in frame S', too. This says that if event P_1 happens before P_2 in a frame of reference S where the two events occur at the same position, then P_1 happens before P_2 in any other frame moving relative to S. **(b)** In frame S the two events occur at different x-coordinates such that $x_3 < x_4$, and events P_3 and P_4 occur at the same time, so $t_3 = t_4$. Hence you can see that $t_3' = \gamma(t_3 - ux_3/c^2)$ is greater than $t_4' = \gamma(t_4 - ux_4/c^2)$, so event P_4 happens before P_3 in frame S'. This says that even though the two events are simultaneous in frame S, they need not be simultaneous in a frame moving relative to S.

37.7 Answer: (ii) Equation (37.27) tells us that the magnitude of momentum of a particle with mass m and speed v is $p = mv/\sqrt{1 - v^2/c^2}$. If v increases by a factor of 2, the numerator mv increases by a factor of 2 *and* the denominator $\sqrt{1 - v^2/c^2}$ decreases. Hence p increases by a factor greater than 2. (Note that in order to double the speed, the initial speed must be less than $c/2$. That's because the speed of light is the ultimate speed limit.)

37.8 Answer: (i) As the proton moves a distance s, the constant force of magnitude F does work $W = Fs$ and increases the kinetic energy by an amount $\Delta K = W = Fs$. This is true no matter what the speed of the proton before moving this distance. Thus the constant force increases the proton's kinetic energy by the same amount during the first meter of travel as during any subsequent meter of travel. (It's true that as the proton approaches the ultimate speed limit of c, the increase in the proton's *speed* is less and less with each subsequent meter of travel. That's not what the question is asking, however.)

Bridging Problem

Answers: (a) $0.268c$ **(b)** 35.6 MeV **(c)** 145 MeV

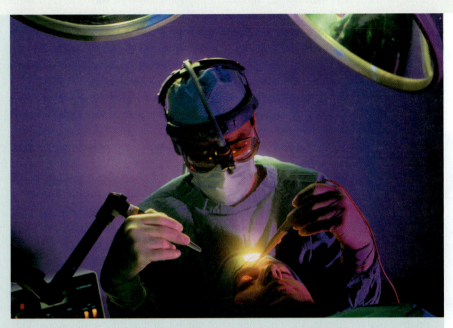

? This plastic surgeon is using two light sources: a headlamp that emits a beam of visible light and a handheld laser that emits infrared light. The light from both sources is emitted in the form of packets of energy called photons. For which source are the photons more energetic: the headlamp or the laser?

LEARNING GOALS

By studying this chapter, you will learn:

- How experiments involving the photoelectric effect and x rays pointed the way to a radical reinterpretation of the nature of light.

- How Einstein's photon picture of light explains the photoelectric effect.

- How experiments with x rays and gamma rays helped confirm the photon picture of light.

- How the wave and particle pictures of light complement each other.

- How the Heisenberg uncertainty principle imposes fundamental limits on what can be measured.

In Chapter 32 we saw how Maxwell, Hertz, and others established firmly that light is an electromagnetic wave. Interference, diffraction, and polarization, discussed in Chapters 35 and 36, further demonstrate this *wave nature* of light.

When we look more closely at the emission, absorption, and scattering of electromagnetic radiation, however, we discover a completely different aspect of light. We find that the energy of an electromagnetic wave is *quantized;* it is emitted and absorbed in particle-like packages of definite energy, called *photons*. The energy of a single photon is proportional to the frequency of the radiation.

We'll find that light and other electromagnetic radiation exhibits *wave–particle duality:* Light acts sometimes like waves and sometimes like particles. Interference and diffraction demonstrate wave behavior, while emission and absorption of photons demonstrate the particle behavior. This radical reinterpretation of light will lead us in the next chapter to no less radical changes in our views of the nature of matter.

38.1 Light Absorbed as Photons: The Photoelectric Effect

A phenomenon that gives insight into the nature of light is the **photoelectric effect,** in which a material emits electrons from its surface when illuminated (Fig. 38.1). To escape from the surface, an electron must absorb enough energy from the incident light to overcome the attraction of positive ions in the material. These attractions constitute a potential-energy barrier; the light supplies the "kick" that enables the electron to escape.

The photoelectric effect has a number of applications. Digital cameras and night-vision scopes use it to convert light energy into an electric signal that is

38.1 The photoelectric effect.

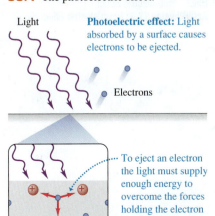

Light

Photoelectric effect: Light absorbed by a surface causes electrons to be ejected.

Electrons

To eject an electron the light must supply enough energy to overcome the forces holding the electron in the material.

38.2 (a) A night-vision scope makes use of the photoelectric effect. Photons entering the scope strike a plate, ejecting electrons that pass through a thin disk in which there are millions of tiny channels. The current through each channel is amplified electronically and then directed toward a screen that glows when hit by electrons. (b) The image formed on the screen, which is a combination of these millions of glowing spots, is thousands of times brighter than the naked-eye view.

(a)

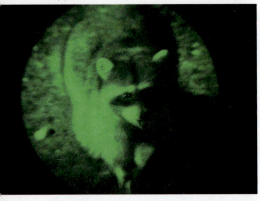

(b)

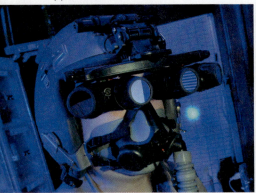

Mastering**PHYSICS**®

PhET: Photoelectric Effect
ActivPhysics 17.3: Photoelectric Effect

reconstructed into an image (Fig. 38.2). On the moon, sunlight striking the surface causes surface dust to eject electrons, leaving the dust particles with a positive charge. The mutual electric repulsion of these charged dust particles causes them to rise above the moon's surface, a phenomenon that was observed from lunar orbit by the Apollo astronauts.

Threshold Frequency and Stopping Potential

In Section 32.1 we explored the wave model of light, which Maxwell formulated two decades before the photoelectric effect was observed. Is the photoelectric effect consistent with this model? Figure 38.3a shows a modern version of one of the experiments that explored this question. Two conducting electrodes are enclosed in an evacuated glass tube and connected by a battery, and the cathode is illuminated. Depending on the potential difference V_{AC} between the two electrodes, electrons emitted by the illuminated cathode (called *photoelectrons*) may travel across to the anode, producing a *photocurrent* in the external circuit. (The tube is evacuated to a pressure of 0.01 Pa or less to minimize collisions between the electrons and gas molecules.)

The illuminated cathode emits photoelectrons with various kinetic energies. If the electric field points toward the cathode, as in Fig. 38.3a, all the electrons are accelerated toward the anode and contribute to the photocurrent. But by reversing the field and adjusting its strength as in Fig. 38.3b, we can prevent the less energetic electrons from reaching the anode. In fact, we can determine the *maximum* kinetic energy K_{max} of the emitted electrons by making the potential of the anode relative to the cathode, V_{AC}, just negative enough so that the current stops. This occurs for $V_{AC} = -V_0$, where V_0 is called the **stopping potential.** As an electron moves from the cathode to the anode, the potential decreases by V_0 and negative work $-eV_0$ is done on the (negatively charged) electron. The most energetic electron leaves the cathode with kinetic energy $K_{max} = \frac{1}{2}mv_{max}^2$ and has zero kinetic energy at the anode. Using the work–energy theorem, we have

$$W_{tot} = -eV_0 = \Delta K = 0 - K_{max} \quad \text{(maximum kinetic energy}$$
$$K_{max} = \frac{1}{2}mv_{max}^2 = eV_0 \quad \text{of photoelectrons)} \tag{38.1}$$

Hence by measuring the stopping potential V_0, we can determine the maximum kinetic energy with which electrons leave the cathode. (We are ignoring any effects due to differences in the materials of the cathode and anode.)

In this experiment, how do we expect the photocurrent to depend on the voltage across the electrodes and on the frequency and intensity of the light? Based on Maxwell's picture of light as an electromagnetic wave, here is what we *would* expect:

Wave-Model Prediction 1: We saw in Section 32.4 that the intensity of an electromagnetic wave depends on its amplitude but not on its frequency. So the photoelectric effect should occur for light of any frequency, and *the magnitude of the photocurrent should not depend on the frequency of the light.*

Wave-Model Prediction 2: It takes a certain minimum amount of energy, called the **work function,** to eject a single electron from a particular surface (see Fig. 38.1). If the light falling on the surface is very faint, some time may elapse before the total energy absorbed by the surface equals the work function. Hence, for faint illumination, *we expect a time delay* between when we switch on the light and when photoelectrons appear.

Wave-Model Prediction 3: Because the energy delivered to the cathode surface depends on the intensity of illumination, *we expect the stopping potential to increase with increasing light intensity.* Since intensity does not depend on frequency, we further expect that *the stopping potential should not depend on the frequency of the light.*

The experimental results proved to be *very* different from these predictions. Here is what was found in the years between 1877 and 1905:

Experimental Result 1: *The photocurrent depends on the light frequency.* For a given material, monochromatic light with a frequency below a minimum **threshold frequency** produces *no* photocurrent, regardless of intensity. For most metals the threshold frequency is in the ultraviolet (corresponding to wavelengths λ between 200 and 300 nm), but for other materials like potassium oxide and cesium oxide it is in the visible spectrum (λ between 380 and 750 nm).

Experimental Result 2: There is *no measurable time delay* between when the light is turned on and when the cathode emits photoelectrons (assuming the frequency of the light exceeds the threshold frequency). This is true no matter how faint the light is.

Experimental Result 3: *The stopping potential does not depend on intensity, but does depend on frequency.* Figure 38.4 shows graphs of photocurrent as a function of potential difference V_{AC} for light of a given frequency and two different intensities. The reverse potential difference $-V_0$ needed to reduce the current to zero is the same for both intensities. The only effect of increasing the intensity is to increase the number of electrons per second and hence the photocurrent i. (The curves level off when V_{AC} is large and positive because at that point all the emitted electrons are being collected by the anode.) If the intensity is held constant but the frequency is increased, the stopping potential also increases. In other words, the greater the light frequency, the higher the energy of the ejected photoelectrons.

These results directly contradict Maxwell's description of light as an electromagnetic wave. A solution to this dilemma was provided by Albert Einstein in 1905. His proposal involved nothing less than a new picture of the nature of light.

Einstein's Photon Explanation

Einstein made the radical postulate that a beam of light consists of small packages of energy called **photons** or *quanta.* This postulate was an extension of an idea developed five years earlier by Max Planck to explain the properties of blackbody radiation, which we discussed in Section 17.7. (We'll explore Planck's ideas in Section 39.5.) In Einstein's picture, the energy E of an individual photon is equal to a constant h times the photon frequency f. From the relationship $f = c/\lambda$ for electromagnetic waves in vacuum, we have

$$E = hf = \frac{hc}{\lambda} \quad \text{(energy of a photon)} \quad (38.2)$$

where h is a universal constant called **Planck's constant.** The numerical value of this constant, to the accuracy known at present, is

$$h = 6.62606896(33) \times 10^{-34} \text{ J} \cdot \text{s}$$

CAUTION **Photons are not "particles" in the usual sense** It's common to envision photons as miniature billiard balls or pellets. While that's a convenient mental picture, it's not very accurate. For one thing, billiard balls and bullets have a rest mass and travel slower than the speed of light c, while photons travel at the speed of light and have *zero* rest mass. For another thing, photons have wave aspects (frequency and wavelength) that are easy to observe. The fact is that the photon concept is a very strange one, and the true nature of photons is difficult to visualize in a simple way. We'll discuss the dual personality of photons in more detail in Section 38.4. ∎

In Einstein's picture, an individual photon arriving at the surface in Fig. 38.1a or 38.2 is absorbed by a single electron. This energy transfer is an all-or-nothing process, in contrast to the continuous transfer of energy in the wave theory of

38.3 An experiment testing whether the photoelectric effect is consistent with the wave model of light.

(a)

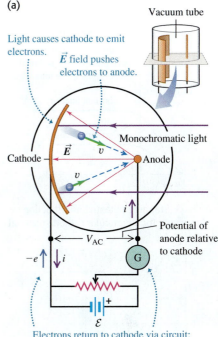

Light causes cathode to emit electrons.

$\vec{E}$ field pushes electrons to anode.

Vacuum tube

Cathode

Monochromatic light

Anode

$\vec{E}$

v

v

$-e$ i

V_{AC}

i

G

Potential of anode relative to cathode

$\mathcal{E}$

Electrons return to cathode via circuit; galvanometer measures current.

(b) We now reverse the electric field so that it tends to repel electrons from the anode. Above a certain field strength, electrons no longer reach the anode.

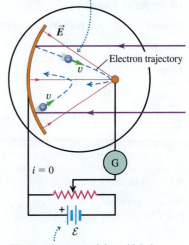

$\vec{E}$

v

v

Electron trajectory

$i = 0$

G

$\mathcal{E}$

The **stopping potential** at which the current ceases has absolute value V_0.

38.4 Photocurrent i for a constant light frequency f as a function of the potential V_{AC} of the anode with respect to the cathode.

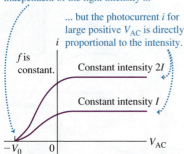

The stopping potential V_0 is independent of the light intensity ...

... but the photocurrent i for large positive V_{AC} is directly proportional to the intensity.

f is constant.

Constant intensity $2I$

Constant intensity I

38.5 Stopping potential as a function of frequency for a particular cathode material.

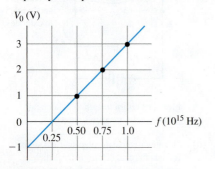

Table 38.1 Work Functions of Several Elements

Element	Work Function (eV)
Aluminum	4.3
Carbon	5.0
Copper	4.7
Gold	5.1
Nickel	5.1
Silicon	4.8
Silver	4.3
Sodium	2.7

38.6 Stopping potential as a function of frequency for two cathode materials having different work functions ϕ.

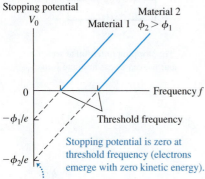

Stopping potential V_0

Material 1

Material 2 $\phi_2 > \phi_1$

Frequency f

$-\phi_1/e$

Threshold frequency

Stopping potential is zero at threshold frequency (electrons emerge with zero kinetic energy).

$-\phi_2/e$

For each material,

$eV = hf - \phi$ or $V_0 = \dfrac{h}{e} - \dfrac{\phi}{e}$

so the plots have same slope h/e but different intercepts $-f/e$ on the vertical axis.

light; the electron gets all of the photon's energy or none at all. The electron can escape from the surface only if the energy it acquires is greater than the work function ϕ. Thus photoelectrons will be ejected only if $hf > \phi$, or $f > \phi/h$. Einstein's postulate therefore explains why the photoelectric effect occurs only for frequencies greater than a minimum threshold frequency. This postulate is also consistent with the observation that greater intensity causes a greater photocurrent (Fig. 38.4). Greater intensity at a particular frequency means a greater number of photons per second absorbed, and thus a greater number of electrons emitted per second and a greater photocurrent.

Einstein's postulate also explains why there is no delay between illumination and the emission of photoelectrons. As soon as photons of sufficient energy strike the surface, electrons can absorb them and be ejected.

Finally, Einstein's postulate explains why the stopping potential for a given surface depends only on the light frequency. Recall that ϕ is the *minimum* energy needed to remove an electron from the surface. Einstein applied conservation of energy to find that the *maximum* kinetic energy $K_{max} = \frac{1}{2}mv_{max}^2$ for an emitted electron is the energy hf gained from a photon minus the work function ϕ:

$$K_{max} = \tfrac{1}{2}mv_{max}^2 = hf - \phi \tag{38.3}$$

Substituting $K_{max} = eV_0$ from Eq. (38.1), we find

$$eV_0 = hf - \phi \quad \text{. (photoelectric effect)} \tag{38.4}$$

Equation (38.4) shows that the stopping potential V_0 increases with increasing frequency f. The intensity doesn't appear in Eq. (38.4), so V_0 is independent of intensity. As a check of Eq. (38.4), we can measure the stopping potential V_0 for each of several values of frequency f for a given cathode material (Fig. 38.5). A graph of V_0 as a function of f turns out to be a straight line, verifying Eq. (38.4), and from such a graph we can determine both the work function ϕ for the material and the value of the quantity h/e. After the electron charge $-e$ was measured by Robert Millikan in 1909, Planck's constant h could also be determined from these measurements.

Electron energies and work functions are usually expressed in electron volts (eV), defined in Section 23.2. To four significant figures,

$$1\ eV = 1.602 \times 10^{-19}\ J$$

To this accuracy, Planck's constant is

$$h = 6.626 \times 10^{-34}\ J \cdot s = 4.136 \times 10^{-15}\ eV \cdot s$$

Table 38.1 lists the work functions of several elements. These values are approximate because they are very sensitive to surface impurities. The greater the work function, the higher the minimum frequency needed to emit photoelectrons (Fig. 38.6).

The photon picture explains a number of other phenomena in which light is absorbed. One example is a *suntan*, which is caused when the energy in sunlight triggers a chemical reaction in skin cells that leads to increased production of the pigment melanin. This reaction can occur only if a specific molecule in the cell absorbs a certain minimum amount of energy. A short-wavelength ultraviolet photon has enough energy to trigger the reaction, but a longer-wavelength visible-light photon does not. Hence ultraviolet light causes tanning, while visible light cannot.

Photon Momentum

Einstein's photon concept applies to *all* regions of the electromagnetic spectrum, including radio waves, x rays, and so on. A photon of any electromagnetic radiation with frequency f and wavelength λ has energy E given by Eq. (38.2).

Furthermore, according to the special theory of relativity, every particle that has energy must also have momentum, even if it has no rest mass. Photons have zero rest mass. As we saw in Eq. (37.40), a particle with zero rest mass and energy E has momentum with magnitude p given by $E = pc$. Thus the wavelength λ of a photon and the magnitude of its momentum p are related simply by

$$p = \frac{E}{c} = \frac{hf}{c} = \frac{h}{\lambda} \qquad \text{(momentum of a photon)} \qquad (38.5)$$

The direction of the photon's momentum is simply the direction in which the electromagnetic wave is moving.

Problem-Solving Strategy 38.1 — Photons

IDENTIFY *the relevant concepts:* The energy and momentum of an individual photon are proportional to the frequency and inversely proportional to the wavelength. Einstein's interpretation of the photoelectric effect is that energy is conserved as a photon ejects an electron from a material surface.

SET UP *the problem:* Identify the target variable. It could be the photon's wavelength λ, frequency f, energy E, or momentum p. If the problem involves the photoelectric effect, the target variable could be the maximum kinetic energy of photoelectrons K_{max}, the stopping potential V_0, or the work function ϕ.

EXECUTE *the solution* as follows:
1. Use Eqs. (38.2) and (38.5) to relate the energy and momentum of a photon to its wavelength and frequency. If the problem involves the photoelectric effect, use Eqs. (38.1), (38.3), and (38.4) to relate the photon frequency, stopping potential, work function, and maximum photoelectron kinetic energy.
2. The electron volt (eV), which we introduced in Section 23.2, is a convenient unit. It is the kinetic energy gained by an electron when it moves freely through an increase of potential of one volt: 1 eV $= 1.602 \times 10^{-19}$ J. If the photon energy E is given in electron volts, use $h = 4.136 \times 10^{-15}$ eV $\cdot$ s; if E is in joules, use $h = 6.626 \times 10^{-34}$ J $\cdot$ s.

EVALUATE *your answer:* In problems involving photons, at first the numbers will be unfamiliar to you and errors will not be obvious. It helps to remember that a visible-light photon with $\lambda = 600$ nm and $f = 5 \times 10^{14}$ Hz has an energy E of about 2 eV, or about 3×10^{-19} J.

Example 38.1 — Laser-pointer photons

A laser pointer with a power output of 5.00 mW emits red light ($\lambda = 650$ nm). (a) What is the magnitude of the momentum of each photon? (b) How many photons does the laser pointer emit each second?

SOLUTION

IDENTIFY and SET UP: This problem involves the ideas of (a) photon momentum and (b) photon energy. In part (a) we'll use Eq. (38.5) and the given wavelength to find the magnitude of each photon's momentum. In part (b), Eq. (38.2) gives the energy per photon, and the power output tells us the energy emitted per second. We can combine these quantities to calculate the number of photons emitted per second.

EXECUTE: (a) We have $\lambda = 650$ nm $= 6.50 \times 10^{-7}$ m, so from Eq. (38.5) the photon momentum is

$$p = \frac{h}{\lambda} = \frac{6.626 \times 10^{-34} \text{ J} \cdot \text{s}}{6.50 \times 10^{-7} \text{ m}}$$
$$= 1.02 \times 10^{-27} \text{ kg} \cdot \text{m/s}$$

(Recall that 1 J $= 1$ kg $\cdot$ m^2/s^2.)

(b) From Eq. (38.2), the energy of a single photon is

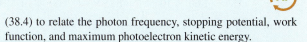

$$E = pc = (1.02 \times 10^{-27} \text{ kg} \cdot \text{m/s})(3.00 \times 10^8 \text{ m/s})$$
$$= 3.06 \times 10^{-19} \text{ J} = 1.91 \text{ eV}$$

The laser pointer emits energy at the rate of 5.00×10^{-3} J/s, so it emits photons at the rate of

$$\frac{5.00 \times 10^{-3} \text{ J/s}}{3.06 \times 10^{-19} \text{ J/photon}} = 1.63 \times 10^{16} \text{ photons/s}$$

EVALUATE: The result in part (a) is very small; a typical oxygen molecule in room-temperature air has 2500 times more momentum. As a check on part (b), we can calculate the photon energy using Eq. (38.2):

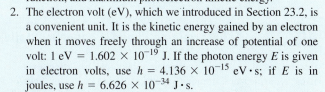

$$E = hf = \frac{hc}{\lambda} = \frac{(6.626 \times 10^{-34} \text{ J} \cdot \text{s})(3.00 \times 10^8 \text{ m/s})}{6.50 \times 10^{-7} \text{ m}}$$
$$= 3.06 \times 10^{-19} \text{ J} = 1.91 \text{ eV}$$

Our result in part (b) shows that a huge number of photons leave the laser pointer each second, each of which has an infinitesimal amount of energy. Hence the discreteness of the photons isn't noticed, and the radiated energy appears to be a continuous flow.

Example 38.2   A photoelectric-effect experiment

While conducting a photoelectric-effect experiment with light of a certain frequency, you find that a reverse potential difference of 1.25 V is required to reduce the current to zero. Find (a) the maximum kinetic energy and (b) the maximum speed of the emitted photoelectrons.

SOLUTION

IDENTIFY and SET UP: The value of 1.25 V is the stopping potential V_0 for this experiment. We'll use this in Eq. (38.1) to find the maximum photoelectron kinetic energy K_{max}, and from this we'll find the maximum photoelectron speed.

EXECUTE: (a) From Eq. (38.1),

$$K_{max} = eV_0 = (1.60 \times 10^{-19}\ \text{C})(1.25\ \text{V}) = 2.00 \times 10^{-19}\ \text{J}$$

(Recall that 1 V = 1 J/C.) In terms of electron volts,

$$K_{max} = eV_0 = e(1.25\ \text{V}) = 1.25\ \text{eV}$$

since the electron volt (eV) is the magnitude of the electron charge e times one volt (1 V).

(b) From $K_{max} = \frac{1}{2}mv_{max}^2$ we get

$$v_{max} = \sqrt{\frac{2K_{max}}{m}} = \sqrt{\frac{2(2.00 \times 10^{-19}\ \text{J})}{9.11 \times 10^{-31}\ \text{kg}}}$$
$$= 6.63 \times 10^5\ \text{m/s}$$

EVALUATE The value of v_{max} is about 0.2% of the speed of light, so we are justified in using the nonrelativistic expression for kinetic energy. (An equivalent justification is that the electron's 1.25-eV kinetic energy is much less than its rest energy $mc^2 = 0.511\ \text{MeV} = 5.11 \times 10^5\ \text{eV}$.)

Example 38.3   Determining ϕ and h experimentally

For a particular cathode material in a photoelectric-effect experiment, you measure stopping potentials $V_0 = 1.0$ V for light of wavelength $\lambda = 600$ nm, 2.0 V for 400 nm, and 3.0 V for 300 nm. Determine the work function ϕ for this material and the implied value of Planck's constant h.

SOLUTION

IDENTIFY and SET UP: This example uses the relationship among stopping potential V_0, frequency f, and work function ϕ in the photoelectric effect. According to Eq. (38.4), a graph of V_0 versus f should be a straight line as in Fig. 38.5 or 38.6. Such a graph is completely determined by its slope and the value at which it intercepts the vertical axis; we will use these to determine the values of the target variables ϕ and h.

EXECUTE: We rewrite Eq. (38.4) as

$$V_0 = \frac{h}{e}f - \frac{\phi}{e}$$

In this form we see that the slope of the line is h/e and the vertical-axis intercept (corresponding to $f = 0$) is $-\phi/e$. The frequencies,

obtained from $f = c/\lambda$ and $c = 3.00 \times 10^8$ m/s, are 0.50×10^{15} Hz, 0.75×10^{15} Hz, and 1.0×10^{15} Hz, respectively. From a graph of these data (see Fig. 38.6), we find

$$-\frac{\phi}{e} = \text{vertical intercept} = -1.0\ \text{V}$$
$$\phi = 1.0\ \text{eV} = 1.6 \times 10^{-19}\ \text{J}$$

and

$$\text{Slope} = \frac{\Delta V_0}{\Delta f} = \frac{3.0\ \text{V} - (-1.0\ \text{V})}{1.00 \times 10^{15}\ \text{s}^{-1} - 0} = 4.0 \times 10^{-15}\ \text{J} \cdot \text{s/C}$$
$$h = \text{slope} \times e = (4.0 \times 10^{-15}\ \text{J} \cdot \text{s/C})(1.60 \times 10^{-19}\ \text{C})$$
$$= 6.4 \times 10^{-34}\ \text{J} \cdot \text{s}$$

EVALUATE: The value of Planck's constant h determined from your experiment differs from the accepted value by only about 3%. The small value $\phi = 1.0$ eV tells us that the cathode surface is not composed solely of one of the elements in Table 38.1.

Application **Sterilizing with High-Energy Photons**
One technique for killing harmful microorganisms is to illuminate them with ultraviolet light with a wavelength shorter than 254 nm. If a photon of such short wavelength strikes a DNA molecule within a microorganism, the energy of the photon is great enough to break the bonds within the molecule. This renders the microorganism unable to grow or reproduce. Such ultraviolet germicidal irradiation is used for medical sanitation, to keep laboratories sterile (as shown here), and to treat both drinking water and wastewater.

Test Your Understanding of Section 38.1 Silicon films become better electrical conductors when illuminated by photons with energies of 1.14 eV or greater, an effect called *photoconductivity*. Which of the following wavelengths of electromagnetic radiation can cause photoconductivity in silicon films? (i) ultraviolet light with $\lambda = 300$ nm; (ii) red light with $\lambda = 600$ nm; (iii) infrared light with $\lambda = 1200$ nm.

38.2   Light Emitted as Photons: X-Ray Production

The photoelectric effect provides convincing evidence that light is *absorbed* in the form of photons. For physicists to accept Einstein's radical photon concept, however, it was also necessary to show that light is *emitted* as photons. An experiment that demonstrates this convincingly is the inverse of the photoelectric effect: Instead of releasing electrons from a surface by shining electromagnetic radiation on it, we cause a surface to emit radiation—specifically, *x rays*—by bombarding it with fast-moving electrons.

X-Ray Photons

X rays were first produced in 1895 by the German physicist Wilhelm Röntgen, using an apparatus similar in principle to the setup shown in Fig. 38.7. Electrons are released from the cathode by *thermionic emission,* in which the escape energy is supplied by heating the cathode to a very high temperature. (As in the photoelectric effect, the minimum energy that an individual electron must be given to escape from the cathode's surface is equal to the work function for the surface. In this case the energy is provided to the electrons by heat rather than by light.) The electrons are then accelerated toward the anode by a potential difference V_{AC}. The bulb is evacuated (residual pressure 10^{-7} atm or less), so the electrons can travel from the cathode to the anode without colliding with air molecules. When V_{AC} is a few thousand volts or more, x rays are emitted from the anode surface.

The anode produces x rays in part simply by slowing the electrons abruptly. (Recall from Section 32.1 that accelerated charges emit electromagnetic waves.) This process is called *bremsstrahlung* (German for "braking radiation"). Because the electrons undergo accelerations of very great magnitude, they emit much of their radiation at short wavelengths in the x-ray range, about 10^{-9} to 10^{-12} m (1 nm to 1 pm). (X-ray wavelengths can be measured quite precisely by crystal diffraction techniques, which we discussed in Section 36.6.) Most electrons are braked by a series of collisions and interactions with anode atoms, so bremsstrahlung produces a continuous spectrum of electromagnetic radiation.

Just as we did for the photoelectric effect in Section 38.1, let's compare what Maxwell's wave theory of electromagnetic radiation would predict about this radiation to what is observed experimentally.

Wave-Model Prediction: The electromagnetic waves produced when an electron slams into the anode should be analogous to the sound waves produced by crashing cymbals together. These waves include sounds of all frequencies. By analogy, the x rays produced by bremsstrahlung should have a spectrum that includes *all* frequencies and hence *all* wavelengths.

Experimental Result: Figure 38.8 shows bremsstrahlung spectra using the same cathode and anode with four different accelerating voltages. We see that *not* all x-ray frequencies and wavelengths are emitted: Each spectrum has a maximum frequency f_{max} and a corresponding minimum wavelength λ_{min}. The greater the potential difference V_{AC}, the higher the maximum frequency and the shorter the minimum wavelength.

The wave model of electromagnetic radiation cannot explain these experimental results. But we can readily understand them using the photon model. An electron has charge $-e$ and gains kinetic energy eV_{AC} when accelerated through a potential increase V_{AC}. The most energetic photon (highest frequency and shortest wavelength) is produced if the electron is braked to a stop all at once when it hits the anode, so that all of its kinetic energy goes to produce one photon; that is,

$$eV_{AC} = hf_{max} = \frac{hc}{\lambda_{min}} \qquad \text{(bremsstrahlung)} \qquad (38.6)$$

(In this equation we neglect the work function of the target anode and the initial kinetic energy of the electrons "boiled off" from the cathode. These energies are very small compared to the kinetic energy eV_{AC} gained due to the potential

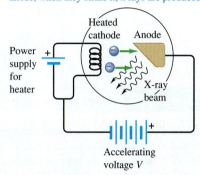

38.7 An apparatus used to produce x rays, similar to Röntgen's 1895 apparatus.

Electrons are emitted thermionically from the heated cathode and are accelerated toward the anode; when they strike it, x rays are produced.

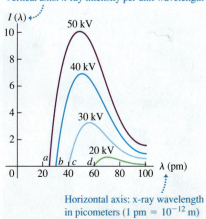

38.8 The continuous spectrum of x rays produced when a tungsten target is struck by electrons accelerated through a voltage V_{AC}. The curves represent different values of V_{AC}; points a, b, c, and d show the minimum wavelength for each voltage.

Vertical axis: x-ray intensity per unit wavelength

Horizontal axis: x-ray wavelength in picometers (1 pm $= 10^{-12}$ m)

difference.) If only a portion of an electron's kinetic energy goes into producing a photon, the photon energy will be less than eV_{AC} and the wavelength will be greater than λ_{min}. As further support for the photon model, the measured values for λ_{min} for different values of eV_{AC} (see Fig. 38.8) agree with Eq. (38.6). Note that according to Eq. (38.6), the maximum frequency and minimum wavelength in the bremsstrahlung process do not depend on the target material; this also agrees with experiment. So we can conclude that the photon picture of electromagnetic radiation is valid for the *emission* as well as the absorption of radiation.

The apparatus shown in Fig. 38.7 can also produce x rays by a second process in which electrons transfer their kinetic energy partly or completely to individual atoms within the target. It turns out that this process not only is consistent with the photon model of electromagnetic radiation, but also provides insight into the structure of atoms. We'll return to this process in Section 41.5.

Example 38.4 **Producing x rays**

Electrons in an x-ray tube accelerate through a potential difference of 10.0 kV before striking a target. If an electron produces one photon on impact with the target, what is the minimum wavelength of the resulting x rays? Find the answer by expressing energies in both SI units and electron volts.

SOLUTION

IDENTIFY and SET UP: To produce an x-ray photon with minimum wavelength and hence maximum energy, all of the electron's kinetic energy must go into producing a single x-ray photon. We'll use Eq. (38.6) to determine the wavelength.

EXECUTE: From Eq. (38.6), using SI units we have

$$\lambda_{min} = \frac{hc}{eV_{AC}} = \frac{(6.626 \times 10^{-34} \text{ J} \cdot \text{s})(3.00 \times 10^8 \text{ m/s})}{(1.602 \times 10^{-19} \text{ C})(10.0 \times 10^3 \text{ V})}$$

$$= 1.24 \times 10^{-10} \text{ m} = 0.124 \text{ nm}$$

Using electron volts, we have

$$\lambda_{min} = \frac{hc}{eV_{AC}} = \frac{(4.136 \times 10^{-15} \text{ eV} \cdot \text{s})(3.00 \times 10^8 \text{ m/s})}{e(10.0 \times 10^3 \text{ V})}$$

$$= 1.24 \times 10^{-10} \text{ m} = 0.124 \text{ nm}$$

In the second calculation, the "e" for the magnitude of the electron charge cancels the "e" in the unit "eV," because the electron volt (eV) is the magnitude of the electron charge e times one volt (1 V).

EVALUATE: To check our result, recall from Example 38.1 that a 1.91-eV photon has a wavelength of 650 nm. Here the electron energy, and therefore the x-ray photon energy, is 10.0×10^3 eV = 10.0 keV, about 5000 times greater than in Example 38.1, and the wavelength is about $\frac{1}{5000}$ as great as in Example 38.1. This makes sense, since wavelength and photon energy are inversely proportional.

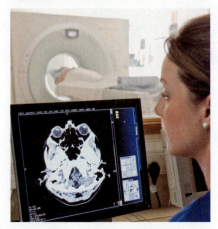

38.9 This radiologist is operating a CT scanner (seen through the window) from a separate room to avoid repeated exposure to x rays.

Applications of X Rays

X rays have many practical applications in medicine and industry. Because x-ray photons are of such high energy, they can penetrate several centimeters of solid matter. Hence they can be used to visualize the interiors of materials that are opaque to ordinary light, such as broken bones or defects in structural steel. The object to be visualized is placed between an x-ray source and an electronic detector (like that used in a digital camera) or a piece of photographic film. The darker an area in the image recorded by such a detector, the greater the radiation exposure. Bones are much more effective x-ray absorbers than soft tissue, so bones appear as light areas. A crack or air bubble allows greater transmission and shows as a dark area.

A widely used and vastly improved x-ray technique is *computed tomography;* the corresponding instrument is called a *CT scanner.* The x-ray source produces a thin, fan-shaped beam that is detected on the opposite side of the subject by an array of several hundred detectors in a line. Each detector measures absorption along a thin line through the subject. The entire apparatus is rotated around the subject in the plane of the beam, and the changing photon-counting rates of the detectors are recorded digitally. A computer processes this information and reconstructs a picture of absorption over an entire cross section of the subject (see Fig. 38.9). Differences in absorption as small as 1% or less can be detected with CT scans, and tumors and other anomalies that are much too small to be seen with older x-ray techniques can be detected.

X rays cause damage to living tissues. As x-ray photons are absorbed in tissues, their energy breaks molecular bonds and creates highly reactive free radicals (such as neutral H and OH), which in turn can disturb the molecular structure of proteins and especially genetic material. Young and rapidly growing cells are particularly susceptible, which is why x rays are useful for selective destruction of cancer cells. Conversely, however, a cell may be damaged by radiation but survive, continue dividing, and produce generations of defective cells; thus x rays can *cause* cancer.

Even when the organism itself shows no apparent damage, excessive exposure to x rays can cause changes in the organism's reproductive system that will affect its offspring. A careful assessment of the balance between risks and benefits of radiation exposure is essential in each individual case.

Test Your Understanding of Section 38.2   In the apparatus shown in Fig. 38.7, suppose you increase the number of electrons that are emitted from the cathode per second while keeping the potential difference V_{AC} the same. How will this affect the intensity I and minimum wavelength λ_{min} of the emitted x rays? (i) I and λ_{min} will both increase; (ii) I will increase but λ_{min} will be unchanged; (iii) I will increase but λ_{min} will decrease; (iv) I will remain the same but λ_{min} will decrease; (v) none of these.   ❚

38.3 Light Scattered as Photons: Compton Scattering and Pair Production

The final aspect of light that we must test against Einstein's photon model is its behavior after the light is produced and before it is eventually absorbed. We can do this by considering the *scattering* of light. As we discussed in Section 33.6, scattering is what happens when light bounces off particles such as molecules in the air.

Compton Scattering

Let's see what Maxwell's wave model and Einstein's photon model predict for how light behaves when it undergoes scattering by a single electron, such as an individual electron within an atom.

Wave-Model Prediction: In the wave description, scattering would be a process of absorption and re-radiation. Part of the energy of the light wave would be absorbed by the electron, which would oscillate in response to the oscillating electric field of the wave. The oscillating electron would act like a miniature antenna (see Section 32.1), re-radiating its acquired energy as *scattered* waves in a variety of directions. The frequency at which the electron oscillates would be the same as the frequency of the incident light, and the re-radiated light would have the same frequency as the oscillations of the electron. So, *in the wave model, the scattered light and incident light have the same frequency and same wavelength*.

Photon-Model Prediction: In the photon model we imagine the scattering process as a collision of two *particles,* the incident photon and an electron that is initially at rest (Fig. 38.10a). The incident photon would give up part of its energy and momentum to the electron, which recoils as a result of this impact. The scattered photon that remains can fly off at a variety of angles ϕ with respect to the incident direction, but it has less energy and less momentum than the incident photon (Fig. 38.10b). The energy and momentum of a photon are given by $E = hf = hc/\lambda$ (Eq. 38.2) and $p = hf/c = h/\lambda$ (Eq. 38.5). Therefore, *in the photon model, the scattered light has a lower frequency f and longer wavelength λ than the incident light*.

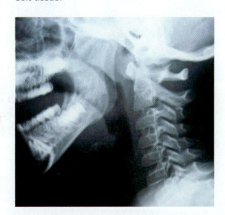

38.10 The photon model of light scattering by an electron.

(a) Before collision: The target electron is at rest.

Incident photon: wavelength λ, momentum $\vec{p}$ Target electron (at rest)

(b) After collision: The angle between the directions of the scattered photon and the incident photon is ϕ.

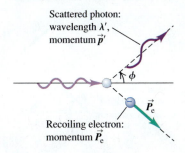

Scattered photon: wavelength λ', momentum $\vec{p}'$

ϕ

$\vec{P}_e$

Recoiling electron: momentum $\vec{P}_e$

38.11 A Compton-effect experiment.

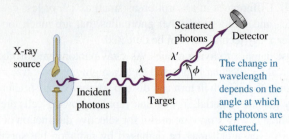

The change in wavelength depends on the angle at which the photons are scattered.

The definitive experiment that tested these predictions of the wave and photon models was carried out in 1922 by the American physicist Arthur H. Compton. In his experiment Compton aimed a beam of x rays at a solid target and measured the wavelength of the radiation scattered from the target (Fig. 38.11). He discovered that some of the scattered radiation has smaller frequency (longer wavelength) than the incident radiation and that the change in wavelength depends on the angle through which the radiation is scattered. This is precisely what the photon model predicts for light scattered from electrons in the target, a process that is now called **Compton scattering.**

Specifically, if the scattered radiation emerges at an angle ϕ with respect to the incident direction, as shown in Fig. 38.11, and if λ and λ' are the wavelengths of the incident and scattered radiation, respectively, Compton found that

$$\lambda' - \lambda = \frac{h}{mc}(1 - \cos\phi) \qquad \text{(Compton scattering)} \qquad (38.7)$$

where m is the electron rest mass. In other words, λ' is greater than λ. The quantity h/mc that appears in Eq. (38.7) has units of length. Its numerical value is

$$\frac{h}{mc} = \frac{6.626 \times 10^{-34} \text{ J} \cdot \text{s}}{(9.109 \times 10^{-31} \text{ kg})(2.998 \times 10^8 \text{ m/s})} = 2.426 \times 10^{-12} \text{ m}$$

Compton showed that Einstein's photon theory, combined with the principles of conservation of energy and conservation of momentum, provides a beautifully clear explanation of his experimental results. We outline the derivation below. The electron recoil energy may be in the relativistic range, so we have to use the relativistic energy–momentum relationships, Eqs. (37.39) and (37.40). The incident photon has momentum $\vec{p}$, with magnitude p and energy pc. The scattered photon has momentum $\vec{p}'$, with magnitude p' and energy $p'c$. The electron is initially at rest, so its initial momentum is zero and its initial energy is its rest energy mc^2. The final electron momentum $\vec{P}_e$ has magnitude P_e, and the final electron energy is given by $E_e^2 = (mc^2)^2 + (P_e c)^2$. Then energy conservation gives us the relationship

$$pc + mc^2 = p'c + E_e$$

Rearranging, we find

$$(pc - p'c + mc^2)^2 = E_e^2 = (mc^2)^2 + (P_e c)^2 \qquad (38.8)$$

We can eliminate the electron momentum $\vec{P}_e$ from Eq. (38.8) by using momentum conservation. From Fig. 38.12 we see that $\vec{p} = \vec{p}' + \vec{P}_e$, or

$$\vec{P}_e = \vec{p} - \vec{p}' \qquad (38.9)$$

38.12 Vector diagram showing conservation of momentum in Compton scattering.

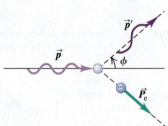

Conservation of momentum during Compton scattering

By taking the scalar product of each side of Eq. (38.9) with itself, we find

$$P_e^2 = p^2 + p'^2 - 2pp'\cos\phi \qquad (38.10)$$

We now substitute this expression for P_e^2 into Eq. (38.8) and multiply out the left side. We divide out a common factor c^2; several terms cancel, and when the resulting equation is divided through by (pp'), the result is

$$\frac{mc}{p'} - \frac{mc}{p} = 1 - \cos\phi \qquad (38.11)$$

Finally, we substitute $p' = h/\lambda'$ and $p = h/\lambda$, then multiply by h/mc to obtain Eq. (38.7).

When the wavelengths of x rays scattered at a certain angle are measured, the curve of intensity per unit wavelength as a function of wavelength has two peaks (Fig. 38.13). The longer-wavelength peak represents Compton scattering. The shorter-wavelength peak, labeled λ_0, is at the wavelength of the incident x rays and corresponds to x-ray scattering from tightly bound electrons. In such scattering processes the entire atom must recoil, so the m in Eq. (38.7) is the mass of the entire atom rather than of a single electron. The resulting wavelength shifts are negligible.

38.13 Intensity as a function of wavelength for photons scattered at an angle of 135° in a Compton-scattering experiment.

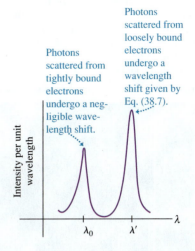

Photons scattered from tightly bound electrons undergo a negligible wavelength shift.

Photons scattered from loosely bound electrons undergo a wavelength shift given by Eq. (38.7).

Example 38.5 **Compton scattering**

You use 0.124-nm x-ray photons in a Compton-scattering experiment. (a) At what angle is the wavelength of the scattered x rays 1.0% longer than that of the incident x rays? (b) At what angle is it 0.050% longer?

SOLUTION

IDENTIFY and SET UP: We'll use the relationship between scattering angle and wavelength shift in the Compton effect. In each case our target variable is the angle ϕ (see Fig. 38.10b). We solve for ϕ using Eq. (38.7).

EXECUTE: (a) In Eq. (38.7) we want $\Delta\lambda = \lambda' - \lambda$ to be 1.0% of 0.124 nm, so $\Delta\lambda = 0.00124$ nm $= 1.24 \times 10^{-12}$ m. Using the value $h/mc = 2.426 \times 10^{-12}$ m, we find

$$\Delta\lambda = \frac{h}{mc}(1 - \cos\phi)$$

$$\cos\phi = 1 - \frac{\Delta\lambda}{h/mc} = 1 - \frac{1.24 \times 10^{-12} \text{ m}}{2.426 \times 10^{-12} \text{ m}} = 0.4889$$

$$\phi = 60.7°$$

(b) For $\Delta\lambda$ to be 0.050% of 0.124 nm, or 6.2×10^{-14} m,

$$\cos\phi = 1 - \frac{6.2 \times 10^{-14} \text{ m}}{2.426 \times 10^{-12} \text{ m}} = 0.9744$$

$$\phi = 13.0°$$

EVALUATE: Our results show that smaller scattering angles give smaller wavelength shifts. Thus in a grazing collision the photon energy loss and the electron recoil energy are smaller than when the scattering angle is larger. This is just what we would expect for an elastic collision, whether between a photon and an electron or between two billiard balls.

Pair Production

Another effect that can be explained only with the photon picture involves *gamma rays,* the shortest-wavelength and highest-frequency variety of electromagnetic radiation. If a gamma-ray photon of sufficiently short wavelength is fired at a target, it may not scatter. Instead, as depicted in Fig. 38.14, it may disappear completely and be replaced by two new particles: an electron and a **positron** (a particle that has the same rest mass m as an electron but has a positive charge $+e$ rather than the negative charge $-e$ of the electron). This process, called **pair production,** was first observed by the physicists Patrick Blackett and Giuseppe Occhialini in 1933. The electron and positron have to be produced in pairs in order to conserve electric charge: The incident photon has zero charge, and the electron–positron pair has net charge $(-e) + (+e) = 0$. Enough energy must be available to account for the rest energy $2mc^2$ of the two particles. To four significant figures, this minimum energy is

38.14 (a) Photograph of bubble-chamber tracks of electron–positron pairs that are produced when 300-MeV photons strike a lead sheet. A magnetic field directed out of the photograph made the electrons (e^-) and positrons (e^+) curve in opposite directions. (b) Diagram showing the pair-production process for two of the gamma-ray photons (γ).

(a)

Electron–positron pair

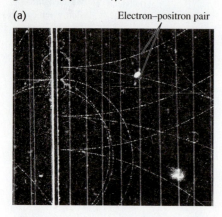

(b)

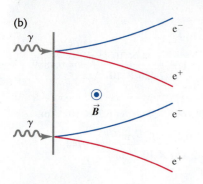

$$E_{min} = 2mc^2 = 2(9.109 \times 10^{-31} \text{ kg})(2.998 \times 10^8 \text{ m/s})^2$$
$$= 1.637 \times 10^{-13} \text{ J} = 1.022 \text{ MeV}$$

Thus the photon must have at least this much energy to produce an electron–positron pair. From Eq. (38.2), $E = hc/\lambda$, the photon wavelength has to be shorter than

$$\lambda_{max} = \frac{hc}{E_{min}} = \frac{(6.626 \times 10^{-34} \text{ J} \cdot \text{s})(2.998 \times 10^8 \text{ m/s})}{1.637 \times 10^{-13} \text{ J}}$$
$$= 1.213 \times 10^{-12} \text{ m} = 1.213 \times 10^{-3} \text{ nm} = 1.213 \text{ pm}$$

This is a very short wavelength, about $\frac{1}{1000}$ as large as the x-ray wavelengths that Compton used in his scattering experiments. (The requisite minimum photon energy is actually a bit higher than 1.022 MeV, so the photon wavelength must be a bit shorter than 1.213 pm. The reason is that when the incident photon encounters an atomic nucleus in the target, some of the photon energy goes into the kinetic energy of the recoiling nucleus.) Just as for the photoelectric effect, the wave model of electromagnetic radiation cannot explain why pair production occurs only when very short wavelengths are used.

The inverse process, *electron–positron pair annihilation*, occurs when a positron and an electron collide. Both particles disappear, and two (or occasionally three) photons can appear, with total energy of at least $2m_ec^2 = 1.022$ MeV. Decay into a *single* photon is impossible because such a process could not conserve both energy and momentum. It's easiest to analyze this annihilation process in the frame of reference called the *center-of-momentum system*, in which the total momentum is zero. It is the relativistic generalization of the center-of-mass system that we discussed in Section 8.5.

Example 38.6   Pair annihilation

An electron and a positron, initially far apart, move toward each other with the same speed. They collide head-on, annihilating each other and producing two photons. Find the energies, wavelengths, and frequencies of the photons if the initial kinetic energies of the electron and positron are (a) both negligible and (b) both 5.000 MeV. The electron rest energy is 0.511 MeV.

SOLUTION

IDENTIFY and SET UP: Just as in the elastic collisions we studied in Chapter 8, both momentum and energy are conserved in pair annihilation. The electron and positron are initially far apart, so the initial electric potential energy is zero and the initial energy is the sum of the particle kinetic and rest energies. The final energy is the sum of the photon energies. The total initial momentum is zero; the total momentum of the two photons must likewise be zero. We find the photon energy E using conservation of energy, conservation of momentum, and the relationship $E = pc$ (see Section 38.1). We then calculate the wavelengths and frequencies using $E = hc/\lambda = hf$.

EXECUTE: If the total momentum of the two photons is to be zero, their momenta must have equal magnitudes p and opposite directions. From $E = pc = hc/\lambda = hf$, the two photons must also have the same energy E, wavelength λ, and frequency f.

Before the collision the energy of each electron is $K + mc^2$, where K is its kinetic energy and $mc^2 = 0.511$ MeV. Conservation of energy then gives

$$(K + mc^2) + (K + mc^2) = E + E$$

Hence the energy of each photon is $E = K + mc^2$.

(a) In this case the electron kinetic energy K is negligible compared to its rest energy mc^2, so each photon has energy $E = mc^2 = 0.511$ MeV. The corresponding photon wavelength and frequency are

$$\lambda = \frac{hc}{E} = \frac{(4.136 \times 10^{-15} \text{ eV} \cdot \text{s})(3.00 \times 10^8 \text{ m/s})}{0.511 \times 10^6 \text{ eV}}$$
$$= 2.43 \times 10^{-12} \text{ m} = 2.43 \text{ pm}$$
$$f = \frac{E}{h} = \frac{0.511 \times 10^6 \text{ eV}}{4.136 \times 10^{-15} \text{ eV} \cdot \text{s}} = 1.24 \times 10^{20} \text{ Hz}$$

(b) In this case $K = 5.000$ MeV, so each photon has energy $E = 5.000$ MeV + 0.511 MeV = 5.511 MeV. Proceeding as in part (a), you can show that the photon wavelength is 0.2250 pm and the frequency is 1.333×10^{21} Hz.

EVALUATE: As a check, recall from Example 38.1 that a 650-nm visible-light photon has energy 1.91 eV and frequency 4.62×10^{14} Hz. The photon energy in part (a) is about 2.5×10^5 times greater. As expected, the photon's wavelength is shorter and its frequency higher than those for a visible-light photon by the same factor. You can check the results for part (b) in the same way.

Test Your Understanding of Section 38.3 If you used visible-light photons in the experiment shown in Fig. 38.11, would the photons undergo a wavelength shift due to the scattering? If so, is it possible to detect the shift with the human eye?

38.4 Wave–Particle Duality, Probability, and Uncertainty

We have studied many examples of the behavior of light and other electromagnetic radiation. Some, including the interference and diffraction effects described in Chapters 35 and 36, demonstrate conclusively the *wave* nature of light. Others, the subject of the present chapter, point with equal force to the *particle* nature of light. At first glance these two aspects seem to be in direct conflict. How can light be a wave and a particle at the same time?

We can find the answer to this apparent wave–particle conflict in the **principle of complementarity,** first stated by the Danish physicist Niels Bohr in 1928. The wave descriptions and the particle descriptions are complementary. That is, we need both to complete our model of nature, but we will never need to use both at the same time to describe a single part of an occurrence.

Diffraction and Interference in the Photon Picture

Let's start by considering again the diffraction pattern for a single slit, which we analyzed in Sections 36.2 and 36.3. Instead of recording the pattern on a digital camera chip or photographic film, we use a detector called a *photomultiplier* that can actually detect individual photons. Using the setup shown in Fig. 38.15, we place the photomultiplier at various positions for equal time intervals, count the photons at each position, and plot out the intensity distribution.

We find that, on average, the distribution of photons agrees with our predictions from Section 36.3. At points corresponding to the maxima of the pattern, we count many photons; at minimum points, we count almost none; and so on. The graph of the counts at various points gives the same diffraction pattern that we predicted with Eq. (36.7).

But suppose we now reduce the intensity to such a low level that only a few photons per second pass through the slit. We now record a series of discrete strikes, each representing a single photon. While we *cannot predict* where any given photon will strike, over time the accumulating strikes build up the familiar diffraction pattern we expect for a wave. To reconcile the wave and particle aspects of this pattern, we have to regard the pattern as a *statistical* distribution that tells us how many photons, on average, go to each spot. Equivalently, the pattern tells us the *probability* that any individual photon will land at a given spot. If we shine our faint light beam on a two-slit apparatus, we get an analogous result (Fig. 38.16). Again we can't predict exactly where an individual photon will go; the interference pattern is a statistical distribution.

How does the principle of complementarity apply to these diffraction and interference experiments? The wave description, not the particle description, explains the single- and double-slit patterns. But the particle description, not the wave description, explains why the photomultiplier records discrete packages of energy. The two descriptions complete our understanding of the results. For instance, suppose we consider an individual photon and ask how it knows "which way to go" when passing through the slit. This question seems like a conundrum, but that is because it is framed in terms of a *particle* description—whereas it is the *wave* nature of light that determines the distribution of photons. Conversely,

38.15 Single-slit diffraction pattern of light observed with a movable photomultiplier. The curve shows the intensity distribution predicted by the wave picture. The photon distribution is shown by the numbers of photons counted at various positions.

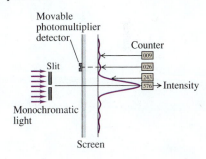

38.16 These images record the positions where individual photons in a two-slit interference experiment strike the screen. As more photons reach the screen, a recognizable interference pattern appears.

After 21 photons reach the screen

After 1000 photons reach the screen

After 10,000 photons reach the screen

the fact that the photomultiplier detects faint light as a sequence of individual "spots" can't be explained in wave terms.

Probability and Uncertainty

Although photons have energy and momentum, they are nonetheless very different from the particle model we used for Newtonian mechanics in Chapters 4 through 8. The Newtonian particle model treats an object as a point mass. We can describe the location and state of motion of such a particle at any instant with three spatial coordinates and three components of momentum, and we can then predict the particle's future motion. This model doesn't work at all for photons, however: We *cannot* treat a photon as a point object. This is because there are fundamental limitations on the precision with which we can simultaneously determine the position and momentum of a photon. Many aspects of a photon's behavior can be stated only in terms of *probabilities*. (In Chapter 39 we will find that the non-Newtonian ideas we develop for photons in this section also apply to particles such as electrons.)

To get more insight into the problem of measuring a photon's position and momentum simultaneously, let's look again at the single-slit diffraction of light (Fig. 38.17). Suppose the wavelength λ is much less than the slit width a. Then most (85%) of the photons go into the central maximum of the diffraction pattern, and the remainder go into other parts of the pattern. We use θ_1 to denote the angle between the central maximum and the first minimum. Using Eq. (36.2) with $m = 1$, we find that θ_1 is given by $\sin \theta_1 = \lambda/a$. Since we assume $\lambda = a$, it follows that θ_1 is very small, $\sin \theta_1$ is very nearly equal to θ_1 (in radians), and

$$\theta_1 = \frac{\lambda}{a} \qquad (38.12)$$

Even though the photons all have the same initial state of motion, they don't all follow the same path. We can't predict the exact trajectory of any individual photon from knowledge of its initial state; we can only describe the *probability* that an individual photon will strike a given spot on the screen. This fundamental indeterminacy has no counterpart in Newtonian mechanics.

Furthermore, there are fundamental *uncertainties* in both the position and the momentum of an individual particle, and these uncertainties are related inseparably. To clarify this point, let's go back to Fig. 38.17. A photon that strikes the screen at the outer edge of the central maximum, at angle θ_1, must have a component of momentum p_y in the y-direction, as well as a component p_x in the x-direction, despite the fact that initially the beam was directed along the x-axis. From the geometry of the situation the two components are related by $p_y/p_x = \tan \theta_1$. Since θ_1 is small, we may use the approximation $\tan \theta_1 = \theta_1$, and

38.17 Interpreting single-slit diffraction in terms of photon momentum.

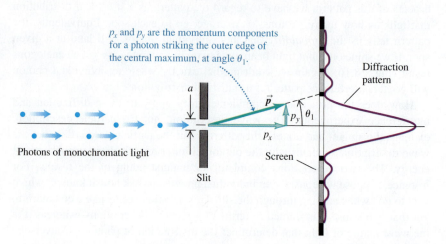

$$p_y = p_x \theta_1 \qquad (38.13)$$

Substituting Eq. (38.12), $\theta_1 = \lambda/a$, into Eq. (38.13) gives

$$p_y = p_x \frac{\lambda}{a} \qquad (38.14)$$

Equation (38.14) says that for the 85% of the photons that strike the detector within the central maximum (that is, at angles between $-\lambda/a$ and $+\lambda/a$), the y-component of momentum is spread out over a range from $-p_x\lambda/a$ to $+p_x\lambda/a$. Now let's consider *all* the photons that pass through the slit and strike the screen. Again, they may hit above or below the center of the pattern, so their component p_y may be positive or negative. However the symmetry of the diffraction pattern shows us the average value $(p_y)_{av} = 0$. There will be an *uncertainty* Δp_y in the y-component of momentum at least as great as $p_x\lambda/a$. That is,

$$\Delta p_y \geq p_x \frac{\lambda}{a} \qquad (38.15)$$

The narrower the slit width a, the broader is the diffraction pattern and the greater is the uncertainty in the y-component of momentum p_y.

The photon wavelength λ is related to the momentum p_x by Eq. (38.5), which we can rewrite as $\lambda = h/p_x$. Using this relationship in Eq. (38.15) and simplifying, we find

$$\Delta p_y \geq p_x \frac{h}{p_x a} = \frac{h}{a}$$

$$\Delta p_y a \geq h \qquad (38.16)$$

What does Eq. (38.16) mean? The slit width a represents an uncertainty in the y-component of the *position* of a photon as it passes through the slit. We don't know exactly *where* in the slit each photon passes through. So both the y-position and the y-component of momentum have uncertainties, and the two uncertainties are related by Eq. (38.16). We can reduce the *momentum* uncertainty Δp_y only by reducing the width of the diffraction pattern. To do this, we have to increase the slit width a, which increases the *position* uncertainty. Conversely, when we *decrease* the position uncertainty by narrowing the slit, the diffraction pattern broadens and the corresponding momentum uncertainty *increases*.

You may protest that it doesn't seem to be consistent with common sense for a photon not to have a definite position and momentum. We reply that what we call *common sense* is based on familiarity gained through experience. Our usual experience includes very little contact with the microscopic behavior of particles like photons. Sometimes we have to accept conclusions that violate our intuition when we are dealing with areas that are far removed from everyday experience.

The Uncertainty Principle

In more general discussions of uncertainty relationships, the uncertainty of a quantity is usually described in terms of the statistical concept of *standard deviation,* which is a measure of the spread or dispersion of a set of numbers around their average value. Suppose we now begin to describe uncertainties in this way [neither Δp_y nor a in Eq. (38.16) is a standard deviation]. If a coordinate x has an uncertainty Δx and if the corresponding momentum component p_x has an uncertainty Δp_x, then those standard-deviation uncertainties are found to be related in general by the inequality

$$\Delta x \, \Delta p_x \geq \hbar/2 \qquad \text{(Heisenberg uncertainty principle for position and momentum)} \qquad (38.17)$$

In this expression the quantity $\hbar$ (pronounced "h-bar") is Planck's constant divided by 2π:

$$\hbar = \frac{h}{2\pi} = 1.054571628(53) \times 10^{-34}\ \text{J} \cdot \text{s}$$

We will use this quantity frequently to avoid writing a lot of factors of 2π in later equations.

CAUTION **h versus h-bar** It's common for students to plug in the value of h when what they really wanted was $\hbar = h/2\pi$, or vice versa. Be careful not to make the same mistake, or you'll find yourself wondering why your answer is off by a factor of 2π! ❚

Equation (38.17) is one form of the **Heisenberg uncertainty principle,** first discovered by the German physicist Werner Heisenberg (1901–1976). It states that, in general, it is impossible to simultaneously determine both the position and the momentum of a particle with arbitrarily great precision, as classical physics would predict. Instead, the uncertainties in the two quantities play complementary roles, as we have described. Figure 38.18 shows the relationship between the two uncertainties. Our derivation of Eq. (38.16), a less refined form of the uncertainty principle given by Eq. (38.17), shows that this principle has its roots in the wave aspect of photons. We will see in Chapter 39 that electrons and other subatomic particles also have a wave aspect, and the same uncertainty principle applies to them as well.

It is tempting to suppose that we could get greater precision by using more sophisticated detectors of position and momentum. This turns out not to be possible. To detect a particle, the detector must *interact* with it, and this interaction unavoidably changes the state of motion of the particle, introducing uncertainty about its original state. For example, we could imagine placing an electron at a certain point in the middle of the slit in Fig. 38.17. If the photon passes through the middle, we would see the electron recoil. We would then know that the photon passed through that point in the slit, and we would be much more certain about the x-coordinate of the photon. However, the collision between the photon and the electron would change the photon momentum, giving us greater uncertainty in the value of that momentum. A more detailed analysis of such hypothetical experiments shows that the uncertainties we have described are fundamental and intrinsic. They *cannot* be circumvented *even in principle* by any experimental technique, no matter how sophisticated.

There is nothing special about the x-axis. In a three-dimensional situation with coordinates (x, y, z) there is an uncertainty relationship for each coordinate and its corresponding momentum component: $\Delta x \Delta p_x \geq \hbar/2$, $\Delta y \Delta p_y \geq \hbar/2$, and $\Delta z \Delta p_z \geq \hbar/2$. However, the uncertainty in one coordinate is *not* related to the uncertainty in a different component of momentum. For example, Δx is not related directly to Δp_y.

Waves and Uncertainty

Here's an alternative way to understand the Heisenberg uncertainty principle in terms of the properties of waves. Consider a sinusoidal electromagnetic wave propagating in the positive x-direction with its electric field polarized in the y-direction. If the wave has wavelength λ, frequency f, and amplitude A, we can write the wave function as

$$E_y(x, t) = A \sin(kx - \omega t) \tag{38.18}$$

In this expression the wave number is $k = 2\pi/\lambda$ and the angular frequency is $\omega = 2\pi f$. We can think of the wave function in Eq. (38.18) as a description of a photon with a definite wavelength and a definite frequency. In terms of k and ω we can express the momentum and energy of the photon as

$$p_x = \frac{h}{\lambda} = \frac{h}{2\pi} \frac{2\pi}{\lambda} = \hbar k \qquad \text{(photon momentum in terms of wave number)} \tag{38.19a}$$

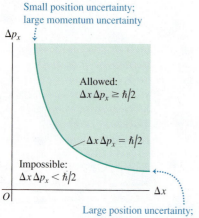

38.18 The Heisenberg uncertainty principle for position and momentum components. It is impossible for the product $\Delta x \Delta p_x$ to be less than $\hbar/2 = h/4\pi$.

Small position uncertainty; large momentum uncertainty

Allowed: $\Delta x \Delta p_x \geq \hbar/2$

$\Delta x \Delta p_x = \hbar/2$

Impossible: $\Delta x \Delta p_x < \hbar/2$

Large position uncertainty; small momentum uncertainty

$$E = hf = \frac{h}{2\pi}\, 2\pi f = \hbar\omega \quad \text{(photon energy in terms of angular frequency)} \quad \text{(38.19b)}$$

Using Eqs. (38.19) in Eq. (38.18), we can rewrite our photon wave equation as

$$E_y(x, t) = A\sin[(p_x x - Et)/\hbar] \quad \text{(wave function for a photon with } x\text{-momentum } p_x \text{ and energy } E) \quad \text{(38.20)}$$

Since this wave function has a definite value of x-momentum p_x, there is *no* uncertainty in the value of this quantity: $\Delta p_x = 0$. The Heisenberg uncertainty principle, Eq. (38.17), says that $\Delta x \Delta p_x \geq \hbar/2$. If Δp_x is zero, then Δx must be infinite. Indeed, the wave described by Eq. (38.20) extends along the entire x-axis and has the same amplitude everywhere. The price we pay for knowing the photon's momentum precisely is that we have no idea *where* the photon is!

In practical situations we always have *some* idea where a photon is. To describe this situation, we need a wave function that is more localized in space. We can create one by superimposing two or more sinusoidal functions. To keep things simple, we'll consider only waves propagating in the positive x-direction. For example, let's add together two sinusoidal wave functions like those in Eqs. (38.18) and (38.20), but with slightly different wavelengths and frequencies and hence slightly different values p_{x1} and p_{x2} of x-momentum and slightly different values E_1 and E_2 of energy. The total wave function is

$$E_y(x, t) = A_1\sin[(p_{1x}x - E_1 t)/\hbar] + A_2\sin[(p_{2x}x - E_2 t)/\hbar] \quad \text{(38.21)}$$

Consider what this wave function looks like at a particular instant of time, say $t = 0$. At this instant Eq. (38.21) becomes

$$E_y(x, t = 0) = A_1\sin(p_{1x}x/\hbar) + A_2\sin(p_{2x}x/\hbar) \quad \text{(38.22)}$$

Figure 38.19a is a graph of the individual wave functions at $t = 0$ for the case $A_2 = -A_1$, and Fig. 38.19b graphs the combined wave function $E_y(x, t = 0)$ given by Eq. (38.22). We saw something very similar to Fig. 38.19b in our discussion of beats in Section 16.7: When we superimposed two sinusoidal waves with slightly different frequencies (see Fig. 16.24), the resulting wave exhibited amplitude variations not present in the original waves. In the same way, a photon represented by the wave function in Eq. (38.21) is most likely to be found in the regions where the wave function's amplitude is greatest. That is, the photon is *localized*. However, the photon's momentum no longer has a definite value because we began with two different x-momentum values, p_{x1} and p_{x2}. This agrees with the Heisenberg uncertainty principle: By decreasing the uncertainty in the photon's position, we have increased the uncertainty in its momentum.

38.19 (a) Two sinusoidal waves with slightly different wave numbers k and hence slightly different values of momentum $p_x = \hbar k$ shown at one instant of time. (b) The superposition of these waves has a momentum equal to the average of the two individual values of momentum. The amplitude varies, giving the total wave a lumpy character not possessed by either individual wave.

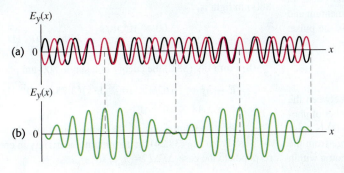

Uncertainty in Energy

Our discussion of combining waves also shows that there is an uncertainty principle that involves *energy* and *time*. To see why this is so, imagine measuring the combined wave function described by Eq. (38.21) at a certain position, say $x = 0$, over a period of time. At $x = 0$, the wave function from Eq. (38.21) becomes

$$E_y(x, t) = A_1 \sin(-E_1 t/\hbar) + A_2 \sin(-E_2 t/\hbar)$$
$$= -A_1 \sin(E_1 t/\hbar) - A_2 \sin(E_2 t/\hbar) \qquad (38.23)$$

What we measure at $x = 0$ is a combination of two oscillating electric fields with slightly different angular frequencies $\omega_1 = E_1/\hbar$ and $\omega_2 = E_2/\hbar$. This is exactly the phenomenon of beats that we discussed in Section 16.7 (compare Fig. 16.24). The amplitude of the combined field rises and falls, so the photon described by this field is localized in *time* as well as in position. The photon is most likely to be found at the times when the amplitude is large. The price we pay for localizing the photon in time is that the wave does not have a definite energy. By contrast, if the photon is described by a sinusoidal wave like that in Eq. (38.20) that *does* have a definite energy E but that has the same amplitude at all times, we have no idea when the photon will appear at $x = 0$. So the better we know the photon's energy, the less certain we are of when we will observe the photon.

Just as for the momentum–position uncertainty principle, we can write a mathematical expression for the uncertainty principle that relates energy and time. In fact, except for an overall minus sign, Eq. (38.23) is identical to Eq. (38.22) if we replace the x-momentum p_x by energy E and the position x by time t. This tells us that in the momentum–position uncertainty relation, Eq. (38.17), we can replace the momentum uncertainty Δp_x with the energy uncertainty ΔE and replace the position uncertainty Δx with the time uncertainty Δt. The result is

$$\Delta t\,\Delta E \geq \hbar/2 \qquad \text{(Heisenberg uncertainty principle for energy and time)} \qquad (38.24)$$

In practice, any real photon has a limited spatial extent and hence passes any point in a limited amount of time. The following example illustrates how this affects the momentum and energy of the photon.

Example 38.7   **Ultrashort laser pulses and the uncertainty principle**

Many varieties of lasers emit light in the form of pulses rather than a steady beam. A tellurium–sapphire laser can produce light at a wavelength of 800 nm in ultrashort pulses that last only 4.00×10^{-15} s (4.00 femtoseconds, or 4.00 fs). The energy in a single pulse produced by one such laser is $2.00\ \mu\text{J} = 2.00 \times 10^{-6}$ J, and the pulses propagate in the positive x-direction. Find (a) the frequency of the light; (b) the energy and minimum energy uncertainty of a single photon in the pulse; (c) the minimum frequency uncertainty of the light in the pulse; (d) the spatial length of the pulse, in meters and as a multiple of the wavelength; (e) the momentum and minimum momentum uncertainty of a single photon in the pulse; and (f) the approximate number of photons in the pulse.

SOLUTION

IDENTIFY and SET UP: It's important to distinguish between the light pulse as a whole (which contains a very large number of photons) and an individual photon within the pulse. The 5.00-fs pulse duration represents the time it takes the pulse to emerge from the laser; it is also the time *uncertainty* for an individual photon within

the pulse, since we don't know when during the pulse that photon emerges. Similarly, the position uncertainty of a photon is the spatial length of the pulse, since a given photon could be found anywhere within the pulse. To find our target variables, we'll use the relationships for photon energy and momentum from Section 38.1 and the two Heisenberg uncertainty principles, Eqs. (38.17) and (38.24).

EXECUTE: (a) From the relationship $c = \lambda f$, the frequency of 800-nm light is

$$f = \frac{c}{\lambda} = \frac{3.00 \times 10^8 \text{ m/s}}{8.00 \times 10^{-7} \text{ m}} = 3.75 \times 10^{14} \text{ Hz}$$

(b) From Eq. (38.2) the energy of a single 800-nm photon is

$$E = hf = (6.626 \times 10^{-34} \text{ J} \cdot \text{s})(3.75 \times 10^{14} \text{ Hz})$$
$$= 2.48 \times 10^{-19} \text{ J}$$

The time uncertainty equals the pulse duration, $\Delta t = 4.00 \times 10^{-15}$ s. From Eq. (38.24) the minimum uncertainty in energy corresponds to the case $\Delta t\,\Delta E = \hbar/2$, so

$$\Delta E = \frac{\hbar}{2\Delta t} = \frac{1.055 \times 10^{-34} \text{ J} \cdot \text{s}}{2(4.00 \times 10^{-15} \text{ s})} = 1.32 \times 10^{-20} \text{ J}$$

This is 5.3% of the photon energy $E = 2.48 \times 10^{-19}$ J, so the energy of a given photon is uncertain by at least 5.3%. The uncertainty could be greater, depending on the shape of the pulse.

(c) From the relationship $f = E/h$, the minimum frequency uncertainty is

$$\Delta f = \frac{\Delta E}{h} = \frac{1.32 \times 10^{-20} \text{ J}}{6.626 \times 10^{-34} \text{ J} \cdot \text{s}} = 1.99 \times 10^{13} \text{ Hz}$$

This is 5.3% of the frequency $f = 3.75 \times 10^{14}$ Hz we found in part (a). Hence these ultrashort pulses do not have a definite frequency; the average frequency of many such pulses will be 3.75×10^{14} Hz, but the frequency of any individual pulse can be anywhere from 5.3% higher to 5.3% lower.

(d) The spatial length Δx of the pulse is the distance that the front of the pulse travels during the time $\Delta t = 4.00 \times 10^{-15}$ s it takes the pulse to emerge from the laser:

$$\Delta x = c\Delta t = (3.00 \times 10^8 \text{ m/s})(4.00 \times 10^{-15} \text{ s})$$

$$= 1.20 \times 10^{-6} \text{ m}$$

$$\Delta x = \frac{1.20 \times 10^{-6} \text{ m}}{8.00 \times 10^{-7} \text{ m/wavelength}} = 1.50 \text{ wavelengths}$$

This justifies the term *ultrashort*. The pulse is less than two wavelengths long!

(e) From Eq. (38.5), the momentum of an average photon in the pulse is

$$p_x = \frac{E}{c} = \frac{2.48 \times 10^{-19} \text{ J}}{3.00 \times 10^8 \text{ m/s}} = 8.28 \times 10^{-28} \text{ kg} \cdot \text{m/s}$$

The spatial uncertainty is $\Delta x = 1.20 \times 10^{-6}$ m. From Eq. (38.17) minimum momentum uncertainty corresponds to $\Delta x \, \Delta p_x = \hbar/2$, so

$$\Delta p_x = \frac{\hbar}{2\Delta x} = \frac{1.055 \times 10^{-34} \text{ J} \cdot \text{s}}{2(1.20 \times 10^{-6} \text{ m})} = 4.40 \times 10^{-29} \text{ kg} \cdot \text{m/s}$$

This is 5.3% of the average photon momentum p_x. An individual photon within the pulse can have a momentum that is 5.3% greater or less than the average.

(f) To estimate the number of photons in the pulse, we divide the total pulse energy by the average photon energy:

$$\frac{2.00 \times 10^{-6} \text{ J/pulse}}{2.48 \times 10^{-19} \text{ J/photon}} = 8.06 \times 10^{12} \text{ photons/pulse}$$

The energy of an individual photon is uncertain, so this is the *average* number of photons per pulse.

EVALUATE: The percentage uncertainties in energy and momentum are large because this laser pulse is so short. If the pulse were longer, both Δt and Δx would be greater and the corresponding uncertainties in photon energy and photon momentum would be smaller.

Our calculation in part (f) shows an important distinction between photons and other kinds of particles. In principle it is possible to make an exact count of the number of electrons, protons, and neutrons in an object such as this book. If you repeated the count, you would get the same answer as the first time. By contrast, if you counted the number of photons in a laser pulse you would *not* necessarily get the same answer every time! The uncertainty in photon energy means that on each count there could be a different number of photons whose individual energies sum to 2.00×10^{-6} J. That's yet another of the many strange properties of photons.

Test Your Understanding of Section 38.4 Through which of the following angles is a photon of wavelength λ most likely to be deflected after passing through a slit of width a? Assume that λ is much less than a. (i) $\theta = \lambda/a$; (ii) $\theta = 3\lambda/2a$; (iii) $\theta = 2\lambda/a$; (iv) $\theta = 3\lambda/a$; (v) not enough information given to decide.

Photons: Electromagnetic radiation behaves as both waves and particles. The energy in an electromagnetic wave is carried in units called photons. The energy E of one photon is proportional to the wave frequency f and inversely proportional to the wavelength λ, and is proportional to a universal quantity h called Planck's constant. The momentum of a photon has magnitude E/c. (See Example 38.1.)

$$E = hf = \frac{hc}{\lambda} \qquad (38.2)$$

$$p = \frac{E}{c} = \frac{hf}{c} = \frac{h}{\lambda} \qquad (38.5)$$

The photoelectric effect: In the photoelectric effect, a surface can eject an electron by absorbing a photon whose energy hf is greater than or equal to the work function ϕ of the material. The stopping potential V_0 is the voltage required to stop a current of ejected electrons from reaching an anode. (See Examples 38.2 and 38.3.)

$$eV_0 = hf - \phi \qquad (38.4)$$

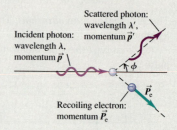

Photon production, photon scattering, and pair production: X rays can be produced when electrons accelerated to high kinetic energy across a potential increase V_{AC} strike a target. The photon model explains why the maximum frequency and minimum wavelength produced are given by Eq. (38.6). (See Example 38.4.) In Compton scattering a photon transfers some of its energy and momentum to an electron with which it collides. For free electrons (mass m), the wavelengths of incident and scattered photons are related to the photon scattering angle ϕ by Eq. (38.7). (See Example 38.5.) In pair production a photon of sufficient energy can disappear and be replaced by an electron–positron pair. In the inverse process, an electron and a positron can annihilate and be replaced by a pair of photons. (See Example 38.6.)

$$eV_{AC} = hf_{\max} = \frac{hc}{\lambda_{\min}} \qquad (38.6)$$
(bremsstrahlung)

$$\lambda' - \lambda = \frac{h}{mc}(1 - \cos\phi) \qquad (38.7)$$
(Compton scattering)

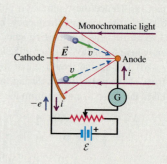

The Heisenberg uncertainty principle: It is impossible to determine both a photon's position and its momentum at the same time to arbitrarily high precision. The precision of such measurements for the x-components is limited by the Heisenberg uncertainty principle, Eq. (38.17); there are corresponding relationships for the y- and z-components. The uncertainty ΔE in the energy of a state that is occupied for a time Δt is given by Eq. (38.24). In these expressions, $\hbar = h/2\pi$. (See Example 38.7.)

$$\Delta x \, \Delta p_x \geq \hbar/2 \qquad (38.17)$$
(Heisenberg uncertainty principle for position and momentum)

$$\Delta t \, \Delta E \geq \hbar/2 \qquad (38.24)$$
(Heisenberg uncertainty principle for energy and time)

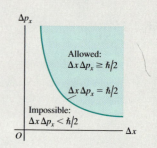

BRIDGING PROBLEM | Compton Scattering and Electron Recoil

An incident x-ray photon is scattered from a free electron that is initially at rest. The photon is scattered straight back at an angle of 180° from its initial direction. The wavelength of the scattered photon is 0.0830 nm. (a) What is the wavelength of the incident photon? (b) What are the magnitude of the momentum and the speed of the electron after the collision? (c) What is the kinetic energy of the electron after the collision?

SOLUTION GUIDE

See MasteringPhysics® study area for a Video Tutor solution.

IDENTIFY and SET UP

1. In this problem a photon is scattered by an electron initially at rest. In Section 38.3 you learned how to relate the wavelengths of the incident and scattered photons; in this problem you must also find the momentum, speed, and kinetic energy of the recoiling electron. You can find these because momentum and energy are conserved in the collision.
2. Which key equation can be used to find the incident photon wavelength? What is the photon scattering angle ϕ in this problem?

EXECUTE

3. Use the equation you selected in step 2 to find the wavelength of the incident photon.
4. Use momentum conservation and your result from step 3 to find the momentum of the recoiling electron. (*Hint:* All of the momentum vectors are along the same line, but not all point in the same direction. Be careful with signs.)
5. Find the speed of the recoiling electron from your result in step 4. (*Hint:* Assume that the electron is nonrelativistic, so you can use the relationship between momentum and speed from Chapter 8. This is acceptable if the speed of the electron is less than about 0.1*c*. Is it?)
6. Use your result from step 4 or step 5 to find the electron kinetic energy.

EVALUATE

7. You can check your answer in step 6 by finding the difference between the energies of the incident and scattered photons. Is your result consistent with conservation of energy?

Problems

For instructor-assigned homework, go to www.masteringphysics.com

•, ••, •••: Problems of increasing difficulty. **CP**: Cumulative problems incorporating material from earlier chapters. **CALC**: Problems requiring calculus. **BIO**: Biosciences problems.

DISCUSSION QUESTIONS

Q38.1 In what ways do photons resemble other particles such as electrons? In what ways do they differ? Do photons have mass? Do they have electric charge? Can they be accelerated? What mechanical properties do they have?

Q38.2 There is a certain probability that a single electron may simultaneously absorb *two* identical photons from a high-intensity laser. How would such an occurrence affect the threshold frequency and the equations of Section 38.1? Explain.

Q38.3 According to the photon model, light carries its energy in packets called quanta or photons. Why then don't we see a series of flashes when we look at things?

Q38.4 Would you expect effects due to the photon nature of light to be generally more important at the low-frequency end of the electromagnetic spectrum (radio waves) or at the high-frequency end (x rays and gamma rays)? Why?

Q38.5 During the photoelectric effect, light knocks electrons out of metals. So why don't the metals in your home lose their electrons when you turn on the lights?

Q38.6 Most black-and-white photographic film (with the exception of some special-purpose films) is less sensitive to red light than blue light and has almost no sensitivity to infrared. How can these properties be understood on the basis of photons?

Q38.7 Human skin is relatively insensitive to visible light, but ultraviolet radiation can cause severe burns. Does this have anything to do with photon energies? Explain.

Q38.8 Explain why Fig. 38.4 shows that most photoelectrons have kinetic energies less than $hf - \phi$, and also explain how these smaller kinetic energies occur.

Q38.9 In a photoelectric-effect experiment, the photocurrent i for large positive values of V_{AC} has the same value no matter what the light frequency f (provided that f is higher than the threshold frequency f_0). Explain why.

Q38.10 In an experiment involving the photoelectric effect, if the intensity of the incident light (having frequency higher than the threshold frequency) is reduced by a factor of 10 without changing anything else, which (if any) of the following statements about this process will be true? (a) The number of photoelectrons will most likely be reduced by a factor of 10. (b) The maximum kinetic energy of the ejected photoelectrons will most likely be reduced by a factor of 10. (c) The maximum speed of the ejected photoelectrons will most likely be reduced by a factor of 10. (d) The maximum speed of the ejected photoelectrons will most likely be reduced by a factor of $\sqrt{10}$. (e) The time for the first photoelectron to be ejected will be increased by a factor of 10.

Q38.11 The materials called *phosphors* that coat the inside of a fluorescent lamp convert ultraviolet radiation (from the mercury-vapor discharge inside the tube) into visible light. Could one also make a phosphor that converts visible light to ultraviolet? Explain.

Q38.12 In a photoelectric-effect experiment, which of the following will increase the maximum kinetic energy of the photoelectrons? (a) Use light of greater intensity; (b) use light of higher frequency; (c) use light of longer wavelength; (d) use a metal surface with a larger work function. In each case justify your answer.

Q38.13 A photon of frequency f undergoes Compton scattering from an electron at rest and scatters through an angle ϕ. The frequency of the scattered photon is f'. How is f' related to f? Does your answer depend on ϕ? Explain.

Q38.14 Can Compton scattering occur with protons as well as electrons? For example, suppose a beam of x rays is directed at a target of liquid hydrogen. (Recall that the nucleus of hydrogen consists of a single proton.) Compared to Compton scattering with electrons, what similarities and differences would you expect? Explain.

Q38.15 Why must engineers and scientists shield against x-ray production in high-voltage equipment?

Q38.16 In attempting to reconcile the wave and particle models of light, some people have suggested that the photon rides up and down on the crests and troughs of the electromagnetic wave. What things are *wrong* with this description?

Q38.17 Some lasers emit light in pulses that are only 10^{-12} s in duration. The length of such a pulse is $(3 \times 10^8 \text{ m/s})(10^{-12} \text{ s}) = 3 \times 10^{-4}$ m $= 0.3$ mm. Can pulsed laser light be as monochromatic as light from a laser that emits a steady, continuous beam? Explain.

EXERCISES

Section 38.1 Light Absorbed as Photons: The Photoelectric Effect

38.1 •• (a) A proton is moving at a speed much slower than the speed of light. It has kinetic energy K_1 and momentum p_1. If the momentum of the proton is doubled, so $p_2 = 2p_1$, how is its new kinetic energy K_2 related to K_1? (b) A photon with energy E_1 has momentum p_1. If another photon has momentum p_2 that is twice p_1, how is the energy E_2 of the second photon related to E_1?

38.2 • **BIO** **Response of the Eye.** The human eye is most sensitive to green light of wavelength 505 nm. Experiments have found that when people are kept in a dark room until their eyes adapt to the darkness, a *single* photon of green light will trigger receptor cells in the rods of the retina. (a) What is the frequency of this photon? (b) How much energy (in joules and electron volts) does it deliver to the receptor cells? (c) To appreciate what a small amount of energy this is, calculate how fast a typical bacterium of mass 9.5×10^{-12} g would move if it had that much energy.

38.3 • A photon of green light has a wavelength of 520 nm. Find the photon's frequency, magnitude of momentum, and energy. Express the energy in both joules and electron volts.

38.4 • **BIO** A laser used to weld detached retinas emits light with a wavelength of 652 nm in pulses that are 20.0 ms in duration. The average power during each pulse is 0.600 W. (a) How much energy is in each pulse in joules? In electron volts? (b) What is the energy of one photon in joules? In electron volts? (c) How many photons are in each pulse?

38.5 • A 75-W light source consumes 75 W of electrical power. Assume all this energy goes into emitted light of wavelength 600 nm. (a) Calculate the frequency of the emitted light. (b) How many photons per second does the source emit? (c) Are the answers to parts (a) and (b) the same? Is the frequency of the light the same thing as the number of photons emitted per second? Explain.

38.6 • A photon has momentum of magnitude 8.24×10^{-28} kg $\cdot$ m/s. (a) What is the energy of this photon? Give your answer in joules and in electron volts. (b) What is the wavelength of this photon? In what region of the electromagnetic spectrum does it lie?

38.7 • The graph in Fig. E38.7 shows the stopping potential as a function of the frequency of the incident light falling on a metal surface. (a) Find the photoelectric work function for this metal. (b) What value of Planck's constant does the graph yield? (c) Why does the graph *not* extend below the x-axis? (d) If a different metal were used, which characteristics of the graph would you expect to be the same and which ones would be different?

Figure **E38.7**

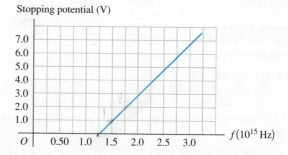

Stopping potential (V)

38.8 • The photoelectric threshold wavelength of a tungsten surface is 272 nm. Calculate the maximum kinetic energy of the electrons ejected from this tungsten surface by ultraviolet radiation of frequency 1.45×10^{15} Hz. Express the answer in electron volts.

38.9 •• A clean nickel surface is exposed to light of wavelength 235 nm. What is the maximum speed of the photoelectrons emitted from this surface? Use Table 38.1.

38.10 •• What would the minimum work function for a metal have to be for visible light (380–750 nm) to eject photoelectrons?

38.11 •• When ultraviolet light with a wavelength of 400.0 nm falls on a certain metal surface, the maximum kinetic energy of the emitted photoelectrons is measured to be 1.10 eV. What is the maximum kinetic energy of the photoelectrons when light of wavelength 300.0 nm falls on the same surface?

38.12 •• The photoelectric work function of potassium is 2.3 eV. If light having a wavelength of 250 nm falls on potassium, find (a) the stopping potential in volts; (b) the kinetic energy in electron volts of the most energetic electrons ejected; (c) the speed of these electrons.

38.13 •• When ultraviolet light with a wavelength of 254 nm falls on a clean copper surface, the stopping potential necessary to stop emission of photoelectrons is 0.181 V. (a) What is the photoelectric threshold wavelength for this copper surface? (b) What is the work function for this surface, and how does your calculated value compare with that given in Table 38.1?

Section 38.2 Light Emitted as Photons: X-Ray Production

38.14 • The cathode-ray tubes that generated the picture in early color televisions were sources of x rays. If the acceleration voltage in a television tube is 15.0 kV, what are the shortest-wavelength x rays produced by the television? (Modern televisions contain shielding to stop these x rays.)

38.15 • Protons are accelerated from rest by a potential difference of 4.00 kV and strike a metal target. If a proton produces one photon on impact, what is the minimum wavelength of the resulting x rays? How does your answer compare to the minimum wavelength if 4.00-keV electrons are used instead? Why do x-ray tubes use electrons rather than protons to produce x rays?

38.16 •• (a) What is the minimum potential difference between the filament and the target of an x-ray tube if the tube is to produce x rays with a wavelength of 0.150 nm? (b) What is the shortest wavelength produced in an x-ray tube operated at 30.0 kV?

Section 38.3 Light Scattered as Photons: Compton Scattering and Pair Production

38.17 • An x ray with a wavelength of 0.100 nm collides with an electron that is initially at rest. The x ray's final wavelength is 0.110 nm. What is the final kinetic energy of the electron?

38.18 • X rays are produced in a tube operating at 18.0 kV. After emerging from the tube, x rays with the minimum wavelength produced strike a target and are Compton-scattered through an angle of 45.0°. (a) What is the original x-ray wavelength? (b) What is the wavelength of the scattered x rays? (c) What is the energy of the scattered x rays (in electron volts)?

38.19 •• X rays with initial wavelength 0.0665 nm undergo Compton scattering. What is the longest wavelength found in the scattered x rays? At which scattering angle is this wavelength observed?

38.20 • A beam of x rays with wavelength 0.0500 nm is Compton-scattered by the electrons in a sample. At what angle from the incident beam should you look to find x rays with a wavelength of (a) 0.0542 nm; (b) 0.0521 nm; (c) 0.0500 nm?

38.21 •• If a photon of wavelength 0.04250 nm strikes a free electron and is scattered at an angle of 35.0° from its original direction, find (a) the change in the wavelength of this photon; (b) the wavelength of the scattered light; (c) the change in energy of the photon (is it a loss or a gain?); (d) the energy gained by the electron.

38.22 •• A photon scatters in the backward direction ($\phi = 180°$) from a free proton that is initially at rest. What must the wavelength of the incident photon be if it is to undergo a 10.0% change in wavelength as a result of the scattering?

38.23 •• X rays with an initial wavelength of 0.900×10^{-10} m undergo Compton scattering. For what scattering angle is the wavelength of the scattered x rays greater by 1.0% than that of the incident x rays?

38.24 •• A photon with wavelength $\lambda = 0.1385$ nm scatters from an electron that is initially at rest. What must be the angle between the direction of propagation of the incident and scattered photons if the speed of the electron immediately after the collision is 8.90×10^6 m/s?

38.25 • An electron and a positron are moving toward each other and each has speed $0.500c$ in the lab frame. (a) What is the kinetic energy of each particle? (b) The e^+ and e^- meet head-on and annihilate. What is the energy of each photon that is produced? (c) What is the wavelength of each photon? How does the wavelength compare to the photon wavelength when the initial kinetic energy of the e^+ and e^- is negligibly small (see Example 38.6)?

Section 38.4 Wave–Particle Duality, Probability, and Uncertainty

38.26 • A laser produces light of wavelength 625 nm in an ultrashort pulse. What is the minimum duration of the pulse if the minimum uncertainty in the energy of the photons is 1.0%?

38.27 • An ultrashort pulse has a duration of 9.00 fs and produces light at a wavelength of 556 nm. What are the momentum and momentum uncertainty of a single photon in the pulse?

38.28 • A horizontal beam of laser light of wavelength 585 nm passes through a narrow slit that has width 0.0620 mm. The intensity of the light is measured on a vertical screen that is 2.00 m from the slit. (a) What is the minimum uncertainty in the vertical component of the momentum of each photon in the beam after the

photon has passed through the slit? (b) Use the result of part (a) to estimate the width of the central diffraction maximum that is observed on the screen.

PROBLEMS

38.29 • **Exposing Photographic Film.** The light-sensitive compound on most photographic films is silver bromide, AgBr. A film is "exposed" when the light energy absorbed dissociates this molecule into its atoms. (The actual process is more complex, but the quantitative result does not differ greatly.) The energy of dissociation of AgBr is 1.00×10^5 J/mol. For a photon that is just able to dissociate a molecule of silver bromide, find (a) the photon energy in electron volts; (b) the wavelength of the photon; (c) the frequency of the photon. (d) What is the energy in electron volts of a photon having a frequency of 100 MHz? (e) Light from a firefly can expose photographic film, but the radiation from an FM station broadcasting 50,000 W at 100 MHz cannot. Explain why this is so.

38.30 •• (a) If the average frequency emitted by a 200-W light bulb is 5.00×10^{14} Hz, and 10.0% of the input power is emitted as visible light, approximately how many visible-light photons are emitted per second? (b) At what distance would this correspond to 1.00×10^{11} visible-light photons per square centimeter per second if the light is emitted uniformly in all directions?

38.31 • When a certain photoelectric surface is illuminated with light of different wavelengths, the following stopping potentials are observed:

Wavelength (nm)	Stopping potential (V)
366	1.48
405	1.15
436	0.93
492	0.62
546	0.36
579	0.24

Plot the stopping potential on the vertical axis against the frequency of the light on the horizontal axis. Determine (a) the threshold frequency; (b) the threshold wavelength; (c) the photoelectric work function of the material (in electron volts); (d) the value of Planck's constant h (assuming that the value of e is known).

38.32 • A 2.50-W beam of light of wavelength 124 nm falls on a metal surface. You observe that the maximum kinetic energy of the ejected electrons is 4.16 eV. Assume that each photon in the beam ejects a photoelectron. (a) What is the work function (in electron volts) of this metal? (b) How many photoelectrons are ejected each second from this metal? (c) If the power of the light beam, but not its wavelength, were reduced by half, what would be the answer to part (b)? (d) If the wavelength of the beam, but not its power, were reduced by half, what would be the answer to part (b)?

38.33 •• **CP BIO Removing Vascular Lesions.** A pulsed dye laser emits light of wavelength 585 nm in 450-μs pulses. Because this wavelength is strongly absorbed by the hemoglobin in the blood, the method is especially effective for removing various types of blemishes due to blood, such as port-wine–colored birthmarks. To get a reasonable estimate of the power required for such laser surgery, we can model the blood as having the same specific heat and heat of vaporization as water (4190 J/kg·K, 2.256×10^6 J/kg). Suppose that each pulse must remove 2.0 μg of blood by evaporating it, starting at 33°C. (a) How much energy must each pulse deliver to the blemish? (b) What must be the power output of this laser? (c) How many photons does each pulse deliver to the blemish?

38.34 • The photoelectric work functions for particular samples of certain metals are as follows: cesium, 2.1 eV; copper, 4.7 eV;

potassium, 2.3 eV; and zinc, 4.3 eV. (a) What is the threshold wavelength for each metal surface? (b) Which of these metals could *not* emit photoelectrons when irradiated with visible light (380–750 nm)?

38.35 •• An incident x-ray photon of wavelength 0.0900 nm is scattered in the backward direction from a free electron that is initially at rest. (a) What is the magnitude of the momentum of the scattered photon? (b) What is the kinetic energy of the electron after the photon is scattered?

38.36 •• **CP** A photon with wavelength $\lambda = 0.0900$ nm is incident on an electron that is initially at rest. If the photon scatters in the backward direction, what is the magnitude of the linear momentum of the electron just after the collision with the photon?

38.37 •• **CP** A photon with wavelength $\lambda = 0.1050$ nm is incident on an electron that is initially at rest. If the photon scatters at an angle of $60.0°$ from its original direction, what are the magnitude and direction of the linear momentum of the electron just after the collision with the photon?

38.38 •• **CP** An x-ray tube is operating at voltage V and current I. (a) If only a fraction p of the electric power supplied is converted into x rays, at what rate is energy being delivered to the target? (b) If the target has mass m and specific heat c (in $J/kg \cdot K$), at what average rate would its temperature rise if there were no thermal losses? (c) Evaluate your results from parts (a) and (b) for an x-ray tube operating at 18.0 kV and 60.0 mA that converts 1.0% of the electric power into x rays. Assume that the 0.250-kg target is made of lead ($c = 130$ J/kg·K). (d) What must the physical properties of a practical target material be? What would be some suitable target elements?

38.39 •• Nuclear fusion reactions at the center of the sun produce gamma-ray photons with energies of about 1 MeV (10^6 eV). By contrast, what we see emanating from the sun's surface are visible-light photons with wavelengths of about 500 nm. A simple model that explains this difference in wavelength is that a photon undergoes Compton scattering many times—in fact, about 10^{26} times, as suggested by models of the solar interior—as it travels from the center of the sun to its surface. (a) Estimate the increase in wavelength of a photon in an average Compton-scattering event. (b) Find the angle in degrees through which the photon is scattered in the scattering event described in part (a). (*Hint:* A useful approximation is $\cos\phi \approx 1 - \phi^2/2$, which is valid for $\phi \ll 1$. Note that ϕ is in radians in this expression.) (c) It is estimated that a photon takes about 10^6 years to travel from the core to the surface of the sun. Find the average distance that light can travel within the interior of the sun without being scattered. (This distance is roughly equivalent to how far you could see if you were inside the sun and could survive the extreme temperatures there. As your answer shows, the interior of the sun is *very* opaque.)

38.40 •• (a) Derive an expression for the total shift in photon wavelength after two successive Compton scatterings from electrons at rest. The photon is scattered by an angle θ_1 in the first scat-

tering and by θ_2 in the second. (b) In general, is the total shift in wavelength produced by two successive scatterings of an angle $\theta/2$ the same as by a single scattering of θ? If not, are there any specific values of θ, other than $\theta = 0°$, for which the total shifts are the same? (c) Use the result of part (a) to calculate the total wavelength shift produced by two successive Compton scatterings of $30.0°$ each. Express your answer in terms of h/mc. (d) What is the wavelength shift produced by a single Compton scattering of $60.0°$? Compare to the answer in part (c).

38.41 •• A photon with wavelength 0.1100 nm collides with a free electron that is initially at rest. After the collision the wavelength is 0.1132 nm. (a) What is the kinetic energy of the electron after the collision? What is its speed? (b) If the electron is suddenly stopped (for example, in a solid target), all of its kinetic energy is used to create a photon. What is the wavelength of this photon?

38.42 •• An x-ray photon is scattered from a free electron (mass m) at rest. The wavelength of the scattered photon is λ', and the final speed of the struck electron is v. (a) What was the initial wavelength λ of the photon? Express your answer in terms of λ', v, and m. (*Hint:* Use the relativistic expression for the electron kinetic energy.) (b) Through what angle ϕ is the photon scattered? Express your answer in terms of λ, λ', and m. (c) Evaluate your results in parts (a) and (b) for a wavelength of 5.10×10^{-3} nm for the scattered photon and a final electron speed of 1.80×10^8 m/s. Give ϕ in degrees.

38.43 •• (a) Calculate the maximum increase in photon wavelength that can occur during Compton scattering. (b) What is the energy (in electron volts) of the lowest-energy x-ray photon for which Compton scattering could result in doubling the original wavelength?

CHALLENGE PROBLEM

38.44 ••• Consider Compton scattering of a photon by a *moving* electron. Before the collision the photon has wavelength λ and is moving in the $+x$-direction, and the electron is moving in the $-x$-direction with total energy E (including its rest energy mc^2). The photon and electron collide head-on. After the collision, both are moving in the $-x$-direction (that is, the photon has been scattered by $180°$). (a) Derive an expression for the wavelength λ' of the scattered photon. Show that if $E \gg mc^2$, where m is the rest mass of the electron, your result reduces to

$$\lambda' = \frac{hc}{E}\left(1 + \frac{m^2c^4\lambda}{4hcE}\right)$$

(b) A beam of infrared radiation from a CO_2 laser ($\lambda = 10.6$ μm) collides head-on with a beam of electrons, each of total energy $E = 10.0$ GeV (1 GeV $= 10^9$ eV). Calculate the wavelength λ' of the scattered photons, assuming a $180°$ scattering angle. (c) What kind of scattered photons are these (infrared, microwave, ultraviolet, etc.)? Can you think of an application of this effect?

Answers

Test Your Understanding Questions

38.1 Answers: (i) and (ii) From Eq. (38.2), a photon of energy $E = 1.14$ eV has wavelength $\lambda = hc/E = (4.136 \times 10^{-15}$ eV$)$ $\cdot (3.00 \times 10^8$ m/s$)/(1.14$ eV$) = 1.09 \times 10^{-6}$ m $= 1090$ nm. This is in the infrared part of the spectrum. Since wavelength is inversely proportional to photon energy, the *minimum* photon energy of 1.14 eV corresponds to the *maximum* wavelength that causes photoconductivity in silicon. Thus the wavelength must be 1090 nm or less.

38.2 Answer: (ii) Equation (38.6) shows that the minimum wavelength of x rays produced by bremsstrahlung depends on the potential difference V_{AC} but does *not* depend on the rate at which electrons strike the anode. Each electron produces at most one photon, so increasing the number of electrons per second causes an increase in the number of x-ray photons emitted per second (that is, the x-ray intensity).

38.3 Answers: yes, no Equation (38.7) shows that the wavelength shift $\Delta \lambda = \lambda' - \lambda$ depends only on the photon scattering angle ϕ, not on the wavelength of the incident photon. So a visible-light photon scattered through an angle ϕ undergoes the same wavelength shift as an x-ray photon. Equation (38.7) also shows that this shift is of the order of $h/mc = 2.426 \times 10^{-12}$ m $=$ 0.002426 nm. This is a few percent of the wavelength of x rays (see Example 38.5), so the effect is noticeable in x-ray scattering. However, h/mc is a tiny fraction of the wavelength of visible light (between 380 and 750 nm). The human eye cannot distinguish such minuscule differences in wavelength (that is, differences in color).

38.4 Answer: (ii) There is *zero* probability that a photon will be deflected by one of the angles where the diffraction pattern has zero intensity. These angles are given by $a \sin \theta = m\lambda$ with $m = \pm 1$, $\pm 2, \pm 3, \ldots$. Since λ is much less than a, we can write these angles as $\theta = m\lambda/a = \pm\lambda/a, \pm 2\lambda/a, \pm 3\lambda/a, \ldots$. These values include answers (i), (iii), and (iv), so it is impossible for a photon to be deflected through any of these angles. The intensity is not zero at $\theta = 3\lambda/2a$ (located between two zeros in the diffraction pattern), so there is some probability that a photon will be deflected through this angle.

Bridging Problem

Answers: (a) 0.0781 nm
(b) 1.65×10^{-23} kg $\cdot$ m/s, 1.81×10^7 m/s
(c) 1.49×10^{-16} J

39 PARTICLES BEHAVING AS WAVES

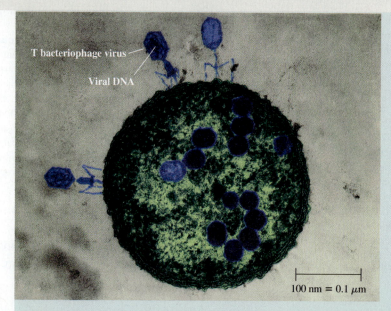

T bacteriophage virus

Viral DNA

100 nm = 0.1 μm

? Viruses (shown in blue) have landed on an *E. coli* bacterium and injected their DNA, converting the bacterium into a virus factory. This false-color image was made using a beam of electrons rather than a light beam. What properties of electrons make them useful for imaging such fine details?

LEARNING GOALS

By studying this chapter, you will learn:

• De Broglie's proposal that electrons, protons, and other particles can behave like waves.

• How electron diffraction experiments provided evidence for de Broglie's ideas.

• How electron microscopes can provide much higher magnification than visible-light microscopes.

• How physicists discovered the atomic nucleus.

• How Bohr's model of electron orbits explained the spectra of hydrogen and hydrogenlike atoms.

• How a laser operates.

• How the idea of energy levels, coupled with the photon model of light, explains the spectrum of light emitted by a hot, opaque object.

• What the uncertainty principle tells us about the nature of the atom.

In Chapter 38 we discovered one aspect of nature's wave–particle duality: Light and other electromagnetic radiation act sometimes like waves and sometimes like particles. Interference and diffraction demonstrate wave behavior, while emission and absorption of photons demonstrate particle behavior.

If light waves can behave like particles, can the particles of matter behave like waves? As we will discover, the answer is a resounding yes. Electrons can be made to interfere and diffract just like other kinds of waves. We will see that the wave nature of electrons is not merely a laboratory curiosity: It is the fundamental reason why atoms, which according to classical physics should be profoundly unstable, are able to exist. In this chapter we'll use the wave nature of matter to help us understand the structure of atoms, the operating principles of a laser, and the curious properties of the light emitted by a heated, glowing object. Without the wave picture of matter, there would be no way to explain these phenomena.

In Chapter 40 we'll introduce an even more complete wave picture of matter called *quantum mechanics*. Through the remainder of this book we'll use the ideas of quantum mechanics to understand the nature of molecules, solids, atomic nuclei, and the fundamental particles that are the building blocks of our universe.

39.1 Electron Waves

In 1924 a French physicist and nobleman, Prince Louis de Broglie (pronounced "de broy"; Fig. 39.1), made a remarkable proposal about the nature of matter. His reasoning, freely paraphrased, went like this: Nature loves symmetry. Light is dualistic in nature, behaving in some situations like waves and in others like particles. If nature is symmetric, this duality should also hold for matter. Electrons and protons, which we usually think of as *particles*, may in some situations behave like *waves*.

If a particle acts like a wave, it should have a wavelength and a frequency. De Broglie postulated that a free particle with rest mass m, moving with nonrelativistic speed v, should have a wavelength λ related to its momentum $p = mv$ in exactly the same way as for a photon, as expressed by Eq. (38.5): $\lambda = h/p$. The **de Broglie wavelength** of a particle is then

$$\lambda = \frac{h}{p} = \frac{h}{mv} \qquad \text{(de Broglie wavelength of a particle)} \qquad (39.1)$$

where h is Planck's constant. If the particle's speed is an appreciable fraction of the speed of light c, we use Eq. (37.27) to replace mv in Eq. (39.1) with $\gamma mv = mv/\sqrt{1 - v^2/c^2}$. The frequency f, according to de Broglie, is also related to the particle's energy E in the same way as for a photon—namely,

$$E = hf \qquad (39.2)$$

Thus in de Broglie's hypothesis, the relationships of wavelength to momentum and of frequency to energy are exactly the same for free particles as for photons.

CAUTION **Not all photon equations apply to particles with mass** Be careful when applying the relationship $E = hf$ to particles with nonzero rest mass, such as electrons and protons. Unlike a photon, they do *not* travel at speed c, so the equations $f = c/\lambda$ and $E = pc$ do *not* apply to them!

Observing the Wave Nature of Electrons

De Broglie's proposal was a bold one, made at a time when there was no direct experimental evidence that particles have wave characteristics. But within a few years of de Broglie's publication of his ideas, they were resoundingly verified by a diffraction experiment with electrons. This experiment was analogous to those we described in Section 36.6, in which atoms in a crystal act as a three-dimensional diffraction grating for x rays. An x-ray beam is strongly reflected when it strikes a crystal at an angle that gives constructive interference among the waves scattered from the various atoms in the crystal. These interference effects demonstrate the *wave* nature of x rays.

In 1927 the American physicists Clinton Davisson and Lester Germer, working at the Bell Telephone Laboratories, were studying the surface of a piece of nickel by directing a beam of *electrons* at the surface and observing how many electrons bounced off at various angles. Figure 39.2 shows an experimental setup like theirs. Like many ordinary metals, the sample was *polycrystalline*: It consisted of many randomly oriented microscopic crystals bonded together. As a result, the electron beam reflected diffusely, like light bouncing off a rough surface (see Fig. 33.6b), with a smooth distribution of intensity as a function of the angle θ.

39.1 Louis-Victor de Broglie, the seventh Duke de Broglie (1892–1987), broke with family tradition by choosing a career in physics rather than as a diplomat. His revolutionary proposal that particles have wave characteristics—for which de Broglie won the 1929 Nobel Prize in physics—was published in his doctoral thesis.

Mastering**PHYSICS**

PhET: Davisson-Germer: Electron Diffraction
ActivPhysics 17.5: Electron Interference

39.2 An apparatus similar to that used by Davisson and Germer to discover electron diffraction.

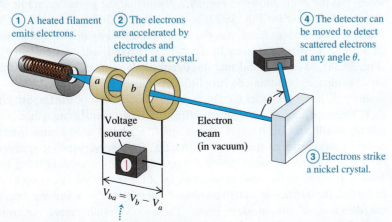

① A heated filament emits electrons.

② The electrons are accelerated by electrodes and directed at a crystal.

④ The detector can be moved to detect scattered electrons at any angle θ.

③ Electrons strike a nickel crystal.

Voltage source

Electron beam (in vacuum)

$V_{ba} = V_b - V_a$

$V_{ba} > 0$, so electrons speed up in moving from a to b.

39.3 (a) Intensity of the scattered electron beam in Fig. 39.2 as a function of the scattering angle θ. (b) Electron waves scattered from two adjacent atoms interfere constructively when $d\sin\theta = m\lambda$. In the case shown here, $\theta = 50°$ and $m = 1$.

(a) This peak in the intensity of scattered electrons is due to constructive interference between electron waves scattered by different surface atoms.

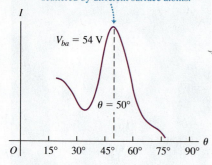

(b) If the scattered waves are in phase, there is a peak in the intensity of scattered electrons.

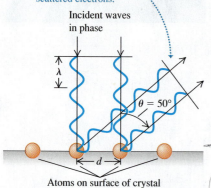

Atoms on surface of crystal

39.4 X-ray and electron diffraction. The upper half of the photo shows the diffraction pattern for 71-pm x rays passing through aluminum foil. The lower half, with a different scale, shows the diffraction pattern for 600-eV electrons from aluminum. The similarity shows that electrons undergo the same kind of diffraction as x rays.

Top: x-ray diffraction

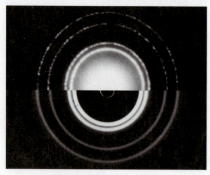

Bottom: electron diffraction

During the experiment an accident occurred that permitted air to enter the vacuum chamber, and an oxide film formed on the metal surface. To remove this film, Davisson and Germer baked the sample in a high-temperature oven, almost hot enough to melt it. Unknown to them, this had the effect of creating large regions within the nickel with crystal planes that were continuous over the width of the electron beam. From the perspective of the electrons, the sample looked like a *single* crystal of nickel.

When the observations were repeated with this sample, the results were quite different. Now strong maxima in the intensity of the reflected electron beam occurred at specific angles (Fig. 39.3a), in contrast to the smooth variation of intensity with angle that Davisson and Germer had observed before the accident. The angular positions of the maxima depended on the accelerating voltage V_{ba} used to produce the electron beam. Davisson and Germer were familiar with de Broglie's hypothesis, and they noticed the similarity of this behavior to x-ray diffraction. This was not the effect they had been looking for, but they immediately recognized that the electron beam was being *diffracted*. They had discovered a very direct experimental confirmation of the wave hypothesis.

Davisson and Germer could determine the speeds of the electrons from the accelerating voltage, so they could compute the de Broglie wavelength from Eq. (39.1). If an electron is accelerated from rest at point a to point b through a potential increase $V_{ba} = V_b - V_a$ as shown in Fig. 39.2, the work done on the electron eV_{ba} equals its kinetic energy K. Using $K = (\frac{1}{2})mv^2 = p^2/2m$ for a nonrelativistic particle, we have

$$eV_{ba} = \frac{p^2}{2m} \qquad p = \sqrt{2meV_{ba}}$$

We substitute this into Eq. (39.1), the expression for the de Broglie wavelength of the electron:

$$\lambda = \frac{h}{p} = \frac{h}{\sqrt{2meV_{ba}}} \qquad \text{(de Broglie wavelength of an electron)} \quad (39.3)$$

The greater the accelerating voltage V_{ba}, the shorter the wavelength of the electron.

To predict the angles at which strong reflection occurs, note that the electrons were scattered primarily by the planes of atoms near the surface of the crystal. Atoms in a surface plane are arranged in rows, with a distance d that can be measured by x-ray diffraction techniques. These rows act like a reflecting diffraction grating; the angles at which strong reflection occurs are the same as for a grating with center-to-center distance d between its slits (Fig. 39.3b). From Eq. (36.13) the angles of maximum reflection are given by

$$d\sin\theta = m\lambda \qquad (m = 1, 2, 3, \dots) \qquad (39.4)$$

where θ is the angle shown in Fig. 39.2. (Note that the geometry in Fig. 39.3b is different from that for Fig. 36.22, so Eq. (39.4) is different from Eq. (36.16).) Davisson and Germer found that the angles predicted by this equation, using the de Broglie wavelength given by Eq. (39.3), agreed with the observed values (Fig. 39.3a). Thus the accidental discovery of **electron diffraction** was the first direct evidence confirming de Broglie's hypothesis.

In 1928, just a year after the Davisson–Germer discovery, the English physicist G. P. Thomson carried out electron-diffraction experiments using a thin, polycrystalline, metallic foil as a target. Debye and Sherrer had used a similar technique several years earlier to study x-ray diffraction from polycrystalline specimens. In these experiments the beam passes *through* the target rather than being reflected from it. Because of the random orientations of the individual microscopic crystals in the foil, the diffraction pattern consists of intensity maxima forming rings around the direction of the incident beam. Thomson's results again confirmed the de Broglie relationship. Figure 39.4 shows both x-ray and electron diffraction patterns for a polycrystalline aluminum foil. (G. P. Thomson was the son of J. J. Thomson, who 31 years earlier discovered the electron. Davisson and the younger Thomson shared the 1937 Nobel Prize in physics for their discoveries.)

Additional diffraction experiments were soon carried out in many laboratories using not only electrons but also various ions and low-energy neutrons. All of these are in agreement with de Broglie's bold predictions. Thus the wave nature of particles, so strange in 1924, became firmly established in the years that followed.

Problem-Solving Strategy 39.1 **Wavelike Properties of Particles**

IDENTIFY *the relevant concepts:* Particles have wavelike properties. A particle's (de Broglie) wavelength is inversely proportional to its momentum, and its frequency is proportional to its energy.

SET UP *the problem:* Identify the target variables and decide which equations you will use to calculate them.

EXECUTE *the solution as follows:*
1. Use Eq. (39.1) to relate a particle's momentum p to its wavelength λ; use Eq. (39.2) to relate its energy E to its frequency f.
2. Nonrelativistic kinetic energy may be expressed as either $K = \frac{1}{2}mv^2$ or (because $p = mv$) $K = p^2/2m$. The latter form is useful in calculations involving the de Broglie wavelength.
3. You may express energies in either joules or electron volts, using $h = 6.626 \times 10^{-34}$ J·s or $h = 4.136 \times 10^{-15}$ eV·s as appropriate.

EVALUATE *your answer:* To check numerical results, it helps to remember some approximate orders of magnitude. Here's a partial list:

Size of an atom: 10^{-10} m = 0.1 nm

Mass of an atom: 10^{-26} kg

Mass of an electron: $m = 10^{-30}$ kg; $mc^2 = 0.511$ MeV

Electron charge magnitude: 10^{-19} C

kT at room temperature: $\frac{1}{40}$ eV

Difference between energy levels of an atom (to be discussed in Section 39.3): 1 to 10 eV

Speed of an electron in the Bohr model of a hydrogen atom (to be discussed in Section 39.3): 10^6 m/s

---

Example 39.1 **An electron-diffraction experiment**

In an electron-diffraction experiment using an accelerating voltage of 54 V, an intensity maximum occurs for $\theta = 50°$ (see Fig. 39.3a). X-ray diffraction indicates that the atomic spacing in the target is $d = 2.18 \times 10^{-10}$ m = 0.218 nm. The electrons have negligible kinetic energy before being accelerated. Find the electron wavelength.

$$\lambda = \frac{6.626 \times 10^{-34} \text{ J} \cdot \text{s}}{\sqrt{2(9.109 \times 10^{-31} \text{ kg})(1.602 \times 10^{-19} \text{ C})(54 \text{ V})}}$$
$$= 1.7 \times 10^{-10} \text{ m} = 0.17 \text{ nm}$$

Alternatively, using Eq. (39.4) and assuming $m = 1$,

$$\lambda = d \sin\theta = (2.18 \times 10^{-10} \text{ m}) \sin 50° = 1.7 \times 10^{-10} \text{ m}$$

SOLUTION

IDENTIFY, SET UP, and EXECUTE: We'll determine λ from both de Broglie's equation, Eq. (39.3), and the diffraction equation, Eq. (39.4). From Eq. (39.3),

EVALUATE: The two numbers agree within the accuracy of the experimental results, which gives us an excellent check on our calculations. Note that this electron wavelength is less than the spacing between the atoms.

Example 39.2 **Energy of a thermal neutron**

Find the speed and kinetic energy of a neutron ($m = 1.675 \times 10^{-27}$ kg) with de Broglie wavelength $\lambda = 0.200$ nm, a typical interatomic spacing in crystals. Compare this energy with the average translational kinetic energy of an ideal-gas molecule at room temperature ($T = 20°C = 293$ K).

$K = \frac{1}{2}mv^2$. We'll use Eq. (18.16) to find the average kinetic energy of a gas molecule.

EXECUTE: From Eq. (39.1), the neutron speed is

$$v = \frac{h}{\lambda m} = \frac{6.626 \times 10^{-34} \text{ J} \cdot \text{s}}{(0.200 \times 10^{-9} \text{ m})(1.675 \times 10^{-27} \text{ kg})}$$
$$= 1.98 \times 10^3 \text{ m/s}$$

SOLUTION

The neutron kinetic energy is

IDENTIFY and SET UP: This problem uses the relationships between particle speed and wavelength, between particle speed and kinetic energy, and between gas temperature and the average kinetic energy of a gas molecule. We'll find the neutron speed v using Eq. (39.1) and from that calculate the neutron kinetic energy

$$K = \frac{1}{2}mv^2 = \frac{1}{2}(1.675 \times 10^{-27} \text{ kg})(1.98 \times 10^3 \text{ m/s})^2$$
$$= 3.28 \times 10^{-21} \text{ J} = 0.0205 \text{ eV}$$

Continued

From Eq. (18.16), the average translational kinetic energy of an ideal-gas molecule at $T = 293$ K is

$$\tfrac{1}{2}m(v^2)_{av} = \tfrac{3}{2}kT = \tfrac{3}{2}(1.38 \times 10^{-23}\ \text{J/K})(293\ \text{K})$$
$$= 6.07 \times 10^{-21}\ \text{J} = 0.0379\ \text{eV}$$

The two energies are comparable in magnitude, which is why a neutron with kinetic energy in this range is called a *thermal neutron*.

Diffraction of thermal neutrons is used to study crystal and molecular structure in the same way as x-ray diffraction. Neutron diffraction has proved to be especially useful in the study of large organic molecules.

EVALUATE: Note that the calculated neutron speed is much less than the speed of light. This justifies our use of the nonrelativistic form of Eq. (39.1).

De Broglie Waves and the Macroscopic World

If the de Broglie picture is correct and matter has wave aspects, you might wonder why we don't see these aspects in everyday life. As an example, we know that waves diffract when sent through a single slit. Yet when we walk through a doorway (a kind of single slit), we don't worry about our body diffracting!

The principal reason we don't see these effects on human scales is that Planck's constant h has such a minuscule value. As a result, the de Broglie wavelengths of even the smallest ordinary objects that you can see are extremely small, and the wave effects are unimportant. For instance, what is the wavelength of a falling grain of sand? If the grain's mass is 5×10^{-10} kg and its diameter is 0.07 mm $= 7 \times 10^{-5}$ m, it will fall in air with a terminal speed of about 0.4 m/s. The magnitude of its momentum is then $p = mv = (5 \times 10^{-10}\ \text{kg}) \times (0.4\ \text{m/s}) = 2 \times 10^{-10}$ kg $\cdot$ m/s. The de Broglie wavelength of this falling sand grain is then

$$\lambda = \frac{h}{p} = \frac{6.626 \times 10^{-34}\ \text{J} \cdot \text{s}}{2 \times 10^{-10}\ \text{kg} \cdot \text{m/s}} = 3 \times 10^{-24}\ \text{m}$$

Not only is this wavelength far smaller than the diameter of the sand grain, but it's also far smaller than the size of a typical atom (about 10^{-10} m). A more massive, faster-moving object would have an even larger momentum and an even smaller de Broglie wavelength. The effects of such tiny wavelengths are so small that they are never noticed in daily life.

The Electron Microscope

The **electron microscope** offers an important and interesting example of the interplay of wave and particle properties of electrons. An electron beam can be used to form an image of an object in much the same way as a light beam. A ray of light can be bent by reflection or refraction, and an electron trajectory can be bent by an electric or magnetic field. Rays of light diverging from a point on an object can be brought to convergence by a converging lens or concave mirror, and electrons diverging from a small region can be brought to convergence by electric and/or magnetic fields.

The analogy between light rays and electrons goes deeper. The *ray* model of geometric optics is an approximate representation of the more general *wave* model. Geometric optics (ray optics) is valid whenever interference and diffraction effects can be neglected. Similarly, the model of an electron as a point particle following a line trajectory is an approximate description of the actual behavior of the electron; this model is useful when we can neglect effects associated with the wave nature of electrons.

How is an electron microscope superior to an optical microscope? The *resolution* of an optical microscope is limited by diffraction effects, as we discussed in Section 36.7. Since an optical microscope uses wavelengths around 500 nm, it can't resolve objects smaller than a few hundred nanometers, no matter how carefully its lenses are made. The resolution of an electron microscope is similarly limited by the wavelengths of the electrons, but these wavelengths may be many thousands of times smaller than wavelengths of visible light. As a result, the useful magnification of an electron microscope can be thousands of times greater than that of an optical microscope.

Note that the ability of the electron microscope to form a magnified image *does not* depend on the wave properties of electrons. Within the limitations of the Heisenberg uncertainty principle (which we'll discuss in Section 39.6), we can compute the electron trajectories by treating them as classical charged particles under the action of electric and magnetic forces. Only when we talk about *resolution* do the wave properties become important.

Example 39.3   An electron microscope

In an electron microscope, the nonrelativistic electron beam is formed by a setup similar to the electron gun used in the Davisson–Germer experiment (see Fig. 39.2). The electrons have negligible kinetic energy before they are accelerated. What accelerating voltage is needed to produce electrons with wavelength 10 pm = 0.010 nm (roughly 50,000 times smaller than typical visible-light wavelengths)?

SOLUTION

IDENTIFY, SET UP, and EXECUTE: We can use the same concepts we used to understand the Davisson–Germer experiment. The accelerating voltage is the quantity V_{ba} in Eq. (39.3). Rewrite this equation to solve for V_{ba}:

$$V_{ba} = \frac{h^2}{2me\lambda^2}$$

$$= \frac{(6.626 \times 10^{-34} \text{ J} \cdot \text{s})^2}{2(9.109 \times 10^{-31} \text{ kg})(1.602 \times 10^{-19} \text{ C})(10 \times 10^{-12} \text{ m})^2}$$

$$= 1.5 \times 10^4 \text{ V} = 15{,}000 \text{ V}$$

EVALUATE: It is easy to attain 15-kV accelerating voltages from 120-V or 240-V line voltage using a step-up transformer (Section 31.6) and a rectifier (Section 31.1). The accelerated electrons have kinetic energy 15 keV; since the electron rest energy is 0.511 MeV = 511 keV, these electrons are indeed nonrelativistic.

Types of Electron Microscope

Figure 39.5 shows the design of a *transmission electron microscope,* in which electrons actually pass through the specimen being studied. The specimen to be viewed can be no more than 10 to 100 nm thick so the electrons are not slowed appreciably as they pass through. The electrons used in a transmission electron microscope are emitted from a hot cathode and accelerated by a potential difference, typically 40 to 400 kV. They then pass through a condensing "lens" that uses magnetic fields to focus the electrons into a parallel beam before they pass through the specimen. The beam then passes through two more magnetic lenses: an objective lens that forms an intermediate image of the specimen and a projection lens that produces a final real image of the intermediate image. The objective and projection lenses play the roles of the objective and eyepiece lenses, respectively, of a compound optical microscope (see Section 34.8). The final image is projected onto a fluorescent screen for viewing or photographing. The entire apparatus, including the specimen, must be enclosed in a vacuum container; otherwise, electrons would scatter off air molecules and muddle the image. The image that opens this chapter was made with a transmission electron microscope.

We might think that when the electron wavelength is 0.01 nm (as in Example 39.3), the resolution would also be about 0.01 nm. In fact, it is seldom better than 0.1 nm, in part because the focal length of a magnetic lens depends on the electron speed, which is never exactly the same for all electrons in the beam.

An important variation is the *scanning electron microscope.* The electron beam is focused to a very fine line and scanned across the specimen. The beam knocks additional electrons off the specimen wherever it hits. These ejected electrons are collected by an anode that is kept at a potential a few hundred volts positive with respect to the specimen. The current of ejected electrons flowing to the collecting anode varies as the microscope beam sweeps across the specimen. The varying strength of the current is then used to create a "map" of the scanned specimen, and this map forms a greatly magnified image of the specimen.

This scheme has several advantages. The specimen can be thick because the beam does not need to pass through it. Also, the knock-off electron production depends on the *angle* at which the beam strikes the surface. Thus scanning electron

39.5 Schematic diagram of a transmission electron microscope (TEM).

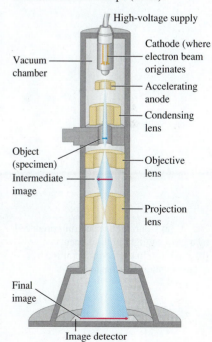

High-voltage supply

Cathode (where electron beam originates)

Vacuum chamber

Accelerating anode

Condensing lens

Object (specimen)

Objective lens

Intermediate image

Projection lens

Final image

Image detector

39.6 This scanning electron microscope image shows *Escherichia coli* bacteria crowded into a stoma, or respiration opening, on the surface of a lettuce leaf. (False color has been added.) If not washed off before the lettuce is eaten, these bacteria can be a health hazard. The *transmission* electron micrograph that opens this chapter shows a greatly magnified view of the surface of an *E. coli* bacterium.

E. coli bacteria

5 μm

micrographs have an appearance that is much more three-dimensional than conventional visible-light micrographs (Fig. 39.6). The resolution is typically of the order of 10 nm, not as good as a transmission electron microscope but still much finer than the best optical microscopes.

Test Your Understanding of Section 39.1   (a) A proton has a slightly smaller mass than a neutron. Compared to the neutron described in Example 39.2, would a proton of the same wavelength have (i) more kinetic energy; (ii) less kinetic energy; or (iii) the same kinetic energy? (b) Example 39.1 shows that to give electrons a wavelength of 1.7×10^{-10} m, they must be accelerated from rest through a voltage of 54 V and so acquire a kinetic energy of 54 eV. Does a photon of this same energy also have a wavelength of 1.7×10^{-10} m? ❙

39.2 The Nuclear Atom and Atomic Spectra

Every neutral atom contains at least one electron. How does the wave aspect of electrons affect atomic structure? As we will see, it is crucial for understanding not only the structure of atoms but also how they interact with light. Historically, the quest to understand the nature of the atom was intimately linked with both the idea that electrons have wave characteristics and the notion that light has particle characteristics. Before we explore how these ideas shaped atomic theory, it's useful to look at what was known about atoms—as well as what remained mysterious—by the first decade of the twentieth century.

Line Spectra

Everyone knows that heated materials emit light, and that different materials emit different kinds of light. The coils of a toaster glow red when in operation, the flame of a match has a characteristic yellow color, and the flame from a gas range is a distinct blue. To analyze these different types of light, we can use a prism or a diffraction grating to separate the various wavelengths in a beam of light into a spectrum. If the light source is a hot solid (such as the filament of an incandescent light bulb) or liquid, the spectrum is *continuous;* light of all wavelengths is present (Fig. 39.7a). But if the source is a heated *gas,* such as the neon in an advertising sign or the sodium vapor formed when table salt is thrown into a campfire, the spectrum includes only a few colors in the form of isolated sharp parallel lines (Fig. 39.7b). (Each "line" is an image of the spectrograph slit, deviated through an angle that depends on the wavelength of the light forming that image; see Section 36.5.) A spectrum of this sort is called an **emission line spectrum,** and the lines are called **spectral lines.** Each spectral line corresponds to a definite wavelength and frequency.

39.7 (a) Continuous spectrum produced by a glowing light bulb filament. (b) Emission line spectrum emitted by a lamp containing a heated gas.

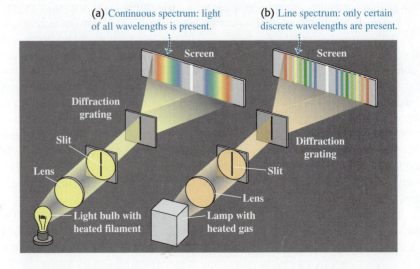

(a) Continuous spectrum: light of all wavelengths is present.

(b) Line spectrum: only certain discrete wavelengths are present.

Screen

Screen

Diffraction grating

Slit

Lens

Light bulb with heated filament

Diffraction grating

Slit

Lens

Lamp with heated gas

39.8 The emission line spectra of several kinds of atoms and molecules. No two are alike. Note that the spectrum of water vapor (H_2O) is similar to that of hydrogen (H_2), but there are important differences that make it straightforward to distinguish these two spectra.

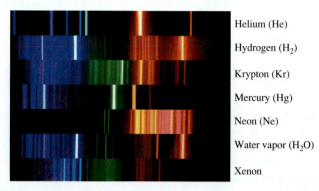

Helium (He)

Hydrogen (H_2)

Krypton (Kr)

Mercury (Hg)

Neon (Ne)

Water vapor (H_2O)

Xenon

It was discovered early in the 19th century that each element in its gaseous state has a unique set of wavelengths in its line spectrum. The spectrum of hydrogen always contains a certain set of wavelengths; mercury produces a different set, neon still another, and so on (Fig. 39.8). Scientists find the use of spectra to identify elements and compounds to be an invaluable tool. For instance, astronomers have detected the spectra from more than 100 different molecules in interstellar space, including some that are not found naturally on earth.

While a *heated* gas selectively *emits* only certain wavelengths, a *cool* gas selectively *absorbs* certain wavelengths. If we pass white (continuous-spectrum) light through a gas and look at the *transmitted* light with a spectrometer, we find a series of dark lines corresponding to the wavelengths that have been absorbed (Fig. 39.9). This is called an **absorption line spectrum.** What's more, a given kind of atom or molecule absorbs the *same* characteristic set of wavelengths when it's cool as it emits when heated. Hence scientists can use absorption line spectra to identify substances in the same manner that they use emission line spectra.

As useful as emission line spectra and absorption line spectra are, they presented a quandary to scientists: *Why* does a given kind of atom emit and absorb only certain very specific wavelengths? To answer this question, we need to have a better idea of what the inside of an atom is like. We know that atoms are much smaller than the wavelengths of visible light, so there is no hope of actually *seeing* an atom using that light. But we can still describe how the mass and electric charge are distributed throughout the volume of the atom.

Here's where things stood in 1910. In 1897 the English physicist J. J. Thomson (Nobel Prize 1906) had discovered the electron and measured its charge-to-mass ratio e/m. By 1909, the American physicist Robert Millikan (Nobel Prize 1923) had made the first measurements of the electron charge $-e$. These and other experiments showed that almost all the mass of an atom had to be associated with the *positive* charge, not with the electrons. It was also known that the overall size of atoms is of the order of 10^{-10} m and that all atoms except hydrogen contain more than one electron.

In 1910 the best available model of atomic structure was one developed by Thomson. He envisioned the atom as a sphere of some as yet unidentified positively charged substance, within which the electrons were embedded like raisins in cake. This model offered an explanation for line spectra. If the atom collided with another atom, as in a heated gas, each electron would oscillate around its equilibrium position with a characteristic frequency and emit electromagnetic radiation with that frequency. If the atom were illuminated with light of many frequencies, each electron would selectively absorb only light whose frequency matched the electron's natural oscillation frequency. (This is the phenomenon of resonance that we discussed in Section 14.8.)

39.9 The absorption line spectrum of the sun. (The spectrum "lines" read from left to right and from top to bottom, like text on a page.) The spectrum is produced by the sun's relatively cool atmosphere, which absorbs photons from deeper, hotter layers. The absorption lines thus indicate what kinds of atoms are present in the solar atmosphere.

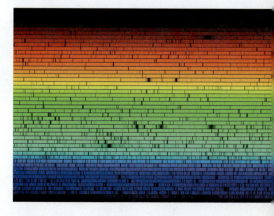

39.10 Born in New Zealand, Ernest Rutherford (1871–1937) spent his professional life in England and Canada. Before carrying out the experiments that established the existence of atomic nuclei, he shared (with Frederick Soddy) the 1908 Nobel Prize in chemistry for showing that radioactivity results from the disintegration of atoms.

MasteringPHYSICS®

PhET: Rutherford Scattering
ActivPhysics 19.1: Particle Scattering

Rutherford's Exploration of the Atom

The first experiments designed to test Thomson's model by probing the interior structure of the atom were carried out in 1910–1911 by Ernest Rutherford (Fig. 39.10) and two of his students, Hans Geiger and Ernest Marsden, at the University of Manchester in England. These experiments consisted of shooting a beam of charged particles at thin foils of various elements and observing how the foil deflected the particles.

The particle accelerators now in common use in laboratories had not yet been invented, and Rutherford's projectiles were *alpha particles* emitted from naturally radioactive elements. The nature of these alpha particles was not completely understood, but it was known that they are ejected from unstable nuclei with speeds of the order of 10^7 m/s, are positively charged, and can travel several centimeters through air or 0.1 mm or so through solid matter before they are brought to rest by collisions.

Figure 39.11 is a schematic view of Rutherford's experimental setup. A radioactive substance at the left emits alpha particles. Thick lead screens stop all particles except those in a narrow beam. The beam passes through the foil target (consisting of gold, silver, or copper) and strikes screens coated with zinc sulfide, creating a momentary flash, or *scintillation*. Rutherford and his students counted the numbers of particles deflected through various angles.

The atoms in a metal foil are packed together like marbles in a box (not spaced apart). Because the particle beam passes through the foil, the alpha particles must pass through the interior of atoms. Within an atom, the charged alpha particle will interact with the electrons and the positive charge. (Because the *total* charge of the atom is zero, alpha particles feel little electrical force outside an atom.) An electron has about 7300 times less mass than an alpha particle, so momentum considerations indicate that the atom's electrons cannot appreciably deflect the alpha particle—any more than a swarm of gnats deflects a tossed pebble. Any deflection will be due to the positively charged material that makes up almost all of the atom's mass.

In the Thomson model, the positive charge and the negative electrons are distributed through the whole atom. Hence the electric field inside the atom should be quite small, and the electric force on an alpha particle that enters the atom should be quite weak. The maximum deflection to be expected is then only a few degrees (Fig. 39.12a). The results of the Rutherford experiments were *very* different

39.11 The Rutherford scattering experiments investigated what happens to alpha particles fired at a thin gold foil. The results of this experiment helped reveal the structure of atoms.

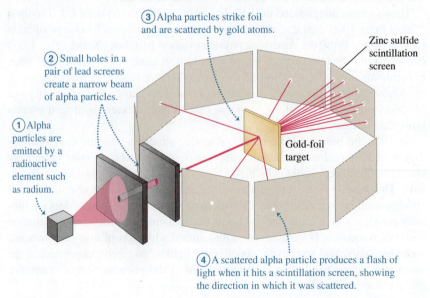

③ Alpha particles strike foil and are scattered by gold atoms.

Zinc sulfide scintillation screen

② Small holes in a pair of lead screens create a narrow beam of alpha particles.

① Alpha particles are emitted by a radioactive element such as radium.

Gold-foil target

④ A scattered alpha particle produces a flash of light when it hits a scintillation screen, showing the direction in which it was scattered.

from the Thomson prediction. Some alpha particles were scattered by nearly 180°—that is, almost straight backward (Fig. 39.12b). Rutherford later wrote:

> **It was quite the most incredible event that ever happened to me in my life. It was almost as incredible as if you had fired a 15-inch shell at a piece of tissue paper and it came back and hit you.**

Clearly the Thomson model was wrong and a new model was needed. Suppose the positive charge, instead of being distributed through a sphere with atomic dimensions (of the order of 10^{-10} m), is all concentrated in a much *smaller* volume. Then it would act like a point charge down to much smaller distances. The maximum electric field repelling the alpha particle would be much larger, and the amazing large-angle scattering that Rutherford observed could occur. Rutherford developed this model and called the concentration of positive charge the **nucleus.** He again computed the numbers of particles expected to be scattered through various angles. Within the accuracy of his experiments, the computed and measured results agreed, down to distances of the order of 10^{-14} m. His experiments therefore established that the atom does have a nucleus—a very small, very dense structure, no larger than 10^{-14} m in diameter. The nucleus occupies only about 10^{-12} of the total volume of the atom or less, but it contains *all* the positive charge and at least 99.95% of the total mass of the atom.

Figure 39.13 shows a computer simulation of alpha particles with a kinetic energy of 5.0 MeV being scattered from a gold nucleus of radius 7.0×10^{-15} m (the actual value) and from a nucleus with a hypothetical radius ten times larger. In the second case there is *no* large-angle scattering. The presence of large-angle scattering in Rutherford's experiments thus attested to the small size of the nucleus.

Later experiments showed that all nuclei are composed of positively charged protons (discovered in 1918) and electrically neutral neutrons (discovered in 1930). For example, the gold atoms in Rutherford's experiments have 79 protons and 118 neutrons. In fact, an alpha particle is itself the nucleus of a helium atom, with two protons and two neutrons. It is much more massive than an electron but only about 2% as massive as a gold nucleus, which helps explain why alpha particles are scattered by gold nuclei but not by electrons.

39.12 A comparison of Thomson's and Rutherford's models of the atom.

(a) Thomson's model of the atom: An alpha particle is scattered through only a small angle.

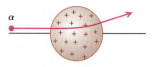

(b) Rutherford's model of the atom: An alpha particle can be scattered through a large angle by the compact, positively charged nucleus (not drawn to scale).

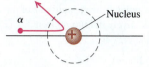

39.13 Computer simulation of scattering of 5.0-MeV alpha particles from a gold nucleus. Each curve shows a possible alpha-particle trajectory. **(a)** The scattering curves match Rutherford's experimental data if a radius of 7.0×10^{-15} m is assumed for a gold nucleus. **(b)** A model with a much larger radius for the gold nucleus does not match the data.

(a) A gold nucleus with radius 7.0×10^{-15} m gives large-angle scattering.

(b) A nucleus with 10 times the radius of the nucleus in **(a)** shows *no* large-scale scattering.

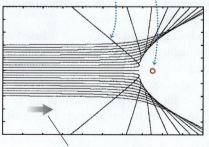
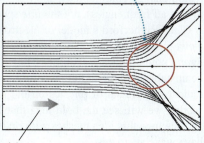

Motion of incident 5.0-MeV alpha particles

Example 39.4   A Rutherford experiment

An alpha particle (charge $2e$) is aimed directly at a gold nucleus (charge $79e$). What minimum initial kinetic energy must the alpha particle have to approach within 5.0×10^{-14} m of the center of the gold nucleus before reversing direction? Assume that the gold nucleus, which has about 50 times the mass of an alpha particle, remains at rest.   *Continued*

SOLUTION

IDENTIFY: The repulsive electric force exerted by the gold nucleus makes the the alpha particle slow to a halt as it approaches, then reverse direction. This force is conservative, so the total mechanical energy (kinetic energy of the alpha particle plus electric potential energy of the system) is conserved.

SET UP: Let point 1 be the initial position of the alpha particle, very far from the gold nucleus, and let point 2 be 5.0×10^{-14} m from the center of the gold nucleus. Our target variable is the kinetic energy K_1 of the alpha particle at point 1 that allows it to reach point 2 with $K_2 = 0$. To find this we'll use the law of conservation of energy and Eq. (23.9) for electric potential energy, $U = qq_0/4\pi\epsilon_0 r$.

EXECUTE: At point 1 the separation r of the alpha particle and gold nucleus is effectively infinite, so from Eq. (23.9) $U_1 = 0$. At point 2 the potential energy is

$$U_2 = \frac{1}{4\pi\epsilon_0} \frac{qq_0}{r}$$

$$= (9.0 \times 10^9 \text{ N} \cdot \text{m}^2/\text{C}^2) \frac{(2)(79)(1.60 \times 10^{-19} \text{ C})^2}{5.0 \times 10^{-14} \text{ m}}$$

$$= 7.3 \times 10^{-13} \text{ J} = 4.6 \times 10^6 \text{ eV} = 4.6 \text{ MeV}$$

By energy conservation $K_1 + U_1 = K_2 + U_2$, so $K_1 = K_2 + U_2 - U_1 = 0 + 4.6 \text{ MeV} - 0 = 4.6 \text{ MeV}$. Thus, to approach within 5.0×10^{-14} m, the alpha particle must have initial kinetic energy $K_1 = 4.6$ MeV.

EVALUATE: Alpha particles emitted from naturally occurring radioactive elements typically have energies in the range 4 to 6 MeV. For example, the common isotope of radium, ^{226}Ra, emits an alpha particle with energy 4.78 MeV.

Was it valid to assume that the gold nucleus remains at rest? To find out, note that when the alpha particle stops momentarily, all of its initial momentum has been transferred to the gold nucleus. An alpha particle has a mass $m_\alpha = 6.64 \times 10^{-27}$ kg; if its initial kinetic energy $K_1 = \frac{1}{2}mv_1^2$ is 7.3×10^{-13} J, you can show that its initial speed is $v_1 = 1.5 \times 10^7$ m/s and its initial momentum is $p_1 = m_\alpha v_1 = 9.8 \times 10^{-20}$ kg·m/s. A gold nucleus (mass $m_{\text{Au}} = 3.27 \times 10^{-25}$ kg) with this much momentum has a much slower speed $v_{\text{Au}} = 3.0 \times 10^5$ m/s and kinetic energy $K_{\text{Au}} = \frac{1}{2}mv_{\text{Au}}^2 = 1.5 \times 10^{-14}$ J = 0.092 MeV. This *recoil kinetic energy* of the gold nucleus is only 2% of the total energy in this situation, so we are justified in neglecting it.

The Failure of Classical Physics

Rutherford's discovery of the atomic nucleus raised a serious question: What prevented the negatively charged electrons from falling into the positively charged nucleus due to the strong electrostatic attraction? Rutherford suggested that perhaps the electrons *revolve* in orbits about the nucleus, just as the planets revolve around the sun.

But according to classical electromagnetic theory, any accelerating electric charge (either oscillating or revolving) radiates electromagnetic waves. An example is the radiation from an oscillating point charge that we depicted in Fig. 32.3 (Section 32.1). An electron orbiting inside an atom would always have a centripetal acceleration toward the nucleus, and so should be emitting radiation *at all times*. The energy of an orbiting electron should therefore decrease continuously, its orbit should become smaller and smaller, and it should spiral into the nucleus within a fraction of a second (Fig. 39.14). Even worse, according to classical theory the *frequency* of the electromagnetic waves emitted should equal the frequency of revolution. As the electrons radiated energy, their angular speeds would change continuously, and they would emit a *continuous* spectrum (a mixture of all frequencies), not the *line* spectrum actually observed.

Thus Rutherford's model of electrons orbiting the nucleus, which is based on Newtonian mechanics and classical electromagnetic theory, makes three entirely *wrong* predictions about atoms: They should emit light continuously, they should be unstable, and the light they emit should have a continuous spectrum. Clearly a radical reappraisal of physics on the scale of the atom was needed. In the next section we will see the bold idea that led to an new understanding of the atom, and see how this idea meshes with de Broglie's no less bold notion that electrons have wave attributes.

39.14 Classical physics makes predictions about the behavior of atoms that do not match reality.

ACCORDING TO CLASSICAL PHYSICS:
- An orbiting electron is accelerating, so it should radiate electromagnetic waves.
- The waves would carry away energy, so the electron should lose energy and spiral inward.
- The electron's angular speed would increase as its orbit shrank, so the frequency of the radiated waves should increase.

Thus, classical physics says that atoms should collapse within a fraction of a second and should emit light with a continuous spectrum as they do so.

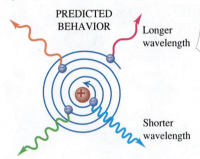

PREDICTED BEHAVIOR

Longer wavelength

Shorter wavelength

IN FACT:
- Atoms are stable.
- They emit light only when excited, and only at specific frequencies (as a line spectrum).

Test Your Understanding of Section 39.2 Suppose you repeated Rutherford's scattering experiment using a thin sheet of solid hydrogen in place of the gold foil. (Hydrogen is a solid at temperatures below 14.0 K.) The nucleus of a hydrogen atom is a single proton, with about one-fourth the mass of an alpha particle. Compared to the original experiment with gold foil, would you expect the alpha particles in this experiment to undergo (i) more large-angle scattering; (ii) the same amount of large-angle scattering; or (iii) less large-angle scattering?

39.3 Energy Levels and the Bohr Model of the Atom

In 1913 a young Danish physicist working with Ernest Rutherford at the University of Manchester made a revolutionary proposal to explain both the stability of atoms and their emission and absorption line spectra. The physicist was Niels Bohr (Fig. 39.15), and his innovation was to combine the photon concept that we introduced in Chapter 38 with a fundamentally new idea: The energy of an atom can have only certain particular values. His hypothesis represented a clean break from 19th-century ideas.

Photon Emission and Absorption by Atoms

Bohr's reasoning went like this. The emission line spectrum of an element tells us that atoms of that element emit photons with only certain specific frequencies f and hence certain specific energies $E = hf$. During the emission of a photon, the internal energy of the atom changes by an amount equal to the energy of the photon. Therefore, said Bohr, each atom must be able to exist with only certain specific values of internal energy. Each atom has a set of possible **energy levels.** An atom can have an amount of internal energy equal to any one of these levels, but it *cannot* have an energy *intermediate* between two levels. All isolated atoms of a given element have the same set of energy levels, but atoms of different elements have different sets.

Suppose an atom is raised, or *excited*, to a high energy level. (In a hot gas this happens when fast-moving atoms undergo inelastic collisions with each other or with the walls of the gas container. In an electric discharge tube, such as those used in a neon light fixture, atoms are excited by collisions with fast-moving electrons.) According to Bohr, an excited atom can make a *transition* from one energy level to a lower level by emitting a photon with energy equal to the energy *difference* between the initial and final levels (Fig. 39.16). If E_i is the initial energy of the atom before such a transition, E_f is its final energy after the transition, and the photon's energy is $hf = hc/\lambda$, then conservation of energy gives

$$hf = \frac{hc}{\lambda} = E_i - E_f \quad \text{(energy of emitted photon)} \quad (39.5)$$

For example, an excited lithium atom emits red light with wavelength $\lambda = 671$ nm. The corresponding photon energy is

$$E = \frac{hc}{\lambda} = \frac{(6.63 \times 10^{-34} \text{ J} \cdot \text{s})(3.00 \times 10^{8} \text{ m/s})}{671 \times 10^{-9} \text{ m}}$$

$$= 2.96 \times 10^{-19} \text{ J} = 1.85 \text{ eV}$$

This photon is emitted during a transition like that shown in Fig. 39.16 between two levels of the atom that differ in energy by $E_i - E_f = 1.85$ eV.

The emission line spectra shown in Fig. 39.8 show that many different wavelengths are emitted by each atom. Hence each kind of atom must have a number of energy levels, with different spacings in energy between them. Each wavelength in the spectrum corresponds to a transition between two specific energy levels of the atom.

CAUTION **Producing a line spectrum** The lines of an emission line spectrum, such as the helium spectrum shown at the top of Fig. 39.8, are *not* all produced by a single atom. The sample of helium gas that produced the spectrum in Fig. 39.8 contained a large number of helium atoms; these were excited in an electric discharge tube to various energy levels. The spectrum of the gas shows the light emitted from all the different transitions that occurred in different atoms of the sample.

The observation that atoms are stable means that each atom has a *lowest* energy level, called the **ground level.** Levels with energies greater than the

39.15 Niels Bohr (1885–1962) was a young postdoctoral researcher when he proposed the novel idea that the energy of an atom could have only certain discrete values. He won the 1922 Nobel Prize in physics for these ideas. Bohr went on to make seminal contributions to nuclear physics and to become a passionate advocate for the free exchange of scientific ideas among all nations.

39.16 An excited atom emitting a photon.

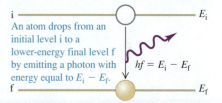

An atom drops from an initial level i to a lower-energy final level f by emitting a photon with energy equal to $E_i - E_f$.

$hf = E_i - E_f$

Mastering PHYSICS

PhET: Models of the Hydrogen Atom
ActivPhysics 18.1: The Bohr Model

ground level are called **excited levels.** An atom in an excited level, called an *excited atom*, can make a transition into the ground level by emitting a photon as in Fig. 39.16. But since there are no levels below the ground level, an atom in the ground level cannot lose energy and so cannot emit a photon.

Collisions are not the only way that an atom's energy can be raised from one level to a higher level. If an atom initially in the lower energy level in Fig. 39.16 is struck by a photon with just the right amount of energy, the photon can be *absorbed* and the atom will end up in the higher level (Fig. 39.17). As an example, we previously mentioned two levels in the lithium atom with an energy difference of 1.85 eV. For a photon to be absorbed and excite the atom from the lower level to the higher one, the photon must have an energy of 1.85 eV and a wavelength of 671 nm. In other words, an atom *absorbs* the same wavelengths that it *emits.* This explains the correspondence between an element's emission line spectrum and its absorption line spectrum that we described in Section 39.2.

Note that a lithium atom *cannot* absorb a photon with a slightly longer wavelength (say, 672 nm) or one with a slightly shorter wavelength (say, 670 nm). That's because these photons have, respectively, slightly too little or slightly too much energy to raise the atom's energy from one level to the next, and an atom cannot have an energy that's intermediate between levels. This explains why absorption line spectra have distinct dark lines (see Fig. 39.9): Atoms can absorb only photons with specific wavelengths.

An atom that's been excited into a high energy level, either by photon absorption or by collisions, does not stay there for long. After a short time, called the *lifetime* of the level (typically around 10^{-8} s), the excited atom will emit a photon and make a transition into a lower excited level or the ground level. A cool gas that's illuminated by white light to make an *absorption* line spectrum thus also produces an *emission* line spectrum when viewed from the side, since when the atoms de-excite they emit photons in all directions (Fig. 39.18). To keep a gas of atoms glowing, you have to continually provide energy to the gas in order to re-excite atoms so that they can emit more photons. If you turn off the energy supply (for example, by turning off the electric current through a neon light fixture, or by shutting off the light source in Fig. 39.18), the atoms drop back into their ground levels and cease to emit light.

By working backward from the observed emission line spectrum of an element, physicists can deduce the arrangement of energy levels in an atom of that element. As an example, Fig. 39.19a shows some of the energy levels for a sodium atom. You may have noticed the yellow-orange light emitted by sodium

39.17 An atom absorbing a photon. (Compare with Fig. 39.16.)

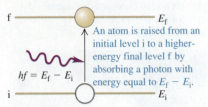

An atom is raised from an initial level i to a higher-energy final level f by absorbing a photon with energy equal to $E_f - E_i$.

$$hf = E_f - E_i$$

39.18 When a beam of white light with a continuous spectrum passes through a cool gas, the transmitted light has an absorption spectrum. The absorbed light energy excites the gas and causes it to emit light of its own, which has an emission spectrum.

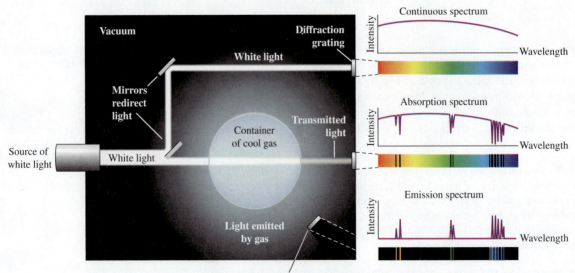

39.19 (a) Energy levels of the sodium atom relative to the ground level. Numbers on the lines between levels are wavelengths of the light emitted or absorbed during transitions between those levels. The column labels, such as $^2S_{1/2}$, refer to some quantum states of the atom. (b) When a sodium compound is placed in a flame, sodium atoms are excited into the lowest excited levels. As they drop back to the ground level, the atoms emit photons of yellow-orange light with wavelengths 589.0 and 589.6 nm.

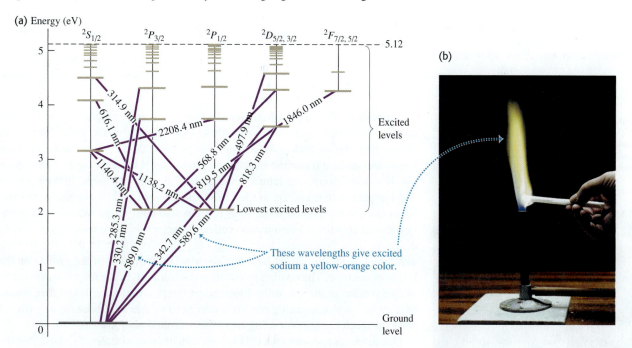

(a) Energy (eV)

vapor street lights. Sodium atoms emit this characteristic yellow-orange light with wavelengths 589.0 and 589.6 nm when they make transitions from the two closely spaced levels labeled *lowest excited levels* to the ground level. A standard test for the presence of sodium compounds is to look for this yellow-orange light from a sample placed in a flame (Fig. 39.19b).

Example 39.5 Emission and absorption spectra

A hypothetical atom (Fig. 39.20a) has energy levels at 0.00 eV (the ground level), 1.00 eV, and 3.00 eV. (a) What are the frequencies and wavelengths of the spectral lines this atom can emit when excited? (b) What wavelengths can this atom absorb if it is in its ground level?

SOLUTION

IDENTIFY and SET UP: Energy is conserved when a photon is emitted or absorbed. In each transition the photon energy is equal to the difference between the energies of the levels involved in the transition.

EXECUTE: (a) The possible energies of emitted photons are 1.00 eV, 2.00 eV, and 3.00 eV. For 1.00 eV, Eq. (39.2) gives

$$f = \frac{E}{h} = \frac{1.00 \text{ eV}}{4.136 \times 10^{-15} \text{ eV} \cdot \text{s}} = 2.42 \times 10^{14} \text{ Hz}$$

For 2.00 eV and 3.00 eV, $f = 4.84 \times 10^{14}$ Hz and 7.25×10^{14} Hz, respectively. For 1.00-eV photons,

$$\lambda = \frac{c}{f} = \frac{3.00 \times 10^8 \text{ m/s}}{2.42 \times 10^{14} \text{ Hz}} = 1.24 \times 10^{-6} \text{ m} = 1240 \text{ nm}$$

39.20 (a) Energy-level diagram for the hypothetical atom, showing the possible transitions for emission from excited levels and for absorption from the ground level. (b) Emission spectrum of this hypothetical atom.

(a)

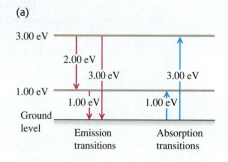

(b)

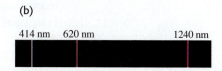

Continued

This is in the infrared region of the spectrum (Fig. 39.20b). For 2.00 eV and 3.00 eV, the wavelengths are 620 nm (red) and 414 nm (violet), respectively.

(b) From the ground level, only a 1.00-eV or a 3.00-eV photon can be absorbed (Fig. 39.20a); a 2.00-eV photon cannot be absorbed because the atom has no energy level 2.00 eV above the ground level. Passing light from a hot solid through a gas of these hypothetical atoms (almost all of which would be in the ground state if the gas were cool) would yield a continuous spectrum with dark absorption lines at 1240 nm and 414 nm.

EVALUATE: Note that if a gas of these atoms were at a sufficiently high temperature, collisions would excite a number of atoms into the 1.00-eV energy level. Such excited atoms *can* absorb 2.00-eV photons, as Fig. 39.20a shows, and an absorption line at 620 nm would appear in the spectrum. Thus the observed spectrum of a given substance depends on its energy levels and its temperature.

Application **Fish Fluorescence**
When illuminated by blue light, this tropical lizardfish (family *Synodontidae*) fluoresces and emits longer-wavelength green light. The fluorescence may be a sexual signal or a way for the fish to camouflage itself among coral (which also have a green fluorescence).

Suppose we take a gas of the hypothetical atoms in Example 39.5 and illuminate it with violet light of wavelength 414 nm. Atoms in the ground level can absorb this photon and make a transition to the 3.00-eV level. Some of these atoms will make a transition back to the ground level by emitting a 414-nm photon. But other atoms will return to the ground level in two steps, first emitting a 620-nm photon to transition to the 1.00-eV level, then a 1240-nm photon to transition back to the ground level. Thus this gas will emit longer-wavelength radiation than it absorbs, a phenomenon called *fluorescence*. For example, the electric discharge in a fluorescent lamp causes the mercury vapor in the tube to emit ultraviolet radiation. This radiation is absorbed by the atoms of the coating on the inside of the tube. The coating atoms then re-emit light in the longer-wavelength, visible portion of the spectrum. Fluorescent lamps are more efficient than incandescent lamps in converting electrical energy to visible light because they do not waste as much energy producing (invisible) infrared photons.

Our discussion of energy levels and spectra has concentrated on *atoms*, but the same ideas apply to *molecules*. Figure 39.8 shows the emission line spectra of two molecules, hydrogen (H_2) and water (H_2O). Just as for sodium or other atoms, physicists can work backward from these molecular spectra and deduce the arrangement of energy levels for each kind of molecule. We'll return to molecules and molecular structure in Chapter 42.

The Franck–Hertz Experiment: Are Energy Levels Real?

Are atomic energy levels real, or just a convenient fiction that helps us to explain spectra? In 1914, the German physicists James Franck and Gustav Hertz answered this question when they found direct experimental evidence for the existence of atomic energy levels.

Franck and Hertz studied the motion of electrons through mercury vapor under the action of an electric field. They found that when the electron kinetic energy was 4.9 eV or greater, the vapor emitted ultraviolet light of wavelength 250 nm. Suppose mercury atoms have an excited energy level 4.9 eV above the ground level. An atom can be raised to this level by collision with an electron; it later decays back to the ground level by emitting a photon. From the photon formula $E = hc/\lambda$, the wavelength of the photon should be

$$\lambda = \frac{hc}{E} = \frac{(4.136 \times 10^{-15} \text{ eV} \cdot \text{s})(3.00 \times 10^8 \text{ m/s})}{4.9 \text{ eV}}$$
$$= 2.5 \times 10^{-7} \text{ m} = 250 \text{ nm}$$

This is equal to the wavelength that Franck and Hertz measured, which demonstrates that this energy level actually exists in the mercury atom. Similar experiments with other atoms yield the same kind of evidence for atomic energy levels. Franck and Hertz shared the 1925 Nobel Prize in physics for their research.

Electron Waves and the Bohr Model of Hydrogen

Bohr's hypothesis established the relationship between atomic spectra and energy levels. By itself, however, it provided no general principles for *predicting* the energy levels of a particular atom. Bohr addressed this problem for the case of the simplest atom, hydrogen, which has just one electron. Let's look at the ideas behind the **Bohr model** of the hydrogen atom.

Bohr postulated that each energy level of a hydrogen atom corresponds to a specific *stable* circular orbit of the electron around the nucleus. In a break with classical physics, Bohr further postulated that an electron in such an orbit does *not* radiate. Instead, an atom radiates energy only when an electron makes a transition from an orbit of energy E_i to a different orbit with lower energy E_f, emitting a photon of energy $hf = E_i - E_f$ in the process.

As a result of a rather complicated argument that related the angular frequency of the light emitted to the angular speed of the electron in highly excited energy levels, Bohr found that the magnitude of the electron's angular momentum is *quantized;* that is, this magnitude must be an integral multiple of $h/2\pi$. (Because $1 \text{ J} = 1 \text{ kg} \cdot \text{m}^2/\text{s}^2$, the SI units of Planck's constant h, J · s, are the same as the SI units of angular momentum, usually written as $\text{kg} \cdot \text{m}^2/\text{s}$.) Let's number the orbits by an integer n, where $n = 1, 2, 3, \ldots$, and call the radius of orbit n r_n and the speed of the electron in that orbit v_n. The value of n for each orbit is called the **principal quantum number** for the orbit. From Section 10.5, Eq. (10.28), the magnitude of the angular momentum of an electron of mass m in such an orbit is $L_n = mv_n r_n$ (Fig. 39.21). So Bohr's argument led to

$$L_n = mv_n r_n = n\frac{h}{2\pi} \quad \text{(quantization of angular momentum)} \quad (39.6)$$

Instead of going through Bohr's argument to justify Eq. (39.6), we can use de Broglie's picture of electron waves. Rather then visualizing the orbiting electron as a particle moving around the nucleus in a circular path, think of it as a sinusoidal *standing wave* with wavelength λ that extends around the circle. A standing wave on a string transmits no energy (see Section 15.7), and electrons in Bohr's orbits radiate no energy. For the wave to "come out even" and join onto itself smoothly, the circumference of this circle must include some *whole number* of wavelengths, as Fig. 39.22 suggests. Hence for an orbit with radius r_n and circumference $2\pi r_n$, we must have $2\pi r_n = n\lambda_n$, where λ_n is the wavelength and $n = 1, 2, 3, \ldots$. According to the de Broglie relationship, Eq. (39.1), the wavelength of a particle with rest mass m moving with nonrelativistic speed v_n is $\lambda_n = h/mv_n$. Combining $2\pi r_n = n\lambda_n$ and $\lambda_n = h/mv_n$, we find $2\pi r_n = nh/mv_n$ or

$$mv_n r_n = n\frac{h}{2\pi}$$

This is the same as Bohr's result, Eq. (39.6). Thus a wave picture of the electron leads naturally to the quantization of the electron's angular momentum.

Now let's consider a model of the hydrogen atom that is Newtonian in spirit but incorporates this quantization assumption (Fig. 39.23). This atom consists of a single electron with mass m and charge $-e$ in a circular orbit around a single proton with charge $+e$. The proton is nearly 2000 times as massive as the electron, so we can assume that the proton does not move. We learned in Section 5.4 that when a particle with mass m moves with speed v_n in a circular orbit with radius r_n, its centripetal (inward) acceleration is v_n^2/r_n. According to Newton's second law, a radially inward net force with magnitude $F = mv_n^2/r_n$ is needed to cause this acceleration. We discussed in Section 12.4 how the gravitational attraction provides that inward force for satellite orbits. In hydrogen the force F is provided by the electrical attraction between the positive proton and the negative electron. From Coulomb's law, Eq. (21.2),

$$F = \frac{1}{4\pi\epsilon_0}\frac{e^2}{r_n^2}$$

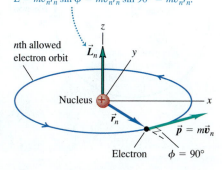

39.21 Calculating the angular momentum of an electron in a circular orbit around an atomic nucleus.

Angular momentum $\vec{L}_n$ of orbiting electron is perpendicular to plane of orbit (since we take origin to be at nucleus) and has magnitude $L = mv_n r_n \sin\phi = mv_n r_n \sin 90° = mv_n r_n$.

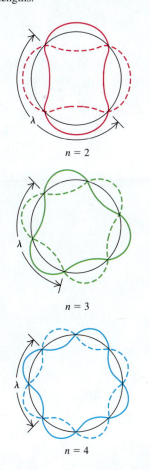

39.22 These diagrams show the idea of fitting a standing electron wave around a circular orbit. For the wave to join onto itself smoothly, the circumference of the orbit must be an integral number n of wavelengths.

$n = 2$

$n = 3$

$n = 4$

39.23 The Bohr model of the hydrogen atom.

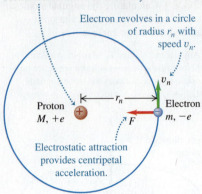

Proton is assumed to be stationary.

Electron revolves in a circle of radius r_n with speed v_n.

Proton $M, +e$

r_n

v_n

Electron $m, -e$

F

Electrostatic attraction provides centripetal acceleration.

Hence Newton's second law states that

$$\frac{1}{4\pi\epsilon_0}\frac{e^2}{r_n^2} = \frac{mv_n^2}{r_n} \tag{39.7}$$

When we solve Eqs. (39.6) and (39.7) simultaneously for r_n and v_n, we get

$$r_n = \epsilon_0 \frac{n^2 h^2}{\pi m e^2} \qquad \text{(orbital radii in the Bohr model)} \tag{39.8}$$

$$v_n = \frac{1}{\epsilon_0}\frac{e^2}{2nh} \qquad \text{(orbital speeds in the Bohr model)} \tag{39.9}$$

Equation (39.8) shows that the orbit radius r_n is proportional to n^2, so the smallest orbit radius corresponds to $n = 1$. We'll denote this minimum radius, called the *Bohr radius,* as a_0:

$$a_0 = \epsilon_0 \frac{h^2}{\pi m e^2} \qquad \text{(Bohr radius)} \tag{39.10}$$

Then we can rewrite Eq. (39.8) as

$$r_n = n^2 a_0 \tag{39.11}$$

The permitted orbits have radii a_0, $4a_0$, $9a_0$, and so on.

You can find the numerical values of the quantities on the right-hand side of Eq. (39.10) in Appendix F. Using these values, we find that the radius a_0 of the smallest Bohr orbit is

$$a_0 = \frac{(8.854 \times 10^{-12}\ \text{C}^2/\text{N}\cdot\text{m}^2)(6.626 \times 10^{-34}\ \text{J}\cdot\text{s})^2}{\pi(9.109 \times 10^{-31}\ \text{kg})(1.602 \times 10^{-19}\ \text{C})^2}$$

$$= 5.29 \times 10^{-11}\ \text{m}$$

This gives an atomic diameter of about 10^{-10} m = 0.1 nm, which is consistent with atomic dimensions estimated by other methods.

Equation (39.9) shows that the orbital speed v_n is proportional to $1/n$. Hence the greater the value of n, the larger the orbital radius of the electron and the slower its orbital speed. (We saw the same relationship between orbital radius and speed for satellite orbits in Section 13.4.) We leave it to you to calculate the speed in the $n = 1$ orbit, which is the greatest possible speed of the electron in the hydrogen atom (see Exercise 39.29); the result is $v_1 = 2.19 \times 10^6$ m/s. This is less than 1% of the speed of light, so relativistic considerations aren't significant.

Hydrogen Energy Levels in the Bohr Model

We can now use Eqs. (39.8) and (39.9) to find the kinetic and potential energies K_n and U_n when the electron is in the orbit with quantum number n:

$$K_n = \tfrac{1}{2}mv_n^2 = \frac{1}{\epsilon_0^2}\frac{me^4}{8n^2 h^2} \qquad \begin{array}{l}\text{(kinetic energies in} \\ \text{the Bohr model)}\end{array} \tag{39.12}$$

$$U_n = -\frac{1}{4\pi\epsilon_0}\frac{e^2}{r_n} = -\frac{1}{\epsilon_0^2}\frac{me^4}{4n^2 h^2} \qquad \begin{array}{l}\text{(potential energies in} \\ \text{the Bohr model)}\end{array} \tag{39.13}$$

The potential energy has a negative sign because we have taken the electric potential energy to be zero when the electron is infinitely far from the nucleus. We are interested only in the *differences* in energy between orbits, so the reference position doesn't matter. The total energy E_n is the sum of the kinetic and potential energies:

$$E_n = K_n + U_n = -\frac{1}{\epsilon_0^2}\frac{me^4}{8n^2 h^2} \qquad \begin{array}{l}\text{(total energies in the} \\ \text{Bohr model)}\end{array} \tag{39.14}$$

39.24 Two ways to represent the energy levels of the hydrogen atom and the transitions between them. Note that the radius of the *n*th permitted orbit is actually n^2 times the radius of the $n = 1$ orbit.

(a) Permitted orbits of an electron in the Bohr model of a hydrogen atom (not to scale). Arrows indicate the transitions responsible for some of the lines of various series.

(b) Energy-level diagram for hydrogen, showing some transitions corresponding to the various series

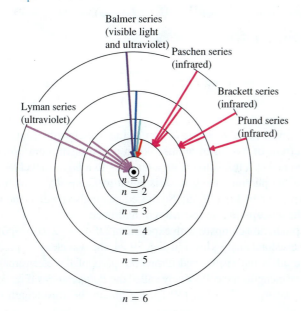

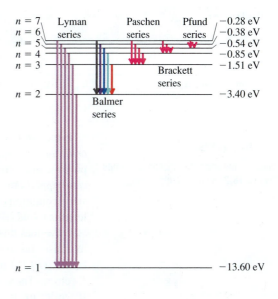

Since E_n in Eq. (39.14) has a different value for each n, you can see that this equation gives the *energy levels* of the hydrogen atom in the Bohr model. Each distinct orbit corresponds to a distinct energy level.

Figure 39.24 depicts the orbits and energy levels. We label the possible energy levels of the atom by values of the quantum number n. For each value of n there are corresponding values of orbit radius r_n, speed v_n, angular momentum $L_n = nh/2\pi$, and total energy E_n. The energy of the atom is least when $n = 1$ and E_n has its most negative value. This is the *ground level* of the hydrogen atom; it is the level with the smallest orbit, of radius a_0. For $n = 2, 3, \ldots$, the absolute value of E_n is smaller and the energy is progressively larger (less negative).

Figure 39.24 also shows some of the possible transitions from one electron orbit to an orbit of lower energy. Consider a transition from orbit n_U (for "upper") to a smaller orbit n_L (for "lower"), with $n_L < n_U$—or, equivalently, from *level* n_U to a lower *level* n_L. Then the energy hc/λ of the emitted photon of wavelength λ is equal to $E_{n_U} - E_{n_L}$. Before we use this relationship to solve for λ, it's convenient to rewrite Eq. (39.14) for the energies as

$$E_n = -\frac{hcR}{n^2}, \quad \text{where} \quad R = \frac{me^4}{8\epsilon_0^2 h^3 c} \quad \begin{array}{l}\text{(total energies in}\\ \text{the Bohr model)}\end{array} \quad \text{(39.15)}$$

The quantity R in Eq. (39.15) is called the **Rydberg constant** (named for the Swedish physicist Johannes Rydberg, who did pioneering work on the hydrogen spectrum). When we substitute the numerical values of the fundamental physical constants m, c, e, h, and ϵ_0, all of which can be determined quite independently of the Bohr theory, we find that $R = 1.097 \times 10^7 \text{ m}^{-1}$. Now we solve for the wavelength of the photon emitted in a transition from level n_U to level n_L:

$$\frac{hc}{\lambda} = E_{n_U} - E_{n_L} = \left(-\frac{hcR}{n_U^2}\right) - \left(-\frac{hcR}{n_L^2}\right) = hcR\left(\frac{1}{n_L^2} - \frac{1}{n_U^2}\right)$$

$$\frac{1}{\lambda} = R\left(\frac{1}{n_L^2} - \frac{1}{n_U^2}\right) \quad \begin{array}{l}\text{(hydrogen wavelengths in the}\\ \text{Bohr model, } n_L < n_U)\end{array} \quad \text{(39.16)}$$

39.25 The Balmer series of spectral lines for atomic hydrogen. You can see these same lines in the spectrum of *molecular* hydrogen (H₂) shown in Fig. 39.8, as well as additional lines that are present only when two hydrogen atoms are combined to make a molecule.

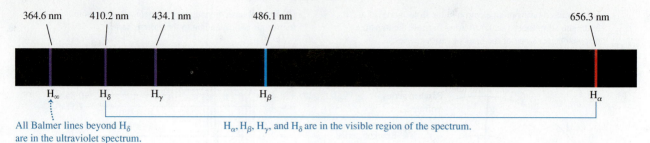

All Balmer lines beyond H_δ are in the ultraviolet spectrum.

H_α, H_β, H_γ, and H_δ are in the visible region of the spectrum.

Mastering PHYSICS

PhET: Neon Lights and Other Discharge Lamps
ActivPhysics 18.2: Spectroscopy

Equation (39.16) is a *theoretical prediction* of the wavelengths found in the *emission* line spectrum of hydrogen atoms. When a hydrogen atom *absorbs* a photon, an electron makes a transition from a level n_L to a *higher* level n_U. This can happen only if the photon energy hc/λ is equal to $E_{n_U} - E_{n_L}$, which is the same condition expressed by Eq. (39.16). So this equation also predicts the wavelengths found in the *absorption* line spectrum of hydrogen.

How does this prediction compare with experiment? If $n_L = 2$, corresponding to transitions to the second energy level in Fig. 39.24, the wavelengths predicted by Eq. (39.16) are all in the visible and ultraviolet parts of the electromagnetic spectrum. These wavelengths are collectively called the *Balmer series* (Fig. 39.25). If we let $n_L = 2$ and $n_U = 3$ in Eq. (39.16) we obtain the wavelength of the H_α line:

$$\frac{1}{\lambda} = (1.097 \times 10^7 \text{ m}^{-1})\left(\tfrac{1}{4} - \tfrac{1}{9}\right) \quad \text{or} \quad \lambda = 656.3 \text{ nm}$$

With $n_L = 2$ and $n_U = 4$ we obtain the wavelength of the H_β line, and so on. With $n_L = 2$ and $n_U = \infty$ we obtain the shortest wavelength in the series, $\lambda = 364.6$ nm. These theoretical predictions are within 0.1% of the observed hydrogen wavelengths! This close agreement provides very strong and direct confirmation of Bohr's theory.

The Bohr model also predicts many other wavelengths in the hydrogen spectrum, as Fig. 39.24 shows. The observed wavelengths of all of these series, each of which is named for its discoverer, match the predicted values with the same percent accuracy as for the Balmer series. The *Lyman series* of spectral lines is caused by transitions between the ground level and the excited levels, corresponding to $n_L = 1$ and $n_U = 2, 3, 4, \ldots$ in Eq. (39.16). The energy difference between the ground level and any of the excited levels is large, so the emitted photons have wavelengths in the ultraviolet part of the electromagnetic spectrum. Transitions among the higher energy levels involve a much smaller energy difference, so the photons emitted in these transitions have little energy and long, infrared wavelengths. That's the case for both the *Brackett series* ($n_L = 3$ and $n_U = 4, 5, 6, \ldots$, corresponding to transitions between the third and higher energy levels) and the *Pfund series* ($n_L = 4$ and $n_U = 5, 6, 7, \ldots$, with transitions between the fourth and higher energy levels).

Figure 39.24 shows only transitions in which a hydrogen atom loses energy and a photon is emitted. But as we discussed previously, the wavelengths of those photons that an atom can *absorb* are the same as those that it can emit. For example, a hydrogen atom in the $n = 2$ level can absorb a 656.3-nm photon and end up in the $n = 3$ level.

One additional test of the Bohr model is its predicted value of the *ionization energy* of the hydrogen atom. This is the energy required to remove the electron completely from the atom. Ionization corresponds to a transition from the ground level ($n = 1$) to an infinitely large orbit radius ($n = \infty$), so the energy that must be added to the atom is $E_\infty - E_1 = 0 - E_1 = -E_1$ (recall that E_1 is negative).

Substituting the constants from Appendix F into Eq. (39.15) gives an ionization energy of 13.606 eV. The ionization energy can also be measured directly; the result is 13.60 eV. These two values agree within 0.1%.

Example 39.6   Exploring the Bohr model

Find the kinetic, potential, and total energies of the hydrogen atom in the first excited level, and find the wavelength of the photon emitted in a transition from that level to the ground level.

SOLUTION

IDENTIFY and SET UP: This problem uses the ideas of the Bohr model. We use simplified versions of Eqs. (39.12), (39.13), and (39.14) to find the energies of the atom, and Eq. (39.16), $hc/\lambda = E_{n_U} - E_{n_L}$, to find the photon wavelength λ in a transition from $n_U = 2$ (the first excited level) to $n_L = 1$ (the ground level).

EXECUTE: We could evaluate Eqs. (39.12), (39.13), and (39.14) for the nth level by substituting the values of m, e, ϵ_0, and h. But we can simplify the calculation by comparing with Eq. (39.15), which shows that the constant $me^4/8\epsilon_0^2 h^2$ that appears in Eqs. (39.12), (39.13), and (39.14) is equal to hcR:

$$\frac{me^4}{8\epsilon_0^2 h^2} = hcR$$

$$= (6.626 \times 10^{-34} \text{ J} \cdot \text{s})(2.998 \times 10^8 \text{ m/s})$$
$$\times (1.097 \times 10^7 \text{ m}^{-1})$$

$$= 2.179 \times 10^{-18} \text{ J} = 13.60 \text{ eV}$$

This allows us to rewrite Eqs. (39.12), (39.13), and (39.14) as

$$K_n = \frac{13.60 \text{ eV}}{n^2} \qquad U_n = \frac{-27.20 \text{ eV}}{n^2} \qquad E_n = \frac{-13.60 \text{ eV}}{n^2}$$

For the first excited level ($n = 2$), we have $K_2 = 3.40$ eV, $U_2 = -6.80$ eV, and $E_2 = -3.40$ eV. For the ground level ($n = 1$), $E_1 = -13.60$ eV. The energy of the emitted photon is then $E_2 - E_1 = -3.40$ eV $- (-13.60 \text{ eV}) = 10.20$ eV, and

$$\lambda = \frac{hc}{E_2 - E_1} = \frac{(4.136 \times 10^{-15} \text{ eV} \cdot \text{s})(3.00 \times 10^8 \text{ m/s})}{10.20 \text{ eV}}$$

$$= 1.22 \times 10^{-7} \text{ m} = 122 \text{ nm}$$

This is the wavelength of the Lyman-alpha (L_α) line, the longest-wavelength line in the Lyman series of ultraviolet lines in the hydrogen spectrum (see Fig. 39.24).

EVALUATE: The total mechanical energy for any level is negative and is equal to one-half the potential energy. We found the same energy relationship for Newtonian satellite orbits in Section 12.4. The situations are similar because both the electrostatic and gravitational forces are inversely proportional to $1/r^2$.

Nuclear Motion and the Reduced Mass of an Atom

The Bohr model is so successful that we can justifiably ask why its predictions for the wavelengths and ionization energy of hydrogen differ from the measured values by about 0.1%. The explanation is that we assumed that the nucleus (a proton) remains at rest. However, as Fig. 39.26 shows, the proton and electron *both* revolve in circular orbits about their common center of mass (see Section 8.5). It turns out that we can take this motion into account very simply by using in Bohr's equations not the electron rest mass m but a quantity called the **reduced mass** m_r of the system. For a system composed of two bodies of masses m_1 and m_2, the reduced mass is

$$m_r = \frac{m_1 m_2}{m_1 + m_2} \tag{39.17}$$

For ordinary hydrogen we let m_1 equal m and m_2 equal the proton mass, $m_p = 1836.2m$. Thus the proton–electron system of ordinary hydrogen has a reduced mass of

$$m_r = \frac{m(1836.2m)}{m + 1836.2m} = 0.99946m$$

When this value is used instead of the electron mass m in the Bohr equations, the predicted values agree very well with the measured values.

In an atom of deuterium, also called *heavy hydrogen*, the nucleus is not a single proton but a proton and a neutron bound together to form a composite body called the *deuteron*. The reduced mass of the deuterium atom turns out to be $0.99973m$. Equations (39.15) and (39.16) (with m replaced by m_r) show that all wavelengths are inversely proportional to m_r. Thus the wavelengths

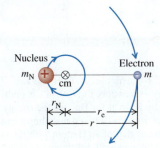

39.26 The nucleus and the electron both orbit around their common center of mass. The distance r_N has been exaggerated for clarity; for ordinary hydrogen it actually equals $r_e/1836.2$.

of the deuterium spectrum should be those of hydrogen divided by $(0.99973m)/(0.99946m) = 1.00027$. This is a small effect but well within the precision of modern spectrometers. This small wavelength shift led the American scientist Harold Urey to the discovery of deuterium in 1932, an achievement that earned him the 1934 Nobel Prize in chemistry.

Hydrogenlike Atoms

We can extend the Bohr model to other one-electron atoms, such as singly ionized helium (He^+), doubly ionized lithium (Li^{2+}), and so on. Such atoms are called *hydrogenlike* atoms. In such atoms, the nuclear charge is not e but Ze, where Z is the *atomic number,* equal to the number of protons in the nucleus. The effect in the previous analysis is to replace e^2 everywhere by Ze^2. In particular, the orbital radii r_n given by Eq. (39.8) become smaller by a factor of Z, and the energy levels E_n given by Eq. (39.14) are multiplied by Z^2. We invite you to verify these statements. The reduced-mass correction in these cases is even less than 0.1% because the nuclei are more massive than the single proton of ordinary hydrogen. Figure 39.27 compares the energy levels for H and for He^+, which has $Z = 2$.

Atoms of the alkali metals (at the far left-hand side of the periodic table; see Appendix D) have one electron outside a core consisting of the nucleus and the inner electrons, with net core charge $+e$. These atoms are approximately hydrogenlike, especially in excited levels. Physicists have studied alkali atoms in which the outer electron has been excited into a very large orbit with $n = 1000$. In accordance with Eq. (39.8), the radius of such a *Rydberg atom* with $n = 1000$ is $n^2 = 10^6$ times the Bohr radius, or about 0.05 mm—about the same size as a small grain of sand.

Although the Bohr model predicted the energy levels of the hydrogen atom correctly, it raised as many questions as it answered. It combined elements of classical physics with new postulates that were inconsistent with classical ideas. The model provided no insight into what happens during a transition from one orbit to another; the angular speeds of the electron motion were not in general the angular frequencies of the emitted radiation, a result that is contrary to classical electrodynamics. Attempts to extend the model to atoms with two or more electrons were not successful. An electron moving in one of Bohr's circular orbits forms a current loop and should produce a magnetic dipole moment (see Section 27.7). However, a hydrogen atom in its ground level has *no* magnetic moment due to orbital motion. In Chapters 40 and 41 we will find that an even more radical departure from classical concepts was needed before the understanding of atomic structure could progress further.

39.27 Energy levels of H and He^+. The energy expression, Eq. (39.14), is multiplied by $Z^2 = 4$ for He^+, so the energy of an He^+ ion with a given n is almost exactly four times that of an H atom with the same n. (There are small differences of the order of 0.05% because the reduced masses are slightly different.)

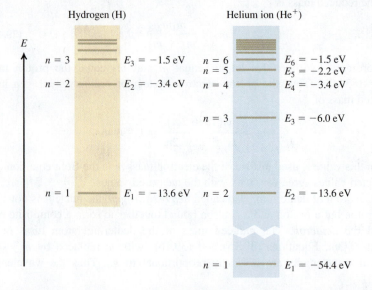

Test Your Understanding of Section 39.3   Consider the possible transitions between energy levels in a He$^+$ ion. For which of these transitions in He$^+$ will the wavelength of the emitted photon be nearly the same as one of the wavelengths emitted by excited H atoms? (i) $n = 2$ to $n = 1$; (ii) $n = 3$ to $n = 2$; (iii) $n = 4$ to $n = 3$; (iv) $n = 4$ to $n = 2$; (v) more than one of these; (vi) none of these. |

39.4 The Laser

The **laser** is a light source that produces a beam of highly coherent and very nearly monochromatic light as a result of cooperative emission from many atoms. The name "laser" is an acronym for "light amplification by stimulated emission of radiation." We can understand the principles of laser operation from what we have learned about atomic energy levels and photons. To do this we'll have to introduce two new concepts: *stimulated emission* and *population inversion*.

MasteringPHYSICS

PhET: Lasers

Spontaneous and Stimulated Emission

Consider a gas of atoms in a transparent container. Each atom is initially in its ground level of energy E_g and also has an excited level of energy E_{ex}. If we shine light of frequency f on the container, an atom can absorb one of the photons provided the photon energy $E = hf$ equals the energy difference $E_{ex} - E_g$ between the levels. Figure 39.28a shows this process, in which three atoms A each absorb a photon and go into the excited level. Some time later, the excited atoms (which we denote as A*) return to the ground level by each emitting a photon with the same frequency as the one originally absorbed (Fig. 39.28b). This process is called **spontaneous emission.** The direction and phase of the spontaneously emitted photons are random.

In **stimulated emission** (Fig. 39.28c), each incident photon encounters a previously excited atom. A kind of resonance effect induces each atom to emit a second photon with the same frequency, direction, phase, and polarization as the incident photon, which is not changed by the process. For each atom there is one photon before a stimulated emission and two photons after—thus the name *light amplification*. Because the two photons have the same phase, they emerge together as *coherent* radiation. The laser makes use of stimulated emission to produce a beam consisting of a large number of such coherent photons.

To discuss stimulated emission from atoms in excited levels, we need to know something about how many atoms are in each of the various energy levels. First, we need to make the distinction between the terms *energy level* and *state*. A system may have more than one way to attain a given energy level; each different way is a different **state.** For instance, there are two ways of putting an ideal unstretched spring in a given energy level. Remembering that the spring potential energy is $U = \frac{1}{2}kx^2$, we could compress the spring by $x = -b$ or we could stretch it by $x = +b$ to get the same $U = \frac{1}{2}kb^2$. The Bohr model had only one state in each energy level, but we will find in Chapter 41 that the hydrogen atom (Fig. 39.24b) actually has two states in its -13.60-eV ground level, eight states in its -3.40-eV first excited level, and so on.

The Maxwell–Boltzmann distribution function (see Section 18.5) determines the number of atoms in a given state in a gas. The function tells us that when the gas is in thermal equilibrium at absolute temperature T, the number n_i of atoms in a state with energy E_i equals $Ae^{-E_i/kT}$, where k is Boltzmann's constant and A is another constant determined by the total number of atoms in the gas. (In Section 18.5, E was the kinetic energy $\frac{1}{2}mv^2$ of a gas molecule; here we're talking about the internal energy of an atom.) Because of the negative exponent, fewer atoms are in higher-energy states, as we should expect. If E_g is a ground-state energy and E_{ex} is the energy of an excited state, then the ratio of numbers of atoms in the two states is

$$\frac{n_{ex}}{n_g} = \frac{Ae^{-E_{ex}/kT}}{Ae^{-E_g/kT}} = e^{-(E_{ex}-E_g)/kT} \qquad (39.18)$$

39.28 Three processes in which atoms interact with light.

(a) Absorption

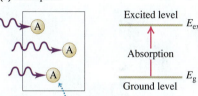

(b) Spontaneous emission

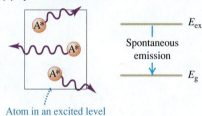

(c) Stimulated emission

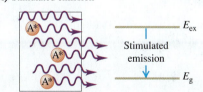

For example, suppose $E_{ex} - E_g = 2.0\,\text{eV} = 3.2 \times 10^{-19}\,\text{J}$, the energy of a 620-nm visible-light photon. At $T = 3000\,\text{K}$ (the temperature of the filament in an incandescent light bulb),

$$\frac{E_{ex} - E_g}{kT} = \frac{3.2 \times 10^{-19}\,\text{J}}{(1.38 \times 10^{-23}\,\text{J/K})(3000\,\text{K})} = 7.73$$

and

$$e^{-(E_{ex} - E_g)/kT} = e^{-7.73} = 0.00044$$

That is, the fraction of atoms in a state 2.0 eV above a ground state is extremely small, even at this high temperature. The point is that at any reasonable temperature there aren't enough atoms in excited states for any appreciable amount of stimulated emission from these states to occur. Rather, a photon emitted by one of the rare excited atoms will almost certainly be absorbed by an atom in the ground state rather than encountering another excited atom.

Enhancing Stimulated Emission: Population Inversions

To make a laser, we need to promote stimulated emission by increasing the number of atoms in excited states. Can we do that simply by illuminating the container with radiation of frequency $f = E/h$ corresponding to the energy difference $E = E_{ex} - E_g$, as in Fig. 39.28a? Some of the atoms absorb photons of energy E and are raised to the excited state, and the population ratio n_{ex}/n_g momentarily increases. But because n_g is originally so much larger than n_{ex}, an enormously intense beam of light would be required to momentarily increase n_{ex} to a value comparable to n_g. The rate at which energy is *absorbed* from the beam by the n_g ground-state atoms far exceeds the rate at which energy is added to the beam by stimulated emission from the relatively rare (n_{ex}) excited atoms.

We need to create a *nonequilibrium* situation in which the number of atoms in a higher-energy state is greater than the number in a lower-energy state. Such a situation is called a **population inversion.** Then the rate of energy radiation by stimulated emission can *exceed* the rate of absorption, and the system will act as a net *source* of radiation with photon energy E. It turns out that we can achieve a population inversion by starting with atoms that have the right kinds of excited states. Figure 39.29a shows an energy-level diagram for such an atom with a ground state and *three* excited states of energies E_1, E_2, and E_3. A laser that uses a material with energy levels like these is called a *four-level laser.* For the laser action to work, the states of energies E_1 and E_3 must have ordinary short lifetimes of about 10^{-8} s, while the state of energy E_2 must have an unusually long lifetime of 10^{-3} s or so. Such a long-lived **metastable state** can occur if, for instance, there are restrictions imposed by conservation of angular momentum that hinder photon emission from this state. (We'll discuss these restrictions in Chapter 41.) The metastable state is the one that we want to populate.

To produce a population inversion, we *pump* the material to excite the atoms out of the ground state into the states of energies E_1, E_2, and E_3 (Fig. 39.29b). If the atoms are in a gas, this pumping can be done by inserting two electrodes into the gas container. When a burst of sufficiently high voltage is applied to the electrodes, an electric discharge occurs. Collisions between ionized atoms and electrons carrying the discharge current then excite the atoms to various energy states. Within about 10^{-8} s the atoms that are excited to states E_1 and E_3 undergo spontaneous photon emission, so these states end up depopulated. But atoms "pile up" in the metastable state with energy E_2. The number of atoms in the metastable state is *less* than the number in the ground state, but is *much greater* than in the nearly unoccupied state of energy E_1. Hence there is a population inversion of state E_2 relative to state E_1 (Fig. 39.29c). You can see why we need

39.29 (a), (b), (c) Stages in the operation of a four-level laser. (d) The light emitted by atoms making spontaneous transitions from state E_2 to state E_1 is reflected between mirrors, so it continues to stimulate emission and gives rise to coherent light. One mirror is partially transmitting and allows the high-intensity light beam to escape.

(a) Before pumping   **(b)** Just after pumping   **(c)** About 10^{-8} s after pumping

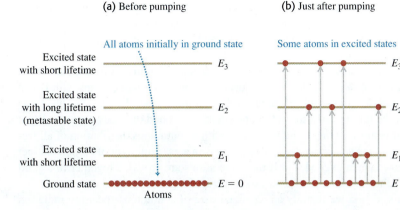

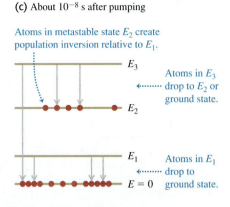

(d) Schematic of gas laser

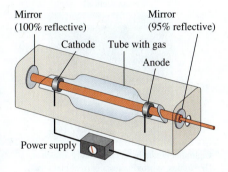

the two levels E_1 and E_3: atoms that undergo spontaneous emission from the E_3 level help to populate the E_2 level, and the presence of the E_1 level makes a population inversion possible.

Over the next 10^{-3} s, some of the atoms in the long-lived metastable state E_2 transition to state E_1 by spontaneous emission. The emitted photons of energy $hf = E_2 - E_1$ are sent back and forth through the gas many times by a pair of parallel mirrors (Fig. 39.29d), so that they can *stimulate* emission from as many of the atoms in state E_2 as possible. The net result of all these processes is a beam of light of frequency f that can be quite intense, has parallel rays, is highly monochromatic, and is spatially *coherent* at all points within a given cross section—that is, a laser beam. One of the mirrors is partially transparent, so a portion of the beam emerges.

What we've described is a *pulsed* laser that produces a burst of coherent light every time the atoms are pumped. Pulsed lasers are used in LASIK eye surgery (an acronym for *laser-assisted in situ keratomileusis*) to reshape the cornea and correct for nearsightedness, farsightedness, or astigmatism. In a *continuous* laser, such as those found in the barcode scanners used at retail checkout counters, energy is supplied to the atoms continuously (for instance, by having the power supply in Fig. 39.29d provide a steady voltage to the electrodes) and a steady beam of light emerges from the laser. For such a laser the pumping must be intense enough to sustain the population inversion, so that the rate at which atoms are added to level E_2 through pumping equals the rate at which atoms in this level emit a photon and transition to level E_1.

Since a special arrangement of energy levels is needed for laser action, it's not surprising that only certain materials can be used to make a laser. Some types of laser use a solid, transparent material such as neodymium glass rather than a gas. The most common kind of laser—used in laser printers (Section 21.2), laser pointers, and to read the data on the disc in a DVD player or Blu-ray player—is a *semiconductor laser,* which doesn't use atomic energy levels at all. As we'll discuss in Chapter 42, these lasers instead use the energy levels of electrons that are free to roam throughout the volume of the semiconductors.

Test Your Understanding of Section 39.4 An ordinary neon light fixture like those used in advertising signs emits red light of wavelength 632.8 nm. Neon atoms are also used in a helium–neon laser (a type of gas laser). The light emitted by a neon light fixture is (i) spontaneous emission; (ii) stimulated emission; (iii) both spontaneous and stimulated emission.

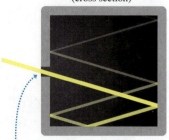

MasteringPHYSICS®

PhET: Blackbody Spectrum
PhET: The Greenhouse Effect

39.30 A hollow box with a small aperture behaves like a blackbody. When the box is heated, the electromagnetic radiation that emerges from the aperture has a blackbody spectrum.

Hollow box with small aperture
(cross section)

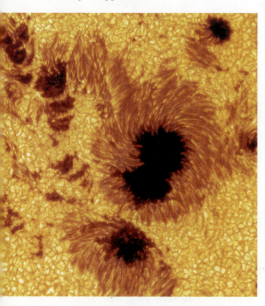

Light that enters box is eventually absorbed. Hence box approximates a perfect blackbody.

39.31 This close-up view of the sun's surface shows two dark sunspots. Their temperature is about 4000 K, while the surrounding solar material is at $T =$ 5800 K. From the Stefan–Boltzmann law, the intensity from a given area of sunspot is only $(4000 \text{ K}/5800 \text{ K})^4 = 0.23$ as great as the intensity from the same area of the surrounding material—which is why sunspots appear dark.

39.5 Continuous Spectra

Emission line spectra come from matter in the gaseous state, in which the atoms are so far apart that interactions between them are negligible and each atom behaves as an isolated system. By contrast, a heated solid or liquid (in which atoms are close to each other) nearly always emits radiation with a *continuous* distribution of wavelengths rather than a line spectrum.

Here's an analogy that suggests why there is a difference. A tuning fork emits sound waves of a single definite frequency (a pure tone) when struck. But if you tightly pack a suitcase full of tuning forks and then shake the suitcase, the proximity of the tuning forks to each other affects the sound that they produce. What you hear is mostly noise, which is sound with a continuous distribution of all frequencies. In the same manner, isolated atoms in a gas emit light of certain distinct frequencies when excited, but if the same atoms are crowded together in a solid or liquid they produce a continuous spectrum of light.

In this section we'll study an idealized case of continuous-spectrum radiation from a hot, dense object. Just as was the case for the emission line spectrum of light from an atom, we'll find that we can understand the continuous spectrum only if we use the ideas of energy levels and photons.

In the same way that an atom's emission spectrum has the same lines as its absorption spectrum, the ideal surface for *emitting* light with a continuous spectrum is one that also *absorbs* all wavelengths of electromagnetic radiation. Such an ideal surface is called a *blackbody* because it would appear perfectly black when illuminated; it would reflect no light at all. The continuous-spectrum radiation that a blackbody emits is called **blackbody radiation.** Like a perfectly frictionless incline or a massless rope, a perfect blackbody does not exist but is nonetheless a useful idealization.

A good approximation to a blackbody is a hollow box with a small aperture in one wall (Fig. 39.30). Light that enters the aperture will eventually be absorbed by the walls of the box, so the box is a nearly perfect absorber. Conversely, when we heat the box, the light that emanates from the aperture is nearly ideal blackbody radiation with a continuous spectrum.

By 1900 blackbody radiation had been studied extensively, and three characteristics had been established. First, the total intensity I (the average rate of radiation of energy per unit surface area or average power per area) emitted from the surface of an ideal radiator is proportional to the fourth power of the absolute temperature (Fig. 39.31). We studied this relationship in Section 17.7 during our study of heat-transfer mechanisms. This total intensity I emitted at absolute temperature T is given by the **Stefan–Boltzmann law:**

$$I = \sigma T^4 \qquad \text{(Stefan–Boltzmann law for a blackbody)} \qquad (39.19)$$

where σ is a fundamental physical constant called the *Stefan–Boltzmann constant.* In SI units,

$$\sigma = 5.670400(40) \times 10^{-8} \frac{\text{W}}{\text{m}^2 \cdot \text{K}^4}$$

Second, the intensity is not uniformly distributed over all wavelengths. Its distribution can be measured and described by the intensity per wavelength interval $I(\lambda)$, called the *spectral emittance.* Thus $I(\lambda) \, d\lambda$ is the intensity corresponding to wavelengths in the interval from λ to $\lambda + d\lambda$. The *total* intensity I, given by Eq. (39.19), is the *integral* of the distribution function $I(\lambda)$ over all wavelengths, which equals the area under the $I(\lambda)$ versus λ curve:

$$I = \int_0^\infty I(\lambda) \, d\lambda \qquad (39.20)$$

CAUTION **Spectral emittance vs. intensity** Although we use the symbol $I(\lambda)$ for spectral emittance, keep in mind that spectral emittance is *not* the same thing as intensity I. Intensity is power per unit area, with units W/m^2. Spectral emittance is power per unit area *per unit wavelength interval*, with units W/m^3.

Figure 39.32 shows the measured spectral emittances $I(\lambda)$ for blackbody radiation at three different temperatures. Each has a peak wavelength λ_m at which the emitted intensity per wavelength interval is largest. Experiment shows that λ_m is inversely proportional to T, so their product is constant. This observation is called the **Wien displacement law.** The experimental value of the constant is 2.90×10^{-3} m·K:

$$\lambda_m T = 2.90 \times 10^{-3} \text{ m·K} \qquad \text{(Wien displacement law)} \qquad (39.21)$$

As the temperature rises, the peak of $I(\lambda)$ becomes higher and shifts to shorter wavelengths. Yellow light has shorter wavelengths than red light, so a body that glows yellow is hotter and brighter than one of the same size that glows red.

Third, experiments show that the *shape* of the distribution function is the same for all temperatures. We can make a curve for one temperature fit any other temperature by simply changing the scales on the graph.

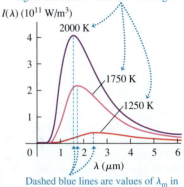

39.32 These graphs show the spectral emittance $I(\lambda)$ for radiation from a blackbody at three different temperatures.

As the temperature increases, the peak of the spectral emittance curve becomes higher and shifts to shorter wavelengths.

Dashed blue lines are values of λ_m in Eq. (39.21) for each temperature.

Rayleigh and the "Ultraviolet Catastrophe"

During the last decade of the 19th century, many attempts were made to derive these empirical results about blackbody radiation from basic principles. In one attempt, the English physicist Lord Rayleigh considered the light enclosed within a rectangular box like that shown in Fig. 39.30. Such a box, he reasoned, has a series of possible *normal modes* for electromagnetic waves, as we discussed in Section 32.5. It also seemed reasonable to assume that the distribution of energy among the various modes would be given by the equipartition principle (see Section 18.4), which had been used successfully in the analysis of heat capacities.

Including both the electric- and magnetic-field energies, Rayleigh assumed that the total energy of each normal mode was equal to kT. Then by computing the *number* of normal modes corresponding to a wavelength interval $d\lambda$, Rayleigh calculated the expected distribution of wavelengths in the radiation within the box. Finally, he computed the predicted intensity distribution $I(\lambda)$ for the radiation emerging from the hole. His result was quite simple:

$$I(\lambda) = \frac{2\pi ckT}{\lambda^4} \qquad \text{(Rayleigh's calculation)} \qquad (39.22)$$

At large wavelengths this formula agrees quite well with the experimental results shown in Fig. 39.32, but there is serious disagreement at small wavelengths. The experimental curves in Fig. 39.32 fall toward zero at small λ. By contrast, Rayleigh's prediction in Eq. (39.22) goes in the opposite direction, approaching infinity at $1/\lambda^4$, a result that was called in Rayleigh's time the "ultraviolet catastrophe." Even worse, the integral of Eq. (39.22) over all λ is infinite, indicating an infinitely large *total* radiated intensity. Clearly, something is wrong.

Planck and the Quantum Hypothesis

Finally, in 1900, the German physicist Max Planck succeeded in deriving a function, now called the **Planck radiation law,** that agreed very well with experimental intensity distribution curves. In his derivation he made what seemed at the time to be a crazy assumption. Planck assumed that electromagnetic oscillators (electrons) in the walls of Rayleigh's box vibrating at a frequency f could have only certain values of energy equal to nhf, where $n = 0, 1, 2, 3, \ldots$ and h turned out to be the constant that now bears Planck's name. These oscillators were in equilibrium with the electromagnetic waves in the box, so they both emitted and absorbed light. His assumption gave quantized energy levels and said that the energy in each normal mode was also a multiple of hf. This was in sharp contrast to Rayleigh's point of view that each normal mode could have any amount of energy.

Planck was not comfortable with this quantum hypothesis; he regarded it as a calculational trick rather than a fundamental principle. In a letter to a friend, he called it "an act of desperation" into which he was forced because "a theoretical explanation had to be found at any cost, whatever the price." But five years later, Einstein identified the energy change hf between levels as the energy of a photon to explain the photoelectric effect (see Section 38.1), and other evidence quickly mounted. By 1915 there was little doubt about the validity of the quantum concept and the existence of photons. By discussing atomic spectra *before* continuous spectra, we have departed from the historical order of things. The credit for inventing the concept of quantization of energy levels goes to Planck, even though he didn't believe it at first. He received the 1918 Nobel Prize in physics for his achievements.

Figure 39.33 shows energy-level diagrams for two of the oscillators that Planck envisioned in the walls of the rectangular box, one with a low frequency and the other with a high frequency. The spacing in energy between adjacent levels is hf. This spacing is small for the low-frequency oscillator that emits and absorbs photons of low frequency f and long wavelength $\lambda = c/f$. The energy spacing is greater for the high-frequency oscillator, which emits high-frequency photons of short wavelength.

According to Rayleigh's picture, both of these oscillators have the same amount of energy kT and are equally effective at emitting radiation. In Planck's model, however, the high-frequency oscillator is very ineffective as a source of light. To see why, we can use the ideas from Section 39.4 about the populations of various energy states. If we consider all the oscillators of a given frequency f in a box at temperature T, the number of oscillators that have energy nhf is $Ae^{-nhf/kT}$. The ratio of the number of oscillators in the first excited state ($n = 1$, energy hf) to the number of oscillators in the ground state ($n = 0$, energy zero) is

$$\frac{n_1}{n_0} = \frac{Ae^{-hf/kT}}{Ae^{-(0)/kT}} = e^{-hf/kT} \tag{39.23}$$

Let's evaluate Eq. (39.23) for $T = 2000$ K, one of the temperatures shown in Fig. 39.32. At this temperature $kT = 2.76 \times 10^{-20}$ J $= 0.172$ eV. For an oscillator that emits photons of wavelength $\lambda = 3.00\ \mu$m, we can show that $hf = hc/\lambda = 0.413$ eV; for a higher-frequency oscillator that emits photons of wavelength $\lambda = 0.500\ \mu$m, $hf = hc/\lambda = 2.48$ eV. For these two cases Eq. (39.23) gives

$$\frac{n_1}{n_0} = e^{-hf/kT} = 0.0909 \text{ for } \lambda = 3.00\ \mu\text{m}$$

$$\frac{n_1}{n_0} = e^{-hf/kT} = 5.64 \times 10^{-7} \text{ for } \lambda = 0.500\ \mu\text{m}$$

The value for $\lambda = 3.00\ \mu$m means that of all the oscillators that can emit light at this wavelength, 0.0909 of them—about one in 11—are in the first excited state. These excited oscillators can each emit a 3.00-μm photon and contribute it to the radiation inside the box. Hence we would expect that this radiation would be rather plentiful in the spectrum of radiation from a 2000 K blackbody. By contrast, the value for $\lambda = 0.500\ \mu$m means that only 5.64×10^{-7} (about one in two million) of the oscillators that can emit this wavelength are in the first excited

39.33 Energy levels for two of the oscillators that Planck envisioned in the walls of a blackbody like that shown in Fig. 39.30. The spacing between adjacent energy levels for each oscillator is hf, which is smaller for the low-frequency oscillator.

Low-frequency oscillator	High-frequency oscillator
$12hf$	$2hf$
$11hf$	
$10hf$	
$9hf$	
$8hf$	
$7hf$	
$6hf$	hf
$5hf$	
$4hf$	
$3hf$	
$2hf$	
$1hf$	
hf	
0	0

state. An oscillator can't emit if it's in the ground state, so the amount of radiation in the box at this wavelength is *tremendously* suppressed compared to Rayleigh's prediction. That's why the spectral emittance curve for 2000 K in Fig. 39.32 has such a low value at $\lambda = 0.500$ μm and shorter wavelengths. So Planck's quantum hypothesis provided a natural way to suppress the spectral emittance of a blackbody at short wavelengths, and hence averted the ultraviolet catastrophe that plagued Rayleigh's calculations.

We won't go into all the details of Planck's derivation of the spectral emittance. Here is his result:

$$I(\lambda) = \frac{2\pi hc^2}{\lambda^5(e^{hc/\lambda kT} - 1)} \quad \text{(Planck radiation law)} \quad (39.24)$$

where h is Planck's constant, c is the speed of light, k is Boltzmann's constant, T is the absolute temperature, and λ is the wavelength. This function turns out to agree well with experimental emittance curves such as those in Fig. 39.32.

The Planck radiation law also contains the Wien displacement law and the Stefan–Boltzmann law as consequences. To derive the Wien law, we find the value of λ at which $I(\lambda)$ is maximum by taking the derivative of Eq. (39.24) and setting it equal to zero. We leave it to you to fill in the details; the result is

$$\lambda_m = \frac{hc}{4.965kT} \quad (39.25)$$

To obtain this result, you have to solve the equation

$$5 - x = 5e^{-x} \quad (39.26)$$

The root of this equation, found by trial and error or more sophisticated means, is 4.965 to four significant figures. You should evaluate the constant $hc/4.965k$ and show that it agrees with the experimental value of 2.90×10^{-3} m·K given in Eq. (39.21).

We can obtain the Stefan–Boltzmann law for a blackbody by integrating Eq. (39.24) over all λ to find the *total* radiated intensity (see Problem 39.67). This is not a simple integral; the result is

$$I = \int_0^\infty I(\lambda)\,d\lambda = \frac{2\pi^5 k^4}{15c^2h^3}T^4 = \sigma T^4 \quad (39.27)$$

in agreement with Eq. (39.19). Our result in Eq. (39.27) also shows that the constant σ in that law can be expressed as a combination of other fundamental constants:

$$\sigma = \frac{2\pi^5 k^4}{15c^2h^3} \quad (39.28)$$

You should substitute the values of k, c, and h from Appendix F and verify that you obtain the value $\sigma = 5.6704 \times 10^{-8}$ W/m²·K⁴ for the Stefan–Boltzmann constant.

The Planck radiation law, Eq. (39.24), looks so different from the unsuccessful Rayleigh expression, Eq. (39.22), that it may seem unlikely that they would agree at large values of λ. But when λ is large, the exponent in the denominator of Eq. (39.24) is very small. We can then use the approximation $e^x \approx 1 + x$ (for $x \ll 1$). You should verify that when this is done, the result approaches Eq. (39.22), showing that the two expressions do agree in the limit of very large λ. We also note that the Rayleigh expression does not contain h. At very long wavelengths (very small photon energies), quantum effects become unimportant.

Example 39.7 Light from the sun

To a good approximation, the sun's surface is a blackbody with a surface temperature of 5800 K. (We are ignoring the absorption produced by the sun's atmosphere, shown in Fig. 39.9.) (a) At what wavelength does the sun emit most strongly? (b) What is the total radiated power per unit surface area?

SOLUTION

IDENTIFY and SET UP: Our target variables are the peak-intensity wavelength λ_m and the radiated power per area I. Hence we'll use the Wien displacement law, Eq. (39.21) (which relates λ_m to the blackbody temperature T), and the Stefan–Boltzmann law, Eq. (39.19) (which relates I to T).

EXECUTE: (a) From Eq. (39.21),

$$\lambda_m = \frac{2.90 \times 10^{-3} \text{ m} \cdot \text{K}}{T} = \frac{2.90 \times 10^{-3} \text{ m} \cdot \text{K}}{5800 \text{ K}}$$

$$= 0.500 \times 10^{-6} \text{ m} = 500 \text{ nm}$$

(b) From Eq. (39.19),

$$I = \sigma T^4 = (5.67 \times 10^{-8} \text{ W/m}^2 \cdot \text{K}^4)(5800 \text{ K})^4$$

$$= 6.42 \times 10^7 \text{ W/m}^2 = 64.2 \text{ MW/m}^2$$

EVALUATE: The 500-nm wavelength found in part (a) is near the middle of the visible spectrum. This should not be a surprise: The human eye evolved to take maximum advantage of natural light.

The enormous value $I = 64.2 \text{ MW/m}^2$ found in part (b) is the intensity at the *surface* of the sun, a sphere of radius 6.96×10^8 m. When this radiated energy reaches the earth, 1.50×10^{11} m away, the intensity has decreased by the factor $[(6.96 \times 10^8 \text{ m})/(1.50 \times 10^{11} \text{ m})]^2 = 2.15 \times 10^{-5}$ to the still-impressive 1.4 kW/m^2.

Example 39.8 A slice of sunlight

Find the power per unit area radiated from the sun's surface in the wavelength range 600.0 to 605.0 nm.

SOLUTION

IDENTIFY and SET UP: This question concerns the power emitted by a blackbody over a narrow range of wavelengths, and so involves the spectral emittance $I(\lambda)$ given by the Planck radiation law, Eq. (39.24). This requires that we find the area under the $I(\lambda)$ curve between 600.0 and 605.0 nm. We'll *approximate* this area as the product of the height of the curve at the median wavelength $\lambda = 602.5$ nm and the width of the interval, $\Delta\lambda = 5.0$ nm. From Example 39.7, $T = 5800$ K.

EXECUTE: To obtain the height of the $I(\lambda)$ curve at $\lambda = 602.5$ nm $= 6.025 \times 10^{-7}$ m, we first evaluate the quantity $hc/\lambda kT$ in Eq. (39.24) and then substitute the result into Eq. (39.24):

$$\frac{hc}{\lambda kT} = \frac{(6.626 \times 10^{-34} \text{ J} \cdot \text{s})(2.998 \times 10^8 \text{ m/s})}{(6.025 \times 10^{-7} \text{m})(1.381 \times 10^{-23} \text{ J/K})(5800 \text{ K})} = 4.116$$

$$I(\lambda) = \frac{2\pi(6.626 \times 10^{-34} \text{ J} \cdot \text{s})(2.998 \times 10^8 \text{ m/s})^2}{(6.025 \times 10^{-7} \text{ m})^5(e^{4.116} - 1)}$$

$$= 7.81 \times 10^{13} \text{ W/m}^3$$

The intensity in the 5.0-nm range from 600.0 to 605.0 nm is then approximately

$$I(\lambda)\Delta\lambda = (7.81 \times 10^{13} \text{ W/m}^3)(5.0 \times 10^{-9} \text{ m})$$

$$= 3.9 \times 10^5 \text{ W/m}^2 = 0.39 \text{ MW/m}^2$$

EVALUATE: In part (b) of Example 39.7, we found the power radiated per unit area by the sun at *all* wavelengths to be $I = 64.2 \text{ MW/m}^2$; here we have found that the power radiated per unit area in the wavelength range from 600 to 605 nm is $I(\lambda)\Delta\lambda = 0.39 \text{ MW/m}^2$, about 0.6% of the total.

Test Your Understanding of Section 39.5 (a) Does a blackbody at 2000 K emit x rays? (b) Does it emit radio waves? ❙

39.6 The Uncertainty Principle Revisited

The discovery of the dual wave–particle nature of matter forces us to reevaluate the kinematic language we use to describe the position and motion of a particle. In classical Newtonian mechanics we think of a particle as a point. We can describe its location and state of motion at any instant with three spatial coordinates and three components of velocity. But because matter also has a wave aspect, when we look at the behavior on a small enough scale—comparable to the de Broglie wavelength of the particle—we can no longer use the Newtonian description. Certainly no Newtonian particle would undergo diffraction like electrons do (Section 39.1).

39.34 (a) A two-slit interference experiment for electrons. (b) The interference pattern after 28, 1000, and 10,000 electrons.

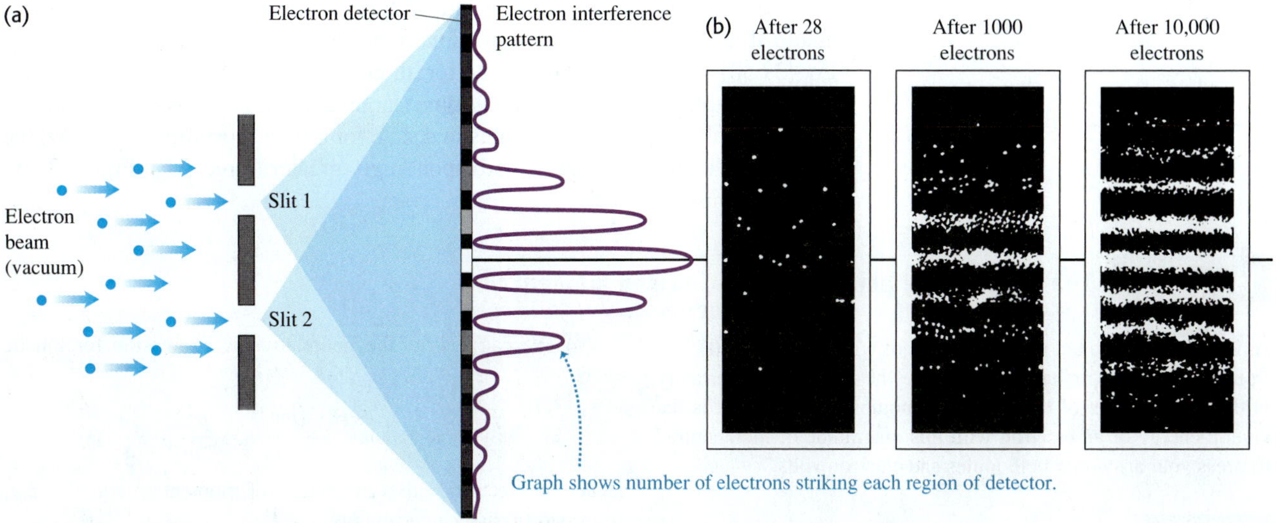

(a) Electron detector — Electron interference pattern

Electron beam (vacuum)

Slit 1

Slit 2

Graph shows number of electrons striking each region of detector.

(b) After 28 electrons

After 1000 electrons

After 10,000 electrons

To demonstrate just how non-Newtonian the behavior of matter can be, let's look at an experiment involving the two-slit interference of electrons (Fig. 39.34). We aim an electron beam at two parallel slits, just as we did for light in Section 38.4. (The electron experiment has to be done in vacuum so that the electrons don't collide with air molecules.) What kind of pattern appears on the detector on the other side of the slits? The answer is: *exactly the same* kind of interference pattern we saw for photons in Section 38.4! Moreover, the principle of complementarity, which we introduced in Section 38.4, tells us that we cannot apply the wave and particle models simultaneously to describe any single element of this experiment. Thus we *cannot* predict exactly where in the pattern (a wave phenomenon) any individual electron (a particle) will land. We can't even ask which slit an individual electron passes through. If we tried to look at where the electrons were going by shining a light on them—that is, by scattering photons off them—the electrons would recoil, which would modify their motions so that the two-slit interference pattern would not appear.

CAUTION   **Electron two-slit interference is not interference between two electrons** It's a common misconception that the pattern in Fig. 39.34b is due to the interference between *two* electron waves, each representing an electron passing through one slit. To show that this cannot be the case, we can send just one electron at a time through the apparatus. It makes no difference; we end up with the same interference pattern. In a sense, each electron wave interferes with itself. ❚

The Heisenberg Uncertainty Principles for Matter

Just as electrons and photons show the same behavior in a two-slit interference experiment, electrons and other forms of matter obey the same Heisenberg uncertainty principles as photons do:

$$\Delta x \Delta p_x \geq \hbar/2$$
$$\Delta y \Delta p_y \geq \hbar/2$$
$$\Delta z \Delta p_z \geq \hbar/2$$

(Heisenberg uncertainty principle for position and momentum)   (39.29)

$$\Delta t \Delta E \geq \hbar/2$$

(Heisenberg uncertainty principle for energy and time interval)   (39.30)

In these equations $\hbar = h/2\pi = 1.055 \times 10^{-34}$ J·s. The uncertainty principle for energy and time interval has a direct application to energy levels. We have assumed that each energy level in an atom has a very definite energy. However, Eq. (39.30) says that this is not true for all energy levels. A system that remains in a metastable state for a very long time (large Δt) can have a very well-defined energy (small ΔE), but if it remains in a state for only a short time (small Δt) the uncertainty in energy must be correspondingly greater (large ΔE). Figure 39.35 illustrates this idea.

Example 39.9   The uncertainty principle: position and momentum

An electron is confined within a region of width 5.000×10^{-11} m (roughly the Bohr radius). (a) Estimate the minimum uncertainty in the x-component of the electron's momentum. (b) What is the kinetic energy of an electron with this magnitude of momentum? Express your answer in both joules and electron volts.

SOLUTION

IDENTIFY and SET UP: This problem uses the Heisenberg uncertainty principle for position and momentum and the relationship between a particle's momentum and its kinetic energy. The electron could be anywhere within the region, so we take $\Delta x = 5.000 \times 10^{-11}$ m as its position uncertainty. We then find the momentum uncertainty Δp_x using Eq. (39.29) and the kinetic energy using the relationships $p = mv$ and $K = \frac{1}{2}mv^2$.

EXECUTE: (a) From Eqs. (39.29), for a given value of Δx, the uncertainty in momentum is minimum when the product $\Delta x \Delta p_x$ equals $\hbar$. Hence

$$\Delta p_x = \frac{\hbar}{2\Delta x} = \frac{1.055 \times 10^{-34} \text{ J·s}}{2(5.000 \times 10^{-11} \text{ m})} = 1.055 \times 10^{-24} \text{ J·s/m}$$
$$= 1.055 \times 10^{-24} \text{ kg·m/s}$$

(b) We can rewrite the nonrelativistic expression for kinetic energy as

$$K = \frac{1}{2}mv^2 = \frac{(mv)^2}{2m} = \frac{p^2}{2m}$$

Hence an electron with a magnitude of momentum equal to Δp_x from part (a) has kinetic energy

$$K = \frac{p^2}{2m} = \frac{(1.055 \times 10^{-24} \text{ kg·m/s})^2}{2(9.11 \times 10^{-31} \text{ kg})}$$
$$= 6.11 \times 10^{-19} \text{ J} = 3.81 \text{ eV}$$

EVALUATE: This energy is typical of electron energies in atoms. This agreement suggests that the uncertainty principle is deeply involved in atomic structure.

A similar calculation explains why electrons in atoms do not fall into the nucleus. If an electron were confined to the interior of a nucleus, its position uncertainty would be $\Delta x \approx 10^{-14}$ m. This would give the electron a momentum uncertainty about 5000 times greater than that of the electron in this example, and a kinetic energy so great that the electron would immediately be ejected from the nucleus.

Example 39.10   The uncertainty principle: energy and time

A sodium atom in one of the states labeled "Lowest excited levels" in Fig. 39.19a remains in that state, on average, for 1.6×10^{-8} s before it makes a transition to the ground level, emitting a photon with wavelength 589.0 nm and energy 2.105 eV. What is the uncertainty in energy of that excited state? What is the wavelength spread of the corresponding spectral line?

SOLUTION

IDENTIFY and SET UP: This problem uses the Heisenberg uncertainty principle for energy and time interval and the relationship between photon energy and wavelength. The average time that the atom spends in this excited state is equal to Δt in Eq. (39.30). We find the minimum uncertainty in the energy of the excited level by replacing the $\geq$ sign in Eq. (39.30) with an equals sign and solving for ΔE.

EXECUTE: From Eq. (39.30),

$$\Delta E = \frac{\hbar}{2\Delta t} = \frac{1.055 \times 10^{-34} \text{ J·s}}{2(1.6 \times 10^{-8} \text{ s})}$$
$$= 3.3 \times 10^{-27} \text{ J} = 2.1 \times 10^{-8} \text{ eV}$$

The atom remains in the ground level indefinitely, so that level has *no* associated energy uncertainty. The fractional uncertainty of the *photon* energy is therefore

$$\frac{\Delta E}{E} = \frac{2.1 \times 10^{-8} \text{ eV}}{2.105 \text{ eV}} = 1.0 \times 10^{-8}$$

You can use some simple calculus and the relation $E = hc/\lambda$ to show that $\Delta\lambda/\lambda \approx \Delta E/E$, so that the corresponding spread in wavelength, or "width," of the spectral line is approximately

$$\Delta\lambda = \lambda\frac{\Delta E}{E} = (589.0 \text{ nm})(1.0 \times 10^{-8}) = 0.0000059 \text{ nm}$$

EVALUATE: This irreducible uncertainty $\Delta\lambda$ is called the *natural line width* of this particular spectral line. Though very small, it is within the limits of resolution of present-day spectrometers. Ordinarily, the natural line width is much smaller than the line width arising from other causes such as the Doppler effect and collisions among the rapidly moving atoms.

The Uncertainty Principle and the Limits of the Bohr Model

We saw in Section 39.3 that the Bohr model of the hydrogen atom was tremendously successful. However, the Heisenberg uncertainty principle for position and momentum shows that this model *cannot* be a correct description of how an electron in an atom behaves. Figure 39.22 shows that in the Bohr model as interpreted by de Broglie, an electron wave moves in a plane around the nucleus. Let's call this the xy-plane, so the z-axis is perpendicular to the plane. Hence the Bohr model says that an electron is always found at $z = 0$, and its z-momentum p_z is always zero (the electron does not move out of the xy-plane). But this implies that there are *no* uncertainties in either z or p_z, so $\Delta z = 0$ and $\Delta p_z = 0$. This directly contradicts Eq. (39.29), which says that the product $\Delta z \Delta p_z$ must be greater than or equal to $\hbar$.

This conclusion isn't too surprising, since the electron in the Bohr model is a mix of particle and wave ideas (the electron moves in an orbit like a miniature planet, but has a wavelength). To get an accurate picture of how electrons behave inside an atom and elsewhere, we need a description that is based *entirely* on the electron's wave properties. Our goal in Chapter 40 will be to develop this description, which we call *quantum mechanics*. To do this we'll introduce the *Schrödinger equation*, the fundamental equation that describes the dynamics of matter waves. This equation, as we will see, is as fundamental to quantum mechanics as Newton's laws are to classical mechanics or as Maxwell's equations are to electromagnetism.

Test Your Understanding of Section 39.6   Rank the following situations according to the uncertainty in x-momentum, from largest to smallest. The mass of the proton is 1836 times the mass of the electron. (i) an electron whose x-coordinate is known to within 2×10^{-15} m; (ii) an electron whose x-coordinate is known to within 4×10^{-15} m; (iii) a proton whose x-coordinate is known to within 2×10^{-15} m; (iv) a proton whose x-coordinate is known to within 4×10^{-15} m. ❙

39.35 The longer the lifetime Δt of a state, the smaller is its spread in energy (shown by the width of the energy levels).

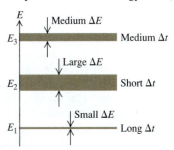

De Broglie waves and electron diffraction: Electrons and other particles have wave properties. A particle's wavelength depends on its momentum in the same way as for photons. A nonrelativistic electron accelerated from rest through a potential difference V_{ba} has a wavelength given by Eq. (39.3). Electron microscopes use the very small wavelengths of fast-moving electrons to make images with resolution thousands of times finer than is possible with visible light. (See Examples 39.1–39.3.)

$$\lambda = \frac{h}{p} = \frac{h}{mv} \qquad (39.1)$$

$$E = hf \qquad (39.2)$$

$$\lambda = \frac{h}{p} = \frac{h}{\sqrt{2meV_{ba}}} \qquad (39.3)$$

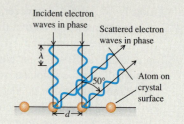

The nuclear atom: The Rutherford scattering experiments show that most of an atom's mass and all of its positive charge are concentrated in a tiny, dense nucleus at the center of the atom. (See Example 39.4.)

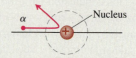

Atomic line spectra and energy levels: The energies of atoms are quantized: They can have only certain definite values, called energy levels. When an atom makes a transition from an energy level E_i to a lower level E_f, it emits a photon of energy $E_i - E_f$. The same photon can be absorbed by an atom in the lower energy level, which excites the atom to the upper level. (See Example 39.5.)

$$hf = \frac{hc}{\lambda} = E_i - E_f \qquad (39.5)$$

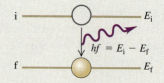

The Bohr model: In the Bohr model of the hydrogen atom, the permitted values of angular momentum are integral multiples of $h/2\pi$. The integer multiplier n is called the principal quantum number for the level. The orbital radii are proportional to n^2 and the orbital speeds are proportional to $1/n$. The energy levels of the hydrogen atom are given by Eq. (39.15), where R is the Rydberg constant. (See Example 39.6.)

$$L_n = mv_n r_n = n\frac{h}{2\pi}$$
$$(n = 1, 2, 3, \dots) \qquad (39.6)$$

$$r_n = \epsilon_0 \frac{n^2 h^2}{\pi m e^2} = n^2 a_0 \qquad (39.8)$$

$$= n^2 (5.29 \times 10^{-11}\ \text{m}) \qquad (39.10)$$

$$v_n = \frac{1}{\epsilon_0}\frac{e^2}{2nh} = \frac{2.19 \times 10^6\ \text{m/s}}{n} \qquad (39.9)$$

$$E_n = -\frac{hcR}{n^2} = -\frac{13.60\ \text{eV}}{n^2} \qquad (39.15)$$

$$(n = 1, 2, 3, \dots)$$

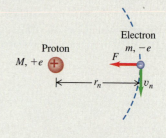

The laser: The laser operates on the principle of stimulated emission, by which many photons with identical wavelength and phase are emitted. Laser operation requires a nonequilibrium condition called a population inversion, in which more atoms are in a higher-energy state than are in a lower-energy state.

Blackbody radiation: The total radiated intensity (average power radiated per area) from a blackbody surface is proportional to the fourth power of the absolute temperature T. The quantity $\sigma = 5.67 \times 10^{-8}\ \text{W/m}^2 \cdot \text{K}^4$ is called the Stefan–Boltzmann constant. The wavelength λ_m at which a blackbody radiates most strongly is inversely proportional to T. The Planck radiation law gives the spectral emittance $I(\lambda)$ (intensity per wavelength interval in blackbody radiation). (See Examples 39.7 and 39.8.)

$$I = \sigma T^4$$
(Stefan–Boltzmann law) $\qquad (39.19)$

$$\lambda_m T = 2.90 \times 10^{-3}\ \text{m} \cdot \text{K}$$
(Wien displacement law) $\qquad (39.21)$

$$I(\lambda) = \frac{2\pi hc^2}{\lambda^5 (e^{hc/\lambda kT} - 1)}$$
(Planck radiation law) $\qquad (39.24)$

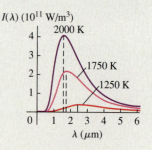

The Heisenberg uncertainty principle for particles: The same uncertainty considerations that apply to photons also apply to particles such as electrons. The uncertainty ΔE in the energy of a state that is occupied for a time Δt is given by Eq. (39.30). (See Examples 39.9 and 39.10.)

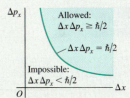

$$\Delta x \Delta p_x \geq \hbar/2$$
$$\Delta y \Delta p_y \geq \hbar/2 \qquad \text{(Heisenberg uncertainty principle for position and momentum)} \qquad (39.29)$$
$$\Delta z \Delta p_z \geq \hbar/2$$

$$\Delta t \, \Delta E \geq \hbar/2 \qquad \text{(Heisenberg uncertainty principle for energy and time interval)} \qquad (39.30)$$

BRIDGING PROBLEM | Hot Stars and Hydrogen Clouds

Figure 39.36 shows a cloud, or *nebula,* of glowing hydrogen in interstellar space. The atoms in this cloud are excited by short-wavelength radiation emitted by the bright blue stars at the center of the nebula. (a) The blue stars act as blackbodies and emit light with a continuous spectrum. What is the wavelength at which a star with a surface temperature of 15,100 K (about $2\frac{1}{2}$ times the surface temperature of the sun) has the maximum spectral emittance? In what region of the electromagnetic spectrum is this? (b) Figure 39.32 shows that most of the energy radiated by a blackbody is at wavelengths between about one half and three times the wavelength of maximum emittance. If a hydrogen atom near the star in part (a) is initially in the ground level, what is the principal quantum number of the highest energy level to which it could be excited by a photon in this wavelength range? (c) The red color of the nebula is primarily due to hydrogen atoms making a transition from $n = 3$ to $n = 2$ and emitting photons of wavelength 656.3 nm. In the Bohr model as interpreted by de Broglie, what are the *electron* wavelengths in the $n = 2$ and $n = 3$ levels?

SOLUTION GUIDE

See MasteringPhysics® study area for a Video Tutor solution.

39.36 The Rosette Nebula.

IDENTIFY and SET UP

1. To solve this problem you need to use your knowledge of both blackbody radiation (Section 39.5) and the Bohr model of the hydrogen atom (Section 39.3).
2. In part (a) the target variable is the wavelength at which the star emits most strongly; in part (b) the target variable is a principal quantum number, and in part (c) it is the de Broglie wavelength of an electron in the $n = 2$ and $n = 3$ Bohr orbits (see Fig. 39.24). Select the equations you will need to find the target variables. (*Hint:* In Section 39.5 you learned how to find the energy change involved in a transition between two given levels of a hydrogen atom. Part (b) is a variation on this: You are to find the final level in a transition that starts in the $n = 1$ level and involves the absorption of a photon of a given wavelength and hence a given energy.)

EXECUTE

3. Use the Wien displacement law to find the wavelength at which the star has maximum spectral emittance. In what part of the electromagnetic spectrum is this wavelength?
4. Use your result from step 3 to find the range of wavelengths in which the star radiates most of its energy. Which end of this range corresponds to a photon with the greatest energy?
5. Write an expression for the wavelength of a photon that must be absorbed to cause an electron transition from the ground level ($n = 1$) to a higher level n. Solve for the value of n that corresponds to the highest-energy photon in the range you calculated in step 4. (*Hint:* Remember than n must be an integer.)
6. Find the electron wavelengths that correspond to the $n = 2$ and $n = 3$ orbits shown in Fig. 39.22.

EVALUATE

7. Check your result in step 5 by calculating the wavelength needed to excite a hydrogen atom from the ground level into the level *above* the highest-energy level that you found in step 5. Is it possible for light in the range of wavelengths you found in step 4 to excite hydrogen atoms from the ground level into this level?
8. How do the electron wavelengths you found in step 6 compare to the wavelength of a *photon* emitted in a transition from the $n = 3$ level to the $n = 2$ level?

Problems

•, ••, •••: Problems of increasing difficulty. **CP**: Cumulative problems incorporating material from earlier chapters. **CALC**: Problems requiring calculus. **BIO**: Biosciences problems.

DISCUSSION QUESTIONS

Q39.1 If a proton and an electron have the same speed, which has the longer de Broglie wavelength? Explain.

Q39.2 If a proton and an electron have the same kinetic energy, which has the longer de Broglie wavelength? Explain.

Q39.3 Does a photon have a de Broglie wavelength? If so, how is it related to the wavelength of the associated electromagnetic wave? Explain.

Q39.4 When an electron beam goes through a very small hole, it produces a diffraction pattern on a screen, just like that of light. Does this mean that an electron spreads out as it goes through the hole? What does this pattern mean?

Q39.5 Galaxies tend to be strong emitters of Lyman-α photons (from the $n = 2$ to $n = 1$ transition in atomic hydrogen). But the intergalactic medium—the very thin gas between the galaxies—tends to *absorb* Lyman-α photons. What can you infer from these observations about the temperature in these two environments? Explain.

Q39.6 A doubly ionized lithium atom (Li^{++}) is one that has had two of its three electrons removed. The energy levels of the remaining single-electron ion are closely related to those of the hydrogen atom. The nuclear charge for lithium is $\pm 3e$ instead of just $+e$. How are the energy levels related to those of hydrogen? How is the *radius* of the ion in the ground level related to that of the hydrogen atom? Explain.

Q39.7 The emission of a photon by an isolated atom is a recoil process in which momentum is conserved. Thus Eq. (39.5) should include a recoil kinetic energy K_r for the atom. Why is this energy negligible in that equation?

Q39.8 How might the energy levels of an atom be measured directly—that is, without recourse to analysis of spectra?

Q39.9 Elements in the gaseous state emit line spectra with well-defined wavelengths. But hot solid bodies always emit a continuous spectrum—that is, a continuous smear of wavelengths. Can you account for this difference?

Q39.10 As a body is heated to a very high temperature and becomes self-luminous, the apparent color of the emitted radiation shifts from red to yellow and finally to blue as the temperature increases. Why does the color shift? What other changes in the character of the radiation occur?

Q39.11 The peak-intensity wavelength of red dwarf stars, which have surface temperatures around 3000 K, is about 1000 nm, which is beyond the visible spectrum. So why are we able to see these stars, and why do they appear red?

Q39.12 You have been asked to design a magnet system to steer a beam of 54-eV electrons like those described in Example 39.1 (Section 39.1). The goal is to be able to direct the electron beam to a specific target location with an accuracy of ± 1.0 mm. In your design, do you need to take the wave nature of electrons into account? Explain.

Q39.13 Why go through the expense of building an electron microscope for studying very small objects such as organic molecules? Why not just use extremely short electromagnetic waves, which are much cheaper to generate?

Q39.14 Which has more total energy: a hydrogen atom with an electron in a high shell (large n) or in a low shell (small n)? Which is moving faster: the high-shell electron or the low-shell electron? Is there a contradiction here? Explain.

Q39.15 Does the uncertainty principle have anything to do with marksmanship? That is, is the accuracy with which a bullet can be aimed at a target limited by the uncertainty principle? Explain.

Q39.16 Suppose a two-slit interference experiment is carried out using an electron beam. Would the same interference pattern result if one slit at a time is uncovered instead of both at once? If not, why not? Doesn't each electron go through one slit or the other? Or does every electron go through both slits? Discuss the latter possibility in light of the principle of complementarity.

Q39.17 Equation (39.30) states that the energy of a system can have uncertainty. Does this mean that the principle of conservation of energy is no longer valid? Explain.

Q39.18 Laser light results from transitions from long-lived metastable states. Why is it more monochromatic than ordinary light?

Q39.19 Could an electron-diffraction experiment be carried out using three or four slits? Using a grating with many slits? What sort of results would you expect with a grating? Would the uncertainty principle be violated? Explain.

Q39.20 As the lower half of Fig. 39.4 shows, the diffraction pattern made by electrons that pass through aluminum foil is a series of concentric rings. But if the aluminum foil is replaced by a single crystal of aluminum, only certain points on these rings appear in the pattern. Explain.

Q39.21 Why can an electron microscope have greater magnification than an ordinary microscope?

Q39.22 When you check the air pressure in a tire, a little air always escapes; the process of making the measurement changes the quantity being measured. Think of other examples of measurements that change or disturb the quantity being measured.

EXERCISES

Section 39.1 Electron Waves

39.1 • (a) An electron moves with a speed of 4.70×10^6 m/s. What is its de Broglie wavelength? (b) A proton moves with the same speed. Determine its de Broglie wavelength.

39.2 •• For crystal diffraction experiments (discussed in Section 39.1), wavelengths on the order of 0.20 nm are often appropriate. Find the energy in electron volts for a particle with this wavelength if the particle is (a) a photon; (b) an electron; (c) an alpha particle ($m = 6.64 \times 10^{-27}$ kg).

39.3 • An electron has a de Broglie wavelength of 2.80×10^{-10} m. Determine (a) the magnitude of its momentum and (b) its kinetic energy (in joules and in electron volts).

39.4 •• **Wavelength of an Alpha Particle.** An alpha particle ($m = 6.64 \times 10^{-27}$ kg) emitted in the radioactive decay of uranium-238 has an energy of 4.20 MeV. What is its de Broglie wavelength?

39.5 • In the Bohr model of the hydrogen atom, what is the de Broglie wavelength for the electron when it is in (a) the $n = 1$

level and (b) the $n = 4$ level? In each case, compare the de Broglie wavelength to the circumference $2\pi r_n$ of the orbit.

39.6 • (a) A nonrelativistic free particle with mass m has kinetic energy K. Derive an expression for the de Broglie wavelength of the particle in terms of m and K. (b) What is the de Broglie wavelength of an 800-eV electron?

39.7 • **Why Don't *We* Diffract?** (a) Calculate the de Broglie wavelength of a typical person walking through a doorway. Make reasonable approximations for the necessary quantities. (b) Will the person in part (a) exhibit wavelike behavior when walking through the "single slit" of a doorway? Why?

39.8 •• What is the de Broglie wavelength for an electron with speed (a) $v = 0.480c$ and (b) $v = 0.960c$? (*Hint:* Use the correct relativistic expression for linear momentum if necessary.)

39.9 • (a) If a photon and an electron each have the same energy of 20.0 eV, find the wavelength of each. (b) If a photon and an electron each have the same wavelength of 250 nm, find the energy of each. (c) You want to study an organic molecule that is about 250 nm long using either a photon or an electron microscope. Approximately what wavelength should you use, and which probe, the electron or the photon, is likely to damage the molecule the least?

39.10 • How fast would an electron have to move so that its de Broglie wavelength is 1.00 mm?

39.11 • **Wavelength of a Bullet.** Calculate the de Broglie wavelength of a 5.00-g bullet that is moving at 340 m/s. Will the bullet exhibit wavelike properties?

39.12 •• Find the wavelengths of a photon and an electron that have the same energy of 25 eV. (*Note:* The energy of the electron is its kinetic energy.)

39.13 •• (a) What accelerating potential is needed to produce electrons of wavelength 5.00 nm? (b) What would be the energy of photons having the same wavelength as these electrons? (c) What would be the wavelength of photons having the same energy as the electrons in part (a)?

39.14 •• Through what potential difference must electrons be accelerated so they will have (a) the same wavelength as an x ray of wavelength 0.150 nm and (b) the same energy as the x ray in part (a)?

39.15 • (a) Approximately how fast should an electron move so it has a wavelength that makes it useful to measure the distance between adjacent atoms in typical crystals (about 0.10 nm)? (b) What is the kinetic energy of the electron in part (a)? (c) What would be the energy of a photon of the same wavelength as the electron in part (b)? (d) Which would make a more effective probe of small-scale structures: electrons or photons? Why?

39.16 •• **CP** A beam of electrons is accelerated from rest through a potential difference of 0.100 kV and then passes through a thin slit. The diffracted beam shows its first diffraction minima at $\pm 11.5°$ from the original direction of the beam when viewed far from the slit. (a) Do we need to use relativity formulas? How do you know? (b) How wide is the slit?

39.17 •• A beam of neutrons that all have the same energy scatters from atoms that have a spacing of 0.0910 nm in the surface plane of a crystal. The $m = 1$ intensity maximum occurs when the angle θ in Fig. 39.2 is 28.6°. What is the kinetic energy (in electron volts) of each neutron in the beam?

39.18 • A beam of 188-eV electrons is directed at normal incidence onto a crystal surface as shown in Fig. 39.3b. The $m = 2$ intensity maximum occurs at an angle $\theta = 60.6°$. (a) What is the spacing between adjacent atoms on the surface? (b) At what other angle or angles is there an intensity maximum? (c) For what electron energy (in electron volts) would the $m = 1$ intensity maximum occur at $\theta = 60.6°$? For this energy, is there an $m = 2$ intensity maximum? Explain.

39.19 • A CD-ROM is used instead of a crystal in an electron-diffraction experiment. The surface of the CD-ROM has tracks of tiny pits with a uniform spacing of 1.60 μm. (a) If the speed of the electrons is 1.26×10^4 m/s, at which values of θ will the $m = 1$ and $m = 2$ intensity maxima appear? (b) The scattered electrons in these maxima strike at normal incidence a piece of photographic film that is 50.0 cm from the CD-ROM. What is the spacing on the film between these maxima?

39.20 • (a) In an electron microscope, what accelerating voltage is needed to produce electrons with wavelength 0.0600 nm? (b) If protons are used instead of electrons, what accelerating voltage is needed to produce protons with wavelength 0.0600 nm? (*Hint:* In each case the initial kinetic energy is negligible.)

39.21 •• You want to study a biological specimen by means of a wavelength of 10.0 nm, and you have a choice of using electromagnetic waves or an electron microscope. (a) Calculate the ratio of the energy of a 10.0-nm-wavelength photon to the kinetic energy of a 10.0-nm-wavelength electron. (b) In view of your answer to part (a), which would be less damaging to the specimen you are studying: photons or electrons?

Section 39.2 The Nuclear Atom and Atomic Spectra

39.22 •• **CP** A 4.78-MeV alpha particle from a ^{226}Ra decay makes a head-on collision with a uranium nucleus. A uranium nucleus has 92 protons. (a) What is the distance of closest approach of the alpha particle to the center of the nucleus? Assume that the uranium nucleus remains at rest and that the distance of closest approach is much greater than the radius of the uranium nucleus. (b) What is the force on the alpha particle at the instant when it is at the distance of closest approach?

39.23 • A beam of alpha particles is incident on a target of lead. A particular alpha particle comes in "head-on" to a particular lead nucleus and stops 6.50×10^{-14} m away from the center of the nucleus. (This point is well outside the nucleus.) Assume that the lead nucleus, which has 82 protons, remains at rest. The mass of the alpha particle is 6.64×10^{-27} kg. (a) Calculate the electrostatic potential energy at the instant that the alpha particle stops. Express your result in joules and in MeV. (b) What initial kinetic energy (in joules and in MeV) did the alpha particle have? (c) What was the initial speed of the alpha particle?

Section 39.3 Energy Levels and the Bohr Model of the Atom

39.24 • The silicon–silicon single bond that forms the basis of the mythical silicon-based creature the Horta has a bond strength of 3.80 eV. What wavelength of photon would you need in a (mythical) phasor disintegration gun to destroy the Horta?

39.25 •• A hydrogen atom is in a state with energy -1.51 eV. In the Bohr model, what is the angular momentum of the electron in the atom, with respect to an axis at the nucleus?

39.26 • A hydrogen atom initially in the ground level absorbs a photon, which excites it to the $n = 4$ level. Determine the wavelength and frequency of the photon.

39.27 • A triply ionized beryllium ion, Be^{3+} (a beryllium atom with three electrons removed), behaves very much like a hydrogen atom except that the nuclear charge is four times as great. (a) What is the ground-level energy of Be^{3+}? How does this compare to the ground-level energy of the hydrogen atom? (b) What is the ionization energy of Be^{3+}? How does this compare to the ionization energy of the hydrogen atom? (c) For the hydrogen atom, the wavelength of the photon emitted in the $n = 2$ to $n = 1$ transition is 122 nm (see Example 39.6). What is the wavelength of the photon emitted when a Be^{3+} ion undergoes this transition? (d) For a given value of n, how does the radius of an orbit in Be^{3+} compare to that for hydrogen?

39.28 •• (a) Show that, as n gets very large, the energy levels of the hydrogen atom get closer and closer together in energy. (b) Do the radii of these energy levels also get closer together?

39.29 • (a) Using the Bohr model, calculate the speed of the electron in a hydrogen atom in the $n = 1$, 2, and 3 levels. (b) Calculate the orbital period in each of these levels. (c) The average lifetime of the first excited level of a hydrogen atom is 1.0×10^{-8} s. In the Bohr model, how many orbits does an electron in the $n = 2$ level complete before returning to the ground level?

39.30 • **CP** The energy-level scheme for the hypothetical one-electron element Searsium is shown in Fig. E39.30. The potential energy is taken to be zero for an electron at an infinite distance from the nucleus. (a) How much energy (in electron volts) does it take to ionize an electron from the ground level? (b) An 18-eV photon is absorbed by a Searsium atom in its ground level. As the atom returns to its ground level, what possible energies can the emitted photons have? Assume that there can be transitions between all pairs of levels. (c) What will happen if a photon with an energy of 8 eV strikes a Searsium atom in its ground level? Why? (d) Photons emitted in the Searsium transitions $n = 3 \rightarrow n = 2$ and $n = 3 \rightarrow n = 1$ will eject photoelectrons from an unknown metal, but the photon emitted from the transition $n = 4 \rightarrow n = 3$ will not. What are the limits (maximum and minimum possible values) of the work function of the metal?

Figure **E39.30**

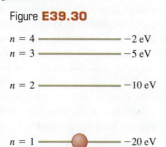

$n = 4$ ——————— -2 eV
$n = 3$ ——————— -5 eV

$n = 2$ ——————— -10 eV

$n = 1$ ——————— -20 eV

39.31 • In a set of experiments on a hypothetical one-electron atom, you measure the wavelengths of the photons emitted from transitions ending in the ground state ($n = 1$), as shown in the energy-level diagram in Fig. E39.31. You also observe that it takes 17.50 eV to ionize this atom. (a) What is the energy of the atom in each of the levels ($n = 1$, $n = 2$, etc.) shown in the figure? (b) If an electron made a transition from the $n = 4$ to the $n = 2$ level, what wavelength of light would it emit?

Figure **E39.31**

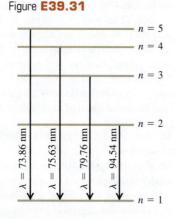

$n = 5$
$n = 4$
$n = 3$
$n = 2$
$n = 1$

$\lambda = 73.86$ nm
$\lambda = 75.63$ nm
$\lambda = 79.76$ nm
$\lambda = 94.54$ nm

39.32 • Find the longest and shortest wavelengths in the Lyman and Paschen series for hydrogen. In what region of the electromagnetic spectrum does each series lie?

39.33 • (a) An atom initially in an energy level with $E = -6.52$ eV absorbs a photon that has wavelength 860 nm. What is the internal energy of the atom after it absorbs the photon? (b) An atom initially in an energy level with $E = -2.68$ eV emits a photon that has wavelength 420 nm. What is the internal energy of the atom after it emits the photon?

39.34 •• Use Balmer's formula to calculate (a) the wavelength, (b) the frequency, and (c) the photon energy for the H_γ line of the Balmer series for hydrogen.

Section 39.4 The Laser

39.35 • **BIO** **Laser Surgery.** Using a mixture of CO_2, N_2, and sometimes He, CO_2 lasers emit a wavelength of 10.6 μm. At power outputs of 0.100 kW, such lasers are used for surgery. How many photons per second does a CO_2 laser deliver to the tissue during its use in an operation?

39.36 • **BIO** **Removing Birthmarks.** Pulsed dye lasers emit light of wavelength 585 nm in 0.45-ms pulses to remove skin blemishes such as birthmarks. The beam is usually focused onto a circular spot 5.0 mm in diameter. Suppose that the output of one such laser is 20.0 W. (a) What is the energy of each photon, in eV? (b) How many photons per square millimeter are delivered to the blemish during each pulse?

39.37 • How many photons per second are emitted by a 7.50-mW CO_2 laser that has a wavelength of 10.6 μm?

39.38 • **BIO** **PRK Surgery.** Photorefractive keratectomy (PRK) is a laser-based surgical procedure that corrects near- and farsightedness by removing part of the lens of the eye to change its curvature and hence focal length. This procedure can remove layers 0.25 μm thick using pulses lasting 12.0 ns from a laser beam of wavelength 193 nm. Low-intensity beams can be used because each individual photon has enough energy to break the covalent bonds of the tissue. (a) In what part of the electromagnetic spectrum does this light lie? (b) What is the energy of a single photon? (c) If a 1.50-mW beam is used, how many photons are delivered to the lens in each pulse?

39.39 • A large number of neon atoms are in thermal equilibrium. What is the ratio of the number of atoms in a $5s$ state to the number in a $3p$ state at (a) 300 K; (b) 600 K; (c) 1200 K? The energies of these states, relative to the ground state, are $E_{5s} = 20.66$ eV and $E_{3p} = 18.70$ eV. (d) At any of these temperatures, the rate at which a neon gas will spontaneously emit 632.8-nm radiation is quite low. Explain why.

39.40 • Figure 39.19a shows the energy levels of the sodium atom. The two lowest excited levels are shown in columns labeled $^2P_{3/2}$ and $^2P_{1/2}$. Find the ratio of the number of atoms in a $^2P_{3/2}$ state to the number in a $^2P_{1/2}$ state for a sodium gas in thermal equilibrium at 500 K. In which state are more atoms found?

Section 39.5 Continuous Spectra

39.41 •• A 100-W incandescent light bulb has a cylindrical tungsten filament 30.0 cm long, 0.40 mm in diameter, and with an emissivity of 0.26. (a) What is the temperature of the filament? (b) For what wavelength does the spectral emittance of the bulb peak? (c) Incandescent light bulbs are not very efficient sources of visible light. Explain why this is so.

39.42 • Determine λ_m, the wavelength at the peak of the Planck distribution, and the corresponding frequency f, at these temperatures: (a) 3.00 K; (b) 300 K; (c) 3000 K.

39.43 • Radiation has been detected from space that is characteristic of an ideal radiator at $T = 2.728$ K. (This radiation is a relic of the Big Bang at the beginning of the universe.) For this temperature, at what wavelength does the Planck distribution peak? In what part of the electromagnetic spectrum is this wavelength?

39.44 • The shortest visible wavelength is about 400 nm. What is the temperature of an ideal radiator whose spectral emittance peaks at this wavelength?

39.45 •• Two stars, both of which behave like ideal blackbodies, radiate the same total energy per second. The cooler one has a surface temperature T and a diameter 3.0 times that of the hotter star. (a) What is the temperature of the hotter star in terms of T? (b) What is the ratio of the peak-intensity wavelength of the hot star to the peak-intensity wavelength of the cool star?

39.46 • **Sirius B.** The brightest star in the sky is Sirius, the Dog Star. It is actually a binary system of two stars, the smaller one (Sirius B) being a white dwarf. Spectral analysis of Sirius B indicates that its surface temperature is 24,000 K and that it radiates energy at a total rate of 1.0×10^{25} W. Assume that it behaves like an ideal blackbody. (a) What is the total radiated intensity of Sirius B? (b) What is the peak-intensity wavelength? Is this wavelength visible to humans? (c) What is the radius of Sirius B? Express your answer in kilometers and as a fraction of our sun's radius. (d) Which star radiates more *total* energy per second, the hot Sirius B or the (relatively) cool sun with a surface temperature of 5800 K? To find out, calculate the ratio of the total power radiated by our sun to the power radiated by Sirius B.

39.47 •• **Blue Supergiants.** A typical blue supergiant star (the type that explodes and leaves behind a black hole) has a surface temperature of 30,000 K and a visual luminosity 100,000 times that of our sun. Our sun radiates at the rate of 3.86×10^{26} W. (Visual luminosity is the total power radiated at visible wavelengths.) (a) Assuming that this star behaves like an ideal blackbody, what is the principal wavelength it radiates? Is this light visible? Use your answer to explain why these stars are blue. (b) If we assume that the power radiated by the star is also 100,000 times that of our sun, what is the radius of this star? Compare its size to that of our sun, which has a radius of 6.96×10^5 km. (c) Is it really correct to say that the visual luminosity is proportional to the total power radiated? Explain.

Section 39.6 The Uncertainty Principle Revisited

39.48 • A pesky 1.5-mg mosquito is annoying you as you attempt to study physics in your room, which is 5.0 m wide and 2.5 m high. You decide to swat the bothersome insect as it flies toward you, but you need to estimate its speed to make a successful hit. (a) What is the maximum uncertainty in the horizontal position of the mosquito? (b) What limit does the Heisenberg uncertainty principle place on your ability to know the horizontal velocity of this mosquito? Is this limitation a serious impediment to your attempt to swat it?

39.49 • By extremely careful measurement, you determine the x-coordinate of a car's center of mass with an uncertainty of only $1.00 \ \mu m$. The car has a mass of 1200 kg. (a) What is the minimum uncertainty in the x-component of the velocity of the car's center of mass as prescribed by the Heisenberg uncertainty principle? (b) Does the uncertainty principle impose a practical limit on our ability to make simultaneous measurements of the positions and velocities of ordinary objects like cars, books, and people? Explain.

39.50 • A 10.0-g marble is gently placed on a horizontal tabletop that is 1.75 m wide. (a) What is the maximum uncertainty in the horizontal position of the marble? (b) According to the Heisenberg uncertainty principle, what is the minimum uncertainty in the horizontal velocity of the marble? (c) In light of your answer to part (b), what is the longest time the marble could remain on the table? Compare this time to the age of the universe, which is approximately 14 billion years. (*Hint:* Can you know that the horizontal velocity of the marble is *exactly* zero?)

39.51 • A scientist has devised a new method of isolating individual particles. He claims that this method enables him to detect simultaneously the position of a particle along an axis with a standard deviation of 0.12 nm and its momentum component along this axis with a standard deviation of 3.0×10^{-25} kg·m/s. Use the Heisenberg uncertainty principle to evaluate the validity of this claim.

39.52 • (a) The x-coordinate of an electron is measured with an uncertainty of 0.20 mm. What is the x-component of the electron's velocity, v_x, if the minimum percentage uncertainty in a simultaneous measurement of v_x is 1.0%? (b) Repeat part (a) for a proton.

39.53 • An atom in a metastable state has a lifetime of 5.2 ms. What is the uncertainty in energy of the metastable state?

39.54 • (a) The uncertainty in the y-component of a proton's position is 2.0×10^{-12} m. What is the minimum uncertainty in a simultaneous measurement of the y-component of the proton's velocity? (b) The uncertainty in the z-component of an electron's velocity is 0.250 m/s. What is the minimum uncertainty in a simultaneous measurement of the z-coordinate of the electron?

PROBLEMS

39.55 •• The negative muon has a charge equal to that of an electron but a mass that is 207 times as great. Consider a hydrogenlike atom consisting of a proton and a muon. (a) What is the reduced mass of the atom? (b) What is the ground-level energy (in electron volts)? (c) What is the wavelength of the radiation emitted in the transition from the $n = 2$ level to the $n = 1$ level?

39.56 • An atom with mass m emits a photon of wavelength λ. (a) What is the recoil speed of the atom? (b) What is the kinetic energy K of the recoiling atom? (c) Find the ratio K/E, where E is the energy of the emitted photon. If this ratio is much less than unity, the recoil of the atom can be neglected in the emission process. Is the recoil of the atom more important for small or large atomic masses? For long or short wavelengths? (d) Calculate K (in electron volts) and K/E for a hydrogen atom (mass 1.67×10^{-27} kg) that emits an ultraviolet photon of energy 10.2 eV. Is recoil an important consideration in this emission process?

39.57 • (a) What is the smallest amount of energy in electron volts that must be given to a hydrogen atom initially in its ground level so that it can emit the H_α line in the Balmer series? (b) How many different possibilities of spectral-line emissions are there for this atom when the electron starts in the $n = 3$ level and eventually ends up in the ground level? Calculate the wavelength of the emitted photon in each case.

39.58 • A large number of hydrogen atoms are in thermal equilibrium. Let n_2/n_1 be the ratio of the number of atoms in an $n = 2$ excited state to the number of atoms in an $n = 1$ ground state. At what temperature is n_2/n_1 equal to (a) 10^{-12}; (b) 10^{-8}; (c) 10^{-4}? (d) Like the sun, other stars have continuous spectra with dark absorption lines (see Fig. 39.9). The absorption takes place in the star's atmosphere, which in all stars is composed primarily of hydrogen. Explain why the Balmer absorption lines are relatively weak in stars with low atmospheric temperatures such as the sun (atmosphere temperature 5800 K) but strong in stars with higher atmospheric temperatures.

39.59 ••• A sample of hydrogen atoms is irradiated with light with wavelength 85.5 nm, and electrons are observed leaving the gas. (a) If each hydrogen atom were initially in its ground level, what would be the maximum kinetic energy in electron volts of these photoelectrons? (b) A few electrons are detected with energies as much as 10.2 eV greater than the maximum kinetic energy calculated in part (a). How can this be?

39.60 • **CP Bohr Orbits of a Satellite.** A 20.0-kg satellite circles the earth once every 2.00 h in an orbit having a radius of 8060 km. (a) Assuming that Bohr's angular-momentum result ($L = nh/2\pi$) applies to satellites just as it does to an electron in the hydrogen atom, find the quantum number n of the orbit of the satellite. (b) Show from Bohr's angular momentum result and Newton's law of gravitation that the radius of an earth-satellite orbit is directly proportional to the square of the quantum number, $r = kn^2$, where k is the constant of proportionality. (c) Using the result from part (b), find the distance between the orbit of the satellite in this problem and its next "allowed" orbit. (Calculate a numerical value.) (d) Comment on the possibility of observing the separation of the two adjacent orbits. (e) Do quantized and classical orbits correspond for this satellite? Which is the "correct" method for calculating the orbits?

39.61 •• **The Red Supergiant Betelgeuse.** The star Betelgeuse has a surface temperature of 3000 K and is 600 times the diameter of our sun. (If our sun were that large, we would be inside it!) Assume that it radiates like an ideal blackbody. (a) If Betelgeuse were to radiate all of its energy at the peak-intensity wavelength, how many photons per second would it radiate? (b) Find the ratio of the power radiated by Betelgeuse to the power radiated by our sun (at 5800 K).

39.62 •• **CP** Light from an ideal spherical blackbody 15.0 cm in diameter is analyzed using a diffraction grating having 3850 lines/cm. When you shine this light through the grating, you observe that the peak-intensity wavelength forms a first-order bright fringe at $\pm 11.6°$ from the central bright fringe. (a) What is the temperature of the blackbody? (b) How long will it take this sphere to radiate 12.0 MJ of energy?

39.63 What must be the temperature of an ideal blackbody so that photons of its radiated light having the peak-intensity wavelength can excite the electron in the Bohr-model hydrogen atom from the ground state to the third excited state?

39.64 • **CP** An ideal spherical blackbody 24.0 cm in diameter is maintained at 225°C by an internal electrical heater and is immersed in a very large open-faced tank of water that is kept boiling by the energy radiated by the sphere. You can neglect any heat transferred by conduction and convection. Consult Table 17.4 as needed. (a) At what rate, in g/s, is water evaporating from the tank? (b) If a physics-wise thermophile organism living in the hot water is observing this process, what will it measure for the peak-intensity (i) wavelength and (ii) frequency of the electromagnetic waves emitted by the sphere?

39.65 ••• When a photon is emitted by an atom, the atom must recoil to conserve momentum. This means that the photon and the recoiling atom share the transition energy. (a) For an atom with mass m, calculate the correction $\Delta\lambda$ due to recoil to the wavelength of an emitted photon. Let λ be the wavelength of the photon if recoil is not taken into consideration. (*Hint:* The correction is very small, as Problem 39.56 suggests, so $|\Delta\lambda|/\lambda \ll 1$. Use this fact to obtain an approximate but very accurate expression for $\Delta\lambda$.) (b) Evaluate the correction for a hydrogen atom in which an electron in the nth level returns to the ground level. How does the answer depend on n?

39.66 •• **An Ideal Blackbody.** A large cavity with a very small hole and maintained at a temperature T is a good approximation to an ideal radiator or blackbody. Radiation can pass into or out of the cavity only through the hole. The cavity is a perfect absorber, since any radiation incident on the hole becomes trapped inside the cavity. Such a cavity at 200°C has a hole with area 4.00 mm². How

long does it take for the cavity to radiate 100 J of energy through the hole?

39.67 •• **CALC** (a) Write the Planck distribution law in terms of the frequency f, rather than the wavelength λ, to obtain $I(f)$. (b) Show that

$$\int_0^\infty I(\lambda)\, d\lambda = \frac{2\pi^5 k^4}{15c^2h^3}T^4$$

where $I(\lambda)$ is the Planck distribution formula of Eq. (39.24). (*Hint:* Change the integration variable from λ to f. You will need to use the following tabulated integral:

$$\int_0^\infty \frac{x^3}{e^{\alpha x} - 1}\, dx = \frac{1}{240}\left(\frac{2\pi}{\alpha}\right)^4$$

(c) The result of part (b) is I and has the form of the Stefan–Boltzmann law, $I = \sigma T^4$ (Eq. 39.19). Evaluate the constants in part (b) to show that σ has the value given in Section 39.5.

39.68 •• **CP** A beam of 40-eV electrons traveling in the $+x$-direction passes through a slit that is parallel to the y-axis and 5.0 μm wide. The diffraction pattern is recorded on a screen 2.5 m from the slit. (a) What is the de Broglie wavelength of the electrons? (b) How much time does it take the electrons to travel from the slit to the screen? (c) Use the width of the central diffraction pattern to calculate the uncertainty in the y-component of momentum of an electron just after it has passed through the slit. (d) Use the result of part (c) and the Heisenberg uncertainty principle (Eq. 39.29 for y) to estimate the minimum uncertainty in the y-coordinate of an electron just after it has passed through the slit. Compare your result to the width of the slit.

39.69 • (a) What is the energy of a photon that has wavelength 0.10 μm? (b) Through approximately what potential difference must electrons be accelerated so that they will exhibit wave nature in passing through a pinhole 0.10 μm in diameter? What is the speed of these electrons? (c) If protons rather than electrons were used, through what potential difference would protons have to be accelerated so they would exhibit wave nature in passing through this pinhole? What would be the speed of these protons?

39.70 • **CP** Electrons go through a single slit 150 nm wide and strike a screen 24.0 cm away. You find that at angles of $\pm 20.0°$ from the center of the diffraction pattern, no electrons hit the screen but electrons hit at all points closer to the center. (a) How fast were these electrons moving when they went through the slit? (b) What will be the next larger angles at which no electrons hit the screen?

39.71 •• **CP** A beam of electrons is accelerated from rest and then passes through a pair of identical thin slits that are 1.25 nm apart. You observe that the first double-slit interference dark fringe occurs at $\pm 18.0°$ from the original direction of the beam when viewed on a distant screen. (a) Are these electrons relativistic? How do you know? (b) Through what potential difference were the electrons accelerated?

39.72 •• **CP** A beam of protons and a beam of alpha particles (of mass 6.64×10^{-27} kg and charge $+2e$) are accelerated from rest through the same potential difference and pass through identical circular holes in a very thin, opaque film. When viewed far from the hole, the diffracted proton beam forms its first dark ring at 15° with respect to its original direction. When viewed similarly, at what angle will the alpha particle form its first dark ring?

39.73 •• **CP** An electron beam and a photon beam pass through identical slits. On a distant screen, the first dark fringe occurs at the same angle for both of the beams. The electron speeds are much slower than that of light. (a) Express the energy of a photon in terms of the kinetic energy K of one of the electrons. (b) Which is greater, the energy of a photon or the kinetic energy of an electron?

39.74 • **CP** Coherent light is passed through two narrow slits whose separation is 40.0 μm. The second-order bright fringe in the interference pattern is located at an angle of 0.0300 rad. If electrons are used instead of light, what must the kinetic energy (in electron volts) of the electrons be if they are to produce an interference pattern for which the second-order maximum is also at 0.0300 rad?

39.75 • **BIO** What is the de Broglie wavelength of a red blood cell, with mass 1.00×10^{-11} g, that is moving with a speed of 0.400 cm/s? Do we need to be concerned with the wave nature of the blood cells when we describe the flow of blood in the body?

39.76 • Calculate the energy in electron volts of (a) an electron that has de Broglie wavelength 400 nm and (b) a photon that has wavelength 400 nm.

39.77 • High-speed electrons are used to probe the interior structure of the atomic nucleus. For such electrons the expression $\lambda = h/p$ still holds, but we must use the relativistic expression for momentum, $p = mv/\sqrt{1 - v^2/c^2}$. (a) Show that the speed of an electron that has de Broglie wavelength λ is

$$v = \frac{c}{\sqrt{1 + (mc\lambda/h)^2}}$$

(b) The quantity h/mc equals 2.426×10^{-12} m. (As we saw in Section 38.3, this same quantity appears in Eq. (38.7), the expression for Compton scattering of photons by electrons.) If λ is small compared to h/mc, the denominator in the expression found in part (a) is close to unity and the speed v is very close to c. In this case it is convenient to write $v = (1 - \Delta)c$ and express the speed of the electron in terms of Δ rather than v. Find an expression for Δ valid when $\lambda \ll h/mc$. [*Hint:* Use the binomial expansion $(1 + z)^n = 1 + nz + n(n - 1)z^2/2 + \cdots$, valid for the case $|z| < 1$.] (c) How fast must an electron move for its de Broglie wavelength to be 1.00×10^{-15} m, comparable to the size of a proton? Express your answer in the form $v = (1 - \Delta)c$, and state the value of Δ.

39.78 • Suppose that the uncertainty of position of an electron is equal to the radius of the $n = 1$ Bohr orbit for hydrogen. Calculate the simultaneous minimum uncertainty of the corresponding momentum component, and compare this with the magnitude of the momentum of the electron in the $n = 1$ Bohr orbit. Discuss your results.

39.79 • **CP** (a) A particle with mass m has kinetic energy equal to three times its rest energy. What is the de Broglie wavelength of this particle? (*Hint:* You must use the relativistic expressions for momentum and kinetic energy: $E^2 = (pc)^2 + (mc^2)^2$ and $K = E - mc^2$.) (b) Determine the numerical value of the kinetic energy (in MeV) and the wavelength (in meters) if the particle in part (a) is (i) an electron and (ii) a proton.

39.80 • **Proton Energy in a Nucleus.** The radii of atomic nuclei are of the order of 5.0×10^{-15} m. (a) Estimate the minimum uncertainty in the momentum of a proton if it is confined within a nucleus. (b) Take this uncertainty in momentum to be an estimate of the magnitude of the momentum. Use the relativistic relationship between energy and momentum, Eq. (37.39), to obtain an estimate of the kinetic energy of a proton confined within a nucleus. (c) For a proton to remain bound within a nucleus, what must the magnitude of the (negative) potential energy for a proton be within the nucleus? Give your answer in eV and in MeV. Compare to the potential energy for an electron in a hydrogen atom, which has a magnitude of a few tens of eV. (This shows why the

interaction that binds the nucleus together is called the "strong nuclear force.")

39.81 • **Electron Energy in a Nucleus.** The radii of atomic nuclei are of the order of 5.0×10^{-15} m. (a) Estimate the minimum uncertainty in the momentum of an electron if it is confined within a nucleus. (b) Take this uncertainty in momentum to be an estimate of the magnitude of the momentum. Use the relativistic relationship between energy and momentum, Eq. (37.39), to obtain an estimate of the kinetic energy of an electron confined within a nucleus. (c) Compare the energy calculated in part (b) to the magnitude of the Coulomb potential energy of a proton and an electron separated by 5.0×10^{-15} m. On the basis of your result, could there be electrons within the nucleus? (*Note:* It is interesting to compare this result to that of Problem 39.80.)

39.82 • In a TV picture tube the accelerating voltage is 15.0 kV, and the electron beam passes through an aperture 0.50 mm in diameter to a screen 0.300 m away. (a) Calculate the uncertainty in the component of the electron's velocity perpendicular to the line between aperture and screen. (b) What is the uncertainty in position of the point where the electrons strike the screen? (c) Does this uncertainty affect the clarity of the picture significantly? (Use nonrelativistic expressions for the motion of the electrons. This is fairly accurate and is certainly adequate for obtaining an estimate of uncertainty effects.)

39.83 • The neutral pion (π^0) is an unstable particle produced in high-energy particle collisions. Its mass is about 264 times that of the electron, and it exists for an average lifetime of 8.4×10^{-17} s before decaying into two gamma-ray photons. Using the relationship $E = mc^2$ between rest mass and energy, find the uncertainty in the mass of the particle and express it as a fraction of the mass.

39.84 • **Quantum Effects in Daily Life?** A 1.25-mg insect flies through a 4.00-mm-diameter hole in an ordinary window screen. The thickness of the screen is 0.500 mm. (a) What should be the approximate wavelength and speed of the insect for her to show wave behavior as she goes through the hole? (b) At the speed found in part (a), how long would it take the insect to pass through the 0.500-mm thickness of the hole in the screen? Compare this time to the age of the universe (about 14 billion years). Would you expect to see "insect diffraction" in daily life?

39.85 • **Doorway Diffraction.** If your wavelength were 1.0 m, you would undergo considerable diffraction in moving through a doorway. (a) What must your speed be for you to have this wavelength? (Assume that your mass is 60.0 kg.) (b) At the speed calculated in part (a), how many years would it take you to move 0.80 m (one step)? Will you notice diffraction effects as you walk through doorways?

39.86 • **Atomic Spectra Uncertainties.** A certain atom has an energy level 2.58 eV above the ground level. Once excited to this level, the atom remains in this level for 1.64×10^{-7} s (on average) before emitting a photon and returning to the ground level. (a) What is the energy of the photon (in electron volts)? What is its wavelength (in nanometers)? (b) What is the smallest possible uncertainty in energy of the photon? Give your answer in electron volts. (c) Show that $|\Delta E/E| = |\Delta\lambda/\lambda|$ if $|\Delta\lambda/\lambda| \ll 1$. Use this to calculate the magnitude of the smallest possible uncertainty in the wavelength of the photon. Give your answer in nanometers.

39.87 •• You intend to use an electron microscope to study the structure of some crystals. For accurate resolution, you want the electron wavelength to be 1.00 nm. (a) Are these electrons relativistic? How do you know? (b) What accelerating potential is needed? (c) What is the kinetic energy of the electrons you are using? To see if it is great enough to damage the crystals you are studying, compare it to the potential energy of a typical NaCl molecule, which is about 6.0 eV. (d) If you decided to use electromagnetic waves as your probe, what energy should their photons have to provide the same resolution as the electrons? Would this energy damage the crystal?

39.88 •• For x rays with wavelength 0.0300 nm, the $m = 1$ intensity maximum for a crystal occurs when the angle θ in Fig. 39.2 is $35.8°$. At what angle θ does the $m = 1$ maximum occur when a beam of 4.50-keV electrons is used instead? Assume that the electrons also scatter from the atoms in the surface plane of this same crystal.

39.89 •• **CP** Electron diffraction can also take place when there is interference between electron waves that scatter from atoms on the surface of a crystal and waves that scatter from atoms in the next plane below the surface, a distance d from the surface (see Fig. 36.23c). (a) Find an equation for the angles θ at which there is an intensity maximum for electron waves of wavelength λ. (b) The spacing between crystal planes in a certain metal is 0.091 nm. If 71.0-eV electrons are used, find the angle at which there is an intensity maximum due to interference between scattered waves from adjacent crystal planes. The angle is measured as shown in Fig. 36.23c. (c) The actual angle of the intensity maximum is slightly different from your result in part (b). The reason is the work function ϕ of the metal (see Section 38.1), which changes the electron potential energy by $-e\phi$ when it moves from vacuum into the metal. If the effect of the work function is taken into account, is the angle of the intensity maximum larger or smaller than the value found in part (b)? Explain.

39.90 •• A certain atom has an energy level 3.50 eV above the ground state. When excited to this state, it remains 4.0 μs, on the average, before emitting a photon and returning to the ground state. (a) What is the energy of the photon? What is its wavelength? (b) What is the smallest possible uncertainty in energy of the photon?

39.91 •• **BIO** **Structure of a Virus.** To investigate the structure of extremely small objects, such as viruses, the wavelength of the probing wave should be about one-tenth the size of the object for sharp images. But as the wavelength gets shorter, the energy of a photon of light gets greater and could damage or destroy the object being studied. One alternative is to use electron matter waves instead of light. Viruses vary considerably in size, but 50 nm is not unusual. Suppose you want to study such a virus, using a wave of wavelength 5.00 nm. (a) If you use light of this wavelength, what would be the energy (in eV) of a single photon? (b) If you use an electron of this wavelength, what would be its kinetic energy (in eV)? Is it now clear why matter waves (such as in the electron microscope) are often preferable to electromagnetic waves for studying microscopic objects?

39.92 •• **CALC** **Zero-Point Energy.** Consider a particle with mass m moving in a potential $U = \frac{1}{2}kx^2$, as in a mass–spring system. The total energy of the particle is $E = p^2/2m + \frac{1}{2}kx^2$. Assume that p and x are approximately related by the Heisenberg uncertainty principle, so $px \approx h$. (a) Calculate the minimum possible value of the energy E, and the value of x that gives this minimum E. This lowest possible energy, which is not zero, is called the

zero-point energy. (b) For the x calculated in part (a), what is the ratio of the kinetic to the potential energy of the particle?

39.93 •• **CALC** A particle with mass m moves in a potential $U(x) = A|x|$, where A is a positive constant. In a simplified picture, quarks (the constituents of protons, neutrons, and other particles, as will be described in Chapter 44) have a potential energy of interaction of approximately this form, where x represents the separation between a pair of quarks. Because $U(x) \rightarrow \infty$ as $x \rightarrow \infty$, it's not possible to separate quarks from each other (a phenomenon called *quark confinement*). (a) Classically, what is the force acting on this particle as a function of x? (b) Using the uncertainty principle as in Problem 39.92, determine approximately the zero-point energy of the particle.

39.94 •• Imagine another universe in which the value of Planck's constant is 0.0663 J · s, but in which the physical laws and all other physical constants are the same as in our universe. In this universe, two physics students are playing catch. They are 12 m apart, and one throws a 0.25-kg ball directly toward the other with a speed of 6.0 m/s. (a) What is the uncertainty in the ball's horizontal momentum, in a direction perpendicular to that in which it is being thrown, if the student throwing the ball knows that it is located within a cube with volume 125 cm³ at the time she throws it? (b) By what horizontal distance could the ball miss the second student?

CHALLENGE PROBLEMS

39.95 ••• (a) Show that in the Bohr model, the frequency of revolution of an electron in its circular orbit around a stationary hydrogen nucleus is $f = me^4/4\epsilon_0^2 n^3 h^3$. (b) In classical physics, the frequency of revolution of the electron is equal to the frequency of the radiation that it emits. Show that when n is very large, the frequency of revolution does indeed equal the radiated frequency calculated from Eq. (39.5) for a transition from $n_1 = n + 1$ to $n_2 = n$. (This illustrates Bohr's *correspondence principle,* which is often used as a check on quantum calculations. When n is small, quantum physics gives results that are very different from those of classical physics. When n is large, the differences are not significant, and the two methods then "correspond." In fact, when Bohr first tackled the hydrogen atom problem, he sought to determine f as a function of n such that it would correspond to classical results for large n.)

39.96 ••• **CP CALC** You have entered a contest in which the contestants drop a marble with mass 20.0 g from the roof of a building onto a small target 25.0 m below. From uncertainty considerations, what is the typical distance by which you will miss the target, given that you aim with the highest possible precision? (*Hint:* The uncertainty Δx_f in the x-coordinate of the marble when it reaches the ground comes in part from the uncertainty Δx_i in the x-coordinate initially and in part from the initial uncertainty in v_x. The latter gives rise to an uncertainty Δv_x in the horizontal motion of the marble as it falls. The values of Δx_i and Δv_x are related by the uncertainty principle. A small Δx_i gives rise to a large Δv_x, and vice versa. Find the value of Δx_i that gives the smallest total uncertainty in x at the ground. Ignore any effects of air resistance.)

Answers

Chapter Opening Question

The smallest detail visible in an image is comparable to the wavelength used to make the image. Electrons can easily be given a large momentum p and hence a short wavelength $\lambda = h/p$, and so can be used to resolve extremely fine details. (See Section 39.1.)

Test Your Understanding Questions

39.1 Answers: (a) (i), (b) no From Example 39.2, the speed of a particle is $v = h/\lambda m$ and the kinetic energy is $K = \frac{1}{2}mv^2 = (m/2)(h/\lambda m)^2 = h^2/2\lambda^2 m$. This shows that for a given wavelength, the kinetic energy is inversely proportional to the mass. Hence the proton, with a smaller mass, has more kinetic energy than the neutron. For part (b), the energy of a photon is $E = hf$, and the frequency of a photon is $f = c/\lambda$. Hence $E = hc/\lambda$ and $\lambda = hc/E = (4.136 \times 10^{-15}\text{ eV}\cdot\text{s})(2.998 \times 10^8\text{ m/s})/(54\text{ eV}) = 2.3 \times 10^{-8}$ m. This is more than 100 times greater than the wavelength of an electron of the same energy. While both photons and electrons have wavelike properties, they have different relationships between their energy and momentum and hence between their frequency and wavelength.

39.2 Answer: (iii) Because the alpha particle is more massive, it won't bounce back in even a head-on collision with a proton that's initially at rest, any more than a bowling ball would when colliding with a Ping-Pong ball at rest (see Fig. 8.22b). Thus there would be *no* large-angle scattering in this case. Rutherford saw large-angle scattering in his experiment because gold nuclei are more massive than alpha particles (see Fig. 8.22a).

39.3 Answer: (iv) Figure 39.27 shows that many (though *not* all) of the energy levels of He$^+$ are the same as those of H. Hence photons emitted during transitions between corresponding pairs of levels in He$^+$ and H have the same energy E and the same wavelength $\lambda = hc/E$. An H atom that drops from the $n = 2$ level to the $n = 1$ level emits a photon of energy 10.20 eV and wavelength 122 nm (see Example 39.6); a He$^+$ ion emits a photon of the same energy and wavelength when it drops from the $n = 4$ level to the $n = 2$ level. Inspecting Fig. 39.27 will show you that every even-numbered level in He$^+$ matches a level in H, while none of the odd-numbered He$^+$ levels do. The first three He$^+$ transitions given in the question ($n = 2$ to $n = 1$, $n = 3$ to $n = 2$, and $n = 4$ to $n = 3$) all involve an odd-numbered level, so none of their wavelengths match a wavelength emitted by H atoms.

39.4 Answer: (i) In a neon light fixture, a large potential difference is applied between the ends of a neon-filled glass tube. This ionizes some of the neon atoms, allowing a current of electrons to flow through the gas. Some of the neon atoms are struck by fast-moving electrons, making them transition to an excited level. From this level the atoms undergo *spontaneous* emission, as depicted in Fig. 39.28b, and emit 632.8-nm photons in the process. No population inversion occurs and the photons are not trapped by mirrors as shown in Fig. 39.29d, so there is no stimulated emission. Hence there is no laser action.

39.5 Answers: (a) yes, (b) yes The Planck radiation law, Eq. (39.24), shows that an ideal blackbody emits radiation at *all* wavelengths: The spectral emittance $I(\lambda)$ is equal to zero only for $\lambda = 0$ and in the limit $\lambda \to \infty$. So a blackbody at 2000 K does indeed emit both x rays and radio waves. However, Fig. 39.32 shows that the spectral emittance for this temperature is very low for wavelengths much shorter than 1 μm (including x rays) and for wavelengths much longer than a few μm (including radio waves). Hence such a blackbody emits very little in the way of x rays or radio waves.

39.6 Answer: (i) and (iii) (tie), (ii) and (iv) (tie) According to the Heisenberg uncertainty principle, the smaller the uncertainty Δx in the x-coordinate, the greater the uncertainty Δp_x in the x-momentum. The relationship between Δx and Δp_x does not depend on the mass of the particle, and so is the same for a proton as for an electron.

Bridging Problem

Answers: (a) 192 nm; ultraviolet **(b)** $n = 4$

(c) $\lambda_2 = 0.665$ nm, $\lambda_3 = 0.997$ nm

LEARNING GOALS

By studying this chapter, you will learn:

- About the wave function that describes the behavior of a particle and the Schrödinger equation that this function must satisfy.

- How to calculate the wave functions and energy levels for a particle confined to a box.

- How to analyze the quantum-mechanical behavior of a particle in a potential well.

- How quantum mechanics makes it possible for particles to go where Newtonian mechanics says they cannot.

- How to use quantum mechanics to analyze a harmonic oscillator.

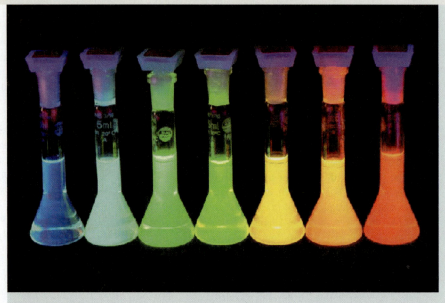

? These containers hold solutions of microscopic semiconductor particles of different sizes. The particles glow when exposed to ultraviolet light; the smallest particles glow blue and the largest particles glow red. Why?

I n Chapter 39 we found that particles can behave like waves. In fact, it turns out that we can use the wave picture to completely describe the behavior of a particle. This approach, called *quantum mechanics,* is the key to understanding the behavior of matter on the molecular, atomic, and nuclear scales. In this chapter we'll see how to find the *wave function* of a particle by solving the *Schrödinger equation,* which is as fundamental to quantum mechanics as Newton's laws are to mechanics or as Maxwell's equations are to electromagnetism.

We'll begin with a quantum-mechanical analysis of a *free particle* that moves along a straight line without being acted on by forces of any kind. We'll then consider particles that are acted on by forces and are trapped in *bound states,* just as electrons are bound within an atom. We'll see that solving the Schrödinger equation automatically gives the possible energy levels for the system.

Besides energies, solving the Schrödinger equation gives us the probabilities of finding a particle in various regions. One surprising result is that there is a nonzero probability that microscopic particles will pass through thin barriers, even though such a process is forbidden by Newtonian mechanics.

In this chapter we'll consider the Schrödinger equation for one-dimensional motion only. In Chapter 41 we'll see how to extend this equation to three-dimensional problems such as the hydrogen atom. The hydrogen-atom wave functions will in turn form the foundation for our analysis of more complex atoms, of the periodic table of the elements, of x-ray energy levels, and of other properties of atoms.

MasteringPHYSICS®

PhET: Fourier: Making Waves

40.1 Wave Functions and the One-Dimensional Schrödinger Equation

We have now seen compelling evidence that on an atomic or subatomic scale, an object such as an electron cannot be described simply as a classical, Newtonian point particle. Instead, we must take into account its *wave* characteristics. In the

Bohr model of the hydrogen atom (Section 39.3) we tried to have it both ways: We pictured the electron as a classical particle in a circular orbit around the nucleus, and used the de Broglie relation between particle momentum and wavelength to explain why only orbits of certain radii are allowed. As we saw in Section 39.6, however, the Heisenberg uncertainty principle tells us that a hybrid description of this kind can't be wholly correct. In this section we'll explore how to describe the state of a particle by using *only* the language of waves. This new description, called **quantum mechanics,** will replace the classical scheme of describing the state of a particle by its coordinates and velocity components.

Our new quantum-mechanical scheme for describing a particle has a lot in common with the language of classical wave motion. In Section 15.3 of Chapter 15, we described transverse waves on a string by specifying the position of each point in the string at each instant of time by means of a *wave function y(x, t)* that represents the displacement from equilibrium, at time *t*, of a point on the string at a distance *x* from the origin (Fig. 40.1). Once we know the wave function for a particular wave motion, we know everything there is to know about the motion. For example, we can find the velocity and acceleration of any point on the string at any time. We worked out specific forms for these functions for *sinusoidal* waves, in which each particle undergoes simple harmonic motion.

We followed a similar pattern for sound waves in Chapter 16. The wave function *p(x, t)* for a wave traveling along the *x*-direction represented the pressure variation at any point *x* and any time *t*. We used this language once more in Section 32.3, where we used *two* wave functions to describe the electric and magnetic fields in an electromagnetic wave.

Thus it's natural to use a wave function as the central element of our new language of quantum mechanics. The customary symbol for this wave function is the Greek letter psi, Ψ or ψ. In general, we'll use an uppercase Ψ to denote a function of all the space coordinates and time, and a lowercase ψ for a function of the space coordinates only—*not* of time. Just as the wave function $y(x, t)$ for mechanical waves on a string provides a complete description of the motion, so the wave function $\Psi(x, y, z, t)$ for a particle contains all the information that can be known about the particle.

> **CAUTION** **Particle waves vs. mechanical waves** Unlike for mechanical waves on a string or sound waves in air, the wave function for a particle is *not* a mechanical wave that needs some material medium in order to propagate. The wave function describes the particle, but we cannot define the function itself in terms of anything material. We can only describe how it is related to physically observable effects. ▌

40.1 These children are talking over a cup-and-string "telephone." The displacement of the string is completely described by a wave function $y(x, t)$. In an analogous way, a particle is completely described by a quantum-mechanical wave function $\Psi(x, y, z, t)$.

Waves in One Dimension

The wave function of a particle depends in general on all three dimensions of space. For simplicity, however, we'll begin our study of these functions by considering *one-dimensional* motion, in which a particle of mass *m* moves parallel to the *x*-axis and the wave function Ψ depends on the coordinate *x* and the time *t* only. (In the same way, we studied one-dimensional kinematics in Chapter 2 before going on to study two- and three-dimensional motion in Chapter 3.)

What does a one-dimensional quantum-mechanical wave look like, and what determines its properties? We can answer this question by first recalling the properties of a wave on a string. We saw in Section 15.3 that any wave function $y(x, t)$ that describes a wave on a string must satisfy the *wave equation:*

$$\frac{\partial^2 y(x, t)}{\partial x^2} = \frac{1}{v^2} \frac{\partial^2 y(x, t)}{\partial t^2} \qquad \text{(wave equation for waves on a string)} \qquad (40.1)$$

In Eq. (40.1) *v* is the speed of the wave, which is the same no matter what the wavelength. As an example, consider the following wave function for a wave of wavelength λ and frequency f moving in the positive x-direction along a string:

$$y(x, t) = A\cos(kx - \omega t) + B\sin(kx - \omega t) \qquad \text{(sinusoidal wave on a string)} \qquad (40.2)$$

Here $k = 2\pi/\lambda$ is the *wave number* and $\omega = 2\pi f$ is the *angular frequency*. (We used these same quantities for mechanical waves in Chapter 15 and electromagnetic waves in Chapter 32.) The quantities A and B are constants that determine the amplitude and phase of the wave. The expression in Eq. (40.2) is a valid wave function if and only if it satisfies the wave equation, Eq. (40.1). To check this, take the first and second derivatives of $y(x, t)$ with respect to x and take the first and second derivatives with respect to t:

$$\frac{\partial y(x, t)}{\partial x} = -kA\sin(kx - \omega t) + kB\cos(kx - \omega t) \qquad (40.3a)$$

$$\frac{\partial^2 y(x, t)}{\partial x^2} = -k^2 A\cos(kx - \omega t) - k^2 B\sin(kx - \omega t) \qquad (40.3b)$$

$$\frac{\partial y(x, t)}{\partial t} = \omega A\sin(kx - \omega t) - \omega B\cos(kx - \omega t) \qquad (40.3c)$$

$$\frac{\partial^2 y(x, t)}{\partial t^2} = -\omega^2 A\cos(kx - \omega t) - \omega^2 B\sin(kx - \omega t) \qquad (40.3d)$$

If we substitute Eqs. (40.3b) and (40.3d) into the wave equation, Eq. (40.1), we get

$$-k^2 A\cos(kx - \omega t) - k^2 B\sin(kx - \omega t)$$

$$= \frac{1}{v^2}[-\omega^2 A\cos(kx - \omega t) - \omega^2 B\sin(kx - \omega t)] \qquad (40.4)$$

For Eq. 40.4 to be satisfied at all coordinates x and all times t, the coefficients of the cosine function must be the same on both sides of the equation, and likewise for the coefficients of the sine function. You can see that both of these conditions will be satisfied if

$$k^2 = \frac{\omega^2}{v^2} \qquad \text{or} \qquad \omega = vk \qquad \text{(waves on a string)} \qquad (40.5)$$

From the definitions of angular frequency ω and wave number k, Eq. (40.5) is equivalent to

$$2\pi f = v\frac{2\pi}{\lambda} \qquad \text{or} \qquad v = \lambda f \qquad \text{(waves on a string)}$$

This equation is just the familiar relationship among wave speed, wavelength, and frequency for waves on a string. So our calculation shows that Eq. (40.2) is a valid wave function for waves on a string for any values of A and B, provided that ω and k are related by Eq. (40.5).

What we need is a quantum-mechanical version of the wave equation, Eq. (40.1), valid for particle waves. We expect this equation to involve partial derivatives of the wave function $\Psi(x, t)$ with respect to x and with respect to t. However, this new equation *cannot* be the same as Eq. (40.1) for waves on a string because the relationship between ω and k is different. We can show this by considering a **free particle,** one that experiences no force at all as it moves along the x-axis. For such a particle the potential energy $U(x)$ has the same value for all x (recall from Chapter 7 that $F_x = -dU(x)/dx$, so zero force means the potential energy has zero derivative). For simplicity let $U = 0$ for all x. Then the energy of the free particle is equal to its kinetic energy, which we can express in terms of its momentum p:

$$E = \tfrac{1}{2}mv^2 = \frac{m^2 v^2}{2m} = \frac{(mv)^2}{2m} = \frac{p^2}{2m} \qquad \text{(energy of a free particle)} \qquad (40.6)$$

The de Broglie relations that we introduced in Section 39.1 tell us that the energy E is proportional to the angular frequency ω and the momentum p is proportional to the wave number:

$$E = hf = \frac{h}{2\pi}2\pi f = \hbar\omega \qquad (40.7a)$$

$$p = \frac{h}{\lambda} = \frac{h}{2\pi}\frac{2\pi}{\lambda} = \hbar k \qquad (40.7b)$$

Remember that $\hbar = h/2\pi$. If we substitute Eqs. (40.7) into Eq. (40.6), we find that the relationship between ω and k for a free particle is

$$\hbar\omega = \frac{\hbar^2 k^2}{2m} \qquad \text{(free particle)} \qquad (40.8)$$

Equation (40.8) is *very* different from the corresponding relationship for waves on a string, Eq. (40.5): The angular frequency ω for particle waves is proportional to the *square* of the wave number, while for waves on a string ω is directly proportional to k. Our task is therefore to construct a quantum-mechanical version of the wave equation whose free-particle solutions satisfy Eq. (40.8).

We'll attack this problem by assuming a sinusoidal wave function $\Psi(x, t)$ of the same form as Eq. (40.2) for a sinusoidal wave on a string. For a wave on a string, Eq. (40.2) represents a wave of wavelength $\lambda = 2\pi/k$ and frequency $f = \omega/2\pi$ propagating in the positive x-direction. By analogy, our sinsuoidal wave function $\Psi(x, t)$ represents a free particle of mass m, momentum $p = \hbar k$, and energy $E = \hbar\omega$ moving in the positive x-direction:

$$\Psi(x, t) = A\cos(kx - \omega t) + B\sin(kx - \omega t) \qquad \begin{array}{l}\text{(sinusoidal wave}\\ \text{function representing}\\ \text{a free particle)}\end{array} \quad (40.9)$$

The wave number k and angular frequency ω in Eq. (40.9) must satisfy Eq. (40.8). If you look at Eq. (40.3b), you'll see that taking the second derivative of $\Psi(x, t)$ in Eq. (40.9) with respect to x gives us $\Psi(x, t)$ multiplied by $-k^2$. Hence if we multiply $\partial^2\Psi(x, t)/\partial x^2$ by $-\hbar^2/2m$, we get

$$-\frac{\hbar^2}{2m}\frac{\partial^2\Psi(x, t)}{\partial x^2} = -\frac{\hbar^2}{2m}[-k^2 A\cos(kx - \omega t) - k^2 B\sin(kx - \omega t)]$$

$$= \frac{\hbar^2 k^2}{2m}[A\cos(kx - \omega t) + B\sin(kx - \omega t)] \qquad (40.10)$$

$$= \frac{\hbar^2 k^2}{2m}\Psi(x, t)$$

Equation (40.10) suggests that $(-\hbar^2/2m)\partial^2\Psi(x, t)/\partial x^2$ should be one side of our quantum-mechanical wave equation, with the other side equal to $\hbar\omega\Psi(x, t)$ in order to satisfy Eq. (40.8). If you look at Eq. (40.3c), you'll see that taking the *first* time derivative of $\Psi(x, t)$ in Eq. (40.9) brings out a factor of ω. So we'll make the educated guess that the right-hand side of our quantum-mechanical wave equation involves $\hbar = h/2\pi$ times $\partial\Psi(x, t)/\partial t$. So our tentative equation is

$$-\frac{\hbar^2}{2m}\frac{\partial^2\Psi(x, t)}{\partial x^2} = C\hbar\frac{\partial\Psi(x, t)}{\partial t} \qquad (40.11)$$

At this point we include a constant C as a "fudge factor" to make sure that everything turns out right. Now let's substitute the wave function from Eq. (40.9) into Eq. (40.11). From Eq. (40.10) and Eq. (40.3c), we get

$$\frac{\hbar^2 k^2}{2m}[A\cos(kx - \omega t) + B\sin(kx - \omega t)]$$

$$= C\hbar\omega[A\sin(kx - \omega t) - B\cos(kx - \omega t)] \qquad (40.12)$$

From Eq. (40.8), $\hbar\omega = \hbar^2 k^2/2m$, so we can cancel these factors on the two sides of Eq. (40.12). What remains is

$$A\cos(kx - \omega t) + B\sin(kx - \omega t) = CA\sin(kx - \omega t)$$

$$- CB\cos(kx - \omega t) \qquad (40.13)$$

As in our discussion above of the wave equation for waves on a string, in order for Eq. (40.13) to be satisfied for all values of x and all values of t, the coefficients of the cosine function must be the same on both sides of the equation, and likewise for the coefficients of the sine function. Hence we have the following relationships among the coefficients A and B in Eq. (40.9) and the coefficient C in Eq. (40.11):

$$A = -CB \qquad (40.14a)$$

$$B = CA \qquad (40.14b)$$

If we use Eq. (40.14b) to eliminate B from Eq. (40.14a), we get $A = -C^2 A$, which means that $C^2 = -1$. Thus C is equal to the *imaginary* number $i = \sqrt{-1}$, and Eq. (40.11) becomes

$$-\frac{\hbar^2}{2m}\frac{\partial^2 \Psi(x, t)}{\partial x^2} = i\hbar\frac{\partial \Psi(x, t)}{\partial t} \qquad \text{(one-dimensional Schrödinger equation for a free particle)} \qquad (40.15)$$

Equation (40.15) is the one-dimensional **Schrödinger equation** for a free particle, developed in 1926 by the Austrian physicist Erwin Schrödinger (Fig. 40.2). The presence of the imaginary number i in Eq. (40.15) means that the solutions to the Schrödinger equation are complex quantities, with a real part and an imaginary part. (The imaginary part of $\Psi(x, t)$ is a real function multiplied by the imaginary number $i = \sqrt{-1}$.) An example is our free-particle wave function from Eq. (40.9). Since we found $C = i$ in Eqs. (40.14), it follows from Eq. (40.14b) that $B = iA$. Then Eq. (40.9) becomes

$$\Psi(x, t) = A[\cos(kx - \omega t) + i\sin(kx - \omega t)] \qquad \text{(sinusoidal wave function representing a free particle)} \qquad (40.16)$$

The real part of $\Psi(x, t)$ is $\text{Re}\,\Psi(x, t) = A\cos(kx - \omega t)$ and the imaginary part is $\text{Im}\,\Psi(x, t) = A\sin(kx - \omega t)$. Figure 40.3 graphs the real and imaginary parts of $\Psi(x, t)$ at $t = 0$, so $\Psi(x, 0) = A\cos kx + iA\sin kx$.

We can rewrite Eq. (40.16) using *Euler's formula*, which states that for any angle θ,

$$e^{i\theta} = \cos\theta + i\sin\theta$$

$$e^{-i\theta} = \cos(-\theta) + i\sin(-\theta) = \cos\theta - i\sin\theta \qquad (40.17)$$

Thus our sinusoidal free-particle wave function becomes

40.2 Erwin Schrödinger (1887–1961) developed the equation that bears his name in 1926, an accomplishment for which he shared (with the British physicist P. A. M. Dirac) the 1933 Nobel Prize in physics. His grave marker is adorned with a large letter ψ.

$$\Psi(x, t) = Ae^{i(kx - \omega t)} = Ae^{ikx}e^{-i\omega t} \qquad \text{(sinusoidal wave function representing a free particle)} \qquad (40.18)$$

If k is positive in Eq. (40.16), the wave function represents a free particle moving in the positive x-direction with momentum $p = \hbar k$ and energy $E = \hbar\omega = \hbar^2 k^2/2m$. If k is negative, the momentum and hence the motion are in the negative x-direction. (With a negative value of k, the wavelength is $\lambda = 2\pi/|k|$).

Interpreting the Wave Function

The complex nature of the wave function for a free particle makes this function challenging to interpret. (We certainly haven't needed imaginary numbers before this point to describe real physical phenomena.) Here's how to think about this function: $\Psi(x, t)$ describes the *distribution* of a particle in space, just as the wave functions for an electromagnetic wave describe the distribution of the electric and magnetic fields. When we worked out interference and diffraction patterns in Chapters 35 and 36, we found that the intensity I of the radiation at any point in a pattern is proportional to the square of the electric-field magnitude—that is, to E^2. In the photon interpretation of interference and diffraction (see Section 38.4), the intensity at each point is proportional to the number of photons striking around that point or, alternatively, to the *probability* that any individual photon will strike around the point. Thus the square of the electric-field magnitude at each point is proportional to the probability of finding a photon around that point.

In exactly the same way, the square of the wave function of a particle at each point tells us about the probability of finding the particle around that point. More precisely, we should say the square of the *absolute value* of the wave function, $|\Psi|^2$. This is necessary because, as we have seen, the wave function is a complex quantity with real and imaginary parts.

For a particle that can move only along the x-direction, the quantity $|\Psi(x, t)|^2 dx$ is the probability that the particle will be found at time t at a coordinate in the range from x to $x + dx$. The particle is most likely to be found in regions where $|\Psi|^2$ is large, and so on. This interpretation, first made by the German physicist Max Born (Fig. 40.4), requires that the wave function Ψ be *normalized*. That is, the integral of $|\Psi(x, t)|^2 dx$ over all possible values of x must equal exactly 1. In other words, the probability is exactly 1, or 100%, that the particle is *somewhere*.

> **CAUTION** **Interpreting** $|\Psi|^2$ Note that $|\Psi(x, t)|^2$ itself is *not* a probability. Rather, $|\Psi(x, t)|^2 dx$ is the probability of finding the particle between position x and position $x + dx$ at time t. If the length dx is made smaller, it becomes less likely that the particle will be found within that length, so the probability decreases. A better name for $|\Psi(x, t)|^2$ is the **probability distribution function,** since it describes how the probability of finding the particle at different locations is distributed over space. Another common name for $|\Psi(x, t)|^2$ is the *probability density*.

We can use the probability interpretation of $|\Psi|^2$ to get a better understanding of Eq. (40.18), the wave function for a free particle. This function describes a particle that has a definite momentum $p = \hbar k$ in the x-direction and *no* uncertainty in momentum: $\Delta p_x = 0$. The Heisenberg uncertainty principle for position and momentum, Eq. (39.29), says that $\Delta x \Delta p_x \geq \hbar/2$. If Δp_x is zero, then Δx must be infinite, and we have no idea whatsoever where along the x-axis the particle can be found. (We saw a similar result for photons in Section 38.4.) We can show this by calculating the probability distribution function $|\Psi(x, t)|^2$. This is the product of Ψ and its *complex conjugate* Ψ^*. To find the complex conjugate of a complex number, we simply replace all i with $-i$. For example, the complex conjugate of $c = a + ib$, where a and b are real, is $c^* = a - ib$, so $|c|^2 = c^*c = (a + ib)(a - ib) = a^2 + b^2$ (recall that $i^2 = -1$). The complex conjugate of Eq. (40.18) is

$$\Psi^*(x, t) = A^* e^{-i(kx - \omega t)} = A^* e^{-ikx} e^{i\omega t}$$

(We have to allow for the possibility that the coefficient A is itself a complex number.) Hence the probability distribution function is

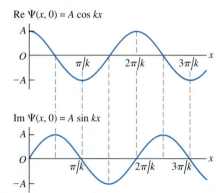

40.3 The spatial wave function $\Psi(x, t) = A e^{i(kx - \omega t)}$ for a free particle of definite momentum $p = \hbar k$ is a complex function: It has both a real part and an imaginary part. These are graphed here as functions of x for $t = 0$.

Re $\Psi(x, 0) = A \cos kx$

Im $\Psi(x, 0) = A \sin kx$

40.4 In 1926, the German physicist Max Born (1882–1970) devised the interpretation that $|\Psi|^2$ is the probability distribution function for a particle that is described by the wave function Ψ. He also coined the term "quantum mechanics" (in the original German, *Quantenmechanik*). For his contributions, Born shared (with Walther Bothe) the 1954 Nobel Prize in physics.

$$|\Psi(x, t)|^2 = \Psi^*(x, t)\Psi(x, t) = (A^* e^{-ikx} e^{i\omega t})(A e^{ikx} e^{-i\omega t})$$
$$= A^* A e^0 = |A|^2$$

The probability distribution function doesn't depend on position, which says that we are equally likely to find the particle *anywhere* along the *x*-axis! Mathematically, this is because the wave function $\Psi(x, t) = A e^{i(kx - \omega t)} = A[\cos(kx - \omega t) + i\sin(kx - \omega t)]$ is a sinusoidal function that extends all the way from $x = -\infty$ to $x = +\infty$ with the same amplitude A. This also means that the wave function can't be normalized: The integral of $|\Psi(x, t)|^2$ over all space would be infinite for any value of A.

Note also that the wave function in Eq. (40.18) describes a particle with a definite energy $E = \hbar\omega$, so there is zero uncertainty in energy: $\Delta E = 0$. The Heisenberg uncertainty principle for energy and time interval, $\Delta t \, \Delta E \geq \hbar$ [Eq. (39.30)], tells us that the time uncertainty Δt for this particle is infinite. In other words, we can have no idea *when* the particle will pass a given point on the *x*-axis. That also agrees with our result $|\Psi(x, t)|^2 = |A|^2$; the probability distribution function has the same value at all times.

Since we always have some idea of where a particle is, the wave function given in Eq. (40.18) isn't a realistic description. In our study of light in Section 38.4, we saw that we can make a wave function that's more *localized* in space by superimposing two or more sinusoidal functions. (This would be a good time to review that section.) As an illustration, let's calculate $|\Psi(x, t)|^2$ for a wave function of this kind.

Example 40.1   **A localized free-particle wave function**

The wave function $\Psi(x, t) = A e^{i(k_1 x - \omega_1 t)} + A e^{i(k_2 x - \omega_2 t)}$ is a superposition of *two* free-particle wave functions of the form given by Eq. (40.18). Both k_1 and k_2 are positive. (a) Show that this wave function satisfies the Schrödinger equation for a free particle of mass m. (b) Find the probability distribution function for $\Psi(x, t)$.

SOLUTION

IDENTIFY and SET UP: The wave functions $A e^{i(k_1 x - \omega_1 t)}$ and $A e^{i(k_2 x - \omega_2 t)}$ both represent particles moving in the positive *x*-direction, but with different momenta and kinetic energies: $p_1 = \hbar k_1$ and $E_1 = \hbar\omega_1 = \hbar^2 k_1^2/2m$ for the first function, $p_2 = \hbar k_2$ and $E_2 = \hbar\omega_2 = \hbar^2 k_2^2/2m$ for the second function. To test whether a superposition of these is also a valid wave function for a free particle, we'll see whether our function $\Psi(x, t)$ satisfies the free-particle Schrödinger equation, Eq. (40.15). It's useful to remember the derivatives of the exponential function: $(d/du)e^{au} = a e^{au}$ and $(d^2/du^2)e^{au} = a^2 e^{au}$. The probability distribution function $|\Psi(x, t)|^2$ is the product of $\Psi(x, t)$ and its complex conjugate.

EXECUTE: (a) If we substitute $\Psi(x, t)$ into Eq. (40.15), the left-hand side of the equation is

$$-\frac{\hbar^2}{2m}\frac{\partial^2 \Psi(x, t)}{\partial x^2} = -\frac{\hbar^2}{2m}\frac{\partial^2 (A e^{i(k_1 x - \omega_1 t)} + A e^{i(k_2 x - \omega_2 t)})}{\partial x^2}$$
$$= -\frac{\hbar^2}{2m}[(ik_1)^2 A e^{i(k_1 x - \omega_1 t)} + (ik_2)^2 A e^{i(k_2 x - \omega_2 t)}]$$
$$= \frac{\hbar^2 k_1^2}{2m} A e^{i(k_1 x - \omega_1 t)} + \frac{\hbar^2 k_2^2}{2m} A e^{i(k_2 x - \omega_2 t)}$$

The right-hand side is

$$i\hbar\frac{\partial \Psi(x, t)}{\partial t} = i\hbar\frac{\partial (A e^{i(k_1 x - \omega_1 t)} + A e^{i(k_2 x - \omega_2 t)})}{\partial t}$$
$$= i\hbar[(-i\omega_1)A e^{i(k_1 x - \omega_1 t)} + (-i\omega_2)A e^{i(k_2 x - \omega_2 t)}]$$
$$= \hbar\omega_1 A e^{i(k_1 x - \omega_1 t)} + \hbar\omega_2 A e^{i(k_2 x - \omega_2 t)}$$

The two sides *are* equal, provided that $\hbar\omega_1 = \hbar^2 k_1^2/2m$ and $\hbar\omega_2 = \hbar^2 k_2^2/2m$. These are just the relationships that we noted above. So we conclude that $\Psi(x, t) = A e^{i(k_1 x - \omega_1 t)} + A e^{i(k_2 x - \omega_2 t)}$ is a valid free-particle wave function. In general, if we take any two wave functions that are solutions of the Schrödinger equation and then make a superposition of these to create a third wave function $\Psi(x, t)$, then $\Psi(x, t)$ is also a solution of the Schrödinger equation.

(b) The complex conjugate of $\Psi(x, t)$ is

$$\Psi^*(x, t) = A^* e^{-i(k_1 x - \omega_1 t)} + A^* e^{-i(k_2 x - \omega_2 t)}$$

Hence

$$|\Psi(x, t)|^2$$
$$= \Psi^*(x, t)\Psi(x, t)$$
$$= (A^* e^{-i(k_1 x - \omega_1 t)} + A^* e^{-i(k_2 x - \omega_2 t)})(A e^{i(k_1 x - \omega_1 t)} + A e^{i(k_2 x - \omega_2 t)})$$
$$= A^* A \begin{bmatrix} e^{-i(k_1 x - \omega_1 t)} e^{i(k_1 x - \omega_1 t)} + e^{-i(k_2 x - \omega_2 t)} e^{i(k_2 x - \omega_2 t)} \\ + e^{-i(k_1 x - \omega_1 t)} e^{i(k_2 x - \omega_2 t)} + e^{-i(k_2 x - \omega_2 t)} e^{i(k_1 x - \omega_1 t)} \end{bmatrix}$$
$$= |A|^2 [e^0 + e^0 + e^{i[(k_2 - k_1)x - (\omega_2 - \omega_1)t]} + e^{-i[(k_2 - k_1)x - (\omega_2 - \omega_1)t]}]$$

To simplify this expression, recall that $e^0 = 1$. From Euler's formula, $e^{i\theta} = \cos\theta + i\sin\theta$ and $e^{-i\theta} = \cos\theta - i\sin\theta$, so $e^{i\theta} + e^{-i\theta} = 2\cos\theta$. Hence

$$|\Psi(x, t)|^2 = |A|^2 \{2 + 2\cos[(k_2 - k_1)x - (\omega_2 - \omega_1)t]\}$$
$$= 2|A|^2 \{1 + \cos[(k_2 - k_1)x - (\omega_2 - \omega_1)t]\}$$

EVALUATE: Figure 40.5 is a graph of the probability distribution function $|\Psi(x, t)|^2$ at $t = 0$. The value of $|\Psi(x, t)|^2$ varies between 0 and $4|A|^2$; probabilities can never be negative! The particle has become *somewhat* localized: The particle is most likely to be found near a point where $|\Psi(x, t)|^2$ is maximum (where the functions $Ae^{i(k_1x-\omega_1t)}$ and $Ae^{i(k_2x-\omega_2t)}$ interfere constructively) and is very unlikely to be found near a point where $|\Psi(x, t)|^2 = 0$ (where $Ae^{i(k_1x-\omega_1t)}$ and $Ae^{i(k_2x-\omega_2t)}$ interfere destructively). This is very similar to the phenomenon of beats for sound waves (see Section 16.7).

Note also that the probability distribution function is not stationary, but moves in the positive x-direction like the particle that it represents. To see this, recall from Section 15.3 that a sinusoidal wave given by $y(x, t) = A\cos(kx - \omega t)$ moves in the positive x-direction with velocity $v = \omega/k$; since $|\Psi(x, t)|^2$ includes a term $\cos[(k_2 - k_1)x - (\omega_2 - \omega_1)t]$, the probability distribution moves at a velocity $v_{av} = (\omega_2 - \omega_1)/(k_2 - k_1)$. The subscript "av" reminds us that v_{av} represents the *average* value of the particle's velocity.

The price we pay for localizing the particle somewhat is that, unlike a particle represented by Eq. (40.18), it no longer has either a definite momentum or a definite energy. That's consistent with the Heisenberg uncertainty principles: If we decrease the uncertainties about where a particle is and when it passes a certain point, the uncertainties in its momentum and energy must increase.

The average momentum of the particle is $p_{av} = (\hbar k_2 + \hbar k_1)/2$, the average of the momenta associated with the free-particle wave functions we added to create $\Psi(x, t)$. This corresponds to the particle having an average velocity $v_{av} = p_{av}/m = (\hbar k_2 + \hbar k_1)/2m$. Can you show that this is equal to the expression $v_{av} = (\omega_2 - \omega_1)/(k_2 - k_1)$ that we found above?

40.5 The probability distribution function at $t = 0$ for $\Psi(x, t) = Ae^{i(k_1x-\omega_1t)} + Ae^{i(k_2x-\omega_2t)}$.

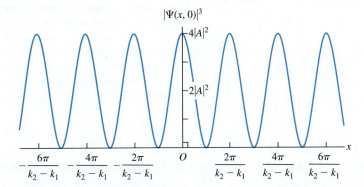

Wave Packets

The wave function that we examined in Example 40.1 is not very well localized: The probability distribution function still extends from $x = -\infty$ to $x = +\infty$. Hence this wave function can't be normalized, either. To make a wave function that's more highly localized, imagine superposing two additional sinusoidal waves with different wave numbers and amplitudes so as to reinforce alternate maxima of $|\Psi(x, t)|^2$ in Fig. 40.5 and cancel out the in-between ones. Finally, if we superpose waves with a very large number of different wave numbers, we can construct a wave with only *one* maximum of $|\Psi(x, t)|^2$ (Fig. 40.6). Then, finally, we have something that begins to look like both a particle and a wave. It is a particle in the sense that it is localized in space; if we look from a distance, it may look like a point. But it also has a periodic structure that is characteristic of a wave.

A localized wave pulse like that shown in Fig. 40.6 is called a **wave packet.** We can represent a wave packet by an expression such as

$$\Psi(x, t) = \int_{-\infty}^{\infty} A(k)e^{i(kx-\omega t)}\, dk \qquad (40.19)$$

This integral represents a superposition of a very large number of waves, each with a different wave number k and angular frequency $\omega = \hbar k^2/2m$, and each with an amplitude $A(k)$ that depends on k.

There is an important relationship between the two functions $\Psi(x, t)$ and $A(k)$, which we show qualitatively in Fig. 40.7. If the function $A(k)$ is sharply peaked, as in Fig. 40.7a, we are superposing only a narrow range of wave numbers. The resulting wave pulse is then relatively broad (Fig. 40.7b). But if we use

40.6 Superposing a large number of sinusoidal waves with different wave numbers and appropriate amplitudes can produce a wave pulse that has a wavelength $\lambda_{av} = 2\pi/k_{av}$ and is localized within a region of space of length Δx. This localized pulse has aspects of both particle and wave.

(a) Real part of the wave function at time t

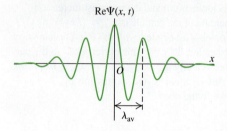

$\text{Re}\,\Psi(x, t)$

λ_{av}

(b) Imaginary part of the wave function at time t

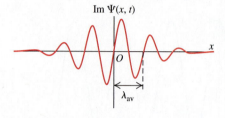

$\text{Im}\,\Psi(x, t)$

λ_{av}

(c) Probability distribution function at time t

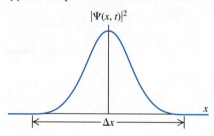

$|\Psi(x, t)|^2$

Δx

a wider range of wave numbers, so that the function $A(k)$ is broader (Fig. 40.7c), then the wave pulse is more narrowly localized (Fig. 40.7d). This is simply the uncertainty principle in action. A narrow range of k means a narrow range of $p_x = \hbar k$ and thus a small Δp_x; the result is a relatively large Δx. A broad range of k corresponds to a large Δp_x, and the resulting Δx is smaller. You can see that the uncertainty principle for position and momentum, $\Delta x \Delta p_x \geq \hbar/2$, is really just a consequence of the properties of integrals like Eq. (40.19).

CAUTION **Matter waves versus light waves in vacuum** We can regard both a wave packet that represents a particle and a short pulse of light from a laser as superpositions of waves of different wave numbers and angular frequencies. An important difference is that the speed of light in vacuum is the same for all wavelengths λ and hence all wave numbers $k = 2\pi/\lambda$, but the speed of a matter wave is *different* for different wavelengths. You can see this from the formula for the speed of the wave crests in a periodic wave, $v = \lambda f = \omega/k$. For a matter wave, $\omega = \hbar k^2/2m$, so $v = \hbar k/2m = h/2m\lambda$. Hence matter waves with longer wavelengths and smaller wave numbers travel more slowly than those with short wavelengths and large wave numbers. (This shouldn't be too surprising. The de Broglie relations that we learned in Section 39.1 tell us that shorter wavelength corresponds to greater momentum and hence a greater speed.) Since the individual sinusoidal waves that make up a wave packet travel at different speeds, the shape of the packet changes as it moves. That's why we've specified the time for which the wave packets in Figs. 40.6 and 40.7 are drawn; at later times, the packets become more spread out. By contrast, a pulse of light waves in vacuum retains the same shape at all times because all of its constituent sinusoidal waves travel together at the same speed. ▮

The One-Dimensional Schrödinger Equation with Potential Energy

The one-dimensional Schrödinger equation that we presented in Eq. (40.15) is valid only for free particles, for which the potential energy function is zero: $U(x) = 0$. But for an electron within an atom, a proton within an atomic nucleus, and many other real situations, the potential energy plays an important role. To study the behavior of matter waves in these situations, we need a version of the Schrödinger equation that describes a particle moving in the presence of a nonzero potential energy function $U(x)$. This equation is

$$-\frac{\hbar^2}{2m}\frac{\partial^2 \Psi(x, t)}{\partial x^2} + U(x)\Psi(x, t) = i\hbar\frac{\partial \Psi(x, t)}{\partial t} \qquad \text{(general one-dimensional Schrödinger equation)} \qquad (40.20)$$

Note that if $U(x) = 0$, Eq. (40.20) reduces to the free-particle Schrödinger equation given in Eq. (40.15).

Here's the motivation behind Eq. (40.20). If $\Psi(x, t)$ is a sinusoidal wave function for a free particle, $\Psi(x, t) = Ae^{i(kx-\omega t)} = Ae^{ikx}e^{-i\omega t}$, the derivative terms in Eq. (40.20) become

$$-\frac{\hbar^2}{2m}\frac{\partial^2 \Psi(x, t)}{\partial x^2} = -\frac{\hbar^2}{2m}\frac{\partial^2}{\partial x^2}(Ae^{ikx}e^{-i\omega t}) = -\frac{\hbar^2}{2m}(ik)^2(Ae^{ikx}e^{-i\omega t})$$

$$= \frac{\hbar^2 k^2}{2m}\Psi(x, t)$$

$$i\hbar\frac{\partial \Psi(x, t)}{\partial t} = i\hbar\frac{\partial}{\partial t}(Ae^{ikx}e^{-i\omega t}) = i\hbar(-i\omega)(Ae^{ikx}e^{-i\omega t}) = \hbar\omega\Psi(x, t)$$

In these expressions $(\hbar^2 k^2/2m)\Psi(x, t)$ is just the kinetic energy $K = p^2/2m = \hbar^2 k^2/2m$ multiplied by the wave function, and $\hbar\omega\Psi(x, t)$ is the total energy $E = \hbar\omega$ multiplied by the wave function. So for a wave function of this kind, Eq. (40.20) says that kinetic energy times $\Psi(x, t)$ plus potential energy times $\Psi(x, t)$ equals total energy times $\Psi(x, t)$. That's equivalent to the statement in classical physics that the sum of kinetic energy and potential energy equals total mechanical energy: $K + U = E$.

The observations we've just made certainly aren't a *proof* that Eq. (40.20) is correct. The real reason we know this equation *is* correct is that it works: Predictions made with this equation agree with experimental results. In the remaining sections of this chapter we'll apply Eq. (40.20) to several physical situations, each with a different form of the function $U(x)$.

Stationary States

We saw in our discussion of wave packets that any free-particle wave function can be built up as a superposition of sinusoidal wave functions of the form $\Psi(x, t) = Ae^{ikx}e^{-i\omega t}$. Each such sinusoidal wave function corresponds to a state of definite energy $E = \hbar\omega = \hbar^2 k^2/2m$ and definite angular frequency $\omega = E/\hbar$, so we can rewrite these functions as $\Psi(x, t) = Ae^{ikx}e^{-iEt/\hbar}$. If the potential energy function $U(x)$ is nonzero, these sinusoidal wave functions do not satisfy the Schrödinger equation, Eq. (40.20), and so these functions cannot be the basic "building blocks" of more complicated wave functions. However, we can still write the wave function for a state of definite energy E in the form

$$\Psi(x, t) = \psi(x)e^{-iEt/\hbar} \qquad \text{(time-dependent wave function for a state of definite energy)} \qquad (40.21)$$

That is, the wave function $\Psi(x, t)$ for a state of definite energy is the product of a *time-independent* wave function $\psi(x)$ and a factor $e^{-iEt/\hbar}$. (For the free-particle sinusoidal wave function, $\psi(x) = Ae^{ikx}$.) States of definite energy are of tremendous importance in quantum mechanics. For example, for each energy level in a hydrogen atom (Section 39.3) there is a specific wave function. It is possible for an atom to be in a state that does not have a definite energy. The wave function for any such state can be written as a combination of definite-energy wave functions, in precisely the same way that a free-particle wave packet can be written as a superposition of sinusoidal wave functions of definite energy as in Eq. (40.19).

A state of definite energy is commonly called a **stationary state.** To see where this name comes from, let's multiply Eq. (40.21) by its complex conjugate to find the probabilty distribution function $|\Psi|^2$:

$$|\Psi(x, t)|^2 = \Psi^*(x, t)\Psi(x, t) = [\psi^*(x)e^{+iEt/\hbar}][\psi(x)e^{-iEt/\hbar}]$$
$$= \psi^*(x)\psi(x)e^{(+iEt/\hbar)+(-iEt/\hbar)} = |\psi(x)|^2 e^0 \qquad (40.22)$$
$$= |\psi(x)|^2$$

Since $|\psi(x)|^2$ does not depend on time, Eq. (40.22) shows that the same must be true for the probability distribution function $|\Psi(x, t)|^2$. This justifies the term "stationary state" for a state of definite energy.

CAUTION **A stationary state does not mean a stationary particle** The name *stationary state* may lead you to think that the particle is not in motion if it is described by such a wave function. That's not the case. It's the *probability distribution* (that is, the relative likelihood of finding the particle at various positions), not the particle itself, that's stationary.

The Schrödinger equation, Eq. (40.20), becomes quite a bit simpler for stationary states. To see this, we substitute Eq. (40.21) into Eq. (40.20):

$$-\frac{\hbar^2}{2m}\frac{\partial^2[\psi(x)e^{-iEt/\hbar}]}{\partial x^2} + U(x)\psi(x)e^{-iEt/\hbar} = i\hbar\frac{\partial[\psi(x)e^{-iEt/\hbar}]}{\partial t}$$

40.7 How varying the function $A(k)$ in the wave-packet expression, Eq. (40.19), changes the character of the wave function $\Psi(x, t)$ (shown here at a specific time $t = 0$).

(a)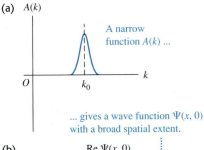
A narrow function $A(k)$ ...

... gives a wave function $\Psi(x, 0)$ with a broad spatial extent.

(b)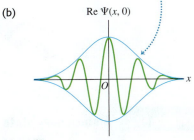

(c)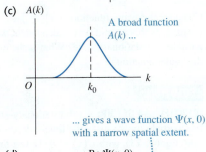
A broad function $A(k)$ ...

... gives a wave function $\Psi(x, 0)$ with a narrow spatial extent.

(d)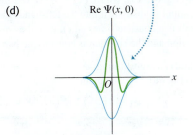

Mastering PHYSICS

PhET: Quantum Tunneling and Wave Packets
ActivPhysics 17.7: Wave Packets

The derivative on the first term on the left-hand side is with respect to x, so the factor of $e^{-iEt/\hbar}$ comes outside of the derivative. Now we take the derivative with respect to t on the right-hand side of the equation:

$$-\frac{\hbar^2}{2m}\frac{d^2\psi(x)}{dx^2}e^{-iEt/\hbar} + U(x)\psi(x)e^{-iEt/\hbar} = i\hbar\left(\frac{-iE}{\hbar}\right)[\psi(x)e^{-iEt/\hbar}]$$

$$= E\psi(x)e^{-iEt/\hbar}$$

If we divide both sides of this equation by $e^{-iEt/\hbar}$, we get

$$-\frac{\hbar^2}{2m}\frac{d^2\psi(x)}{dx^2} + U(x)\psi(x) = E\psi(x) \qquad \text{(time-independent Schrödinger equation)} \qquad (40.23)$$

This is called the **time-independent Schrödinger equation.** The time-dependent factor $e^{-iEt/\hbar}$ does not appear, and Eq. (40.23) is an equation that involves only the time-independent wave function $\psi(x)$. We'll devote much of this chapter to solving this equation to find the definite-energy, stationary-state wave functions $\psi(x)$ and the corresponding values of E—that is, the energies of the allowed levels—for different physical situations.

Example 40.2   A stationary state

Consider the wave function $\psi(x) = A_1e^{ikx} + A_2e^{-ikx}$, where k is positive. Is this a valid time-independent wave function for a free particle in a stationary state? What is the energy corresponding to this wave function?

SOLUTION

IDENTIFY and SET UP: A valid stationary-state wave function for a free particle must satisfy the time-independent Schrödinger equation, Eq. (40.23), with $U(x) = 0$. To test the given function $\psi(x)$, we simply substitute it into the left-hand side of the equation. If the result is a constant times $\psi(x)$, then the wave function is indeed a solution and the constant is equal to the particle energy E.

EXECUTE: Substituting $\psi(x) = A_1e^{ikx} + A_2e^{-ikx}$ and $U(x) = 0$ into Eq. (40.23), we obtain

$$-\frac{\hbar^2}{2m}\frac{d^2\psi(x)}{dx^2} = -\frac{\hbar^2}{2m}\frac{d^2(A_1e^{ikx} + A_2e^{-ikx})}{dx^2}$$

$$= -\frac{\hbar^2}{2m}[(ik)^2A_1e^{ikx} + (-ik)^2A_2e^{ikx}]$$

$$= \frac{\hbar^2k^2}{2m}(A_1e^{ikx} + A_2e^{-ikx}) = \frac{\hbar^2k^2}{2m}\psi(x)$$

The result is a constant times $\psi(x)$, so this $\psi(x)$ is indeed a valid stationary-state wave function for a free particle. Comparing with Eq. (40.23) shows that the constant on the right-hand side is the particle energy: $E = \hbar^2k^2/2m$.

EVALUATE: Note that $\psi(x)$ is a *superposition* of two different wave functions: one function (A_1e^{ikx}) that represents a particle with magnitude of momentum $p = \hbar k$ moving in the positive x-direction, and one function (A_2e^{-ikx}) that represents a particle with the same magnitude of momentum moving in the negative x-direction. So while the combined wave function $\psi(x)$ represents a stationary state with a definite energy, this state does *not* have a definite momentum. We'll see in Section 40.2 that such a wave function can represent a *standing wave,* and we'll explore situations in which such standing matter waves can arise.

Test Your Understanding of Section 40.1   Does a wave packet given by Eq. (40.19) represent a stationary state?

40.2   Particle in a Box

An important problem in quantum mechanics is how to use the time-independent Schrödinger equation, Eq. (40.23), to determine the possible energy levels and the corresponding wave functions for various systems. The fundamental problem is then the following: For a given potential energy function $U(x)$, what are the possible stationary-state wave functions $\psi(x)$, and what are the corresponding energies E?

In Section 40.1 we solved this problem for the case $U(x) = 0$, corresponding to a *free* particle. The allowed wave functions and corresponding energies are

$$\psi(x) = Ae^{ikx} \quad E = \frac{\hbar^2 k^2}{2m} \quad \text{(free particle)} \quad (40.24)$$

The wave number k is equal to $2\pi/\lambda$, where λ is the wavelength. We found that k can have any real value, so the energy E of a free particle can have any value from zero to infinity. Furthermore, the particle can be found with equal probability at any value of x from $-\infty$ to $+\infty$.

Now let's look at a simple model in which a particle is *bound* so that it cannot escape to infinity, but rather is confined to a restricted region of space. Our system consists of a particle confined between two rigid walls separated by a distance L (Fig. 40.8). The motion is purely one dimensional, with the particle moving along the x-axis only and the walls at $x = 0$ and $x = L$. The potential energy corresponding to the rigid walls is infinite, so the particle cannot escape; between the walls, the potential energy is zero (Fig. 40.9). This situation is often described as a **"particle in a box."** This model might represent an electron that is free to move within a long, straight molecule or along a very thin wire.

Wave Functions for a Particle in a Box

To solve the Schrödinger equation for this system, we begin with some restrictions on the particle's stationary-state wave function $\psi(x)$. Because the particle is confined to the region $0 \le x \le L$, we expect the probability distribution function $|\Psi(x, t)|^2 = |\psi(x)|^2$ and the wave function $\psi(x)$ to be zero outside that region. This agrees with the Schrödinger equation: If the term $U(x)\psi(x)$ in Eq. (40.23) is to be finite, then $\psi(x)$ must be zero where $U(x)$ is infinite.

Furthermore, $\psi(x)$ must be a *continuous* function to be a mathematically well-behaved solution to the Schrödinger equation. This implies that $\psi(x)$ must be zero at the region's boundary, $x = 0$ and $x = L$. These two conditions serve as *boundary conditions* for the problem. They should look familiar, because they are the same conditions that we used to find the normal modes of a vibrating string in Section 15.8 (Fig. 40.10); you should review that discussion.

An additional condition is that to calculate the second derivative $d^2\psi(x)/dx^2$ in Eq. (40.23), the *first* derivative $d\psi(x)/dx$ must also be continuous except at points where the potential energy becomes infinite (as it does at the walls of the box). This is analogous to the requirement that a vibrating string, like those shown in Fig. 40.10, can't have any kinks in it (which would correspond to a discontinuity in the first derivative of the wave function) except at the ends of the string.

We now solve for the wave functions in the region $0 \le x \le L$ subject to the above conditions. In this region $U(x) = 0$, so the wave function in this region must satisfy

$$-\frac{\hbar^2}{2m}\frac{d^2\psi(x)}{dx^2} = E\psi(x) \quad \text{(particle in a box)} \quad (40.25)$$

Equation (40.25) is the *same* Schrödinger equation as for a free particle, so it is tempting to conclude that the wave functions and energies are given by Eq. (40.24). It is true that $\psi(x) = Ae^{ikx}$ satisfies the Schrödinger equation with $U(x) = 0$, is continuous, and has a continuous first derivative $d\psi(x)/dx = ikAe^{ikx}$. However, this wave function does *not* satisfy the boundary conditions that $\psi(x)$ must be zero at $x = 0$ and $x = L$: At $x = 0$ the wave function in Eq. (40.24) is equal to $Ae^0 = A$, and at $x = L$ it is equal to Ae^{ikL}. (These would be equal to zero if $A = 0$, but then the wave function would be zero and there would be no particle at all!)

The way out of this dilemma is to recall Example 40.2 (Section 40.1), in which we found that a more general stationary-state solution to the time-independent Schrödinger equation with $U(x) = 0$ is

$$\psi(x) = A_1 e^{ikx} + A_2 e^{-ikx} \quad (40.26)$$

40.8 The Newtonian view of a particle in a box.

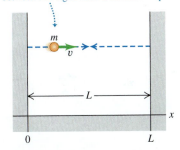

A particle with mass m moves along a straight line at constant speed, bouncing between two rigid walls a distance L apart.

40.9 The potential-energy function for a particle in a box.

The potential energy U is zero in the interval $0 < x < L$ and is infinite everywhere outside this interval.

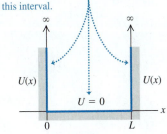

40.10 Normal modes of vibration for a string with length L, held at both ends.

Each end is a node, and there are $n - 1$ additional nodes between the ends.

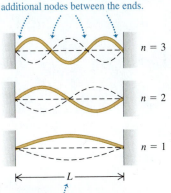

The length is an integral number of half-wavelengths: $L = n\lambda_n/2$.

This wave function is a superposition of two waves: one traveling in the $+x$-direction of amplitude A_1, and one traveling in the $-x$-direction with the same wave number but amplitude A_2. This is analogous to a standing wave on a string (Fig. 40.10), which we can regard as the superposition of two sinusoidal waves propagating in opposite directions (see Section 15.7). The energy that corresponds to Eq. (40.26) is $E = \hbar^2 k^2/2m$, just as for a single wave.

To see whether the wave function given by Eq. (40.26) can satisfy the boundary conditions, let's first rewrite it in terms of sines and cosines using Euler's formula, Eq. (40.17):

$$\psi(x) = A_1(\cos kx + i\sin kx) + A_2[\cos(-kx) + i\sin(-kx)]$$

$$= A_1(\cos kx + i\sin kx) + A_2(\cos kx - i\sin kx) \qquad (40.27)$$

$$= (A_1 + A_2)\cos kx + i(A_1 - A_2)\sin kx$$

At $x = 0$ this is equal to $\psi(0) = A_1 + A_2$, which must equal zero if we are to satisfy the boundary condition at that point. Hence $A_2 = -A_1$, and Eq. (40.27) becomes

$$\psi(x) = 2iA_1\sin kx = C\sin kx \qquad (40.28)$$

We have simplified the expression by introducing the constant $C = 2iA_1$. (We'll come back to this constant later.) We can also satisfy the second boundary condition that $\psi = 0$ at $x = L$ by choosing values of k such that $kL = n\pi$ ($n = 1, 2, 3, \dots$). Hence Eq. (40.28) does indeed give the stationary-state wave functions for a particle in a box in the region $0 \leq x \leq L$. (Outside this region, $\psi(x) = 0$.) The possible values of k and the wavelength $\lambda = 2\pi/k$ are

$$k = \frac{n\pi}{L} \quad \text{and} \quad \lambda = \frac{2\pi}{k} = \frac{2L}{n} \quad (n = 1, 2, 3, \dots) \qquad (40.29)$$

Just as for the string in Fig. 40.10, the length L of the region is an integral number of half-wavelengths.

Energy Levels for a Particle in a Box

The possible energy levels for a particle in a box are given by $E = \hbar^2 k^2/2m = p^2/2m$, where $p = \hbar k = (h/2\pi)(2\pi/\lambda) = h/\lambda$ is the magnitude of momentum of a free particle with wave number k and wavelength λ. This makes sense, since inside the region $0 \leq x \leq L$ the potential energy is zero and the energy is all kinetic. For each value of n, there are corresponding values of p, λ, and E; let's call them p_n, λ_n, and E_n. Putting the pieces together, we get

$$p_n = \frac{h}{\lambda_n} = \frac{nh}{2L} \qquad (40.30)$$

and so the energy levels for a particle in a box are

$$E_n = \frac{p_n^2}{2m} = \frac{n^2 h^2}{8mL^2} = \frac{n^2 \pi^2 \hbar^2}{2mL^2} \quad (n = 1, 2, 3, \dots) \qquad \begin{array}{l}\text{(energy levels,} \\ \text{particle in a box)}\end{array} \quad (40.31)$$

Each energy level has its own value of the quantum number n and a corresponding wave function, which we denote by ψ_n. When we replace k in Eq. (40.28) by $n\pi/L$ from Eq. (40.29), we find

$$\psi_n(x) = C\sin\frac{n\pi x}{L} \quad (n = 1, 2, 3, \dots) \qquad (40.32)$$

The energy-level diagram in Fig. 40.11a shows the five lowest levels for a particle in a box. The energy levels are proportional to n^2, so successively higher levels are spaced farther and farther apart. There are an infinite number of levels because the walls are perfectly rigid; even a particle of infinitely great kinetic

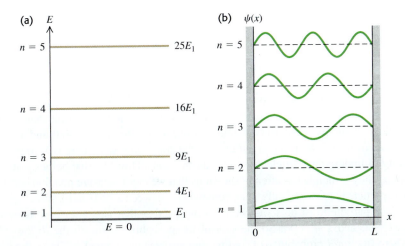

40.11 (a) Energy-level diagram for a particle in a box. Each energy is $n^2 E_1$, where E_1 is the ground-level energy. (b) Wave functions for a particle in a box, with $n = 1, 2, 3, 4,$ and 5. **CAUTION:** The five graphs have been displaced vertically for clarity, as in Fig. 40.10. Each of the horizontal dashed lines represents $\psi = 0$ for the respective wave function.

energy is confined within the box. Figure 40.11b shows graphs of the wave functions $\psi_n(x)$ for $n = 1, 2, 3, 4,$ and 5. Note that these functions look identical to those for a standing wave on a string (see Fig. 40.10).

CAUTION **A particle in a box cannot have zero energy** Note that the energy of a particle in a box *cannot* be zero. Equation (40.31) shows that $E = 0$ would require $n = 0$, but substituting $n = 0$ into Eq. (40.32) gives a zero wave function. Since a particle is described by a *nonzero* wave function, this means that there cannot be a particle with $E = 0$. This is a consequence of the Heisenberg uncertainty principle: A particle in a zero-energy state would have a definite value of momentum (precisely zero), so its position uncertainty would be infinite and the particle could be found anywhere along the x-axis. But this is impossible, since a particle in a box can be found only between $x = 0$ and $x = L$. Hence $E = 0$ is not allowed. By contrast, the allowed stationary-state wave functions with $n = 1, 2, 3, \ldots$ do not represent states of definite momentum (each is an equal mixture of a state of x-momentum $+p_n = nh/2L$ and a state of x-momentum $-p_n = -nh/2L$). Hence each stationary state has a nonzero momentum uncertainty, consistent with having a finite position uncertainty.

Example 40.3   **Electron in an atom-size box**

Find the first two energy levels for an electron confined to a one-dimensional box 5.0×10^{-10} m across (about the diameter of an atom).

SOLUTION

IDENTIFY and SET UP: This problem uses what we have learned in this section about a particle in a box. The first two energy levels correspond to $n = 1$ and $n = 2$ in Eq. (40.31).

EXECUTE: From Eq. (40.31),

$$E_1 = \frac{h^2}{8mL^2} = \frac{(6.626 \times 10^{-34} \text{ J} \cdot \text{s})^2}{8(9.109 \times 10^{-31} \text{ kg})(5.0 \times 10^{-10} \text{ m})^2}$$

$$= 2.4 \times 10^{-19} \text{ J} = 1.5 \text{ eV}$$

$$E_2 = \frac{2^2 h^2}{8mL^2} = 4E_1 = 9.6 \times 10^{-19} \text{ J} = 6.0 \text{ eV}$$

EVALUATE: The difference between the first two energy levels is $E_2 - E_1 = 4.5$ eV. An electron confined to a box is different from an electron bound in an atom, but it is reassuring that this result is of the same order of magnitude as the difference between actual atomic energy levels.

You can also show that for a proton or neutron ($m = 1.67 \times 10^{-27}$ kg) confined to a box 1.1×10^{-14} m across (the width of a medium-sized atomic nucleus), the energies of the first two levels are about a million times larger: $E_1 = 1.7 \times 10^6$ eV $= 1.7$ MeV, $E_2 = 4E_1 = 6.8$ MeV, $E_2 - E_1 = 5.1$ MeV. This suggests why nuclear reactions (which involve transitions between energy levels in nuclei) release so much more energy than chemical reactions (which involve transitions between energy levels of electrons in atoms).

Finally, you can show (see Exercise 40.11) that the energy levels of a billiard ball ($m = 0.2$ kg) confined to a box 1.3 m across—the width of a billiard table—are separated by about 5×10^{-67} J. Quantum effects won't disturb a game of billiards.

Probability and Normalization

Let's look a bit more closely at the wave functions for a particle in a box, keeping in mind the *probability* interpretation of the wave function ψ that we discussed in Section 40.1. In our one-dimensional situation the quantity $|\psi(x)|^2 \, dx$ is proportional

40.12 Graphs of (a) $\psi(x)$ and (b) $|\psi(x)|^2$ for the first three wave functions ($n = 1, 2, 3$) for a particle in a box. The horizontal dashed lines represent $\psi(x) = 0$ and $|\psi(x)|^2 = 0$ for each of the three levels. The value of $|\psi(x)|^2 dx$ at each point is the probability of finding the particle in a small interval dx about the point. As in Fig. 40.11b, the three graphs in each part have been displaced vertically for clarity.

(a) $\psi(x)$

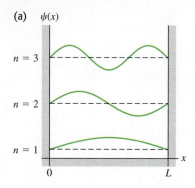

(b) $|\psi(x)|^2$

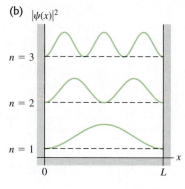

to the probability that the particle will be found within a small interval dx about x. For a particle in a box,

$$|\psi(x)|^2 dx = C^2 \sin^2 \frac{n\pi x}{L} dx$$

Figure 40.12 shows graphs of both $\psi(x)$ and $|\psi(x)|^2$ for $n = 1, 2,$ and 3. Note that not all positions are equally likely. By contrast, in classical mechanics the particle is equally likely to be found at any position between $x = 0$ and $x = L$. We see from Fig. 40.12b that $|\psi(x)|^2 = 0$ at some points, so there is zero probability of finding the particle at exactly these points. Don't let that bother you; the uncertainty principle has already shown us that we can't measure position exactly. The particle is localized only to be somewhere between $x = 0$ and $x = L$.

The particle must be *somewhere* on the x-axis—that is, somewhere between $x = -\infty$ and $x = +\infty$. So the *sum* of the probabilities for all the dx's everywhere (the *total* probability of finding the particle) must equal 1. That's the normalization condition that we discussed in Section 40.1:

$$\int_{-\infty}^{\infty} |\psi(x)|^2 \, dx = 1 \qquad \text{(normalization condition)} \qquad (40.33)$$

A wave function is said to be *normalized* if it has a constant such as C in Eq. (40.32) that is calculated to make the total probability equal 1 in Eq. (40.33). For a normalized wave function, $|\psi(x)|^2 dx$ is not merely proportional to, but *equals*, the probability of finding the particle between the coordinates x and $x + dx$. That's why we call $|\psi(x)|^2$ the probability distribution function. (In Section 40.1 we called $|\Psi(x, t)|^2$ the probability distribution function. For the case of a stationary-state wave function, however, $|\Psi(x, t)|^2$ is equal to $|\psi(x)|^2$.)

Let's normalize the particle-in-a-box wave functions $\psi_n(x)$ given by Eq. (40.32). Since $\psi_n(x)$ is zero except between $x = 0$ and $x = L$, Eq. (40.33) becomes

$$\int_0^L C^2 \sin^2 \frac{n\pi x}{L} \, dx = 1 \qquad (40.34)$$

You can evaluate this integral using the trigonometric identity $\sin^2 \theta = \frac{1}{2}(1 - \cos 2\theta)$; the result is $C^2 L/2$. Thus our probability interpretation of the wave function demands that $C^2 L/2 = 1$, or $C = (2/L)^{1/2}$; the constant C is *not* arbitrary. (This is in contrast to the classical vibrating string problem, in which C represents an amplitude that depends on initial conditions.) Thus the normalized stationary-state wave functions for a particle in a box are

$$\psi_n(x) = \sqrt{\frac{2}{L}} \sin \frac{n\pi x}{L} \qquad (n = 1, 2, 3, \dots) \qquad \text{(particle in a box)} \qquad (40.35)$$

Example 40.4 **A nonsinusoidal wave function?**

(a) Show that $\psi(x) = Ax + B$, where A and B are constants, is a solution of the Schrödinger equation for an $E = 0$ energy level of a particle in a box. (b) What constraints do the boundary conditions at $x = 0$ and $x = L$ place on the constants A and B?

SOLUTION

IDENTIFY and SET UP: To be physically reasonable, a wave function must satisfy both the Schrödinger equation and the appropriate boundary conditions. In part (a) we'll substitute $\psi(x)$ into the

Schrödinger equation for a particle in a box, Eq. (40.25), to determine whether it is a solution. In part (b) we'll see what restrictions on $\psi(x)$ arise from applying the boundary conditions that $\psi(x) = 0$ at $x = 0$ and $x = L$.

EXECUTE: (a) From Eq. (40.25), the Schrödinger equation for an $E = 0$ energy level of a particle in a box is

$$-\frac{\hbar^2}{2m} \frac{d^2\psi(x)}{dx^2} = E\psi(x) = 0$$

in the region $0 \le x \le L$. Differentiating $\psi(x) = Ax + B$ twice with respect to x gives $d^2\psi(x)/dx^2 = 0$, so the left side of the equation is zero, and so $\psi(x) = Ax + B$ is a solution of this Schrödinger equation for $E = 0$. (Note that both $\psi(x)$ and its derivative $d\psi(x)/dx = A$ are continuous functions, as they must be.)

(b) Applying the boundary condition at $x = 0$ gives $\psi(0) = B = 0$, and so $\psi(x) = Ax$. Applying the boundary condition at $x = L$ gives $\psi(L) = AL = 0$, so $A = 0$. Hence $\psi(x) = 0$ both

inside the box ($0 \le x \le L$) *and outside:* There is *zero* probability of finding the particle anywhere with this wave function, and so $\psi(x) = Ax + B$ is *not* a physically valid wave function.

EVALUATE: The moral is that there are many functions that satisfy the Schrödinger equation for a given physical situation, but most of these—including the function considered here—have to be rejected because they don't satisfy the appropriate boundary conditions.

Time Dependence

Finally, we note that the wave functions $\psi_n(x)$ in Eq. (40.35) depend only on the *spatial* coordinate x. Equation (40.21) shows that if $\psi(x)$ is the wave function for a state of definite energy E, the full time-dependent wave function is $\Psi(x, t) = \psi(x)e^{-iEt/\hbar}$. Hence the *time-dependent* stationary-state wave functions for a particle in a box are

$$\Psi_n(x, t) = \sqrt{\frac{2}{L}} \sin\left(\frac{n\pi x}{L}\right)e^{-iE_n t/\hbar} \qquad (n = 1, 2, 3, \dots) \qquad (40.36)$$

In this expression the energies E_n are given by Eq. (40.31). The higher the quantum number n, the greater the angular frequency $\omega_n = E_n/\hbar$ at which the wave function oscillates. Note that since $|e^{-iE_n t/\hbar}|^2 = e^{+iE_n t/\hbar}e^{-iE_n t/\hbar} = e^0 = 1$, the probability distribution function $|\Psi_n(x, t)|^2 = (2/L)\sin^2(n\pi x/L)$ is independent of time and does *not* oscillate. (Remember, this is why we say that these states of definite energy are *stationary*.)

Test Your Understanding of Section 40.2   If a particle in a box is in the nth energy level, what is the average value of its x-component of momentum p_x? (i) $nh/2L$; (ii) $(\sqrt{2}/2)nh/L$; (iii) $(1/\sqrt{2})nh/L$; (iv) $[1/(2\sqrt{2})]nh/L$; (v) zero.

40.3 Potential Wells

A **potential well** is a potential-energy function $U(x)$ that has a minimum. We introduced this term in Section 7.5, and we also used it in our discussion of periodic motion in Chapter 14. In Newtonian mechanics a particle trapped in a potential well can vibrate back and forth with periodic motion. Our first application of the Schrödinger equation, the particle in a box, involved a rudimentary potential well with a function $U(x)$ that is zero within a certain interval and infinite everywhere else. As we mentioned in Section 40.2, this function corresponds to a few situations found in nature, but the correspondence is only approximate.

A better approximation to several actual physical situations is a **finite well**, which is a potential well with straight sides but *finite* height. Figure 40.13 shows a potential-energy function that is zero in the interval $0 \le x \le L$ and has the value U_0 outside this interval. This function is often called a **square-well potential**. It could serve as a simple model of an electron within a metallic sheet with thickness L, moving perpendicular to the surfaces of the sheet. The electron can move freely inside the metal but has to climb a potential-energy barrier with height U_0 to escape from either surface of the metal. The energy U_0 is related to the *work function* that we discussed in Section 38.1 in connection with the photoelectric effect. In three dimensions, a spherical version of a finite well gives an approximate description of the motions of protons and neutrons within a nucleus.

Bound States of a Square-Well Potential

In Newtonian mechanics, the particle is trapped (localized) in a well if the total mechanical energy E is less than U_0. In quantum mechanics, such a trapped state is often called a **bound state.** All states are bound for an infinitely deep well like

MasteringPHYSICS

PhET: Double Wells and Covalent Bonds
PhET: Quantum Bound States
ActivPhysics 20.1: Potential Energy Diagrams
ActivPhysics 20.3: Potential Wells

40.13 A square-well potential.

The potential energy U is zero within the potential well (in the interval $0 \le x \le L$) and has the constant value U_0 outside this interval.

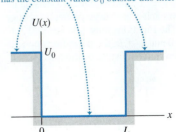

the one we described in Section 40.2. For a finite well like that shown in Fig. 40.13, if E is greater than U_0, the particle is *not* bound.

Let's see how to solve the Schrödinger equation for the bound states of a square-well potential. Our goal is to find the energies and wave functions for which $E < U_0$. The easiest approach is to consider separately the regions where $U = 0$ and where $U = U_0$. Where $U = 0$, the time-independent Schrödinger equation is

$$-\frac{\hbar^2}{2m}\frac{d^2\psi(x)}{dx^2} = E\psi(x) \quad \text{or} \quad \frac{d^2\psi(x)}{dx^2} = -\frac{2mE}{\hbar^2}\psi(x) \quad \text{[40.37]}$$

This is the same as Eq. (40.25) from Section 40.2, which describes a particle in a box. As in Section 40.2, we can express the solutions of this equation as combinations of $\cos kx$ and $\sin kx$, where $E = \hbar^2 k^2/2m$. We can rewrite the relationship between E and k as $k = \sqrt{2mE}/\hbar$. Hence inside the square well ($0 \le x \le L$) we have

$$\psi(x) = A\cos\left(\frac{\sqrt{2mE}}{\hbar}x\right) + B\sin\left(\frac{\sqrt{2mE}}{\hbar}x\right) \quad \text{(inside the well)} \quad \text{[40.38]}$$

where A and B are constants. So far, this looks a lot like the particle-in-a-box analysis in Section 40.2. The difference is that for the square-well potential, the potential energy outside the well is not infinite, so the wave function $\psi(x)$ outside the well is *not* zero.

For the regions outside the well ($x < 0$ and $x > L$) the potential-energy function in the time-independent Schrödinger equation is $U = U_0$:

$$-\frac{\hbar^2}{2m}\frac{d^2\psi(x)}{dx^2} + U_0\psi(x) = E\psi(x) \quad \text{or} \quad \frac{d^2\psi(x)}{dx^2} = \frac{2m(U_0 - E)}{\hbar^2}\psi(x) \quad \text{[40.39]}$$

The quantity $U_0 - E$ is positive, so the solutions of this equation are exponential. Using κ (the Greek letter kappa) to represent the quantity $[2m(U_0 - E)]^{1/2}/\hbar$ and taking κ as positive, we can write the solutions as

$$\psi(x) = Ce^{\kappa x} + De^{-\kappa x} \quad \text{(outside the well)} \quad \text{[40.40]}$$

where C and D are constants with different values in the two regions $x < 0$ and $x > L$. Note that ψ can't be allowed to approach infinity as $x \to +\infty$ or $x \to -\infty$. [If it did, we wouldn't be able to satisfy the normalization condition, Eq. (40.33).] This means that in Eq. (40.40), we must have $D = 0$ for $x < 0$ and $C = 0$ for $x > L$.

Our calculations so far show that the bound-state wave functions for a finite well are sinusoidal inside the well [Eq. (40.38)] and exponential outside it [Eq. (40.40)]. We have to *match* the wave functions inside and outside the well so that they satisfy the boundary conditions that we mentioned in Section 40.2: $\psi(x)$ and $d\psi(x)/dx$ must be continuous at the boundary points $x = 0$ and $x = L$. If the wave function $\psi(x)$ or the slope $d\psi(x)/dx$ were to change discontinuously at a point, the second derivative $d^2\psi(x)/dx^2$ would be *infinite* at that point. That would violate the time-independent Schrödinger equation, Eq. (40.23), which says that at every point $d^2\psi(x)/dx^2$ is proportional to $U - E$. For a finite well $U - E$ is finite everywhere, so $d^2\psi(x)/dx^2$ must also be finite everywhere.

Matching the sinusoidal and exponential functions at the boundary points so that they join smoothly is possible only for certain specific values of the total energy E, so this requirement determines the possible energy levels of the finite square well. There is no simple formula for the energy levels as there was for the infinitely deep well. Finding the levels is a fairly complex mathematical problem that requires solving a transcendental equation by numerical approximation; we won't go into the details. Figure 40.14 shows the general shape of a possible wave function. The most striking features of this wave function are the

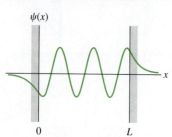

40.14 A possible wave function for a particle in a finite potential well. The function is sinusoidal inside the well ($0 \le x \le L$) and exponential outside it. It approaches zero asymptotically at large $|x|$. The functions must join smoothly at $x = 0$ and $x = L$; the wave function and its derivative must be continuous.

"exponential tails" that extend outside the well into regions that are forbidden by Newtonian mechanics (because in those regions the particle would have negative kinetic energy). We see that there is some probability for finding the particle *outside* the potential well, which would be impossible in classical mechanics. In Section 40.4 we'll discuss an amazing result of this effect.

Example 40.5 Outside a finite well

(a) Show that Eq. (40.40), $\psi(x) = Ce^{\kappa x} + De^{-\kappa x}$, is indeed a solution of the time-independent Schrödinger equation outside a finite well of height U_0. (b) What happens to $\psi(x)$ in the limit $U_0 \to \infty$?

SOLUTION

IDENTIFY and SET UP: In part (a), we try the given function $\psi(x)$ in the time-independent Schrödinger equation for $x < 0$ and for $x > L$, Eq. (40.39). In part (b), we note that in the limit $U_0 \to \infty$ the finite well becomes an *infinite* well, like those we considered in Section 40.2 for a particle in a box. So in this limit the wave functions outside a finite well must reduce to the wave functions outside the box.

EXECUTE: (a) We must show that $\psi(x) = Ce^{\kappa x} + De^{-\kappa x}$ satisfies $d^2\psi(x)/dx^2 = [2m(U_0 - E)/\hbar^2]\psi(x)$. We recall that $(d/du)e^{au} = ae^{au}$ and $(d^2/du^2)e^{au} = a^2e^{au}$; the left-hand side of the Schrödinger equation is then

$$\frac{d^2\psi(x)}{dx^2} = \frac{d^2}{dx^2}(Ce^{\kappa x}) + \frac{d^2}{dx^2}(De^{-\kappa x})$$
$$= C\kappa^2 e^{\kappa x} + D(-\kappa)^2 e^{-\kappa x}$$
$$= \kappa^2(Ce^{\kappa x} + De^{-\kappa x})$$
$$= \kappa^2\psi(x)$$

Since from Eq. (40.40) $\kappa^2 = 2m(U_0 - E)/\hbar^2$, this is equal to the right-hand side of the equation. The equation is satisfied, and $\psi(x)$ is a solution.

(b) As U_0 approaches infinity, κ also approaches infinity. In the region $x < 0$, $\psi(x) = Ce^{\kappa x}$; as $\kappa \to \infty$, $\kappa x \to -\infty$ (since x is negative) and $e^{\kappa x} \to 0$, so the wave function approaches zero for all $x < 0$. Likewise, we can show that the wave function also approaches zero for all $x > L$. This is just what we found in Section 40.2; the wave function for a particle in a box must be zero outside the box.

EVALUATE: Our result in part (b) shows that the infinite square well is a *limiting case* of the finite well. We've seen many cases in Newtonian mechanics where it's important to consider limiting cases (such as Examples 5.11 and 5.13 in Section 5.2). Limiting cases are no less important in quantum mechanics.

Comparing Finite and Infinite Square Wells

Let's continue the comparison of the finite-depth potential well with the infinitely deep well, which we began in Example 40.5. First, because the wave functions for the finite well don't go to zero at $x = 0$ and $x = L$, the wavelength of the sinusoidal part of each wave function is *longer* than it would be with an infinite well. This increase in λ corresponds to a reduced magnitude of momentum $p = h/\lambda$ and therefore a reduced energy. Thus each energy level, including the ground level, is *lower* for a finite well than for an infinitely deep well with the same width.

Second, a well with finite depth U_0 has only a *finite* number of bound states and corresponding energy levels, compared to the *infinite* number for an infinitely deep well. How many levels there are depends on the magnitude of U_0 in comparison with the ground-level energy for the infinitely deep well (IDW), which we call $E_{1-\text{IDW}}$. From Eq. (40.31),

$$E_{1-\text{IDW}} = \frac{\pi^2\hbar^2}{2mL^2} \qquad \text{(ground-level energy, infinitely deep well)} \qquad (40.41)$$

When the well is very deep so U_0 is much larger than $E_{1-\text{IDW}}$, there are many bound states and the energies of the lowest few are nearly the same as the energies for the infinitely deep well. When U_0 is only a few times as large as $E_{1-\text{IDW}}$ there are only a few bound states. (There is always at least *one* bound state, no matter how shallow the well.) As with the infinitely deep well, there is no state with $E = 0$; such a state would violate the uncertainty principle.

40.15 (a) Wave functions for the three bound states for a particle in a finite potential well with depth U_0, for the case $U_0 = 6E_{1-\text{IDW}}$. (Here $E_{1-\text{IDW}}$ is the ground-level energy for an infinite well of the same width.) The horizontal brown line for each wave function corresponds to $\psi = 0$; the vertical placement of these lines indicates the energy of each bound state (compare Fig. 40.11). (b) Energy-level diagram for this system. The energies are expressed both as multiples of $E_{1-\text{IDW}}$ and as fractions of U_0. All energies greater than U_0 are possible; states with $E > U_0$ form a continuum.

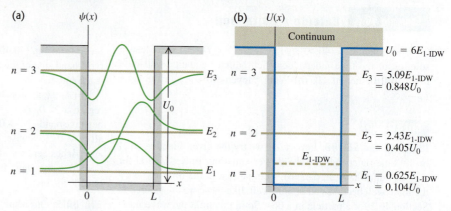

Figure 40.15 shows the case $U_0 = 6E_{1-\text{IDW}}$; for this particular case there are three bound states. In the figure, we express the energy levels both as fractions of the well depth U_0 and as multiples of $E_{1-\text{IDW}}$. Note that if the well were infinitely deep, the lowest three levels, as given by Eq. (40.31), would be $E_{1-\text{IDW}}$, $4E_{1-\text{IDW}}$, and $9E_{1-\text{IDW}}$. Figure 40.15 also shows the wave functions for the three bound states.

It turns out that when U_0 is less than $E_{1-\text{IDW}}$, there is only one bound state. In the limit when U_0 is *much smaller* than $E_{1-\text{IDW}}$ (a very shallow well), the energy of this single state is approximately $E = 0.68U_0$.

Figure 40.16 shows graphs of the probability distributions—that is, the values of $|\psi|^2$—for the wave functions shown in Fig. 40.15a. As with the infinite well, not all positions are equally likely. Unlike the infinite well, there is some probability of finding the particle outside the well in the classically forbidden regions.

There are also states for which E is *greater* than U_0. In these *free-particle states* the particle is not bound but is free to move through all values of x. *Any* energy E greater than U_0 is possible, so the free-particle states form a *continuum* rather than a discrete set of states with definite energy levels. The free-particle wave functions are sinusoidal both inside and outside the well. The wavelength is shorter inside the well than outside, corresponding to greater kinetic energy inside the well than outside it.

Figure 40.17 shows a graphic demonstration of particles in a *two*-dimensional finite potential well. Example 40.6 describes another application of the square-well potential.

40.16 Probability distribution functions $|\psi(x)|^2$ for the square-well wave functions shown in Fig. 40.15. The horizontal brown line for each wave function corresponds to $|\psi|^2 = 0$.

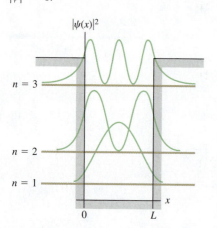

Example 40.6   **An electron in a finite well**

An electron is trapped in a square well 0.50 nm across (roughly five times a typical atomic diameter). (a) Find the ground-level energy $E_{1-\text{IDW}}$ if the well is infinitely deep. (b) Find the energy levels if the actual well depth U_0 is six times the ground-level energy found in part (a). (c) Find the wavelength of the photon emitted when the electron makes a transition from the $n = 2$ level to the $n = 1$ level. In what region of the electromagnetic spectrum does the photon wavelength lie? (d) If the electron is in the $n = 1$ (ground) level and absorbs a photon, what is the minimum photon energy that will free the electron from the well? In what region of the spectrum does the wavelength of this photon lie?

SOLUTION

IDENTIFY and SET UP: Equation (40.41) gives the ground-level energy $E_{1-\text{IDW}}$ for an infinitely deep well, and Fig. 40.15b shows the energies for a square well with $U_0 = 6E_{1-\text{IDW}}$. The energy of the photon emitted or absorbed in a transition is equal to the difference in energy between two levels involved in the transition; the photon wavelength is given by $E = hc/\lambda$ (see Chapter 38).

EXECUTE: (a) From Eq. (40.41),

$$E_{1-\text{IDW}} = \frac{\pi^2 \hbar^2}{2mL^2} = \frac{\pi^2 (1.055 \times 10^{-34}\,\text{J} \cdot \text{s})^2}{2(9.11 \times 10^{-31}\,\text{kg})(0.50 \times 10^{-9}\,\text{m})^2}$$

$$= 2.4 \times 10^{-19}\,\text{J} = 1.5\,\text{eV}$$

(b) We have $U_0 = 6E_{1-\text{IDW}} = 6(1.5\,\text{eV}) = 9.0\,\text{eV}$. We can read off the energy levels from Fig. 40.15b:

$$E_1 = 0.625 E_{1-\text{IDW}} = 0.625(1.5\,\text{eV}) = 0.94\,\text{eV}$$

$$E_2 = 2.43 E_{1-\text{IDW}} = 2.43(1.5\,\text{eV}) = 3.6\,\text{eV}$$

$$E_3 = 5.09 E_{1-\text{IDW}} = 5.09(1.5\,\text{eV}) = 7.6\,\text{eV}$$

(c) The photon energy and wavelength for the $n = 2$ to $n = 1$ transition are

$$E_2 - E_1 = 3.6\,\text{eV} - 0.94\,\text{eV} = 2.7\,\text{eV}$$

$$\lambda = \frac{hc}{E} = \frac{(4.136 \times 10^{-15}\,\text{eV} \cdot \text{s})(3.00 \times 10^8\,\text{m/s})}{2.7\,\text{eV}}$$

$$= 460\,\text{nm}$$

in the blue region of the visible spectrum.

(d) We see from Fig. 40.15b that the minimum energy needed to free the electron from the well from the $n = 1$ level is $U_0 - E_1 = 9.0\,\text{eV} - 0.94\,\text{eV} = 8.1\,\text{eV}$, which is three times the 2.7-eV photon energy found in part (c). Hence the corresponding photon wavelength is one-third of 460 nm, or (to two significant figures) 150 nm, which is in the ultraviolet region of the spectrum.

EVALUATE: As a check, you can also calculate the bound-state energies by using the formulas $E_1 = 0.104U_0$, $E_2 = 0.405U_0$, and $E_3 = 0.848U_0$ given in Fig. 40.15b. As an additional check, note that the first three energy levels of an infinitely deep well of the same width are $E_{1-\text{IDW}} = 1.5\,\text{eV}$, $E_{2-\text{IDW}} = 4E_{1-\text{IDW}} = 6.0\,\text{eV}$, and $E_{3-\text{IDW}} = 9E_{1-\text{IDW}} = 13.5\,\text{eV}$. The energies we found in part (b) are less than these values: As we mentioned earlier, the finite depth of the well lowers the energy levels compared to the levels for an infinitely deep well.

One application of these ideas is to *quantum dots,* which are nanometer-sized particles of a semiconductor such as cadmium selenide (CdSe). An electron within a quantum dot behaves much like a particle in a finite potential well of width L equal to the size of the dot. When quantum dots are illuminated with ultraviolet light, the electrons absorb the ultraviolet photons and are excited into high energy levels, such as the $n = 3$ level described in this example. If the electron returns to the ground level ($n = 1$) in two or more steps (for example, from $n = 3$ to $n = 2$ and from $n = 2$ to $n = 1$), one of the steps will involve emitting a visible-light photon, as we have calculated here. (We described this process of *fluorescence* in Section 39.3.) Increasing the value of L decreases the energies of the levels and hence the spacing between them, and thus decreases the energy and increases the wavelength of the emitted photons. The photograph that opens this chapter shows quantum dots of different sizes in solution: Each emits a characteristic wavelength that depends on the dot size. Quantum dots can be injected into living tissue and their fluorescent glow used as a tracer for biological research and for medicine. They may also be the key to a new generation of lasers and ultrafast computers.

Test Your Understanding of Section 40.3 Suppose that the width of the finite potential well shown in Fig. 40.15 is reduced by one-half. How must the value of U_0 change so that there are still just three bound energy levels whose energies are the fractions of U_0 shown in Fig. 40.15b? U_0 must: (i) increase by a factor of four; (ii) increase by a factor of two; (iii) remain the same; (iv) decrease by a factor of one-half; (v) decrease by a factor of one-fourth.

40.4 Potential Barriers and Tunneling

A **potential barrier** is the opposite of a potential well; it is a potential-energy function with a *maximum*. Figure 40.18 shows an example. In classical Newtonian mechanics, if a particle (such as a roller coaster) is located to the left of the barrier (which might be a hill), and if the total mechanical energy of the system is E_1, the particle cannot move farther to the right than $x = a$. If it did, the potential energy U would be greater than the total energy E and the kinetic energy $K = E - U$ would be negative. This is impossible in classical mechanics since $K = \frac{1}{2}mv^2$ can never be negative.

A quantum-mechanical particle behaves differently: If it encounters a barrier like the one in Fig. 40.18 and has energy less than E_2, it *may* appear on the other side. This phenomenon is called *tunneling*. In quantum-mechanical tunneling, unlike macroscopic, mechanical tunneling, the particle does not actually push through the barrier and loses no energy in the process.

Tunneling Through a Rectangular Barrier

To understand how tunneling can occur, let's look at the potential-energy function $U(x)$ shown in Fig. 40.19. It's like Fig. 40.13 turned upside-down; the potential energy is zero everywhere except in the range $0 \le x \le L$, where it has the value U_0. This might represent a simple model for the potential energy of an

40.17 To make this image, 48 iron atoms (shown as yellow peaks) were placed in a circle on a copper surface. The "elevation" at each point inside the circle indicates the electron density within the circle. The standing-wave pattern is very similar to the probability distribution function for a particle in a one-dimensional finite potential well. (This image was made with a scanning tunneling microscope, discussed in Section 40.4.)

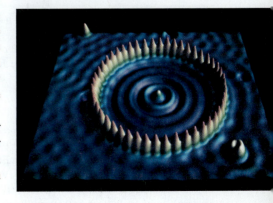

MasteringPHYSICS

PhET: Quantum Tunneling and Wave Packets
ActivPhysics 20.4: Potential Barriers

40.18 A potential-energy barrier. According to Newtonian mechanics, if the total energy of the system is E_1, a particle to the left of the barrier can go no farther than $x = a$. If the total energy is greater than E_2, the particle can pass over the barrier.

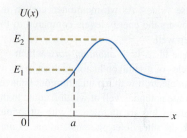

40.19 A rectangular potential-energy barrier with width L and height U_0. According to Newtonian mechanics, if the total energy E is less than U_0, a particle cannot pass over this barrier but is confined to the side where it starts.

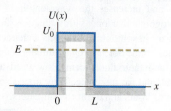

40.20 A possible wave function for a particle tunneling through the potential-energy barrier shown in Fig. 40.19.

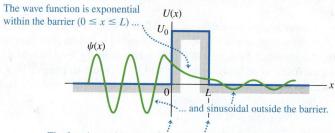

The wave function is exponential within the barrier ($0 \le x \le L$) ...

... and sinusoidal outside the barrier.

The function and its derivative (slope) are continuous at $x = 0$ and $x = L$ so that the sinusoidal and exponential functions join smoothly.

electron in the presence of two slabs of metal separated by an air gap of thickness L. The potential energy is lower within either slab than in the gap between them.

Let's consider solutions of the Schrödinger equation for this potential-energy function for the case in which E is less than U_0. We can use our results from Section 40.3. In the regions $x < 0$ and $x > L$, where $U = 0$, the solution is sinusoidal and is given by Eq. (40.38). Within the barrier ($0 \le x \le L$), $U = U_0$ and the solution is exponential as in Eq. (40.40). Just as with the finite potential well, the functions have to join smoothly at the boundary points $x = 0$ and $x = L$, which means that both $\psi(x)$ and $d\psi(x)/dx$ have to be continuous at these points.

These requirements lead to a wave function like the one shown in Fig. 40.20. The function is *not* zero inside the barrier (the region forbidden by Newtonian mechanics). Even more remarkable, a particle that is initially to the *left* of the barrier has some probability of being found to the *right* of the barrier. How great this probability is depends on the width L of the barrier and the particle's energy E in comparison with the barrier height U_0. The **tunneling probability** T that the particle gets through the barrier is proportional to the square of the ratio of the amplitudes of the sinusoidal wave functions on the two sides of the barrier. These amplitudes are determined by matching wave functions and their derivatives at the boundary points, a fairly involved mathematical problem. When T is much smaller than unity, it is given approximately by

$$T = Ge^{-2\kappa L} \text{ where } G = 16\frac{E}{U_0}\left(1 - \frac{E}{U_0}\right) \text{ and } \kappa = \frac{\sqrt{2m(U_0 - E)}}{\hbar} \quad (40.42)$$

(probability of tunneling)

The probability decreases rapidly with increasing barrier width L. It also depends critically on the energy difference $U_0 - E$, which in Newtonian physics is the additional kinetic energy the particle would need to be able to climb over the barrier.

Example 40.7 Tunneling through a barrier

A 2.0-eV electron encounters a barrier 5.0 eV high. What is the probability that it will tunnel through the barrier if the barrier width is (a) 1.00 nm and (b) 0.50 nm?

SOLUTION

IDENTIFY and SET UP: This problem uses the ideas of tunneling through a rectangular barrier, as in Figs. 40.19 and 40.20. Our target variable is the tunneling probability T in Eq. (40.42), which we evaluate for the given values $E = 2.0$ eV (electron energy), $U = 5.0$ eV (barrier height), $m = 9.11 \times 10^{-31}$ kg (mass of the electron), and $L = 1.00$ nm or 0.50 nm (barrier width).

EXECUTE: First we evaluate G and κ in Eq. (40.42), using $E = 2.0$ eV:

$$G = 16\left(\frac{2.0 \text{ eV}}{5.0 \text{ eV}}\right)\left(1 - \frac{2.0 \text{ eV}}{5.0 \text{ eV}}\right) = 3.8$$

$$U_0 - E = 5.0 \text{ eV} - 2.0 \text{ eV} = 3.0 \text{ eV} = 4.8 \times 10^{-19} \text{ J}$$

$$\kappa = \frac{\sqrt{2(9.11 \times 10^{-31} \text{ kg})(4.8 \times 10^{-19} \text{ J})}}{1.055 \times 10^{-34} \text{ J} \cdot \text{s}} = 8.9 \times 10^9 \text{ m}^{-1}$$

(a) When $L = 1.00$ nm $= 1.00 \times 10^{-9}$ m, $2\kappa L = 2(8.9 \times 10^9 \text{ m}^{-1})(1.00 \times 10^{-9} \text{ m}) = 17.8$ and $T = Ge^{-2\kappa L} = 3.8e^{-17.8} = 7.1 \times 10^{-8}$.

(b) When $L = 0.50$ nm, one-half of 1.00 nm, $2\kappa L$ is one-half of 17.8, or 8.9. Hence $T = 3.8e^{-8.9} = 5.2 \times 10^{-4}$.

EVALUATE: Halving the width of this barrier increases the tunneling probability T by a factor of $(5.2 \times 10^{-4})/(7.1 \times 10^{-8}) = 7.3 \times 10^3$, or nearly ten thousand. The tunneling probability is an *extremely* sensitive function of the barrier width.

(a)

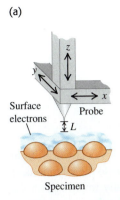

(b)

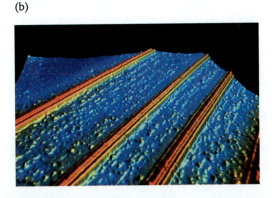

40.21 (a) Schematic diagram of the probe of a scanning tunneling microscope (STM). As the sharp conducting probe is scanned across the surface in the x- and y-directions, it is also moved in the z-direction to maintain a constant tunneling current. The changing position of the probe is recorded and used to construct an image of the surface. (b) This colored STM image shows "quantum wires": thin strips, just 10 atoms wide, of a conductive rare-earth silicide atop a silicon surface. Such quantum wires may one day be the basis of ultraminiaturized circuits.

Applications of Tunneling

Tunneling has a number of practical applications, some of considerable importance. When you twist two copper wires together or close the contacts of a switch, current passes from one conductor to the other despite a thin layer of nonconducting copper oxide between them. The electrons tunnel through this thin insulating layer. A *tunnel diode* is a semiconductor device in which electrons tunnel through a potential barrier. The current can be switched on and off very quickly (within a few picoseconds) by varying the height of the barrier. A *Josephson junction* consists of two superconductors separated by an oxide layer a few atoms (1 to 2 nm) thick. Electron pairs in the superconductors can tunnel through the barrier layer, giving such a device unusual circuit properties. Josephson junctions are useful for establishing precise voltage standards and measuring tiny magnetic fields, and they play a crucial role in the developing field of quantum computing.

The *scanning tunneling microscope* (STM) uses electron tunneling to create images of surfaces down to the scale of individual atoms. An extremely sharp conducting needle is brought very close to the surface, within 1 nm or so (Fig. 40.21a). When the needle is at a positive potential with respect to the surface, electrons can tunnel through the surface potential-energy barrier and reach the needle. As Example 40.7 shows, the tunneling probability and hence the tunneling current are very sensitive to changes in the width L of the barrier (the distance between the surface and the needle tip). In one mode of operation the needle is scanned across the surface and at the same time is moved perpendicular to the surface to maintain a constant tunneling current. The needle motion is recorded, and after many parallel scans, an image of the surface can be reconstructed. Extremely precise control of needle motion, including isolation from vibration, is essential. Figure 40.21b shows an STM image. (Figure 40.17 is also an STM image.)

Tunneling is also of great importance in nuclear physics. A fusion reaction can occur when two nuclei tunnel through the barrier caused by their electrical repulsion and approach each other closely enough for the attractive nuclear force to cause them to fuse. Fusion reactions occur in the cores of stars, including the sun; without tunneling, the sun wouldn't shine. The emission of alpha particles from unstable nuclei such as radium also involves tunneling. An alpha particle is a cluster of two protons and two neutrons (the same as a nucleus of the most common form of helium). Such clusters form naturally within larger atomic nuclei. An alpha particle trying to escape from a nucleus encounters a potential barrier that results from the combined effect of the attractive nuclear force and the electrical repulsion of the remaining part of the nucleus (Fig. 40.22). The alpha particle can escape only by tunneling through this barrier. Depending on the barrier height and width for a given kind of alpha-emitting nucleus, the tunneling probability can be low or high, and the alpha-emitting material will have low or high radioactivity. Recall from Section 39.2 that Ernest Rutherford used alpha particles

Application **Electron Tunneling in Enzymes**
Protein molecules play essential roles as enzymes in living organisms. Enzymes like the one shown here are large molecules, and in many cases their function depends on the ability of electrons to tunnel across the space that separates one part of the molecule from another. Without tunneling, life as we know it would be impossible!

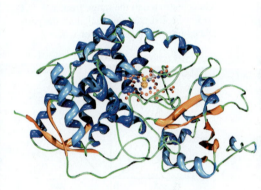

40.22 Approximate potential-energy function for an alpha particle interacting with a nucleus of radius R. If an alpha particle inside the nucleus has energy E greater than zero, it can tunnel through the barrier and escape from the nucleus.

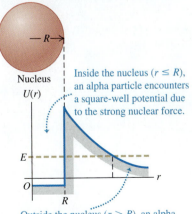

Inside the nucleus ($r \leq R$), an alpha particle encounters a square-well potential due to the strong nuclear force.

Outside the nucleus ($r > R$), an alpha particle experiences a $1/r$ potential due to electrostatic repulsion.

from a radioactive source to discover the atomic nucleus. Although Rutherford did not know it, tunneling by these alpha particles made his experiments possible! We'll learn more about alpha decay in Chapter 43.

Test Your Understanding of Section 40.4 Is it possible for a particle undergoing tunneling to be found *within* the barrier rather than on either side of it? ▮

40.5 The Harmonic Oscillator

Systems that *oscillate* are of tremendous importance in the physical world, from the oscillations of your eardrums in response to a sound wave to the vibrations of the ground caused by an earthquake. Oscillations are equally important on the microscopic scale where quantum effects dominate. The molecules of the air around you can be set into vibration when they collide with each other, the protons and neutrons in an excited atomic nucleus can oscillate in opposite directions, and a microwave oven transfers energy to food by making water molecules in the food flip back and forth. In this section we'll look at the solutions of the Schrödinger equation for the simplest kind of vibrating system, the quantum-mechanical harmonic oscillator.

As we learned in Chapter 14, a **harmonic oscillator** is a particle with mass m that moves along the x-axis under the influence of a conservative force $F_x = -k'x$. The constant k' is called the *force constant*. (In Chapter 14 we used the symbol k for the force constant. In this section we'll use the symbol k' instead to minimize confusion with the wave number $k = 2\pi/\lambda$.) The force is proportional to the particle's displacement x from its equilibrium position, $x = 0$. The corresponding potential-energy function is $U = \frac{1}{2}k'x^2$ (Fig. 40.23). In Newtonian mechanics, when the particle is displaced from equilibrium, it undergoes sinusoidal motion with frequency $f = (1/2\pi)(k'/m)^{1/2}$ and angular frequency $\omega = 2\pi f = (k'/m)^{1/2}$. The amplitude (that is, the maximum displacement from equilibrium) of these Newtonian oscillations is A, which is related to the energy E of the oscillator by $E = \frac{1}{2}k'A^2$.

Let's make an enlightened guess about the energy levels of a quantum-mechanical harmonic oscillator. In classical physics an electron oscillating with angular frequency ω emits electromagnetic radiation with that same angular frequency. It's reasonable to guess that when an excited quantum-mechanical harmonic oscillator with angular frequency $\omega = (k'/m)^{1/2}$ (according to Newtonian mechanics, at least) makes a transition from one energy level to a lower level, it would emit a photon with this same angular frequency ω. The energy of such a photon is $hf = (h/2\pi)(\omega/2\pi) = \hbar\omega$. So we would expect that the spacing between adjacent energy levels of the harmonic oscillator would be

$$hf = \hbar\omega = \hbar\sqrt{\frac{k'}{m}} \tag{40.43}$$

That's the same spacing between energy levels that Planck assumed in deriving his radiation law (see Section 39.5). It was a good assumption; as we'll see, the energy levels are in fact half-integer $\left(\frac{1}{2}, \frac{3}{2}, \frac{5}{2}, \ldots\right)$ multiples of $\hbar\omega$.

Wave Functions, Boundary Conditions, and Energy Levels

We'll begin our quantum-mechanical analysis of the harmonic oscillator by writing down the one-dimensional time-independent Schrödinger equation, Eq. (40.23), with $\frac{1}{2}k'x^2$ in place of U:

$$-\frac{\hbar^2}{2m}\frac{d^2\psi(x)}{dx^2} + \frac{1}{2}k'x^2\psi(x) = E\psi(x) \quad \text{(Schrödinger equation for the harmonic oscillator)} \tag{40.44}$$

The solutions of this equation are wave functions for the physically possible states of the system.

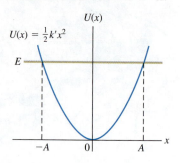

40.23 Potential-energy function for the harmonic oscillator. In Newtonian mechanics the amplitude A is related to the total energy E by $E = \frac{1}{2}k'A^2$, and the particle is restricted to the range from $x = -A$ to $x = A$. In quantum mechanics the particle can be found at $x > A$ or $x < -A$.

In the discussion of square-well potentials in Section 40.2 we found that the energy levels are determined by boundary conditions at the walls of the well. However, the harmonic-oscillator potential has no walls as such; what, then, are the appropriate boundary conditions? Classically, $|x|$ cannot be greater than the amplitude A given by $E = \frac{1}{2}k'A^2$. Quantum mechanics does allow some penetration into classically forbidden regions, but the probability decreases as that penetration increases. Thus the wave functions must approach zero as $|x|$ grows large.

Satisfying the requirement that $\psi(x) \to 0$ as $|x| \to \infty$ is not as trivial as it may seem. To see why this is, let's rewrite Eq. (40.44) in the form

$$\frac{d^2\psi(x)}{dx^2} = \frac{2m}{\hbar^2}\left(\tfrac{1}{2}k'x^2 - E\right)\psi(x) \qquad (40.45)$$

Equation (40.45) shows that when x is large enough (either positive or negative) to make the quantity $\left(\tfrac{1}{2}k'x^2 - E\right)$ positive, the function $\psi(x)$ and its second derivative $d^2\psi(x)/dx^2$ have the same sign. Figure 40.24 shows four possible kinds of behavior of $\psi(x)$ beginning at a point where x is greater than the classical amplitude A, so that $\frac{1}{2}k'x^2 - \frac{1}{2}k'A^2 = \frac{1}{2}k'x^2 - E > 0$. Let's look at these four cases more closely. Note that if $\psi(x)$ is positive as shown in Fig. 40.24, Eq. (40.45) tells us that $d^2\psi(x)/dx^2$ is also positive and the function is *concave upward*. Note also that $d^2\psi(x)/dx^2$ is the rate of change of the *slope* of $\psi(x)$; this will help us understand how our four possible wave functions behave.

- *Curve a:* The slope of $\psi(x)$ is positive at point x. Since $d^2\psi(x)/dx^2 > 0$, the function curves upward increasingly steeply and goes to infinity. This violates the boundary condition that $\psi(x) \to 0$ as $|x| \to \infty$, so this isn't a viable wave function.
- *Curve b:* The slope of $\psi(x)$ is negative at point x, and $d^2\psi(x)/dx^2$ has a large positive value. Hence the slope changes rapidly from negative to positive and keeps on increasing—so, again, the wave function goes to infinity. This wave function isn't viable either.
- *Curve c:* As for curve b, the slope is negative at point x. However, $d^2\psi(x)/dx^2$ now has a *small* positive value, so the slope increases only gradually as $\psi(x)$ decreases to zero and crosses over to negative values. Equation (40.45) tells us that once $\psi(x)$ becomes negative, $d^2\psi(x)/dx^2$ also becomes negative. Hence the curve becomes concave *downward* and heads for *negative* infinity. This wave function, too, fails to satisfy the requirement that $\psi(x) \to 0$ as $|x| \to \infty$ and thus isn't viable.
- *Curve d:* If the slope of $\psi(x)$ at point x is negative, and the positive value of $d^2\psi(x)/dx^2$ at this point is neither too large nor too small, the curve bends just enough to glide in asymptotically to the x-axis. In this case $\psi(x)$, $d\psi(x)/dx$, and $d^2\psi(x)/dx^2$ all approach zero at large x. This case offers the only hope of satisfying the boundary condition that $\psi(x) \to 0$ as $|x| \to \infty$, and it occurs only for certain very special values of the energy E.

This qualitative discussion suggests how the boundary conditions as $|x| \to \infty$ determine the possible energy levels for the quantum-mechanical harmonic oscillator. It turns out that these boundary conditions are satisfied only if the energy E is equal to one of the values E_n, given by the simple formula

$$E_n = \left(n + \tfrac{1}{2}\right)\hbar\sqrt{\frac{k'}{m}} = \left(n + \tfrac{1}{2}\right)\hbar\omega \quad (n = 0, 1, 2, \ldots) \qquad (40.46)$$

(energy levels, harmonic oscillator)

where n is the quantum number identifying each state and energy level. Note that the ground level of energy $E_0 = \frac{1}{2}\hbar\omega$ is denoted by $n = 0$, *not* $n = 1$.

40.24 Possible behaviors of harmonic-oscillator wave functions in the region $\frac{1}{2}k'x^2 > E$. In this region, $\psi(x)$ and $d^2\psi(x)/dx^2$ have the same sign. The curve is concave upward when $d^2\psi(x)/dx^2$ is positive and concave downward when $d^2\psi(x)/dx^2$ is negative.

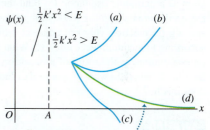

Only curve d, which approaches the x-axis asymptotically for large x, is an acceptable wave function for this system.

40.25 Energy levels for the harmonic oscillator. The spacing between any two adjacent levels is $\Delta E = \hbar\omega$. The energy of the ground level is $E_0 = \frac{1}{2}\hbar\omega$.

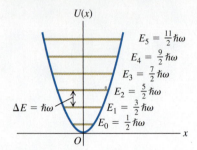

Equation (40.46) confirms our guess [(Eq. 40.43)] that adjacent energy levels are separated by a constant interval of $\hbar\omega = hf$, as Planck assumed in 1900. There are infinitely many levels; this shouldn't be surprising because we are dealing with an infinitely deep potential well. As $|x|$ increases, $U = \frac{1}{2}k'x^2$ increases without bound.

Figure 40.25 shows the lowest six energy levels and the potential-energy function $U(x)$. For each level n, the value of $|x|$ at which the horizontal line representing the total energy E_n intersects $U(x)$ gives the amplitude A_n of the corresponding Newtonian oscillator.

Example 40.8 **Vibration in a crystal**

A sodium atom of mass 3.82×10^{-26} kg vibrates within a crystal. The potential energy increases by 0.0075 eV when the atom is displaced 0.014 nm from its equilibrium position. Treat the atom as a harmonic oscillator. (a) Find the angular frequency of the oscillations according to Newtonian mechanics. (b) Find the spacing (in electron volts) of adjacent vibrational energy levels according to quantum mechanics. (c) What is the wavelength of a photon emitted as the result of a transition from one level to the next lower level? In what region of the electromagnetic spectrum does this lie?

SOLUTION

IDENTIFY and SET UP: We'll find the force constant k' from the expression $U = \frac{1}{2}k'x^2$ for potential energy. We'll then find the angular frequency $\omega = (k'/m)^{1/2}$ and use this in Eq. (40.46) to find the spacing between adjacent energy levels. We'll calculate the wavelength of the emitted photon as in Example 40.6.

EXECUTE: We are given that $U = 0.0075\,\text{eV} = 1.2 \times 10^{-21}$ J when $x = 0.014 \times 10^{-9}$ m, so we can solve $U = \frac{1}{2}k'x^2$ for k':

$$k' = \frac{2U}{x^2} = \frac{2(1.2 \times 10^{-21}\,\text{J})}{(0.014 \times 10^{-9}\,\text{m})^2} = 12.2\,\text{N/m}$$

(a) The Newtonian angular frequency is

$$\omega = \sqrt{\frac{k'}{m}} = \sqrt{\frac{12.2\,\text{N/m}}{3.82 \times 10^{-26}\,\text{kg}}} = 1.79 \times 10^{13}\,\text{rad/s}$$

(b) From Eq. (40.46) and Fig. 40.25, the spacing between adjacent energy levels is

$$\hbar\omega = (1.054 \times 10^{-34}\,\text{J} \cdot \text{s})(1.79 \times 10^{13}\,\text{s}^{-1})$$

$$= 1.88 \times 10^{-21}\,\text{J}\left(\frac{1\,\text{eV}}{1.602 \times 10^{-19}\,\text{J}}\right) = 0.0118\,\text{eV}$$

(c) The energy E of the emitted photon is equal to the energy lost by the oscillator in the transition, 0.0118 eV. Then

$$\lambda = \frac{hc}{E} = \frac{(4.136 \times 10^{-15}\,\text{eV} \cdot \text{s})(3.00 \times 10^8\,\text{m/s})}{0.0118\,\text{eV}}$$

$$= 1.05 \times 10^{-4}\,\text{m} = 105\,\mu\text{m}$$

This photon wavelength is in the infrared region of the spectrum.

EVALUATE: This example shows us that interatomic force constants are a few newtons per meter, about the same as those of household springs or spring-based toys such as the Slinky. It also suggests that we can learn about the vibrations of molecules by measuring the radiation that they emit in transitioning to a lower vibrational state. We will explore this idea further in Chapter 42.

MasteringPHYSICS

ActivPhysics 20.1.6: Potential Energy Diagrams, Question 6

Comparing Quantum and Newtonian Oscillators

The wave functions for the levels $n = 0, 1, 2, \ldots$ of the harmonic oscillator are called *Hermite functions;* they aren't encountered in elementary calculus courses but are well known to mathematicians. Each Hermite function is an exponential function multiplied by a polynomial in x. The harmonic-oscillator wave function corresponding to $n = 0$ and $E = E_0$ (the ground level) is

$$\psi(x) = Ce^{-\sqrt{mk'}\,x^2/2\hbar} \tag{40.47}$$

The constant C is chosen to normalize the function—that is, to make $\int_{-\infty}^{\infty}|\psi|^2\,dx = 1$. (We're using C rather than A as a normalization constant in this section, since we've already appropriated the symbol A to denote the Newtonian amplitude of a harmonic oscillator.) You can find C using the following result from integral tables:

$$\int_{-\infty}^{\infty} e^{-a^2x^2}\,dx = \frac{\sqrt{\pi}}{a}$$

40.26 The first four wave functions for the harmonic oscillator. The amplitude A of a Newtonian oscillator with the same total energy is shown for each. Each wave function penetrates somewhat into the classically forbidden regions $|x| > A$. The total number of finite maxima and minima for each function is $n + 1$, one more than the quantum number.

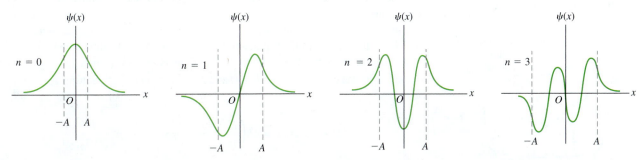

To confirm that $\psi(x)$ as given by Eq. (40.47) really *is* a solution of the Schrödinger equation for the harmonic oscillator, we invite you to calculate the second derivative of this wave function, substitute it into Eq. (40.44), and verify that the equation is satisfied if the energy E is equal to $E_0 = \frac{1}{2}\hbar\omega$ (see Exercise 40.38). It's a little messy, but the result is satisfying and worth the effort.

Figure 40.26 shows the the first four harmonic-oscillator wave functions. Each graph also shows the amplitude A of a Newtonian harmonic oscillator with the same energy—that is, the value of A determined from

$$\tfrac{1}{2}k'A^2 = \left(n + \tfrac{1}{2}\right)\hbar\omega \qquad (40.48)$$

In each case there is some penetration of the wave function into the regions $|x| > A$ that are forbidden by Newtonian mechanics. This is similar to the effect that we noted in Section 40.3 for a particle in a finite square well.

Figure 40.27 shows the probability distributions $|\psi(x)|^2$ for these same states. Each graph also shows the probability distribution determined from Newtonian physics, in which the probability of finding the particle near a randomly chosen point is inversely proportional to the particle's speed at that point. If we average out the wiggles in the quantum-mechanical probability curves, the results for $n > 0$ resemble the Newtonian predictions. This agreement improves with increasing n; Fig. 40.28 shows the classical and quantum-mechanical probability functions for $n = 10$. Notice that the spacing between zeros of $|\psi(x)|^2$ in Fig. 40.28 increases with increasing distance from $x = 0$. This makes sense from the Newtonian perspective: As a particle moves away from $x = 0$, its kinetic energy K and the magnitude p of its momentum both decrease. Thinking quantum-mechanically, this means that the wavelength $\lambda = h/p$ increases, so the spacing between zeros of $\psi(x)$ (and hence of $|\psi(x)|^2$) also increases.

In the Newtonian analysis of the harmonic oscillator the minimum energy is zero, with the particle at rest at its equilibrium position $x = 0$. This is not possible in quantum mechanics; no solution of the Schrödinger equation has $E = 0$ and satisfies the boundary conditions. Furthermore, if there were such a state, it

40.27 Probability distribution functions $|\psi(x)|^2$ for the harmonic-oscillator wave functions shown in Fig. 40.26. The amplitude A of the Newtonian motion with the same energy is shown for each. The blue lines show the corresponding probability distributions for the Newtonian motion. As n increases, the averaged-out quantum-mechanical functions resemble the Newtonian curves more and more.

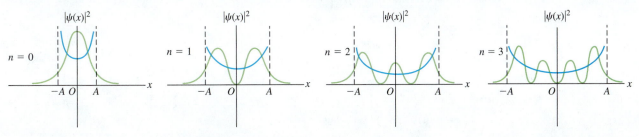

40.28 Newtonian and quantum-mechanical probability distribution functions for a harmonic oscillator for the state $n = 10$. The Newtonian amplitude A is also shown.

The larger the value of n, the more closely the quantum-mechanical probability distribution (green) matches the Newtonian probability distribution (blue).

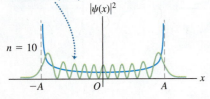

would violate the Heisenberg uncertainty principle because there would be no uncertainty in either position or momentum. The energy must be at least $\frac{1}{2}\hbar\omega$ for the system to conform to the uncertainty principle. To see qualitatively why this is so, consider a Newtonian oscillator with total energy $\frac{1}{2}\hbar\omega$. We can find the amplitude A and the maximum velocity just as we did in Section 14.3. When the particle is at its maximum displacement $(x = \pm A)$ and instantaneously at rest, $K = 0$ and $E = U = \frac{1}{2}k'A^2$. When the particle is at equilibrium $(x = 0)$ and moving at its maximum speed, $U = 0$ and $E = K = \frac{1}{2}mv_{\text{max}}^2$. Setting $E = \frac{1}{2}\hbar\omega$, we find

$$E = \tfrac{1}{2}k'A^2 = \tfrac{1}{2}\hbar\omega = \tfrac{1}{2}\hbar\left(\frac{k'}{m}\right)^{1/2} \quad \text{so} \quad A = \frac{\hbar^{1/2}}{k'^{1/4}m^{1/4}}$$

$$E = \tfrac{1}{2}mv_{\text{max}}^2 = \tfrac{1}{2}k'A^2 \quad \text{so} \quad v_{\text{max}} = A\left(\frac{k'}{m}\right)^{1/2} = \frac{\hbar^{1/2}k'^{1/4}}{m^{3/4}}$$

The maximum *momentum* of the particle is

$$p_{\text{max}} = mv_{\text{max}} = \hbar^{1/2}k'^{1/4}m^{1/4}$$

Here's where the Heisenberg uncertainty principle comes in. It turns out that the uncertainties in the particle's position and momentum (calculated as standard deviations) are, respectively, $\Delta x = A/\sqrt{2} = A/2^{1/2}$ and $\Delta p_x = p_{\text{max}}/\sqrt{2} = p_{\text{max}}/2^{1/2}$. Then the product of the two uncertainties is

$$\Delta x\,\Delta p_x = \left(\frac{\hbar^{1/2}}{2^{1/2}k'^{1/4}m^{1/4}}\right)\left(\frac{\hbar^{1/2}k'^{1/4}m^{1/4}}{2^{1/2}}\right) = \frac{\hbar}{2}$$

This product equals the minimum value allowed by Eq. (39.29), $\Delta x\,\Delta p_x \geq \hbar/2$, and thus satisfies the uncertainty principle. If the energy had been less than $\frac{1}{2}\hbar\omega$, the product $\Delta x\,\Delta p_x$ would have been less than $\hbar/2$, and the uncertainty principle would have been violated.

Even when a potential-energy function isn't precisely parabolic in shape, we may be able to approximate it by the harmonic-oscillator potential for sufficiently small displacements from equilibrium. Figure 40.29 shows a typical potential-energy function for an interatomic force in a molecule. At large separations the curve of $U(r)$ versus r levels off, corresponding to the absence of force at great distances. But the curve is approximately parabolic near the minimum of $U(r)$ (the equilibrium separation of the atoms). Near equilibrium the molecular vibration is approximately simple harmonic with energy levels given by Eq. (40.46), as we assumed in Example 40.8.

40.29 A potential-energy function describing the interaction of two atoms in a diatomic molecule. The distance r is the separation between the centers of the atoms, and the equilibrium separation is $r = r_0$. The energy needed to dissociate the molecule is U_∞.

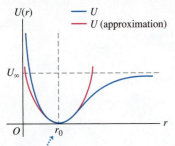

When r is near r_0, the potential-energy curve is approximately parabolic (as shown by the red curve) and the system behaves approximately like a harmonic oscillator.

Test Your Understanding of Section 40.5 A quantum-mechanical system initially in its ground level absorbs a photon and ends up in the first excited state. The system then absorbs a second photon and ends up in the second excited state. For which of the following systems does the second photon have a longer wavelength than the first one? (i) a harmonic oscillator; (ii) a hydrogen atom; (iii) a particle in a box.

Wave functions: The wave function for a particle contains all of the information about that particle. If the particle moves in one dimension in the presence of a potential energy function $U(x)$, the wave function $\Psi(x, t)$ obeys the one-dimensional Schrödinger equation. (For a *free* particle on which no forces act, $U(x) = 0$.) The quantity $|\Psi(x, t)|^2$, called the probability distribution function, determines the relative probability of finding a particle near a given position at a given time. If the particle is in a state of definite energy, called a stationary state, $\Psi(x, t)$ is a product of a function $\psi(x)$ that depends only on spatial coordinates and a function $e^{-iEt/\hbar}$ that depends only on time. For a stationary state, the probability distribution function is independent of time.

A spatial stationary-state wave function $\psi(x)$ for a particle that moves in one dimension in the presence of a potential-energy function $U(x)$ satisfies the time-independent Schrödinger equation. More complex wave functions can be constructed by superposing stationary-state wave functions. These can represent particles that are localized in a certain region, thus representing both particle and wave aspects. (See Examples 40.1 and 40.2.)

$$-\frac{\hbar^2}{2m}\frac{\partial^2\Psi(x, t)}{\partial x^2} + U(x)\Psi(x, t)$$
$$= i\hbar\frac{\partial\Psi(x, t)}{\partial t} \quad (40.20)$$
(general 1-D Schrödinger equation)

$$\Psi(x, t) = \psi(x)e^{-iEt/\hbar} \quad (40.21)$$
(time-dependent wave function for a state of definite energy)

$$-\frac{\hbar^2}{2m}\frac{d^2\psi(x)}{dx^2} + U(x)\psi(x) = E\psi(x)$$
(time-independent Schrödinger equation) $\quad (40.23)$

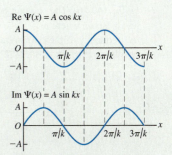

Particle in a box: The energy levels for a particle of mass m in a box (an infinitely deep square potential well) with width L are given by Eq. (40.31). The corresponding normalized stationary-state wave functions of the particle are given by Eq. (40.35). (See Examples 40.3 and 40.4.)

$$E_n = \frac{p_n^2}{2m} = \frac{n^2h^2}{8mL^2} = \frac{n^2\pi^2\hbar^2}{2mL^2}$$
$(n = 1, 2, 3, \dots) \quad (40.31)$

$$\psi_n(x) = \sqrt{\frac{2}{L}}\sin\frac{n\pi x}{L}$$
$(n = 1, 2, 3, \dots) \quad (40.35)$

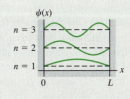

Wave functions and normalization: To be a solution of the Schrödinger equation, the wave function $\psi(x)$ and its derivative $d\psi(x)/dx$ must be continuous everywhere except where the potential-energy function $U(x)$ has an infinite discontinuity. Wave functions are usually normalized so that the total probability of finding the particle somewhere is unity.

$$\int_{-\infty}^{\infty}|\psi(x)|^2\, dx = 1 \quad (40.33)$$
(normalization condition)

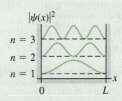

Finite potential well: In a potential well with finite depth U_0, the energy levels are lower than those for an infinitely deep well with the same width, and the number of energy levels corresponding to bound states is finite. The levels are obtained by matching wave functions at the well walls to satisfy the continuity of $\psi(x)$ and $d\psi(x)/dx$. (See Examples 40.5 and 40.6.)

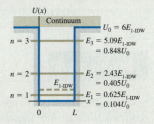

Potential barriers and tunneling: There is a certain probability that a particle will penetrate a potential-energy barrier even though its initial energy is less than the barrier height. This process is called tunneling. (See Example 40.7.)

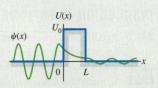

Quantum harmonic oscillator: The energy levels for the harmonic oscillator (for which $U(x) = \frac{1}{2}k'x^2$) are given by Eq. (40.46). The spacing between any two adjacent levels is $\hbar\omega$, where $\omega = \sqrt{k'/m}$ is the oscillation angular frequency of the corresponding Newtonian harmonic oscillator. (See Example 40.8.)

$$E_n = \left(n + \tfrac{1}{2}\right)\hbar\sqrt{\frac{k'}{m}} = \left(n + \tfrac{1}{2}\right)\hbar\omega$$

$$(n = 0, 1, 2, 3, \ldots) \qquad (40.46)$$

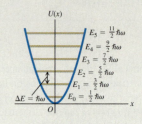

BRIDGING PROBLEM | A Packet in a Box

A particle of mass m in an infinitely deep well has the following wave function in the region from $x = 0$ to $x = L$:

$$\Psi(x, t) = \frac{1}{\sqrt{2}}\psi_1(x)e^{-iE_1t/\hbar} + \frac{1}{\sqrt{2}}\psi_2(x)e^{-iE_2t/\hbar}$$

Here $\psi_1(x)$ and $\psi_2(x)$ are the normalized stationary-state wave functions for the first two levels ($n = 1$ and $n = 2$), given by Eq. (40.35). E_1 and E_2, given by Eq. (40.31), are the energies of these levels. The wave function is zero for $x < 0$ and for $x > L$. (a) Find the probability distribution function for this wave function. (b) Does $\Psi(x, t)$ represent a stationary state of definite energy? How can you tell? (c) Show that the wave function $\Psi(x, t)$ is normalized. (d) Find the angular frequency of oscillation of the probability distribution function. What is the interpretation of this oscillation? (e) Suppose instead that $\Psi(x, t)$ is a combination of the wave functions of the two lowest levels of a finite well of length L and height U_0 equal to six times the energy of the lowest-energy bound state of an infinite well of length L. What would be the angular frequency of the probability distribution function in this case?

SOLUTION GUIDE

See MasteringPhysics® study area for a Video Tutor solution.

IDENTIFY and SET UP

1. In Section 40.1 we saw how to interpret a combination of two free-particle wave functions of different energies. In this problem you need to apply these same ideas to a combination of wave functions for the infinite well (Section 40.2) and the finite well (Section 40.3).

EXECUTE

2. Write down the full time-dependent wave function $\Psi(x, t)$ and its complex conjugate $\Psi^*(x, t)$ using the functions $\psi_1(x)$ and $\psi_2(x)$ from Eq. (40.35). Use these to calculate the probability distribution function, and decide whether or not this function depends on time.

3. To check for normalization, you'll need to verify that when you integrate the probability distribution function from step 2 over all values of x, the integral is equal to 1. [*Hint:* The trigonometric identities $\sin^2\theta = \frac{1}{2}(1 - \cos 2\theta)$ and $\sin\theta\sin\phi = \cos(\theta - \phi) - \cos(\theta + \phi)$ may be helpful.]

4. To find the answer to part (d) you'll need to identify the oscillation angular frequency ω_{osc} in your expression from step 2 for the probability distribution function. To interpret the oscillations, draw graphs of the probability distribution functions at times $t = 0$, $t = T/4$, $t = T/2$, and $t = 3T/4$, where $T = 2\pi/\omega_{osc}$ is the oscillation period of the probability distribution function.

5. For the finite well you do not have simple expressions for the first two stationary-state wave functions $\psi_1(x)$ and $\psi_2(x)$. However, you can still find the oscillation angular frequency ω_{osc}, which is related to the energies E_1 and E_2 in the same way as for the infinite-well case. (Can you see why?)

EVALUATE

6. Why are the factors of $1/\sqrt{2}$ in the wave function $\Psi(x, t)$ important?

7. Why do you suppose the oscillation angular frequency for a finite well is lower than for an infinite well of the same width?

Problems

For instructor-assigned homework, go to www.masteringphysics.com

•, ••, •••: Problems of increasing difficulty. **CP**: Cumulative problems incorporating material from earlier chapters. **CALC**: Problems requiring calculus. **BIO**: Biosciences problems.

DISCUSSION QUESTIONS

Q40.1 If quantum mechanics replaces the language of Newtonian mechanics, why don't we have to use wave functions to describe the motion of macroscopic bodies such as baseballs and cars?

Q40.2 A student remarks that the relationship of ray optics to the more general wave picture is analogous to the relationship of New-

tonian mechanics, with well-defined particle trajectories, to quantum mechanics. Comment on this remark.

Q40.3 As Eq. (40.21) indicates, the time-dependent wave function for a stationary state is a complex number having a real part and an imaginary part. How can this function have any physical meaning, since part of it is *imaginary*?

Q40.4 Why must the wave function of a particle be normalized?

Q40.5 If a particle is in a stationary state, does that mean that the particle is not moving? If a particle moves in empty space with constant momentum $\vec{p}$ and hence constant energy $E = p^2/2m$, is it in a stationary state? Explain your answers.

Q40.6 For the particle in a box, we chose $k = n\pi/L$ with $n = 1, 2, 3, \ldots$ to fit the boundary condition that $\psi = 0$ at $x = L$. However, $n = 0, -1, -2, -3, \ldots$ also satisfy that boundary condition. Why didn't we also choose those values of n?

Q40.7 If ψ is normalized, what is the physical significance of the area under a graph of $|\psi|^2$ versus x between x_1 and x_2? What is the total area under the graph of $|\psi|^2$ when all x are included? Explain.

Q40.8 For a particle in a box, what would the probability distribution function $|\psi|^2$ look like if the particle behaved like a classical (Newtonian) particle? Do the actual probability distributions approach this classical form when n is very large? Explain.

Q40.9 In Chapter 15 we represented a standing wave as a superposition of two waves traveling in opposite directions. Can the wave functions for a particle in a box also be thought of as a combination of two traveling waves? Why or why not? What physical interpretation does this representation have? Explain.

Q40.10 A particle in a box is in the ground level. What is the probability of finding the particle in the right half of the box? (Refer to Fig. 40.12, but don't evaluate an integral.) Is the answer the same if the particle is in an excited level? Explain.

Q40.11 The wave functions for a particle in a box (see Fig. 40.12a) are zero at certain points. Does this mean that the particle can't move past one of these points? Explain.

Q40.12 For a particle confined to an infinite square well, is it correct to say that each state of definite energy is also a state of definite wavelength? Is it also a state of definite momentum? Explain. (*Hint:* Remember that momentum is a vector.)

Q40.13 For a particle in a finite potential well, is it correct to say that each bound state of definite energy is also a state of definite wavelength? Is it a state of definite momentum? Explain.

Q40.14 In Fig. 40.12b, the probability function is zero at the points $x = 0$ and $x = L$, the "walls" of the box. Does this mean that the particle never strikes the walls? Explain.

Q40.15 A particle is confined to a finite potential well in the region $0 < x < L$. How does the area under the graph of $|\psi|^2$ in the region $0 < x < L$ compare to the total area under the graph of $|\psi|^2$ when including all possible x?

Q40.16 Compare the wave functions for the first three energy levels for a particle in a box of width L (see Fig. 40.12a) to the corresponding wave functions for a finite potential well of the same width (see Fig. 40.15a). How does the wavelength in the interval $0 \le x \le L$ for the $n = 1$ level of the particle in a box compare to the corresponding wavelength for the $n = 1$ level of the finite potential well? Use this to explain why E_1 is less than $E_{1-\text{IDW}}$ in the situation depicted in Fig. 40.15b.

Q40.17 It is stated in Section 40.3 that a finite potential well always has at least one bound level, no matter how shallow the well. Does this mean that as $U_0 \rightarrow 0$, $E_1 \rightarrow 0$? Does this violate the Heisenberg uncertainty principle? Explain.

Q40.18 Figure 40.15a shows that the higher the energy of a bound state for a finite potential well, the more the wave function extends outside the well (into the intervals $x < 0$ and $x > L$). Explain why this happens.

Q40.19 In classical (Newtonian) mechanics, the total energy E of a particle can never be less than the potential energy U because the kinetic energy K cannot be negative. Yet in barrier tunneling (see Section 40.4) a particle passes through regions where E is less than U. Is this a contradiction? Explain.

Q40.20 Figure 40.17 shows the scanning tunneling microscope image of 48 iron atoms placed on a copper surface, the pattern indicating the density of electrons on the copper surface. What can you infer about the potential-energy function inside the circle of iron atoms?

Q40.21 Qualitatively, how would you expect the probability for a particle to tunnel through a potential barrier to depend on the height of the barrier? Explain.

Q40.22 The wave function shown in Fig. 40.20 is nonzero for both $x < 0$ and $x > L$. Does this mean that the particle splits into two parts when it strikes the barrier, with one part tunneling through the barrier and the other part bouncing off the barrier? Explain.

Q40.23 The probability distributions for the harmonic oscillator wave functions (see Figs. 40.27 and 40.28) begin to resemble the classical (Newtonian) probability distribution when the quantum number n becomes large. Would the distributions become the same as in the classical case in the limit of very large n? Explain.

Q40.24 In Fig. 40.28, how does the probability of finding a particle in the center half of the region $-A < x < A$ compare to the probability of finding the particle in the outer half of the region? Is this consistent with the physical interpretation of the situation?

Q40.25 Compare the allowed energy levels for the hydrogen atom, the particle in a box, and the harmonic oscillator. What are the values of the quantum number n for the ground level and the second excited level of each system?

Q40.26 Sketch the wave function for the potential-energy well shown in Fig. Q40.26 when E_1 is less than U_0 and when E_3 is greater than U_0.

Figure **Q40.26**

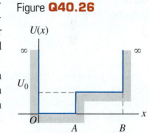

EXERCISES

Section 40.1 Wave Functions and the One-Dimensional Schrödinger Equation

40.1 • An electron is moving as a free particle in the $-x$-direction with momentum that has magnitude 4.50×10^{-24} kg $\cdot$ m/s. What is the one-dimensional time-dependent wave function of the electron?

40.2 • A free particle moving in one dimension has wave function

$$\Psi(x, t) = A\left[e^{i(kx - \omega t)} - e^{i(2kx - 4\omega t)}\right]$$

where k and ω are positive real constants. (a) At $t = 0$ what are the two smallest positive values of x for which the probability function $|\Psi(x, t)|^2$ is a maximum? (b) Repeat part (a) for time $t = 2\pi/\omega$. (c) Calculate v_{av} as the distance the maxima have moved divided by the elapsed time. Compare your result to the expression $v_{av} = (\omega_2 - \omega_1)/(k_2 - k_1)$ from Example 40.1.

40.3 • Consider the free-particle wave function of Example 40.1. Let $k_2 = 3k_1 = 3k$. At $t = 0$ the probability distribution function $|\Psi(x, t)|^2$ has a maximum at $x = 0$. (a) What is the smallest positive value of x for which the probability distribution function has a maximum at time $t = 2\pi/\omega$, where $\omega = \hbar k^2/2m$. (b) From your result in part (a), what is the average speed with which the probability distribution is moving in the $+x$-direction? Compare your result to the expression $v_{av} = (\omega_2 - \omega_1)/(k_2 - k_1)$ from Example 40.1.

40.4 • Consider the free particle of Example 40.1. Show that $v_{av} = (\omega_2 - \omega_1)/(k_2 - k_1)$ can be written as $v_{av} = p_{av}/m$, where $p_{av} = (\hbar k_2 + \hbar k_1)/2$.

40.5 • Consider a wave function given by $\psi(x) = A \sin kx$, where $k = 2\pi/\lambda$ and A is a real constant. (a) For what values of x is there the highest probability of finding the particle described by this wave function? Explain. (b) For which values of x is the probability *zero*? Explain.

40.6 •• Compute $|\Psi|^2$ for $\Psi = \psi \sin \omega t$, where ψ is time independent and ω is a real constant. Is this a wave function for a stationary state? Why or why not?

40.7 • **CALC** Let ψ_1 and ψ_2 be two solutions of Eq. (40.23) with energies E_1 and E_2, respectively, where $E_1 \neq E_2$. Is $\psi = A\psi_1 + B\psi_2$, where A and B are nonzero constants, a solution to Eq. (40.23)? Explain your answer.

40.8 • A particle is described by a wave function $\psi(x) = Ae^{-\alpha x^2}$, where A and α are real, positive constants. If the value of α is increased, what effect does this have on (a) the particle's uncertainty in position and (b) the particle's uncertainty in momentum? Explain your answers.

40.9 • **CALC** **Linear Combinations of Wave Functions.** Let ψ_1 and ψ_2 be two solutions of Eq. (40.23) with the same energy E. Show that $\psi = B\psi_1 + C\psi_2$ is also a solution with energy E, for any values of the constants B and C.

Section 40.2 Particle in a Box

40.10 •• **CALC** A particle moving in one dimension (the x-axis) is described by the wave function

$$\psi(x) = \begin{cases} Ae^{-bx}, & \text{for } x \geq 0 \\ Ae^{bx}, & \text{for } x < 0 \end{cases}$$

where $b = 2.00 \text{ m}^{-1}$, $A > 0$, and the $+x$-axis points toward the right. (a) Determine A so that the wave function is normalized. (b) Sketch the graph of the wave function. (c) Find the probability of finding this particle in each of the following regions: (i) within 50.0 cm of the origin, (ii) on the left side of the origin (can you first guess the answer by looking at the graph of the wave function?), (iii) between $x = 0.500$ m and $x = 1.00$ m.

40.11 • **Ground-Level Billiards.** (a) Find the lowest energy level for a particle in a box if the particle is a billiard ball ($m = 0.20$ kg) and the box has a width of 1.3 m, the size of a billiard table. (Assume that the billiard ball slides without friction rather than rolls; that is, ignore the *rotational* kinetic energy.) (b) Since the energy in part (a) is all kinetic, to what speed does this correspond? How much time would it take at this speed for the ball to move from one side of the table to the other? (c) What is the difference in energy between the $n = 2$ and $n = 1$ levels? (d) Are quantum-mechanical effects important for the game of billiards?

40.12 • A proton is in a box of width L. What must the width of the box be for the ground-level energy to be 5.0 MeV, a typical value for the energy with which the particles in a nucleus are bound? Compare your result to the size of a nucleus—that is, on the order of 10^{-14} m.

40.13 •• Find the width L of a one-dimensional box for which the ground-state energy of an electron in the box equals the absolute value of the ground state of a hydrogen atom.

40.14 •• When a hydrogen atom undergoes a transition from the $n = 2$ to the $n = 1$ level, a photon with $\lambda = 122$ nm is emitted. (a) If the atom is modeled as an electron in a one-dimensional box, what is the width of the box in order for the $n = 2$ to $n = 1$ transition to correspond to emission of a photon of this energy? (b) For a box with the width calculated in part (a), what is the ground-state energy? How does this correspond to the ground-state energy of a hydrogen atom? (c) Do you think a one-dimensional box is a good

model for a hydrogen atom? Explain. (*Hint:* Compare the spacing between adjacent energy levels as a function of n.)

40.15 •• A certain atom requires 3.0 eV of energy to excite an electron from the ground level to the first excited level. Model the atom as an electron in a box and find the width L of the box.

40.16 • An electron in a one-dimensional box has ground-state energy 1.00 eV. What is the wavelength of the photon absorbed when the electron makes a transition to the second excited state?

40.17 • **CALC** Show that the time-dependent wave function given by Eq. (40.35) is a solution to the one-dimensional Schrödinger equation, Eq. (40.23).

40.18 • Recall that $|\psi|^2 \, dx$ is the probability of finding the particle that has normalized wave function $\psi(x)$ in the interval x to $x + dx$. Consider a particle in a box with rigid walls at $x = 0$ and $x = L$. Let the particle be in the ground level and use ψ_n as given in Eq. (40.35). (a) For which values of x, if any, in the range from 0 to L is the probability of finding the particle zero? (b) For which values of x is the probability highest? (c) In parts (a) and (b) are your answers consistent with Fig. 40.12? Explain.

40.19 • Repeat Exercise 40.18 for the particle in the first excited level.

40.20 • **CALC** (a) Show that $\psi = A \sin kx$ is a solution to Eq. (40.25) if $k = \sqrt{2mE}/\hbar$. (b) Explain why this is an acceptable wave function for a particle in a box with rigid walls at $x = 0$ and $x = L$ only if k is an integer multiple of π/L.

40.21 • **CALC** (a) Repeat Exercise 40.20 for $\psi = A \cos kx$. (b) Explain why this cannot be an acceptable wave function for a particle in a box with rigid walls at $x = 0$ and $x = L$ no matter what the value of k.

40.22 • (a) Find the excitation energy from the ground level to the third excited level for an electron confined to a box that has a width of 0.125 nm. (b) The electron makes a transition from the $n = 1$ to $n = 4$ level by absorbing a photon. Calculate the wavelength of this photon.

40.23 • An electron is in a box of width 3.0×10^{-10} m. What are the de Broglie wavelength and the magnitude of the momentum of the electron if it is in (a) the $n = 1$ level; (b) the $n = 2$ level; (c) the $n = 3$ level? In each case how does the wavelength compare to the width of the box?

40.24 •• **CALC** **Normalization of the Wave Function.** Consider a particle moving in one dimension, which we shall call the x-axis. (a) What does it mean for the wave function of this particle to be *normalized*? (b) Is the wave function $\psi(x) = e^{ax}$, where a is a positive real number, normalized? Could this be a valid wave function? (c) If the particle described by the wave function $\psi(x) = Ae^{bx}$, where A and b are positive real numbers, is confined to the range $x \geq 0$, determine A (including its units) so that the wave function is normalized.

Section 40.3 Potential Wells

40.25 • **CALC** (a) Show that $\psi = A \sin kx$, where k is a real (not complex) constant, is *not* a solution of Eq. (40.23) for $U = U_0$ and $E < U_0$. (b) Is this ψ a solution for $E > U_0$?

40.26 •• An electron is moving past the square well shown in Fig. 40.13. The electron has energy $E = 3U_0$. What is the ratio of the de Broglie wavelength of the electron in the region $x > L$ to the wavelength for $0 < x < L$?

40.27 • An electron is bound in a square well of depth $U_0 = 6E_{1-\text{IDW}}$. What is the width of the well if its ground-state energy is 2.00 eV?

40.28 •• An electron is bound in a square well of width 1.50 nm and depth $U_0 = 6E_{1-\text{IDW}}$. If the electron is initially in the ground level and absorbs a photon, what maximum wavelength can the photon have and still liberate the electron from the well?

40.29 • **CALC** Calculate $d^2\psi/dx^2$ for the wave function of Eq. (40.38), and show that the function is a solution of Eq. (40.37).

40.30 •• An electron is bound in a square well with a depth equal to six times the ground-level energy $E_{1-\text{IDW}}$ of an infinite well of the same width. The longest-wavelength photon that is absorbed by the electron has a wavelength of 400.0 nm. Determine the width of the well.

40.31 •• A proton is bound in a square well of width 4.0 fm = 4.0×10^{-15} m. The depth of the well is six times the ground-level energy $E_{1-\text{IDW}}$ of the corresponding infinite well. If the proton makes a transition from the level with energy E_1 to the level with energy E_3 by absorbing a photon, find the wavelength of the photon.

Section 40.4 Potential Barriers and Tunneling

40.32 •• **Alpha Decay.** In a simple model for a radioactive nucleus, an alpha particle ($m = 6.64 \times 10^{-27}$ kg) is trapped by a square barrier that has width 2.0 fm and height 30.0 MeV. (a) What is the tunneling probability when the alpha particle encounters the barrier if its kinetic energy is 1.0 MeV below the top of the barrier (Fig. E40.32)? (b) What is the tunneling probability if the energy of the alpha particle is 10.0 MeV below the top of the barrier?

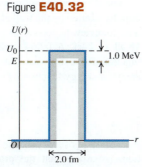

Figure **E40.32**

40.33 • An electron with initial kinetic energy 6.0 eV encounters a barrier with height 11.0 eV. What is the probability of tunneling if the width of the barrier is (a) 0.80 nm and (b) 0.40 nm?

40.34 • An electron with initial kinetic energy 5.0 eV encounters a barrier with height U_0 and width 0.60 nm. What is the transmission coefficient if (a) $U_0 = 7.0$ eV; (b) $U_0 = 9.0$ eV; (c) $U_0 = 13.0$ eV?

40.35 •• An electron is moving past the square barrier shown in Fig. 40.19, but the energy of the electron is *greater* than the barrier height. If $E = 2U_0$, what is the ratio of the de Broglie wavelength of the electron in the region $x > L$ to the wavelength for $0 < x < L$?

40.36 • A proton with initial kinetic energy 50.0 eV encounters a barrier of height 70.0 eV. What is the width of the barrier if the probability of tunneling is 3.0×10^{-3}? How does this compare with the barrier width for an electron with the same energy tunneling through a barrier of the same height with the same probability?

40.37 •• (a) An electron with initial kinetic energy 32 eV encounters a square barrier with height 41 eV and width 0.25 nm. What is the probability that the electron will tunnel through the barrier? (b) A proton with the same kinetic energy encounters the same barrier. What is the probability that the proton will tunnel through the barrier?

Section 40.5 The Harmonic Oscillator

40.38 • **CALC** Show that $\psi(x)$ given by Eq. (40.47) is a solution to Eq. (40.44) with energy $E_0 = \hbar\omega/2$.

40.39 • A wooden block with mass 0.250 kg is oscillating on the end of a spring that has force constant 110 N/m. Calculate the ground-level energy and the energy separation between adjacent levels. Express your results in joules and in electron volts. Are quantum effects important?

40.40 • A harmonic oscillator absorbs a photon of wavelength 8.65×10^{-6} m when it undergoes a transition from the ground state to the first excited state. What is the ground-state energy, in electron volts, of the oscillator?

40.41 • Chemists use infrared absorption spectra to identify chemicals in a sample. In one sample, a chemist finds that light of wavelength 5.8 μm is absorbed when a molecule makes a transition from its ground harmonic oscillator level to its first excited level. (a) Find the energy of this transition. (b) If the molecule can be treated as a harmonic oscillator with mass 5.6×10^{-26} kg, find the force constant.

40.42 •• The ground-state energy of a harmonic oscillator is 5.60 eV. If the oscillator undergoes a transition from its $n = 3$ to $n = 2$ level by emitting a photon, what is the wavelength of the photon?

40.43 • In Section 40.5 it is shown that for the ground level of a harmonic oscillator, $\Delta x \Delta p_x = \hbar/2$. Do a similar analysis for an excited level that has quantum number n. How does the uncertainty product $\Delta x \Delta p_x$ depend on n?

40.44 •• For the ground-level harmonic oscillator wave function $\psi(x)$ given in Eq. (40.47), $|\psi|^2$ has a maximum at $x = 0$. (a) Compute the ratio of $|\psi|^2$ at $x = +A$ to $|\psi|^2$ at $x = 0$, where A is given by Eq. (40.48) with $n = 0$ for the ground level. (b) Compute the ratio of $|\psi|^2$ at $x = +2A$ to $|\psi|^2$ at $x = 0$. In each case is your result consistent with what is shown in Fig. 40.27?

40.45 •• For the sodium atom of Example 40.8, find (a) the ground-state energy, (b) the wavelength of a photon emitted when the $n = 4$ to $n = 3$ transition occurs; (c) the energy difference for any $\Delta n = 1$ transition.

PROBLEMS

40.46 • The discussion in Section 40.1 shows that the wave function $\Psi = \psi e^{-i\omega t}$ is a stationary state, where ψ is time independent and ω is a real (not complex) constant. Consider the wave function $\Psi = \psi_1 e^{-i\omega_1 t} + \psi_2 e^{-i\omega_2 t}$, where ψ_1 and ψ_2 are different time-independent functions and ω_1 and ω_2 are different real constants. Assume that ψ_1 and ψ_2 are real-valued functions, so that $\psi_1^* = \psi_1$ and $\psi_2^* = \psi_2$. Is this Ψ a wave function for a stationary state? Why or why not?

40.47 •• A particle of mass m in a one-dimensional box has the following wave function in the region $x = 0$ to $x = L$:

$$\Psi(x, t) = \frac{1}{\sqrt{2}} \psi_1(x) e^{-iE_1 t/\hbar} + \frac{1}{\sqrt{2}} \psi_3(x) e^{-iE_3 t/\hbar}$$

Here $\psi_1(x)$ and $\psi_3(x)$ are the normalized stationary-state wave functions for the $n = 1$ and $n = 3$ levels, and E_1 and E_3 are the energies of these levels. The wave function is zero for $x < 0$ and for $x > L$. (a) Find the value of the probability distribution function at $x = L/2$ as a function of time. (b) Find the angular frequency at which the probability distribution function oscillates.

40.48 •• **CALC** Consider the wave packet defined by

$$\psi(x) = \int_0^\infty B(k) \cos kx \, dk$$

Let $B(k) = e^{-\alpha^2 k^2}$. (a) The function $B(k)$ has its maximum value at $k = 0$. Let k_h be the value of k at which $B(k)$ has fallen to half its maximum value, and define the width of $B(k)$ as $w_k = k_\text{h}$. In terms of α, what is w_k? (b) Use integral tables to evaluate the integral that gives $\psi(x)$. For what value of x is $\psi(x)$ maximum? (c) Define the width of $\psi(x)$ as $w_x = x_\text{h}$, where x_h is the positive value of x at which $\psi(x)$ has fallen to half its maximum value. Calculate w_x in terms of α. (d) The momentum p is equal to $hk/2\pi$, so the width of B in momentum is $w_p = hw_k/2\pi$. Calculate the product $w_p w_x$ and compare to the Heisenberg uncertainty principle.

40.49 •• **CALC** (a) Using the integral in Problem 40.48, determine the wave function $\psi(x)$ for a function $B(k)$ given by

$$B(k) = \begin{cases} 0 & k < 0 \\ 1/k_0, & 0 \le k \le k_0 \\ 0, & k > k_0 \end{cases}$$

This represents an equal combination of all wave numbers between 0 and k_0. Thus $\psi(x)$ represents a particle with average wave number $k_0/2$, with a total spread or uncertainty in wave number of k_0. We will call this spread the *width* w_k of $B(k)$, so $w_k = k_0$. (b) Graph $B(k)$ versus k and $\psi(x)$ versus x for the case $k_0 = 2\pi/L$, where L is a length. Locate the point where $\psi(x)$ has its maximum value and label this point on your graph. Locate the two points closest to this maximum (one on each side of it) where $\psi(x) = 0$, and define the distance along the x-axis between these two points as w_x, the width of $\psi(x)$. Indicate the distance w_x on your graph. What is the value of w_x if $k_0 = 2\pi/L$? (c) Repeat part (b) for the case $k_0 = \pi/L$. (d) The momentum p is equal to $hk/2\pi$, so the width of B in momentum is $w_p = hw_k/2\pi$. Calculate the product $w_p w_x$ for each of the cases $k_0 = 2\pi/L$ and $k_0 = \pi/L$. Discuss your results in light of the Heisenberg uncertainty principle.

40.50 • **CALC** Show that the wave function $\psi(x) = Ae^{ikx}$ is a solution of Eq. (40.23) for a particle of mass m, in a region where the potential energy is a constant $U_0 < E$. Find an expression for k, and relate it to the particle's momentum and to its de Broglie wavelength.

40.51 •• **CALC** Wave functions like the one in Problem 40.50 can represent free particles moving with velocity $v = p/m$ in the x-direction. Consider a beam of such particles incident on a potential-energy step $U(x) = 0$, for $x < 0$, and $U(x) = U_0 < E$, for $x > 0$. The wave function for $x < 0$ is $\psi(x) = Ae^{ik_1x} + Be^{-ik_1x}$, representing incident and reflected particles, and for $x > 0$ is $\psi(x) = Ce^{ik_2x}$, representing transmitted particles. Use the conditions that both ψ and its first derivative must be continuous at $x = 0$ to find the constants B and C in terms of k_1, k_2, and A.

40.52 • Let ΔE_n be the energy difference between the adjacent energy levels E_n and E_{n+1} for a particle in a box. The ratio $R_n = \Delta E_n/E_n$ compares the energy of a level to the energy separation of the next higher energy level. (a) For what value of n is R_n largest, and what is this largest R_n? (b) What does R_n approach as n becomes very large? How does this result compare to the classical value for this quantity?

40.53 • **Photon in a Dye Laser.** An electron in a long, organic molecule used in a dye laser behaves approximately like a particle in a box with width 4.18 nm. What is the wavelength of the photon emitted when the electron undergoes a transition (a) from the first excited level to the ground level and (b) from the second excited level to the first excited level?

40.54 • **CALC** A particle is in the ground level of a box that extends from $x = 0$ to $x = L$. (a) What is the probability of finding the particle in the region between 0 and $L/4$? Calculate this by integrating $|\psi(x)|^2 \, dx$, where ψ is normalized, from $x = 0$ to $x = L/4$. (b) What is the probability of finding the particle in the region $x = L/4$ to $x = L/2$? (c) How do the results of parts (a) and (b) compare? Explain. (d) Add the probabilities calculated in parts (a) and (b). (e) Are your results in parts (a), (b), and (d) consistent with Fig. 40.12b? Explain.

40.55 •• **CALC** What is the probability of finding a particle in a box of length L in the region between $x = L/4$ and $x = 3L/4$ when the particle is in (a) the ground level and (b) the first excited level? (*Hint:* Integrate $|\psi(x)|^2 \, dx$, where ψ is normalized, between $L/4$ and $3L/4$.) (c) Are your results in parts (a) and (b) consistent with Fig. 40.12b? Explain.

40.56 •• Consider a particle in a box with rigid walls at $x = 0$ and $x = L$. Let the particle be in the ground level. Calculate the probability $|\psi|^2 \, dx$ that the particle will be found in the interval x to $x + dx$ for (a) $x = L/4$; (b) $x = L/2$; (c) $x = 3L/4$.

40.57 •• Repeat Problem 40.56 for a particle in the first excited level.

40.58 •• **CP** A particle is confined within a box with perfectly rigid walls at $x = 0$ and $x = L$. Although the magnitude of the instantaneous force exerted on the particle by the walls is infinite and the time over which it acts is zero, the impulse (that involves a product of force and time) is both finite and quantized. Show that the impulse exerted by the wall at $x = 0$ is $(nh/L)\hat{\imath}$ and that the impulse exerted by the wall at $x = L$ is $-(nh/L)\hat{\imath}$. (*Hint:* You may wish to review Section 8.1.)

40.59 •• **CALC** A fellow student proposes that a possible wave function for a free particle with mass m (one for which the potential-energy function $U(x)$ is zero) is

$$\psi(x) = \begin{cases} e^{+\kappa x}, & x < 0 \\ e^{-\kappa x}, & x \ge 0 \end{cases}$$

where κ is a positive constant. (a) Graph this proposed wave function. (b) Show that the proposed wave function satisfies the Schrödinger equation for $x < 0$ if the energy is $E = -\hbar^2\kappa^2/2m$—that is, if the energy of the particle is *negative*. (c) Show that the proposed wave function also satisfies the Schrödinger equation for $x \ge 0$ with the same energy as in part (b). (d) Explain why the proposed wave function is nonetheless *not* an acceptable solution of the Schrödinger equation for a free particle. (*Hint:* What is the behavior of the function at $x = 0$?) It is in fact impossible for a free particle (one for which $U(x) = 0$) to have an energy less than zero.

40.60 •• The *penetration distance* η in a finite potential well is the distance at which the wave function has decreased to $1/e$ of the wave function at the classical turning point:

$$\psi(x = L + \eta) = \frac{1}{e}\psi(L)$$

The penetration distance can be shown to be

$$\eta = \frac{\hbar}{\sqrt{2m(U_0 - E)}}$$

The probability of finding the particle beyond the penetration distance is nearly zero. (a) Find η for an electron having a kinetic energy of 13 eV in a potential well with $U_0 = 20$ eV. (b) Find η for a 20.0-MeV proton trapped in a 30.0-MeV-deep potential well.

40.61 • **CALC** (a) For the finite potential well of Fig. 40.13, what relationships among the constants A and B of Eq. (40.38) and C and D of Eq. (40.40) are obtained by applying the boundary condition that ψ be continuous at $x = 0$ and at $x = L$? (b) What relationships among A, B, C, and D are obtained by applying the boundary condition that $d\psi/dx$ be continuous at $x = 0$ and at $x = L$?

40.62 • An electron with initial kinetic energy 5.5 eV encounters a square potential barrier with height 10.0 eV. What is the width of the barrier if the electron has a 0.10% probability of tunneling through the barrier?

40.63 •• A particle with mass m and total energy E tunnels through a square barrier of height U_0 and width L. When the transmission coefficient is *not* much less than unity, it is given by

$$T = \left[1 + \frac{(U_0 \sinh \kappa L)^2}{4E(U_0 - E)}\right]^{-1}$$

where $\sinh \kappa L = (e^{\kappa L} - e^{-\kappa L})/2$ is the hyperbolic sine of κL. (a) Show that if $\kappa L \gg 1$, this expression for T approaches Eq. (40.42). (b) Explain why the restriction $\kappa L \gg 1$ in part (a) implies either that the barrier is relatively wide or that the energy E is relatively low compared to U_0. (c) Show that as the particle's incident kinetic energy E approaches the barrier height U_0, T approaches $[1 + (kL/2)^2]^{-1}$, where $k = \sqrt{2mE}/\hbar$ is the wave number of the incident particle. (*Hint:* If $|z| \ll 1$, then $\sinh z \approx z$.)

40.64 • **CP** A harmonic oscillator consists of a 0.020-kg mass on a spring. Its frequency is 1.50 Hz, and the mass has a speed of 0.360 m/s as it passes the equilibrium position. (a) What is the value of the quantum number n for its energy level? (b) What is the difference in energy between the levels E_n and E_{n+1}? Is this difference detectable?

40.65 • For small amplitudes of oscillation the motion of a pendulum is simple harmonic. For a pendulum with a period of 0.500 s, find the ground-level energy and the energy difference between adjacent energy levels. Express your results in joules and in electron volts. Are these values detectable?

40.66 •• Some 164.9-nm photons are emitted in a $\Delta n = 1$ transition within a solid-state lattice. The lattice is modeled as electrons in a box having length 0.500 nm. What transition corresponds to the emitted light?

40.67 •• **CALC** Show that for $\psi(x)$ given by Eq. (40.47), the probability distribution function has a maximum at $x = 0$.

40.68 •• **CALC** (a) Show by direct substitution in the Schrödinger equation for the one-dimensional harmonic oscillator that the wave function $\psi_1(x) = A_1 x e^{-\alpha^2 x^2/2}$, where $\alpha^2 = m\omega/\hbar$, is a solution with energy corresponding to $n = 1$ in Eq. (40.46). (b) Find the normalization constant A_1. (c) Show that the probability density has a minimum at $x = 0$ and maxima at $x = \pm 1/\alpha$, corresponding to the classical turning points for the ground state $n = 0$.

40.69 •• **CP** (a) The wave nature of particles results in the quantum-mechanical situation that a particle confined in a box can assume only wavelengths that result in standing waves in the box, with nodes at the box walls. Use this to show that an electron confined in a one-dimensional box of length L will have energy levels given by

$$E_n = \frac{n^2 h^2}{8mL^2}$$

(*Hint:* Recall that the relationship between the de Broglie wavelength and the speed of a nonrelativistic particle is $mv = h/\lambda$. The energy of the particle is $\frac{1}{2}mv^2$.) (b) If a hydrogen atom is modeled as a one-dimensional box with length equal to the Bohr radius, what is the energy (in electron volts) of the lowest energy level of the electron?

40.70 ••• Consider a potential well defined as $U(x) = \infty$ for $x < 0$, $U(x) = 0$ for $0 < x < L$, and $U(x) = U_0 > 0$ for $x > L$ (Fig. P40.70). Consider a particle with mass m and kinetic energy $E < U_0$ that is trapped in the well. (a) The boundary condition at the infinite wall $(x = 0)$ is $\psi(0) = 0$. What must the form of

the function $\psi(x)$ for $0 < x < L$ be in order to satisfy both the Schrödinger equation and this boundary condition? (b) The wave function must remain finite as $x \rightarrow \infty$. What must the form of the function $\psi(x)$ for $x > L$ be in order to satisfy both the Schrödinger equation and this boundary condition at infinity? (c) Impose the boundary conditions that ψ and $d\psi/dx$ are continuous at $x = L$. Show that the energies of the allowed levels are obtained from solutions of the equation $k \cot kL = -\kappa$, where $k = \sqrt{2mE}/\hbar$ and $\kappa = \sqrt{2m(U_0 - E)}/\hbar$.

Figure **P40.70**

40.71 ••• Section 40.2 considered a box with walls at $x = 0$ and $x = L$. Now consider a box with width L but centered at $x = 0$, so that it extends from $x = -L/2$ to $x = +L/2$ (Fig. P40.71). Note that this box is symmetric about $x = 0$. (a) Consider possible wave functions of the form $\psi(x) = A \sin kx$. Apply the boundary conditions at the wall to obtain the allowed energy levels. (b) Another set of possible wave functions are functions of the form $\psi(x) = A \cos kx$. Apply the boundary conditions at the wall to obtain the allowed energy levels. (c) Compare the energies obtained in parts (a) and (b) to the set of energies given in Eq. (40.31). (d) An odd function f satisfies the condition $f(x) = -f(-x)$, and an even function g satisfies $g(x) = g(-x)$. Of the wave functions from parts (a) and (b), which are even and which are odd?

Figure **P40.71**

CHALLENGE PROBLEMS

40.72 ••• **CALC** **The WKB Approximation.** It can be a challenge to solve the Schrödinger equation for the bound-state energy levels of an arbitrary potential well. An alternative approach that can yield good approximate results for the energy levels is the *WKB approximation* (named for the physicists Gregor Wentzel, Hendrik Kramers, and Léon Brillouin, who pioneered its application to quantum mechanics). The WKB approximation begins from three physical statements: (i) According to de Broglie, the magnitude of momentum p of a quantum-mechanical particle is $p = h/\lambda$. (ii) The magnitude of momentum is related to the kinetic energy K by the relationship $K = p^2/2m$. (iii) If there are no nonconservative forces, then in Newtonian mechanics the energy E for a particle is constant and equal at each point to the sum of the kinetic and potential energies at that point: $E = K + U(x)$, where x is the coordinate. (a) Combine these three relationships to show that the wavelength of the particle at a coordinate x can be written as

$$\lambda(x) = \frac{h}{\sqrt{2m[E - U(x)]}}$$

Thus we envision a quantum-mechanical particle in a potential well $U(x)$ as being like a free particle, but with a wavelength $\lambda(x)$ that is a function of position. (b) When the particle moves into a region of increasing potential energy, what happens to its wavelength? (c) At a point where $E = U(x)$, Newtonian mechanics says that the particle has zero kinetic energy and must be instantaneously at rest. Such a point is called a *classical turning point*, since this is where a Newtonian particle must stop its motion and reverse direction. As an example, an object oscillating in simple harmonic motion with amplitude A moves back and forth between the points $x = -A$ and $x = +A$; each of these is a classical turning point, since there the potential energy $\frac{1}{2}k'x^2$ equals the total energy $\frac{1}{2}k'A^2$. In the WKB expression for $\lambda(x)$, what is the wavelength at a classical turning point? (d) For a particle in a box with length L, the walls of the box are classical turning points (see Fig. 40.8). Furthermore, the number of wavelengths that fit within the box must be a half-integer (see Fig. 40.10), so that $L = (n/2)\lambda$ and hence $L/\lambda = n/2$, where $n = 1, 2, 3, \ldots$. [Note that this is a restatement of Eq. (40.29).] The WKB scheme for finding the allowed bound-state energy levels of an *arbitrary* potential well is an extension of these observations. It demands that for an allowed energy E, there must be a half-integer number of wavelengths between the classical turning points for that energy. Since the wavelength in the WKB approximation is not a constant but depends on x, the number of wavelengths between the classical turning points a and b for a given value of the energy is the integral of $1/\lambda(x)$ between those points:

$$\int_a^b \frac{dx}{\lambda(x)} = \frac{n}{2} \quad (n = 1, 2, 3, \ldots)$$

Using the expression for $\lambda(x)$ you found in part (a), show that the *WKB condition for an allowed bound-state energy* can be written as

$$\int_a^b \sqrt{2m[E - U(x)]}\, dx = \frac{nh}{2} \quad (n = 1, 2, 3, \ldots)$$

(e) As a check on the expression in part (d), apply it to a particle in a box with walls at $x = 0$ and $x = L$. Evaluate the integral and show that the allowed energy levels according to the WKB approximation are the same as those given by Eq. (40.31). (*Hint:* Since the walls of the box are infinitely high, the points $x = 0$ and $x = L$ are classical turning points for *any* energy E. Inside the box, the potential energy is zero.) (f) For the finite square well shown in Fig. 40.13, show that the WKB expression given in part (d) predicts the *same* bound-state energies as for an infinite square well of the same width. (*Hint:* Assume $E < U_0$. Then the classical turning points are at $x = 0$ and $x = L$.) This shows that the WKB approximation does a poor job when the potential-energy function changes discontinuously, as for a finite potential well. In the next two problems we consider situations in which the potential-energy function changes gradually and the WKB approximation is much more useful.

40.73 ••• **CALC** The WKB approximation (see Challenge Problem 40.72) can be used to calculate the energy levels for a harmonic oscillator. In this approximation, the energy levels are the solutions to the equation

$$\int_a^b \sqrt{2m[E - U(x)]}\, dx = \frac{nh}{2} \quad n = 1, 2, 3, \ldots$$

Here E is the energy, $U(x)$ is the potential-energy function, and $x = a$ and $x = b$ are the classical turning points (the points at

which E is equal to the potential energy, so the Newtonian kinetic energy would be zero). (a) Determine the classical turning points for a harmonic oscillator with energy E and force constant k'. (b) Carry out the integral in the WKB approximation and show that the energy levels in this approximation are $E_n = \hbar\omega$, where $\omega = \sqrt{k'/m}$ and $n = 1, 2, 3, \ldots$. (*Hint:* Recall that $\hbar = h/2\pi$. A useful standard integral is

$$\int \sqrt{A^2 - x^2}\, dx = \frac{1}{2}\left[x\sqrt{A^2 - x^2} + A^2 \arcsin\left(\frac{x}{|A|}\right)\right]$$

where arcsin denotes the inverse sine function. Note that the integrand is even, so the integral from $-x$ to x is equal to twice the integral from 0 to x.) (c) How do the approximate energy levels found in part (b) compare with the true energy levels given by Eq. (40.46)? Does the WKB approximation give an underestimate or an overestimate of the energy levels?

40.74 ••• **CALC** Protons, neutrons, and many other particles are made of more fundamental particles called *quarks* and *antiquarks* (the antimatter equivalent of quarks). A quark and an antiquark can form a bound state with a variety of different energy levels, each of which corresponds to a different particle observed in the laboratory. As an example, the ψ particle is a low-energy bound state of a so-called charm quark and its antiquark, with a rest energy of 3097 MeV; the $\psi(2S)$ particle is an excited state of this same quark–antiquark combination, with a rest energy of 3686 MeV. A simplified representation of the potential energy of interaction between a quark and an antiquark is $U(x) = A|x|$, where A is a positive constant and x represents the distance between the quark and the antiquark. You can use the WKB approximation (see Challenge Problem 40.72) to determine the bound-state energy levels for this potential-energy function. In the WKB approximation, the energy levels are the solutions to the equation

$$\int_a^b \sqrt{2m[E - U(x)]}\, dx = \frac{nh}{2} \quad (n = 1, 2, 3, \ldots)$$

Here E is the energy, $U(x)$ is the potential-energy function, and $x = a$ and $x = b$ are the classical turning points (the points at which E is equal to the potential energy, so the Newtonian kinetic energy would be zero). (a) Determine the classical turning points for the potential $U(x) = A|x|$ and for an energy E. (b) Carry out the above integral and show that the allowed energy levels in the WKB approximation are given by

$$E_n = \frac{1}{2m}\left(\frac{3mAh}{4}\right)^{2/3} n^{2/3} \quad (n = 1, 2, 3, \ldots)$$

(*Hint:* The integrand is even, so the integral from $-x$ to x is equal to twice the integral from 0 to x.) (c) Does the difference in energy between successive levels increase, decrease, or remain the same as n increases? How does this compare to the behavior of the energy levels for the harmonic oscillator? For the particle in a box? Can you suggest a simple rule that relates the difference in energy between successive levels to the shape of the potential-energy function?

Answers

Chapter Opening Question

When an electron in one of these particles—called *quantum dots*—makes a transition from an excited level to a lower level, it emits a photon whose energy is equal to the difference in energy between the levels. The smaller the quantum dot, the larger the energy spacing between levels and hence the shorter (bluer) the wavelength of the emitted photons. See Example 40.6 (Section 40.3) for more details.

Test Your Understanding Questions

40.1 Answer: no Equation (40.19) represents a superposition of wave functions with different values of wave number k and hence different values of energy $E = \hbar^2 k^2/2m$. The state that this combined wave function represents is not a state of definite energy, and therefore not a stationary state. Another way to see this is to note that there is a factor $e^{-iEt/\hbar}$ inside the integral in Eq. (40.19), with a different value of E for each value of k. This wave function therefore has a very complicated time dependence, and the probability distribution function $|\Psi(x, t)|^2$ does depend on time.

40.2 Answer: (v) Our derivation of the stationary-state wave functions for a particle in a box shows that they are superpositions of waves propagating in opposite directions, just like a standing wave on a string. One wave has momentum in the positive x-direction, while the other wave has an equal magnitude of momentum in the negative x-direction. The *total* x-component of momentum is zero.

40.3 Answer: (i) The energy levels are arranged as shown in Fig. 40.15b if $U_0 = 6E_{1-\text{IDW}}$, where $E_{1-\text{IDW}} = \pi^2\hbar^2/2mL^2$ is the ground-level energy of an infinite well. If the well width L is reduced to one-half of its initial value, $E_{1-\text{IDW}}$ increases by a factor of four and so U_0 must also increase by a factor of four. The energies E_1, E_2, and E_3 shown in Fig. 40.15b are all specific fractions of U_0, so they will also increase by a factor of four.

40.4 Answer: yes Figure 40.20 shows a possible wave function $\psi(x)$ for tunneling. Since $\psi(x)$ is not zero within the barrier $(0 \leq x \leq L)$, there is some probability that the particle can be found there.

40.5 Answer: (ii) If the second photon has a longer wavelength and hence lower energy than the first photon, the difference in energy between the first and second excited levels must be less than the difference between the ground level and the first excited level. This is the case for the hydrogen atom, for which the energy difference between levels decreases as the energy increases (see Fig. 39.24). By contrast, the energy difference between successive levels increases for a particle in a box (see Fig. 40.11b) and is constant for a harmonic oscillator (see Fig. 40.25).

Bridging Problem

Answers: (a)

$$|\Psi(x, t)|^2 = \frac{1}{L}\left[\sin^2\frac{\pi x}{L} + \sin^2\frac{2\pi x}{L} \right.$$

$$\left. + 2\sin\frac{\pi x}{L}\sin\frac{2\pi x}{L}\cos\left(\frac{(E_2 - E_1)t}{\hbar}\right) \right]$$

(b) no   **(c)** yes   **(d)** $\dfrac{3\pi^2\hbar}{2mL^2}$ **(e)** $\dfrac{0.905\pi^2\hbar}{mL^2}$

41 ATOMIC STRUCTURE

LEARNING GOALS

By studying this chapter, you will learn:

- How to extend quantum-mechanical calculations to three-dimensional problems.

- How to solve the Schrödinger equation for a particle trapped in a cubical box.

- How to describe the states of a hydrogen atom in terms of quantum numbers.

- How magnetic fields affect the orbital motion of atomic electrons.

- How we know that electrons have their own intrinsic angular momentum.

- How to analyze the structure of many-electron atoms.

- How x rays emitted by atoms reveal their inner structure.

? Lithium (with three electrons per atom) is a metal that burns spontaneously in water, while helium (with two electrons per atom) is a gas that undergoes almost no chemical reactions. How can one extra electron make these two elements so dramatically different?

Some physicists claim that all of chemistry is contained in the Schrödinger equation. This is somewhat of an exaggeration, but this equation can teach us a great deal about the chemical behavior of elements and the nature of chemical bonds. It provides insight into the periodic table of the elements and the microscopic basis of magnetism.

In order to learn about the quantum-mechanical structure of atoms, we'll first construct a three-dimensional version of the Schrödinger equation. We'll try this equation out by looking at a three-dimensional version of a particle in a box: a particle confined to a cubical volume.

We'll then see that we can learn a great deal about the structure and properties of *all* atoms from the solutions to the Schrödinger equation for the hydrogen atom. These solutions have quantized values of angular momentum; we don't need to make a separate statement about quantization as we did with the Bohr model. We label the states with a set of quantum numbers, which we'll use later with many-electron atoms as well. We'll find that the electron also has an intrinsic *spin* angular momentum in addition to the orbital angular momentum associated with its motion.

We'll also encounter the exclusion principle, a kind of microscopic zoning ordinance that is the key to understanding many-electron atoms. This principle says that no two electrons in an atom can have the same quantum-mechanical state. Finally, we'll use the principles of this chapter to explain the characteristic x-ray spectra of atoms.

41.1 The Schrödinger Equation in Three Dimensions

We have discussed the Schrödinger equation and its applications only for *one-dimensional* problems, the analog of a Newtonian particle moving along a straight line. The straight-line model is adequate for some applications, but to understand atomic structure, we need a three-dimensional generalization.

It's not difficult to guess what the three-dimensional Schrödinger equation should look like. First, the wave function Ψ is a function of time and all three space coordinates (x, y, z). In general, the potential-energy function also depends on all three coordinates and can be written as $U(x, y, z)$. Next, recall from Section 40.1 that the term $-(\hbar^2/2m)\partial^2\Psi/\partial x^2$ in the one-dimensional Schrödinger equation, Eq. (40.20), is related to the kinetic energy of the particle in the state described by the wave function Ψ. For example, if we insert into this term the wave function $\Psi(x, t) = Ae^{ikx}e^{-i\omega t}$ for a free particle with magnitude of momentum $p = \hbar k$ and kinetic energy $K = p^2/2m$, we obtain $-(\hbar^2/2m)(ik)^2Ae^{ikx}e^{-i\omega t} = (\hbar^2 k^2/2m)Ae^{ikx}e^{-i\omega t} = (p^2/2m)\Psi(x, t) = K\Psi(x, t)$. If the particle can move in three dimensions, its momentum has three components (p_x, p_y, p_z) and its kinetic energy is

$$K = \frac{p_x^2}{2m} + \frac{p_y^2}{2m} + \frac{p_z^2}{2m} \tag{41.1}$$

These observations, taken together, suggest that the correct generalization of the Schrödinger equation to three dimensions is

$$-\frac{\hbar^2}{2m}\left(\frac{\partial^2\Psi(x, y, z, t)}{\partial x^2} + \frac{\partial^2\Psi(x, y, z, t)}{\partial y^2} + \frac{\partial^2\Psi(x, y, z, t)}{\partial z^2}\right)$$

$$+ U(x, y, z)\Psi(x, y, z, t) = i\hbar\frac{\partial\Psi(x, y, z, t)}{\partial t} \tag{41.2}$$

(general three-dimensional Schrödinger equation)

The three-dimensional wave function $\Psi(x, y, z, t)$ has a similar interpretation as in one dimension. The wave function itself is a complex quantity with both a real part and an imaginary part, but $|\Psi(x, y, z, t)|^2$—the square of its absolute value, equal to the product of $\Psi(x, y, z, t)$ and its complex conjugate $\Psi^*(x, y, z, t)$— is real and either positive or zero at every point in space. We interpret $|\Psi(x, y, z, t)|^2\, dV$ as the *probability* of finding the particle within a small volume dV centered on the point (x, y, z) at time t, so $|\Psi(x, y, z, t)|^2$ is the *probability distribution function* in three dimensions. The *normalization condition* on the wave function is that the probability that the particle is *somewhere* in space is exactly 1. Hence the integral of $|\Psi(x, y, z, t)|^2$ over all space must equal 1:

$$\int |\Psi(x, y, z, t)|^2\, dV = 1 \qquad \text{(normalization condition in three dimensions)} \tag{41.3}$$

If the wave function $\Psi(x, y, z, t)$ represents a state of a definite energy E— that is, a stationary state—we can write it as the product of a spatial wave function $\psi(x, y, z)$ and a function of time $e^{-iEt/\hbar}$:

$$\Psi(x, y, z, t) = \psi(x, y, z)e^{-iEt/\hbar} \qquad \text{(time-dependent wave function for a state of definite energy)} \tag{41.4}$$

(Compare this to Eq. (40.21) for a one-dimensional state of definite energy.) If we substitute Eq. (41.4) into Eq. (41.2), the right-hand side of the equation becomes $i\hbar\psi(x, y, z)(-iE/\hbar)e^{-iEt/\hbar} = E\psi(x, y, z)e^{-iEt/\hbar}$. We can then divide both sides by the factor $e^{-iEt/\hbar}$, leaving the *time-independent* Schrödinger equation in three dimensions for a stationary state:

$$-\frac{\hbar^2}{2m}\left(\frac{\partial^2\psi(x, y, z)}{\partial x^2} + \frac{\partial^2\psi(x, y, z)}{\partial y^2} + \frac{\partial^2\psi(x, y, z)}{\partial z^2}\right) + U(x, y, z)\psi(x, y, z)$$

$$= E\psi(x, y, z) \qquad \text{(three-dimensional time-independent} \qquad (41.5)$$
$$\text{Schrödinger equation)}$$

The probability distribution function for a stationary state is just the square of the absolute value of the spatial wave function: $|\psi(x, y, z)e^{-iEt/\hbar}|^2 = \psi^*(x, y, z)e^{+iEt/\hbar}\psi(x, y, z)e^{-iEt/\hbar} = |\psi(x, y, z)|^2$. Note that this doesn't depend on time. (As we discussed in Section 40.1, that's why we call these states *stationary*.) Hence for a stationary state the wave function normalization condition, Eq. (41.3), becomes

$$\int |\psi(x, y, z)|^2 \, dV = 1 \qquad \text{(normalization condition for a} \qquad (41.6)$$
$$\text{stationary state in three dimensions)}$$

We won't pretend that we have *derived* Eqs. (41.2) and (41.5). Like their one-dimensional versions, these equations have to be tested by comparison of their predictions with experimental results. Happily, Eqs. (41.2) and (41.5) both pass this test with flying colors, so we are confident that they *are* the correct equations.

An important topic that we will address in this chapter is the solutions for Eq. (41.5) for the stationary states of the hydrogen atom. The potential-energy function for an electron in a hydrogen atom is *spherically symmetric;* it depends only on the distance $r = (x^2 + y^2 + z^2)^{1/2}$ from the origin of coordinates. To take advantage of this symmetry, it's best to use *spherical coordinates* rather than the Cartesian coordinates (x, y, z) to solve the Schrödinger equation for the hydrogen atom. Before introducing these new coordinates and investigating the hydrogen atom, it's useful to look at the three-dimensional version of the particle in a box that we considered in Section 40.2. Solving this simpler problem will give us insight into the more complicated stationary states found in atomic physics.

Test Your Understanding of Section 41.1   In a certain region of space the potential-energy function for a quantum-mechanical particle is zero. In this region the wave function $\psi(x, y, z)$ for a certain stationary state satisfies $\partial^2\psi/\partial x^2 > 0$, $\partial^2\psi/\partial y^2 > 0$, and $\partial^2\psi/\partial z^2 > 0$. The particle has a definite energy E that is positive. What can you conclude about $\psi(x, y, z)$ in this region? (i) It must be positive; (ii) it must be negative; (iii) it must be zero; (iv) not enough information given to decide. ∎

41.2 Particle in a Three-Dimensional Box

41.1 A particle is confined in a cubical box with walls at $x = 0$, $x = L$, $y = 0$, $y = L$, $z = 0$, and $z = L$.

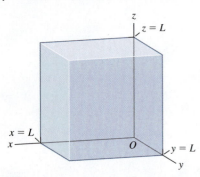

Consider a particle enclosed within a cubical box of side L. This could represent an electron that's free to move anywhere within the interior of a solid metal cube but cannot escape the cube. We'll choose the origin to be at one corner of the box, with the x-, y-, and z-axes along edges of the box. Then the particle is confined to the region $0 \le x \le L$, $0 \le y \le L$, $0 \le z \le L$ (Fig. 41.1). What are the stationary states of this system?

As for the particle in a one-dimensional box that we considered in Section 40.2, we'll say that the potential energy is zero inside the box but infinite outside. Hence the spatial wave function $\psi(x, y, z)$ must be zero outside the box in order that the term $U(x, y, z)\psi(x, y, z)$ in the time-independent Schrödinger equation, Eq. (41.5), not be infinite. Hence the probability distribution function $|\psi(x, y, z)|^2$ is zero outside the box, and there is zero probability that the particle will be found there. Inside the box, the spatial wave function for a stationary state obeys the time-independent Schrödinger equation, Eq. (41.5), with $U(x, y, z) = 0$:

$$-\frac{\hbar^2}{2m}\left(\frac{\partial^2\psi(x, y, z)}{\partial x^2} + \frac{\partial^2\psi(x, y, z)}{\partial y^2} + \frac{\partial^2\psi(x, y, z)}{\partial z^2}\right) = E\psi(x, y, z)$$

(particle in a three-dimensional box)   (41.7)

In order for the wave function to be continuous from the inside to the outside of the box, $\psi(x, y, z)$ must equal zero on the walls. Hence our boundary conditions are that $\psi(x, y, z) = 0$ at $x = 0$, $x = L$, $y = 0$, $y = L$, $z = 0$, and $z = L$.

Guessing a solution to a complicated partial differential equation like Eq. (41.7) seems like quite a challenge. To make progress, recall that we wrote the time-*dependent* wave function for a stationary state as the product of one function that depends only on the spatial coordinates x, y, and z and a second function that depends only on the time t: $\Psi(x, y, z, t) = \psi(x, y, z)e^{-iEt/\hbar}$. In the same way, let's try a technique called *separation of variables:* We'll write the spatial wave function $\psi(x, y, z)$ as a product of one function X that depends only on x, a second function Y that depends only on y, and a third function Z that depends only on z:

$$\psi(x, y, z) = X(x)Y(y)Z(z) \qquad (41.8)$$

If we substitute Eq. (41.8) into Eq. (41.7), we get

$$-\frac{\hbar^2}{2m}\left(Y(y)Z(z)\frac{d^2X(x)}{dx^2} + X(x)Z(z)\frac{d^2Y(y)}{dy^2} + X(x)Y(y)\frac{d^2Z(z)}{dz^2}\right)$$
$$= EX(x)Y(y)Z(z) \qquad (41.9)$$

The partial derivatives in Eq. (41.7) have become ordinary derivatives since they act on functions of a single variable. Now we divide both sides of Eq. (41.9) by the product $X(x)Y(y)Z(z)$:

$$\left(-\frac{\hbar^2}{2m}\frac{1}{X(x)}\frac{d^2X(x)}{dx^2}\right) + \left(-\frac{\hbar^2}{2m}\frac{1}{Y(y)}\frac{d^2Y(y)}{dy^2}\right) + \left(-\frac{\hbar^2}{2m}\frac{1}{Z(z)}\frac{d^2Z(z)}{dz^2}\right) = E$$

(41.10)

The right-hand side of Eq. (41.10) is the energy of the stationary state, which does not and cannot depend on the values of x, y, or z. For this to be true, the left-hand side of the equation must also be independent of the values of x, y, and z. Hence the first term in parentheses on the left-hand side of Eq. (41.10) must equal a constant that doesn't depend on x, the second term in parentheses must equal another constant that doesn't depend on y, and the third term in parentheses must equal a third constant that doesn't depend on z. Let's call these constants E_X, E_Y, and E_Z, respectively. We then have a separate equation for each of the three functions $X(x)$, $Y(y)$, and $Z(z)$:

$$-\frac{\hbar^2}{2m}\frac{d^2X(x)}{dx^2} = E_X X(x) \qquad (41.11a)$$

$$-\frac{\hbar^2}{2m}\frac{d^2Y(y)}{dy^2} = E_Y Y(y) \qquad (41.11b)$$

$$-\frac{\hbar^2}{2m}\frac{d^2Z(z)}{dz^2} = E_Z Z(z) \qquad (41.11c)$$

To satisfy the boundary conditions that $\psi(x, y, z) = X(x)Y(y)Z(z)$ be equal to zero on the walls of the box, we demand that $X(x) = 0$ at $x = 0$ and $x = L$, $Y(y) = 0$ at $y = 0$ and $y = L$, and $Z(z) = 0$ at $z = 0$ and $z = L$.

How can we interpret the three constants E_X, E_Y, and E_Z in Eqs. (41.11)? From Eq. (41.10), they are related to the energy E by

$$E_X + E_Y + E_Z = E \qquad (41.12)$$

Equation (41.12) should remind you of Eq. (41.1) in Section 41.1, which states that the kinetic energy of a particle is the sum of contributions coming from its x-, y-, and z-components of momentum. Hence the constants E_X, E_Y, and E_Z tell us how much of the particle's energy is due to motion along each of the three coordinate axes. (Inside the box the potential energy is zero, so the particle's energy is purely kinetic.)

Equations (41.11) represent an enormous simplification; we've reduced the problem of solving a fairly complex *partial* differential equation with three independent variables to the much simpler problem of solving three separate *ordinary* differential equations with one independent variable each. What's more, each of these ordinary differential equations is just the same as the time-independent Schrödinger equation for a particle in a *one-dimensional* box, Eq. (40.25), and with exactly the same boundary conditions at 0 and L. (The only differences are that some of the quantities are labeled by different symbols.) By comparing with our work in Section 40.2, you can see that the solutions to Eqs. (41.11) are

$$X_{n_X}(x) = C_X \sin\frac{n_X \pi x}{L} \quad (n_X = 1, 2, 3, \dots) \qquad (41.13a)$$

$$Y_{n_Y}(y) = C_Y \sin\frac{n_Y \pi y}{L} \quad (n_Y = 1, 2, 3, \dots) \qquad (41.13b)$$

$$Z_{n_Z}(z) = C_Z \sin\frac{n_Z \pi z}{L} \quad (n_Z = 1, 2, 3, \dots) \qquad (41.13c)$$

where C_X, C_Y, and C_Z are constants. The corresponding values of E_X, E_Y, and E_Z are

$$E_X = \frac{n_X^2 \pi^2 \hbar^2}{2mL^2} \quad (n_X = 1, 2, 3, \dots) \qquad (41.14a)$$

$$E_Y = \frac{n_Y^2 \pi^2 \hbar^2}{2mL^2} \quad (n_Y = 1, 2, 3, \dots) \qquad (41.14b)$$

$$E_Z = \frac{n_Z^2 \pi^2 \hbar^2}{2mL^2} \quad (n_Z = 1, 2, 3, \dots) \qquad (41.14c)$$

There is only one quantum number n for the one-dimensional particle in a box, but *three* quantum numbers n_X, n_Y, and n_Z for the three-dimensional box. If we substitute Eqs. (41.13) back into Eq. (41.8) for the total spatial wave function, $\psi(x, y, z) = X(x)Y(y)Z(z)$, we get the following stationary-state wave functions for a particle in a three-dimensional cubical box:

$$\psi_{n_X, n_Y, n_Z}(x, y, z) = C \sin\frac{n_X \pi x}{L} \sin\frac{n_Y \pi y}{L} \sin\frac{n_Z \pi z}{L}$$

$$(n_X = 1, 2, 3, \dots; n_Y = 1, 2, 3, \dots; n_Z = 1, 2, 3, \dots) \qquad (41.15)$$

where $C = C_X C_Y C_Z$. The value of the constant C is determined by the normalization condition, Eq. (41.6).

In Section 40.2 we saw that the stationary-state wave functions for a particle in a one-dimensional box were analogous to standing waves on a string. In a similar way, the *three*-dimensional wave functions given by Eq. (41.15) are analogous to standing electromagnetic waves in a cubical cavity like the interior of a microwave oven (see Section 32.5). In a microwave oven there are "dead spots" where the wave intensity is zero, corresponding to the nodes of the standing wave. (The rotating platform in a microwave oven ensures even cooking by making sure that no part of the food sits at any "dead spot.") In a similar fashion, the probability distribution function corresponding to Eq. (41.15) can have "dead

41.2 Probability distribution function $|\psi_{n_X,n_Y,n_Z}(x, y, z)|^2$ for (n_X, n_Y, n_Z) equal to **(a)** (2, 1, 1), **(b)** (1, 2, 1), and **(c)** (1, 1, 2). The value of $|\psi|^2$ is proportional to the density of dots. The wave function is zero on the walls of the box and on the midplane of the box, so $|\psi|^2 = 0$ at these locations.

(a) $|\psi_{2,1,1}|^2$

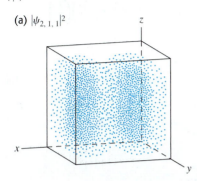

(b) $|\psi_{1,2,1}|^2$

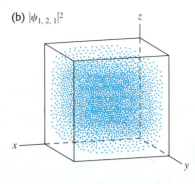

(c) $|\psi_{1,1,2}|^2$

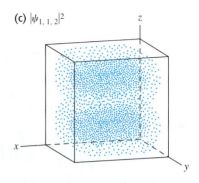

spots" where there is zero probability of finding the particle. As an example, consider the case $(n_X, n_Y, n_Z) = (2, 1, 1)$. From Eq. (41.15), the probability distribution function for this case is

$$|\psi_{2,1,1}(x, y, z)|^2 = |C|^2 \sin^2 \frac{2\pi x}{L} \sin^2 \frac{\pi y}{L} \sin^2 \frac{\pi z}{L}$$

As Fig. 41.2a shows, the probability distribution function is zero on the plane $x = L/2$, where $\sin^2(2\pi x/L) = \sin^2 \pi = 0$. The particle is most likely to be found near where all three of the sine-squared functions are greatest, at $(x, y, z) = (L/4, L/2, L/2)$ or $(x, y, z) = (3L/4, L/2, L/2)$. Figures 41.2b and 41.2c show the similar cases $(n_X, n_Y, n_Z) = (1, 2, 1)$ and $(n_X, n_Y, n_Z) = (1, 1, 2)$. For higher values of the quantum numbers n_X, n_Y, and n_Z there are additional planes on which the probability distribution function equals zero, just as the probability distribution function $|\psi(x)|^2$ for a one-dimensional box has more zeros for higher values of n (see Fig. 40.12).

Example 41.1 Probability in a three-dimensional box

(a) Find the value of the constant C that normalizes the wave function of Eq. (41.15). (b) Find the probability that the particle will be found somewhere in the region $0 \le x \le L/4$ (Fig. 41.3) for the cases (i) $(n_X, n_Y, n_Z) = (1, 2, 1)$, (ii) $(n_X, n_Y, n_Z) = (2, 1, 1)$, and (iii) $(n_X, n_Y, n_Z) = (3, 1, 1)$.

SOLUTION

IDENTIFY and SET UP: Equation (41.6) tells us that to normalize the wave function, we have to choose the value of C so that the

integral of the probability distribution function $|\psi_{n_X, n_Y, n_Z}(x, y, z)|^2$ over the volume within the box equals 1. (The integral is actually over *all* space, but the particle-in-a-box wave functions are zero outside the box.)

The probability of finding the particle within a certain volume within the box equals the integral of the probability distribution function over that volume. Hence in part (b) we'll integrate $|\psi_{n_X, n_Y, n_Z}(x, y, z)|^2$ for the given values of (n_X, n_Y, n_Z) over the volume $0 \le x \le L/4, 0 \le y \le L, 0 \le z \le L$.

EXECUTE: (a) From Eq. (41.15),

$$|\psi_{n_X,n_Y,n_Z}(x, y, z)|^2 = |C|^2 \sin^2 \frac{n_X \pi x}{L} \sin^2 \frac{n_Y \pi y}{L} \sin^2 \frac{n_Z \pi z}{L}$$

Hence the normalization condition is

$$\int |\psi_{n_X,n_Y,n_Z}(x, y, z)|^2 \, dV$$

$$= |C|^2 \int_{x=0}^{x=L} \int_{y=0}^{y=L} \int_{z=0}^{z=L} \sin^2 \frac{n_X \pi x}{L} \sin^2 \frac{n_Y \pi y}{L} \sin^2 \frac{n_Z \pi z}{L} \, dx \, dy \, dz$$

$$= |C|^2 \left(\int_{x=0}^{x=L} \sin^2 \frac{n_X \pi x}{L} \, dx \right) \left(\int_{y=0}^{y=L} \sin^2 \frac{n_Y \pi y}{L} \, dy \right)$$

$$\times \left(\int_{z=0}^{z=L} \sin^2 \frac{n_Z \pi z}{L} \, dz \right) = 1$$

41.3 What is the probability that the particle is in the dark-colored quarter of the box?

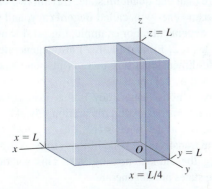

Continued

We can use the identity $\sin^2\theta = \frac{1}{2}(1 - \cos 2\theta)$ and the variable substitution $\theta = n_X\pi x/L$ to show that

$$\int \sin^2\frac{n_X\pi x}{L}\,dx = \frac{L}{2n_X\pi}\left[\frac{n_X\pi x}{L} - \frac{1}{2}\sin\left(\frac{2n_X\pi x}{L}\right)\right]$$

$$= \frac{x}{2} - \frac{L}{4n_X\pi}\sin\left(\frac{2n_X\pi x}{L}\right)$$

If we evaluate this integral between $x = 0$ and $x = L$, the result is $L/2$ (recall that $\sin 0 = 0$ and $\sin 2n_X\pi = 0$ for any integer n_X). The y- and z-integrals each yield the same result, so the normalization condition is

$$|C|^2\left(\frac{L}{2}\right)\left(\frac{L}{2}\right)\left(\frac{L}{2}\right) = |C|^2\left(\frac{L}{2}\right)^3 = 1$$

or $|C|^2 = (2/L)^3$. If we choose C to be real and positive, then $C = (2/L)^{3/2}$.

(b) We have the same y- and z-integrals as in part (a), but now the limits of integration on the x-integral are $x = 0$ and $x = L/4$:

$$P = \int_{0\le x\le L/4}|\psi_{n_X,n_Y,n_Z}|^2\,dV = |C|^2\left(\int_{x=0}^{x=L/4}\sin^2\frac{n_X\pi x}{L}\,dx\right)$$

$$\times\left(\int_{y=0}^{y=L}\sin^2\frac{n_Y\pi y}{L}\,dy\right)\left(\int_{z=0}^{z=L}\sin^2\frac{n_Z\pi z}{L}\,dz\right)$$

The x-integral is

$$\int_{x=0}^{x=L/4}\sin^2\frac{n_X\pi x}{L}\,dx = \left(\frac{x}{2} - \frac{L}{4n_X\pi}\sin\left(\frac{2n_X\pi x}{L}\right)\right)\Bigg|_{x=0}^{x=L/4}$$

$$= \frac{L}{8} - \frac{L}{4n_X\pi}\sin\left(\frac{n_X\pi}{2}\right)$$

Hence the probability of finding the particle somewhere in the region $0 \le x \le L/4$ is

$$P = \left(\frac{2}{L}\right)^3\left(\frac{L}{8} - \frac{L}{4n_X\pi}\sin\left(\frac{n_X\pi}{2}\right)\right)\left(\frac{L}{2}\right)\left(\frac{L}{2}\right)$$

$$= \frac{1}{4} - \frac{1}{2n_X\pi}\sin\left(\frac{n_X\pi}{2}\right)$$

This depends only on the value of n_X, not on n_Y or n_Z. Hence for the three cases we have

(i) $n_X = 1$: $P = \frac{1}{4} - \frac{1}{2(1)\pi}\sin\left(\frac{\pi}{2}\right) = \frac{1}{4} - \frac{1}{2\pi}(1)$

$$= \frac{1}{4} - \frac{1}{2\pi} = 0.091$$

(ii) $n_X = 2$: $P = \frac{1}{4} - \frac{1}{2(2)\pi}\sin\left(\frac{2\pi}{2}\right) = \frac{1}{4} - \frac{1}{4\pi}\sin\pi$

$$= \frac{1}{4} - 0 = 0.250$$

(iii) $n_X = 3$: $P = \frac{1}{4} - \frac{1}{2(3)\pi}\sin\left(\frac{3\pi}{2}\right) = \frac{1}{4} - \frac{1}{6\pi}(-1)$

$$= \frac{1}{4} + \frac{1}{6\pi} = 0.303$$

EVALUATE: You can see why the probabilities in part (b) are different by looking at part (b) of Fig. 40.12, which shows $\sin^2 n_X\pi x/L$ for $n_X = 1, 2,$ and 3. For $n_X = 2$ the area under the curve between $x = 0$ and $x = L/4$ (equal to the integral between these two points) is exactly $\frac{1}{4}$ of the total area between $x = 0$ and $x = L$. For $n_X = 1$ the area between $x = 0$ and $x = L/4$ is less than $\frac{1}{4}$ of the total area, and for $n_X = 3$ it is greater than $\frac{1}{4}$ of the total area.

Energy Levels, Degeneracy, and Symmetry

From Eqs. (41.12) and (41.14), the allowed energies for a particle of mass m in a cubical box of side L are

$$E_{n_X,n_Y,n_Z} = \frac{(n_X^2 + n_Y^2 + n_Z^2)\pi^2\hbar^2}{2mL^2} \qquad (n_X = 1, 2, 3, \ldots; n_Y = 1, 2, 3, \ldots; n_Z = 1, 2, 3, \ldots) \qquad \text{(41.16)}$$

(energy levels, particle in a three-dimensional cubical box)

Figure 41.4 shows the six lowest energy levels given by Eq. (41.16). Note that most energy levels correspond to more than one set of quantum numbers (n_X, n_Y, n_Z) and hence to more than one quantum state. Having two or more distinct quantum states with the same energy is called **degeneracy,** and states with the same energy are said to be **degenerate.** For example, Fig. 41.4 shows that the states $(n_X, n_Y, n_Z) = (2, 1, 1)$, $(1, 2, 1)$, and $(1, 1, 2)$ are degenerate. By comparison, for a particle in a one-dimensional box there is just one state for each energy level (see Fig. 40.11a) and no degeneracy.

The reason the cubical box exhibits degeneracy is that it is *symmetric:* All sides of the box have the same dimensions. As an illustration, Fig. 41.2 shows the probability distribution functions for the three states $(n_X, n_Y, n_Z) = (2, 1, 1)$, $(1, 2, 1)$, and $(1, 1, 2)$. You can transform any one of these three states into a different one by simply rotating the cubical box by 90°. This rotation doesn't change the energy, so the three states are degenerate.

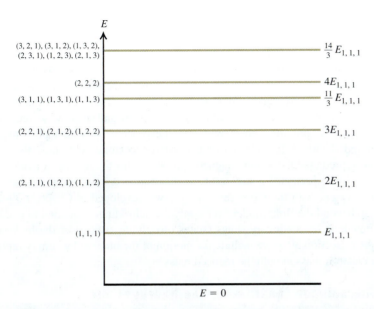

41.4 Energy-level diagram for a particle in a three-dimensional cubical box. We label each level with the quantum numbers of the states (n_X, n_Y, n_Z) with that energy. Several of the levels are degenerate (more than one state has the same energy). The lowest (ground) level, $(n_X, n_Y, n_Z) = (1, 1, 1)$, has energy $E_{1,1,1} = (1^2 + 1^2 + 1^2)\pi^2\hbar^2/2mL^2 = 3\pi^2\hbar^2/2mL^2$; we show the energies of the other levels as multiples of $E_{1,1,1}$.

Since degeneracy is a consequence of symmetry, we can remove the degeneracy by making the box asymmetric. We do this by giving the three sides of the box different lengths L_X, L_Y, and L_Z. If we repeat the steps that we followed to solve the time-independent Schrödinger equation, we find that the energy levels are given by

$$E_{n_X,n_Y,n_Z} = \left(\frac{n_X^2}{L_X^2} + \frac{n_Y^2}{L_Y^2} + \frac{n_Z^2}{L_Z^2}\right)\frac{\pi^2\hbar^2}{2m} \quad (n_X = 1, 2, 3, \ldots; n_Y = 1, 2, 3, \ldots;$$
$$n_Z = 1, 2, 3, \ldots) \quad \text{(41.17)}$$

(energy levels, particle in a three-dimensional box with sides of length L_X, L_Y, and L_Z)

If L_X, L_Y, and L_Z are all different, the states $(n_X, n_Y, n_Z) = (2, 1, 1)$, $(1, 2, 1)$, and $(1, 1, 2)$ have different energies and hence are no longer degenerate. Note that Eq. (41.17) reduces to Eq. (41.16) if the lengths are all the same ($L_X = L_Y = L_Z = L$).

Let's summarize the key differences between the three-dimensional particle in a box and the one-dimensional case that we examined in Section 40.2:

- We can write the wave function for a three-dimensional stationary state as a product of three functions, one for each spatial coordinate. Only a single function of the coordinate x is needed in one dimension.
- In the three-dimensional case, three quantum numbers are needed to describe each stationary state. Only one quantum number is needed in the one-dimensional case.
- Most of the energy levels for the three-dimensional case are degenerate: More than one stationary state has this energy. There is no degeneracy in the one-dimensional case.
- For a stationary state of the three-dimensional case, there are surfaces on which the probability distribution function $|\psi|^2$ is zero. In the one-dimensional case there are positions on the x-axis where $|\psi|^2$ is zero.

We'll see these same features in the following section as we examine a three-dimensional situation that's more realistic than a particle in a box: a hydrogen atom in which a negatively charged electron orbits a positively charged nucleus.

Test Your Understanding of Section 41.2 Rank the following states of a particle in a cubical box of side L in order from highest to lowest energy: (i) $(n_X, n_Y, n_Z) = (2, 3, 2)$; (ii) $(n_X, n_Y, n_Z) = (4, 1, 1)$; (iii) $(n_X, n_Y, n_Z) = (2, 2, 3)$; (iv) $(n_X, n_Y, n_Z) = (1, 3, 3)$.

41.3 The Hydrogen Atom

Let's continue the discussion of the hydrogen atom that we began in Chapter 39. In the Bohr model, electrons move in circular orbits like Newtonian particles, but with quantized values of angular momentum. While this model gave the correct energy levels of the hydrogen atom, as deduced from spectra, it had many conceptual difficulties. It mixed classical physics with new and seemingly contradictory concepts. It provided no insight into the process by which photons are emitted and absorbed. It could not be generalized to atoms with more than one electron. It predicted the wrong magnetic properties for the hydrogen atom. And perhaps most important, its picture of the electron as a localized point particle was inconsistent with the more general view we developed in Chapters 39 and 40. To go beyond the Bohr model, let's apply the Schrödinger equation to find the wave functions for stationary states (states of definite energy) of the hydrogen atom. As in Section 39.3, we include the motion of the nucleus by simply replacing the electron mass m with the reduced mass m_r.

The Schrödinger Equation for the Hydrogen Atom

We discussed the three-dimensional version of the Schrödinger equation in Section 41.1. The potential-energy function is *spherically symmetric:* It depends only on the distance $r = (x^2 + y^2 + z^2)^{1/2}$ from the origin of coordinates:

$$U(r) = -\frac{1}{4\pi\epsilon_0}\frac{e^2}{r} \tag{41.18}$$

41.5 The Schrödinger equation for the hydrogen atom can be solved most readily using spherical coordinates.

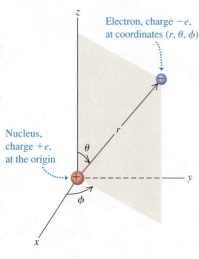

The hydrogen-atom problem is best formulated in spherical coordinates (r, θ, ϕ), shown in Fig. 41.5; the spherically symmetric potential-energy function depends only on r, not on θ or ϕ. The Schrödinger equation with this potential-energy function can be solved exactly; the solutions are combinations of familiar functions. Without going into a lot of detail, we can describe the most important features of the procedure and the results.

First, we find the solutions using the same method of separation of variables that we employed for a particle in a cubical box in Section 41.2. We express the wave function $\psi(r, \theta, \phi)$ as a product of three functions, each one a function of only one of the three coordinates:

$$\psi(r, \theta, \phi) = R(r)\Theta(\theta)\Phi(\phi) \tag{41.19}$$

That is, the function $R(r)$ depends only on r, $\Theta(\theta)$ depends only on θ, and $\Phi(\phi)$ depends only on ϕ. Just as for a particle in a three-dimensional box, when we substitute Eq. (41.19) into the Schrödinger equation, we get three separate ordinary differential equations. One equation involves only r and $R(r)$, a second involves only θ and $\Theta(\theta)$, and a third involves only ϕ and $\Phi(\phi)$:

$$-\frac{\hbar^2}{2m_r r^2}\frac{d}{dr}\left(r^2\frac{dR(r)}{dr}\right) + \left(\frac{\hbar^2 l(l+1)}{2m_r r^2} + U(r)\right)R(r) = ER(r) \tag{41.20a}$$

$$\frac{1}{\sin\theta}\frac{d}{d\theta}\left(\sin\theta\frac{d\Theta(\theta)}{d\theta}\right) + \left(l(l+1) - \frac{m_l^2}{\sin^2\theta}\right)\Theta(\theta) = 0 \tag{41.20b}$$

$$\frac{d^2\Phi(\phi)}{d\phi^2} + m_l^2\Phi(\phi) = 0 \tag{41.20c}$$

In Eqs. (41.20) E is the energy of the stationary state and l and m_l are constants that we'll discuss later. (Be careful! Don't confuse the constant m_l with the reduced mass m_r.)

We won't attempt to solve this set of three equations, but we can describe how it's done. As for the particle in a cubical box, the physically acceptable solutions of these three equations are determined by boundary conditions. The radial function $R(r)$ in Eq. (41.20a) must approach zero at large r, because we are describing *bound states* of the electron that are localized near the nucleus. This is analogous to the requirement that the harmonic-oscillator wave functions (see Section 40.5) must approach zero at large x. The angular functions $\Theta(\theta)$ and $\Phi(\phi)$ in Eqs. (41.20b) and (41.20c) must be *finite* for all relevant values of the angles. For example, there are solutions of the Θ equation that become infinite at $\theta = 0$ and $\theta = \pi$; these are unacceptable, since $\psi(r, \theta, \phi)$ must be normalizable. Furthermore, the angular function $\Phi(\phi)$ in Eq. (41.20c) must be *periodic*. For example, (r, θ, ϕ) and $(r, \theta, \phi + 2\pi)$ describe the same point, so $\Phi(\phi + 2\pi)$ must equal $\Phi(\phi)$.

The allowed radial functions $R(r)$ turn out to be an exponential function $e^{-\alpha r}$ (where α is positive) multiplied by a polynomial in r. The functions $\Theta(\theta)$ are polynomials containing various powers of $\sin\theta$ and $\cos\theta$, and the functions $\Phi(\phi)$ are simply proportional to $e^{im_l\phi}$, where $i = \sqrt{-1}$ and m_l is an integer that may be positive, zero, or negative.

In the process of finding solutions that satisfy the boundary conditions, we also find the corresponding energy levels. We denote the energies of these levels [E in Eq. (41.20a)] by E_n ($n = 1, 2, 3, \ldots$). These turn out to be *identical* to those from the Bohr model, as given by Eq. (39.15), with the electron rest mass m replaced by the reduced mass m_r. Rewriting that equation using $\hbar = h/2\pi$, we have

$$E_n = -\frac{1}{(4\pi\epsilon_0)^2}\frac{m_r e^4}{2n^2\hbar^2} = -\frac{13.60\text{ eV}}{n^2} \qquad \text{(energy levels of hydrogen)} \quad (41.21)$$

As in Section 39.3, we call n the **principal quantum number** for the level of energy E_n.

Equation (41.21) is an important validation of our Schrödinger-equation analysis of the hydrogen atom. The Schrödinger analysis is quite different from the Bohr model, both mathematically and conceptually, yet both yield the same energy-level scheme—a scheme that agrees with the energies determined from spectra. As we will see, the Schrödinger analysis can explain many more aspects of the hydrogen atom than can the Bohr model.

Quantization of Orbital Angular Momentum

The solutions to Eqs. (41.20) that satisfy the boundary conditions mentioned above also have quantized values of *orbital angular momentum*. That is, only certain discrete values of the magnitude and components of orbital angular momentum are permitted. In discussing the Bohr model in Section 39.3, we mentioned that quantization of angular momentum was a result with little fundamental justification. With the Schrödinger equation it appears automatically.

The possible values of the magnitude L of orbital angular momentum $\vec{L}$ are determined by the requirement that the $\Theta(\theta)$ function in Eq. (41.20b) must be finite at $\theta = 0$ and $\theta = \pi$. In a level with energy E_n and principal quantum number n, the possible values of L are

$$L = \sqrt{l(l + 1)}\,\hbar \qquad (l = 0, 1, 2, \ldots, n - 1) \qquad \begin{array}{l}\text{(magnitude of}\\ \text{orbital angular}\\ \text{momentum)}\end{array} \quad (41.22)$$

The *orbital angular-momentum quantum number* l, which is the same l that appears in Eqs. (41.20a) and (41.20b), is called the **orbital quantum number** for short. In the Bohr model, each energy level corresponded to a single value of angular momentum. Equation (41.22) shows that in fact there are n different possible values of L for the nth energy level.

An interesting feature of Eq. (41.22) is that the orbital angular momentum is *zero* for $l = 0$ states. This result disagrees with the Bohr model, in which the electron always moved in a circle of definite radius and L was never zero. The $l = 0$ wave functions ψ depend only on r; for these states, the functions $\Theta(\theta)$ and $\Phi(\phi)$ are constants. Thus the wave functions for $l = 0$ states are spherically symmetric. There is nothing in their probability distribution $|\psi|^2$ to favor one direction over any other, and there is no orbital angular momentum.

The permitted values of the *component* of $\vec{L}$ in a given direction, say the z-component L_z, are determined by the requirement that the $\Phi(\phi)$ function must equal $\Phi(\phi + 2\pi)$. The possible values of L_z are

$$L_z = m_l\hbar \qquad (m_l = 0, \pm 1, \pm 2, \dots, \pm l) \qquad \begin{array}{l}\text{(components of}\\ \text{orbital angular} \qquad [41.23]\\ \text{momentum)}\end{array}$$

The quantum number m_l is the same as that in Eqs. (41.20b) and (41.20c). We see that m_l can be zero or a positive or negative integer up to, but no larger in magnitude than, l. That is, $|m_l| \leq l$. For example, if $l = 1$, m_l can equal 1, 0, or -1. For reasons that will emerge later, we call m_l the *orbital magnetic quantum number*, or **magnetic quantum number** for short.

The component L_z can never be quite as large as L (unless both are zero). For example, when $l = 2$, the largest possible value of m_l is also 2; then Eqs. (41.22) and (41.23) give

$$L = \sqrt{2(2 + 1)}\hbar = \sqrt{6}\hbar = 2.45\hbar$$

$$L_z = 2\hbar$$

Figure 41.6 shows the situation. The minimum value of the angle θ_L between the vector $\vec{L}$ and the z-axis is given by

$$\theta_L = \arccos\frac{L_z}{L} = \arccos\frac{2}{2.45} = 35.3°$$

That $|L_z|$ is always less than L is also required by the uncertainty principle. Suppose we could know the precise *direction* of the orbital angular momentum vector. Then we could let that be the direction of the z-axis, and L_z would equal L. This corresponds to a particle moving in the xy-plane only, in which case the z-component of the linear momentum $\vec{p}$ would be zero with no uncertainty Δp_z. Then the uncertainty principle $\Delta z\Delta p_z \geq \hbar$ requires infinite uncertainty Δz in the coordinate z. This is impossible for a localized state; we conclude that we can't know the direction of $\vec{L}$ precisely. Thus, as we've already stated, the component of $\vec{L}$ in a given direction can never be quite as large as its magnitude L. Also, if we can't know the direction of $\vec{L}$ precisely, we can't determine the components L_x and L_y precisely. Thus we show *cones* of possible directions for $\vec{L}$ in Fig. 41.6b.

You may wonder why we have singled out the z-axis for special attention. There's no fundamental reason for this; the atom certainly doesn't care what coordinate system we use. The point is that we can't determine all three components of orbital angular momentum with certainty, so we arbitrarily pick one as the component we want to measure. When we discuss interactions of the atom with a magnetic field, we will consistently choose the positive z-axis to be in the direction of $\vec{B}$.

Quantum Number Notation

The wave functions for the hydrogen atom are determined by the values of three quantum numbers n, l, and m_l. (Compare this to the particle in a three-dimensional box that we considered in Section 41.2. There, too, three quantum numbers were needed to describe each stationary state.) The energy E_n is determined by the principal quantum number n according to Eq. (41.21). The magnitude of orbital angular momentum is determined by the orbital quantum number l, as in Eq. (41.22). The component of orbital angular momentum in a specified axis direction (customarily the z-axis) is determined by the magnetic quantum number m_l, as in Eq. (41.23). The energy does not depend on the values of l or m_l (Fig. 41.7), so for each energy level E_n given by Eq. (41.21), there is more than one distinct state having the same energy but different quantum numbers. That is, these states are *degenerate,* just like most of the states of a particle in a

41.6 (a) When $l = 2$, the magnitude of the angular momentum vector $\vec{L}$ is $\sqrt{6}\hbar = 2.45\hbar$, but $\vec{L}$ does not have a definite direction. In this semiclassical vector picture, $\vec{L}$ makes an angle of 35.3° with the z-axis when the z-component has its maximum value of $2\hbar$. (b) These cones show the possible directions of $\vec{L}$ for different values of L_z.

(a)

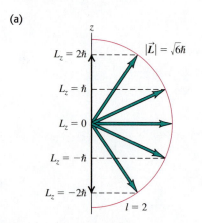

(b)

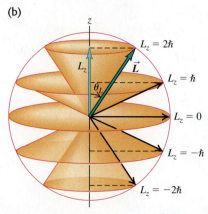

three-dimensional box. As for the three-dimensional box, degeneracy arises because the hydrogen atom is symmetric: If you rotate the atom through any angle, the potential-energy function at a distance r from the nucleus has the same value.

States with various values of the orbital quantum number l are often labeled with letters, according to the following scheme:

$l = 0$: s states

$l = 1$: p states

$l = 2$: d states

$l = 3$: f states

$l = 4$: g states

$l = 5$: h states

and so on alphabetically. This seemingly irrational choice of the letters s, p, d, and f originated in the early days of spectroscopy and has no fundamental significance. In an important form of *spectroscopic notation* that we'll use often, a state with $n = 2$ and $l = 1$ is called a $2p$ state; a state with $n = 4$ and $l = 0$ is a $4s$ state; and so on. Only s states ($l = 0$) are spherically symmetric.

Here's another bit of notation. The radial extent of the wave functions increases with the principal quantum number n, and we can speak of a region of space associated with a particular value of n as a **shell.** Especially in discussions of many-electron atoms, these shells are denoted by capital letters:

$n = 1$: K shell

$n = 2$: L shell

$n = 3$: M shell

$n = 4$: N shell

and so on alphabetically. For each n, different values of l correspond to different *subshells.* For example, the L shell ($n = 2$) contains the $2s$ and $2p$ subshells.

Table 41.1 shows some of the possible combinations of the quantum numbers n, l, and m_l for hydrogen-atom wave functions. The spectroscopic notation and the shell notation for each are also shown.

41.7 The energy for an orbiting satellite such as the Hubble Space Telescope depends on the average distance between the satellite and the center of the earth. It does *not* depend on whether the orbit is circular (with a large orbital angular momentum L) or elliptical (in which case L is smaller). In the same way, the energy of a hydrogen atom does not depend on the orbital angular momentum.

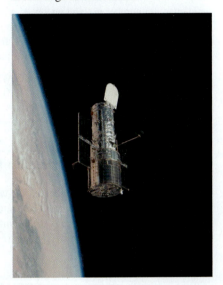

Table 41.1 Quantum States of the Hydrogen Atom

n	l	m_l	Spectroscopic Notation	Shell
1	0	0	$1s$	K
2	0	0	$2s$	L
2	1	$-1, 0, 1$	$2p$	
3	0	0	$3s$	
3	1	$-1, 0, 1$	$3p$	M
3	2	$-2, -1, 0, 1, 2$	$3d$	
4	0	0	$4s$	N

and so on

Problem-Solving Strategy 41.1   Atomic Structure

IDENTIFY *the relevant concepts:* Many problems in atomic structure can be solved simply by reference to the quantum numbers n, l, and m_l that describe the total energy E, the magnitude of the orbital angular momentum $\vec{L}$, the z-component of $\vec{L}$, and other properties of an atom.

SET UP *the problem:* Identify the target variables and choose the appropriate equations, which may include Eqs. (41.21), (41.22), and (41.23).

EXECUTE *the solution* as follows:

1. Be sure you understand the possible values of the quantum numbers n, l, and m_l for the hydrogen atom. They are all integers; n is always greater than zero, l can be zero or positive up to $n - 1$, and m_l can range from $-l$ to l. You should know how to count the number of (n, l, m_l) states in each shell (K, L, M, and so on) and subshell ($3s$, $3p$, $3d$, and so on). Be able to *construct* Table 41.1, not just to write it from memory.

2. Solve for the target variables.

EVALUATE *your answer:* It helps to be familiar with typical magnitudes in atomic physics. For example, the electric potential energy of a proton and electron 0.10 nm apart (typical of atomic dimensions) is about -15 eV. Visible light has wavelengths around 500 nm and frequencies around 5×10^{14} Hz. Problem-Solving Strategy 39.1 (Section 39.1) gives other typical magnitudes.

Example 41.2 **Counting hydrogen states**

How many distinct (n, l, m_l) states of the hydrogen atom with $n = 3$ are there? What are their energies?

SOLUTION

IDENTIFY and SET UP: This problem uses the relationships among the principal quantum number n, orbital quantum number l, magnetic quantum number m_l, and energy of a state for the hydrogen atom. We use the rule that l can have n integer values, from 0 to $n - 1$, and that m_l can have $2l + 1$ values, from $-l$ to l. Equation (41.21) gives the energy of any particular state.

EXECUTE: When $n = 3$, l can be 0, 1, or 2. When $l = 0$, m_l can be only 0 (1 state). When $l = 1$, m_l can be -1, 0, or 1 (3 states). When $l = 2$, m_l can be -2, -1, 0, 1, or 2 (5 states). The total number of

(n, l, m_l) states with $n = 3$ is therefore $1 + 3 + 5 = 9$. (In Section 41.5 we'll find that the total number of $n = 3$ states is in fact twice this, or 18, because of electron spin.)

The energy of a hydrogen-atom state depends only on n, so all 9 of these states have the same energy. From Eq. (41.21),

$$E_3 = \frac{-13.60 \text{ eV}}{3^2} = -1.51 \text{ eV}$$

EVALUATE: For a given value of n, the total number of (n, l, m_l) states turns out to be n^2. In this case $n = 3$ and there are $3^2 = 9$ states. Remember that the ground level of hydrogen has $n = 1$ and $E_1 = -13.6$ eV; the $n = 3$ excited states have a higher (less negative) energy.

Example 41.3 **Angular momentum in an excited level of hydrogen**

Consider the $n = 4$ states of hydrogen. (a) What is the maximum magnitude L of the orbital angular momentum? (b) What is the maximum value of L_z? (c) What is the minimum angle between $\vec{L}$ and the z-axis? Give your answers to parts (a) and (b) in terms of $\hbar$.

SOLUTION

IDENTIFY and SET UP: We again need to relate the principal quantum number n and the orbital quantum number l for a hydrogen atom. We also need to relate the value of l and the magnitude and possible directions of the orbital angular momentum vector. We'll use Eq. (41.22) in part (a) to determine the maximum value of L; then we'll use Eq. (41.23) in part (b) to determine the maximum value of L_z. The angle between $\vec{L}$ and the z-axis is minimum when L_z is maximum (so that $\vec{L}$ is most nearly aligned with the positive z-axis).

EXECUTE: (a) When $n = 4$, the maximum value of the orbital angular-momentum quantum number l is $(n - 1) = (4 - 1) = 3$; from Eq. (41.22),

$$L_{max} = \sqrt{3(3 + 1)}\hbar = \sqrt{12}\hbar = 3.464\hbar$$

(b) For $l = 3$ the maximum value of m_l is 3. From Eq. (41.23),

$$(L_z)_{max} = 3\hbar$$

(c) The *minimum* allowed angle between $\vec{L}$ and the z-axis corresponds to the *maximum* allowed values of L_z and m_l (Fig. 41.6b shows an $l = 2$ example). For the state with $l = 3$ and $m_l = 3$,

$$\theta_{min} = \arccos\frac{(L_z)_{max}}{L} = \arccos\frac{3\hbar}{3.464\hbar} = 30.0°$$

EVALUATE: As a check, you can verify that θ is greater than 30.0° for all states with smaller values of l.

Electron Probability Distributions

Rather than picturing the electron as a point particle moving in a precise circle, the Schrödinger equation gives an electron *probability distribution* surrounding the nucleus. The hydrogen-atom probability distributions are three-dimensional, so they are harder to visualize than the two-dimensional circular orbits of the Bohr model. It's helpful to look at the *radial probability distribution function* $P(r)$—that is, the probability per radial length for the electron to be found at various distances from the proton. From Section 41.1 the probability for finding the electron in a small volume element dV is $|\psi|^2 dV$. (We assume that ψ is normalized in accordance with Eq. (41.6)—that is, that the integral of $|\psi|^2 dV$ over all space equals unity so that there is 100% probability of finding the electron somewhere in the universe.) Let's take as our volume element a thin spherical shell with inner radius r and outer radius $r + dr$. The volume dV of this shell is approximately its area $4\pi r^2$ multiplied by its thickness dr:

$$dV = 4\pi r^2 \, dr \qquad (41.24)$$

We denote by $P(r)\,dr$ the probability of finding the particle within the radial range dr; then, using Eq. (41.24),

$$P(r)\,dr = |\psi|^2\,dV = |\psi|^2 4\pi r^2\,dr \qquad \text{(probability that the electron is between } r \text{ and } r + dr)} \qquad \text{(41.25)}$$

For wave functions that depend on θ and ϕ as well as r, we use the value of $|\psi|^2$ averaged over all angles in Eq. (41.25).

Figure 41.8 shows graphs of $P(r)$ for several hydrogen-atom wave functions. The r scales are labeled in multiples of a, the smallest distance between the electron and the nucleus in the Bohr model:

$$a = \frac{\epsilon_0 h^2}{\pi m_r e^2} = \frac{4\pi\epsilon_0 \hbar^2}{m_r e^2} = 5.29 \times 10^{-11} \text{ m} \qquad \text{(smallest } r, \text{ Bohr model)} \qquad \text{(41.26)}$$

Just as for a particle in a cubical box (see Section 41.2), there are some locations where the probability is zero. These surfaces are planes for a particle in a box; for a hydrogen atom these are spherical surfaces (that is, surfaces of constant r). But again, the uncertainty principle tells us not to worry; we can't localize the electron exactly anyway. Note that for the states having the largest possible l for each n (such as 1s, 2p, 3d, and 4f states), $P(r)$ has a single maximum at n^2a. For these states, the electron is most likely to be found at the distance from the nucleus that is predicted by the Bohr model, $r = n^2a$.

Figure 41.8 shows *radial* probability distribution functions $P(r) = 4\pi r^2|\psi|^2$, which indicate the relative probability of finding the electron within a thin spherical shell of radius r. By contrast, Figs. 41.9 and 41.10 show the *three-dimensional* probability distribution functions $|\psi|^2$, which indicate the relative probability of finding the electron within a small box at a given position. The darker the blue "cloud," the greater the value of $|\psi|^2$. (These are similar to the "clouds" shown in Fig. 41.2.) Figure 41.9 shows cross sections of the spherically symmetric probability clouds for the lowest three s subshells, for which $|\psi|^2$ depends only on the radial coordinate r. Figure 41.10 shows cross sections of the clouds for other electron states for which $|\psi|^2$ depends on both r and θ. For these states the probability distribution function is zero for certain values of θ as well as for certain values of r. In *any* stationary state of the hydrogen atom, $|\psi|^2$ is independent of ϕ.

41.8 Radial probability distribution functions $P(r)$ for several hydrogen-atom wave functions, plotted as functions of the ratio r/a [see Eq. (41.26)]. For each function, the number of maxima is $(n - l)$. The curves for which $l = n - 1$ (1s, 2p, 3d, ...) have only one maximum, located at $r = n^2a$.

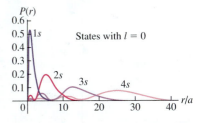

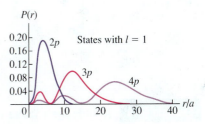

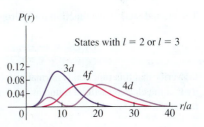

41.9 Three-dimensional probability distribution functions $|\psi|^2$ for the spherically symmetric 1s, 2s, and 3s hydrogen-atom wave functions.

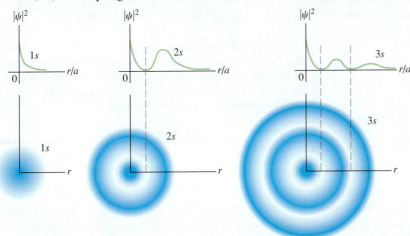

41.10 Cross sections of three-dimensional probability distributions for a few quantum states of the hydrogen atom. They are not to the same scale. Mentally rotate each drawing about the z-axis to obtain the three-dimensional representation of $|\psi|^2$. For example, the $2p, m_l = \pm 1$ probability distribution looks like a fuzzy donut.

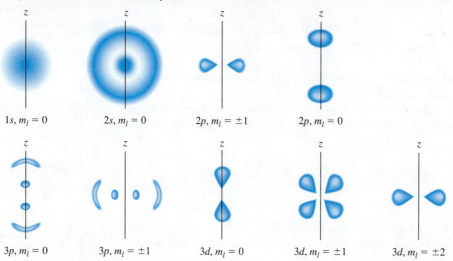

$1s, m_l = 0$ $\qquad$ $2s, m_l = 0$ $\qquad$ $2p, m_l = \pm 1$ $\qquad$ $2p, m_l = 0$

$3p, m_l = 0$ $\qquad$ $3p, m_l = \pm 1$ $\qquad$ $3d, m_l = 0$ $\qquad$ $3d, m_l = \pm 1$ $\qquad$ $3d, m_l = \pm 2$

Example 41.4 A hydrogen wave function

The ground-state wave function for hydrogen (a $1s$ state) is

$$\psi_{1s}(r) = \frac{1}{\sqrt{\pi a^3}} e^{-r/a}$$

(a) Verify that this function is normalized. (b) What is the probability that the electron will be found at a distance less than a from the nucleus?

SOLUTION

IDENTIFY and SET UP: This example is similar to Example 41.1 in Section 41.2. We need to show that this wave function satisfies the condition that the probability of finding the electron *somewhere* is 1. We then need to find the probability that it will be found in the region $r < a$. In part (a) we'll carry out the integral $\int |\psi|^2 \, dV$ over all space; if it is equal to 1, the wave function is normalized. In part (b) we'll carry out the same integral over a spherical volume that extends from the origin (the nucleus) out to a distance a from the nucleus.

EXECUTE: (a) Since the wave function depends only on the radial coordinate r, we can choose our volume elements to be spherical shells of radius r, thickness dr, and volume dV given by Eq. (41.24). We then have

$$\int_{\text{all space}} |\psi_{1s}|^2 \, dV = \int_0^\infty \frac{1}{\pi a^3} e^{-2r/a} (4\pi r^2 \, dr)$$

$$= \frac{4}{a^3} \int_0^\infty r^2 e^{-2r/a} \, dr$$

You can find the following indefinite integral in a table of integrals or by integrating by parts:

$$\int r^2 e^{-2r/a} \, dr = \left(-\frac{ar^2}{2} - \frac{a^2 r}{2} - \frac{a^3}{4} \right) e^{-2r/a}$$

Evaluating this between the limits $r = 0$ and $r = \infty$ is simple; it is zero at $r = \infty$ because of the exponential factor, and at $r = 0$ only the last term in the parentheses survives. Thus the value of the definite integral is $a^3/4$. Putting it all together, we find

$$\int_0^\infty |\psi_{1s}|^2 \, dV = \frac{4}{a^3} \int_0^\infty r^2 e^{-2r/a} \, dr = \frac{4}{a^3} \frac{a^3}{4} = 1$$

The wave function *is* normalized.

(b) To find the probability P that the electron is found within $r < a$, we carry out the same integration but with the limits 0 and a. We'll leave the details to you (Exercise 41.15). From the upper limit we get $-5e^{-2}a^3/4$; the final result is

$$P = \int_0^a |\psi_{1s}|^2 \, 4\pi r^2 \, dr = \frac{4}{a^3} \left(-\frac{5a^3 e^{-2}}{4} + \frac{a^3}{4} \right)$$

$$= (-5e^{-2} + 1) = 1 - 5e^{-2} = 0.323$$

EVALUATE: Our results tell us that in a ground state we expect to find the electron at a distance from the nucleus less than a about $\frac{1}{3}$ of the time and at a greater distance about $\frac{2}{3}$ of the time. It's hard to tell, but in Fig. 41.8, about $\frac{2}{3}$ of the area under the $1s$ curve is at distances greater than a (that is, $r/a > 1$).

Hydrogenlike Atoms

Two generalizations that we discussed with the Bohr model in Section 39.3 are equally valid in the Schrödinger analysis. First, if the "atom" is not composed of a single proton and a single electron, using the reduced mass m_r of the system in Eqs. (41.21) and (41.26) will lead to changes that are substantial for some exotic systems. One example is *positronium,* in which a positron and an electron orbit each other; another is a *muonic atom,* in which the electron is replaced by an unstable particle called a muon that has the same charge as an electron but is 207 times more massive. Second, our analysis is applicable to single-electron ions, such as He^+, Li^{2+}, and so on. For such ions we replace e^2 by Ze^2 in Eqs. (41.21) and (41.26), where Z is the number of protons (the **atomic number**).

Test Your Understanding of Section 41.3 Rank the following states of the hydrogen atom in order from highest to lowest probability of finding the electron in the vicinity of $r = 5a$: (i) $n = 1$, $l = 0$, $m_l = 0$; (ii) $n = 2$, $l = 1$, $m_l = +1$; (iii) $n = 2$, $l = 1$, $m_l = 0$.

41.4 The Zeeman Effect

The **Zeeman effect** is the splitting of atomic energy levels and the associated spectral lines when the atoms are placed in a magnetic field (Fig. 41.11). This effect confirms experimentally the quantization of angular momentum. The discussion in this section, which assumes that the only angular momentum is the *orbital* angular momentum of a single electron, also shows why we call m_l the magnetic quantum number.

Atoms contain charges in motion, so it should not be surprising that magnetic forces cause changes in that motion and in the energy levels. As early as the middle of the 19th century, physicists speculated that the sources of visible light might be vibrating electric charge on an atomic scale. In 1896 the Dutch physicist Pieter Zeeman was the first to show that in the presence of a magnetic field, some spectral lines were split into groups of closely spaced lines (Fig. 41.12). This effect now bears his name.

Magnetic Moment of an Orbiting Electron

Let's begin our analysis of the Zeeman effect by reviewing the concept of *magnetic dipole moment* or *magnetic moment,* introduced in Section 27.7. A plane current loop with vector area $\vec{A}$ carrying current I has a magnetic moment $\vec{\mu}$ given by

$$\vec{\mu} = I\vec{A} \tag{41.27}$$

When a magnetic dipole of moment $\vec{\mu}$ is placed in a magnetic field $\vec{B}$, the field exerts a torque $\vec{\tau} = \vec{\mu} \times \vec{B}$ on the dipole. The potential energy U associated with this interaction is given by Eq. (27.27):

$$U = -\vec{\mu} \cdot \vec{B} \tag{41.28}$$

Now let's use Eqs. (41.27) and (41.28) and the Bohr model to look at the interaction of a hydrogen atom with a magnetic field. The orbiting electron with speed v is equivalent to a current loop with radius r and area πr^2. The average current I is the average charge per unit time that passes a given point of the orbit. This is equal to the charge magnitude e divided by the time T for one revolution, given by $T = 2\pi r/v$. Thus $I = ev/2\pi r$, and from Eq. (41.27) the magnitude μ of the magnetic moment is

$$\mu = IA = \frac{ev}{2\pi r}\pi r^2 = \frac{evr}{2} \tag{41.29}$$

We can also express this in terms of the magnitude L of the orbital angular momentum. From Eq. (10.28) the angular momentum of a particle in a circular orbit is $L = mvr$, so Eq. (41.29) becomes

$$\mu = \frac{e}{2m}L \tag{41.30}$$

41.11 Magnetic effects on the spectrum of sunlight. (a) The slit of a spectrograph is positioned along the black line crossing a portion of a sunspot. (b) The 0.4-T magnetic field in the sunspot (a thousand times greater than the earth's field) splits the middle spectral line into three lines.

(a)

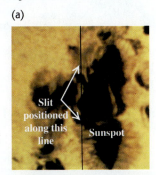

(b)

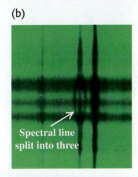

41.12 The normal Zeeman effect. Compare this to the magnetic splitting in the solar spectrum shown in Fig. 41.11b.

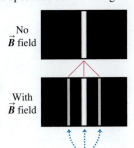

When an excited gas is placed in a magnetic field, the interaction of orbital magnetic moments with the field splits individual spectral lines of the gas into sets of three lines.

The ratio of the magnitude of $\vec{\mu}$ to the magnitude of $\vec{L}$ is $\mu/L = e/2m$ and is called the *gyromagnetic ratio*.

In the Bohr model, $L = nh/2\pi = n\hbar$, where $n = 1, 2, \ldots$. For an $n = 1$ state (a ground state), Eq. (41.30) becomes $\mu = (e/2m)\hbar$. This quantity is a natural unit for magnetic moment; it is called one **Bohr magneton,** denoted by μ_B:

$$\mu_B = \frac{e\hbar}{2m} \qquad \text{(definition of the Bohr magneton)} \qquad (41.31)$$

Evaluating Eq. (41.31) gives

$$\mu_B = 5.788 \times 10^{-5} \text{ eV/T} = 9.274 \times 10^{-24} \text{ J/T or A} \cdot \text{m}^2$$

Note that the units J/T and A $\cdot$ m^2 are equivalent. We defined this quantity previously in Section 28.8.

While the Bohr model suggests that the orbital motion of an atomic electron gives rise to a magnetic moment, this model does *not* give correct predictions about magnetic interactions. As an example, the Bohr model predicts that an electron in a hydrogen-atom ground state has an orbital magnetic moment of magnitude μ_B. But the Schrödinger picture tells us that such a ground-state electron is in an s state with zero angular momentum, so the orbital magnetic moment must be *zero!* To get the correct results, we must describe the states by using Schrödinger wave functions.

It turns out that in the Schrödinger formulation, electrons have the same ratio of μ to L (gyromagnetic ratio) as in the Bohr model—namely, $e/2m$. Suppose the magnetic field $\vec{B}$ is directed along the $+z$-axis. From Eq. (41.28) the interaction energy U of the atom's magnetic moment with the field is

$$U = -\mu_z B \qquad (41.32)$$

where μ_z is the z-component of the vector $\vec{\mu}$.

Now we use Eq. (41.30) to find μ_z, recalling that e is the *magnitude* of the electron charge and that the actual charge is $-e$. Because the electron charge is negative, the orbital angular momentum and magnetic moment vectors are opposite. We find

$$\mu_z = -\frac{e}{2m}L_z \qquad (41.33)$$

For the Schrödinger wave functions, $L_z = m_l\hbar$, with $m_l = 0, \pm 1, \pm 2, \ldots, \pm l$, so Eq. (41.33) becomes

$$\mu_z = -\frac{e}{2m}L_z = -m_l\frac{e\hbar}{2m} \qquad (41.34)$$

CAUTION **Two uses of the symbol m** Be careful not to confuse the electron mass m with the magnetic quantum number m_l. ▌

Finally, we can express the interaction energy, Eq. (41.32), as

$$U = -\mu_z B = m_l\frac{e\hbar}{2m}B \qquad (m_l = 0, \pm 1, \pm 2, \ldots, \pm l)$$

(orbital magnetic interaction energy) $\qquad (41.35)$

In terms of the Bohr magneton $\mu_B = e\hbar/2m$, we can write Eq. (41.35) as

$$U = m_l\mu_B B \qquad \text{(orbital magnetic interaction energy)} \qquad (41.36)$$

The magnetic field shifts the energy of each orbital state by an amount U. The interaction energy U depends on the value of m_l because m_l determines the orientation of the orbital magnetic moment relative to the magnetic field. This dependence is the reason m_l is called the magnetic quantum number.

The values of m_l range from $-l$ to $+l$ in steps of one, so an energy level with a particular value of the orbital quantum number l contains $(2l + 1)$ different orbital states. Without a magnetic field these states all have the same energy; that is, they are degenerate. The magnetic field removes this degeneracy. In the presence of a magnetic field they are split into $2l + 1$ distinct energy levels; adjacent levels differ in energy by $(e\hbar/2m)B = \mu_B B$. We can understand this in terms of the connection between degeneracy and symmetry. With a magnetic field applied along the z-axis, the atom is no longer completely symmetric under rotation: There is a preferred direction in space. By removing the symmetry, we remove the degeneracy of states.

Figure 41.13 shows the effect on the energy levels of hydrogen. Spectral lines corresponding to transitions from one set of levels to another set are correspondingly split and appear as a series of three closely spaced spectral lines replacing a single line. As the following example shows, the splitting of spectral lines is quite small because the value of $\mu_B B$ is small even for substantial magnetic fields.

41.13 This energy-level diagram for hydrogen shows how the levels are split when the electron's orbital magnetic moment interacts with an external magnetic field. The values of m_l are shown adjacent to the various levels. The relative magnitudes of the level splittings are exaggerated for clarity. The $n = 4$ splittings are not shown; can you draw them in?

Example 41.5   An atom in a magnetic field

An atom in a state with $l = 1$ emits a photon with wavelength 600.000 nm as it decays to a state with $l = 0$. If the atom is placed in a magnetic field with magnitude $B = 2.00$ T, what are the shifts in the energy levels and in the wavelength that result from the interaction between the atom's orbital magnetic moment and the magnetic field?

SOLUTION

IDENTIFY and SET UP: This problem concerns the splitting of atomic energy levels by a magnetic field (the Zeeman effect). We use Eq. (41.35) or (41.36) to determine the energy-level shifts. The relationship $E = hc/\lambda$ between the energy and wavelength of a photon then lets us calculate the wavelengths emitted during transitions from the $l = 1$ states to the $l = 0$ state.

EXECUTE: The energy of a 600-nm photon is

$$E = \frac{hc}{\lambda} = \frac{(4.14 \times 10^{-15} \text{ eV} \cdot \text{s})(3.00 \times 10^8 \text{ m/s})}{600 \times 10^{-9} \text{ m}}$$
$$= 2.07 \text{ eV}$$

If there is no external magnetic field, that is the difference in energy between the $l = 0$ and $l = 1$ levels.

With a 2.00-T field present, Eq. (41.36) shows that there is no shift of the $l = 0$ state (which has $m_l = 0$). For the $l = 1$ states, the splitting of levels is given by

$$U = m_l \mu_B B = m_l(5.788 \times 10^{-5} \text{ eV/T})(2.00 \text{ T})$$
$$= m_l(1.16 \times 10^{-4} \text{ eV}) = m_l(1.85 \times 10^{-23} \text{ J})$$

The possible values of m_l for $l = 1$ are -1, 0, and $+1$, and the three corresponding levels are separated by equal intervals of

Continued

1.16×10^{-4} eV. This is a small fraction of the 2.07-eV photon energy:

$$\frac{\Delta E}{E} = \frac{1.16 \times 10^{-4} \text{ eV}}{2.07 \text{ eV}} = 5.60 \times 10^{-5}$$

The possible values of m_l for $l = 1$ are -1, 0, and $+1$, and the three corresponding levels are separated by equal intervals of 1.16×10^{-4} eV. This is a small fraction of the 2.07-eV photon energy:

$$\frac{1.16 \times 10^{-4} \text{ eV}}{2.07 \text{ eV}} = 5.60 \times 10^{-5}$$

The corresponding *wavelength* shifts are approximately $(5.60 \times 10^{-5})(600 \text{ nm}) = 0.034$ nm. The original 600.000-nm line is split into a triplet with wavelengths 599.966, 600.000, and 600.034 nm.

EVALUATE: Even though 2.00 T would be a strong field in most laboratories, the wavelength splittings are extremely small. Nonetheless, modern spectrographs have more than enough chromatic resolving power to measure these splittings (see Section 36.5).

Selection Rules

Figure 41.14 shows what happens to a set of d states ($l = 2$) as the magnetic field increases. With zero field the five states $m_l = -2$, -1, 0, 1, and 2 are degenerate (have the same energy), but the applied field spreads the states out. Figure 41.15 shows the splittings of both the $3d$ and $2p$ states. Equal energy differences $(e\hbar/2m)B = \mu_B B$ separate adjacent levels. In the absence of a magnetic field, a transition from a $3d$ to a $2p$ state would yield a single spectral line with photon energy $E_i - E_f$. With the levels split as shown, it might seem that there are five possible photon energies.

In fact, there are only three possibilities. Not all combinations of initial and final levels are possible because of a restriction associated with conservation of angular momentum. The photon ordinarily carries off one unit ($\hbar$) of angular momentum, which leads to the requirements that in a transition l must change by 1 and m_l must change by 0 or ± 1. These requirements are called **selection rules.** Transitions that obey these rules are called *allowed transitions;* those that don't are *forbidden transitions.* In Fig. 41.15 we show the allowed transitions by solid arrows. You should count the possible transition energies to convince yourself that the nine solid arrows give only three possible energies; the zero-field value $E_i - E_f$, and that value plus or minus $\Delta E = (e\hbar/2m)B = \mu_B B$. Figure 41.12 shows the corresponding spectral lines.

What we have described is called the *normal* Zeeman effect. It is based entirely on the orbital angular momentum of the electron. However, it leaves out a very important consideration: the electron *spin* angular momentum, the subject of the next section.

41.14 This figure shows how the splitting of the energy levels of a d state ($l = 2$) depends on the magnitude B of an external magnetic field, assuming only an orbital magnetic moment.

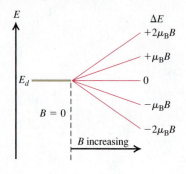

41.15 The cause of the normal Zeeman effect. The magnetic field splits the levels, but selection rules allow transitions with only three different energy changes, giving three different photon frequencies and wavelengths.

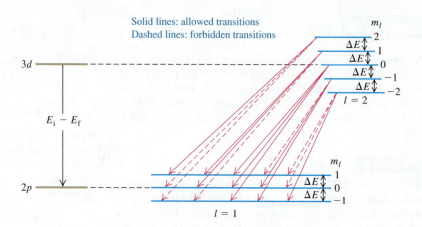

Test Your Understanding of Section 41.4 In this section we assumed that the magnetic field points in the positive z-direction. Would the results be different if the magnetic field pointed in the positive x-direction? ∎

41.5 Electron Spin

Despite the success of the Schrödinger equation in predicting the energy levels of the hydrogen atom, experimental observations indicate that it doesn't tell the whole story of the behavior of electrons in atoms. First, spectroscopists have found magnetic-field splitting into other than the three lines we've explained, sometimes unequally spaced. Before this effect was understood, it was called the *anomalous* Zeeman effect to distinguish it from the "normal" effect discussed in the preceding section. Figure 41.16 shows both kinds of splittings.

Second, some energy levels show splittings that resemble the Zeeman effect even when there is *no* external magnetic field. For example, when the lines in the hydrogen spectrum are examined with a high-resolution spectrograph, some lines are found to consist of sets of closely spaced lines called *multiplets*. Similarly, the orange-yellow line of sodium, corresponding to the transition $4p \rightarrow 3s$ of the outer electron, is found to be a doublet ($\lambda = 589.0$, 589.6 nm), suggesting that the $4p$ level might in fact be two closely spaced levels. The Schrödinger equation in its original form didn't predict any of this.

41.16 Illustrations of the normal and anomalous Zeeman effects for two elements, zinc and sodium. The brackets under each illustration show the "normal" splitting predicted by neglecting the effect of electron spin.

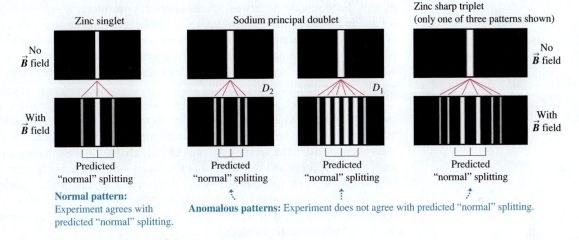

Zinc singlet   Sodium principal doublet   Zinc sharp triplet (only one of three patterns shown)

No $\vec{B}$ field

With $\vec{B}$ field

D_2 D_1

No $\vec{B}$ field

With $\vec{B}$ field

Predicted "normal" splitting   Predicted "normal" splitting   Predicted "normal" splitting   Predicted "normal" splitting

Normal pattern: Experiment agrees with predicted "normal" splitting.

Anomalous patterns: Experiment does not agree with predicted "normal" splitting.

The Stern–Gerlach Experiment

Similar anomalies appeared in 1922 in atomic-beam experiments performed in Germany by Otto Stern and Walter Gerlach. When they passed a beam of neutral atoms through a nonuniform magnetic field (Fig. 41.17), atoms were deflected

PhET: Stern–Gerlach Experiment

41.17 The Stern–Gerlach experiment.

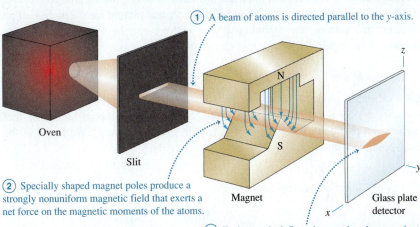

① A beam of atoms is directed parallel to the y-axis.

② Specially shaped magnet poles produce a strongly nonuniform magnetic field that exerts a net force on the magnetic moments of the atoms.

③ Each atom is deflected upward or downward according to the orientation of its magnetic moment.

Oven

Slit

Magnet

Glass plate detector

N

S

z

y

x

according to the orientation of their magnetic moments with respect to the field. These experiments demonstrated the quantization of angular momentum in a very direct way. If there were only orbital angular momentum, the deflections would split the beam into an odd number $(2l + 1)$ of different components. However, some atomic beams were split into an *even* number of components. If we use a different symbol j for an angular momentum quantum number, setting $2j + 1$ equal to an even number gives $j = \frac{1}{2}, \frac{3}{2}, \frac{5}{2}, \ldots$, suggesting a half-integer angular momentum. This can't be understood on the basis of the Bohr model and similar pictures of atomic structure.

In 1925 two graduate students in the Netherlands, Samuel Goudsmidt and George Uhlenbeck, proposed that the electron might have some additional motion. Using a semiclassical model, they suggested that the electron might behave like a spinning sphere of charge instead of a particle. If so, it would have an additional *spin* angular momentum and magnetic moment. If these were quantized in much the same way as *orbital* angular momentum and magnetic moment, they might help to explain the observed energy-level anomalies.

An Analogy for Electron Spin

To introduce the concept of **electron spin,** let's start with an analogy. The earth travels in a nearly circular orbit around the sun, and at the same time it *rotates* on its axis. Each motion has its associated angular momentum, which we call the *orbital* and *spin* angular momentum, respectively. The total angular momentum of the earth is the vector sum of the two. If we were to model the earth as a single point, it would have no moment of inertia about its spin axis and thus no spin angular momentum. But when our model includes the finite size of the earth, spin angular momentum becomes possible.

In the Bohr model, suppose the electron is not just a point charge but a small spinning sphere moving in orbit. Then the electron has not only orbital angular momentum but also spin angular momentum associated with the rotation of its mass about its axis. The sphere carries an electric charge, so the spinning motion leads to current loops and to a magnetic moment, as we discussed in Section 27.7. In a magnetic field, the *spin* magnetic moment has an interaction energy in addition to that of the *orbital* magnetic moment (the normal Zeeman-effect interaction that we discussed in Section 41.4). We should see additional Zeeman shifts due to the spin magnetic moment.

As we mentioned, such shifts *are* indeed observed in precise spectroscopic analysis. This and a variety of other experimental evidence have shown conclusively that the electron *does* have a spin angular momentum and a spin magnetic moment that do not depend on its orbital motion but are intrinsic to the electron itself. The origin of this spin angular momentum is fundamentally quantum-mechanical, so it's not correct to model the electron as a spinning charged sphere. But just as the Bohr model can be a useful conceptual picture for the motion of an electron in an atom, the spinning-sphere analogy can help you visualize the intrinsic spin angular momentum of an electron.

Spin Quantum Numbers

Like orbital angular momentum, the spin angular momentum of an electron (denoted by $\vec{S}$) is found to be quantized. Suppose we have an apparatus that measures a particular component of $\vec{S}$, say the z-component S_z. We find that the only possible values are

$$S_z = \pm\tfrac{1}{2}\hbar \text{ (components of spin angular momentum)} \qquad (41.37)$$

This relationship is reminiscent of the expression $L_z = m_l\hbar$ for the z-component of orbital angular momentum, except that $|S_z|$ is *one-half* of $\hbar$ instead of an *integer* multiple. Equation (41.37) also suggests that the magnitude S of the spin angular momentum is given by an expression analogous to Eq. (41.22) with the orbital quantum number l replaced by the **spin quantum number** $s = \frac{1}{2}$:

$$S = \sqrt{\tfrac{1}{2}\left(\tfrac{1}{2} + 1\right)}\,\hbar = \sqrt{\tfrac{3}{4}}\,\hbar \qquad \text{(magnitude of spin angular momentum)} \qquad (41.38)$$

The electron is often called a "spin $-\frac{1}{2}$ particle."

To visualize the quantized spin of an electron in a hydrogen atom, think of the electron probability distribution function $|\psi|^2$ as a cloud surrounding the nucleus like those shown in Figs. 41.9 and 41.10. Then imagine many tiny spin arrows distributed throughout the cloud, either all with components in the $+z$-direction or all with components in the $-z$-direction. But don't take this picture too seriously.

To label completely the state of the electron in a hydrogen atom, we now need a fourth quantum number m_s to specify the electron spin orientation. In analogy to the orbital magnetic quantum number m_l, we call m_s the **spin magnetic quantum number.** For an electron we give m_s the value $+\frac{1}{2}$ or $-\frac{1}{2}$ to agree with Eq. (41.37):

$$S_z = m_s \hbar \quad \left(m_s = \pm \tfrac{1}{2}\right) \qquad \begin{array}{l}\text{(allowed values of } m_s \text{ and}\\ S_z \text{ for an electron)}\end{array} \qquad (41.39)$$

The spin angular momentum vector $\vec{S}$ can have only two orientations in space relative to the z-axis: *"spin up"* with a z-component of $+\frac{1}{2}\hbar$ and *"spin down"* with a z-component of $-\frac{1}{2}\hbar$.

The z-component of the associated spin magnetic moment (μ_z) turns out to be related to S_z by

$$\mu_z = -(2.00232)\frac{e}{2m}S_z \qquad (41.40)$$

where $-e$ and m are (as usual) the charge and mass of the electron. When the atom is placed in a magnetic field, the interaction energy $-\vec{\mu} \cdot \vec{B}$ of the spin magnetic dipole moment with the field causes further splittings in energy levels and in the corresponding spectral lines.

Equation (41.40) shows that the gyromagnetic ratio for electron spin is approximately *twice* as great as the value $e/2m$ for *orbital* angular momentum and magnetic dipole moment. This result has no classical analog. But in 1928 Paul Dirac developed a relativistic generalization of the Schrödinger equation for electrons. His equation gave a spin gyromagnetic ratio of exactly $2(e/2m)$. It took another two decades to develop the area of physics called *quantum electrodynamics,* abbreviated QED, which predicts the value we've given to "only" six significant figures as 2.00232. QED now predicts a value that agrees with a recent (2006) measurement of 2.00231930436170(152), making QED the most precise theory in all science.

Example 41.6 **Energy of electron spin in a magnetic field**

Calculate the interaction energy for an electron in an $l = 0$ state in a magnetic field with magnitude 2.00 T.

SOLUTION

IDENTIFY and SET UP: For $l = 0$ the electron has zero orbital angular momentum and zero orbital magnetic moment. Hence the only magnetic interaction is that between the $\vec{B}$ field and the spin magnetic moment $\vec{\mu}$. From Eq. (41.28), the interaction energy is $U = -\vec{\mu} \cdot \vec{B}$. As in Section 41.4, we take $\vec{B}$ to be in the positive z-direction so that $U = -\mu_z B$ [Eq. (41.32)]. Equation (41.40) gives μ_z in terms of S_z, and Eq. (41.37) gives S_z.

EXECUTE: Combining Eqs. (41.37) and (41.40), we have

$$\mu_z = -(2.00232)\left(\frac{e}{2m}\right)\left(\pm\tfrac{1}{2}\hbar\right)$$

$$= \mp\tfrac{1}{2}(2.00232)\left(\frac{e\hbar}{2m}\right) = \mp(1.00116)\mu_B$$

$$= \mp(1.00116)(9.274 \times 10^{-24}\,\text{J/T})$$

$$= \mp 9.285 \times 10^{-24}\,\text{J/T}$$

$$= \mp 5.795 \times 10^{-5}\,\text{eV/T}$$

Continued

Then from Eq. (41.32),

$$U = -\mu_z B = \pm(9.285 \times 10^{-24} \text{ J/T})(2.00 \text{ T})$$
$$= \pm 1.86 \times 10^{-23} \text{ J} = \pm 1.16 \times 10^{-4} \text{ eV}$$

The positive value of U and the negative value of μ_z correspond to $S_z = +\frac{1}{2}\hbar$ (spin up); the negative value of U and the positive value of μ_z correspond to $S_z = -\frac{1}{2}\hbar$ (spin down).

EVALUATE: Let's check the *signs* of our results. If the electron is spin down, $\vec{S}$ points generally opposite to $\vec{B}$. Then the magnetic moment $\vec{\mu}$ (which is opposite to $\vec{S}$ because the electron charge is negative) points generally parallel to $\vec{B}$, and μ_z is positive. From Eq. (41.28), $U = -\vec{\mu} \cdot \vec{B}$, the interaction energy is negative if $\vec{\mu}$ and $\vec{B}$ are parallel. Our results show that U is indeed negative in this case. We can similarly confirm that U must be positive and μ_z negative for a spin-up electron.

The red lines in Fig. 41.18 show how the interaction energies for the two spin states vary with the magnetic field magnitude B. The graphs are straight lines because, from Eq. (41.32), U is proportional to B.

41.18 An $l = 0$ level of a single electron is split by interaction of the spin magnetic moment with an external magnetic field. The greater the magnitude B of the magnetic field, the greater the splitting. The quantity 5.795×10^{-5} eV/T is just $(1.00116)\mu_B$.

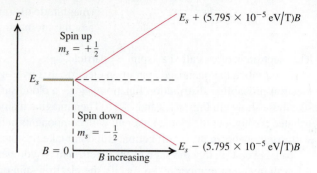

Spin-Orbit Coupling

We mentioned earlier that the spin magnetic dipole moment also gives splitting of energy levels even when there is *no* external field. One cause involves the orbital motion of the electron. In the Bohr model, observers moving with the electron would see the positively charged nucleus revolving around them (just as to earthbound observers the sun seems to be orbiting the earth). This apparent motion of charge causes a magnetic field at the location of the electron, as measured in the electron's moving frame of reference. The resulting interaction with the spin magnetic moment causes a twofold splitting of this level, corresponding to the two possible orientations of electron spin.

Discussions based on the Bohr model can't be taken too seriously, but a similar result can be derived from the Schrödinger equation. The interaction energy U can be expressed in terms of the scalar product of the angular momentum vectors $\vec{L}$ and $\vec{S}$. This effect is called **spin-orbit coupling;** it is responsible for the small energy difference between the two closely spaced, lowest excited levels of sodium shown in Fig. 39.19a and for the corresponding doublet (589.0, 589.6 nm) in the spectrum of sodium.

Example 41.7   **An effective magnetic field**

To six significant figures, the wavelengths of the two spectral lines that make up the sodium doublet are $\lambda_1 = 588.995$ nm and $\lambda_2 = 589.592$ nm. Calculate the effective magnetic field experienced by the electron in the $3p$ levels of the sodium atom.

SOLUTION

IDENTIFY and SET UP: The two lines in the sodium doublet result from transitions from the two $3p$ levels, which are split by spin-orbit coupling, to the $3s$ level, which is *not* split because it has $L = 0$. We picture the spin-orbit coupling as an interaction between the electron spin magnetic moment and an effective magnetic field due to the nucleus. This example is like Example 41.6 in reverse: There we were given B and found the difference between the energies of the two spin states, while here we use the energy difference to find the target variable B. The difference in energy between the two $3p$ levels is equal to the difference in energy between the two $3p$ levels is equal to the difference in energy

between the two photons of the sodium doublet. We use this relationship and the results of Example 41.6 to determine B.

EXECUTE: The energies of the two photons are $E_1 = hc/\lambda_1$ and $E_2 = hc/\lambda_2$. Here $E_1 > E_2$ because $\lambda_1 < \lambda_2$, so the difference in their energies is

$$\Delta E = \frac{hc}{\lambda_1} - \frac{hc}{\lambda_2} = hc\left(\frac{\lambda_2 - \lambda_1}{\lambda_2 \lambda_1}\right)$$
$$= (4.136 \times 10^{-15} \text{ eV} \cdot \text{s})(2.998 \times 10^8 \text{ m/s})$$
$$\times \frac{(589.592 \times 10^{-9} \text{ m}) - (588.995 \times 10^{-9} \text{ m})}{(589.592 \times 10^{-9} \text{ m})(588.995 \times 10^{-9} \text{ m})}$$
$$= 0.00213 \text{ eV} = 3.41 \times 10^{-22} \text{ J}$$

This equals the energy difference between the two $3p$ levels. The spin-orbit interaction raises one level by 1.70×10^{-22} J (one-half of 3.41×10^{-22} J) and lowers the other by 1.70×10^{-22} J. From Example 41.6, the amount each state is raised or lowered is $|U| = (1.00116)\mu_B B$, so

$$B = \left| \frac{U}{(1.00116)\mu_B} \right| = \frac{1.70 \times 10^{-22} \text{ J}}{9.28 \times 10^{-24} \text{ J/T}} = 18.0 \text{ T}$$

EVALUATE: The electron experiences a *very* strong effective magnetic field. To produce a steady, macroscopic field of this magnitude in the laboratory requires state-of-the-art electromagnets.

Combining Orbital and Spin Angular Momenta

The orbital and spin angular momenta ($\vec{L}$ and $\vec{S}$, respectively) can combine in various ways. The vector sum of $\vec{L}$ and $\vec{S}$ is the *total* angular momentum $\vec{J}$:

$$\vec{J} = \vec{L} + \vec{S} \tag{41.41}$$

The possible values of the magnitude J are given in terms of a quantum number j by

$$J = \sqrt{j(j + 1)}\hbar \tag{41.42}$$

We can then have states in which $j = |l \pm \frac{1}{2}|$. The $l + \frac{1}{2}$ states correspond to the case in which the vectors $\vec{L}$ and $\vec{S}$ have parallel z-components; for the $l - \frac{1}{2}$ states, $\vec{L}$ and $\vec{S}$ have antiparallel z-components. For example, when $l = 1$, j can be $\frac{1}{2}$ or $\frac{3}{2}$. In another spectroscopic notation these p states are labeled $^2P_{1/2}$ and $^2P_{3/2}$, respectively. The superscript is the number of possible spin orientations, the letter P (now capitalized) indicates states with $l = 1$, and the subscript is the value of j. We used this scheme to label the energy levels of the sodium atom in Fig. 39.19a.

The various line splittings resulting from magnetic interactions are collectively called *fine structure*. There are also additional, much smaller splittings associated with the fact that the *nucleus* of the atom has a magnetic dipole moment that interacts with the orbital and/or spin magnetic dipole moments of the electrons. These effects are called *hyperfine structure*. For example, the ground level of hydrogen is split into two states, separated by only 5.9×10^{-6} eV. The photon that is emitted in the transitions between these states has a wavelength of 21 cm. Radio astronomers use this wavelength to map clouds of interstellar hydrogen gas that are too cold to emit visible light (Fig. 41.19).

41.19 In a visible-light image (top), these three distant galaxies appear to be unrelated. But in fact these galaxies are connected by immense streamers of hydrogen gas. This is revealed by by the false-color image (bottom) made with a radio telescope tuned to the 21-cm wavelength emitted by hydrogen atoms.

Galaxies in visible light (negative image; galaxies appear dark)

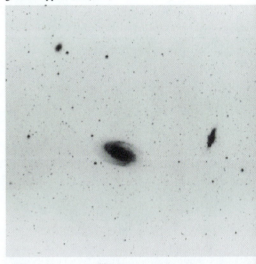

Radio image at wavelength 21 cm

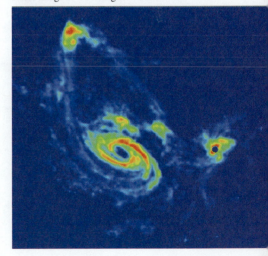

Test Your Understanding of Section 41.5 In which of the following situations is the magnetic moment of an electron perfectly aligned with a magnetic field that points in the positive z-direction? (i) $m_s = +\frac{1}{2}$; (ii) $m_s = -\frac{1}{2}$; (iii) both (i) and (ii); (iv) neither (i) nor (ii).

41.6 Many-Electron Atoms and the Exclusion Principle

So far our analysis of atomic structure has concentrated on the hydrogen atom. That's natural; neutral hydrogen, with only one electron, is the simplest atom. If we can't understand hydrogen, we certainly can't understand anything more complex. But now let's move on to many-electron atoms.

In general, an atom in its normal (electrically neutral) state has Z electrons and Z protons. Recall from Section 41.3 that we call Z the *atomic number*. The total electric charge of such an atom is exactly zero because the neutron has no charge while the proton and electron charges have the same magnitude but opposite sign.

We can apply the Schrödinger equation to this general atom. However, the complexity of the analysis increases very rapidly with increasing Z. Each of the Z electrons interacts not only with the nucleus but also with every other electron.

The wave functions and the potential energy are functions of $3Z$ coordinates, and the equation contains second derivatives with respect to all of them. The mathematical problem of finding solutions of such equations is so complex that it has not been solved exactly even for the neutral helium atom, which has only two electrons.

Fortunately, various approximation schemes are available. The simplest approximation is to ignore all interactions between electrons and consider each electron as moving under the action only of the nucleus (considered to be a point charge). In this approximation the wave function for each individual electron is a function like that for the hydrogen atom, specified by four quantum numbers (n, l, m_l, m_s). The nuclear charge is Ze instead of e, so we replace every factor of e^2 in the wave functions and the energy levels by Ze^2. In particular, the energy levels are given by Eq. (41.21) with e^4 replaced by $Z^2 e^4$:

$$E_n = -\frac{1}{(4\pi\epsilon_0)^2} \frac{m_r Z^2 e^4}{2n^2 \hbar^2} = -\frac{Z^2}{n^2}(13.6\text{ eV}) \qquad (41.43)$$

This approximation is fairly drastic; when there are many electrons, their interactions with each other are as important as the interaction of each with the nucleus. So this model isn't very useful for quantitative predictions.

The Central-Field Approximation

A less drastic and more useful approximation is to think of all the electrons together as making up a charge cloud that is, on average, *spherically symmetric*. We can then think of each individual electron as moving in the total electric field due to the nucleus and this averaged-out cloud of all the other electrons. There is a corresponding spherically symmetric potential-energy function $U(r)$. This picture is called the **central-field approximation;** it provides a useful starting point for understanding atomic structure.

In the central-field approximation we can again deal with one-electron wave functions. The Schrödinger equation differs from the equation for hydrogen only in that the $1/r$ potential-energy function is replaced by a different function $U(r)$. But it turns out that $U(r)$ does not enter the differential equations for $\Theta(\theta)$ and $\Phi(\phi)$, so those angular functions are exactly the same as for hydrogen, and the orbital angular-momentum *states* are also the same as before. The quantum numbers l, m_l, and m_s have the same meanings as before, and Eqs. (41.22) and (41.23) again give the magnitude and z-component of the orbital angular momentum.

The radial wave functions and probabilities are different than for hydrogen because of the change in $U(r)$, so the energy levels are no longer given by Eq. (41.21). We can still label a state using the four quantum numbers (n, l, m_l, m_s). In general, the energy of a state now depends on both n and l, rather than just on n as with hydrogen. The restrictions on values of the quantum numbers are the same as before:

$$n \geq 1 \quad 0 \leq l \leq n-1 \quad |m_l| \leq l \quad m_s = \pm\tfrac{1}{2} \qquad \begin{array}{l}\text{(allowed values of}\\ \text{quantum numbers)}\end{array} \quad (41.44)$$

The Exclusion Principle

To understand the structure of many-electron atoms, we need an additional principle, the *exclusion principle*. To see why this principle is needed, let's consider the lowest-energy state, or *ground state*, of a many-electron atom. In the one-electron states of the central-field model, there is a lowest-energy state (corresponding to an $n = 1$ state of hydrogen). We might expect that in the ground state of a complex atom, *all* the electrons should be in this lowest state. If so, then we should see only gradual changes in physical and chemical properties when we look at the behavior of atoms with increasing numbers of electrons (Z).

Such gradual changes are *not* what is observed. Instead, properties of elements vary widely from one to the next, with each element having its own distinct personality. For example, the elements fluorine, neon, and sodium have 9, 10, and 11 electrons, respectively, per atom. Fluorine $(Z = 9)$ is a *halogen;* it tends strongly to form compounds in which each fluorine atom acquires an extra electron. Sodium $(Z = 11)$ is an *alkali metal;* it forms compounds in which each sodium atom *loses* an electron. Neon $(Z = 10)$ is a *noble gas,* forming no compounds at all. Such observations show that in the ground state of a complex atom the electrons *cannot* all be in the lowest-energy states. But why not?

The key to this puzzle, discovered by the Austrian physicist Wolfgang Pauli (Fig. 41.20) in 1925, is called the **exclusion principle.** This principle states that **no two electrons can occupy the same quantum-mechanical state** in a given system. That is, **no two electrons in an atom can have the same values of all four quantum numbers** (n, l, m_l, m_s). Each quantum state corresponds to a certain distribution of the electron "cloud" in space. Therefore the principle also says, in effect, that no more than two electrons with opposite values of the quantum number m_s can occupy the same region of space. We shouldn't take this last statement too seriously because the electron probability functions don't have sharp, definite boundaries. But the exclusion principle limits the amount by which electron wave functions can overlap. Think of it as the quantum-mechanical analog of a university rule that allows only one student per desk.

CAUTION **The meaning of the exclusion principle** Don't confuse the exclusion principle with the electric repulsion between electrons. While both effects tend to keep electrons within an atom separated from each other, they are very different in character. Two electrons can always be pushed closer together by adding energy to combat electric repulsion; in contrast, *nothing* can overcome the exclusion principle and force two electrons into the same quantum-mechanical state.

Table 41.2 lists some of the sets of quantum numbers for electron states in an atom. It's similar to Table 41.1 (Section 41.3), but we've added the number of states in each subshell and shell. Because of the exclusion principle, the "number of states" is the same as the *maximum* number of electrons that can be found in those states. For each state, m_s can be either $+\frac{1}{2}$ or $-\frac{1}{2}$.

As with the hydrogen wave functions, different states correspond to different spatial distributions; electrons with larger values of n are concentrated at larger distances from the nucleus. Figure 41.8 (Section 41.3) shows this effect. When an atom has more than two electrons, they can't all huddle down in the low-energy $n = 1$ states nearest to the nucleus because there are only two of these states; the exclusion principle forbids multiple occupancy of a state. Some electrons are forced into states farther away, with higher energies. Each value of n corresponds roughly to a region of space around the nucleus in the form of a spherical *shell.* Hence we speak of the K shell as the region that is occupied by the electrons in the $n = 1$ states, the L shell as the region of the $n = 2$ states, and so on. States with the same n but different l form *subshells,* such as the $3p$ subshell.

41.20 The key to understanding the periodic table of the elements was the discovery by Wolfgang Pauli (1900–1958) of the exclusion principle. Pauli received the 1945 Nobel Prize in physics for his accomplishment. This photo shows Pauli (on the left) and Niels Bohr watching the physics of a toy top spinning on the floor—a macroscopic analog of a microscopic electron with spin.

Table 41.2 Quantum States of Electrons in the First Four Shells

n	l	m_l	Spectroscopic Notation	Number of States		Shell
1	0	0	$1s$	2		K
2	0	0	$2s$	2	8	L
2	1	$-1, 0, 1$	$2p$	6		
3	0	0	$3s$	2	18	M
3	1	$-1, 0, 1$	$3p$	6		
3	2	$-2, -1, 0, 1, 2$	$3d$	10		
4	0	0	$4s$	2	32	N
4	1	$-1, 0, 1$	$4p$	6		
4	2	$-2, -1, 0, 1, 2$	$4d$	10		
4	3	$-3, -2, -1, 0, 1, 2, 3$	$4f$	14		

The Periodic Table

We can use the exclusion principle to derive the most important features of the structure and chemical behavior of multielectron atoms, including the periodic table of the elements. Let's imagine constructing a neutral atom by starting with a bare nucleus with Z protons and then adding Z electrons, one by one. To obtain the ground state of the atom as a whole, we fill the lowest-energy electron states (those closest to the nucleus, with the smallest values of n and l) first, and we use successively higher states until all the electrons are in place. The chemical properties of an atom are determined principally by interactions involving the outermost, or *valence,* electrons, so we particularly want to learn how these electrons are arranged.

Let's look at the ground-state electron configurations for the first few atoms (in order of increasing Z). For hydrogen the ground state is $1s$; the single electron is in a state $n = 1$, $l = 0$, $m_l = 0$, and $m_s = \pm\frac{1}{2}$. In the helium atom ($Z = 2$), *both* electrons are in $1s$ states, with opposite spins; one has $m_s = -\frac{1}{2}$ and the other has $m_s = +\frac{1}{2}$. We denote the helium ground state as $1s^2$. (The superscript 2 is not an exponent; the notation $1s^2$ tells us that there are two electrons in the $1s$ subshell. Also, the superscript 1 is understood, as in $2s$.) For helium the K shell is completely filled, and all others are empty. Helium is a noble gas; it has no tendency to gain or lose an electron, and it forms no compounds.

Lithium ($Z = 3$) has three electrons. In its ground state, two are in $1s$ states and one is in a $2s$ state, so we denote the lithium ground state as $1s^2 2s$. On average, the $2s$ electron is considerably farther from the nucleus than are the $1s$ electrons (Fig. 41.21). According to Gauss's law, the *net* charge Q_{encl} attracting the $2s$ electron is nearer to $+e$ than to the value $+3e$ it would have without the two $1s$ electrons present. As a result, the $2s$ electron is loosely bound; only 5.4 eV is required to remove it, compared with the 30.6 eV given by Eq. (41.43) with $Z = 3$ and $n = 2$. In chemical behavior, lithium is an *alkali metal.* It forms ionic compounds in which each lithium atom loses an electron and has a valence of $+1$.

Next is beryllium ($Z = 4$); its ground-state configuration is $1s^2 2s^2$, with its two valence electrons filling the s subshell of the L shell. Beryllium is the first of the *alkaline earth* elements, forming ionic compounds in which the valence of the atoms is $+2$.

Table 41.3 shows the ground-state electron configurations of the first 30 elements. The L shell can hold eight electrons. At $Z = 10$, both the K and L shells are filled, and there are no electrons in the M shell. We expect this to be a particularly stable configuration, with little tendency to gain or lose electrons. This element is neon, a noble gas with no known compounds. The next element after neon is sodium ($Z = 11$), with filled K and L shells and one electron in the M shell. Its "noble-gas-plus-one-electron" structure resembles that of lithium; both are alkali metals. The element *before* neon is fluorine, with $Z = 9$. It has a vacancy in the L shell and has an affinity for an extra electron to fill the shell. Fluorine forms ionic compounds in which it has a valence of -1. This behavior is characteristic of the *halogens* (fluorine, chlorine, bromine, iodine, and astatine), all of which have "noble-gas-minus-one" configurations (Fig. 41.22).

Proceeding down the list, we can understand the regularities in chemical behavior displayed by the **periodic table of the elements** (Appendix D) on the basis of electron configurations. The similarity of elements in each *group* (vertical column) of the periodic table is the result of similarity in outer-electron configuration. All the noble gases (helium, neon, argon, krypton, xenon, and radon) have filled-shell or filled-shell plus filled p subshell configurations. All the alkali metals (lithium, sodium, potassium, rubidium, cesium, and francium) have "noble-gas-plus-one" configurations. All the alkaline earth metals (beryllium, magnesium, calcium, strontium, barium, and radium) have "noble-gas-plus-two" configurations, and, as we just mentioned, all the halogens (fluorine, chlorine, bromine, iodine, and astatine) have "noble-gas-minus-one" structures.

41.21 Schematic representation of the charge distribution in a lithium atom. The nucleus has a charge of $+3e$.

On average, the $2s$ electron is considerably farther from the nucleus than the $1s$ electrons. Therefore, it experiences a net nuclear charge of approximately $+3e - 2e = +e$ (rather than $+3e$).

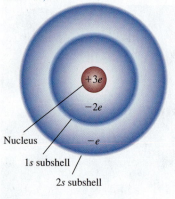

Nucleus

$1s$ subshell

$2s$ subshell

$+3e$

$-2e$

$-e$

41.22 Salt (sodium chloride, NaCl) dissolves readily in water, making seawater salty. This is due to the electron configurations of sodium and chlorine: Sodium can easily lose an electron to form an Na^+ ion, and chlorine can easily gain an electron to form a Cl^- ion. These ions are held in solution because they are attracted to the polar ends of water molecules (see Fig. 21.30a).

Table 41.3 **Ground-State Electron Configurations**

Element	Symbol	Atomic Number (Z)	Electron Configuration
Hydrogen	H	1	$1s$
Helium	He	2	$1s^2$
Lithium	Li	3	$1s^2 2s$
Beryllium	Be	4	$1s^2 2s^2$
Boron	B	5	$1s^2 2s^2 2p$
Carbon	C	6	$1s^2 2s^2 2p^2$
Nitrogen	N	7	$1s^2 2s^2 2p^3$
Oxygen	O	8	$1s^2 2s^2 2p^4$
Fluorine	F	9	$1s^2 2s^2 2p^5$
Neon	Ne	10	$1s^2 2s^2 2p^6$
Sodium	Na	11	$1s^2 2s^2 2p^6 3s$
Magnesium	Mg	12	$1s^2 2s^2 2p^6 3s^2$
Aluminum	Al	13	$1s^2 2s^2 2p^6 3s^2 3p$
Silicon	Si	14	$1s^2 2s^2 2p^6 3s^2 3p^2$
Phosphorus	P	15	$1s^2 2s^2 2p^6 3s^2 3p^3$
Sulfur	S	16	$1s^2 2s^2 2p^6 3s^2 3p^4$
Chlorine	Cl	17	$1s^2 2s^2 2p^6 3s^2 3p^5$
Argon	Ar	18	$1s^2 2s^2 2p^6 3s^2 3p^6$
Potassium	K	19	$1s^2 2s^2 2p^6 3s^2 3p^6 4s$
Calcium	Ca	20	$1s^2 2s^2 2p^6 3s^2 3p^6 4s^2$
Scandium	Sc	21	$1s^2 2s^2 2p^6 3s^2 3p^6 4s^2 3d$
Titanium	Ti	22	$1s^2 2s^2 2p^6 3s^2 3p^6 4s^2 3d^2$
Vanadium	V	23	$1s^2 2s^2 2p^6 3s^2 3p^6 4s^2 3d^3$
Chromium	Cr	24	$1s^2 2s^2 2p^6 3s^2 3p^6 4s 3d^5$
Manganese	Mn	25	$1s^2 2s^2 2p^6 3s^2 3p^6 4s^2 3d^5$
Iron	Fe	26	$1s^2 2s^2 2p^6 3s^2 3p^6 4s^2 3d^6$
Cobalt	Co	27	$1s^2 2s^2 2p^6 3s^2 3p^6 4s^2 3d^7$
Nickel	Ni	28	$1s^2 2s^2 2p^6 3s^2 3p^6 4s^2 3d^8$
Copper	Cu	29	$1s^2 2s^2 2p^6 3s^2 3p^6 4s 3d^{10}$
Zinc	Zn	30	$1s^2 2s^2 2p^6 3s^2 3p^6 4s^2 3d^{10}$

A slight complication occurs with the M and N shells because the $3d$ and $4s$ subshell levels ($n = 3$, $l = 2$, and $n = 4$, $l = 0$, respectively) have similar energies. (We'll discuss in the next subsection why this happens.) Argon ($Z = 18$) has all the $1s$, $2s$, $2p$, $3s$, and $3p$ subshells filled, but in potassium ($Z = 19$) the additional electron goes into a $4s$ energy state rather than a $3d$ state (because the $4s$ state has slightly lower energy).

The next several elements have one or two electrons in the $4s$ subshell and increasing numbers in the $3d$ subshell. These elements are all metals with rather similar chemical and physical properties; they form the first *transition series,* starting with scandium ($Z = 21$) and ending with zinc ($Z = 30$), for which all the $3d$ and $4s$ subshells are filled.

Something similar happens with $Z = 57$ through $Z = 71$, which have one or two electrons in the $6s$ subshell but only partially filled $4f$ and $5d$ subshells. These are the *rare earth* elements; they all have very similar physical and chemical properties. Yet another such series, called the *actinide* series, starts with $Z = 91$.

Screening

We have mentioned that in the central-field picture, the energy levels depend on l as well as n. Let's take sodium ($Z = 11$) as an example. If 10 of its electrons fill its K and L shells, the energies of some of the states for the remaining electron are found experimentally to be

$3s$ states: -5.138 eV

$3p$ states: -3.035 eV

$3d$ states: -1.521 eV

$4s$ states: -1.947 eV

Application **Electron Configurations and Bone Cancer Radiotherapy**
The orange spots in this colored x-ray image are bone cancer tumors. One method of treating bone cancer is to inject a radioactive isotope of strontium (^{89}Sr) into a patient's vein. Strontium is chemically similar to calcium because in both atoms the two outer electrons are in an s state (the structures are $1s^2 2s^2 2p^6 3s^2 3p^6 4s^2 3d^{10} 4p^6 5s^2$ for strontium and $1s^2 2s^2 2p^6 3s^2 3p^6 4s^2$ for calcium). Hence the strontium is readily taken up by the tumors, where calcium turnover is more rapid than in healthy bone. Radiation from the strontium helps to destroy the tumors.

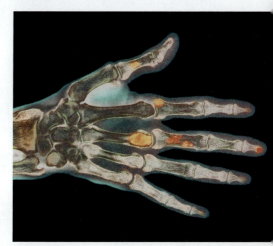

The 3s states are the lowest (most negative); one is the ground state for the 11th electron in sodium. The energy of the 3d states is quite close to the energy of the $n = 3$ state in hydrogen. The surprise is that the 4s state energy is 0.426 eV *below* the 3d state, even though the 4s state has larger n.

We can understand these results using Gauss's law and the radial probability distribution. For any spherically symmetric charge distribution, the electric-field magnitude at a distance r from the center is $Q_{encl}/4\pi\epsilon_0 r^2$, where Q_{encl} is the total charge enclosed within a sphere with radius r. Mentally remove the outer (valence) electron atom from a sodium atom. What you have left is a spherically symmetric collection of 10 electrons (filling the K and L shells) and 11 protons, so $Q_{encl} = -10e + 11e = +e$. If the 11th electron is completely outside this collection of charges, it is attracted by an effective charge of $+e$, not $+11e$. This is a more extreme example of the effect depicted in Fig. 41.21.

This effect is called **screening**; the 10 electrons *screen* 10 of the 11 protons, leaving an effective net charge of $+e$. In general, an electron that spends all its time completely outside a positive charge $Z_{eff}e$ has energy levels given by the hydrogen expression with e^2 replaced by $Z_{eff}e^2$. From Eq. (41.43) this is

$$E_n = -\frac{Z_{eff}^2}{n^2}(13.6 \text{ eV}) \qquad \text{(energy levels with screening)} \qquad (41.45)$$

If the 11th electron in the sodium atom is completely outside the remaining charge distribution, then $Z_{eff} = 1$.

CAUTION **Different equations for different atoms** Equations (41.21), (41.43), and (41.45) all give values of E_n in terms of $(13.6 \text{ eV})/n^2$, but they don't apply in general to the same atoms. Equation (41.21) is *only* for hydrogen. Equation (41.43) is only for the case in which there is no interaction with any other electron (and is thus accurate only when the atom has just one electron). Equation (41.45) is useful when one electron is screened from the nucleus by other electrons. ▌

Now let's use the radial probability functions shown in Fig. 41.8 to explain why the energy of a sodium 3d state is approximately the same as the $n = 3$ value of hydrogen, -1.51 eV. The distribution for the 3d state (for which l has the maximum value $n - 1$) has one peak, and its most probable radius is *outside* the positions of the electrons with $n = 1$ or 2. (Those electrons also are pulled closer to the nucleus than in hydrogen because they are less effectively screened from the positive charge $11e$ of the nucleus.) Thus in sodium a 3d electron spends most of its time well outside the $n = 1$ and $n = 2$ states (the K and L shells). The 10 electrons in these shells screen about ten-elevenths of the charge of the 11 protons, leaving a net charge of about $Z_{eff}e = (1)e$. Then, from Eq. (41.45), the corresponding energy is approximately $-(1)^2(13.6 \text{ eV})/3^2 = -1.51$ eV. This approximation is very close to the experimental value of -1.521 eV.

Looking again at Fig. 41.8, we see that the radial probability density for the 3p state (for which $l = n - 2$) has two peaks and that for the 3s state ($l = n - 3$) has three peaks. For sodium the first small peak in the 3p distribution gives a 3p electron a higher probability (compared to the 3d state) of being *inside* the charge distributions for the electrons in the $n = 2$ states. That is, a 3p electron is less completely screened from the nucleus than is a 3d electron because it spends some of its time within the filled K and L shells. Thus for the 3p electrons, Z_{eff} is greater than unity. From Eq. (41.45) the 3p energy is lower (more negative) than the 3d energy of -1.521 eV. The actual value is -3.035 eV. A 3s electron spends even more time within the inner electron shells than a 3p electron does, giving an even larger Z_{eff} and an even more negative energy.

Example 41.8 Determining Z_{eff} experimentally

The measured energy of a $3s$ state of sodium is -5.138 eV. Calculate the value of Z_{eff}.

SOLUTION

IDENTIFY and SET UP: Sodium has a single electron in the M shell outside filled K and L shells. The ten K and L electrons partially screen the single M electron from the $+11e$ charge of the nucleus; our goal is to determine the extent of this screening. We are given $n = 3$ and $E_n = -5.138$ eV, so we can use Eq. (41.45) to determine Z_{eff}.

EXECUTE: Solving Eq. (41.45) for Z_{eff}, we have

$$Z_{eff}^{2} = -\frac{n^2 E_n}{13.6 \text{ eV}} = -\frac{3^2(-5.138 \text{ eV})}{13.6 \text{ eV}} = 3.40$$

$$Z_{eff} = 1.84$$

EVALUATE: The effective charge attracting a $3s$ electron is $1.84e$. Sodium's 11 protons are screened by an average of $11 - 1.84 = 9.16$ electrons instead of 10 electrons because the $3s$ electron spends some time within the inner (K and L) shells.

Each alkali metal (lithium, sodium, potassium, rubidium, and cesium) has one more electron than the corresponding noble gas (helium, neon, argon, krypton, and xenon). This extra electron is mostly outside the other electrons in the filled shells and subshells. Therefore all the alkali metals behave similarly to sodium.

Example 41.9 Energies for a valence electron

The valence electron in potassium has a $4s$ ground state. Calculate the approximate energy of the $n = 4$ state having the smallest Z_{eff}, and discuss the relative energies of the $4s$, $4p$, $4d$, and $4f$ states.

SOLUTION

IDENTIFY and SET UP: The state with the smallest Z_{eff} is the one in which the valence electron spends the most time outside the inner filled shells and subshells, so that it is most effectively screened from the charge of the nucleus. Once we have determined which state has the smallest Z_{eff}, we can use Eq. (41.45) to determine the energy of this state.

EXECUTE: A $4f$ state has $n = 4$ and $l = 3 = 4 - 1$. Thus it is the state of greatest orbital angular momentum for $n = 4$, and thus the state in which the electron spends the most time outside the electron charge clouds of the inner filled shells and subshells. This makes Z_{eff} for a $4f$ state close to unity. Equation (41.45) then gives

$$E_4 = -\frac{Z_{eff}^2}{n^2}(13.6 \text{ eV}) = -\frac{1}{4^2}(13.6 \text{ eV}) = -0.85 \text{ eV}$$

This approximation agrees with the measured energy of the sodium $4f$ state to the precision given.

An electron in a $4d$ state spends a bit more time within the inner shells, and its energy is therefore a bit more negative (measured to be -0.94 eV). For the same reason, a $4p$ state has an even lower energy (measured to be -2.73 eV) and a $4s$ state has the lowest energy (measured to be -4.339 eV).

EVALUATE: We can extend this analysis to the *singly ionized alkaline earth elements*: Be^+, Mg^+, Ca^+, Sr^+, and Ba^+. For any allowed value of n, the highest-l state ($l = n - 1$) of the one remaining outer electron sees an effective charge of almost $+2e$, so for these states, $Z_{eff} = 2$. A $3d$ state for Mg^+, for example, has an energy of about $-2^2(13.6 \text{ eV})/3^2 = -6.0$ eV.

Test Your Understanding of Section 41.6 If electrons did *not* obey the exclusion principle, would it be easier or more difficult to remove the first electron from sodium? ❙

41.7 X-Ray Spectra

X-ray spectra provide yet another example of the richness and power of the Schrödinger equation and of the model of atomic structure that we derived from it in the preceding section. In Section 38.2 we discussed x-ray production on the basis of the photon concept. With the development of x-ray diffraction techniques (see Section 36.6) by von Laue, Bragg, and others, beginning in 1912, it became possible to measure x-ray wavelengths quite precisely (to within 0.1% or less).

Detailed studies of x-ray spectra showed a continuous spectrum of wavelengths (see Fig. 38.8 in Section 38.2), with minimum wavelength (corresponding to maximum frequency and photon energy) determined by the accelerating voltage V_{AC} in the x-ray tube, according to the relationship derived in Section 38.2 for *bremsstrahlung* processes:

$$\lambda_{min} = \frac{hc}{eV_{AC}} \qquad (41.46)$$

This continuous-spectrum radiation is nearly independent of the target material in the x-ray tube.

Moseley's Law and Atomic Energy Levels

Depending on the accelerating voltage and the target element, we may find sharp peaks superimposed on this continuous spectrum, as in Fig. 41.23. These peaks are at different wavelengths for different elements; they form what is called a *characteristic x-ray spectrum* for each target element. In 1913 the British scientist Henry G. J. Moseley studied these spectra in detail using x-ray diffraction techniques. He found that the most intense short-wavelength line in the characteristic x-ray spectrum from a particular target element, called the K_α line, varied smoothly with that element's atomic number Z (Fig. 41.24). This is in sharp contrast to optical spectra, in which elements with adjacent Z-values have spectra that often bear no resemblance to each other.

Moseley found that the relationship could be expressed in terms of x-ray frequencies f by a simple formula called *Moseley's law*:

$$f = (2.48 \times 10^{15}\ \text{Hz})(Z - 1)^2 \qquad \text{(Moseley's law)} \qquad (41.47)$$

Moseley went far beyond this empirical relationship; he showed how characteristic x-ray spectra could be understood on the basis of energy levels of atoms in the target. His analysis was based on the Bohr model, published in the same year. We will recast it somewhat, using the ideas of atomic structure that we discussed in Section 41.6. First recall that the *outer* electrons of an atom are responsible for optical spectra. Their excited states are usually only a few electron volts above their ground state. In transitions from excited states to the ground state, they usually emit photons in or near the visible region.

Characteristic x rays, by contrast, are emitted in transitions involving the *inner* shells of a complex atom. In an x-ray tube the electrons may strike the target with enough energy to knock electrons out of the inner shells of the target atoms. These inner electrons are much closer to the nucleus than are the electrons in the outer shells; they are much more tightly bound, and hundreds or thousands of electron volts may be required to remove them.

Suppose one electron is knocked out of the K shell. This process leaves a vacancy, which we'll call a *hole*. (One electron remains in the K shell.) The hole can then be filled by an electron falling in from one of the outer shells, such as the $L, M, N, \ldots$ shell. This transition is accompanied by a decrease in the energy of the

41.23 Graph of intensity per unit wavelength as a function of wavelength for x rays produced with an accelerating voltage of 35 kV and a molybdenum target. The curve is a smooth function similar to the bremsstrahlung spectra in Fig. 38.8 (Section 38.2), but with two sharp spikes corresponding to part of the characteristic x-ray spectrum for molybdenum.

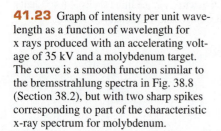
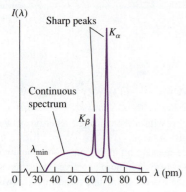

41.24 The square root of Moseley's measured frequencies of the K_α line for 14 elements.

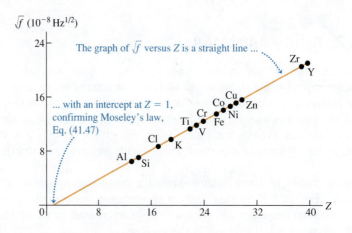

atom (because *less* energy would be needed to remove an electron from an *L*, *M*, *N*, ... shell), and an x-ray photon is emitted with energy equal to this decrease. Each state has definite energy, so the emitted x rays have definite wavelengths; the emitted spectrum is a *line* spectrum.

We can estimate the energy and frequency of K_α x-ray photons using the concept of screening from Section 41.6. A K_α x-ray photon is emitted when an electron in the *L* shell ($n = 2$) drops down to fill a hole in the *K* shell ($n = 1$). As the electron drops down, it is attracted by the *Z* protons in the nucleus screened by the one remaining electron in the *K* shell. We therefore approximate the energy by Eq. (41.45), with $Z_{eff} = Z - 1$, $n_i = 2$, and n_f. The energy before the transition is

$$E_i \approx -\frac{(Z-1)^2}{2^2}(13.6 \text{ eV}) = -(Z-1)^2(3.4 \text{ eV})$$

and the energy after the transition is

$$E_f \approx -\frac{(Z-1)^2}{1^2}(13.6 \text{ eV}) = -(Z-1)^2(13.6 \text{ eV})$$

The energy of the K_α x-ray photon is $E_{K\alpha} = E_i - E_f \approx (Z-1)^2(-3.4 \text{ eV} + 13.6 \text{ eV})$. That is,

$$E_{K\alpha} \approx (Z-1)^2(10.2 \text{ eV}) \qquad (41.48)$$

The frequency of the photon is its energy divided by Planck's constant:

$$f = \frac{E}{h} \approx \frac{(Z-1)^2(10.2 \text{ eV})}{4.136 \times 10^{-15} \text{ eV} \cdot \text{s}} = (2.47 \times 10^{15} \text{ Hz})(Z-1)^2$$

This relationship agrees almost exactly with Moseley's experimental law, Eq. (41.47). Indeed, considering the approximations we have made, the agreement is better than we have a right to expect. But our calculation does show how Moseley's law can be understood on the bases of screening and transitions between energy levels.

The hole in the *K* shell may also be filled by an electron falling from the *M* or *N* shell, assuming that these are occupied. If so, the x-ray spectrum of a large group of atoms of a single element shows a series, named the *K* series, of three lines, called the K_α, K_β, and K_γ lines. These three lines result from transitions in which the *K*-shell hole is filled by an *L*, *M*, or *N* electron, respectively. Figure 41.25 shows the *K* series for tungsten ($Z = 74$), molybdenum ($Z = 42$), and copper ($Z = 29$).

There are other series of x-ray lines, called the *L*, *M*, and *N* series, that are produced after the ejection of electrons from the *L*, *M*, and *N* shells rather than the *K* shell. Electrons in these outer shells are farther away from the nucleus and are not held as tightly as are those in the *K* shell, so removing these outer electrons requires less energy. Hence the x-ray photons that are emitted when these vacancies are filled have lower energy than those in the *K* series.

Application **X Rays in Forensic Science**
When a handgun is fired, a cloud of gunshot residue (GSR) is ejected from the barrel. The x-ray emission spectrum of GSR includes characteristic peaks from lead (Pb), antimony (Sb), and barium (Ba). If a sample taken from a suspect's skin or clothing has an x-ray emission spectrum with these characteristics, it indicates that the suspect recently fired a gun.

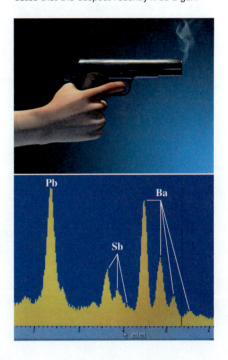

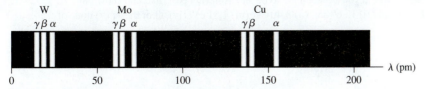

41.25 Wavelengths of the K_α, K_β, and K_γ lines of tungsten (W), molybdenum (Mo), and copper (Cu).

The three lines in each series are called the K_α, K_β, and K_γ lines. The K_α line is produced by the transition of an *L* electron to the vacancy in the *K* shell, the K_β line by an *M* electron, and the K_γ line by an *N* electron.

Example 41.10 Chemical analysis by x-ray emission

You measure the K_α wavelength for an unknown element, obtaining the value 0.0709 nm. What is the element?

SOLUTION

IDENTIFY and SET UP: To determine which element this is, we need to know its atomic number Z. We can find this using Moseley's law, which relates the frequency of an element's K_α x-ray emission line to that element's atomic number Z. We'll use the relationship $f = c/\lambda$ to calculate the frequency for the K_α line, and then use Eq. (41.47) to find the corresponding value of the atomic number Z. We'll then consult the periodic table (Appendix D) to determine which element has this atomic number.

EXECUTE: The frequency is

$$f = \frac{c}{\lambda} = \frac{3.00 \times 10^8 \text{ m/s}}{0.0709 \times 10^{-9} \text{ m}} = 4.23 \times 10^{18} \text{ Hz}$$

Solving Moseley's law for Z, we get

$$Z = 1 + \sqrt{\frac{f}{2.48 \times 10^{15} \text{ Hz}}} = 1 + \sqrt{\frac{4.23 \times 10^{18} \text{ Hz}}{2.48 \times 10^{15} \text{ Hz}}} = 42.3$$

We know that Z has to be an integer; we conclude that $Z = 42$, corresponding to the element molybdenum.

EVALUATE: If you're worried that our calculation did not give an integer for Z, remember that Moseley's law is an empirical relationship. There are slight variations from one atom to another due to differences in the structure of the electron shells. Nonetheless, this example suggests the power of Moseley's law.

Niels Bohr commented that it was Moseley's observations, not the alpha-particle scattering experiments of Rutherford, Geiger, and Marsden (see Section 39.2), that truly convinced physicists that the atom consists of a positive nucleus surrounded by electrons in motion. Unlike Bohr or Rutherford, Moseley did not receive a Nobel Prize for his important work; these awards are given only to living scientists, and Moseley was killed in combat during the First World War.

X-Ray Absorption Spectra

We can also observe x-ray *absorption* spectra. Unlike optical spectra, the absorption wavelengths are usually not the same as those for emission, especially in many-electron atoms, and do not give simple line spectra. For example, the K_α emission line results from a transition from the L shell to a hole in the K shell. The reverse transition doesn't occur in atoms with $Z \geq 10$ because in the atom's ground state, there is no vacancy in the L shell. To be absorbed, a photon must have enough energy to move an electron to an empty state. Since empty states are only a few electron volts in energy below the free-electron continuum, the minimum absorption energies in many-electron atoms are about the same as the minimum energies that are needed to remove an electron from its shell. Experimentally, if we gradually increase the accelerating voltage and hence the maximum photon energy, we observe sudden increases in absorption when we reach these minimum energies. These sudden jumps of absorption are called *absorption edges* (Fig. 41.26).

Characteristic x-ray spectra provide a very useful analytical tool. Satellite-borne x-ray spectrometers are used to study x-ray emission lines from highly excited atoms in distant astronomical sources. X-ray spectra are also used in air-pollution monitoring and in studies of the abundance of various elements in rocks.

41.26 When a beam of x rays is passed through a slab of molybdenum, the extent to which the beam is absorbed depends on the energy E of the x-ray photons. A sharp increase in absorption occurs at the K absorption edge at 20 keV. The increase occurs because photons with energies above this value can excite an electron from the K shell of a molybdenum atom into an empty state.

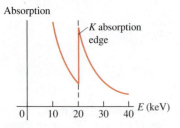

Test Your Understanding of Section 41.7 A beam of photons is passed through a sample of high-temperature atomic hydrogen. At what photon energy would you expect there to be an absorption edge like that shown in Fig. 41.26? (i) 13.60 eV; (ii) 3.40 eV; (iii) 1.51 eV; (iv) all of these; (v) none of these.

Three-dimensional problems: The time-independent Schrödinger equation for three-dimensional problems is given by Eq. (41.5).

$$-\frac{\hbar^2}{2m}\left(\frac{\partial^2\psi(x,y,z)}{\partial x^2} + \frac{\partial^2\psi(x,y,z)}{\partial y^2}\right.$$

$$\left. + \frac{\partial^2\psi(x,y,z)}{\partial z^2}\right) + U(x,y,z)\psi(x,y,z)$$

$$= E\psi(x,y,z)$$

(three-dimensional time-independent Schrödinger equation)   (41.5)

Particle in a three-dimensional box: The wave function for a particle in a cubical box is the product of a function of x only, a function of y only, and a function of z only. Each stationary state is described by three quantum numbers (n_X, n_Y, n_Z). Most of the energy levels given by Eq. (41.16) exhibit degeneracy: More than one quantum state has the same energy. (See Example 41.1.)

$$E_{n_X,n_Y,n_Z} = \frac{(n_X^2 + n_Y^2 + n_Z^2)\pi^2\hbar^2}{2mL^2}$$

$$(n_X = 1, 2, 3, \ldots;$$

$$n_Y = 1, 2, 3, \ldots; n_Z = 1, 2, 3, \ldots)$$

(energy levels, particle in a three-dimensional cubical box)   (41.16)

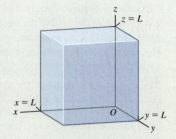

The hydrogen atom: The Schrödinger equation for the hydrogen atom gives the same energy levels as the Bohr model. If the nucleus has charge Ze, there is an additional factor of Z^2 in the numerator of Eq. (41.21). The possible magnitudes L of orbital angular momentum are given by Eq. (41.22), and the possible values of the z-component of orbital angular momentum are given by Eq. (41.23). (See Examples 41.2 and 41.3.)

The probability that an atomic electron is between r and $r + dr$ from the nucleus is $P(r)\,dr$, given by Eq. (41.25). Atomic distances are often measured in units of a, the smallest distance between the electron and the nucleus in the Bohr model. (See Example 41.4.)

$$E_n = -\frac{1}{(4\pi\epsilon_0)^2}\frac{m_r e^4}{2n^2\hbar^2} = -\frac{13.60\text{ eV}}{n^2}$$

(energy levels of hydrogen)   (41.21)

$$L = \sqrt{l(l+1)}\,\hbar$$
$$(l = 0, 1, 2, \ldots, n-1)$$ (41.22)

$$L_z = m_l\hbar$$
$$(m_l = 0, \pm1, \pm2, \ldots, \pm l)$$ (41.23)

$$P(r)\,dr = |\psi|^2\,dV = |\psi|^2\,4\pi r^2\,dr$$ (41.25)

$$a = \frac{\epsilon_0 h^2}{\pi m_r e^2} = \frac{4\pi\epsilon_0 \hbar^2}{m_r e^2}$$

$$= 5.29 \times 10^{-11}\text{ m}$$ (41.26)

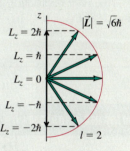

The Zeeman effect: The interaction energy of an electron (mass m) with magnetic quantum number m_l in a magnetic field $\vec{B}$ along the $+z$-direction is given by Eq. (41.35) or (41.36), where $\mu_B = e\hbar/2m$ is called the Bohr magneton. (See Example 41.5.)

$$U = -\mu_z B = m_l\frac{e\hbar}{2m}B = m_l\mu_B B$$

$$(m_l = 0, \pm1, \pm2, \ldots, \pm l)$$

(41.35), (41.36)

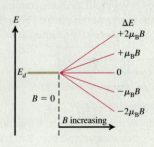

Electron spin: An electron has an intrinsic spin angular momentum of magnitude S, given by Eq. (41.38). The possible values of the z-component of the spin angular momentum are $S_z = m_s\hbar$, where $m_s = \pm\frac{1}{2}$. (See Examples 41.6 and 41.7.)

$$S = \sqrt{\tfrac{1}{2}(\tfrac{1}{2} + 1)}\,\hbar = \sqrt{\tfrac{3}{4}}\,\hbar \quad (41.38)$$

$$S_z = \pm\tfrac{1}{2}\hbar \quad (41.37)$$

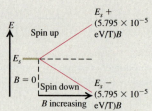

Many-electron atoms: In a hydrogen atom, the quantum numbers n, l, m_l, and m_s of the electron have certain allowed values given by Eq. (41.44). In a many-electron atom, the allowed quantum numbers for each electron are the same as in hydrogen, but the energy levels depend on both n and l because of screening, the partial cancellation of the field of the nucleus by the inner electrons. If the effective (screened) charge attracting an electron is $Z_{eff}e$, the energies of the levels are given approximately by Eq. (41.45). (See Examples 41.8 and 41.9.)

$$n \geq 1 \quad 0 \leq l \leq n - 1$$
$$|m_l| \leq l \quad m_s = \pm\tfrac{1}{2} \quad (41.44)$$

$$E_n = -\frac{Z_{eff}^2}{n^2}(13.6 \text{ eV}) \quad (41.45)$$

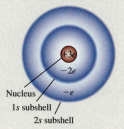

X-ray spectra: Moseley's law states that the frequency of a K_α x ray from a target with atomic number Z is given by Eq. (41.47). Characteristic x-ray spectra result from transitions to a hole in an inner energy level of an atom. (See Example 41.10.)

$$f = (2.48 \times 10^{15} \text{ Hz})(Z - 1)^2 \quad (41.47)$$

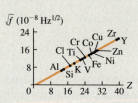

BRIDGING PROBLEM   A Many-Electron Atom in a Box

An atom of titanium (Ti) has 22 electrons and has a radius of 1.47×10^{-10} m. As a simple model of this atom, imagine putting 22 electrons into a cubical box that has the same volume as a titanium atom. (a) What is the length of each side of the box? (b) What will be the configuration of the 22 electrons? (c) Find the energies of each of the levels occupied by the electrons. (Ignore the electric forces that the electrons exert on each other.) (d) You remove one of the electrons from the lowest level. As a result, one of the electrons from the highest occupied level drops into the lowest level to fill the hole, emitting a photon in the process. What is the energy of this photon? How does this compare to the energy of the K_α photon for titanium as predicted by Moseley's law?

SOLUTION GUIDE

See MasteringPhysics® study area for a Video Tutor solution. **MP**

IDENTIFY and SET UP

1. In this problem you'll use ideas from Section 41.2 about a particle in a cubical box. You'll also apply the exclusion principle from Section 41.6 to find the electron configuration of this cubical "atom." The ideas about x-ray spectra from Section 41.7 are also important.
2. The target variables are (a) the dimensions of the box, (b) the electron configurations (like those given in Table 41.3 for real atoms), (c) the occupied energy levels of the cubical box, and (d) the energy of the emitted photon.

EXECUTE

3. Use your knowledge of geometry to find the length of each side of the box.
4. Each electron state is described by four quantum numbers: n_X, n_Y, and n_Z as described in Section 41.2 and the spin magnetic quantum number m_s described in Section 41.5. Use the exclusion principle to determine the quantum numbers of each of the 22 electrons in the "atom." (*Hint:* Figure 41.4 in Section 41.2 shows the first several energy levels of a cubical box relative to the ground level $E_{1,1,1}$.)
5. Use your results from steps 3 and 4 to find the energies of each of the occupied levels.
6. Use your result from step 5 to find the energy of the photon emitted when an electron makes a transition from the highest occupied level to the ground level. Compare this to the energy calculated for titanium using Moseley's law.

EVALUATE·

7. Is this cubical "atom" a useful model for titanium? Why or why not?
8. In this problem you ignored the electric interactions between electrons. To estimate how large these are, find the electrostatic potential energy of two electrons separated by half the length of the box. How does this compare to the energy levels you calculated in step 5? Is it a good approximation to ignore these interactions?

Problems

For instructor-assigned homework, go to www.masteringphysics.com

•, ••, •••: Problems of increasing difficulty. **CP**: Cumulative problems incorporating material from earlier chapters. **CALC**: Problems requiring calculus. **BIO**: Biosciences problems.

DISCUSSION QUESTIONS

Q41.1 Particle A is described by the wave function $\psi(x, y, z)$. Particle B is described by the wave function $\psi(x, y, z)e^{i\phi}$, where ϕ is a real constant. How does the probability of finding particle A within a volume dV around a certain point in space compare with the probability of finding particle B within this same volume?

Q41.2 What are the most significant differences between the Bohr model of the hydrogen atom and the Schrödinger analysis? What are the similarities?

Q41.3 For a body orbiting the sun, such as a planet, comet, or asteroid, is there any restriction on the z-component of its orbital angular momentum such as there is with the z-component of the electron's orbital angular momentum in hydrogen? Explain.

Q41.4 Why is the analysis of the helium atom much more complex than that of the hydrogen atom, either in a Bohr type of model or using the Schrödinger equation?

Q41.5 The Stern–Gerlach experiment is always performed with beams of *neutral* atoms. Wouldn't it be easier to form beams using *ionized* atoms? Why won't this work?

Q41.6 (a) If two electrons in hydrogen atoms have the same principal quantum number, can they have different orbital angular momentum? How? (b) If two electrons in hydrogen atoms have the same orbital angular-momentum quantum number, can they have different principal quantum numbers? How?

Q41.7 In the Stern–Gerlach experiment, why is it essential for the magnetic field to be *inhomogeneous* (that is, nonuniform)?

Q41.8 In the ground state of the helium atom one electron must have "spin down" and the other "spin up." Why?

Q41.9 An electron in a hydrogen atom is in an s level, and the atom is in a magnetic field $\vec{B} = B\hat{k}$. Explain why the "spin up" state $\left(m_s = +\frac{1}{2}\right)$ has a higher energy than the "spin down" state $\left(m_s = -\frac{1}{2}\right)$.

Q41.10 The central-field approximation is more accurate for alkali metals than for transition metals such as iron, nickel, or copper. Why?

Q41.11 Table 41.3 shows that for the ground state of the potassium atom, the outermost electron is in a $4s$ state. What does this tell you about the relative energies of the $3d$ and $4s$ levels for this atom? Explain.

Q41.12 Do gravitational forces play a significant role in atomic structure? Explain.

Q41.13 Why do the transition elements ($Z = 21$ to 30) all have similar chemical properties?

Q41.14 Use Table 41.3 to help determine the ground-state electron configuration of the neutral gallium atom (Ga) as well as the ions Ga$^+$ and Ga$^-$. Gallium has an atomic number of 31.

Q41.15 On the basis of the Pauli exclusion principle, the structure of the periodic table of the elements shows that there must be a fourth quantum number in addition to n, l, and m_l. Explain.

Q41.16 A small amount of magnetic-field splitting of spectral lines occurs even when the atoms are not in a magnetic field. What causes this?

Q41.17 The ionization energies of the alkali metals (that is, the lowest energy required to remove one outer electron when the atom is in its ground state) are about 4 or 5 eV, while those of the noble gases are in the range from 11 to 25 eV. Why is there a difference?

Q41.18 The energy required to remove the $3s$ electron from a sodium atom in its ground state is about 5 eV. Would you expect the energy required to remove an additional electron to be about the same, or more, or less? Why?

Q41.19 What is the "central-field approximation" and why is it only an approximation?

Q41.20 The nucleus of a gold atom contains 79 protons. How does the energy required to remove a $1s$ electron completely from a gold atom compare with the energy required to remove the electron from the ground level in a hydrogen atom? In what region of the electromagnetic spectrum would a photon with this energy for each of these two atoms lie?

Q41.21 (a) Can you show that the orbital angular momentum of an electron in any given direction (e.g., along the z-axis) is *always* less than or equal to its total orbital angular momentum? In which cases would the two be equal to each other? (b) Is the result in part (a) true for a classical object, such as a spinning top or planet?

Q41.22 An atom in its ground level absorbs a photon with energy equal to the K absorption edge. Does absorbing this photon ionize this atom? Explain.

Q41.23 Can a hydrogen atom emit x rays? If so, how? If not, why not?

EXERCISES

Section 41.2 Particle in a Three-Dimensional Box

41.1 • For a particle in a three-dimensional box, what is the degeneracy (number of different quantum states with the same energy) of the following energy levels: (a) $3\pi^2\hbar^2/2mL^2$ and (b) $9\pi^2\hbar^2/2mL^2$?

41.2 • **CP** Model a hydrogen atom as an electron in a cubical box with side length L. Set the value of L so that the volume of the box equals the volume of a sphere of radius $a = 5.29 \times 10^{-11}$ m, the Bohr radius. Calculate the energy separation between the ground and first excited levels, and compare the result to this energy separation calculated from the Bohr model.

41.3 • **CP** A photon is emitted when an electron in a three-dimensional box of side length 8.00×10^{-11} m makes a transition from the $n_X = 2, n_Y = 2, n_Z = 1$ state to the $n_X = 1, n_Y = 1, n_Z = 1$ state. What is the wavelength of this photon?

41.4 • For each of the following states of a particle in a three-dimensional box, at what points is the probability distribution function a maximum: (a) $n_X = 1, n_Y = 1, n_Z = 1$ and (b) $n_X = 2, n_Y = 2, n_Z = 1$?

41.5 •• A particle is in the three-dimensional box of Section 41.1. For the state $n_X = 2, n_Y = 2, n_Z = 1$, for what planes (in addition to the walls of the box) is the probability distribution function zero? Compare this number of planes to the corresponding number of planes where $|\psi|^2$ is zero for the lower-energy state $n_X = 2, n_Y = 1, n_Z = 1$ and for the ground state $n_X = 1, n_Y = 1, n_Z = 1$.

41.6 • What is the energy difference between the two lowest energy levels for a proton in a cubical box with side length 1.00×10^{-14} m, the approximate diameter of a nucleus?

Section 41.3 The Hydrogen Atom

41.7 •• Consider an electron in the N shell. (a) What is the smallest orbital angular momentum it could have? (b) What is the largest orbital angular momentum it could have? Express your answers in terms of $\hbar$ and in SI units. (c) What is the largest orbital angular momentum this electron could have in any chosen direction? Express your answers in terms of $\hbar$ and in SI units. (d) What is the largest spin angular momentum this electron could have in any chosen direction? Express your answers in terms of $\hbar$ and in SI units. (e) For the electron in part (c), what is the ratio of its spin angular momentum in the z-direction to its orbital angular momentum in the z-direction?

41.8 • An electron is in the hydrogen atom with $n = 5$. (a) Find the possible values of L and L_z for this electron, in units of $\hbar$. (b) For each value of L, find all the possible angles between $\vec{L}$ and the z-axis. (c) What are the maximum and minimum values of the magnitude of the angle between $\vec{L}$ and the z-axis?

41.9 • The orbital angular momentum of an electron has a magnitude of 4.716×10^{-34} kg·m^2/s. What is the angular-momentum quantum number l for this electron?

41.10 • Consider states with angular-momentum quantum number $l = 2$. (a) In units of $\hbar$, what is the largest possible value of L_z? (b) In units of $\hbar$, what is the value of L? Which is larger: L or the maximum possible L_z? (c) For each allowed value of L_z, what angle does the vector $\vec{L}$ make with the $+z$-axis? How does the minimum angle for $l = 2$ compare to the minimum angle for $l = 3$ calculated in Example 41.3?

41.11 • Calculate, in units of $\hbar$, the magnitude of the maximum orbital angular momentum for an electron in a hydrogen atom for states with a principal quantum number of 2, 20, and 200. Compare each with the value of $n\hbar$ postulated in the Bohr model. What trend do you see?

41.12 • (a) Make a chart showing all the possible sets of quantum numbers l and m_l for the states of the electron in the hydrogen atom when $n = 5$. How many combinations are there? (b) What are the energies of these states?

41.13 •• (a) How many different $5g$ states does hydrogen have? (b) Which of the states in part (a) has the largest angle between $\vec{L}$ and the z-axis, and what is that angle? (c) Which of the states in part (a) has the smallest angle between $\vec{L}$ and the z-axis, and what is that angle?

41.14 •• CALC (a) What is the probability that an electron in the $1s$ state of a hydrogen atom will be found at a distance less than $a/2$ from the nucleus? (b) Use the results of part (a) and of Example 41.4 to calculate the probability that the electron will be found at distances between $a/2$ and a from the nucleus.

41.15 • CALC In Example 41.4 fill in the missing details that show that $P = 1 - 5e^{-2}$.

41.16 • Show that $\Phi(\phi) = e^{im_l\phi} = \Phi(\phi + 2\pi)$ (that is, show that $\Phi(\phi)$ is periodic with period 2π) if and only if m_l is restricted to the values $0, \pm1, \pm2, \ldots$. (Hint: Euler's formula states that $e^{i\phi} = \cos\phi + i\sin\phi$.)

Section 41.4 The Zeeman Effect

41.17 • A hydrogen atom in a $3p$ state is placed in a uniform external magnetic field $\vec{B}$. Consider the interaction of the magnetic field with the atom's orbital magnetic dipole moment. (a) What field magnitude B is required to split the $3p$ state into multiple lev-els with an energy difference of 2.71×10^{-5} eV between adjacent levels? (b) How many levels will there be?

41.18 • A hydrogen atom is in a d state. In the absence of an external magnetic field the states with different m_l values have (approximately) the same energy. Consider the interaction of the magnetic field with the atom's orbital magnetic dipole moment. (a) Calculate the splitting (in electron volts) of the m_l levels when the atom is put in a 0.400-T magnetic field that is in the $+z$-direction. (b) Which m_l level will have the lowest energy? (c) Draw an energy-level diagram that shows the d levels with and without the external magnetic field.

41.19 • A hydrogen atom in the $5g$ state is placed in a magnetic field of 0.600 T that is in the z-direction. (a) Into how many levels is this state split by the interaction of the atom's orbital magnetic dipole moment with the magnetic field? (b) What is the energy separation between adjacent levels? (c) What is the energy separation between the level of lowest energy and the level of highest energy?

41.20 •• CP A hydrogen atom undergoes a transition from a $2p$ state to the $1s$ ground state. In the absence of a magnetic field, the energy of the photon emitted is 122 nm. The atom is then placed in a strong magnetic field in the z-direction. Ignore spin effects; consider only the interaction of the magnetic field with the atom's orbital magnetic moment. (a) How many different photon wavelengths are observed for the $2p \rightarrow 1s$ transition? What are the m_l values for the initial and final states for the transition that leads to each photon wavelength? (b) One observed wavelength is exactly the same with the magnetic field as without. What are the initial and final m_l values for the transition that produces a photon of this wavelength? (c) One observed wavelength with the field is longer than the wavelength without the field. What are the initial and final m_l values for the transition that produces a photon of this wavelength? (d) Repeat part (c) for the wavelength that is shorter than the wavelength in the absence of the field.

Section 41.5 Electron Spin

41.21 •• CP **Classical Electron Spin.** (a) If you treat an electron as a classical spherical object with a radius of 1.0×10^{-17} m, what angular speed is necessary to produce a spin angular momentum of magnitude $\sqrt{\frac{3}{4}}\hbar$? (b) Use $v = r\omega$ and the result of part (a) to calculate the speed v of a point at the electron's equator. What does your result suggest about the validity of this model?

41.22 •• A hydrogen atom in the $n = 1$, $m_s = -\frac{1}{2}$ state is placed in a magnetic field with a magnitude of 0.480 T in the $+z$-direction. (a) Find the magnetic interaction energy (in electron volts) of the electron with the field. (b) Is there any orbital magnetic dipole moment interaction for this state? Explain. Can there be an orbital magnetic dipole moment interaction for $n \neq 1$?

41.23 • Calculate the energy difference between the $m_s = \frac{1}{2}$ ("spin up") and $m_s = -\frac{1}{2}$ ("spin down") levels of a hydrogen atom in the $1s$ state when it is placed in a 1.45-T magnetic field in the *negative* z-direction. Which level, $m_s = \frac{1}{2}$ or $m_s = -\frac{1}{2}$, has the lower energy?

41.24 • CP The hyperfine interaction in a hydrogen atom between the magnetic dipole moment of the proton and the spin magnetic dipole moment of the electron splits the ground level into two levels separated by 5.9×10^{-6} eV. (a) Calculate the wavelength and frequency of the photon emitted when the atom makes a transition between these states, and compare your answer to the value given at the end of Section 41.5. In what part of the electromagnetic spectrum does this lie? Such photons are emitted by cold hydrogen clouds in interstellar space; by detecting these photons, astronomers can learn about the number and density of such clouds. (b) Calculate the effective magnetic field experienced by the electron in these states (see Fig. 41.18). Compare your result to the effective magnetic field due to the spin-orbit coupling calculated in Example 41.7.

41.25 • A hydrogen atom in a particular orbital angular momentum state is found to have j quantum numbers $\frac{7}{2}$ and $\frac{9}{2}$. What is the letter that labels the value of l for the state?

Section 41.6 Many-Electron Atoms and the Exclusion Principle

41.26 • For germanium (Ge, $Z = 32$), make a list of the number of electrons in each subshell ($1s, 2s, 2p, \ldots$). Use the allowed values of the quantum numbers along with the exclusion principle; do *not* refer to Table 41.3.

41.27 • Make a list of the four quantum numbers n, l, m_l, and m_s for each of the 10 electrons in the ground state of the neon atom. Do *not* refer to Table 41.2 or 41.3.

41.28 •• (a) Write out the ground-state electron configuration ($1s^2$, $2s^2$, ...) for the carbon atom. (b) What element of next-larger Z has chemical properties similar to those of carbon? Give the ground-state electron configuration for this element.

41.29 •• (a) Write out the ground-state electron configuration ($1s^2$, $2s^2$, ...) for the beryllium atom. (b) What element of next-larger Z has chemical properties similar to those of beryllium? Give the ground-state electron configuration of this element. (c) Use the procedure of part (b) to predict what element of next-larger Z than in (b) will have chemical properties similar to those of the element you found in part (b), and give its ground-state electron configuration.

41.30 • For magnesium, the first ionization potential is 7.6 eV. The second ionization potential (additional energy required to remove a second electron) is almost twice this, 15 eV, and the third ionization potential is much larger, about 80 eV. How can these numbers be understood?

41.31 • The $5s$ electron in rubidium (Rb) sees an effective charge of $2.771e$. Calculate the ionization energy of this electron.

41.32 • The energies of the $4s$, $4p$, and $4d$ states of potassium are given in Example 41.9. Calculate Z_{eff} for each state. What trend do your results show? How can you explain this trend?

41.33 • (a) The doubly charged ion N^{2+} is formed by removing two electrons from a nitrogen atom. What is the ground-state electron configuration for the N^{2+} ion? (b) Estimate the energy of the least strongly bound level in the L shell of N^{2+}. (c) The doubly charged ion P^{2+} is formed by removing two electrons from a phosphorus atom. What is the ground-state electron configuration for the P^{2+} ion? (d) Estimate the energy of the least strongly bound level in the M shell of P^{2+}.

41.34 • (a) The energy of the $2s$ state of lithium is -5.391 eV. Calculate the value of Z_{eff} for this state. (b) The energy of the $4s$ state of potassium is -4.339 eV. Calculate the value of Z_{eff} for this state. (c) Compare Z_{eff} for the $2s$ state of lithium, the $3s$ state of sodium (see Example 41.8), and the $4s$ state of potassium. What trend do you see? How can you explain this trend?

41.35 • Estimate the energy of the highest-l state for (a) the L shell of Be^+ and (b) the N shell of Ca^+.

Section 41.7 X-Ray Spectra

41.36 • A K_α x ray emitted from a sample has an energy of 7.46 keV. Of which element is the sample made?

41.37 • Calculate the frequency, energy (in keV), and wavelength of the K_α x ray for the elements (a) calcium (Ca, $Z = 20$); (b) cobalt (Co, $Z = 27$); (c) cadmium (Cd, $Z = 48$).

41.38 •• The energies for an electron in the K, L, and M shells of the tungsten atom are $-69,500$ eV, $-12,000$ eV, and -2200 eV, respectively. Calculate the wavelengths of the K_α and K_β x rays of tungsten.

PROBLEMS

41.39 • In terms of the ground-state energy $E_{1,1,1}$, what is the energy of the highest level occupied by an electron when 10 electrons are placed into a cubical box?

41.40 •• **CALC** A particle in the three-dimensional box of Section 41.2 is in the ground state, where $n_X = n_Y = n_Z = 1$. (a) Calculate the probability that the particle will be found somewhere between $x = 0$ and $x = L/2$. (b) Calculate the probability that the particle will be found somewhere between $x = L/4$ and $x = L/2$. Compare your results to the result of Example 41.1 for the probability of finding the particle in the region $x = 0$ to $x = L/4$.

41.41 •• **CALC** A particle is in the three-dimensional box of Section 41.2. (a) Consider the cubical volume defined by $0 \leq x \leq L/4$, $0 \leq y \leq L/4$, and $0 \leq z \leq L/4$. What fraction of the total volume of the box is this cubical volume? (b) If the particle is in the ground state ($n_X = 1$, $n_Y = 1$, $n_Z = 1$) calculate the probability that the particle will be found in the cubical volume defined in part (a). (c) Repeat the calculation of part (b) when the particle is in the state $n_X = 2$, $n_Y = 1$, $n_Z = 1$.

41.42 •• **CALC** A particle is described by the normalized wave function $\psi(x, y, z) = Axe^{-\alpha x^2}e^{-\beta y^2}e^{-\gamma z^2}$, where A, α, β, and γ are all real, positive constants. The probability that the particle will be found in the infinitesimal volume $dx\, dy\, dz$ centered at the point (x_0, y_0, z_0) is $|\psi(x_0, y_0, z_0)|^2\, dx\, dy\, dz$. (a) At what value of x_0 is the particle most likely to be found? (b) Are there values of x_0 for which the probability of the particle being found is zero? If so, at what x_0?

41.43 •• **CALC** A particle is described by the normalized wave function $\psi(x, y, z) = Ae^{-\alpha(x^2+y^2+z^2)}$, where A and α are real, positive constants. (a) Determine the probability of finding the particle at a distance between r and $r + dr$ from the origin. (*Hint:* See Problem 41.42. Consider a spherical shell centered on the origin with inner radius r and thickness dr.) (b) For what value of r does the probability in part (a) have its maximum value? Is this the same value of r for which $|\psi(x, y, z)|^2$ is a maximum? Explain any differences.

41.44 •• **CP CALC** **A Three-Dimensional Isotropic Harmonic Oscillator.** An isotropic harmonic oscillator has the potential-energy function $U(x, y, z) = \frac{1}{2}k'(x^2 + y^2 + z^2)$. (*Isotropic* means that the force constant k' is the same in all three coordinate directions.) (a) Show that for this potential, a solution to Eq. (41.5) is given by $\psi = \psi_{n_x}(x)\psi_{n_y}(y)\psi_{n_z}(z)$. In this expression, $\psi_{n_x}(x)$ is a solution to the one-dimensional harmonic oscillator Schrödinger equation, Eq. (40.44), with energy $E_{n_x} = \left(n_x + \frac{1}{2}\right)\hbar\omega$. The functions $\psi_{n_y}(y)$ and $\psi_{n_z}(z)$ are analogous one-dimensional wave functions for oscillations in the y- and z-directions. Find the energy associated with this ψ. (b) From your results in part (a) what are the ground-level and first-excited-level energies of the three-dimensional isotropic oscillator? (c) Show that there is only one state (one set of quantum numbers n_x, n_y, and n_z) for the ground level but three states for the first excited level.

41.45 •• **CP CALC** **Three-Dimensional Anisotropic Harmonic Oscillator.** An oscillator has the potential-energy function $U(x, y, z) = \frac{1}{2}k'_1(x^2 + y^2) + \frac{1}{2}k'_2z^2$, where $k'_1 > k'_2$. This oscillator is called *anisotropic* because the force constant is not the same in all three coordinate directions. (a) Find a general expression for the energy levels of the oscillator (see Problem 41.44). (b) From your results in part (a), what are the ground-level and first-excited-level energies of this oscillator? (c) How many states (different sets of quantum numbers n_x, n_y, and n_z) are there for the ground level and for the first excited level? Compare to part (c) of Problem 41.44.

41.46 •• An electron in hydrogen is in the $5f$ state. (a) Find the largest possible value of the z-component of its angular momentum. (b) Show that for the electron in part (a), the corresponding x- and y-components of its angular momentum satisfy the equation $\sqrt{L_x^2 + L_y^2} = \hbar\sqrt{3}$.

41.47 •• (a) Show that the total number of atomic states (including different spin states) in a shell of principal quantum number n is $2n^2$. [*Hint:* The sum of the first N integers $1 + 2 + 3 + \cdots + N$ is equal to $N(N + 1)/2$.] (b) Which shell has 50 states?

41.48 •• (a) What is the lowest possible energy (in electron volts) of an electron in hydrogen if its orbital angular momentum is $\sqrt{12}\hbar$? (b) What are the largest and smallest values of the z-component of the orbital angular momentum (in terms of $\hbar$) for the electron in part (a)? (c) What are the largest and smallest values of the spin angular momentum (in terms of $\hbar$) for the electron in part (a)? (d) What are the largest and smallest values of the orbital angular momentum (in terms of $\hbar$) for an electron in the M shell of hydrogen?

41.49 • Consider an electron in hydrogen having total energy -0.5440 eV. (a) What are the possible values of its orbital angular momentum (in terms of $\hbar$)? (b) What wavelength of light would it take to excite this electron to the next higher shell? Is this photon visible to humans?

41.50 • (a) Show all the distinct states for an electron in the N shell of hydrogen. Include all four quantum numbers. (b) For an f electron in the N shell, what is the largest possible orbital angular momentum and the greatest positive value for the component of this angular momentum along any chosen direction (the z-axis)? What is the magnitude of its spin angular momentum? Express these quantities in units of $\hbar$. (c) For an electron in the d state of the N shell, what are the maximum and minimum angles between its angular momentum vector and any chosen direction (the z-axis)? (d) What is the largest value of the orbital angular momentum for an f electron in the M shell?

41.51 • (a) The energy of an electron in the $4s$ state of sodium is -1.947 eV. What is the effective net charge of the nucleus "seen" by this electron? On the average, how many electrons screen the nucleus? (b) For an outer electron in the $4p$ state of potassium, on the average 17.2 inner electrons screen the nucleus. (i) What is the effective net charge of the nucleus "seen" by this outer electron? (ii) What is the energy of this outer electron?

41.52 • CALC For a hydrogen atom, the probability $P(r)$ of finding the electron within a spherical shell with inner radius r and outer radius $r + dr$ is given by Eq. (41.25). For a hydrogen atom in the $1s$ ground state, at what value of r does $P(r)$ have its maximum value? How does your result compare to the distance between the electron and the nucleus for the $n = 1$ state in the Bohr model, Eq. (41.26)?

41.53 •• CALC Consider a hydrogen atom in the $1s$ state. (a) For what value of r is the potential energy $U(r)$ equal to the total energy E? Express your answer in terms of a. This value of r is called the *classical turning point*, since this is where a Newtonian particle would stop its motion and reverse direction. (b) For r greater than the classical turning point, $U(r) > E$. Classically, the particle cannot be in this region, since the kinetic energy cannot be negative. Calculate the probability of the electron being found in this classically forbidden region.

41.54 • CP **Rydberg Atoms.** *Rydberg atoms* are atoms whose outermost electron is in an excited state with a *very* large principal quantum number. Rydberg atoms have been produced in the laboratory and detected in interstellar space. (a) Why do all neutral Rydberg atoms with the same n value have essentially the same ionization energy, independent of the total number of electrons in the atom? (b) What is the ionization energy for a Rydberg atom with a principal quantum number of 350? What is the radius in the Bohr model of the Rydberg electron's orbit? (c) Repeat part (b) for $n = 650$.

41.55 ••• CALC The wave function for a hydrogen atom in the $2s$ state is

$$\psi_{2s}(r) = \frac{1}{\sqrt{32\pi a^3}}\left(2 - \frac{r}{a}\right)e^{-r/2a}$$

(a) Verify that this function is normalized. (b) In the Bohr model, the distance between the electron and the nucleus in the $n = 2$ state is exactly $4a$. Calculate the probability that an electron in the $2s$ state will be found at a distance less than $4a$ from the nucleus.

41.56 •• CALC The normalized wave function for a hydrogen atom in the $2s$ state is given in Problem 41.55. (a) For a hydrogen atom in the $2s$ state, at what value of r is $P(r)$ maximum? How does your result compare to $4a$, the distance between the electron and the nucleus in the $n = 2$ state of the Bohr model? (b) At what value of r (other than $r = 0$ or $r = \infty$) is $P(r)$ equal to zero, so that the probability of finding the electron at that separation from the nucleus is zero? Compare your result to Fig. 41.9.

41.57 •• (a) For an excited state of hydrogen, show that the smallest angle that the orbital angular momentum vector $\vec{L}$ can have with the z-axis is

$$(\theta_L)_{min} = \arccos\left(\frac{n-1}{\sqrt{n(n-1)}}\right)$$

(b) What is the corresponding expression for $(\theta_L)_{max}$, the largest possible angle between $\vec{L}$ and the z-axis?

41.58 •• (a) If the value of L_z is known, we cannot know either L_x or L_y precisely. But we can know the value of the quantity $\sqrt{L_x^2 + L_y^2}$. Write an expression for this quantity in terms of l, m_l, and $\hbar$. (b) What is the meaning of $\sqrt{L_x^2 + L_y^2}$? (c) For a state of nonzero orbital angular momentum, find the maximum and minimum values of $\sqrt{L_x^2 + L_y^2}$. Explain your results.

41.59 •• CALC The normalized radial wave function for the $2p$ state of the hydrogen atom is $R_{2p} = (1/\sqrt{24a^5})re^{-r/2a}$. After we average over the angular variables, the radial probability function becomes $P(r)\,dr = (R_{2p})^2r^2\,dr$. At what value of r is $P(r)$ for the $2p$ state a maximum? Compare your results to the radius of the $n = 2$ state in the Bohr model.

41.60 •• CP **Stern–Gerlach Experiment.** In a Stern–Gerlach experiment, the deflecting force on the atom is $F_z = -\mu_z(dB_z/dz)$, where μ_z is given by Eq. (41.40) and dB_z/dz is the magnetic-field gradient. In a particular experiment the magnetic-field region is 50.0 cm long; assume the magnetic-field gradient is constant in this region. A beam of silver atoms enters the magnetic field with a speed of 525 m/s. What value of dB_z/dz is required to give a separation of 1.0 mm between the two spin components as they exit the field? (*Note:* The magnetic dipole moment of silver is the same as that for hydrogen, since its valence electron is in an $l = 0$ state.)

41.61 • Consider the transition from a $3d$ to a $2p$ state of hydrogen in an external magnetic field. Assume that the effects of electron spin can be ignored (which is not actually the case) so that the magnetic field interacts only with the orbital angular momentum. Identify each allowed transition by the m_l values of the initial and final states. For each of these allowed transitions, determine the shift of the transition energy from the zero-field value and show that there are three different transition energies.

41.62 •• An atom in a $3d$ state emits a photon of wavelength 475.082 nm when it decays to a $2p$ state. (a) What is the energy (in electron volts) of the photon emitted in this transition? (b) Use the selection rules described in Section 41.4 to find the allowed transitions if the atom is now in an external magnetic field of 3.500 T. Ignore the effects of the electron's spin. (c) For the case in part (b), if the energy of the $3d$ state was originally -8.50000 eV with no magnetic field present, what will be the energies of the states into which it splits in the magnetic field? (d) What are the allowed wavelengths of the light emitted during transition in part (b)?

41.63 •• **CALC** **Spectral Analysis.** While studying the spectrum of a gas cloud in space, an astronomer magnifies a spectral line that results from a transition from a p state to an s state. She finds that the line at 575.050 nm has actually split into three lines, with adjacent lines 0.0462 nm apart, indicating that the gas is in an external magnetic field. (Ignore effects due to electron spin.) What is the strength of the external magnetic field?

41.64 •• A hydrogen atom makes a transition from an $n = 3$ state to an $n = 2$ state (the Balmer H_α line) while in a magnetic field in the $+z$-direction and with magnitude 1.40 T. (a) If the magnetic quantum number is $m_l = 2$ in the initial $(n = 3)$ state and $m_l = 1$ in the final $(n = 2)$ state, by how much is each energy level shifted from the zero-field value? (b) By how much is the wavelength of the H_α line shifted from the zero-field value? Is the wavelength increased or decreased? Disregard the effect of electron spin. [*Hint:* Use the result of Problem 39.86(c).]

41.65 • **CP** A large number of hydrogen atoms in $1s$ states are placed in an external magnetic field that is in the $+z$-direction. Assume that the atoms are in thermal equilibrium at room temperature, $T = 300$ K. According to the Maxwell–Boltzmann distribution (see Section 39.4), what is the ratio of the number of atoms in the $m_s = \frac{1}{2}$ state to the number in the $m_s = -\frac{1}{2}$ state when the magnetic-field magnitude is (a) 5.00×10^{-5} T (approximately the earth's field); (b) 0.500 T; (c) 5.00 T?

41.66 •• **Effective Magnetic Field.** An electron in a hydrogen atom is in the $2p$ state. In a simple model of the atom, assume that the electron circles the proton in an orbit with radius r equal to the Bohr-model radius for $n = 2$. Assume that the speed v of the orbiting electron can be calculated by setting $L = mvr$ and taking L to have the quantum-mechanical value for a $2p$ state. In the frame of the electron, the proton orbits with radius r and speed v. Model the orbiting proton as a circular current loop, and calculate the magnetic field it produces at the location of the electron.

41.67 •• **Weird Universe.** In another universe, the electron is a spin-$\frac{3}{2}$ rather than a spin-$\frac{1}{2}$ particle, but all other physics are the same as in our universe. In this universe, (a) what are the atomic numbers of the lightest two inert gases? (b) What is the ground-state electron configuration of sodium?

41.68 • For an ion with nuclear charge Z and a single electron, the electric potential energy is $-Ze^2/4\pi\epsilon_0 r$ and the expression for the energies of the states and for the normalized wave functions are obtained from those for hydrogen by replacing e^2 by Ze^2. Consider the N^{6+} ion, with seven protons and one electron. (a) What is the ground-state energy in electron volts? (b) What is the ionization energy, the energy required to remove the electron from the N^{6+}

ion if it is initially in the ground state? (c) What is the distance a [given for hydrogen by Eq. (41.26)] for this ion? (d) What is the wavelength of the photon emitted when the N^{6+} ion makes a transition from the $n = 2$ state to the $n = 1$ ground state?

41.69 •• A hydrogen atom in an $n = 2$, $l = 1$, $m_l = -1$ state emits a photon when it decays to an $n = 1$, $l = 0$, $m_l = 0$ ground state. (a) In the absence of an external magnetic field, what is the wavelength of this photon? (b) If the atom is in a magnetic field in the $+z$-direction and with a magnitude of 2.20 T, what is the shift in the wavelength of the photon from the zero-field value? Does the magnetic field increase or decrease the wavelength? Disregard the effect of electron spin. [*Hint:* Use the result of Problem 39.86(c).]

41.70 •• A lithium atom has three electrons, and the $^2S_{1/2}$ ground-state electron configuration is $1s^2 2s$. The $1s^2 2p$ excited state is split into two closely spaced levels, $^2P_{3/2}$ and $^2P_{1/2}$, by the spin-orbit interaction (see Example 41.7 in Section 41.5). A photon with wavelength 67.09608 μm is emitted in the $^2P_{3/2} \rightarrow {}^2S_{1/2}$ transition, and a photon with wavelength 67.09761 μm is emitted in the $^2P_{1/2} \rightarrow {}^2S_{1/2}$ transition. Calculate the effective magnetic field seen by the electron in the $1s^2 2p$ state of the lithium atom. How does your result compare to that for the $3p$ level of sodium found in Example 41.7?

41.71 • Estimate the minimum and maximum wavelengths of the characteristic x rays emitted by (a) vanadium $(Z = 23)$ and (b) rhenium $(Z = 45)$. Discuss any approximations that you make.

41.72 •• **CP** **Electron Spin Resonance.** Electrons in the lower of two spin states in a magnetic field can absorb a photon of the right frequency and move to the higher state. (a) Find the magnetic-field magnitude B required for this transition in a hydrogen atom with $n = 1$ and $l = 0$ to be induced by microwaves with wavelength λ. (b) Calculate the value of B for a wavelength of 3.50 cm.

CHALLENGE PROBLEMS

41.73 ••• Each of $2N$ electrons (mass m) is free to move along the x-axis. The potential-energy function for each electron is $U(x) = \frac{1}{2}k'x^2$, where k' is a positive constant. The electric and magnetic interactions between electrons can be ignored. Use the exclusion principle to show that the minimum energy of the system of $2N$ electrons is $\hbar N^2 \sqrt{k'/m}$. (*Hint:* See Section 40.5 and the hint given in Problem 41.47.)

41.74 ••• **CP** Consider a simple model of the helium atom in which two electrons, each with mass m, move around the nucleus (charge $+2e$) in the same circular orbit. Each electron has orbital angular momentum $\hbar$ (that is, the orbit is the smallest-radius Bohr orbit), and the two electrons are always on opposite sides of the nucleus. Ignore the effects of spin. (a) Determine the radius of the orbit and the orbital speed of each electron. [*Hint:* Follow the procedure used in Section 39.3 to derive Eqs. (39.8) and (39.9). Each electron experiences an attractive force from the nucleus and a repulsive force from the other electron.] (b) What is the total kinetic energy of the electrons? (c) What is the potential energy of the system (the nucleus and the two electrons)? (d) In this model, how much energy is required to remove both electrons to infinity? How does this compare to the experimental value of 79.0 eV?

41.75 ••• **CALC** Repeat the calculation of Problem 41.53 for a one-electron ion with nuclear charge Z. (See Problem 41.68.) How does the probability of the electron being found in the classically forbidden region depend on Z?

Answers

Chapter Opening Question

Helium is inert because it has a filled K shell, while sodium is very reactive because its third electron is loosely bound in an L shell. See Section 41.6 for more details.

Test Your Understanding Questions

41.1 Answer: (iv) If $U(x, y, z) = 0$ in a certain region of space, we can rewrite the time-independent Schrödinger equation [Eq. (41.5)] for that region as $\partial^2\psi/\partial x^2 + \partial^2\psi/\partial y^2 + \partial^2\psi/\partial z^2 = (-2mE/\hbar^2)\psi$. We are told that all of the second derivatives of $\psi(x, y, z)$ are positive in this region, so the left-hand side of this equation is positive. Hence the right-hand side $(-2mE/\hbar^2)\psi$ must also be positive. Since $E > 0$, the quantity $-2mE/\hbar^2$ is negative, and so $\psi(x, y, z)$ must be negative.

41.2 Answer: (iv), (ii), (i) and (iii) (tie) Equation (41.16) shows that the energy levels for a cubical box are proportional to the quantity $n_X^2 + n_Y^2 + n_Z^2$. Hence ranking in order of this quantity is the same as ranking in order of energy. For the four cases we are given, we have (i) $n_X^2 + n_Y^2 + n_Z^2 = 2^2 + 3^2 + 2^2 = 17$; (ii) $n_X^2 + n_Y^2 + n_Z^2 = 4^2 + 1^2 + 1^2 = 18$; (iii) $n_X^2 + n_Y^2 + n_Z^2 = 2^2 + 2^2 + 3^2 = 17$; and (iv) $n_X^2 + n_Y^2 + n_Z^2 = 1^2 + 3^2 + 3^2 = 19$. The states $(n_X, n_Y, n_Z) = (2, 3, 2)$ and $(n_X, n_Y, n_Z) = (2, 2, 3)$ have the same energy (they are degenerate).

41.3 Answer: (ii) and (iii) (tie), (i) An electron in a state with principal quantum number n is most likely to be found at $r = n^2a$. This result is independent of the values of the quantum numbers l and m_l. Hence an electron with $n = 2$ (most likely to be found at $r = 4a$) is more likely to be found near $r = 5a$ than an electron with $n = 1$ (most likely to be found at $r = a$).

41.4 Answer: no All that matters is the component of the electron's orbital magnetic moment along the direction of $\vec{B}$. We called this quantity μ_z in Eq. (41.32) because we *defined* the positive z-axis to be in the direction of $\vec{B}$. In reality, the names of the axes are entirely arbitrary.

41.5 Answer: (iv) For the magnetic moment to be perfectly aligned with the z-direction, the z-component of the spin vector $\vec{S}$ would have to have the same absolute value as $\vec{S}$. However, the possible values of S_z are $\pm\frac{1}{2}\hbar$ [Eq. (41.37)], while the magnitude of the spin vector is $S = \sqrt{\frac{3}{4}}\hbar$ [Eq. (41.38)]. Hence $\vec{S}$ can never be perfectly aligned with any one direction in space.

41.6 Answer: more difficult If there were no exclusion principle, all 11 electrons in the sodium atom would be in the level of lowest energy (the 1s level) and the configuration would be $1s^{11}$. Consequently, it would be more difficult to remove the first electron. (In a real sodium atom the valence electron is in a screened 3s state, which has a comparatively high energy.)

41.7 Answer: (iv) An absorption edge appears if the photon energy is just high enough to remove an electron in a given energy level from the atom. In a sample of high-temperature hydrogen we expect to find atoms whose electron is in the ground level $(n = 1)$, the first excited level $(n = 2)$, and the second excited level $(n = 3)$. From Eq. (41.21) these levels have energies $E_n = (-13.60 \text{ eV})/n^2 = -13.60 \text{ eV}, -3.40 \text{ eV}$, and -1.51 eV (see Fig. 38.9b).

Bridging Problem

Answers: (a) 2.37×10^{-10} m

(b) Values of (n_X, n_Y, n_Z, m_s) for the 22 electrons: $(1, 1, 1, +\frac{1}{2})$, $(1, 1, 1, -\frac{1}{2})$, $(2, 1, 1, +\frac{1}{2})$, $(2, 1, 1, -\frac{1}{2})$, $(1, 2, 1, +\frac{1}{2})$, $(1, 2, 1, -\frac{1}{2})$, $(1, 1, 2, +\frac{1}{2})$, $(1, 1, 2, -\frac{1}{2})$, $(2, 2, 1, +\frac{1}{2})$, $(2, 2, 1, -\frac{1}{2})$, $(2, 1, 2, +\frac{1}{2})$, $(2, 1, 2, -\frac{1}{2})$, $(1, 2, 2, +\frac{1}{2})$, $(1, 2, 2, -\frac{1}{2})$, $(3, 1, 1, +\frac{1}{2})$, $(3, 1, 1, -\frac{1}{2})$, $(1, 3, 1, +\frac{1}{2})$, $(1, 3, 1, -\frac{1}{2})$, $(1, 1, 3, +\frac{1}{2})$, $(1, 1, 3, -\frac{1}{2})$, $(2, 2, 2, +\frac{1}{2})$, $(2, 2, 2, -\frac{1}{2})$

(c) 20.1 eV, 40.2 eV, 60.3 eV, 73.7 eV, and 80.4 eV

(d) 60.3 eV versus 4.52×10^3 eV

MOLECULES AND CONDENSED MATTER

42

? This false-color image of Venus shows its thick, cloud-shrouded atmosphere, which is 96.5% carbon dioxide (CO_2). The atmosphere raises the surface temperature of Venus to 735 K ($462°C = 863°F$)—hotter even than Mercury, the closest planet to the sun. What property of CO_2 molecules makes them a potent agent for raising Venus's temperature?

LEARNING GOALS

By studying this chapter, you will learn:

- The various types of bonds that hold atoms together.

- How the rotational and vibrational dynamics of molecules are revealed by molecular spectra.

- How and why atoms form into crystalline structures.

- How to use the energy-band concept to explain the electrical properties of solids.

- A simple model for metals that explains many of their physical properties.

- How the character of a semiconductor can be radically transformed by adding small amounts of an impurity.

- Some of the technological applications of semiconductor devices.

- Why certain materials become superconductors at low temperature.

In Chapter 41 we discussed the structure and properties of isolated atoms. But such atoms are the exception; usually we find atoms combined to form molecules or more extended structures we call condensed matter (liquid or solid). It's the attractive forces between atoms, called molecular bonds, that cause them to combine. In this chapter we'll study several kinds of bonds as well as the energy levels and spectra associated with diatomic molecules. We will see that just as atoms have quantized energies determined by the quantum-mechanical state of their electrons, so molecules have quantized energies determined by their rotational and vibrational states.

The same physical principles behind molecular bonds also apply to the study of condensed matter, in which various types of bonding occur. We'll explore the concept of energy bands and see how it helps us understand the properties of solids. Then we'll look more closely at the properties of a special class of solids called semiconductors. Devices using semiconductors are found in every radio, TV, pocket calculator, and computer used today; they have revolutionized the entire field of electronics during the past half-century.

42.1 Types of Molecular Bonds

We can use our discussion of atomic structure in Chapter 41 as a basis for exploring the nature of *molecular bonds,* the interactions that hold atoms together to form stable structures such as molecules and solids.

42.1 When the separation r between two oppositely charged ions is large, the potential energy $U(r)$ is proportional to $1/r$ as for point charges and the force is attractive. As r decreases, the charge clouds of the two atoms overlap and the force becomes less attractive. If r is less than the equilibrium separation r_0, the force is repulsive.

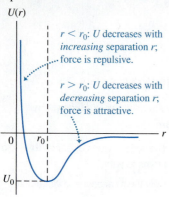

$r < r_0$: U decreases with increasing separation r; force is repulsive.

$r > r_0$: U decreases with decreasing separation r; force is attractive.

Ionic Bonds

The **ionic bond** is an interaction between oppositely charged *ionized* atoms. The most familiar example is sodium chloride (NaCl), in which the sodium atom gives its one $3s$ electron to the chlorine atom, filling the vacancy in the $3p$ subshell of chlorine.

Let's look at the energy balance in this transaction. Removing the $3s$ electron from a neutral sodium atom requires 5.138 eV of energy; this is called the *ionization energy* of sodium. The neutral chlorine atom can attract an extra electron into the vacancy in the $3p$ subshell, where it is incompletely screened by the other electrons and therefore is attracted to the nucleus. This state has 3.613 eV lower energy than a neutral chlorine atom and a distant free electron; 3.613 eV is the magnitude of the *electron affinity* of chlorine. Thus creating the well-separated Na$^+$ and Cl$^-$ ions requires a net investment of only 5.138 eV − 3.613 eV = 1.525 eV. When the two oppositely charged ions are brought together by their mutual attraction, the magnitude of their negative potential energy is determined by how closely they can approach each other. This in turn is limited by the exclusion principle, which forbids extensive overlap of the electron clouds of the two ions. As the distance decreases, the exclusion principle distorts the charge clouds, so the ions no longer interact like point charges and the interaction eventually becomes repulsive (Fig. 42.1).

The minimum electric potential energy for NaCl turns out to be −5.7 eV at a separation of 0.24 nm. The net energy released in creating the ions and letting them come together to the equilibrium separation of 0.24 nm is 5.7 eV − 1.525 eV = 4.2 eV. Thus, if the kinetic energy of the ions is neglected, 4.2 eV is the *binding energy* of the NaCl molecule, the energy that is needed to dissociate the molecule into separate neutral atoms.

Ionic bonds can involve more than one electron per atom. For instance, alkaline earth elements form ionic compounds in which an atom loses *two* electrons; an example is magnesium chloride, or Mg^{2+}(Cl$^-$)$_2$. Ionic bonds that involve a loss of more than two electrons are relatively rare. Instead, a different kind of bond, the *covalent* bond, comes into operation. We'll discuss this type of bond below.

Example 42.1 | Electric potential energy of the NaCl molecule

Find the electric potential energy of an Na$^+$ ion and a Cl$^-$ ion separated by 0.24 nm. Consider the ions as point charges.

SOLUTION

IDENTIFY and SET UP: Equation (23.9) in Section 23.1 tells us that the electric potential energy of two point charges q and q_0 separated by a distance r is $U = qq_0/4\pi\epsilon_0 r$.

EXECUTE: We have $q = +e$ (for Na$^+$), $q_0 = -e$ (for Cl$^-$), and $r = 0.24$ nm $= 0.24 \times 10^{-9}$ m. From Eq. (23.9),

$$U = -\frac{1}{4\pi\epsilon_0}\frac{e^2}{r_0} = -(9.0 \times 10^9 \text{ N} \cdot \text{m}^2/\text{C}^2)\frac{(1.6 \times 10^{-19} \text{ C})^2}{0.24 \times 10^{-9} \text{ m}}$$

$$= -9.6 \times 10^{-19}\text{J} = -6.0 \text{ eV}$$

EVALUATE: This result agrees fairly well with the observed value of −5.7 eV. The reason for the difference is that when the two ions are at their equilibrium separation of 0.24 nm, the outer regions of their electron clouds overlap. Hence the two ions don't behave exactly like point charges.

Mastering PHYSICS

PhET: Double Wells and Covalent Bonds

Covalent Bonds

The **covalent bond** is characterized by a more egalitarian participation of the two atoms than occurs with the ionic bond. The simplest covalent bond is found in the hydrogen molecule, a structure containing two protons and two electrons. As the separate atoms (Fig. 42.2a) come together, the electron wave functions are distorted and become more concentrated in the region between the two protons (Fig. 42.2b). The net attraction of the electrons for each proton more than balances the repulsion of the two protons and of the two electrons.

The attractive interaction is then supplied by a *pair* of electrons, one contributed by each atom, with charge clouds that are concentrated primarily in the region between the two atoms. The energy of the covalent bond in the hydrogen molecule H$_2$ is −4.48 eV.

As we saw in Chapter 41, the exclusion principle permits two electrons to occupy the same region of space (that is, to be in the same spatial quantum state) only when they have opposite spins. When the spins are parallel, the exclusion principle forbids the molecular state that would be most favorable from energy considerations (with both electrons in the region between atoms). Opposite spins are an essential requirement for a covalent bond, and no more than two electrons can participate in such a bond.

However, an atom with several electrons in its outermost shell can form several covalent bonds. The bonding of carbon and hydrogen atoms, of central importance in organic chemistry, is an example. In the *methane* molecule (CH_4) the carbon atom is at the center of a regular tetrahedron, with a hydrogen atom at each corner. The carbon atom has four electrons in its *L* shell, and each of these four electrons forms a covalent bond with one of the four hydrogen atoms (Fig. 42.3). Similar patterns occur in more complex organic molecules.

Because of the role played by the exclusion principle, covalent bonds are highly directional. In the methane molecule the wave function for each of carbon's four valence electrons is a combination of the 2*s* and 2*p* wave functions called a *hybrid wave function*. The probability distribution for each one has a lobe protruding toward a corner of a tetrahedron. This symmetric arrangement minimizes the overlap of wave functions for the electron pairs, minimizing their repulsive potential energy.

Ionic and covalent bonds represent two extremes in molecular bonding, but there is no sharp division between the two types. Often there is a *partial* transfer of one or more electrons from one atom to another. As a result, many molecules that have dissimilar atoms have electric dipole moments—that is, a preponderance of positive charge at one end and of negative charge at the other. Such molecules are called *polar* molecules. Water molecules have large electric dipole moments; these are responsible for the exceptionally large dielectric constant of liquid water (see Sections 24.4 and 24.5).

van der Waals Bonds

Ionic and covalent bonds, with typical bond energies of 1 to 5 eV, are called *strong bonds*. There are also two types of weaker bonds. One of these, the **van der Waals bond,** is an interaction between the electric dipole moments of atoms or molecules; typical energies are 0.1 eV or less. The bonding of water molecules in the liquid and solid states results partly from dipole–dipole interactions.

No atom has a permanent electric dipole moment, nor do many molecules. However, fluctuating charge distributions can lead to fluctuating dipole moments; these in turn can induce dipole moments in neighboring structures. Overall, the resulting dipole–dipole interaction is attractive, giving a weak bonding of atoms or molecules. The interaction potential energy drops off very quickly with distance *r* between molecules, usually as $1/r^6$. The liquefaction and solidification of the inert gases and of molecules such as H_2, O_2, and N_2 are due to induced-dipole van der Waals interactions. Not much thermal-agitation energy is needed to break these weak bonds, so such substances usually exist in the liquid and solid states only at very low temperatures.

Hydrogen Bonds

In the other type of weak bond, the **hydrogen bond,** a proton (H^+ ion) gets between two atoms, polarizing them and attracting them by means of the induced dipoles. This bond is unique to hydrogen-containing compounds because only hydrogen has a singly ionized state with no remaining electron cloud; the hydrogen ion is a bare proton, much smaller than any other singly ionized atom. The bond energy is usually less than 0.5 eV. The hydrogen bond is responsible for the cross-linking of long-chain organic molecules such as polyethylene (used in plastic bags). Hydrogen bonding also plays a role in the structure of ice.

42.2 Covalent bond in a hydrogen molecule.

(a) Separate hydrogen atoms

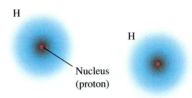

Individual H atoms are usually widely separated and do not interact.

(b) H_2 molecule

Covalent bond: the charge clouds for the two electrons with opposite spins are concentrated in the region between the nuclei.

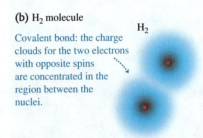

42.3 Schematic diagram of the methane (CH_4) molecule. The carbon atom is at the center of a regular tetrahedron and forms four covalent bonds with the hydrogen atoms at the corners. Each covalent bond includes two electrons with opposite spins, forming a charge cloud that is concentrated between the carbon atom and a hydrogen atom.

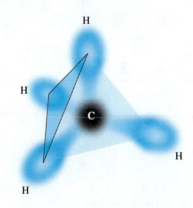

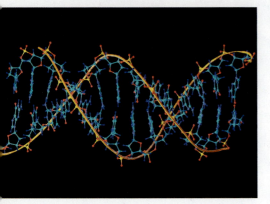

42.4 A diatomic molecule modeled as two point masses m_1 and m_2 separated by a distance r_0. The distances of the masses from the center of mass are r_1 and r_2, where $r_1 + r_2 = r_0$.

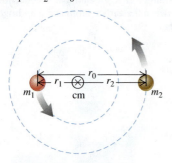

42.5 The ground level and first four excited rotational energy levels for a diatomic molecule. The levels are not equally spaced.

E

$10\hbar^2/I$ —————— $l = 4$

$6\hbar^2/I$ —————— $l = 3$

$3\hbar^2/I$ —————— $l = 2$

$\hbar^2/I$ —————— $l = 1$
0 —————— $l = 0$

All these bond types hold the atoms together in *solids* as well as in molecules. Indeed, a solid is in many respects a giant molecule. Still another type of bonding, the *metallic bond,* comes into play in the structure of metallic solids. We'll return to this subject in Section 42.3.

Test Your Understanding of Section 42.1 If electrons obeyed the exclusion principle but did *not* have spin, how many electrons could participate in a covalent bond? (i) one; (ii) two; (iii) three; (iv) more than three.

42.2 Molecular Spectra

Molecules have energy levels that are associated with rotational motion of a molecule as a whole and with vibrational motion of the atoms relative to each other. Just as transitions between energy levels in atoms lead to atomic spectra, transitions between rotational and vibrational levels in molecules lead to *molecular spectra.*

Rotational Energy Levels

In this discussion we'll concentrate mostly on *diatomic* molecules, to keep things as simple as possible. In Fig. 42.4 we picture a diatomic molecule as a rigid dumbbell (two point masses m_1 and m_2 separated by a constant distance r_0) that can *rotate* about axes through its center of mass, perpendicular to the line joining them. What are the energy levels associated with this motion?

We showed in Section 10.5 that when a rigid body rotates with angular speed ω about a perpendicular axis through its center of mass, the magnitude L of its angular momentum is given by Eq. (10.28), $L = I\omega$, where I is its moment of inertia about that symmetry axis. Its kinetic energy is given by Eq. (9.17), $K = \frac{1}{2}I\omega^2$. Combining these two equations, we find $K = L^2/2I$. There is no potential energy U, so the kinetic energy K is equal to the total mechanical energy E:

$$E = \frac{L^2}{2I} \tag{42.1}$$

Zero potential energy means that U does not depend on the angular coordinate of the molecule. But the potential-energy function U for the hydrogen atom (see Section 41.3) also has no dependence on angular coordinates. Thus the angular solutions to the Schrödinger equation for rigid-body rotation are the same as for the hydrogen atom, and the angular momentum is quantized in the same way. As in Eq. (41.21),

$$L = \sqrt{l(l + 1)}\,\hbar \qquad (l = 0, 1, 2, \dots) \tag{42.2}$$

Combining Eqs. (42.1) and (42.2), we obtain the *rotational energy levels:*

$$E_l = l(l + 1)\frac{\hbar^2}{2I} \qquad (l = 0, 1, 2, \dots) \qquad \text{(rotational energy levels, diatomic molecule)} \tag{42.3}$$

Figure 42.5 is an energy-level diagram showing these rotational levels. The ground level has zero quantum number l, corresponding to zero angular momentum (no rotation and zero rotational energy E). The spacing of adjacent levels increases with increasing l.

We can express the moment of inertia I in Eqs. (42.1) and (42.3) in terms of the *reduced mass* m_r of the molecule:

$$m_r = \frac{m_1 m_2}{m_1 + m_2} \tag{42.4}$$

We introduced the reduced mass in Section 39.3 to accommodate the finite nuclear mass of the hydrogen atom. In Fig. 42.4 the distances r_1 and r_2 are the distances from the center of mass to the nuclei of the atoms. By definition of the center of mass, $m_1 r_1 = m_2 r_2$, and the figure also shows that $r_0 = r_1 + r_2$. Solving these equations for r_1 and r_2, we find

MasteringPHYSICS

PhET: The Greenhouse Effect

$$r_1 = \frac{m_2}{m_1 + m_2} r_0 \qquad r_2 = \frac{m_1}{m_1 + m_2} r_0 \qquad (42.5)$$

The moment of inertia is $I = m_1 r_1^2 + m_2 r_2^2$; substituting Eq. (42.5), we find

$$I = m_1 \frac{m_2^2}{(m_1 + m_2)^2} r_0^2 + m_2 \frac{m_1^2}{(m_1 + m_2)^2} r_0^2 = \frac{m_1 m_2}{m_1 + m_2} r_0^2 \quad \text{or}$$

$$I = m_r r_0^2 \text{ (moment of inertia of a diatomic molecule)} \qquad (42.6)$$

The reduced mass enables us to reduce this two-body problem to an equivalent one-body problem (a particle of mass m_r moving around a circle with radius r_0), just as we did with the hydrogen atom. Indeed, the only difference between this problem and the hydrogen atom is the difference in the radial forces. To conserve angular momentum and account for the angular momentum of the emitted or absorbed photon, the allowed transitions between rotational states must satisfy the same selection rule that we discussed in Section 41.4 for allowed transitions between the states of an atom: l must change by exactly one unit, that is, $\Delta l = \pm 1$.

Example 42.2 Rotational spectrum of carbon monoxide

The two nuclei in the carbon monoxide (CO) molecule are 0.1128 nm apart. The mass of the most common carbon atom is 1.993×10^{-26} kg; that of the most common oxygen atom is 2.656×10^{-26} kg. (a) Find the energies of the lowest three rotational energy levels of CO. Express your results in meV (1 meV $= 10^{-3}$ eV). (b) Find the wavelength of the photon emitted in the transition from the $l = 2$ to the $l = 1$ level.

SOLUTION

IDENTIFY and SET UP: This problem uses the ideas developed in this section about the rotational energy levels of molecules. We are given the distance r_0 between the atoms and their masses m_1 and m_2. We find the reduced mass m_r using Eq. (42.4), the moment of inertia I using Eq. (42.6), and the energies E_l using Eq. (42.3). The energy E of the emitted photon is equal to the difference in energy between the $l = 2$ and $l = 1$ levels. (This transition obeys the $\Delta l = \pm 1$ selection rule, since $\Delta l = 1 - 2 = -1$.) We determine the photon wavelength using $E = hc/\lambda$.

EXECUTE: (a) From Eqs. (42.4) and (42.6), the reduced mass and moment of inertia of the CO molecule are:

$$m_r = \frac{m_1 m_2}{m_1 + m_2}$$

$$= \frac{(1.993 \times 10^{-26} \text{ kg})(2.656 \times 10^{-26} \text{ kg})}{(1.993 \times 10^{-26} \text{ kg}) + (2.656 \times 10^{-26} \text{ kg})}$$

$$= 1.139 \times 10^{-26} \text{ kg}$$

$$I = m_r r_0^2$$

$$= (1.139 \times 10^{-26} \text{ kg})(0.1128 \times 10^{-9} \text{ m})^2$$

$$= 1.449 \times 10^{-46} \text{ kg} \cdot \text{m}^2$$

The rotational levels are given by Eq. (42.3):

$$E_l = l(l + 1)\frac{\hbar^2}{2I} = l(l + 1)\frac{(1.0546 \times 10^{-34} \text{ J} \cdot \text{s})^2}{2(1.449 \times 10^{-46} \text{ kg} \cdot \text{m}^2)}$$

$$= l(l + 1)(3.838 \times 10^{-23} \text{ J}) = l(l + 1)0.2395 \text{ meV}$$

(1 meV $= 10^{-3}$ eV.) Substituting $l = 0, 1, 2$, we find

$$E_0 = 0 \qquad E_1 = 0.479 \text{ meV} \qquad E_2 = 1.437 \text{ meV}$$

(b) The photon energy and wavelength are:

$$E = E_2 - E_1 = 0.958 \text{ meV}$$

$$\lambda = \frac{hc}{E} = \frac{(4.136 \times 10^{-15} \text{ eV} \cdot \text{s})(3.00 \times 10^8 \text{ m/s})}{0.958 \times 10^{-3} \text{ eV}}$$

$$= 1.29 \times 10^{-3} \text{ m} = 1.29 \text{ mm}$$

EVALUATE: The differences between the first few rotational energy levels of CO are very small (about 1 meV $= 10^{-3}$ eV) compared to the differences between atomic energy levels (typically a few eV). Hence a photon emitted by a CO molecule in a transition from the $l = 2$ to the $l = 1$ level has very low energy and a very long wavelength compared to the visible light emitted by excited atoms. Photon wavelengths for rotational transitions in molecules are typically in the microwave and far infrared regions of the spectrum.

In this example we were given the equilibrium separation between the atoms, also called the *bond length*, and we used it to calculate one of the wavelengths emitted by excited CO molecules. In actual experiments, scientists work this problem backward: By measuring the long-wavelength emissions of a sample of diatomic molecules, they determine the moment of inertia of the molecule and hence the bond length.

42.6 A diatomic molecule modeled as two point masses m_1 and m_2 connected by a spring with force constant k'.

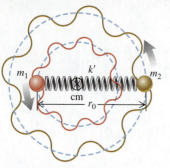

42.7 The ground level and first three excited vibrational levels for a diatomic molecule, assuming small displacements from equilibrium so we can treat the oscillations as simple harmonic. (Compare Fig. 40.25.)

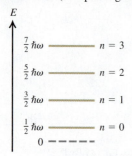

42.8 Energy-level diagram for vibrational and rotational energy levels of a diatomic molecule. For each vibrational level (n) there is a series of more closely spaced rotational levels (l). Several transitions corresponding to a single band in a band spectrum are shown. These transitions obey the selection rule $\Delta l = \pm 1$.

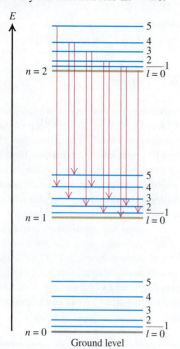

Vibrational Energy Levels

Molecules are never completely rigid. In a more realistic model of a diatomic molecule we represent the connection between atoms not as a rigid rod but as a spring (Fig. 42.6). Then, in addition to rotating, the atoms of the molecule can *vibrate* about their equilibrium positions along the line joining them. For small oscillations the restoring force can be taken as proportional to the displacement from the equilibrium separation r_0 (like a spring that obeys Hooke's law with a force constant k'), and the system is a harmonic oscillator. We discussed the quantum-mechanical harmonic oscillator in Section 40.5. The energy levels are given by Eq. (40.46) with the mass m replaced by the reduced mass m_r:

$$E_n = \left(n + \tfrac{1}{2}\right)\hbar\omega = \left(n + \tfrac{1}{2}\right)\hbar\sqrt{\frac{k'}{m_r}} \qquad (n = 0, 1, 2, \dots)$$

(vibrational energy levels of a diatomic molecule) (42.7)

This represents a series of levels equally spaced in energy, with an energy separation of

$$\Delta E = \hbar\omega = \hbar\sqrt{\frac{k'}{m_r}} \qquad (42.8)$$

Figure 42.7 is an energy-level diagram showing these vibrational levels. As an example, for the carbon monoxide molecule of Example 42.2 the spacing $\hbar\omega$ between vibrational energy levels is 0.2690 eV. From Eq. (42.8) this corresponds to a force constant of 1.90×10^3 N/m, which is a fairly loose spring. (To stretch a macroscopic spring with this value of k' by 1.0 cm would require a force of only 19 N, or about 4 lb.) Force constants for diatomic molecules are typically about 100 to 2000 N/m.

CAUTION **Watch out for k, k', and K** As in Section 40.5 we're again using k' for the force constant, this time to minimize confusion with Boltzmann's constant k, the gas constant per molecule (introduced in Section 18.3). Besides the quantities k and k', we also use the absolute temperature unit 1 K = 1 kelvin. ▮

Rotation and Vibration Combined

Visible-light photons have energies between 1.65 eV and 3.26 eV. The 0.2690-eV energy difference between vibrational levels for carbon monoxide corresponds to a photon of wavelength 4.613 μm, in the infrared region of the spectrum. This is much closer to the visible region than is the photon in the rotational transition in Example 42.2. In general the energy differences for molecular *vibration* are much smaller than those that produce atomic spectra, but much larger than the energy differences for molecular *rotation*.

When we include *both* rotational and vibrational energies, the energy levels for our diatomic molecule are

$$E_{nl} = l(l + 1)\frac{\hbar^2}{2I} + \left(n + \tfrac{1}{2}\right)\hbar\sqrt{\frac{k'}{m_r}} \qquad (42.9)$$

Figure 42.8 shows the energy-level diagram. For each value of n there are many values of l, forming a series of closely spaced levels.

The red arrows in Fig. 42.8 show several possible transitions in which a molecule goes from a level with $n = 2$ to a level with $n = 1$ by emitting a photon. As we mentioned above, these transitions must obey the selection rule $\Delta l = \pm 1$ in order to conserve angular momentum. An additional selection rule states that if the vibrational level changes, the vibrational quantum number n in Eq. (42.9) must increase by 1 ($\Delta n = 1$) if a photon is absorbed or decrease by 1 ($\Delta n = -1$) if a photon is emitted.

42.9 A typical molecular band spectrum.

As an illustration of these selection rules, Fig. 42.8 shows that a molecule in the $n = 2$, $l = 4$ level can emit a photon and drop into the $n = 1$, $l = 5$ level ($\Delta n = -1$, $\Delta l = +1$) or the $n = 1$, $l = 3$ level ($\Delta n = -1$, $\Delta l = -1$), but is forbidden from making a $\Delta n = -1$, $\Delta l = 0$ transition into the $n = 1$, $l = 4$ level.

Transitions between states with various pairs of n-values give different series of spectrum lines, and the resulting spectrum has a series of *bands*. Each band corresponds to a particular vibrational transition, and each individual line in a band represents a particular rotational transition, with the selection rule $\Delta l = \pm 1$. Figure 42.9 shows a typical *band spectrum*.

All molecules can have excited states of the *electrons* in addition to the rotational and vibrational states that we have described. In general, these lie at higher energies than the rotational and vibrational states, and there is no simple rule relating them. When there is a transition between electronic states, the $\Delta n = \pm 1$ selection rule for the vibrational levels no longer holds.

Example 42.3 **Vibration-rotation spectrum of carbon monoxide**

Consider again the CO molecule of Example 42.2. Find the wavelength of the photon emitted by a CO molecule when its vibrational energy changes and its rotational energy is (a) initially zero and (b) finally zero.

SOLUTION

IDENTIFY and SET UP: We need to use the selection rules for the vibrational and rotational transitions of a diatomic molecule. Since a photon is emitted as the vibrational energy changes, the selection rule $\Delta n = -1$ tells us that the vibrational quantum number n decreases by 1 in both parts (a) and (b). In part (a) the initial value of l is zero; the selection rule $\Delta l = \pm 1$ tells us that the *final* value of l is 1, so the rotational energy increases in this case. In part (b) the *final* value of l is zero; $\Delta l = \pm 1$ then tells us that the *initial* value of l is 1, and the rotational energy decreases.

In each case the energy E of the emitted photon is the difference between the initial and final energies of the molecule, accounting for the change in both vibrational and rotational energy. In part (a) E equals the difference $\hbar \omega$ between adjacent vibrational energy levels *minus* the rotational energy that the molecule *gains*; in part (b) E equals $\hbar \omega$ *plus* the rotational energy that the molecule *loses*. Example 42.2 tells us that the difference between the $l = 0$ and $l = 1$ rotational energy levels is 0.479 meV = 0.000479 eV, and we learned above that the

vibrational energy-level separation for CO is $\hbar \omega = 0.2690$ eV. We use $E = hc/\lambda$ to determine the corresponding wavelengths (our target variables).

EXECUTE: (a) In this transition the CO molecule loses $\hbar \omega = 0.2690$ eV of vibrational energy and gains 0.000479 eV of rotational energy. Hence the energy E that goes into the emitted photon equals 0.2690 eV *less* 0.000479 eV, or 0.2685 eV. The photon wavelength is

$$\lambda = \frac{hc}{E} = \frac{(4.136 \times 10^{-15} \text{ eV} \cdot \text{s})(2.998 \times 10^{8} \text{ m/s})}{0.2685 \text{ eV}}$$

$$= 4.618 \times 10^{-6} \text{ m} = 4.618 \ \mu\text{m}$$

(b) Now the CO molecule loses $\hbar \omega = 0.2690$ eV of vibrational energy and also loses 0.000479 eV of rotational energy, so the energy that goes into the photon is $E = 0.2690$ eV $+ 0.000479$ eV $= 0.2695$ eV. The wavelength is

$$\lambda = \frac{hc}{E} = \frac{(4.136 \times 10^{-15} \text{ eV} \cdot \text{s})(2.998 \times 10^{8} \text{ m/s})}{0.2695 \text{ eV}}$$

$$= 4.601 \times 10^{-6} \text{ m} = 4.601 \ \mu\text{m}$$

EVALUATE: In part (b) the molecule loses more energy than it does in part (a), so the emitted photon must have greater energy and a shorter wavelength. That's just what our results show.

Complex Molecules

We can apply these same principles to more complex molecules. A molecule with three or more atoms has several different kinds or *modes* of vibratory motion. Each mode has its own set of energy levels, related to its frequency by Eq. (42.7).

42.10 The carbon dioxide molecule can vibrate in three different modes. For clarity, the atoms are not shown to scale: The separation between atoms is actually comparable to their diameters.

(a) Bending mode

(b) Symmetric stretching mode

(c) Asymmetric stretching mode

In nearly all cases the associated radiation lies in the infrared region of the electromagnetic spectrum.

Infrared spectroscopy has proved to be an extremely valuable analytical tool. It provides information about the strength, rigidity, and length of molecular bonds and the structure of complex molecules. Also, because every molecule (like every atom) has its characteristic spectrum, infrared spectroscopy can be used to identify unknown compounds.

One molecule that can readily absorb and emit infrared radiation is carbon dioxide (CO_2). Figure 42.10 shows the three possible modes of vibration of a CO_2 molecule. A number of transitions are possible between excited levels of the same vibrational mode as well as between levels of different vibrational modes. The energy differences are less than 1 eV in all of these transitions, and so involve infrared photons of wavelength longer than 1 μm. Hence a gas of CO_2 can readily absorb light at a number of different infrared wavelengths. This makes CO_2 a very effective greenhouse gas (see Section 17.7) even on earth, where carbon dioxide is just 0.04% of the atmosphere by volume. On Venus, however, the atmosphere has more than 90 times the total mass of our atmosphere and is almost entirely CO_2. The resulting greenhouse effect is tremendous: The surface temperature on Venus is more than 400 kelvins greater than what it would be if the planet had no atmosphere at all.

Test Your Understanding of Section 42.2 A rotating diatomic molecule emits a photon when it makes a transition from level l to level $l - 1$. If the value of l increases, does the wavelength of the emitted photon (i) increase, (ii) decrease, or (iii) remain unchanged?

42.3 Structure of Solids

The term *condensed matter* includes both solids and liquids. In both states, the interactions between atoms or molecules are strong enough to give the material a definite volume that changes relatively little with applied stress. In condensed matter, adjacent atoms attract one another until their outer electron charge clouds begin to overlap significantly. Thus the distances between adjacent atoms in condensed matter are about the same as the diameters of the atoms themselves, typically 0.1 to 0.5 nm. Also, when we speak of the distances between atoms, we mean the center-to-center (nucleus-to-nucleus) distances.

Ordinarily, we think of a liquid as a material that can flow and of a solid as a material with a definite shape. However, if you heat a horizontal glass rod in the flame of a burner, you'll find that the rod begins to sag (flow) more and more easily as its temperature rises. Glass has no definite transition from solid to liquid, and no definite melting point. On this basis, we can consider glass at room temperature as being an extremely viscous liquid. Tar and butter show similar behavior.

What is the microscopic difference between materials like glass or butter and solids like ice or copper, which do have definite melting points? Ice and copper are examples of *crystalline solids* in which the atoms have *long-range order,* a recurring pattern of atomic positions that extends over many atoms. This pattern is called the *crystal structure.* In contrast, glass at room temperature is an example of an *amorphous* solid, one that has no long-range order, but only *short-range order* (correlations between neighboring atoms or molecules). Liquids also have only short-range order. The boundaries between crystalline solid, amorphous solid, and liquid may be sometimes blurred. Some solids, crystalline when perfect, can form with so many imperfections in their structure that they have almost no long-range order. Conversely, some liquid crystals (organic compounds composed of cylindrical molecules that tend to line up parallel to each other) have a fairly high degree of long-range order.

Nearly everything we know about crystal structure has been learned from diffraction experiments, initially with x rays and later with electrons and neutrons.

42.11 Portions of some common types of crystal lattices.

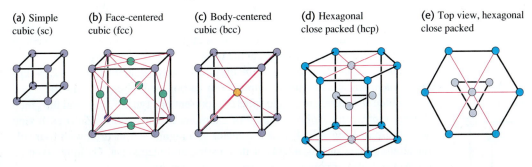

(a) Simple cubic (sc)

(b) Face-centered cubic (fcc)

(c) Body-centered cubic (bcc)

(d) Hexagonal close packed (hcp)

(e) Top view, hexagonal close packed

A typical distance between atoms is of the order of 0.1 nm. You can show that 12.4-keV x rays, 150-eV electrons, and 0.0818-eV neutrons all have wavelengths $\lambda = 0.1$ nm.

Crystal Lattices and Structures

A *crystal lattice* is a repeating pattern of mathematical points that extends throughout space. There are 14 general types of such patterns; Fig. 42.11 shows small portions of some common examples. The *simple cubic lattice* (sc) has a lattice point at each corner of a cubic array (Fig. 42.11a). The *face-centered cubic lattice* (fcc) is like the simple cubic but with an additional lattice point at the center of each cube face (Fig. 42.11b). The *body-centered cubic lattice* (bcc) is like the simple cubic but with an additional point at the center of each cube (Fig. 42.11c). The *hexagonal close-packed lattice* has layers of lattice points in hexagonal patterns, each hexagon made up of six equilateral triangles (Figs. 42.11d and 42.11e).

> **CAUTION** **A perfect crystal lattice is infinitely large** Figure 42.11 shows just enough lattice points so you can easily visualize the pattern; the lattice, a mathematical abstraction, extends throughout space. Thus the lattice points shown repeat endlessly in all directions.

   In a crystal structure, a single atom or a group of atoms is associated with each lattice point. The group may contain the same or different kinds of atoms. This atom or group of atoms is called a *basis*. Thus a complete description of a crystal structure includes both the lattice and the basis. We initially consider *perfect crystals,* or *ideal single crystals,* in which the crystal structure extends uninterrupted throughout space.

   The bcc and fcc structures are two common simple crystal structures. The alkali metals have a bcc structure—that is, a bcc lattice with a basis of one atom at each lattice point. Each atom in a bcc structure has eight nearest neighbors (Fig. 42.12a). The elements Al, Ca, Cu, Ag, and Au have an fcc structure—that is, an fcc lattice with a basis of one atom at each lattice point. Each atom in an fcc structure has 12 nearest neighbors (Fig. 42.12b).

   Figure 42.13 shows a representation of the structure of sodium chloride (NaCl, ordinary salt). It may look like a simple cubic structure, but it isn't. The sodium and chloride ions each form an fcc structure, so we can think roughly of the sodium chloride structure as being composed of two interpenetrating fcc structures. More correctly, the sodium chloride crystal structure of Fig. 42.13 has an fcc lattice with one chloride ion at each lattice point and one sodium ion half a cube length above it. That is, its basis consists of one chloride and one sodium ion.

   Another example is the *diamond structure;* it's called that because it is the crystal structure of carbon in the diamond form. It's also the crystal structure of silicon, germanium, and gray tin. The diamond lattice is fcc; the basis consists of one atom at each lattice point and a second *identical* atom displaced a quarter of a cube length in each of the three cube-edge directions. Figure 42.14 will help you

42.12 (a) The bcc *structure* is composed of a bcc *lattice* with a basis of one atom for each lattice point. (b) The fcc *structure* is composed of an fcc *lattice* with a basis of one atom for each lattice point. These structures repeat precisely to make up perfect crystals.

(a) The bcc structure

(b) The fcc structure

42.13 Representation of part of the sodium chloride crystal structure. The distances between ions are exaggerated.

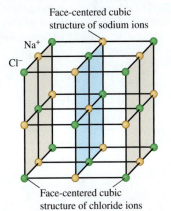

Face-centered cubic structure of sodium ions

Na^+

Cl^-

Face-centered cubic structure of chloride ions

42.14 The diamond structure, shown as two interpenetrating face-centered cubic structures with distances between atoms exaggerated. Relative to the corresponding green atom, each purple atom is shifted up, back, and to the left by a distance $a/4$.

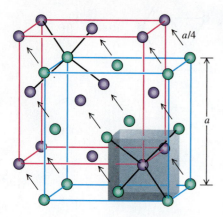

visualize this. The shaded volume in Fig. 42.14 shows the bottom right front eighth of the basic cube; the four atoms at alternate corners of this cube are at the corners of a regular tetrahedron, and there is an additional atom at the center. Thus each atom in the diamond structure is at the center of a regular tetrahedron with four nearest-neighbor atoms at the corners.

In the diamond structure, both the purple and green spheres in Fig. 42.14 represent *identical* atoms—for example, both carbon or both silicon. In the cubic zinc sulfide structure, the purple spheres represent one type of atom and the green spheres represent a *different* type. For example, in zinc sulfide (ZnS) each zinc atom (purple in Fig. 42.14) is at the center of a regular tetrahedron with four sulfur atoms (green in Fig. 42.14) at its corners, and vice versa. Gallium arsenide (GaAs) and similar compounds have this same structure.

Bonding in Solids

The forces that are responsible for the regular arrangement of atoms in a crystal are the same as those involved in molecular bonds, plus one additional type. Not surprisingly, *ionic* and *covalent* molecular bonds are found in ionic and covalent crystals, respectively. The most familiar *ionic crystals* are the alkali halides, such as ordinary salt (NaCl). The positive sodium ions and the negative chloride ions occupy alternate positions in a cubic arrangement (see Fig. 42.13). The attractive forces are the familiar Coulomb's-law forces between charged particles. These forces have no preferred direction, and the arrangement in which the material crystallizes is partly determined by the relative sizes of the two ions. Such a structure is *stable* in the sense that it has lower total energy than the separated ions (see the following example). The negative potential energies of pairs of opposite charges are greater in absolute value than the positive energies of pairs of like charges because the pairs of unlike charges are closer together, on average.

Example 42.4 Potential energy of an ionic crystal

Imagine a one-dimensional ionic crystal consisting of a very large number of alternating positive and negative ions with charges e and $-e$, with equal spacing a along a line. Show that the total interaction potential energy is negative, which means that such a "crystal" is stable.

SOLUTION

IDENTIFY and SET UP: We treat each ion as a point charge and use our results from Section 23.1 for the electric potential energy of a collection of point charges. Equations (23.10) and (23.11) tell us to consider the electric potential energy U of each pair of charges. The total potential energy of the system is the sum of the values of U for every possible pair; we take the number of pairs to be infinite.

EXECUTE: Let's pick an ion somewhere in the middle of the line and add the potential energies of its interactions with all the ions to one side of it. From Eq. (23.11), that sum is

$$\sum U = -\frac{e^2}{4\pi\epsilon_0}\frac{1}{a} + \frac{e^2}{4\pi\epsilon_0}\frac{1}{2a} - \frac{e^2}{4\pi\epsilon_0}\frac{1}{3a} + \cdots$$

$$= -\frac{e^2}{4\pi\epsilon_0 a}\left(1 - \tfrac{1}{2} + \tfrac{1}{3} - \tfrac{1}{4} + \cdots\right)$$

You may notice that the series in parentheses resembles the Taylor series for the function $\ln(1 + x)$:

$$\ln(1 + x) = x - \frac{x^2}{2} + \frac{x^3}{3} - \frac{x^4}{4} + \cdots$$

When $x = 1$ this becomes the series in parentheses above, so

$$\sum U = -\frac{e^2}{4\pi\epsilon_0 a}\ln 2$$

This is certainly a negative quantity. The atoms on the other side of the ion we're considering make an equal contribution to the potential energy. And if we include the potential energies of all pairs of atoms, the sum is certainly negative.

EVALUATE: We conclude that this one-dimensional ionic "crystal" is stable: It has lower energy than the zero electric potential energy that is obtained when all the ions are infinitely far apart from each other.

Types of Crystals

Carbon, silicon, germanium, and tin in the diamond structure are simple examples of *covalent crystals*. These elements are in Group IV of the periodic table, meaning that each atom has four electrons in its outermost shell. Each atom forms a covalent bond with each of four adjacent atoms at the corners of a tetrahedron (Fig. 42.14). These bonds are strongly directional because of the asymmetric electron distributions dictated by the exclusion principle, and the result is the tetrahedral diamond structure.

A third crystal type, less directly related to the chemical bond than are ionic or covalent crystals, is the **metallic crystal.** In this structure, one or more of the outermost electrons in each atom become detached from the parent atom (leaving a positive ion) and are free to move through the crystal. These electrons are not localized near the individual ions. The corresponding electron wave functions extend over many atoms.

Thus we can picture a metallic crystal as an array of positive ions immersed in a sea of freed electrons whose attraction for the positive ions holds the crystal together (Fig. 42.15). These electrons also give metals their high electrical and thermal conductivities. This sea of electrons has many of the properties of a gas, and indeed we speak of the *electron-gas model* of metallic solids. The simplest version of this model is the *free-electron model,* which ignores interactions with the ions completely (except at the surface). We'll return to this model in Section 42.5.

In a metallic crystal the freed electrons are not localized but are shared among *many* atoms. This gives a bonding that is neither localized nor strongly directional. The crystal structure is determined primarily by considerations of *close packing*—that is, the maximum number of atoms that can fit into a given volume. The two most common metallic crystal lattices are the face-centered cubic and hexagonal close-packed (see Figs. 42.11b, 42.11d, and 42.11e). In structures composed of these lattices with a basis of one atom, each atom has 12 nearest neighbors.

As we mentioned in Section 42.1, van der Waals interactions and hydrogen bonding also play a role in the structure of some solids. In polyethylene and similar polymers, covalent bonding of atoms forms long-chain molecules, and hydrogen bonding forms cross-links between adjacent chains. In solid water, both van der Waals forces and hydrogen bonds are significant in determining the crystal structures of ice.

Our discussion has centered on perfect crystals, or ideal single crystals. Real crystals show a variety of departures from this idealized structure. Materials are often *polycrystalline,* composed of many small single crystals bonded together at *grain boundaries.* There may be *point defects* within a single crystal: *Interstitial* atoms may occur in places where they do not belong, and there may be *vacancies,* positions that should be occupied by an atom but are not. A point defect of particular interest in semiconductors, which we will discuss in Section 42.6, is the *substitutional impurity,* a foreign atom replacing a regular atom (for example, arsenic in a silicon crystal).

There are several basic types of extended defects called *dislocations.* One type is the *edge dislocation,* shown schematically in Fig. 42.16, in which one plane of atoms slips relative to another. The mechanical properties of metallic crystals are influenced strongly by the presence of dislocations. The ductility and malleability of some metals depend on the presence of dislocations that can move through the crystal during plastic deformations. Solid-state physicists often point out that the biggest extended defect of all, present in *all* real crystals, is the surface of the material with its dangling bonds and abrupt change in potential energy.

42.15 In a metallic solid, one or more electrons are detached from each atom and are free to wander around the crystal, forming an "electron gas." The wave functions for these electrons extend over many atoms. The positive ions vibrate around fixed locations in the crystal.

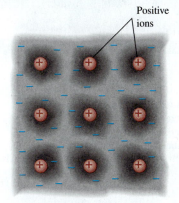

Positive ions

42.16 An edge dislocation in two dimensions. In three dimensions an edge dislocation would look like an extra plane of atoms slipped partway into the crystal.

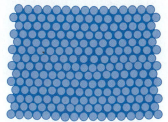

The irregularity is seen most easily by viewing the figure from various directions at a grazing angle with the page.

Test Your Understanding of Section 42.3 If a is the distance in an $NaCl$ crystal from an Na^+ ion to one of its nearest-neighbor Cl^- ions, what is the distance from an Na^+ ion to one of its *next-to-nearest*-neighbor Cl^- ions? (i) $a\sqrt{2}$; (ii) $a\sqrt{3}$; (iii) $2a$; (iv) none of these.

42.17 The concept of energy bands was first developed by the Swiss-American physicist Felix Bloch (1905–1983) in his doctoral thesis. Our modern understanding of electrical conductivity stems from that landmark work. Bloch's work in nuclear physics brought him (along with Edward Purcell) the 1952 Nobel Prize in physics.

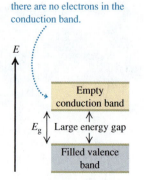

42.18 Origin of energy bands in a solid. (a) As the distance r between atoms decreases, the energy levels spread into bands. The vertical line at r_0 shows the actual atomic spacing in the crystal. (b) Symbolic representation of energy bands.

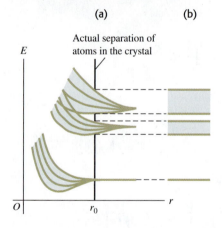

42.4 Energy Bands

The **energy-band** concept, introduced in 1928 (Fig. 42.17), is a great help in understanding several properties of solids. To introduce the idea, suppose we have a large number N of identical atoms, far enough apart that their interactions are negligible. Every atom has the same energy-level diagram. We can draw an energy-level diagram for the *entire system.* It looks just like the diagram for a single atom, but the exclusion principle, applied to the entire system, permits each state to be occupied by N electrons instead of just one.

Now we begin to push the atoms uniformly closer together. Because of the electrical interactions and the exclusion principle, the wave functions begin to distort, especially those of the outer, or *valence,* electrons. The corresponding energies also shift, some upward and some downward, by varying amounts, as the valence electron wave functions become less localized and extend over more and more atoms. Thus the valence states that formerly gave the *system* a state with a sharp energy level that could accommodate N electrons now give a *band* containing N closely spaced levels (Fig. 42.18). Ordinarily, N is very large, somewhere near the order of Avogadro's number (10^{24}), so we can accurately treat the levels as forming a *continuous* distribution of energies within a band. Between adjacent energy bands are gaps or forbidden regions where there are *no* allowed energy levels. The inner electrons in an atom are affected much less by nearby atoms than are the valence electrons, and their energy levels remain relatively sharp.

Insulators, Semiconductors, and Conductors

The nature of the energy bands determines whether the material is an electrical insulator, a semiconductor, or a conductor. In particular, what matters are the extent to which the states in each band are occupied and the spacing, or *energy gap,* between adjacent bands. A crucial factor is the exclusion principle (see Section 41.6), which states that only one electron can occupy a given quantum-mechanical state.

In an *insulator* at absolute zero temperature, the highest band that is completely filled, called the **valence band,** is also the highest band that has *any* electrons in it. The next higher band, called the **conduction band,** is completely empty; there are no electrons in its states (Fig. 42.19a). Imagine what happens if an electric field is applied to a material of this kind. To move in response to the field, an electron would have to go into a different quantum state with a slightly different energy. It can't do that, however, because all the neighboring states are already occupied. The only way such an electron can move is to jump across the energy gap into the conduction band, where there are plenty of nearby unoccupied states. At any temperature above absolute zero

42.19 Three types of energy-band structure.

(a) In an insulator at absolute zero, there are no electrons in the conduction band.

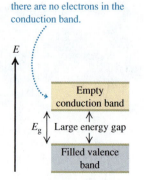

(b) A semiconductor has the same band structure as an insulator but a smaller gap between the valence and conduction bands.

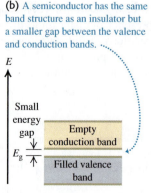

(c) A conductor has a partially filled conduction band.

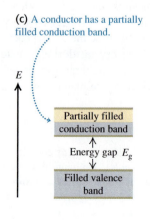

there is some probability this jump can happen, because an electron can gain energy from thermal motion. In an insulator, however, the energy gap between the valence and conduction bands can be 5 eV or more, and that much thermal energy is not ordinarily available. Hence little or no current flows in response to an applied electric field, and the electric conductivity (Section 25.2) is low. The thermal conductivity (Section 17.7), which also depends on mobile electrons, is likewise low.

We saw in Section 24.4 that an insulator becomes a conductor if it is subjected to a large enough electric field; this is called *dielectric breakdown.* If the electric field is of order 10^{10} V/m, there is a potential difference of a few volts over a distance comparable to atomic sizes. In this case the field can do enough work on a valence electron to boost it across the energy gap and into the conduction band. (In practice dielectric breakdown occurs for fields much less than 10^{10} V/m, because imperfections in the structure of an insulator provide some more accessible energy states *within* the energy gap.)

As in an insulator, a *semiconductor* at absolute zero has an empty conduction band above the full valence band. The difference is that in a semiconductor the energy gap between these bands is relatively small and electrons can more readily jump into the conduction band (Fig. 42.19b). As the temperature of a semiconductor increases, the population in the conduction band increases very rapidly, as does the electric conductivity. For example, in a semiconductor near room temperature with an energy gap of 1 eV, the number of conduction electrons doubles when the temperature rises by just 10°C. We will use the concept of energy bands to explore semiconductors in more depth in Section 42.6.

In a *conductor* such as a metal, there are electrons in the conduction band even at absolute zero (Fig. 42.19c). The metal sodium is an example. An analysis of the atomic energy-level diagram for sodium (see Fig. 39.19a) shows that for an isolated sodium atom, the six lowest excited states (all 3*p* states) are about 2.1 eV above the two 3*s* ground states. In solid sodium, however, the atoms are so close together that the 3*s* and 3*p bands* spread out and overlap into a single band. Each sodium atom contributes one electron to the band, leaving an Na^+ ion behind. Each atom also contributes eight *states* to that band (two 3*s*, six 3*p*), so the band is only one-eighth occupied. We call this structure a *conduction* band because it is only partially occupied. Electrons near the top of the filled portion of the band have many adjacent unoccupied states available, and they can easily gain or lose small amounts of energy in response to an applied electric field. Therefore these electrons are mobile, giving solid sodium its high electrical and thermal conductivity. A similar description applies to other conducting materials.

Mastering**PHYSICS**

PhET: Band Structure
PhET: Conductivity

Example 42.5 **Photoconductivity in germanium**

At room temperature, pure germanium has an almost completely filled valence band separated by a 0.67-eV gap from an almost completely empty conduction band. It is a poor electrical conductor, but its conductivity increases greatly when it is irradiated with electromagnetic waves of a certain maximum wavelength. What is that wavelength?

SOLUTION

IDENTIFY and SET UP: The conductivity of a semiconductor increases greatly when electrons are excited from the valence band into the conduction band. In germanium, the excitation occurs when an electron absorbs a photon with an energy of at least $E_{min} = 0.67$ eV. From the relationship $E = hc/\lambda$, the *maximum* wavelength λ_{max} (our target variable) corresponds to this *minimum* photon energy.

EXECUTE: The wavelength of a photon with energy $E_{min} = 0.67$ eV is

$$\lambda_{max} = \frac{hc}{E_{min}} = \frac{(4.136 \times 10^{-15} \text{ eV} \cdot \text{s})(3.00 \times 10^8 \text{ m/s})}{0.67 \text{ eV}}$$
$$= 1.9 \times 10^{-6} \text{ m} = 1.9 \ \mu\text{m} = 1900 \text{ nm}$$

EVALUATE: This wavelength is in the infrared part of the spectrum, so visible-light photons (which have shorter wavelength and higher energy) will also induce conductivity in germanium. As we'll see in Section 42.7, semiconductor crystals are widely used as photocells and for many other applications.

Test Your Understanding of Section 42.4 One type of thermometer works by measuring the temperature-dependent electrical resistivity of a sample. Which of the following types of material displays the greatest change in resistivity for a given temperature change? (i) insulator; (ii) semiconductor; (iii) resistor.

42.5 Free-Electron Model of Metals

Studying the energy states of electrons in metals can give us a lot of insight into their electrical and magnetic properties, the electron contributions to heat capacities, and other behavior. As we discussed in Section 42.3, one of the distinguishing features of a metal is that one or more valence electrons are detached from their home atom and can move freely within the metal, with wave functions that extend over many atoms.

The **free-electron model** assumes that these electrons are completely free inside the material, that they don't interact at all with the ions or with each other, but that there are infinite potential-energy barriers at the surfaces. The idea is that a typical electron moves so rapidly within the metal that it "sees" the effect of the ions and other electrons as a uniform potential-energy function, whose value we can choose to be zero.

We can represent the surfaces of the metal by the same cubical box that we analyzed in Section 41.2 (the three-dimensional version of the particle in a box studied in Section 40.2). If the box has sides of length L (Fig. 42.20), the energies of the stationary states (quantum states of definite energy) are

$$E_{n_X, n_Y, n_Z} = \frac{(n_X^2 + n_Y^2 + n_Z^2)\pi^2\hbar^2}{2mL^2} \quad \begin{array}{l}(n_X = 1, 2, 3, \ldots; n_Y = 1, \\ 2, 3, \ldots; n_Z = 1, 2, 3, \ldots)\end{array} \tag{42.10}$$

Each state is labeled by the three positive-integer quantum numbers (n_X, n_Y, n_Z).

Density of States

Later we'll need to know the *number dn* of quantum states that have energies in a given range dE. The number of states per unit energy range dn/dE is called the **density of states,** denoted by $g(E)$. We'll begin by working out an expression for $g(E)$. Think of a three-dimensional space with coordinates (n_X, n_Y, n_Z) (Fig. 42.21). The radius n_{rs} of a sphere centered at the origin in that space is $n_{rs} = (n_X^2 + n_Y^2 + n_Z^2)^{1/2}$. Each point with integer coordinates in that space represents one spatial quantum state. Thus each point corresponds to one unit of volume in the space, and the total number of points with integer coordinates inside a sphere equals the volume of the sphere, $\frac{4}{3}\pi n_{rs}^3$. Because all our n's are positive, we must take only one *octant* of the sphere, with $\frac{1}{8}$ the total volume, or $(\frac{1}{8})(\frac{4}{3}\pi n_{rs}^3) = \frac{1}{6}\pi n_{rs}^3$. The particles are electrons, so each point corresponds to *two* states with opposite spin components ($m_s = \pm\frac{1}{2}$), and the total number n of electron states corresponding to points inside the octant is twice $\frac{1}{6}\pi n_{rs}^3$, or

$$n = \frac{\pi n_{rs}^3}{3} \tag{42.11}$$

The energy E of states at the surface of the sphere can be expressed in terms of n_{rs}. Equation (42.10) becomes

$$E = \frac{n_{rs}^2\pi^2\hbar^2}{2mL^2} \tag{42.12}$$

We can combine Eqs. (42.11) and (42.12) to get a relationship between E and n that doesn't contain n_{rs}. We'll leave the details as an exercise (Exercise 42.24); the result is

$$n = \frac{(2m)^{3/2}VE^{3/2}}{3\pi^2\hbar^3} \tag{42.13}$$

42.20 A cubical box with side length L. We studied this three-dimensional version of the infinite square well in Section 41.2. The energy levels for a particle in this box are given by Eq. (42.10).

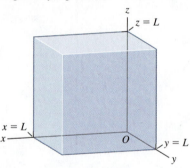

42.21 The allowed values of n_X, n_Y, and n_Z are positive integers for the electron states in the free-electron gas model. Including spin, there are two states for each unit volume in n space.

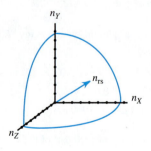

where $V = L^3$ is the volume of the box. Equation (42.13) gives the total number of states with energies of E or less.

To get the number of states dn in an energy interval dE, we treat n and E as continuous variables and take differentials of both sides of Eq. (42.13). We get

$$dn = \frac{(2m)^{3/2}VE^{1/2}}{2\pi^2\hbar^3}\, dE \qquad (42.14)$$

The density of states $g(E)$ is equal to dn/dE, so from Eq. (42.14) we get

$$g(E) = \frac{(2m)^{3/2}V}{2\pi^2\hbar^3}E^{1/2} \qquad \text{(density of states, free-electron model)} \qquad (42.15)$$

Fermi–Dirac Distribution

Now we need to know how the electrons are distributed among the various quantum states at any given temperature. The Maxwell–Boltzmann distribution states that the average number of particles in a state of energy E is proportional to $e^{-E/kT}$ (see Sections 18.5 and 39.4). However, there are two very important reasons why it wouldn't be right to use the Maxwell–Boltzmann distribution. The first reason is the exclusion principle. At absolute zero the Maxwell–Boltzmann function predicts that *all* the electrons would go into the two ground states of the system, with $n_X = n_Y = n_Z = 1$ and $m_s = \pm\frac{1}{2}$. But the exclusion principle allows only one electron in each state. At absolute zero the electrons can fill up the lowest *available* states, but there's not enough room for *all* of them to go into the lowest states. Thus a reasonable guess as to the shape of the distribution would be Fig. 42.22. At absolute zero temperature the states are filled up to some value E_{F0}, and all states above this value are empty.

The second reason we can't use the Maxwell–Boltzmann distribution is more subtle. That distribution assumes that we are dealing with *distinguishable* particles. It might seem that we could put a tag on each electron and know which is which. But overlapping electrons in a system such as a metal are *indistinguishable*. Suppose we have two electrons; a state in which the first is in energy level E_1 and the second is in level E_2 is not distinguishable from a state in which the two electrons are reversed, because we can't tell which electron is which.

The statistical distribution function that emerges from the exclusion principle and the indistinguishability requirement is called (after its inventors) the **Fermi–Dirac distribution.** Because of the exclusion principle, the probability that a particular state with energy E is occupied by an electron is the same as $f(E)$, the fraction of states with that energy that are occupied:

$$f(E) = \frac{1}{e^{(E-E_F)/kT} + 1} \qquad \text{(Fermi–Dirac distribution)} \qquad (42.16)$$

The energy E_F is called the **Fermi energy** or the *Fermi level;* we'll discuss its significance below. We use E_{F0} for its value at absolute zero ($T = 0$) and E_F for other temperatures. We can accurately let $E_F = E_{F0}$ for metals because the Fermi energy does not change much with temperature for solid conductors. However, it is not safe to assume that $E_F = E_{F0}$ for semiconductors, in which the Fermi energy usually does change with temperature.

Figure 42.23 shows graphs of Eq. (42.16) for three temperatures. The trend of this function as kT approaches zero confirms our guess. When $E = E_F$, the exponent is zero and $f(E_F) = \frac{1}{2}$. That is, the probability is $\frac{1}{2}$ that a state at the Fermi energy contains an electron. Alternatively, at $E = E_F$, half the states are filled (and half are empty).

42.22 The probability distribution for occupation of free-electron energy states at absolute zero.

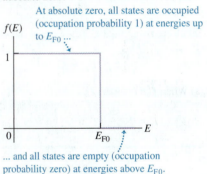

At absolute zero, all states are occupied (occupation probability 1) at energies up to E_{F0} ...

... and all states are empty (occupation probability zero) at energies above E_{F0}.

42.23 Graphs of the Fermi–Dirac distribution function for various values of kT, assuming that the Fermi energy E_F is independent of the temperature T.

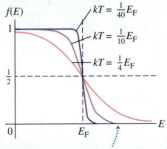

As T increases, more and more of the electrons are excited to states with energy $E > E_F$.

For $E < E_F$ the exponent is negative, and $f(E) > \frac{1}{2}$. For $E > E_F$ the exponent is positive, and $f(E) < \frac{1}{2}$. The shape depends on the ratio E_F/kT. At $T \ll E_F/k$ this ratio is very large. Then for $E < E_F$ the curve very quickly approaches 1, and for $E > E_F$ it quickly approaches zero. When T is larger, the changes are more gradual. When T is zero, all the states up to the Fermi level E_{F0} are filled, and all states above that level are empty (Fig. 42.22).

Example 42.6 Probabilities in the free-electron model

For free electrons in a solid, at what energy is the probability that a particular state is occupied equal to (a) 0.01 and (b) 0.99?

SOLUTION

IDENTIFY and SET UP: This problem asks us to explore the Fermi–Dirac distribution. Equation (42.16) gives the occupation probability $f(E)$ for a given energy E. If we solve this equation for E, we get an expression for the energy that corresponds to a given occupation probability—which is just what we need to solve this problem.

EXECUTE: Using Eq. (42.16), you can show that

$$E = E_F + kT \ln\left(\frac{1}{f(E)} - 1\right)$$

(a) When $f(E) = 0.01$,

$$E = E_F + kT \ln\left(\frac{1}{0.01} - 1\right) = E_F + 4.6kT$$

The probability that a state $4.6kT$ above the Fermi level is occupied is only 0.01, or 1%.

(b) When $f(E) = 0.99$,

$$E = E_F + kT \ln\left(\frac{1}{0.99} - 1\right) = E_F - 4.6kT$$

The probability that a state $4.6kT$ below the Fermi level is occupied is 0.99, or 99%.

EVALUATE: At very low temperatures, $4.6kT$ is much less than E_F. Then the occupation probability of levels even slightly below E_F is nearly 1 (100%), and that for levels even slightly above E_F is nearly zero (see Fig. 42.23). In general, if the probability is P that a state with an energy ΔE *above* E_F is occupied, then the probability is $1 - P$ that a state ΔE *below* E_F is occupied. We leave the proof to you (Problem 42.50).

Electron Concentration and Fermi Energy

Equation (42.16) gives the probability that any specific state with energy E is occupied at a temperature T. To get the actual number of electrons in any energy range dE, we have to multiply this probability by the number dn of states in that range $g(E)\, dE$. Thus the number dN of electrons with energies in the range dE is

$$dN = g(E)f(E)\, dE = \frac{(2m)^{3/2}VE^{1/2}}{2\pi^2\hbar^3} \frac{1}{e^{(E-E_F)/kT} + 1}\, dE \quad (42.17)$$

The Fermi energy E_F is determined by the total number N of electrons; at any temperature the electron states are filled up to a point at which all electrons are accommodated. At absolute zero there is a simple relationship between E_{F0} and N. All states below E_{F0} are filled; in Eq. (42.13) we set n equal to the total number of electrons N and E to the Fermi energy at absolute zero E_{F0}:

$$N = \frac{(2m)^{3/2}VE_{F0}^{3/2}}{3\pi^2\hbar^3} \quad (42.18)$$

Solving Eq. (42.18) for E_{F0}, we get

$$E_{F0} = \frac{3^{2/3}\pi^{4/3}\hbar^2}{2m}\left(\frac{N}{V}\right)^{2/3} \quad (42.19)$$

The quantity N/V is the number of free electrons per unit volume. It is called the *electron concentration* and is usually denoted by n.

If we replace N/V with n, Eq. (42.19) becomes

$$E_{F0} = \frac{3^{2/3}\pi^{4/3}\hbar^2 n^{2/3}}{2m} \quad (42.20)$$

CAUTION **Electron concentration and number of electrons** Don't confuse the electron concentration n with any quantum number n. Furthermore, the number of states is *not* in general the same as the total number of electrons N.

Example 42.7 The Fermi energy in copper

At low temperatures, copper has a free-electron concentration $n = 8.45 \times 10^{28}$ m^{-3}. Using the free-electron model, find the Fermi energy for solid copper, and find the speed of an electron with a kinetic energy equal to the Fermi energy.

SOLUTION

IDENTIFY and SET UP: This problem uses the relationship between Fermi energy and free-electron concentration. Because copper is a solid conductor, its Fermi energy changes very little with temperature and we can use the expression for the Fermi energy at absolute zero, Eq. (42.20). We'll find the *Fermi speed* v_F that corresponds to kinetic energy E_F using the nonrelativistic formula $E_F = \frac{1}{2}mv_F^2$.

EXECUTE: Using the given value of n, we solve for E_F and v_F:

$$E_F = \frac{3^{2/3}\pi^{4/3}(1.055 \times 10^{-34} \text{ J} \cdot \text{s})^2(8.45 \times 10^{28} \text{ m}^{-3})^{2/3}}{2(9.11 \times 10^{-31} \text{ kg})}$$

$$= 1.126 \times 10^{-18} \text{ J} = 7.03 \text{ eV}$$

$$v_F = \sqrt{\frac{2E_F}{m}} = \sqrt{\frac{2(1.126 \times 10^{-18} \text{ J})}{9.11 \times 10^{-31} \text{ kg}}} = 1.57 \times 10^6 \text{ m/s}$$

EVALUATE: Our values of E_F and v_F are within the ranges of typical values for metals, 1.6–14 eV and 0.8–2.2 $\times$ 10^6 m/s, respectively. Note that the calculated Fermi speed is far less than the speed of light $c = 3.00 \times 10^8$ m/s, which justifies our use of the nonrelativistic formula $\frac{1}{2}mv_F^2 = E_F$.

Our calculated Fermi energy is much larger than kT at ordinary temperatures. (At room temperature $T = 20°C = 293$ K, the quantity kT equals $(1.381 \times 10^{-23}$ J/K$)(293$ K$) = 4.04 \times 10^{-21}$ J $= 0.0254$ eV.) So it is a good approximation to take almost all the states below E_F as completely full and almost all those above E_F as completely empty (see Fig. 42.22).

We can also use Eq. (42.15) to find $g(E)$ if E and V are known. You can show that if $E = 7.03$ eV and $V = 1$ cm^3, $g(E)$ is about 2×10^{22} states/eV. This huge number shows why we were justified in treating n and E as continuous variables in our density-of-states derivation.

Average Free-Electron Energy

We can calculate the *average* free-electron energy in a metal at absolute zero by using the same ideas that we used to find E_{F0}. From Eq. (42.17) the number dN of electrons with energies in the range dE is $g(E)f(E)\,dE$. The energy of these electrons is $E\,dN = Eg(E)f(E)\,dE$. At absolute zero we substitute $f(E) = 1$ from $E = 0$ to $E = E_{F0}$ and $f(E) = 0$ for all other energies. Therefore the total energy E_{tot} of all the N electrons is

$$E_{tot} = \int_0^{E_{F0}} Eg(E)(1)\,dE + \int_{E_{F0}}^{\infty} Eg(E)(0)\,dE = \int_0^{E_{F0}} Eg(E)\,dE$$

The simplest way to evaluate this expression is to compare Eqs. (42.15) and (42.19), noting that

$$g(E) = \frac{3NE^{1/2}}{2E_{F0}^{3/2}}$$

Substituting this expression into the integral and using $E_{av} = E_{tot}/N$, we get

$$E_{av} = \frac{3}{2E_{F0}^{3/2}} \int_0^{E_{F0}} E^{3/2}\,dE = \tfrac{3}{5}E_{F0} \qquad (42.21)$$

At absolute zero the average free-electron energy equals $\frac{3}{5}$ of the Fermi energy.

Example 42.8 Free-electron gas versus ideal gas

(a) Find the average energy of the free electrons in copper at absolute zero (see Example 42.7). (b) What would be the average kinetic energy of electrons if they behaved like an ideal gas at room temperature, 20°C (see Section 18.3)? What would be the speed of an electron with this kinetic energy? Compare these ideal-gas values with the (correct) free-electron values.

Continued

SOLUTION

IDENTIFY and SET UP: Free electrons in a metal behave like a kind of gas. In part (a) we use Eq. (42.21) to determine the average kinetic energy of free electrons in terms of the Fermi energy at absolute zero, which we know for copper from Example 42.7. In part (b) we treat electrons as an ideal gas at room temperature: Eq. (18.16) then gives the average kinetic energy per electron as $E_{av} = \frac{3}{2}kT$, and $E_{av} = \frac{1}{2}mv^2$ gives the corresponding electron speed v.

EXECUTE: (a) From Example 42.7, the Fermi energy in copper at absolute zero is 1.126×10^{-18} J = 7.03 eV. According to Eq. (42.21), the average energy is $\frac{3}{5}$ of this, or 6.76×10^{-19} J = 4.22 eV.

(b) In Example 42.7 we found that $kT = 4.04 \times 10^{-21}$ J = 0.0254 eV at room temperature $T = 20°C = 293$ K. If electrons behaved like an ideal gas at this temperature, the average kinetic energy per electron would be $\frac{3}{2}$ of this, or 6.07×10^{-21} J = 0.0379 eV. The speed of an electron with this kinetic energy would be

$$v = \sqrt{\frac{2E_{av}}{m}} = \sqrt{\frac{2(6.07 \times 10^{-21}\,\text{J})}{9.11 \times 10^{-31}\,\text{kg}}} = 1.15 \times 10^5 \text{ m/s}$$

EVALUATE: The ideal-gas model predicts an average energy that is about 1% of the value given by the free-electron model, and a speed that is about 7% of the free-electron Fermi speed $v_F = 1.57 \times 10^6$ m/s that we found in Example 42.7. Thus temperature plays a *very* small role in determining the properties of electrons in metals; their average energies are determined almost entirely by the exclusion principle.

A similar analysis allows us to determine the contributions of electrons to the heat capacities of a solid metal. If there is one conduction electron per atom, the principle of equipartition of energy (see Section 18.4) would predict that the kinetic energies of these electrons contribute $3R/2$ to the molar heat capacity at constant volume C_V. But when kT is much smaller than E_F, which is usually the situation in metals, only those few electrons near the Fermi level can find empty states and change energy appreciably when the temperature changes. The number of such electrons is proportional to kT/E_F, so we expect that the electron molar heat capacity at constant volume is proportional to $(kT/E_F)(3R/2) = (3kT/2E_F)R$. A more detailed analysis shows that the actual electron contribution to C_V for a solid metal is $(\pi^2 kT/2E_F)R$, not far from our prediction. You can verify that if $T = 293$ K and $E_F = 7.03$ eV, the electron contribution to C_V is $0.018R$, which is only 1.2% of the (incorrect) $3R/2$ prediction of the equipartition principle. Because the electron contribution is so small, the overall heat capacity of most solid metals is due primarily to vibration of the atoms in the crystal structure (see Fig. 18.18 in Section 18.4).

Test Your Understanding of Section 42.5 An ideal gas obeys the relationship $pV = nRT$ (see Section 18.1). That is, for a given volume V and a number of moles n, as the temperature T decreases, the pressure p decreases proportionally and tends to zero as T approaches absolute zero. Is this also true of the free-electron gas in a solid metal?

PhET: Semiconductors
PhET: Conductivity

42.6 Semiconductors

A **semiconductor** has an electrical resistivity that is intermediate between those of good conductors and of good insulators. The tremendous importance of semiconductors in present-day electronics stems in part from the fact that their electrical properties are very sensitive to very small concentrations of impurities. We'll discuss the basic concepts using the semiconductor elements silicon (Si) and germanium (Ge) as examples.

Silicon and germanium are in Group IV of the periodic table. Both have four electrons in the outermost atomic subshells ($3s^2 3p^2$ for silicon, $4s^2 4p^2$ for germanium), and both crystallize in the covalently bonded diamond structure discussed in Section 42.3 (see Fig. 42.14). Because all four of the outer electrons are involved in the bonding, at absolute zero the band structure (see Section 42.4) has a completely empty conduction band (see Fig. 42.19b). As we discussed in Section 42.4, at very low temperatures electrons cannot jump from the filled valence band into the conduction band. This property makes these materials insulators at very low temperatures; their electrons have no nearby states available into which they can move in response to an applied electric field.

However, in semiconductors the energy gap E_g between the valence and conduction bands is small in comparison to the gap of 5 eV or more for many insulators; room-temperature values are 1.12 eV for silicon and only 0.67 eV for germanium. Thus even at room temperature a substantial number of electrons can gain enough energy to jump the gap to the conduction band, where they are dissociated from their parent atoms and are free to move about the crystal. The number of these electrons increases rapidly with temperature.

Example 42.9 | **Jumping a band gap**

Consider a material with the band structure described above, with its Fermi energy in the middle of the gap (Fig. 42.24). Find the probability that a state at the bottom of the conduction band is occupied at $T = 300$ K, and compare that with the probability at $T = 310$ K, for band gaps of (a) 0.200 eV; (b) 1.00 eV; (c) 5.00 eV.

SOLUTION

IDENTIFY and SET UP: The Fermi–Dirac distribution function gives the probability that a state of energy E is occupied at temperature T. Figure 42.24 shows that the state of interest at the bottom of the conduction band has an energy $E = E_F + E_g/2$ that is greater than the Fermi energy E_F, with $E - E_F = E_g/2$.

42.24 Band structure of a semiconductor. At absolute zero a completely filled valence band is separated by a narrow energy gap E_g of 1 eV or so from a completely empty conduction band. At ordinary temperatures, a number of electrons are excited to the conduction band.

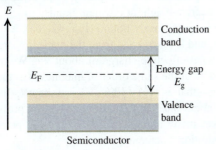

Semiconductor

Figure 42.23 shows that the higher the temperature, the larger the fraction of electrons with energies greater than the Fermi energy.

EXECUTE: (a) When $E_g = 0.200$ eV,

$$\frac{E - E_F}{kT} = \frac{E_g}{2} \frac{0.100 \text{ eV}}{(8.617 \times 10^{-5} \text{ eV/K})(300 \text{ K})} = 3.87$$

$$f(E) = \frac{1}{e^{3.87} + 1} = 0.0205$$

For $T = 310$ K, the exponent is 3.74 and $f(E) = 0.0231$, a 13% increase in probability for a temperature rise of 10 K.

(b) For $E_g = 1.00$ eV, both exponents are five times as large as in part (a), namely 19.3 and 18.7; the values of $f(E)$ are 4.0×10^{-9} and 7.4×10^{-9}. In this case the (low) probability nearly doubles with a temperature rise of 10 K.

(c) For $E_g = 5.0$ eV, the exponents are 96.7 and 93.6; the values of $f(E)$ are 1.0×10^{-42} and 2.3×10^{-41}. The (extremely low) probability increases by a factor of 23 for a 10 K temperature rise.

EVALUATE: This example illustrates two important points. First, the probability of finding an electron in a state at the bottom of the conduction band is extremely sensitive to the width of the band gap. At room temperature, the probability is about 2% for a 0.200-eV gap, a few in a thousand million for a 1.00-eV gap, and essentially zero for a 5.00-eV gap. (Pure diamond, with a 5.47-eV band gap, has essentially no electrons in the conduction band and is an excellent insulator.) Second, for any given band gap the probability depends strongly on temperature, and even more strongly for large gaps than for small ones.

In principle, we could continue the calculation in Example 42.9 to find the actual density $n = N/V$ of electrons in the conduction band at any temperature. To do this, we would have to evaluate the integral $\int g(E)f(E)\, dE$ from the bottom of the conduction band to its top. First we would need to know the density of states function $g(E)$. It wouldn't be correct to use Eq. (42.15) because the energy-level structure and the density of states for real solids are more complex than those for the simple free-electron model. However, there are theoretical methods for predicting what $g(E)$ should be near the bottom of the conduction band, and such calculations have been carried out. Once we know n, we can *begin* to determine the resistivity of the material (and its temperature dependence) using the analysis of Section 25.2, which you may want to review. But next we'll see that the electrons in the conduction band don't tell the whole story about conduction in semiconductors.

Holes

When an electron is removed from a covalent bond, it leaves a vacancy behind. An electron from a neighboring atom can move into this vacancy, leaving the neighbor with the vacancy. In this way the vacancy, called a **hole,** can travel through the material and serve as an additional current carrier. It's like describing the motion of a bubble in a liquid. In a pure, or *intrinsic*, semiconductor, valence-band holes and conduction-band electrons are always present in equal numbers. When an electric field is applied, they move in opposite directions (Fig. 42.25). Thus a hole in the valence band behaves like a positively charged particle, even though the moving charges in that band are electrons. The conductivity that we

42.25 Motion of electrons in the conduction band and of holes in the valence band of a semiconductor under the action of an applied electric field $\vec{E}$.

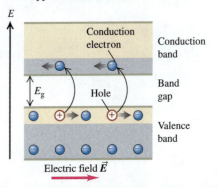

just described for a pure semiconductor is called *intrinsic conductivity.* Another kind of conductivity, to be discussed in the next subsection, is due to impurities.

An analogy helps to picture conduction in an intrinsic semiconductor. The valence band at absolute zero is like a floor of a parking garage that's filled bumper to bumper with cars (which represent electrons). No cars can move because there is nowhere for them to go. But if one car is moved to the vacant floor above, it can move freely, just as electrons can move freely in the conduction band. Also, the empty space that it leaves permits cars to move on the nearly filled floor, thereby moving the empty space just as holes move in the normally filled valence band.

Impurities

Suppose we mix into melted germanium ($Z = 32$) a small amount of arsenic ($Z = 33$), the next element after germanium in the periodic table. This deliberate addition of impurity elements is called *doping.* Arsenic is in Group V; it has *five* valence electrons. When one of these electrons is removed, the remaining electron structure is essentially identical to that of germanium. The only difference is that it is smaller; the arsenic nucleus has a charge of $+33e$ rather than $+32e$, and it pulls the electrons in a little more. An arsenic atom can comfortably take the place of a germanium atom as a substitutional impurity. Four of its five valence electrons form the necessary nearest-neighbor covalent bonds.

The fifth valence electron is very loosely bound (Fig. 42.26a); it doesn't participate in the covalent bonds, and it is screened from the nuclear charge of $+33e$ by the 32 electrons, leaving a net effective charge of about $+e$. We might guess that the binding energy would be of the same order of magnitude as the energy of the $n = 4$ level in hydrogen—that is, $(\frac{1}{4})^2(13.6\text{ eV}) = 0.85\text{ eV}$. In fact, it is much smaller than this, only about 0.01 eV, because the electron probability distribution actually extends over many atomic diameters and the polarization of intervening atoms provides additional screening.

The energy level of this fifth electron corresponds in the band picture to an isolated energy level lying in the gap, about 0.01 eV below the bottom of the conduction band (Fig. 42.26b). This level is called a *donor level,* and the impurity atom that is responsible for it is simply called a *donor.* All Group V elements, including N, P, As, Sb, and Bi, can serve as donors. At room temperature, kT is about 0.025 eV. This is substantially greater than 0.01 eV, so at ordinary temperatures, most electrons can gain enough energy to jump from donor levels into the conduction band, where they are free to wander through the material. The remaining ionized donor stays at its site in the structure and does not participate in conduction.

Example 42.9 shows that at ordinary temperatures and with a band gap of 1.0 eV, only a very small fraction (of the order of 10^{-9}) of the states at the bottom of the conduction band in a pure semiconductor contain electrons to participate in intrinsic conductivity. Thus we expect the conductivity of such a semiconductor to be about 10^{-9} as great as that of good metallic conductors, and measurements bear out this prediction. However, a concentration of donors as small as one part in 10^8 can increase the conductivity so drastically that conduction due to impurities becomes by far the dominant mechanism. In this case the conductivity is due almost entirely to *negative* charge (electron) motion. We call the material an **n-type semiconductor,** with *n*-type impurities.

Adding atoms of an element in Group III (B, Al, Ga, In, Tl), with only *three* valence electrons, has an analogous effect. An example is gallium ($Z = 31$); as a substitutional impurity in germanium, the gallium atom would like to form four covalent bonds, but it has only three outer electrons. It can, however, steal an electron from a neighboring germanium atom to complete the required four covalent bonds (Fig. 42.27a). The resulting atom has the same electron configuration as Ge but is somewhat larger because gallium's nuclear charge is smaller, $+31e$ instead of $+32e$.

42.26 An *n*-type semiconductor.

(a) A donor (*n*-type) impurity atom has a fifth valence electron that does not participate in the covalent bonding and is very loosely bound.

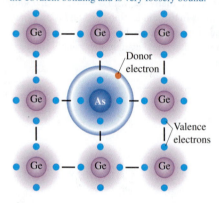

(b) Energy-band diagram for an *n*-type semiconductor at a low temperature. One donor electron has been excited from the donor levels into the conduction band.

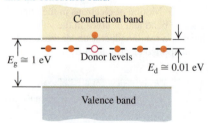

This theft leaves the neighboring atom with a *hole,* or missing electron. The hole acts as a positive charge that can move through the crystal just as with intrinsic conductivity. The stolen electron is bound to the gallium atom in a level called an *acceptor level* about 0.01 eV above the top of the valence band (Fig. 42.27b). The gallium atom, called an *acceptor,* thus accepts an electron to complete its desire for four covalent bonds. This extra electron gives the previously neutral gallium atom a net charge of $-e$. The resulting gallium ion is *not* free to move. In a semiconductor that is doped with acceptors, we consider the conductivity to be almost entirely due to *positive* charge (hole) motion. We call the material a **p-type semiconductor,** with *p*-type impurities. Some semiconductors are doped with *both* n- and p-type impurities. Such materials are called *compensated* semiconductors.

> **CAUTION** **The meaning of "p-type" and "n-type"** Saying that a material is a *p*-type semiconductor does *not* mean that the material has a positive charge; ordinarily, it would be neutral. Rather, it means that its *majority carriers* of current are positive holes (and therefore its *minority carriers* are negative electrons). The same idea holds for an *n*-type semiconductor; ordinarily, it will *not* have a negative charge, but its majority carriers are negative electrons. ▌

We can verify the assertion that the current in n- and p-type semiconductors really *is* carried by electrons and holes, respectively, by using the Hall effect (see Section 27.9). The sign of the Hall emf is opposite in the two cases. Hall-effect devices constructed from semiconductor materials are used in probes to measure magnetic fields and the currents that cause those fields.

Test Your Understanding of Section 42.6 Would there be any advantage to adding *n*-type or *p*-type impurities to copper?

42.7 Semiconductor Devices

Semiconductor devices play an indispensable role in contemporary electronics. In the early days of radio and television, transmitting and receiving equipment relied on vacuum tubes, but these have been almost completely replaced in the last six decades by solid-state devices, including transistors, diodes, integrated circuits, and other semiconductor devices. The only surviving vacuum tubes in consumer electronics are the picture tubes in older TV receivers and computer monitors; these are rapidly being replaced by flat-screen displays.

One simple semiconductor device is the *photocell* (Fig. 42.28). When a thin slab of semiconductor is irradiated with an electromagnetic wave whose photons have at least as much energy as the band gap between the valence and conduction bands, an electron in the valence band can absorb a photon and jump to the conduction band, where it and the hole it left behind contribute to the conductivity (see Example 42.5 in Section 42.4). The conductivity therefore increases with wave intensity, thus increasing the current I in the photocell circuit of Fig. 42.28. Hence the ammeter reading indicates the intensity of the light.

Detectors for charged particles operate on the same principle. An external circuit applies a voltage across a semiconductor. An energetic charged particle passing through the semiconductor collides inelastically with valence electrons, exciting them from the valence to the conduction band and creating pairs of holes and conduction electrons. The conductivity increases momentarily, causing a pulse of current in the external circuit. Solid-state detectors are widely used in nuclear and high-energy physics research.

The p-n Junction

In many semiconductor devices the essential principle is the fact that the conductivity of the material is controlled by impurity concentrations, which can be varied

42.27 A *p*-type semiconductor.

(a) An acceptor (*p*-type) impurity atom has only three valence electrons, so it can borrow an electron from a neighboring atom. The resulting hole is free to move about the crystal.

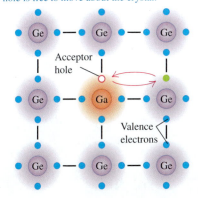

(b) Energy-band diagram for a *p*-type semiconductor at a low temperature. One acceptor level has accepted an electron from the valence band, leaving a hole behind.

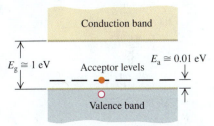

42.28 A semiconductor photocell in a circuit. The more intense the light falling on the photocell, the greater the conductivity of the photocell and the greater the current measured by the ammeter (A).

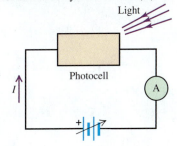

42.29 (a) A semiconductor *p-n* junction in a circuit. (b) Graph showing the asymmetric current–voltage relationship. The curve is described by Eq. (42.22).

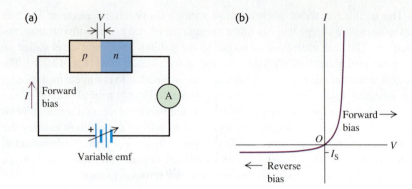

within wide limits from one region of a device to another. An example is the ***p-n* junction** at the boundary between one region of a semiconductor with *p*-type impurities and another region containing *n*-type impurities. One way of fabricating a *p-n* junction is to deposit some *n*-type material on the *very* clean surface of some *p*-type material. (We can't just stick *p*- and *n*-type pieces together and expect the junction to work properly because of the impossibility of matching their surfaces at the atomic level.)

When a *p-n* junction is connected to an external circuit, as in Fig. 42.29a, and the potential difference $V_p - V_n = V$ across the junction is varied, the current *I* varies as shown in Fig. 42.29b. In striking contrast to the symmetric behavior of resistors that obey Ohm's law and give a straight line on an *I–V* graph, a *p-n* junction conducts much more readily in the direction from *p* to *n* than the reverse. Such a (mostly) one-way device is called a **diode rectifier.** Later we'll discuss a simple model of *p-n* junction behavior that predicts a current–voltage relationship in the form

$$I = I_S(e^{eV/kT} - 1) \qquad \text{(current through a } p\text{-}n \text{ junction)} \qquad (42.22)$$

In the exponent, $e = 1.602 \times 10^{-19}$ C is the quantum of charge, *k* is Boltzmann's constant, and *T* is absolute temperature.

CAUTION **Two different uses of *e*** In $e^{eV/kT}$ the base of the exponent also uses the symbol *e*, standing for the base of the natural logarithms, 2.71828 . . . . This *e* is quite different from $e = 1.602 \times 10^{-19}$ C in the exponent. ▌

Equation (42.22) is valid for both positive and negative values of *V*; note that *V* and *I* always have the same sign. As *V* becomes very negative, *I* approaches the value $-I_S$. The magnitude I_S (always positive) is called the *saturation current*.

Currents Through a *p-n* Junction

We can understand the behavior of a *p-n* junction diode qualitatively on the basis of the mechanisms for conductivity in the two regions. Suppose, as in Fig. 42.29a, you connect the positive terminal of the battery to the *p* region and the negative terminal to the *n* region. Then the *p* region is at higher potential than the *n* region, corresponding to positive *V* in Eq. (42.22), and the resulting electric field is in the direction *p* to *n*. This is called the *forward* direction, and the positive potential difference is called *forward bias*. Holes, plentiful in the *p* region, flow easily across the junction into the *n* region, and free electrons, plentiful in the *n* region, easily flow into the *p* region; these movements of charge constitute a *forward* current. Connecting the battery with the opposite polarity gives *reverse bias,* and the field tends to push electrons from *p* to *n* and holes from *n* to *p*. But there are very few free electrons in the *p* region and very few holes in the *n* region. As a result, the current in the *reverse* direction is much smaller than that with the same potential difference in the forward direction.

42.30 A *p-n* junction in equilibrium, with no externally applied field or potential difference. The generation and recombination currents exactly balance. The Fermi energy E_F is the same on both sides of the junction. The excess positive and negative charges on the *n* and *p* sides produce an electric field $\vec{E}$ in the direction shown.

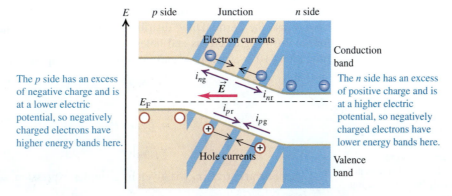

The *p* side has an excess of negative charge and is at a lower electric potential, so negatively charged electrons have higher energy bands here.

The *n* side has an excess of positive charge and is at a higher electric potential, so negatively charged electrons have lower energy bands here.

Suppose you have a box with a barrier separating the left and right sides: You fill the left side with oxygen gas and the right side with nitrogen gas. What happens if the barrier leaks? Oxygen diffuses to the right, and nitrogen diffuses to the left. A similar diffusion occurs across a *p-n* junction. First consider the equilibrium situation with no applied voltage (Fig. 42.30). The many holes in the *p* region act like a hole gas that diffuses across the junction into the *n* region. Once there, the holes recombine with some of the many free electrons. Similarly, electrons diffuse from the *n* region to the *p* region and fall into some of the many holes there. The hole and electron diffusion currents lead to a net positive charge in the *n* region and a net negative charge in the *p* region, causing an electric field in the direction from *n* to *p* at the junction. The potential energy associated with this field raises the electron energy levels in the *p* region relative to the same levels in the *n* region.

There are four currents across the junction, as shown. The diffusion processes lead to *recombination currents* of holes and electrons, labeled i_{pr} and i_{nr} in Fig. 42.30. At the same time, electron–hole pairs are generated in the junction region by thermal excitation. The electric field described above sweeps these electrons and holes out of the junction; electrons are swept opposite the field to the *n* side, and holes are swept in the same direction as the field to the *p* side. The corresponding currents, called *generation currents,* are labeled i_{pg} and i_{ng}. At equilibrium the magnitudes of the generation and recombination currents are equal:

$$|i_{pg}| = |i_{pr}| \quad \text{and} \quad |i_{ng}| = |i_{nr}| \qquad (42.23)$$

In thermal equilibrium the Fermi energy is the same at each point across the junction.

Now we apply a forward bias—that is, a positive potential difference *V* across the junction. A forward bias *decreases* the electric field in the junction region. It also decreases the difference between the energy levels on the *p* and *n* sides (Fig. 42.31)

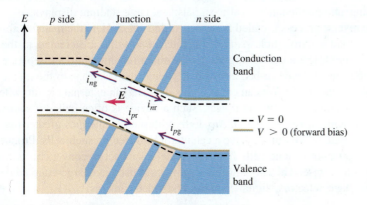

42.31 A *p-n* junction under forward-bias conditions. The potential difference between *p* and *n* regions is reduced, as is the electric field within the junction. The recombination currents increase but the generation currents are nearly constant, causing a net current from left to right. (Compare Fig. 42.30.)

by an amount $\Delta E = -eV$. It becomes easier for the electrons in the n region to climb the potential-energy hill and diffuse into the p region and for the holes in the p region to diffuse into the n region. This effect increases both recombination currents by the Maxwell–Boltzmann factor $e^{-\Delta E/kT} = e^{eV/kT}$. (We don't have to use the Fermi–Dirac distribution because most of the available states for the diffusing electrons and holes are empty, so the exclusion principle has little effect.) The generation currents don't change appreciably, so the net hole current is

$$
\begin{aligned}
i_{ptot} &= i_{pr} - |i_{pg}| \\
&= |i_{pg}|e^{eV/kT} - |i_{pg}| \\
&= |i_{pg}|(e^{eV/kT} - 1)
\end{aligned}
\tag{42.24}
$$

The net electron current i_{ntot} is given by a similar expression, so the total current $I = i_{ptot} + i_{ntot}$ is

$$
I = I_S(e^{eV/kT} - 1)
\tag{42.25}
$$

in agreement with Eq. (42.22). We can repeat this entire discussion for reverse bias (negative V and I) with the same result. Therefore Eq. (42.22) is valid for both positive and negative values.

Several effects make the behavior of practical p-n junction diodes more complex than this simple analysis predicts. One effect, *avalanche breakdown,* occurs under large reverse bias. The electric field in the junction is so great that the carriers can gain enough energy between collisions to create electron–hole pairs during inelastic collisions. The electrons and holes then gain energy and collide to form more pairs, and so on. (A similar effect occurs in dielectric breakdown in insulators, discussed in Section 42.4.)

A second type of breakdown begins when the reverse bias becomes large enough that the top of the valence band in the p region is just higher in energy than the bottom of the conduction band in the n region (Fig. 42.32). If the junction region is thin enough, the probability becomes large that electrons can *tunnel* from the valence band of the p region to the conduction band of the n region. This process is called *Zener breakdown.* It occurs in Zener diodes, which are used for voltage regulation and protection against voltage surges.

Semiconductor Devices and Light

A *light-emitting diode (LED)* is a p-n junction diode that emits light. When the junction is forward biased, many holes are pushed from their p region to the junction region, and many electrons are pushed from their n region to the junction region. In the junction region the electrons fall into holes (recombine). In recombining, the electron can emit a photon with energy approximately equal to the band gap. This energy (and therefore the photon wavelength and the color of the light) can be varied by using materials with different band gaps. Light-emitting diodes are very energy-efficient light sources and have many applications, including automobile lamps, traffic signals, and large stadium displays.

The reverse process is called the *photovoltaic effect.* Here the material absorbs photons, and electron–hole pairs are created. Pairs that are created in the p-n junction, or close enough to migrate to it without recombining, are separated by the electric field we described above that sweeps the electrons to the n side and the holes to the p side. We can connect this device to an external circuit, where it becomes a source of emf and power. Such a device is often called a *solar cell,* although sunlight isn't required. *Any* light with photon energies greater than the band gap will do. You might have a calculator powered by such cells. Production of low-cost photovoltaic cells for large-scale solar energy conversion is a very active field of research. The same basic physics is used in charge-coupled device (CCD) image detectors, digital cameras, and video cameras.

42.32 Under reverse-bias conditions the potential-energy difference between the p and n sides of a junction is greater than at equilibrium. If this difference is great enough, the bottom of the conduction band on the n side may actually be below the top of the valence band on the p side.

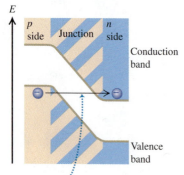

If a p-n junction under reverse bias is thin enough, electrons can tunnel from the valence band to the conduction band (a process called Zener breakdown).

Application **Swallow This Semiconductor Device**

This tiny capsule—designed to be swallowed by a patient—contains a miniature camera with a CCD light detector, plus six LEDs to illuminate the subject. The capsule radios high-resolution images to an external recording unit as it passes painlessly through the patient's stomach and intestines. This technique makes it possible to examine the small intestine, which is not readily accessible with conventional endoscopy.

Transistors

A *bipolar junction transistor* includes two *p-n* junctions in a "sandwich" configuration, which may be either *p-n-p* or *n-p-n*. Figure 42.33 shows such a *p-n-p* transistor. The three regions are called the emitter, base, and collector, as shown. When there is no current in the left loop of the circuit, there is only a very small current through the resistor R because the voltage across the base–collector junction is in the reverse direction. But when a forward bias is applied between emitter and base, as shown, most of the holes traveling from emitter to base travel *through* the base (which is typically both narrow and lightly doped) to the second junction, where they come under the influence of the collector-to-base potential difference and flow on through the collector to give an increased current to the resistor.

In this way the current in the collector circuit is *controlled* by the current in the emitter circuit. Furthermore, V_c may be considerably larger than V_e, so the *power* dissipated in R may be much larger than the power supplied to the emitter circuit by the battery V_e. Thus the device functions as a *power amplifier*. If the potential drop across R is greater than V_e, it may also be a voltage amplifier.

In this configuration the *base* is the common element between the "input" and "output" sides of the circuit. Another widely used arrangement is the *common-emitter* circuit, shown in Fig. 42.34. In this circuit the current in the collector side of the circuit is much larger than that in the base side, and the result is current amplification.

The *field-effect transistor* (Fig. 42.35) is an important type. In one variation a slab of *p*-type silicon is made with two *n*-type regions on the top, called the *source* and the *drain;* a metallic conductor is fastened to each. A third electrode called the *gate* is separated from the slab, source, and drain by an insulating layer of SiO_2. When there is no charge on the gate and a potential difference of either polarity is applied between the source and the drain, there is very little current because one of the *p-n* junctions is reverse biased.

Now we place a positive charge on the gate. With dimensions of the order of 10^{-6} m, it takes little charge to provide a substantial electric field. Thus there is very little current into or out of the gate. There aren't many free electrons in the *p*-type material, but there are some, and the effect of the field is to attract them toward the positive gate. The resulting greatly enhanced concentration of electrons near the gate (and between the two junctions) permits current to flow between the source and the drain. The current is very sensitive to the gate charge and potential, and the device functions as an amplifier. The device just described is called an *enhancement-type MOSFET* (metal-oxide-semiconductor field-effect transistor).

Integrated Circuits

A further refinement in semiconductor technology is the *integrated circuit*. By successively depositing layers of material and etching patterns to define current paths, we can combine the functions of several MOSFETs, capacitors, and resistors on a single square of semiconductor material that may be only a few millimeters on a side. An elaboration of this idea leads to *large-scale integrated circuits*. The resulting integrated circuit chips are the heart of all pocket calculators and present-day computers, large and small (Fig. 42.36).

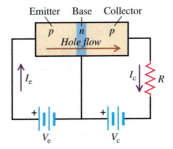

42.33 Schematic diagram of a *p-n-p* transistor and circuit.

- When $V_e = 0$, the current is very small.
- When a potential V_e is applied between emitter and base, holes travel from the emitter to the base.
- When V_c is sufficiently large, most of the holes continue into the collector.

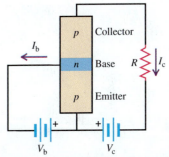

42.34 A common-emitter circuit.

- When $V_b = 0$, I_c is very small, and most of the voltage V_c appears across the base–collector junction.
- As V_b increases, the base–collector potential decreases, and more holes can diffuse into the collector; thus, I_c increases. Ordinarily, I_c is much larger than I_b.

42.35 A field-effect transistor. The current from source to drain is controlled by the potential difference between the source and the drain and by the charge on the gate; no current flows through the gate.

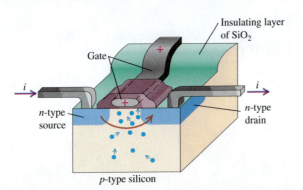

42.36 An integrated circuit chip the size of your thumb can contain millions of transistors.

The first semiconductor devices were invented in 1947. Since then, they have completely revolutionized the electronics industry through miniaturization, reliability, speed, energy usage, and cost. They have found applications in communications, computer systems, control systems, and many other areas. In transforming these areas, they have changed, and continue to change, human civilization itself.

Test Your Understanding of Section 42.7 Suppose a negative charge is placed on the gate of the MOSFET shown in Fig. 42.35. Will a substantial current flow between the source and the drain? ❙

42.8 Superconductivity

Superconductivity is the complete disappearance of all electrical resistance at low temperatures. We described this property at the end of Section 25.2 and the magnetic properties of type-I and type-II superconductors in Section 29.8. In this section we'll relate superconductivity to the structure and energy-band model of a solid.

Although superconductivity was discovered in 1911, it was not well understood on a theoretical basis until 1957. In that year, the American physicists John Bardeen, Leon Cooper, and Robert Schrieffer published the theory of superconductivity, now called the BCS theory, that was to earn them the Nobel Prize in physics in 1972. (It was Bardeen's second Nobel Prize; he shared his first for his work on the development of the transistor.) The key to the BCS theory is an interaction between *pairs* of conduction electrons, called *Cooper pairs,* caused by an interaction with the positive ions of the crystal. Here's a rough qualitative picture of what happens. A free electron exerts attractive forces on nearby positive ions, pulling them slightly closer together. The resulting slight concentration of positive charge then exerts an attractive force on another free electron with momentum opposite to the first. At ordinary temperatures this electron-pair interaction is very small in comparison to energies of thermal motion, but at very low temperatures it becomes significant.

Bound together this way, the pairs of electrons cannot *individually* gain or lose very small amounts of energy, as they would ordinarily be able to do in a partly filled conduction band. Their pairing gives an energy gap in the allowed electron quantum levels, and at low temperatures there is not enough collision energy to jump this gap. Therefore the electrons can move freely through the crystal without any energy exchange through collisions—that is, with zero resistance.

Researchers have not yet reached a consensus on whether some modification of the BCS theory can explain the properties of the high-T_C superconductors that have been discovered since 1986. There *is* evidence for pairing, but of a different sort than for conventional superconductors. Furthermore, the original pairing mechanism of the BCS theory seems too weak to explain the high transition temperatures and critical fields of these new superconductors.

Molecular bonds and molecular spectra: The principal types of molecular bonds are ionic, covalent, van der Waals, and hydrogen bonds. In a diatomic molecule the rotational energy levels are given by Eq. (42.3), where I is the moment of inertia of the molecule, m_r is its reduced mass, and r_0 is the distance between the two atoms. The vibrational energy levels are given by Eq. (42.7), where k' is the effective force constant of the interatomic force. (See Examples 42.1–42.3.)

$$E_l = l(l+1)\frac{\hbar^2}{2I} \quad (l = 0, 1, 2, \dots)$$

$$(42.3)$$

$$I = m_r r_0^2 \qquad (42.6)$$

$$m_r = \frac{m_1 m_2}{m_1 + m_2} \qquad (42.4)$$

$$E_n = \left(n + \tfrac{1}{2}\right)\hbar\omega = \left(n + \tfrac{1}{2}\right)\hbar\sqrt{\frac{k'}{m_r}}$$

$$(n = 0, 1, 2, \dots) \qquad (42.7)$$

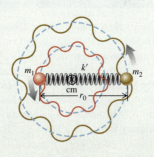

Solids and energy bands: Interatomic bonds in solids are of the same types as in molecules plus one additional type, the metallic bond. Associating the basis with each lattice point gives the crystal structure. (See Example 42.4.)

When atoms are bound together in condensed matter, their outer energy levels spread out into bands. At absolute zero, insulators and conductors have a completely filled valence band separated by an energy gap from an empty conduction band. Conductors, including metals, have partially filled conduction bands. (See Example 42.5.)

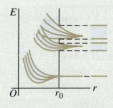

Free-electron model of metals: In the free-electron model of the behavior of conductors, the electrons are treated as completely free particles within the conductor. In this model the density of states is given by Eq. (42.15). The probability that an energy state of energy E is occupied is given by the Fermi–Dirac distribution, Eq. (42.16), which is a consequence of the exclusion principle. In Eq. (42.16), E_F is the Fermi energy. (See Examples 42.6–42.8.)

$$g(E) = \frac{(2m)^{3/2}V}{2\pi^2\hbar^3}E^{1/2} \qquad (42.15)$$

$$f(E) = \frac{1}{e^{(E-E_F)/kT} + 1} \qquad (42.16)$$

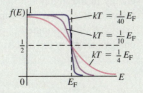

Semiconductors: A semiconductor has an energy gap of about 1 eV between its valence and conduction bands. Its electrical properties may be drastically changed by the addition of small concentrations of donor impurities, giving an n-type semiconductor, or acceptor impurities, giving a p-type semiconductor. (See Example 42.9.)

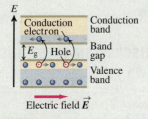

Semiconductor devices: Many semiconductor devices, including diodes, transistors, and integrated circuits, use one or more p-n junctions. The current–voltage relationship for an ideal p-n junction diode is given by Eq. (42.22).

$$I = I_S\left(e^{eV/kT} - 1\right) \qquad (42.22)$$

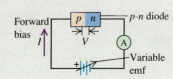

BRIDGING PROBLEM · Detecting Infrared Photons

At 80 K, the band gap in the semiconductor indium antimonide (InSb) is 0.230 eV. A photon emitted by a hydrogen fluoride (HF) molecule undergoing a vibration-rotation transition from $(n = 1, l = 0)$ to $(n = 0, l = 1)$ is absorbed by an electron at the top of the valence band of InSb. (a) How far above the top of the band gap (in eV) is the final state of the electron? (b) What is the probability that the final state was already occupied? The vibration frequency for HF is 1.24×10^{14} Hz, the mass of a hydrogen atom is 1.67×10^{-27} kg, the mass of a fluorine atom is 3.15×10^{-26} kg, and the equilibrium distance between the two nuclei is 0.092 nm. Assume that the Fermi energy for InSb is in the middle of the gap.

SOLUTION GUIDE

See MasteringPhysics® study area for a Video Tutor solution.

IDENTIFY and SET UP

1. This problem involves what you learned about molecular transitions in Section 42.2, about the Fermi–Dirac distribution in Section 42.5, and about semiconductors in Section 42.6.
2. Equation (42.9) gives the combined vibrational-rotational energy in the initial and final molecular states. The difference between the initial and final molecular energies equals the energy E of the emitted photon, which is in turn equal to the energy gained by the InSb valence electron when it absorbs that photon. The probability that the final state is occupied is given by the Fermi–Dirac distribution, Eq. (42.16).

EXECUTE

3. Before you can use Eq. (42.9), you'll first need to use the data given to calculate the moment of inertia I and the quantity $\hbar\omega$ for the HF molecule. (*Hint:* Be careful not to confuse frequency f and angular frequency ω.)
4. Use your results from step 3 to calculate the initial and final energies of the HF molecule. (*Hint:* Does the vibrational energy increase or decrease? What about the rotational energy?)
5. Use your result from step 4 to find the energy imparted to the InSb electron. Determine the final energy of this electron relative to the bottom of the conduction band.
6. Use your result from step 5 to determine the probability that the InSb final state is already occupied.

EVALUATE

7. Is the molecular transition of the HF molecule allowed? Which is larger: the vibrational energy change or the rotational energy change?
8. Is it likely that the excited InSb electron will be blocked from entering a state in the conduction band?

Problems

For instructor-assigned homework, go to www.masteringphysics.com

•, ••, •••: Problems of increasing difficulty. **CP:** Cumulative problems incorporating material from earlier chapters. **CALC:** Problems requiring calculus. **BIO:** Biosciences problems.

DISCUSSION QUESTIONS

Q42.1 Ionic bonds result from the electrical attraction of oppositely charged particles. Are other types of molecular bonds also electrical in nature, or is some other interaction involved? Explain.

Q42.2 In ionic bonds, an electron is transferred from one atom to another and thus no longer "belongs" to the atom from which it came. Are there similar transfers of ownership of electrons with other types of molecular bonds? Explain.

Q42.3 Van der Waals bonds occur in many molecules, but hydrogen bonds occur only with materials that contain hydrogen. Why is this type of bond unique to hydrogen?

Q42.4 The bonding of gallium arsenide (GaAs) is said to be 31% ionic and 69% covalent. Explain.

Q42.5 The H_2^+ molecule consists of two hydrogen nuclei and a single electron. What kind of molecular bond do you think holds this molecule together? Explain.

Q42.6 The moment of inertia for an axis through the center of mass of a diatomic molecule calculated from the wavelength emitted in an $l = 19 \rightarrow l = 18$ transition is different from the moment of inertia calculated from the wavelength of the photon emitted in an $l = 1 \rightarrow l = 0$ transition. Explain this difference. Which transition corresponds to the larger moment of inertia?

Q42.7 Analysis of the photon absorption spectrum of a diatomic molecule shows that the vibrational energy levels for small values of n are very nearly equally spaced but the levels for large n are not equally spaced. Discuss the reason for this observation. Do you expect the adjacent levels to move closer together or farther apart as n increases? Explain.

Q42.8 Discuss the differences between the rotational and vibrational energy levels of the deuterium ("heavy hydrogen") molecule D_2 and those of the ordinary hydrogen molecule H_2. A deuterium atom has twice the mass of an ordinary hydrogen atom.

Q42.9 Various organic molecules have been discovered in interstellar space. Why were these discoveries made with radio telescopes rather than optical telescopes?

Q42.10 The air you are breathing contains primarily nitrogen (N_2) and oxygen (O_2). Many of these molecules are in excited rotational energy levels ($l = 1, 2, 3, \ldots$), but almost all of them are in the vibrational ground level ($n = 0$). Explain this difference between the rotational and vibrational behaviors of the molecules.

Q42.11 In what ways do atoms in a diatomic molecule behave as though they were held together by a spring? In what ways is this a poor description of the interaction between the atoms?

Q42.12 Individual atoms have discrete energy levels, but certain solids (which are made up of only individual atoms) show energy bands and gaps. What causes the solids to behave so differently from the atoms of which they are composed?

Q42.13 What factors determine whether a material is a conductor of electricity or an insulator? Explain.

Q42.14 Ionic crystals are often transparent, whereas metallic crystals are always opaque. Why?

Q42.15 Speeds of molecules in a gas vary with temperature, whereas speeds of electrons in the conduction band of a metal are nearly independent of temperature. Why are these behaviors so different?

Q42.16 Use the band model to explain how it is possible for some materials to undergo a semiconductor-to-metal transition as the temperature or pressure varies.

Q42.17 An isolated zinc atom has a ground-state electron configuration of filled 1s, 2s, 2p, 3s, 3p, and 4s subshells. How can zinc be a conductor if its valence subshell is full?

Q42.18 The assumptions of the *free-electron model* of metals may seem contrary to reason, since electrons exert powerful electrical forces on each other. Give some reasons why these assumptions actually make physical sense.

Q42.19 Why are materials that are good thermal conductors also good electrical conductors? What kinds of problems does this pose for the design of appliances such as clothes irons and electric heaters? Are there materials that do not follow this general rule?

Q42.20 What is the essential characteristic for an element to serve as a donor impurity in a semiconductor such as Si or Ge? For it to serve as an acceptor impurity? Explain.

Q42.21 There are several methods for removing electrons from the surface of a semiconductor. Can holes be removed from the surface? Explain.

Q42.22 A student asserts that silicon and germanium become good insulators at very low temperatures and good conductors at very high temperatures. Do you agree? Explain your reasoning.

Q42.23 The electrical conductivities of most metals decrease gradually with increasing temperature, but the intrinsic conductivity of semiconductors always *increases* rapidly with increasing temperature. What causes the difference?

Q42.24 How could you make compensated silicon that has twice as many acceptors as donors?

Q42.25 For electronic devices such as amplifiers, what are some advantages of transistors compared to vacuum tubes? What are some disadvantages? Are there any situations in which vacuum tubes *cannot* be replaced by solid-state devices? Explain your reasoning.

Q42.26 Why does tunneling limit the miniaturization of MOSFETs?

Q42.27 The saturation current I_S for a p-n junction, Eq. (42.22), depends strongly on temperature. Explain why.

EXERCISES

Section 42.1 Types of Molecular Bonds

42.1 • If the energy of the H_2 covalent bond is -4.48 eV, what wavelength of light is needed to break that molecule apart? In what part of the electromagnetic spectrum does this light lie?

42.2 • **An Ionic Bond.** (a) Calculate the electric potential energy for a K^+ ion and a Br^- ion separated by a distance of 0.29 nm, the equilibrium separation in the KBr molecule. Treat the ions as point charges. (b) The ionization energy of the potassium atom is 4.3 eV. Atomic bromine has an electron affinity of 3.5 eV. Use these data and the results of part (a) to estimate the binding energy of the KBr molecule. Do you expect the actual binding energy to be higher or lower than your estimate? Explain your reasoning.

42.3 • We know from Chapter 18 that the average kinetic energy of an ideal-gas atom or molecule at Kelvin temperature T is $\frac{3}{2}kT$. For what value of T does this energy correspond to (a) the bond energy of the van der Waals bond in He_2 $(7.9 \times 10^{-4}$ eV) and (b) the bond energy of the covalent bond in H_2 (4.48 eV)? (c) The kinetic energy in a collision between molecules can go into dissociating one or both molecules, provided the kinetic energy is higher than the bond energy. At room temperature (300 K), is it likely that He_2 molecules will remain intact after a collision? What about H_2 molecules? Explain.

42.4 •• Light of wavelength 3.10 mm strikes and is absorbed by a molecule. Is this process most likely to alter the rotational, vibrational, or atomic energy levels of the molecule? Explain your reasoning. (b) If the light in part (a) had a wavelength of 207 nm, which energy levels would it most likely affect? Explain.

42.5 • For the H_2 molecule the equilibrium spacing of the two protons is 0.074 nm. The mass of a hydrogen atom is 1.67×10^{-27} kg. Calculate the wavelength of the photon emitted in the rotational transition $l = 2$ to $l = 1$.

42.6 • (a) A molecule decreases its vibrational energy by 0.250 eV by giving up a photon of light. What wavelength of light does it give up during this process, and in what part of the electromagnetic spectrum does that wavelength of light lie? (b) An atom decreases its energy by 8.50 eV by giving up a photon of light. What wavelength of light does it give up during this process, and in what part of the electromagnetic spectrum does that wavelength of light lie? (c) A molecule decreases its rotational energy by 3.20×10^{-3} eV by giving up a photon of light. What wavelength of light does it give up during this process, and in what part of the electromagnetic spectrum does that wavelength of light lie?

Section 42.2 Molecular Spectra

42.7 • A hypothetical NH molecule makes a rotational-level transition from $l = 3$ to $l = 1$ and gives off a photon of wavelength 1.780 nm in doing so. What is the separation between the two atoms in this molecule if we model them as point masses? The mass of hydrogen is 1.67×10^{-27} kg, and the mass of nitrogen is 2.33×10^{-26} kg.

42.8 • The water molecule has an $l = 1$ rotational level 1.01×10^{-5} eV above the $l = 0$ ground level. Calculate the wavelength and frequency of the photon absorbed by water when it undergoes a rotational-level transition from $l = 0$ to $l = 1$. The magnetron oscillator in a microwave oven generates microwaves with a frequency of 2450 MHz. Does this make sense, in view of the frequency you calculated in this problem? Explain.

42.9 • In Example 42.2 the moment of inertia for CO was calculated using Eq. (42.6). (a) In CO, how far is each atom from the center of mass of the molecule? (b) Use $I = m_1 r_1^2 + m_2 r_2^2$ to calculate the moment of inertia of CO about an axis through the center of mass and perpendicular to the line joining the centers of the two atoms. Does your result agree with the value obtained in Example 42.2?

42.10 • Two atoms of cesium (Cs) can form a Cs_2 molecule. The equilibrium distance between the nuclei in a Cs_2 molecule is 0.447 nm. Calculate the moment of inertia about an axis through the center of mass of the two nuclei and perpendicular to the line joining them. The mass of a cesium atom is 2.21×10^{-25} kg.

42.11 •• **CP** The rotational energy levels of CO are calculated in Example 42.2. If the energy of the rotating molecule is described by the classical expression $K = \frac{1}{2}I\omega^2$, for the $l = 1$ level what are (a) the angular speed of the rotating molecule; (b) the linear speed of each atom (use the result of Exercise 42.9); (c) the rotational period (the time for one rotation)?

42.12 • If a sodium chloride (NaCl) molecule could undergo an $n \rightarrow n - 1$ vibrational transition with no change in rotational quantum number, a photon with wavelength 20.0 μm would be emitted. The mass of a sodium atom is 3.82×10^{-26} kg, and the mass of a chlorine atom is 5.81×10^{-26} kg. Calculate the force constant k' for the interatomic force in NaCl.

42.13 • A lithium atom has mass 1.17×10^{-26} kg, and a hydrogen atom has mass 1.67×10^{-27} kg. The equilibrium separation between the two nuclei in the LiH molecule is 0.159 nm. (a) What is the difference in energy between the $l = 3$ and $l = 4$ rotational levels? (b) What is the wavelength of the photon emitted in a transition from the $l = 4$ to the $l = 3$ level?

42.14 •• When a hypothetical diatomic molecule having atoms 0.8860 nm apart undergoes a rotational transition from the $l = 2$ state to the next lower state, it gives up a photon having energy 8.841×10^{-4} eV. When the molecule undergoes a vibrational transition from one energy state to the next lower energy state, it gives up 0.2560 eV. Find the force constant of this molecule.

42.15 • (a) Show that the energy difference between rotational levels with angular-momentum quantum numbers l and $l - 1$ is $l\hbar^2/I$. (b) In terms of l, $\hbar$, and I, what is the frequency of the photon emitted in the pure rotation transition $l \rightarrow l - 1$?

42.16 • The vibrational and rotational energies of the CO molecule are given by Eq. (42.9). Calculate the wavelength of the photon absorbed by CO in each of the following vibration–rotation transitions: (a) $n = 0$, $l = 1 \rightarrow n = 1$, $l = 2$; (b) $n = 0$, $l = 2 \rightarrow n = 1$, $l = 1$; (c) $n = 0$, $l = 3 \rightarrow n = 1$, $l = 2$.

Section 42.3 Structure of Solids

42.17 • **Density of NaCl.** The spacing of adjacent atoms in a crystal of sodium chloride is 0.282 nm. The mass of a sodium atom is 3.82×10^{-26} kg, and the mass of a chlorine atom is 5.89×10^{-26} kg. Calculate the density of sodium chloride.

42.18 • Potassium bromide (KBr) has a density of 2.75×10^3 kg/m^3 and the same crystal structure as NaCl. The mass of a potassium atom is 6.49×10^{-26} kg, and the mass of a bromine atom is 1.33×10^{-25} kg. (a) Calculate the average spacing between adjacent atoms in a KBr crystal. (b) How does the value calculated in part (a) compare with the spacing in NaCl (see Exercise 42.17)? Is the relationship between the two values qualitatively what you would expect? Explain.

Section 42.4 Energy Bands

42.19 • The maximum wavelength of light that a certain silicon photocell can detect is 1.11 μm. (a) What is the energy gap (in electron volts) between the valence and conduction bands for this photocell? (b) Explain why pure silicon is opaque.

42.20 • The gap between valence and conduction bands in diamond is 5.47 eV. (a) What is the maximum wavelength of a photon that can excite an electron from the top of the valence band into the conduction band? In what region of the electromagnetic spectrum does this photon lie? (b) Explain why pure diamond is transparent and colorless. (c) Most gem diamonds have a yellow color. Explain how impurities in the diamond can cause this color.

42.21 • The gap between valence and conduction bands in silicon is 1.12 eV. A nickel nucleus in an excited state emits a gamma-ray photon with wavelength 9.31×10^{-4} nm. How many electrons can be excited from the top of the valence band to the bottom of the conduction band by the absorption of this gamma ray?

Section 42.5 Free-Electron Model of Metals

42.22 • Calculate v_{rms} for free electrons with average kinetic energy $\frac{3}{2}kT$ at a temperature of 300 K. How does your result compare to the speed of an electron with a kinetic energy equal to the Fermi energy of copper, calculated in Example 42.7? Why is there such a difference between these speeds?

42.23 • Calculate the density of states $g(E)$ for the free-electron model of a metal if $E = 7.0$ eV and $V = 1.0$ cm^3. Express your answer in units of states per electron volt.

42.24 • Supply the details in the derivation of Eq. (42.13) from Eqs. (42.11) and (42.12).

42.25 • CP Silver has a Fermi energy of 5.48 eV. Calculate the electron contribution to the molar heat capacity at constant volume of silver, C_V, at 300 K. Express your result (a) as a multiple of R and (b) as a fraction of the actual value for silver, $C_V = 25.3$ J/mol · K. (c) Is the value of C_V due principally to the electrons? If not, to what is it due? (*Hint:* See Section 18.4.)

42.26 • The Fermi energy of sodium is 3.23 eV. (a) Find the average energy E_{av} of the electrons at absolute zero. (b) What is the speed of an electron that has energy E_{av}? (c) At what Kelvin temperature T is kT equal to E_F? (This is called the *Fermi temperature* for the metal. It is approximately the temperature at which molecules in a classical ideal gas would have the same kinetic energy as the fastest-moving electron in the metal.)

42.27 •• For a solid metal having a Fermi energy of 8.500 eV, what is the probability, at room temperature, that a state having an energy of 8.520 eV is occupied by an electron?

Section 42.6 Semiconductors

42.28 • Pure germanium has a band gap of 0.67 eV. The Fermi energy is in the middle of the gap. (a) For temperatures of 250 K, 300 K, and 350 K, calculate the probability $f(E)$ that a state at the bottom of the conduction band is occupied. (b) For each temperature in part (a), calculate the probability that a state at the top of the valence band is empty.

42.29 • Germanium has a band gap of 0.67 eV. Doping with arsenic adds donor levels in the gap 0.01 eV below the bottom of the conduction band. At a temperature of 300 K, the probability is 4.4×10^{-4} that an electron state is occupied at the bottom of the conduction band. Where is the Fermi level relative to the conduction band in this case?

Section 42.7 Semiconductor Devices

42.30 •• (a) Suppose a piece of very pure germanium is to be used as a light detector by observing, through the absorption of photons, the increase in conductivity resulting from generation of electron–hole pairs. If each pair requires 0.67 eV of energy, what is the maximum wavelength that can be detected? In what portion of the spectrum does it lie? (b) What are the answers to part (a) if the material is silicon, with an energy requirement of 1.14 eV per pair, corresponding to the gap between valence and conduction bands in that element?

42.31 • CP At a temperature of 290 K, a certain p-n junction has a saturation current $I_S = 0.500$ mA. (a) Find the current at this temperature when the voltage is (i) 1.00 mV, (ii) −1.00 mV, (iii) 100 mV, and (iv) −100 mV. (b) Is there a region of applied voltage where the diode obeys Ohm's law?

42.32 • For a certain p-n junction diode, the saturation current at room temperature (20°C) is 0.750 mA. What is the resistance of this diode when the voltage across it is (a) 85.0 mV and (b) −50.0 mV?

42.33 •• (a) A forward-bias voltage of 15.0 mV produces a positive current of 9.25 mA through a *p-n* junction at 300 K. What does the positive current become if the forward-bias voltage is reduced to 10.0 mV? (b) For reverse-bias voltages of −15.0 mV and −10.0 mV, what is the reverse-bias negative current?

42.34 •• A *p-n* junction has a saturation current of 3.60 mA. (a) At a temperature of 300 K, what voltage is needed to produce a positive current of 40.0 mA? (b) For a voltage equal to the negative of the value calculated in part (a), what is the negative current?

PROBLEMS

42.35 •• A hypothetical diatomic molecule of oxygen (mass = 2.656×10^{-26} kg) and hydrogen (mass = 1.67×10^{-27} kg) emits a photon of wavelength 2.39 μm when it makes a transition from one vibrational state to the next lower state. If we model this molecule as two point masses at opposite ends of a massless spring, (a) what is the force constant of this spring, and (b) how many vibrations per second is the molecule making?

42.36 • When a diatomic molecule undergoes a transition from the $l = 2$ to the $l = 1$ rotational state, a photon with wavelength 63.8 μm is emitted. What is the moment of inertia of the molecule for an axis through its center of mass and perpendicular to the line connecting the nuclei?

42.37 •• **CP** (a) The equilibrium separation of the two nuclei in an NaCl molecule is 0.24 nm. If the molecule is modeled as charges $+e$ and $-e$ separated by 0.24 nm, what is the electric dipole moment of the molecule (see Section 21.7)? (b) The measured electric dipole moment of an NaCl molecule is 3.0×10^{-29} C·m. If this dipole moment arises from point charges $+q$ and $-q$ separated by 0.24 nm, what is q? (c) A definition of the *fractional ionic character* of the bond is q/e. If the sodium atom has charge $+e$ and the chlorine atom has charge $-e$, the fractional ionic character would be equal to 1. What is the actual fractional ionic character for the bond in NaCl? (d) The equilibrium distance between nuclei in the hydrogen iodide (HI) molecule is 0.16 nm, and the measured electric dipole moment of the molecule is 1.5×10^{-30} C·m. What is the fractional ionic character for the bond in HI? How does your answer compare to that for NaCl calculated in part (c)? Discuss reasons for the difference in these results.

42.38 • The binding energy of a potassium chloride molecule (KCl) is 4.43 eV. The ionization energy of a potassium atom is 4.3 eV, and the electron affinity of chlorine is 3.6 eV. Use these data to estimate the equilibrium separation between the two atoms in the KCl molecule. Explain why your result is only an estimate and not a precise value.

42.39 • (a) For the sodium chloride molecule (NaCl) discussed at the beginning of Section 42.1, what is the maximum separation of the ions for stability if they may be regarded as point charges? That is, what is the largest separation for which the energy of an Na$^+$ ion and a Cl$^-$ ion, calculated in this model, is lower than the energy of the two separate atoms Na and Cl? (b) Calculate this distance for the potassium bromide molecule, described in Exercise 42.2.

42.40 •• The rotational spectrum of HCl contains the following wavelengths (among others): 60.4 μm, 69.0 μm, 80.4 μm, 96.4 μm, and 120.4 μm. Use this spectrum to find the moment of inertia of the HCl molecule about an axis through the center of mass and perpendicular to the line joining the two nuclei.

42.41 • (a) Use the result of Problem 42.40 to calculate the equilibrium separation of the atoms in an HCl molecule. The mass of a chlorine atom is 5.81×10^{-26} kg, and the mass of a hydrogen

atom is 1.67×10^{-27} kg. (b) The value of l changes by ± 1 in rotational transitions. What is the value of l for the upper level of the transition that gives rise to each of the wavelengths listed in Problem 42.40? (c) What is the longest-wavelength line in the rotational spectrum of HCl? (d) Calculate the wavelengths of the emitted light for the corresponding transitions in the deuterium chloride (DCl) molecule. In this molecule the hydrogen atom in HCl is replaced by an atom of deuterium, an isotope of hydrogen with a mass of 3.34×10^{-27} kg. Assume that the equilibrium separation between the atoms is the same as for HCl.

42.42 • When a NaF molecule makes a transition from the $l = 3$ to the $l = 2$ rotational level with no change in vibrational quantum number or electronic state, a photon with wavelength 3.83 mm is emitted. A sodium atom has mass 3.82×10^{-26} kg, and a fluorine atom has mass 3.15×10^{-26} kg. Calculate the equilibrium separation between the nuclei in a NaF molecule. How does your answer compare with the value for NaCl given in Section 42.1? Is this result reasonable? Explain.

42.43 •• **CP** Consider a gas of diatomic molecules (moment of inertia I) at an absolute temperature T. If E_g is a ground-state energy and E_{ex} is the energy of an excited state, then the Maxwell–Boltzmann distribution (see Section 39.4) predicts that the ratio of the numbers of molecules in the two states is

$$\frac{n_{ex}}{n_g} = e^{-(E_{ex} - E_g)/kT}$$

(a) Explain why the ratio of the number of molecules in the *l*th rotational energy *level* to the number of molecules in the ground ($l = 0$) rotational level is

$$\frac{n_l}{n_0} = (2l + 1)e^{-[l(l+1)\hbar^2]/2IkT}$$

(*Hint:* For each value of l, how many states are there with different values of m_l?) (b) Determine the ratio n_l/n_0 for a gas of CO molecules at 300 K for the cases (i) $l = 1$; (ii) $l = 2$; (iii) $l = 10$; (iv) $l = 20$; (v) $l = 50$. The moment of inertia of the CO molecule is given in Example 42.2 (Section 42.2). (c) Your results in part (b) show that as l is increased, the ratio n_l/n_0 first increases and then decreases. Explain why.

42.44 •• Our galaxy contains numerous *molecular clouds,* regions many light-years in extent in which the density is high enough and the temperature low enough for atoms to form into molecules. Most of the molecules are H$_2$, but a small fraction of the molecules are carbon monoxide (CO). Such a molecular cloud in the constellation Orion is shown in Fig. P42.44. The left-hand image was made with an ordinary visible-light telescope; the right-hand image shows the molecular cloud in Orion as imaged with a radio telescope tuned to a wavelength emitted by CO in a rotational transition. The different colors in the radio image indicate regions of the cloud that are moving either toward us (blue) or away from us (red) relative to the motion of the cloud as a whole, as determined by the Doppler shift of the radiation. (Since a

Figure **P42.44**

molecular cloud has about 10,000 hydrogen molecules for each CO molecule, it might seem more reasonable to tune a radio telescope to emissions from H_2 than to emissions from CO. Unfortunately, it turns out that the H_2 molecules in molecular clouds do not radiate in either the radio or visible portions of the electromagnetic spectrum.) (a) Using the data in Example 42.2 (Section 42.2), calculate the energy and wavelength of the photon emitted by a CO molecule in an $l = 1 \rightarrow l = 0$ rotational transition. (b) As a rule, molecules in a gas at temperature T will be found in a certain excited rotational energy level provided the energy of that level is no higher than kT (see Problem 42.43). Use this rule to explain why astronomers can detect radiation from CO in molecular clouds even though the typical temperature of a molecular cloud is a very low 20 K.

42.45 • **Spectral Lines from Isotopes.** The equilibrium separation for NaCl is 0.2361 nm. The mass of a sodium atom is 3.8176×10^{-26} kg. Chlorine has two stable isotopes, ^{35}Cl and ^{37}Cl, that have different masses but identical chemical properties. The atomic mass of ^{35}Cl is 5.8068×10^{-26} kg, and the atomic mass of ^{37}Cl is 6.1384×10^{-26} kg. (a) Calculate the wavelength of the photon emitted in the $l = 2 \rightarrow l = 1$ and $l = 1 \rightarrow l = 0$ transitions for Na^{35}Cl. (b) Repeat part (a) for Na^{37}Cl. What are the differences in the wavelengths for the two isotopes?

42.46 • When an OH molecule undergoes a transition from the $n = 0$ to the $n = 1$ vibrational level, its internal vibrational energy increases by 0.463 eV. Calculate the frequency of vibration and the force constant for the interatomic force. (The mass of an oxygen atom is 2.66×10^{-26} kg, and the mass of a hydrogen atom is 1.67×10^{-27} kg.)

42.47 • The force constant for the internuclear force in a hydrogen molecule (H_2) is $k' = 576$ N/m. A hydrogen atom has mass 1.67×10^{-27} kg. Calculate the zero-point vibrational energy for H_2 (that is, the vibrational energy the molecule has in the $n = 0$ ground vibrational level). How does this energy compare in magnitude with the H_2 bond energy of -4.48 eV?

42.48 • Suppose the hydrogen atom in HF (see the Bridging Problem for this chapter) is replaced by an atom of deuterium, an isotope of hydrogen with a mass of 3.34×10^{-27} kg. The force constant is determined by the electron configuration, so it is the same as for the normal HF molecule. (a) What is the vibrational frequency of this molecule? (b) What wavelength of light corresponds to the energy difference between the $n = 1$ and $n = 0$ levels? In what region of the spectrum does this wavelength lie?

42.49 • The hydrogen iodide (HI) molecule has equilibrium separation 0.160 nm and vibrational frequency 6.93×10^{13} Hz. The mass of a hydrogen atom is 1.67×10^{-27} kg, and the mass of an iodine atom is 2.11×10^{-25} kg. (a) Calculate the moment of inertia of HI about a perpendicular axis through its center of mass. (b) Calculate the wavelength of the photon emitted in each of the following vibration–rotation transitions: (i) $n = 1$, $l = 1 \rightarrow n = 0$, $l = 0$; (ii) $n = 1$, $l = 2 \rightarrow n = 0$, $l = 1$; (iii) $n = 2$, $l = 2 \rightarrow n = 1$, $l = 3$.

42.50 •• Prove this statement: For free electrons in a solid, if a state that is at an energy ΔE above E_F has probability P of being occupied, then the probability is $1 - P$ that a state at an energy ΔE below E_F is occupied.

42.51 •• Compute the Fermi energy of potassium by making the simple approximation that each atom contributes one free electron. The density of potassium is 851 kg/m^3, and the mass of a single potassium atom is 6.49×10^{-26} kg.

42.52 •• Hydrogen is found in two naturally occurring isotopes; normal hydrogen (containing a single proton in its nucleus) and deu-

terium (having a proton and a neutron). Assuming that both molecules are the same size and that the proton and neutron have the same mass (which is almost the case), find the ratio of (a) the energy of any given rotational state in a diatomic hydrogen molecule to the energy of the same state in a diatomic deuterium molecule and (b) the energy of any given vibrational state in hydrogen to the same state in deuterium (assuming that the force constant is the same for both molecules). Why is it physically reasonable that the force constant would be the same for hydrogen and deuterium molecules?

42.53 ••• Metallic lithium has a bcc crystal structure. Each unit cell is a cube of side length $a = 0.35$ nm. (a) For a bcc lattice, what is the number of atoms per unit volume? Give your answer in terms of a. (*Hint:* How many atoms are there per unit cell?) (b) Use the result of part (a) to calculate the zero-temperature Fermi energy E_{F0} for metallic lithium. Assume there is one free electron per atom.

42.54 •• **CALC** The one-dimensional calculation of Example 42.4 (Section 42.3) can be extended to three dimensions. For the three-dimensional fcc NaCl lattice, the result for the potential energy of a pair of Na$^+$ and Cl$^-$ ions due to the electrostatic interaction with all of the ions in the crystal is $U = -\alpha e^2/4\pi\epsilon_0 r$, where $\alpha = 1.75$ is the *Madelung constant*. Another contribution to the potential energy is a repulsive interaction at small ionic separation r due to overlap of the electron clouds. This contribution can be represented by A/r^8, where A is a positive constant, so the expression for the total potential energy is

$$U_{\text{tot}} = -\frac{\alpha e^2}{4\pi\epsilon_0 r} + \frac{A}{r^8}$$

(a) Let r_0 be the value of the ionic separation r for which U_{tot} is a minimum. Use this definition to find an equation that relates r_0 and A, and use this to write U_{tot} in terms of r_0. For NaCl, $r_0 = 0.281$ nm. Obtain a numerical value (in electron volts) of U_{tot} for NaCl. (b) The quantity $-U_{\text{tot}}$ is the energy required to remove a Na$^+$ ion and a Cl$^-$ ion from the crystal. Forming a pair of neutral atoms from this pair of ions involves the release of 5.14 eV (the ionization energy of Na) and the expenditure of 3.61 eV (the electron affinity of Cl). Use the result of part (a) to calculate the energy required to remove a pair of neutral Na and Cl atoms from the crystal. The experimental value for this quantity is 6.39 eV; how well does your calculation agree?

42.55 •• **CALC** Consider a system of N free electrons within a volume V. Even at absolute zero, such a system exerts a pressure p on its surroundings due to the motion of the electrons. To calculate this pressure, imagine that the volume increases by a small amount dV. The electrons will do an amount of work $p\,dV$ on their surroundings, which means that the total energy E_{tot} of the electrons will change by an amount $dE_{\text{tot}} = -p\,dV$. Hence $p = -dE_{\text{tot}}/dV$. (a) Show that the pressure of the electrons at absolute zero is

$$p = \frac{3^{2/3}\pi^{4/3}\hbar^2}{5m}\left(\frac{N}{V}\right)^{5/3}$$

(b) Evaluate this pressure for copper, which has a free-electron concentration of 8.45×10^{28} m^{-3}. Express your result in pascals and in atmospheres. (c) The pressure you found in part (b) is extremely high. Why, then, don't the electrons in a piece of copper simply explode out of the metal?

42.56 •• **CALC** When the pressure p on a material increases by an amount Δp, the volume of the material will change from V to $V + \Delta V$, where ΔV is negative. The *bulk modulus B* of the material is defined to be the ratio of the pressure change Δp to the absolute value $|\Delta V/V|$ of the fractional volume change. The greater the bulk modulus, the greater the pressure increase required for a given fractional volume change, and the more incompressible the material (see Section 11.4). Since $\Delta V < 0$, the bulk modulus can be written as $B = -\Delta p/(\Delta V/V_0)$. In the limit that the pressure and volume changes are very small, this becomes

$$B = -V\frac{dp}{dV}$$

(a) Use the result of Problem 42.55 to show that the bulk modulus for a system of N free electrons in a volume V at low temperatures is $B = \frac{5}{3}p$. (*Hint:* The quantity p in the expression $B = -V(dp/dV)$ is the *external* pressure on the system. Can you explain why this is equal to the *internal* pressure of the system itself, as found in Problem 42.55?) (b) Evaluate the bulk modulus for the electrons in copper, which has a free-electron concentration of 8.45×10^{28} m^{-3}. Express your result in pascals. (c) The actual bulk modulus of copper is 1.4×10^{11} Pa. Based on your result in part (b), what fraction of this is due to the free electrons in copper? (This result shows that the free electrons in a metal play a major role in making the metal resistant to compression.) What do you think is responsible for the remaining fraction of the bulk modulus?

42.57 •• In the discussion of free electrons in Section 42.5, we assumed that we could ignore the effects of relativity. This is not a safe assumption if the Fermi energy is greater than about $\frac{1}{100}mc^2$ (that is, more than about 1% of the rest energy of an electron). (a) Assume that the Fermi energy at absolute zero, as given by Eq. (42.19), is equal to $\frac{1}{100}mc^2$. Show that the electron concentration is

$$\frac{N}{V} = \frac{2^{3/2}m^3c^3}{3000\pi^2\hbar^3}$$

and determine the numerical value of N/V. (b) Is it a good approximation to ignore relativistic effects for electrons in a metal such as copper, for which the electron concentration is 8.45×10^{28} m^{-3}? Explain. (c) A *white dwarf star* is what is left behind by a star like the sun after it has ceased to produce energy by nuclear reactions. (Our own sun will become a white dwarf star in another 6×10^9 years or so.) A typical white dwarf has mass 2×10^{30} kg (comparable to the sun) and radius 6000 km (comparable to that of the earth). The gravitational attraction of different parts of the white dwarf for each other tends to compress the star; what prevents it from compressing is the pressure of free electrons within the star (see Problem 42.55). Estimate the electron concentration within a typical white dwarf star using the following assumptions: (i) the white dwarf star is made of carbon, which has a mass per atom of

1.99×10^{-26} kg; and (ii) all six of the electrons from each carbon atom are able to move freely throughout the star. (d) Is it a good approximation to ignore relativistic effects in the structure of a white dwarf star? Explain.

42.58 •• **CP** A variable DC battery is connected in series with a 125-Ω resistor and a *p-n* junction diode that has a saturation current of 0.625 mA at room temperature (20°C). When a voltmeter across the 125-Ω resistor reads 35.0 V, what are (a) the voltage across the diode and (b) the resistance of the diode?

CHALLENGE PROBLEMS

42.59 ••• **CP** Van der Waals bonds arise from the interaction between two permanent or induced electric dipole moments in a pair of atoms or molecules. (a) Consider two identical dipoles, each consisting of charges $+q$ and $-q$ separated by a distance d and oriented as shown in Fig. P42.59a. Calculate the electric potential energy, expressed in terms of the electric dipole moment $p = qd$, for the situation where $r \gg d$. Is the interaction attractive or repulsive, and how does this potential energy vary with r, the separation between the centers of the two dipoles? (b) Repeat part (a) for the orientation of the dipoles shown in Fig. P42.59b. The dipole interaction is more complicated when we have to average over the relative orientations of the two dipoles due to thermal motion or when the dipoles are induced rather than permanent.

Figure **P42.59**

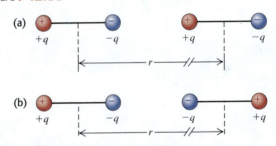

42.60 ••• **CP CALC** (a) Consider the hydrogen molecule (H_2) to be a simple harmonic oscillator with an equilibrium spacing of 0.074 nm, and estimate the vibrational energy-level spacing for H_2. The mass of a hydrogen atom is 1.67×10^{-27} kg. (*Hint:* Estimate the force constant by equating the change in Coulomb repulsion of the protons, when the atoms move slightly closer together than r_0, to the "spring" force. That is, assume that the chemical binding force remains approximately constant as r is decreased slightly from r_0.) (b) Use the results of part (a) to calculate the vibrational energy-level spacing for the deuterium molecule, D_2. Assume that the spring constant is the same for D_2 as for H_2. The mass of a deuterium atom is 3.34×10^{-27} kg.

Answers

Chapter Opening Question ?

Venus must radiate energy into space at the same rate that it receives energy in the form of sunlight. However, carbon dioxide (CO_2) molecules in the atmosphere absorb infrared radiation

emitted by the surface of Venus and re-emit it toward the ground. To compensate for this and to maintain the balance between emitted and received energy, the surface temperature of Venus and hence the rate of blackbody radiation from the surface both increase.

Test Your Understanding Questions

42.1 Answer: (i) The exclusion principle states that only one electron can be in a given state. Real electrons have spin, so two electrons (one spin up, one spin down) can be in a given *spatial* state and hence two can participate in a given covalent bond between two atoms. If electrons obeyed the exclusion principle but did not have spin, that state of an electron would be completely described by its spatial distribution and only *one* electron could participate in a covalent bond. (We will learn in Chapter 44 that this situation is wholly imaginary: There are subatomic particles without spin, but they do *not* obey the exclusion principle.)

42.2 Answer: (ii) Figure 42.5 shows that the difference in energy between adjacent rotational levels increases with increasing l. Hence, as l increases, the energy E of the emitted photon increases and the wavelength $\lambda = hc/E$ decreases.

42.3 Answer: (ii) In Fig. 42.13 let a be the distance between adjacent Na^+ and Cl^- ions. This figure shows that the Cl^- ion that is the next nearest neighbor to a Na^+ ion is on the opposite corner of a cube of side a. The distance between these two ions is $\sqrt{a^2 + a^2 + a^2} = \sqrt{3a^2} = a\sqrt{3}$.

42.4 Answer: (ii) A small temperature change causes a substantial increase in the population of electrons in a semiconductor's conduction band and a comparably substantial increase in conductivity. The conductivity of conductors and insulators varies more gradually with temperature.

42.5 Answer: no The kinetic-molecular model of an ideal gas (see Section 18.3) shows that the gas pressure is proportional to the average translational kinetic energy E_{av} of the particles that make up the gas. In a classical ideal gas, E_{av} is directly proportional to the average temperature T, so the pressure decreases as T decreases. In a free-electron gas, the average kinetic energy per electron is *not* related simply to T; as Example 42.8 shows, for the free-electron gas in a metal, E_{av} is almost completely a consequence of the exclusion principle at room temperature and colder. Hence the pressure of a free-electron gas in a solid metal does *not* change appreciably between room temperature and absolute zero.

42.6 Answer: no Pure copper is already an excellent conductor since it has a partially filled conduction band (Fig. 42.19c). Furthermore, copper forms a metallic crystal (Fig. 42.15) as opposed to the covalent crystals of silicon or germanium, so the scheme of using an impurity to donate or accept an electron does not work for copper. In fact, adding impurities to copper *decreases* the conductivity because an impurity tends to scatter electrons, impeding the flow of current.

42.7 Answer: no A negative charge on the gate will repel, not attract, electrons in the *p*-type silicon. Hence the electron concentration in the region between the two *p-n* junctions will be made even smaller. With so few charge carriers present in this region, very little current will flow between the source and the drain.

Bridging Problem

Answers: **(a)** 0.278 eV
(b) 1.74×10^{-25}

NUCLEAR PHYSICS

? This sculpture of a wooly mammoth, just 3.7 cm (1.5 in.) in length, was carved from a mammoth's ivory tusk by an artist who lived in southwestern Germany 35,000 years ago. What physical principles make it possible to date biological specimens such as these?

LEARNING GOALS

By studying this chapter, you will learn:

- Some key properties of atomic nuclei, including radii, densities, spins, and magnetic moments.

- How the binding energy of a nucleus depends on the numbers of protons and neutrons that it contains.

- The most important ways in which unstable nuclei undergo radioactive decay.

- How the decay rate of a radioactive substance depends on time.

- Some of the biological hazards and medical uses of radiation.

- How to analyze some important types of nuclear reactions.

- What happens in a nuclear fission chain reaction, and how it can be controlled.

- The sequence of nuclear reactions that allow the sun and stars to shine.

D uring the past century, applications of nuclear physics have had enormous effects on humankind, some beneficial, some catastrophic. Many people have strong opinions about applications such as bombs and reactors. Ideally, those opinions should be based on understanding, not on prejudice or emotion, and we hope this chapter will help you to reach that ideal.

Every atom contains at its center an extremely dense, positively charged *nucleus,* which is much smaller than the overall size of the atom but contains most of its total mass. We will look at several important general properties of nuclei and of the nuclear force that holds them together. The stability or instability of a particular nucleus is determined by the competition between the attractive nuclear force among the protons and neutrons and the repulsive electrical interactions among the protons. Unstable nuclei *decay,* transforming themselves spontaneously into other nuclei by a variety of processes. Nuclear reactions can also be induced by impact on a nucleus of a particle or another nucleus. Two classes of reactions of special interest are *fission* and *fusion.* We could not survive without the energy released by one nearby fusion reactor, our sun.

43.1 Properties of Nuclei

As we described in Section 39.2, Rutherford found that the nucleus is tens of thousands of times smaller in radius than the atom itself. Since Rutherford's initial experiments, many additional scattering experiments have been performed, using high-energy protons, electrons, and neutrons as well as alpha particles (helium-4 nuclei). These experiments show that we can model a nucleus as a sphere with a radius R that depends on the total number of *nucleons* (neutrons and protons) in the nucleus. This number is called the **nucleon number** A. The radii of most nuclei are represented quite well by the equation

$$R = R_0 A^{1/3} \quad \text{(radius of a nucleus)} \tag{43.1}$$

where R_0 is an experimentally determined constant:

$$R_0 = 1.2 \times 10^{-15} \text{ m} = 1.2 \text{ fm}$$

The nucleon number A in Eq. (43.1) is also called the **mass number** because it is the nearest whole number to the mass of the nucleus measured in unified atomic mass units (u). (The proton mass and the neutron mass are both approximately 1 u.) The best current conversion factor is

$$1 \text{ u} = 1.660538782(83) \times 10^{-27} \text{ kg}$$

In Section 43.2 we'll discuss the masses of nuclei in more detail. Note that when we speak of the masses of nuclei and particles, we mean their *rest* masses.

Nuclear Density

The volume V of a sphere is equal to $4\pi R^3/3$, so Eq. (43.1) shows that the *volume* of a nucleus is proportional to A. Dividing A (the approximate mass in u) by the volume gives us the approximate density and cancels out A. Thus *all nuclei have approximately the same density*. This fact is of crucial importance in understanding nuclear structure.

Example 43.1   Calculating nuclear properties

The most common kind of iron nucleus has mass number $A = 56$. Find the radius, approximate mass, and approximate density of the nucleus.

SOLUTION

IDENTIFY and SET UP: Equation (43.1) tells us how the nuclear radius R depends on the mass number A. The mass of the nucleus in atomic mass units is approximately equal to the value of A, and the density ρ is mass divided by volume.

EXECUTE: The radius and approximate mass are

$$R = R_0 A^{1/3} = (1.2 \times 10^{-15} \text{ m})(56)^{1/3}$$

$$= 4.6 \times 10^{-15} \text{ m} = 4.6 \text{ fm}$$

$$m \approx (56 \text{ u})(1.66 \times 10^{-27} \text{ kg/u}) = 9.3 \times 10^{-26} \text{ kg}$$

The volume V of the nucleus (which we treat as a sphere of radius R) and its density ρ are

$$V = \tfrac{4}{3}\pi R^3 = \tfrac{4}{3}\pi R_0^3 A = \tfrac{4}{3}\pi(4.6 \times 10^{-15} \text{ m})^3$$

$$= 4.1 \times 10^{-43} \text{ m}^3$$

$$\rho = \frac{m}{V} \approx \frac{9.3 \times 10^{-26} \text{ kg}}{4.1 \times 10^{-43} \text{ m}^3} = 2.3 \times 10^{17} \text{ kg/m}^3$$

EVALUATE: As we mentioned above, *all* nuclei have approximately this same density. The density of solid iron is about 7000 kg/m³; the iron nucleus is more than 10^{13} times as dense as iron in bulk. Such densities are also found in *neutron stars*, which are similar to gigantic nuclei made almost entirely of neutrons. A 1-cm cube of material with this density would have a mass of 2.3×10^{11} kg, or 230 million metric tons!

Nuclides and Isotopes

The building blocks of the nucleus are the proton and the neutron. In a neutral atom, the nucleus is surrounded by one electron for every proton in the nucleus. We introduced these particles in Section 21.1; we'll recount the discovery of the neutron and proton in Chapter 44. The masses of these particles are

Proton: $m_p = 1.007276 \text{ u} = 1.672622 \times 10^{-27} \text{ kg}$

Neutron: $m_n = 1.008665 \text{ u} = 1.674927 \times 10^{-27} \text{ kg}$

Electron: $m_e = 0.000548580 \text{ u} = 9.10938 \times 10^{-31} \text{ kg}$

The number of protons in a nucleus is the **atomic number** Z. The number of neutrons is the **neutron number** N. The nucleon number or mass number A is the sum of the number of protons Z and the number of neutrons N:

$$A = Z + N \qquad (43.2)$$

A single nuclear species having specific values of both Z and N is called a **nuclide.** Table 43.1 lists values of A, Z, and N for some nuclides. The electron structure of an atom, which is responsible for its chemical properties, is determined by the charge Ze of the nucleus. The table shows some nuclides that have the same Z but different N. These nuclides are called **isotopes** of that element; they have different masses because they have different numbers of neutrons in their nuclei. A familiar example is chlorine (Cl, $Z = 17$). About 76% of chlorine nuclei have $N = 18$; the other 24% have $N = 20$. Different isotopes of an element usually have slightly different physical properties such as melting and boiling temperatures and diffusion rates. The two common isotopes of uranium with $A = 235$ and 238 are usually separated industrially by taking advantage of the different diffusion rates of gaseous uranium hexafluoride (UF_6) containing the two isotopes.

Table 43.1 also shows the usual notation for individual nuclides: the symbol of the element, with a pre-subscript equal to Z and a pre-superscript equal to the mass number A. The general format for an element El is ^A_ZEl. The isotopes of chlorine mentioned above, with $A = 35$ and 37, are written $^{35}_{17}\text{Cl}$ and $^{37}_{17}\text{Cl}$ and pronounced "chlorine-35" and "chlorine-37," respectively. This name of the element determines the atomic number Z, so the pre-subscript Z is sometimes omitted, as in ^{35}Cl.

Table 43.2 gives the masses of some common atoms, including their electrons. Note that this table gives masses of *neutral* atoms (with Z electrons) rather than masses of *bare* nuclei, because it is much more difficult to measure masses of bare nuclei with high precision. The mass of a neutral carbon-12 atom is exactly 12 u; that's how the unified atomic mass unit is defined. The masses of other atoms are *approximately* equal to A atomic mass units, as we stated earlier. In fact, the atomic masses are *less* than the sum of the masses of their parts (the Z protons, the Z electrons, and the N neutrons). We'll explain this very important mass difference in the next section.

Table 43.1 Compositions of Some Common Nuclides

Z = atomic number (number of protons)

N = neutron number

$A = Z + N$ = mass number (total number of nucleons)

Nucleus	Z	N	$A = Z + N$
^1_1H	1	0	1
^2_1H	1	1	2
^4_2He	2	2	4
^6_3Li	3	3	6
^7_3Li	3	4	7
^9_4Be	4	5	9
$^{10}_5\text{B}$	5	5	10
$^{11}_5\text{B}$	5	6	11
$^{12}_6\text{C}$	6	6	12
$^{13}_6\text{C}$	6	7	13
$^{14}_7\text{N}$	7	7	14
$^{16}_8\text{O}$	8	8	16
$^{23}_{11}\text{Na}$	11	12	23
$^{65}_{29}\text{Cu}$	29	36	65
$^{200}_{80}\text{Hg}$	80	120	200
$^{235}_{92}\text{U}$	92	143	235
$^{238}_{92}\text{U}$	92	146	238

Table 43.2 **Neutral Atomic Masses for Some Light Nuclides**

Element and Isotope	Atomic Number, Z	Neutron Number, N	Atomic Mass (u)	Mass Number, A
Hydrogen ($^{1}_{1}$H)	1	0	1.007825	1
Deuterium ($^{2}_{1}$H)	1	1	2.014102	2
Tritium ($^{3}_{1}$H)	1	2	3.016049	3
Helium ($^{3}_{2}$He)	2	1	3.016029	3
Helium ($^{4}_{2}$He)	2	2	4.002603	4
Lithium ($^{6}_{3}$Li)	3	3	6.015122	6
Lithium ($^{7}_{3}$Li)	3	4	7.016004	7
Beryllium ($^{9}_{4}$Be)	4	5	9.012182	9
Boron ($^{10}_{5}$B)	5	5	10.012937	10
Boron ($^{11}_{5}$B)	5	6	11.009305	11
Carbon ($^{12}_{6}$C)	6	6	12.000000	12
Carbon ($^{13}_{6}$C)	6	7	13.003355	13
Nitrogen ($^{14}_{7}$N)	7	7	14.003074	14
Nitrogen ($^{15}_{7}$N)	7	8	15.000109	15
Oxygen ($^{16}_{8}$O)	8	8	15.994915	16
Oxygen ($^{17}_{8}$O)	8	9	16.999132	17
Oxygen ($^{18}_{8}$O)	8	10	17.999160	18

Source: A. H. Wapstra and G. Audi, *Nuclear Physics* **A595**, 4 (1995).

Nuclear Spins and Magnetic Moments

Like electrons, protons and neutrons are also spin-$\frac{1}{2}$ particles with spin angular momenta given by the same equations as in Section 41.5. The magnitude of the spin angular momentum $\vec{S}$ of a nucleon is

$$S = \sqrt{\tfrac{1}{2}\left(\tfrac{1}{2} + 1\right)}\hbar = \sqrt{\tfrac{3}{4}}\hbar \tag{43.3}$$

and the z-component is

$$S_z = \pm\tfrac{1}{2}\hbar \tag{43.4}$$

In addition to the spin angular momentum of the nucleons, there may be *orbital* angular momentum associated with their motions within the nucleus. The orbital angular momentum of the nucleons is quantized in the same way as that of electrons in atoms.

The *total* angular momentum $\vec{J}$ of the nucleus is the vector sum of the individual spin and orbital angular momenta of all the nucleons. It has magnitude

$$J = \sqrt{j(j + 1)}\hbar \tag{43.5}$$

and z-component

$$J_z = m_j\hbar \qquad (m_j = -j, -j + 1, \ldots, j - 1, j) \tag{43.6}$$

When the total number of nucleons A is *even*, j is an integer; when it is *odd*, j is a half-integer. All nuclides for which both Z and N are even have $J = 0$, which suggests that pairing of particles with opposite spin components may be an important consideration in nuclear structure. The total nuclear angular momentum quantum number j is usually called the *nuclear spin*, even though in general it refers to a combination of the orbital and spin angular momenta of the nucleons that make up the nucleus.

Associated with nuclear angular momentum is a *magnetic moment*. When we discussed *electron* magnetic moments in Section 41.4, we introduced the Bohr magneton $\mu_B = e\hbar/2m_e$ as a natural unit of magnetic moment. We found that the magnitude of the z-component of the electron-spin magnetic moment is almost exactly equal to μ_B; that is, $|\mu_{sz}|_{\text{electron}} \approx \mu_B$. In discussing *nuclear* magnetic moments, we can define an analogous quantity, the **nuclear magneton** μ_n:

$$\mu_n = \frac{e\hbar}{2m_p} = 5.05078 \times 10^{-27} \, \text{J/T} = 3.15245 \times 10^{-8} \, \text{eV/T} \qquad \text{(43.7)}$$

(nuclear magneton)

where m_p is the proton mass. Because the proton mass m_p is 1836 times larger than the electron mass m_e, the nuclear magneton μ_n is 1836 times smaller than the Bohr magneton μ_B.

We might expect the magnitude of the z-component of the spin magnetic moment of the proton to be approximately μ_n. Instead, it turns out to be

$$|\mu_{sz}|_{\text{proton}} = 2.7928\mu_n \qquad \text{(43.8)}$$

Even more surprising, the neutron, which has zero charge, has a spin magnetic moment; its z-component has magnitude

$$|\mu_{sz}|_{\text{neutron}} = 1.9130\mu_n \qquad \text{(43.9)}$$

The proton has a positive charge; as expected, its spin magnetic moment $\vec{\mu}$ is parallel to its spin angular momentum $\vec{S}$. However, $\vec{\mu}$ and $\vec{S}$ are opposite for a neutron, as would be expected for a *negative* charge distribution. These *anomalous* magnetic moments arise because the proton and neutron aren't really fundamental particles but are made of simpler particles called *quarks*. We'll discuss quarks in some detail in Chapter 44.

The magnetic moment of an entire nucleus is typically a few nuclear magnetons. When a nucleus is placed in an external magnetic field $\vec{B}$, there is an interaction energy $U = -\vec{\mu} \cdot \vec{B} = -\mu_z B$ just as with atomic magnetic moments. The components of the magnetic moment in the direction of the field μ_z are quantized, so a series of energy levels results from this interaction.

Example 43.2 Proton spin flips

Protons are placed in a 2.30-T magnetic field that points in the positive z-direction. (a) What is the energy difference between states with the z-component of proton spin angular momentum parallel and antiparallel to the field? (b) A proton can make a transition from one of these states to the other by emitting or absorbing a photon with the appropriate energy. Find the frequency and wavelength of such a photon.

SOLUTION

IDENTIFY and SET UP: The proton is a spin-$\frac{1}{2}$ particle with a magnetic moment $\vec{\mu}$ in the same direction as its spin $\vec{S}$, so its energy depends on the orientation of its spin relative to an applied magnetic field $\vec{B}$. If the z-component of $\vec{S}$ is aligned with $\vec{B}$, then μ_z is equal to the positive value given in Eq. (43.8). If the z-component of $\vec{S}$ is opposite $\vec{B}$, then μ_z is the negative of this value. The interaction energy in either case is $U = -\mu_z B$; the difference between these energies is our target variable in part (a). We find the photon frequency and wavelength using $E = hf = hc/\lambda$.

EXECUTE: (a) When the z-components of $\vec{S}$ and $\vec{\mu}$ are parallel to $\vec{B}$, the interaction energy is

$$U = -|\mu_z|B = -(2.7928)(3.152 \times 10^{-8}\,\text{eV/T})(2.30\,\text{T})$$
$$= -2.025 \times 10^{-7}\,\text{eV}$$

When the z-components of $\vec{S}$ and $\vec{\mu}$ are antiparallel to the field, the energy is $+2.025 \times 10^{-7}\,\text{eV}$. Hence the energy *difference* between the states is

$$\Delta E = 2(2.025 \times 10^{-7}\,\text{eV}) = 4.05 \times 10^{-7}\,\text{eV}$$

(b) The corresponding photon frequency and wavelength are

$$f = \frac{\Delta E}{h} = \frac{4.05 \times 10^{-7}\,\text{eV}}{4.136 \times 10^{-15}\,\text{eV}\cdot\text{s}} = 9.79 \times 10^7\,\text{Hz} = 97.9\,\text{MHz}$$

$$\lambda = \frac{c}{f} = \frac{3.00 \times 10^8\,\text{m/s}}{9.79 \times 10^7\,\text{s}^{-1}} = 3.06\,\text{m}$$

EVALUATE: This frequency is in the middle of the FM radio band. When a hydrogen specimen is placed in a 2.30-T magnetic field and irradiated with radio waves of this frequency, proton *spin flips* can be detected by the absorption of energy from the radiation.

43.1 Magnetic resonance imaging (MRI).

(a)

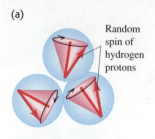

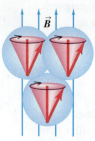

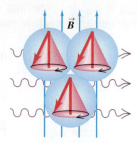

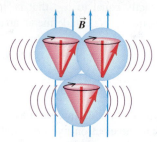

Protons, the nuclei of hydrogen atoms in the tissue under study, normally have random spin orientations.

In the presence of a strong magnetic field, the spins become aligned with a component parallel to $\vec{B}$.

A brief radio signal causes the spins to flip orientation.

As the protons realign with the $\vec{B}$ field, they emit radio waves that are picked up by sensitive detectors.

(b) Since $\vec{B}$ has a different value at different locations in the tissue, the radio waves from different locations have different frequencies. This makes it possible to construct an image.

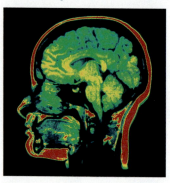

(c) An electromagnet used for MRI

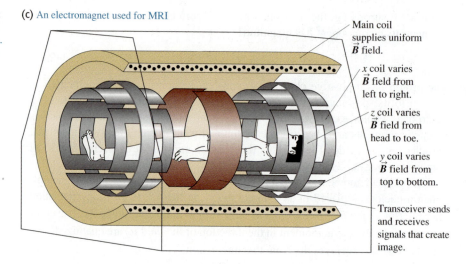

Main coil supplies uniform $\vec{B}$ field.

x coil varies $\vec{B}$ field from left to right.

z coil varies $\vec{B}$ field from head to toe.

y coil varies $\vec{B}$ field from top to bottom.

Transceiver sends and receives signals that create image.

PhET: Simplified MRI

Nuclear Magnetic Resonance and MRI

Spin-flip experiments of the sort referred to in Example 43.2 are called *nuclear magnetic resonance* (NMR). They have been carried out with many different nuclides. Frequencies and magnetic fields can be measured very precisely, so this technique permits precise measurements of nuclear magnetic moments. An elaboration of this basic idea leads to *magnetic resonance imaging* (MRI), a noninvasive imaging technique that discriminates among various body tissues on the basis of the differing environments of protons in the tissues (Fig. 43.1).

The magnetic moment of a nucleus is also the *source* of a magnetic field. In an atom the interaction of an electron's magnetic moment with the field of the nucleus's magnetic moment causes additional splittings in atomic energy levels and spectra. We called this effect *hyperfine structure* in Section 41.5. Measurements of the hyperfine structure may be used to directly determine the nuclear spin.

Test Your Understanding of Section 43.1   (a) By what factor must the mass number of a nucleus increase to double its volume? (i) $\sqrt[3]{2}$; (ii) $\sqrt{2}$; (iii) 2; (iv) 4; (v) 8. (b) By what factor must the mass number increase to double the radius of the nucleus? (i) $\sqrt[3]{2}$; (ii) $\sqrt{2}$; (iii) 2; (iv) 4; (v) 8.

43.2 Nuclear Binding and Nuclear Structure

Because energy must be added to a nucleus to separate it into its individual protons and neutrons, the total rest energy E_0 of the separated nucleons is greater than the rest energy of the nucleus. The energy that must be added to separate the nucleons is called the **binding energy** E_B; it is the magnitude of the energy by which the nucleons are bound together. Thus the rest energy of the nucleus is $E_0 - E_B$. Using the equivalence of rest mass and energy (see Section 37.8), we see that the total mass of the nucleons is always greater than the mass of the nucleus by an amount E_B/c^2 called the *mass defect*. The binding energy for a nucleus containing Z protons and N neutrons is defined as

$$E_B = (ZM_H + Nm_n - {}_Z^A M)c^2 \quad \text{(nuclear binding energy)} \quad (43.10)$$

where ${}_Z^A M$ is the mass of the *neutral* atom containing the nucleus, the quantity in the parentheses is the mass defect, and $c^2 = 931.5 \text{ MeV/u}$. Note that Eq. (43.10) does not include Zm_p, the mass of Z protons. Rather, it contains ZM_H, the mass of Z protons and Z electrons combined as Z neutral ${}_1^1 H$ atoms, to balance the Z electrons included in ${}_Z^A M$, the mass of the neutral atom.

The simplest nucleus is that of hydrogen, a single proton. Next comes the nucleus of ${}_1^2 H$, the isotope of hydrogen with mass number 2, usually called *deuterium*. Its nucleus consists of a proton and a neutron bound together to form a particle called the *deuteron*. By using values from Table 43.2 in Eq. (43.10), we find that the binding energy of the deuteron is

$$E_B = (1.007825 \text{ u} + 1.008665 \text{ u} - 2.014102 \text{ u})(931.5 \text{ MeV/u})$$
$$= 2.224 \text{ MeV}$$

This much energy would be required to pull the deuteron apart into a proton and a neutron. An important measure of how tightly a nucleus is bound is the *binding energy per nucleon*, E_B/A. At $(2.224 \text{ MeV})/(2 \text{ nucleons}) = 1.112 \text{ MeV}$ per nucleon, ${}_1^2 H$ has the lowest binding energy per nucleon of all nuclides.

Application **Deuterium and Heavy Water Toxicity**

A crucial step in plant and animal cell division is the formation of a spindle, which separates the two sets of daughter chromosomes. If a plant is given only heavy water—in which one or both of the hydrogen atoms in an H_2O molecule are replaced with a deuterium atom—cell division stops and the plant stops growing. The reason is that deuterium is more massive than ordinary hydrogen, so the O–H bond in heavy water has a slightly different binding energy and heavy water has slightly different properties as a solvent. The biochemical reactions that occur during cell division are very sensitive to these solvent properties, so a spindle never forms and the cell cannot reproduce.

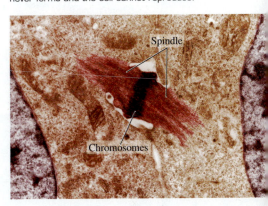

Spindle

Chromosomes

Problem-Solving Strategy 43.1   Nuclear Properties

IDENTIFY *the relevant concepts:* The key properties of a nucleus are its mass, radius, binding energy, mass defect, binding energy per nucleon, and angular momentum.

SET UP *the problem:* Once you have identified the target variables, assemble the equations needed to solve the problem. A relatively small number of equations from this section and Section 43.1 are all you need.

EXECUTE *the solution:* Solve for the target variables. Binding-energy calculations using Eq. (43.10) often involve subtracting two nearly equal quantities. To get enough precision in the difference, you may need to carry as many as nine significant figures, if that many are available.

EVALUATE *your answer:* It's useful to be familiar with the following benchmark magnitudes. Protons and neutrons are about 1840 times as massive as electrons. Nuclear radii are of the order of 10^{-15} m. The electric potential energy of two protons in a nucleus is roughly 10^{-13} J or 1 MeV, so nuclear interaction energies are typically a few MeV rather than a few eV as with atoms. The binding energy per nucleon is about 1% of the nucleon rest energy. (The ionization energy of the hydrogen atom is only 0.003% of the electron's rest energy.) Angular momenta are determined only by the value of $\hbar$, so they are of the same order of magnitude in both nuclei and atoms. Nuclear magnetic moments, however, are about a factor of 1000 *smaller* than those of electrons in atoms because nuclei are so much more massive than electrons.

Example 43.3   The most strongly bound nuclide

Find the mass defect, the total binding energy, and the binding energy per nucleon of ${}_{28}^{62}\text{Ni}$, which has the highest binding energy per nucleon of all nuclides (Fig. 43.2). The neutral atomic mass of ${}_{28}^{62}\text{Ni}$ is 61.928349 u.

SOLUTION

IDENTIFY and SET UP: The mass defect ΔM is the difference between the mass of the nucleus and the combined mass of its constituent nucleons. The binding energy E_B is this quantity multiplied by c^2, and the binding energy per nucleon is E_B divided by the mass number A. We use Eq. (43.10), $\Delta M = ZM_H + Nm_n - {}_Z^A M$, to determine both the mass defect and the binding energy.

EXECUTE: With $Z = 28$, $M_H = 1.007825 \text{ u}$, $N = A - Z = 62 - 28 = 34$, $m_n = 1.008665 \text{ u}$, and ${}_Z^A M = 61.928349 \text{ u}$, Eq. (43.10) gives $\Delta M = 0.585361 \text{ u}$. The binding energy is then

$$E_B = (0.585361 \text{ u})(931.5 \text{ MeV/u}) = 545.3 \text{ MeV}$$

Continued

The binding energy *per nucleon* is $E_B/A = (545.3 \text{ MeV})/62$, or 8.795 MeV per nucleon.

EVALUATE: Our result means that it would take a minimum of 545.3 MeV to pull a $^{62}_{28}\text{Ni}$ completely apart into 28 protons and 34 neutrons. The mass defect of $^{62}_{28}\text{Ni}$ is about 1% of the atomic (or the nuclear) mass. The binding energy is therefore about 1% of the rest energy of the nucleus, and the binding energy per nucleon is about 1% of the rest energy of a nucleon. Note that the mass defect is more than half the mass of a nucleon, which suggests how tightly bound nuclei are.

Nearly all stable nuclides, from the lightest to the most massive, have binding energies in the range of 7–9 MeV per nucleon. Figure 43.2 is a graph of binding energy per nucleon as a function of the mass number A. Note the spike at $A = 4$, showing the unusually large binding energy per nucleon of the ^4_2He nucleus (alpha particle) relative to its neighbors. To explain this curve, we must consider the interactions among the nucleons.

The Nuclear Force

The force that binds protons and neutrons together in the nucleus, despite the electrical repulsion of the protons, is an example of the *strong interaction* that we mentioned in Section 5.5. In the context of nuclear structure, this interaction is called the *nuclear force*. Here are some of its characteristics. First, it does not depend on charge; neutrons as well as protons are bound, and the binding is the same for both. Second, it has short range, of the order of nuclear dimensions—that is, 10^{-15} m. (Otherwise, the nucleus would grow by pulling in additional protons and neutrons.) But within its range, the nuclear force is much stronger than electrical forces; otherwise, the nucleus could never be stable. It would be nice if we could write a simple equation like Newton's law of gravitation or Coulomb's law for this force, but physicists have yet to fully determine its dependence on the separation r. Third, the nearly constant density of nuclear matter and the nearly constant binding energy per nucleon of larger nuclides show that a particular nucleon cannot interact simultaneously with *all* the other nucleons in a nucleus, but only with those few in its immediate vicinity. This is different from electrical forces; *every* proton in the nucleus repels every other one. This limited number of interactions is called *saturation;* it is analogous to covalent bonding in molecules and solids. Finally, the nuclear force favors binding of *pairs* of protons or neutrons with opposite spins and of *pairs of pairs*—that is, a pair of protons and a pair of neutrons, each pair having opposite spins. Hence the alpha particle (two protons and two neutrons) is an exceptionally stable nucleus for its mass number. We'll see other evidence for pairing effects in nuclei in the

43.2 Approximate binding energy per nucleon as a function of mass number A (the total number of nucleons) for stable nuclides.

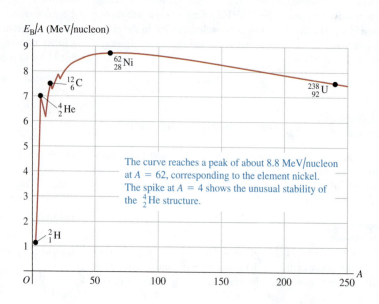

The curve reaches a peak of about 8.8 MeV/nucleon at $A = 62$, corresponding to the element nickel. The spike at $A = 4$ shows the unusual stability of the ^4_2He structure.

next subsection. (In Section 42.8 we described an analogous pairing that binds opposite-spin electrons in Cooper pairs in the BCS theory of superconductivity.)

The analysis of nuclear structure is more complex than the analysis of many-electron atoms. Two different kinds of interactions are involved (electrical and nuclear), and the nuclear force is not yet completely understood. Even so, we can gain some insight into nuclear structure by the use of simple models. We'll discuss briefly two rather different but successful models, the *liquid-drop model* and the *shell model.*

The Liquid-Drop Model

The **liquid-drop model,** first proposed in 1928 by the Russian physicist George Gamow and later expanded on by Niels Bohr, is suggested by the observation that all nuclei have nearly the same density. The individual nucleons are analogous to molecules of a liquid, held together by short-range interactions and surface-tension effects. We can use this simple picture to derive a formula for the estimated total binding energy of a nucleus. We'll include five contributions:

1. We've remarked that nuclear forces show *saturation;* an individual nucleon interacts only with a few of its nearest neighbors. This effect gives a binding-energy term that is proportional to the number of nucleons. We write this term as $C_1 A$, where C_1 is an experimentally determined constant.

2. The nucleons on the surface of the nucleus are less tightly bound than those in the interior because they have no neighbors outside the surface. This decrease in the binding energy gives a *negative* energy term proportional to the surface area $4\pi R^2$. Because R is proportional to $A^{1/3}$, this term is proportional to $A^{2/3}$; we write it as $-C_2 A^{2/3}$, where C_2 is another constant.

3. Every one of the Z protons repels every one of the $(Z - 1)$ other protons. The total repulsive electric potential energy is proportional to $Z(Z - 1)$ and inversely proportional to the radius R and thus to $A^{1/3}$. This energy term is negative because the nucleons are less tightly bound than they would be without the electrical repulsion. We write this correction as $-C_3 Z(Z - 1)/A^{1/3}$.

4. To be in a stable, low-energy state, the nucleus must have a balance between the energies associated with the neutrons and with the protons. This means that N is close to Z for small A and N is greater than Z (but not too much greater) for larger A. We need a negative energy term corresponding to the difference $|N - Z|$. The best agreement with observed binding energies is obtained if this term is proportional to $(N - Z)^2/A$. If we use $N = A - Z$ to express this energy in terms of A and Z, this correction is $-C_4(A - 2Z)^2/A$.

5. Finally, the nuclear force favors *pairing* of protons and of neutrons. This energy term is positive (more binding) if both Z and N are even, negative (less binding) if both Z and N are odd, and zero otherwise. The best fit to the data occurs with the form $\pm C_5 A^{-4/3}$ for this term.

The total estimated binding energy E_B is the sum of these five terms:

$$E_B = C_1 A - C_2 A^{2/3} - C_3 \frac{Z(Z - 1)}{A^{1/3}} - C_4 \frac{(A - 2Z)^2}{A} \pm C_5 A^{-4/3} \qquad \text{(43.11)}$$

$$\text{(nuclear binding energy)}$$

The constants C_1, C_2, C_3, C_4, and C_5, chosen to make this formula best fit the observed binding energies of nuclides, are

$$C_1 = 15.75 \text{ MeV}$$
$$C_2 = 17.80 \text{ MeV}$$
$$C_3 = 0.7100 \text{ MeV}$$
$$C_4 = 23.69 \text{ MeV}$$
$$C_5 = 39 \text{ MeV}$$

The constant C_1 is the binding energy per nucleon due to the saturated nuclear force. This energy is almost 16 MeV per nucleon, about double the *total* binding energy per nucleon in most nuclides.

If we estimate the binding energy E_B using Eq. (43.11), we can solve Eq. (43.10) to use it to estimate the mass of any neutral atom:

$$_Z^A M = ZM_H + Nm_n - \frac{E_B}{c^2} \qquad \text{(semiempirical mass formula)} \qquad (43.12)$$

Equation (43.12) is called the *semiempirical mass formula*. The name is apt; it is *empirical* in the sense that the C's have to be determined empirically (experimentally), yet it does have a sound theoretical basis.

Example 43.4   Estimating binding energy and mass

For the nuclide $_{28}^{62}$Ni of Example 43.3, (a) calculate the five terms in the binding energy and the total estimated binding energy, and (b) find the neutral atomic mass using the semiempirical mass formula.

SOLUTION

IDENTIFY and SET UP: We use the liquid-drop model of the nucleus and its five contributions to the binding energy, as given by Eq. (43.11), to calculate the total binding energy E_B. We then use Eq. (43.12) to find the neutral atomic mass $_{28}^{62}M$.

EXECUTE: (a) With $Z = 28$, $A = 62$, and $N = 34$, the five terms in Eq. (43.11) are

1. $C_1 A = (15.75 \text{ MeV})(62) = 976.5 \text{ MeV}$

2. $-C_2 A^{2/3} = -(17.80 \text{ MeV})(62)^{2/3} = -278.8 \text{ MeV}$

3. $-C_3 \dfrac{Z(Z-1)}{A^{1/3}} = -(0.7100 \text{ MeV}) \dfrac{(28)(27)}{(62)^{1/3}}$

   $= -135.6 \text{ MeV}$

4. $-C_4 \dfrac{(A-2Z)^2}{A} = -(23.69 \text{ MeV}) \dfrac{(62-56)^2}{62}$

   $= -13.8 \text{ MeV}$

5. $+C_5 A^{-4/3} = (39 \text{ MeV})(62)^{-4/3} = 0.2 \text{ MeV}$

The pairing correction (term 5) is by far the smallest of all the terms; it is positive because both Z and N are even. The sum of all five terms is the total estimated binding energy, $E_B = 548.5 \text{ MeV}$.

(b) We use $E_B = 548.5 \text{ MeV}$ in Eq. (43.12):

$$_{28}^{62}M = 28(1.007825 \text{ u}) + 34(1.008665 \text{ u})$$

$$- \frac{548.5 \text{ MeV}}{931.5 \text{ MeV/u}} = 61.925 \text{ u}$$

EVALUATE: The binding energy of $_{28}^{62}$Ni calculated in part (a) is only about 0.6% larger than the true value of 545.3 MeV found in Example 43.3, and the mass calculated in part (b) is only about 0.005% smaller than the measured value of 61.928349 u. The semiempirical mass formula can be quite accurate!

The liquid-drop model and the mass formula derived from it are quite successful in correlating nuclear masses, and we will see later that they are a great help in understanding decay processes of unstable nuclides. Some other aspects of nuclei, such as angular momentum and excited states, are better approached with different models.

The Shell Model

The **shell model** of nuclear structure is analogous to the central-field approximation in atomic physics (see Section 41.6). We picture each nucleon as moving in a potential that represents the averaged-out effect of all the other nucleons. This may not seem to be a very promising approach; the nuclear force is very strong, very short range, and therefore strongly distance dependent. However, in some respects, this model turns out to work fairly well.

The potential-energy function for the nuclear force is the same for protons as for neutrons. Figure 43.3a shows a reasonable assumption for the shape of this function: a spherical version of the square-well potential we discussed in Section 40.3. The corners are somewhat rounded because the nucleus doesn't have a sharply defined surface. For protons there is an additional potential energy associated with electrical repulsion. We consider each proton to interact with a sphere of uniform charge density, with radius R and total charge $(Z - 1)e$. Figure 43.3b shows the nuclear, electric, and total potential energies for a proton as functions of the distance r from the center of the nucleus.

In principle, we could solve the Schrödinger equation for a proton or neutron moving in such a potential. For any spherically symmetric potential energy, the angular-momentum states are the same as for the electrons in the central-field approximation in atomic physics. In particular, we can use the concept of *filled shells and subshells* and their relationship to stability. In atomic structure we found that the values $Z = 2, 10, 18, 36, 54,$ and 86 (the atomic numbers of the noble gases) correspond to particularly stable electron arrangements.

A comparable effect occurs in nuclear structure. The numbers are different because the potential-energy function is different and the nuclear spin-orbit interaction is much stronger and of opposite sign than in atoms, so the sub-shells fill up in a different order from those for electrons in an atom. It is found that when the number of neutrons *or* the number of protons is 2, 8, 20, 28, 50, 82, or 126, the resulting structure is unusually stable—that is, has an unusually high binding energy. (Nuclides with $Z = 126$ have not been observed in nature.) These numbers are called *magic numbers.* Nuclides in which Z is a magic number tend to have an above-average number of stable isotopes. There are several *doubly magic* nuclides for which both Z and N are magic, including

$$\quad {}^{4}_{2}\text{He} \quad {}^{16}_{8}\text{O} \quad {}^{40}_{20}\text{Ca} \quad {}^{48}_{20}\text{Ca} \quad {}^{208}_{82}\text{Pb}$$

All these nuclides have substantially higher binding energy per nucleon than do nuclides with neighboring values of N or Z. They also all have zero nuclear spin. The magic numbers correspond to filled-shell or -subshell configurations of nucleon energy levels with a relatively large jump in energy to the next allowed level.

Test Your Understanding of Section 43.2 Rank the following nuclei in order from largest to smallest value of the binding energy per nucleon. (i) ${}^{4}_{2}\text{He}$; (ii) ${}^{52}_{24}\text{Cr}$; (iii) ${}^{152}_{62}\text{Sm}$; (iv) ${}^{200}_{80}\text{Hg}$; (v) ${}^{252}_{92}\text{Cf}$. ❙

43.3 Nuclear Stability and Radioactivity

Among about 2500 known nuclides, fewer than 300 are stable. The others are unstable structures that decay to form other nuclides by emitting particles and electromagnetic radiation, a process called **radioactivity.** The time scale of these decay processes ranges from a small fraction of a microsecond to billions of years. The *stable* nuclides are shown by dots on the graph in Fig. 43.4, where the neutron number N and proton number (or atomic number) Z for each nuclide are plotted. Such a chart is called a *Segrè chart,* after its inventor, the Italian-American physicist Emilio Segrè (1905–1989).

Each blue line perpendicular to the line $N = Z$ represents a specific value of the mass number $A = Z + N$. Most lines of constant A pass through only one or two stable nuclides; that is, there is usually a very narrow range of stability for a given mass number. The lines at $A = 20$, $A = 40$, $A = 60$, and $A = 80$ are examples. In four cases these lines pass through *three* stable nuclides—namely, at $A = 96, 124, 130,$ and 136.

Four stable nuclides have both odd Z and odd N:

$$\quad {}^{2}_{1}\text{H} \quad {}^{6}_{3}\text{Li} \quad {}^{10}_{5}\text{B} \quad {}^{14}_{7}\text{N}$$

These are called *odd-odd nuclides.* The absence of other odd-odd nuclides shows the influence of pairing. Also, there is *no* stable nuclide with $A = 5$ or $A = 8$. The doubly magic ${}^{4}_{2}\text{He}$ nucleus, with a pair of protons and a pair of neutrons, has no interest in accepting a fifth particle into its structure. Collections of eight nucleons decay to smaller nuclides, with a ${}^{8}_{4}\text{Be}$ nucleus immediately splitting into two ${}^{4}_{2}\text{He}$ nuclei.

43.3 Approximate potential-energy functions for a nucleon in a nucleus. The approximate nuclear radius is R.

(a) The potential energy U_{nuc} due to the nuclear force is the same for protons and neutrons. For neutrons, it is the *total* potential energy.

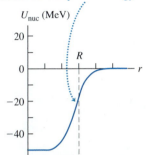

(b) For protons, the total potential energy U_{tot} is the sum of the nuclear (U_{nuc}) and electric (U_{el}) potential energies.

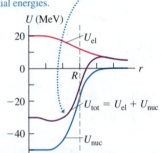

43.4 Segrè chart showing neutron number and proton number for stable nuclides.

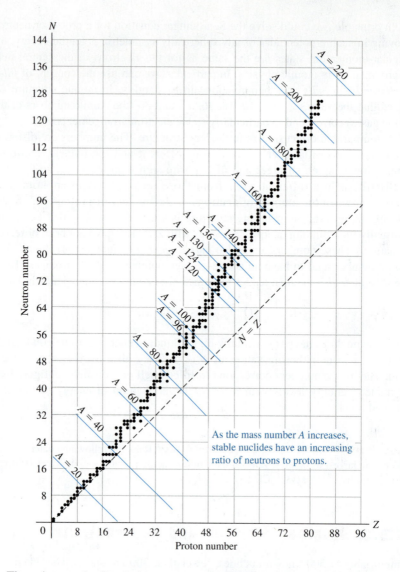

The points on the Segrè chart representing stable nuclides define a rather narrow stability region. For low mass numbers, the numbers of protons and neutrons are approximately equal, $N \approx Z$. The ratio N/Z increases gradually with A, up to about 1.6 at large mass numbers, because of the increasing influence of the electrical repulsion of the protons. Points to the right of the stability region represent nuclides that have too many protons relative to neutrons to be stable. In these cases, repulsion wins, and the nucleus comes apart. To the left are nuclides with too many neutrons relative to protons. In these cases the energy associated with the neutrons is out of balance with that associated with the protons, and the nuclides decay in a process that converts neutrons to protons. The graph also shows that no nuclide with $A > 209$ or $Z > 83$ is stable. A nucleus is unstable if it is too big. Note that there is no stable nuclide with $Z = 43$ (technetium) or 61 (promethium).

Alpha Decay

Nearly 90% of the 2500 known nuclides are *radioactive;* they are not stable but decay into other nuclides. When unstable nuclides decay into different nuclides, they usually emit alpha (α) or beta (β) particles. An **alpha particle** is a ^{4}He nucleus, two protons and two neutrons bound together, with total spin zero. Alpha emission occurs principally with nuclei that are too large to be stable. When a nucleus emits an alpha particle, its N and Z values each decrease by 2 and A decreases by 4, moving it closer to stable territory on the Segrè chart.

PhET: Alpha Decay

43.5 Alpha decay of the unstable radium nuclide $^{226}_{88}$Ra.

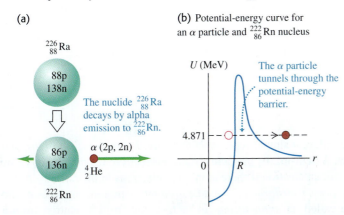

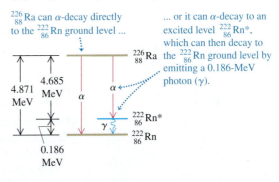

(c) Energy-level diagram for the system

A familiar example of an alpha emitter is radium, $^{226}_{88}$Ra (Fig. 43.5a). The speed of the emitted alpha particle, determined from the curvature of its path in a transverse magnetic field, is about 1.52×10^7 m/s. This speed, although large, is only 5% of the speed of light, so we can use the nonrelativistic kinetic-energy expression $K = \frac{1}{2}mv^2$:

$$K = \frac{1}{2}(6.64 \times 10^{-27} \text{ kg})(1.52 \times 10^7 \text{ m/s})^2 = 7.67 \times 10^{-13} \text{ J} = 4.79 \text{ MeV}$$

Alpha particles are always emitted with definite kinetic energies, determined by conservation of momentum and energy. Because of their charge and mass, alpha particles can travel only several centimeters in air, or a few tenths or hundredths of a millimeter through solids, before they are brought to rest by collisions.

Some nuclei can spontaneously decay by emission of α particles because energy is released in their alpha decay. You can use conservation of mass-energy to show that

> **alpha decay is possible whenever the mass of the original neutral atom is greater than the sum of the masses of the final neutral atom and the neutral helium-4 atom.**

In alpha decay, the α particle tunnels through a potential-energy barrier, as Fig. 43.5b shows. You may want to review the discussion of tunneling in Section 40.4.

Example 43.5   Alpha decay of radium

Show that the α-emission process $^{226}_{88}$Ra $\rightarrow$ $^{222}_{86}$Rn + $^4_2$He is energetically possible, and calculate the kinetic energy of the emitted α particle. The neutral atomic masses are 226.025403 u for $^{226}_{88}$Ra and 222.017571 u for $^{222}_{86}$Rn.

SOLUTION

IDENTIFY and SET UP: Alpha emission is possible if the mass of the $^{226}_{88}$Ra atom is greater than the sum of the atomic masses of $^{222}_{86}$Rn and $^4_2$He. The mass difference between the initial radium atom and the final radon and helium atoms corresponds (through $E = mc^2$) to the energy E released in the decay. Because momentum is conserved as well as energy, *both* the alpha particle and the $^{222}_{86}$Rn atom are in motion after the decay; we will have to account for this in determining the kinetic energy of the alpha particle.

EXECUTE: From Table 43.2, the mass of the $^4_2$He atom is 4.002603 u. The difference in mass between the original nucleus and the decay products is

$$226.025403 \text{ u} - (222.017571 \text{ u} + 4.002603 \text{ u}) = +0.005229 \text{ u}$$

Since this is positive, α decay is energetically possible. The energy equivalent of this mass difference is

$$E = (0.005229 \text{ u})(931.5 \text{ MeV/u}) = 4.871 \text{ MeV}$$

Thus we expect the decay products to emerge with total kinetic energy 4.871 MeV. Momentum is also conserved; if the parent $^{226}_{88}$Ra nucleus is at rest, the daughter $^{222}_{86}$Rn nucleus and the α particle have momenta of equal magnitude p but opposite direction. Kinetic

Continued

energy is $K = \frac{1}{2}mv^2 = p^2/2m$: Since p is the same for the two particles, the kinetic energy divides inversely as their masses. Hence the α particle gets $222/(222 + 4)$ of the total, or 4.78 MeV.

EVALUATE: Experiment shows that $^{226}_{88}$Ra does undergo alpha decay, and the observed α-particle energy is 4.78 MeV. You can check your results by verifying that the alpha particle and the $^{222}_{86}$Rn nucleus produced in the decay have the same magnitude of momentum $p = mv$. You can calculate the speed v of each of the decay products from its respective kinetic energy. You'll find that the alpha particle moves at a sprightly $0.0506c = 1.52 \times 10^7$ m/s; if momentum is conserved, you should find that the $^{222}_{86}$Rn nucleus moves $\frac{4}{222}$ as fast. Does it?

Beta Decay

There are three different simple types of *beta decay: beta-minus, beta-plus,* and *electron capture.* A **beta-minus particle** (β^-) is an electron. It's not obvious how a nucleus can emit an electron if there aren't any electrons in the nucleus. Emission of a β^- involves *transformation* of a neutron into a proton, an electron, and a third particle called an *antineutrino.* In fact, if you freed a neutron from a nucleus, it would decay into a proton, an electron, and an antineutrino in an average time of about 15 minutes.

Beta particles can be identified and their speeds can be measured with techniques that are similar to the Thomson experiments we described in Section 27.5. The speeds of beta particles range up to 0.9995 of the speed of light, so their motion is highly relativistic. They are emitted with a continuous spectrum of energies. This would not be possible if the only two particles were the β^- and the recoiling nucleus, since energy and momentum conservation would then require a definite speed for the β^-. Thus there must be a *third* particle involved. From conservation of charge, it must be neutral, and from conservation of angular momentum, it must be a spin-$\frac{1}{2}$ particle.

This third particle is an antineutrino, the *antiparticle* of a **neutrino.** The symbol for a neutrino is ν_e (the Greek letter nu). Both the neutrino and the antineutrino have zero charge and zero (or very small) mass and therefore produce very little observable effect when passing through matter. Both evaded detection until 1953, when Frederick Reines and Clyde Cowan succeeded in observing the antineutrino directly. We now know that there are at least three varieties of neutrinos, each with its corresponding antineutrino; one is associated with beta decay and the other two are associated with the decay of two unstable particles, the muon and the tau particle. We'll discuss these particles in more detail in Chapter 44. The antineutrino that is emitted in β^- decay is denoted as $\bar{\nu}_e$. The basic process of β^- decay is

$$n \rightarrow p + \beta^- + \bar{\nu}_e \qquad (43.13)$$

Beta-minus decay usually occurs with nuclides for which the neutron-to-proton ratio N/Z is too large for stability. In β^- decay, N decreases by 1, Z increases by 1, and A doesn't change. You can use conservation of mass-energy to show that

beta-minus decay can occur whenever the mass of the original neutral atom is larger than that of the final atom.

Example 43.6 Why cobalt-60 is a beta-minus emitter

The nuclide $^{60}_{27}$Co, an odd-odd unstable nucleus, is used in medical and industrial applications of radiation. Show that it is unstable relative to β^- decay. The atomic masses you need are 59.933822 u for $^{60}_{27}$Co and 59.930791 u for $^{60}_{28}$Ni.

SOLUTION

IDENTIFY and SET UP: Beta-minus decay is possible if the mass of the original neutral atom is greater than that of the final atom.

We must first identify the nuclide that will result if $^{60}_{27}$Co undergoes β^- decay and then compare its neutral atomic mass to that of $^{60}_{27}$Co.

EXECUTE: In the presumed β^- decay of $^{60}_{27}$Co, Z increases by 1 from 27 to 28 and A remains at 60, so the final nuclide is $^{60}_{28}$Ni. The neutral atomic mass of $^{60}_{27}$Co is greater than that of $^{60}_{28}$Ni by 0.003031 u, so β^- decay *can* occur.

EVALUATE: With three decay products in β^- decay—the $^{60}_{28}\text{Ni}$ nucleus, the electron, and the antineutrino—the energy can be shared in many different ways that are consistent with conservation of energy and momentum. It's impossible to predict precisely how the energy will be shared for the decay of a particular $^{60}_{27}\text{Co}$ nucleus. By contrast, in alpha decay there are just two decay products, and their energies and momenta are determined uniquely (see Example 43.5).

We have noted that β^- decay occurs with nuclides that have too large a neutron-to-proton ratio N/Z. Nuclides for which N/Z is too *small* for stability can emit a *positron,* the electron's antiparticle, which is identical to the electron but with positive charge. (We'll discuss the positron in more detail in Chapter 44.) The basic process, called *beta-plus decay* (β^+), is

$$\text{p} \rightarrow \text{n} + \beta^+ + \nu_\text{e} \qquad (43.14)$$

where β^+ is a positron and ν_e is the electron neutrino.

> **Beta-plus decay can occur whenever the mass of the original neutral atom is at least two electron masses larger than that of the final atom.**

You can show this using conservation of mass-energy.

The third type of beta decay is *electron capture.* There are a few nuclides for which β^+ emission is not energetically possible but in which an orbital electron (usually in the K shell) can combine with a proton in the nucleus to form a neutron and a neutrino. The neutron remains in the nucleus and the neutrino is emitted. The basic process is

$$\text{p} + \beta^- \rightarrow \text{n} + \nu_\text{e} \qquad (43.15)$$

You can use conservation of mass-energy to show that

> **electron capture can occur whenever the mass of the original neutral atom is larger than that of the final atom.**

In all types of beta decay, A remains constant. However, in beta-plus decay and electron capture, N increases by 1 and Z decreases by 1 as the neutron–proton ratio increases toward a more stable value. The reaction of Eq. (43.15) also helps to explain the formation of a neutron star, mentioned in Example 43.1.

CAUTION **Beta decay inside and outside nuclei** The beta-decay reactions given by Eqs. (43.13), (43.14), and (43.15) occur *within* a nucleus. Although the decay of a neutron outside the nucleus proceeds through the reaction of Eq. (43.13), the reaction of Eq. (43.14) is forbidden by conservation of mass-energy for a proton outside the nucleus. The reaction of Eq. (43.15) can occur outside the nucleus only with the addition of some extra energy, as in a collision. ❚

Example 43.7   Why cobalt-57 is not a beta-plus emitter

The nuclide $^{57}_{27}\text{Co}$ is an odd-even unstable nucleus. Show that it cannot undergo β^+ decay, but that it *can* decay by electron capture. The atomic masses you need are 56.936296 u for $^{57}_{27}\text{Co}$ and 56.935399 u for $^{57}_{26}\text{Fe}$.

SOLUTION

IDENTIFY and SET UP: Beta-plus decay is possible if the mass of the original neutral atom is greater than that of the final atom plus two electron masses (0.001097 u). Electron capture is possible if the mass of the original atom is greater than that of the final atom. We must first identify the nuclide that will result if $^{57}_{27}\text{Co}$ undergoes β^+ decay or electron capture and then find the corresponding mass difference.

EXECUTE: The original nuclide is $^{57}_{27}\text{Co}$. In both the presumed β^+ decay and electron capture, Z decreases by 1 from 27 to 26, and A remains at 57, so the final nuclide is $^{57}_{26}\text{Fe}$. Its mass is less than that of $^{57}_{27}\text{Co}$ by 0.000897 u, a value smaller than 0.001097 u (two electron masses), so β^+ decay *cannot* occur. However, the mass of the original atom is greater than the mass of the final atom, so electron capture *can* occur.

EVALUATE: In electron capture there are just two decay products, the final nucleus and the emitted neutrino. As in alpha decay (Example 43.5) but unlike in β^- decay (Example 43.6), the decay products of electron capture have unique energies and momenta. In Section 43.4 we'll see how to relate the probability that electron capture will occur to the *half-life* of this nuclide.

Gamma Decay

The energy of internal motion of a nucleus is quantized. A typical nucleus has a set of allowed energy levels, including a *ground state* (state of lowest energy) and several *excited states*. Because of the great strength of nuclear interactions, excitation energies of nuclei are typically of the order of 1 MeV, compared with a few eV for atomic energy levels. In ordinary physical and chemical transformations the nucleus always remains in its ground state. When a nucleus is placed in an excited state, either by bombardment with high-energy particles or by a radioactive transformation, it can decay to the ground state by emission of one or more photons called **gamma rays** or *gamma-ray photons,* with typical energies of 10 keV to 5 MeV. This process is called *gamma* (γ) *decay.* For example, alpha particles emitted from ^{226}Ra have two possible kinetic energies, either 4.784 MeV or 4.602 MeV. Including the recoil energy of the resulting ^{222}Rn nucleus, these correspond to a total released energy of 4.871 MeV or 4.685 MeV, respectively. When an alpha particle with the smaller energy is emitted, the ^{222}Rn nucleus is left in an excited state. It then decays to its ground state by emitting a gamma-ray photon with energy

$$(4.871 - 4.685)\,\text{MeV} = 0.186\,\text{MeV}$$

A photon with this energy is observed during this decay (Fig. 43.5c).

CAUTION γ **decay vs.** α **and** β **decay** In both α and β decay, the Z value of a nucleus changes and the nucleus of one element becomes the nucleus of a different element. In γ decay, the element does *not* change; the nucleus merely goes from an excited state to a less excited state. ▌

Natural Radioactivity

Many radioactive elements occur in nature. For example, you are very slightly radioactive because of unstable nuclides such as carbon-14 and potassium-40 that are present throughout your body. The study of natural radioactivity began in 1896, one year after Röntgen discovered x rays. Henri Becquerel discovered a radiation from uranium salts that seemed similar to x rays. Intensive investigation in the following two decades by Marie and Pierre Curie, Ernest Rutherford, and many others revealed that the emissions consist of positively and negatively charged particles and neutral rays; they were given the names *alpha, beta,* and *gamma* because of their differing penetration characteristics.

The decaying nucleus is usually called the *parent nucleus;* the resulting nucleus is the *daughter nucleus.* When a radioactive nucleus decays, the daughter nucleus may also be unstable. In this case a *series* of successive decays occurs until a stable configuration is reached. Several such series are found in nature. The most abundant radioactive nuclide found on earth is the uranium isotope ^{238}U, which undergoes a series of 14 decays, including eight α emissions and six β^- emissions, terminating at a stable isotope of lead, ^{206}Pb (Fig. 43.6).

Radioactive decay series can be represented on a Segrè chart, as in Fig. 43.7. The neutron number N is plotted vertically, and the atomic number Z is plotted horizontally. In alpha emission, both N and Z decrease by 2. In β^- emission, N decreases by 1 and Z increases by 1. The decays can also be represented in equation form; the first two decays in the series are written as

$$^{238}\text{U} \rightarrow {}^{234}\text{Th} + \alpha$$
$$^{234}\text{Th} \rightarrow {}^{234}\text{Pa} + \beta^- + \bar{\nu}_e$$

or more briefly as

$$^{238}\text{U} \xrightarrow{\alpha} {}^{234}\text{Th}$$
$$^{234}\text{Th} \xrightarrow{\beta^-} {}^{234}\text{Pa}$$

43.6 Earthquakes are caused in part by the radioactive decay of ^{238}U in the earth's interior. These decays release energy that helps to produce convection currents in the earth's interior. Such currents drive the motions of the earth's crust, including the sudden sharp motions that we call earthquakes (like the one that caused this damage).

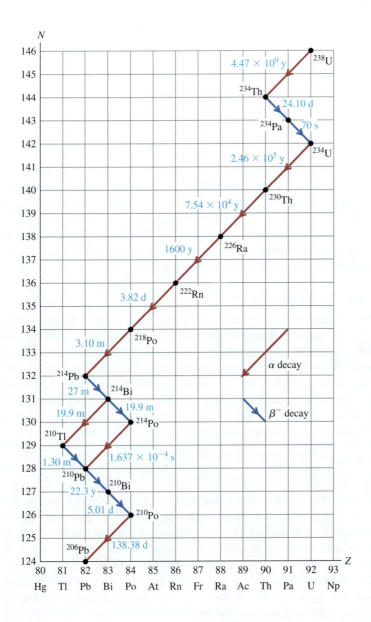

43.7 Segrè chart showing the uranium ^{238}U decay series, terminating with the stable nuclide ^{206}Pb. The times are half-lives (discussed in the next section), given in years (y), days (d), hours (h), minutes (m), or seconds (s).

In the second process, the beta decay leaves the daughter nucleus ^{234}Pa in an excited state, from which it decays to the ground state by emitting a gamma-ray photon. An excited state is denoted by an asterisk, so we can represent the γ emission as

$$^{234}Pa^* \rightarrow \,^{234}Pa + \gamma$$

or

$$^{234}Pa^* \xrightarrow{\gamma} \,^{234}Pa$$

An interesting feature of the ^{238}U decay series is the branching that occurs at ^{214}Bi. This nuclide decays to ^{210}Pb by emission of an α and a β^-, which can occur in either order. We also note that the series includes unstable isotopes of several elements that also have stable isotopes, including thallium (Tl), lead (Pb), and bismuth (Bi). The unstable isotopes of these elements that occur in the ^{238}U series all have too many neutrons to be stable.

Many other decay series are known. Two of these occur in nature, one starting with the uncommon isotope ^{235}U and ending with ^{207}Pb, the other starting with thorium (^{232}Th) and ending with ^{208}Pb.

43.4 Activities and Half-Lives

Suppose you need to dispose of some radioactive waste that contains a certain number of nuclei of a particular radioactive nuclide. If no more are produced, that number decreases in a simple manner as the nuclei decay. This decrease is a statistical process; there is no way to predict when any individual nucleus will decay. No change in physical or chemical environment, such as chemical reactions or heating or cooling, greatly affects most decay rates. The rate varies over an extremely wide range for different nuclides.

Radioactive Decay Rates

Let $N(t)$ be the (very large) number of radioactive nuclei in a sample at time t, and let $dN(t)$ be the (negative) change in that number during a short time interval dt. (We'll use $N(t)$ to minimize confusion with the neutron number N.) The number of decays during the interval dt is $-dN(t)$. The rate of change of $N(t)$ is the negative quantity $dN(t)/dt$; thus $-dN(t)/dt$ is called the *decay rate* or the **activity** of the specimen. The larger the number of nuclei in the specimen, the more nuclei decay during any time interval. That is, the activity is directly proportional to $N(t)$; it equals a constant λ multiplied by $N(t)$:

$$-\frac{dN(t)}{dt} = \lambda N(t) \qquad (43.16)$$

The constant λ is called the **decay constant,** and it has different values for different nuclides. A large value of λ corresponds to rapid decay; a small value corresponds to slower decay. Solving Eq. (43.16) for λ shows us that λ is the ratio of the number of decays per time to the number of remaining radioactive nuclei; λ can then be interpreted as the *probability per unit time* that any individual nucleus will decay.

The situation is reminiscent of a discharging capacitor, which we studied in Section 26.4. Equation (43.16) has the same form as the negative of Eq. (26.15), with q and $1/RC$ replaced by $N(t)$ and λ. Then we can make the same substitutions in Eq. (26.16), with the initial number of nuclei $N(0) = N_0$, to find the exponential function:

$$N(t) = N_0 e^{-\lambda t} \qquad \text{(number of remaining nuclei)} \qquad (43.17)$$

Figure 43.8 is a graph of this function, showing the number of remaining nuclei $N(t)$ as a function of time.

The **half-life** $T_{1/2}$ is the time required for the number of radioactive nuclei to decrease to one-half the original number N_0. Then half of the remaining radioactive nuclei decay during a second interval $T_{1/2}$, and so on. The numbers remaining after successive half-lives are $N_0/2$, $N_0/4$, $N_0/8$, ...

To get the relationship between the half-life $T_{1/2}$ and the decay constant λ, we set $N(t)/N_0 = \frac{1}{2}$ and $t = T_{1/2}$ in Eq. (43.17), obtaining

$$\tfrac{1}{2} = e^{-\lambda T_{1/2}}$$

We take logarithms of both sides and solve for $T_{1/2}$:

$$T_{1/2} = \frac{\ln 2}{\lambda} = \frac{0.693}{\lambda} \qquad (43.18)$$

43.8 The number of nuclei in a sample of a radioactive element as a function of time. The sample's activity has an exponential decay curve with the same shape.

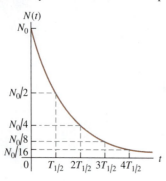

The mean lifetime T_{mean}, generally called the *lifetime,* of a nucleus or unstable particle is proportional to the half-life $T_{1/2}$:

$$T_{mean} = \frac{1}{\lambda} = \frac{T_{1/2}}{\ln 2} = \frac{T_{1/2}}{0.693} \quad \begin{array}{l} \text{(lifetime } T_{mean}, \text{ decay} \\ \text{constant } \lambda, \text{ and half-life } T_{1/2}) \end{array} \quad \text{(43.19)}$$

In particle physics the life of an unstable particle is usually described by the lifetime, not the half-life.

Because the activity $-dN(t)/dt$ at any time equals $\lambda N(t)$, Eq. (43.17) tells us that the activity also depends on time as $e^{-\lambda t}$. Thus the graph of activity versus time has the same shape as Fig. 43.8. Also, after successive half-lives, the activity is one-half, one-fourth, one-eighth, and so on of the original activity.

CAUTION **A half-life may not be enough** It is sometimes implied that any radioactive sample will be safe after a half-life has passed. That's wrong. If your radioactive waste initially has ten times too much activity for safety, it is not safe after one half-life, when it still has five times too much. Even after three half-lives it still has 25% more activity than is safe. The number of radioactive nuclei and the activity approach zero only as t approaches infinity. ▮

A common unit of activity is the **curie,** abbreviated Ci, which is defined to be 3.70×10^{10} decays per second. This is approximately equal to the activity of one gram of radium. The SI unit of activity is the *becquerel,* abbreviated Bq. One becquerel is one decay per second, so

$$1 \text{ Ci} = 3.70 \times 10^{10} \text{ Bq} = 3.70 \times 10^{10} \text{ decays/s}$$

Example 43.8 **Activity of ^{57}Co**

The isotope ^{57}Co decays by electron capture to ^{57}Fe with a half-life of 272 d. The ^{57}Fe nucleus is produced in an excited state, and it almost instantaneously emits gamma rays that we can detect. (a) Find the mean lifetime and decay constant for ^{57}Co. (b) If the activity of a ^{57}Co radiation source is now 2.00 μCi, how many ^{57}Co nuclei does the source contain? (c) What will be the activity after one year?

SOLUTION

IDENTIFY and SET UP: This problem uses the relationships among decay constant λ, lifetime T_{mean}, and activity $-dN(t)/dt$. In part (a) we use Eq. (43.19) to find λ and T_{mean} from $T_{1/2}$. In part (b), we use Eq. (43.16) to calculate the number of nuclei $N(t)$ from the activity. Finally, in part (c) we use Eqs. (43.16) and (43.17) to find the activity after one year.

EXECUTE: (a) It's convenient to convert the half-life to seconds:

$$T_{1/2} = (272 \text{ d})(86{,}400 \text{ s/d}) = 2.35 \times 10^7 \text{ s}$$

From Eq. (43.19), the mean lifetime and decay constant are

$$T_{mean} = \frac{T_{1/2}}{\ln 2} = \frac{2.35 \times 10^7 \text{ s}}{0.693} = 3.39 \times 10^7 \text{ s} = 392 \text{ days}$$

$$\lambda = \frac{1}{T_{mean}} = 2.95 \times 10^{-8} \text{s}^{-1}$$

(b) The activity $-dN(t)/dt$ is given as 2.00 μCi, so

$$-\frac{dN(t)}{dt} = 2.00 \text{ } \mu\text{Ci} = (2.00 \times 10^{-6})(3.70 \times 10^{10} \text{s}^{-1})$$

$$= 7.40 \times 10^4 \text{ decays/s}$$

From Eq. (43.16) this is equal to $\lambda N(t)$, so we find

$$N(t) = -\frac{dN(t)/dt}{\lambda} = \frac{7.40 \times 10^4 \text{s}^{-1}}{2.95 \times 10^{-8} \text{s}^{-1}} = 2.51 \times 10^{12} \text{ nuclei}$$

If you feel we're being too cavalier about the "units" decays and nuclei, you can use decays/(nucleus · s) as the unit for λ.

(c) From Eq. (43.17) the number $N(t)$ of nuclei remaining after one year $(3.156 \times 10^7 \text{ s})$ is

$$N(t) = N_0 e^{-\lambda t} = N_0 e^{-(2.95 \times 10^{-8} \text{s}^{-1})(3.156 \times 10^7 \text{s})} = 0.394 N_0$$

The number of nuclei has decreased to 0.394 of the original number. Equation (43.16) says that the activity is proportional to the number of nuclei, so the activity has decreased by this same factor to $(0.394)(2.00 \text{ } \mu\text{Ci}) = 0.788 \text{ } \mu\text{Ci}$.

EVALUATE: The number of nuclei found in part (b) is equivalent to 4.17×10^{-12} mol, with a mass of 2.38×10^{-10} g. This is a far smaller mass than even the most sensitive balance can measure.

After one 272-day half-life, the number of ^{57}Co nuclei has decreased to $N_0/2$; after $2(272 \text{ d}) = 544 \text{ d}$, it has decreased to $N_0/2^2 = N_0/4$. This result agrees with our answer to part (c), which says that after 365 d the number of nuclei is between $N_0/2$ and $N_0/4$.

Radioactive Dating

An interesting application of radioactivity is the dating of archaeological and geological specimens by measuring the concentration of radioactive isotopes. The most familiar example is *carbon dating*. The unstable isotope ^{14}C, produced during nuclear reactions in the atmosphere that result from cosmic-ray bombardment, gives a small proportion of ^{14}C in the CO_2 in the atmosphere. Plants that obtain their carbon from this source contain the same proportion of ^{14}C as the atmosphere. When a plant dies, it stops taking in carbon, and its ^{14}C β^- decays to ^{14}N with a half-life of 5730 years. By measuring the proportion of ^{14}C in the remains, we can determine how long ago the organism died.

One difficulty with radiocarbon dating is that the ^{14}C concentration in the atmosphere changes over long time intervals. Corrections can be made on the basis of other data such as measurements of tree rings that show annual growth cycles. Similar radioactive techniques are used with other isotopes for dating geological specimens. Some rocks, for example, contain the unstable potassium isotope ^{40}K, a beta emitter that decays to the stable nuclide ^{40}Ar with a half-life of 2.4×10^8 y. The age of the rock can be determined by comparing the concentrations of ^{40}K and ^{40}Ar.

Example 43.9 Radiocarbon dating

Before 1900 the activity per unit mass of atmospheric carbon due to the presence of ^{14}C averaged about 0.255 Bq per gram of carbon. (a) What fraction of carbon atoms were ^{14}C? (b) In analyzing an archaeological specimen containing 500 mg of carbon, you observe 174 decays in one hour. What is the age of the specimen, assuming that its activity per unit mass of carbon when it died was that average value of the air?

SOLUTION

IDENTIFY and SET UP: The key idea is that the present-day activity of a biological sample containing ^{14}C is related to both the elapsed time since it stopped taking in atmospheric carbon and its activity at that time. We use Eqs. (43.16) and (43.17) to solve for the age t of the specimen. In part (a) we determine the number of ^{14}C atoms $N(t)$ from the activity $-dN(t)/dt$ using Eq. (43.16). We find the total number of carbon atoms in 500 mg by using the molar mass of carbon (12.011 g/mol, given in Appendix D), and we use the result to calculate the fraction of carbon atoms that are ^{14}C. The activity decays at the same rate as the number of ^{14}C nuclei; we use this and Eq. (43.17) to solve for the age t of the specimen.

EXECUTE: (a) To use Eq. (43.16), we must first find the decay constant λ from Eq. (43.18):

$$T_{1/2} = 5730 \text{ y} = (5730 \text{ y})(3.156 \times 10^7 \text{ s/y}) = 1.808 \times 10^{11} \text{ s}$$

$$\lambda = \frac{\ln 2}{T_{1/2}} = \frac{0.693}{1.808 \times 10^{11} \text{ s}} = 3.83 \times 10^{-12} \text{ s}^{-1}$$

Then, from Eq. (43.16),

$$N(t) = \frac{-dN/dt}{\lambda} = \frac{0.255 \text{ s}^{-1}}{3.83 \times 10^{-12} \text{ s}^{-1}} = 6.65 \times 10^{10} \text{ atoms}$$

The *total* number of C atoms in 1 gram (1/12.011 mol) is $(1/12.011)(6.022 \times 10^{23}) = 5.01 \times 10^{22}$. The ratio of ^{14}C atoms to all C atoms is

$$\frac{6.65 \times 10^{10}}{5.01 \times 10^{22}} = 1.33 \times 10^{-12}$$

Only four carbon atoms in every 3×10^{12} are ^{14}C.

(b) Assuming that the activity per gram of carbon in the specimen when it died ($t = 0$) was 0.255 Bq/g = $(0.255 \text{ s}^{-1} \cdot \text{g}^{-1})(3600 \text{ s/h}) = 918 \text{ h}^{-1} \cdot \text{g}^{-1}$, the activity of 500 mg of carbon then was $(0.500 \text{ g})(918 \text{ h}^{-1} \cdot \text{g}^{-1}) = 459 \text{ h}^{-1}$. The observed activity now, at time t, is 174 h^{-1}. Since the activity is proportional to the number of radioactive nuclei, the activity ratio $174/459 = 0.379$ equals the number ratio $N(t)/N_0$.

Now we solve Eq. (43.17) for t and insert values for $N(t)/N_0$ and λ:

$$t = \frac{\ln(N(t)/N_0)}{-\lambda} = \frac{\ln 0.379}{-3.83 \times 10^{-12} \text{ s}^{-1}} = 2.53 \times 10^{11} \text{ s} = 8020 \text{ y}$$

EVALUATE: After 8020 y the ^{14}C activity has decreased from 459 to 174 decays per hour. The specimen died and stopped taking CO_2 out of the air about 8000 years ago.

Radiation in the Home

A serious health hazard in some areas is the accumulation in houses of ^{222}Rn, an inert, colorless, odorless radioactive gas. Looking at the ^{238}U decay chain in Fig. 43.7, we see that the half-life of ^{222}Rn is 3.82 days. If so, why not just move out of the house for a while and let it decay away? The answer is that ^{222}Rn is continuously being *produced* by the decay of ^{226}Ra, which is found in minute quantities in the rocks and soil on which some houses are built. It's a dynamic equilibrium situation, in which the rate of production equals the rate of decay. The reason ^{222}Rn is a bigger hazard than the other elements in the ^{238}U decay series is that it's a gas. During its short half-life of 3.82 days it can migrate from the soil into your house. If a ^{222}Rn nucleus decays in your lungs, it emits a damaging α particle and its daughter nucleus ^{218}Po, which is *not* chemically inert and is likely to stay in your lungs until it decays, emits another damaging α particle and so on down the ^{238}U decay series.

How much of a hazard is radon? Although reports indicate values as high as 3500 pCi/L, the average activity per volume in the air inside American homes due to ^{222}Rn is about 1.5 pCi/L (over a thousand decays each second in an average-sized room). *If* your environment has this level of activity, it has been estimated that a lifetime exposure would reduce your life expectancy by about 40 days. For comparison, smoking one pack of cigarettes per day reduces life expectancy by 6 years, and it is estimated that the average emission from all the nuclear power plants in the world reduces life expectancy by anywhere from 0.01 day to 5 days. These figures include catastrophes such as the 1986 nuclear reactor disaster at Chernobyl, for which the *local* effect on life expectancy is much greater.

Test Your Understanding of Section 43.4   Which sample contains a greater number of nuclei: a 5.00-μCi sample of ^{240}Pu (half-life 6560 y) or a 4.45-μCi sample of ^{243}Am (half-life 7370 y)? (i) the ^{240}Pu sample; (ii) the ^{243}Am sample; (iii) both have the same number of nuclei.

43.5 Biological Effects of Radiation

The above discussion of radon introduced the interaction of radiation with living organisms, a topic of vital interest and importance. Under *radiation* we include radioactivity (alpha, beta, gamma, and neutrons) and electromagnetic radiation such as x rays. As these particles pass through matter, they lose energy, breaking molecular bonds and creating ions—hence the term *ionizing radiation*. Charged particles interact directly with the electrons in the material. X rays and γ rays interact by the photoelectric effect, in which an electron absorbs a photon and breaks loose from its site, or by Compton scattering (see Section 38.3). Neutrons cause ionization indirectly through collisions with nuclei or absorption by nuclei with subsequent radioactive decay of the resulting nuclei.

These interactions are extremely complex. It is well known that excessive exposure to radiation, including sunlight, x rays, and all the nuclear radiations, can destroy tissues. In mild cases it results in a burn, as with common sunburn. Greater exposure can cause very severe illness or death by a variety of mechanisms, including massive destruction of tissue cells, alterations of genetic material, and destruction of the components in bone marrow that produce red blood cells.

Calculating Radiation Doses

Radiation dosimetry is the quantitative description of the effect of radiation on living tissue. The *absorbed dose* of radiation is defined as the energy delivered to the tissue per unit mass. The SI unit of absorbed dose, the joule per kilogram, is called the *gray* (Gy); 1 Gy = 1 J/kg. Another unit is the *rad*, defined as 0.01 J/kg:

$$1 \text{ rad} = 0.01 \text{ J/kg} = 0.01 \text{ Gy}$$

Absorbed dose by itself is not an adequate measure of biological effect because equal energies of different kinds of radiation cause different extents of biological effect. This variation is described by a numerical factor called the **relative biological effectiveness (RBE),** also called the *quality factor* (QF), of each specific radiation. X rays with 200 keV of energy are defined to have an

Table 43.3 Relative Biological Effectiveness (RBE) for Several Types of Radiation

Radiation	RBE (Sv/Gy or rem/rad)
X rays and γ rays	1
Electrons	1.0–1.5
Slow neutrons	3–5
Protons	10
α particles	20
Heavy ions	20

RBE of unity, and the effects of other radiations can be compared experimentally. Table 43.3 shows approximate values of RBE for several radiations. All these values depend somewhat on the kind of tissue in which the radiation is absorbed and on the energy of the radiation.

The biological effect is described by the product of the absorbed dose and the RBE of the radiation; this quantity is called the *biologically equivalent dose,* or simply the equivalent dose. The SI unit of equivalent dose for humans is the sievert (Sv):

$$\text{Equivalent dose (Sv)} = \text{RBE} \times \text{Absorbed dose (Gy)} \qquad (43.20)$$

A more common unit, corresponding to the rad, is the rem (an abbreviation of *röntgen equivalent for man*):

$$\text{Equivalent dose (rem)} = \text{RBE} \times \text{Absorbed dose (rad)} \qquad (43.21)$$

Thus the unit of the RBE is 1 Sv/Gy or 1 rem/rad, and 1 rem = 0.01 Sv.

Example 43.10 Dose from a medical x ray

During a diagnostic x-ray examination a 1.2-kg portion of a broken leg receives an equivalent dose of 0.40 mSv. (a) What is the equivalent dose in mrem? (b) What is the absorbed dose in mrad and in mGy? (c) If the x-ray energy is 50 keV, how many x-ray photons are absorbed?

SOLUTION

IDENTIFY and SET UP: We are asked to relate the equivalent dose (the biological effect of the radiation, measured in sieverts or rems) to the absorbed dose (the energy absorbed per mass, measured in grays or rads). In part (a) we use the conversion factor 1 rem = 0.01 Sv for equivalent dose. Table 43.3 gives the RBE for x rays; we use this value in part (b) to determine the absorbed dose using Eqs. (43.20) and (43.21). Finally, in part (c) we use the mass and the definition of absorbed dose to find the total energy absorbed and the total number of photons absorbed.

EXECUTE: (a) The equivalent dose in mrem is

$$\frac{0.40 \text{ mSv}}{0.01 \text{ Sv/rem}} = 40 \text{ mrem}$$

(b) For x rays, RBE = 1 rem/rad or 1 Sv/Gy, so the absorbed dose is

$$\frac{40 \text{ mrem}}{1 \text{ rem/rad}} = 40 \text{ mrad}$$

$$\frac{0.40 \text{ mSv}}{1 \text{ Sv/Gy}} = 0.40 \text{ mGy} = 4.0 \times 10^{-4} \text{ J/kg}$$

(c) The total energy absorbed is

$$(4.0 \times 10^{-4} \text{ J/kg})(1.2 \text{ kg}) = 4.8 \times 10^{-4} \text{ J} = 3.0 \times 10^{15} \text{ eV}$$

The number of x-ray photons is

$$\frac{3.0 \times 10^{15} \text{ eV}}{5.0 \times 10^{4} \text{ eV/photon}} = 6.0 \times 10^{10} \text{ photons}$$

EVALUATE: The absorbed dose is relatively large because x rays have a low RBE. If the ionizing radiation had been a beam of α particles, for which RBE = 20, the absorbed dose needed for an equivalent dose of 0.40 mSv would be only 0.020 mGy, corresponding to a smaller total absorbed energy of 2.4×10^{-5} J.

Radiation Hazards

Here are a few numbers for perspective. To convert from Sv to rem, simply multiply by 100. An ordinary chest x-ray exam delivers about 0.20–0.40 mSv to about 5 kg of tissue. Radiation exposure from cosmic rays and natural radioactivity in soil, building materials, and so on is of the order of 2–3 mSv per year at sea level and twice that at an elevation of 1500 m (5000 ft). A whole-body dose of up to about 0.20 Sv causes no immediately detectable effect. A short-term whole-body dose of 5 Sv or more usually causes death within a few days or weeks. A localized dose of 100 Sv causes complete destruction of the exposed tissues.

The long-term hazards of radiation exposure in causing various cancers and genetic defects have been widely publicized, and the question of whether there is any "safe" level of radiation exposure has been hotly debated. U.S. government regulations are based on a maximum *yearly* exposure, from all except natural resources, of 2 to 5 mSv. Workers with occupational exposure to radiation are permitted 50 mSv per year. Recent studies suggest that these limits are too high and that even extremely small exposures carry hazards, but it is very difficult to gather reliable statistics on the effects of low doses. It has become clear that any use of x rays for medical diagnosis should be preceded by a very careful estimation of the relationship of risk to possible benefit.

Another sharply debated question is that of radiation hazards from nuclear power plants. The radiation level from these plants is *not* negligible. However, to make a meaningful evaluation of hazards, we must compare these levels with the alternatives, such as coal-powered plants. The health hazards of coal smoke are serious and well documented, and the natural radioactivity in the smoke from a coal-fired power plant is believed to be roughly 100 times as great as that from a properly operating nuclear plant with equal capacity. But the comparison is not this simple; the possibility of a nuclear accident and the very serious problem of safe disposal of radioactive waste from nuclear plants must also be considered. It is clearly impossible to eliminate *all* hazards to health. Our goal should be to try to take a rational approach to the problem of *minimizing* the hazard from all sources. Figure 43.9 shows one estimate of the various sources of radiation exposure for the U.S. population. Ionizing radiation is a two-edged sword; it poses very serious health hazards, yet it also provides many benefits to humanity, including the diagnosis and treatments of disease and a wide variety of analytical techniques.

Beneficial Uses of Radiation

Radiation is widely used in medicine for intentional selective destruction of tissue such as tumors. The hazards are considerable, but if the disease would be fatal without treatment, any hazard may be preferable. Artificially produced isotopes are often used as radiation sources. Such isotopes have several advantages over naturally radioactive isotopes. They may have shorter half-lives and correspondingly greater activity. Isotopes can be chosen that emit the type and energy of radiation desired. Some artificial isotopes have been replaced by photon and electron beams from linear accelerators.

Nuclear medicine is an expanding field of application. Radioactive isotopes have virtually the same electron configurations and resulting chemical behavior as stable isotopes of the same element. But the location and concentration of radioactive isotopes can easily be detected by measurements of the radiation they emit. A familiar example is the use of radioactive iodine for thyroid studies. Nearly all the iodine ingested is either eliminated or stored in the thyroid, and the body's chemical reactions do not discriminate between the unstable isotope ^{131}I and the stable isotope ^{127}I. A minute quantity of ^{131}I is fed or injected into the patient, and the speed with which it becomes concentrated in the thyroid provides a measure of thyroid function. The half-life is 8.02 days, so there are no long-lasting radiation hazards. By use of more sophisticated scanning detectors, one can also obtain a "picture" of the thyroid, which shows enlargement and other abnormalities. This procedure, a type of *autoradiography*, is comparable to photographing the glowing filament of an incandescent light bulb by using the light emitted by the filament itself. If this process discovers cancerous thyroid nodules, they can be destroyed by much larger quantities of ^{131}I.

Another useful nuclide for nuclear medicine is technetium-99 (^{99}Tc), which is formed in an excited state by the β^- decay of molybdenum (^{99}Mo). The technetium then decays to its ground state by emitting a γ-ray photon with energy 143 keV. The half-life is 6.01 hours, unusually long for γ emission. (The ground state of ^{99}Tc is also unstable, with a half-life of 2.11×10^5 y; it decays by β^- emission to the stable ruthenium nuclide ^{99}Ru.) The chemistry of technetium is such that it can readily be attached to organic molecules that are taken up by various organs of the body. A small quantity of such technetium-bearing molecules is injected into a patient, and a scanning detector or *gamma camera* is used to produce an image, or *scintigram*, that reveals which parts of the body take up these γ-emitting molecules. This technique, in which ^{99}Tc acts as a radioactive *tracer*, plays an important role in locating cancers, embolisms, and other pathologies (Fig. 43.10).

43.9 Contribution of various sources to the total average radiation exposure in the U.S. population, expressed as percentages of the total.

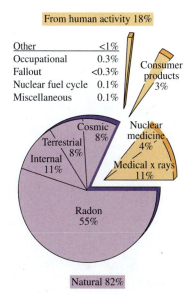

From human activity 18%

Other	<1%
Occupational	0.3%
Fallout	<0.3%
Nuclear fuel cycle	0.1%
Miscellaneous	0.1%

Consumer products 3%

Cosmic 8%

Terrestrial 8%

Internal 11%

Nuclear medicine 4%

Medical x rays 11%

Radon 55%

Natural 82%

43.10 This colored scintigram shows where a chemical containing radioactive ^{99}Tc was taken up by a patient's lungs. The orange color in the lung on the left indicates strong γ-ray emission by the ^{99}Tc, which shows that the chemical was able to pass into this lung through the bloodstream. The lung on the right shows weaker emission, indicating the presence of an embolism (a blood clot or other obstruction in an artery) that is restricting the flow of blood to this lung.

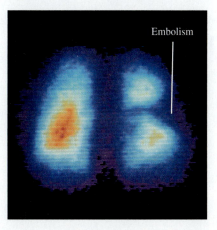

Embolism

Tracer techniques have many other applications. Tritium (^{3}H), a radioactive hydrogen isotope, is used to tag molecules in complex organic reactions; radioactive tags on pesticide molecules, for example, can be used to trace their passage through food chains. In the world of machinery, radioactive iron can be used to study piston-ring wear. Laundry detergent manufacturers have even tested the effectiveness of their products using radioactive dirt.

Many direct effects of radiation are also useful, such as strengthening polymers by cross-linking, sterilizing surgical tools, dispersing of unwanted static electricity in the air, and intentionally ionizing the air in smoke detectors. Gamma rays are also being used to sterilize and preserve some food products.

Test Your Understanding of Section 43.5 Alpha particles have 20 times the relative biological effectiveness of 200-keV x rays. Which would be better to use to radiate tissue deep inside the body? (i) a beam of alpha particles; (ii) a beam of 200-keV x rays; (iii) both are equally effective.

43.6 Nuclear Reactions

In the preceding sections we studied the decay of unstable nuclei, especially spontaneous emission of an α or β particle, sometimes followed by γ emission. Nothing needs to be done to initiate this decay, and nothing can be done to control it. This section examines some *nuclear reactions,* rearrangements of nuclear components that result from a bombardment by a particle rather than a spontaneous natural process. Rutherford suggested in 1919 that a massive particle with sufficient kinetic energy might be able to penetrate a nucleus. The result would be either a new nucleus with greater atomic number and mass number or a decay of the original nucleus. Rutherford bombarded nitrogen (^{14}N) with α particles and obtained an oxygen (^{17}O) nucleus and a proton:

$$ {}^4_2\text{He} + {}^{14}_7\text{N} \rightarrow {}^{17}_8\text{O} + {}^1_1\text{H} \qquad (43.22) $$

Rutherford used alpha particles from naturally radioactive sources. In Chapter 44 we'll describe some of the particle accelerators that are now used to initiate nuclear reactions.

Nuclear reactions are subject to several *conservation laws.* The classical conservation principles for charge, momentum, angular momentum, and energy (including rest energies) are obeyed in all nuclear reactions. An additional conservation law, not anticipated by classical physics, is conservation of the total number of nucleons. The numbers of protons and neutrons need not be conserved separately; in β decay, neutrons and protons change into one another. We'll study the basis of the conservation of nucleon number in Chapter 44.

When two nuclei interact, charge conservation requires that the sum of the initial atomic numbers must equal the sum of the final atomic numbers. Because of conservation of nucleon number, the sum of the initial mass numbers must also equal the sum of the final mass numbers. In general, these are *not* elastic collisions, and the total initial mass does *not* equal the total final mass.

Reaction Energy

The difference between the masses before and after the reaction corresponds to the **reaction energy,** according to the mass–energy relationship $E = mc^2$. If initial particles A and B interact to produce final particles C and D, the reaction energy Q is defined as

$$ Q = (M_A + M_B - M_C - M_D)c^2 \qquad \text{(reaction energy)} \qquad (43.23) $$

To balance the electrons, we use the neutral atomic masses in Eq. (43.23). That is, we use the mass of $^1_1$H for a proton, $^2_1$H for a deuteron, $^4_2$He for an α particle, and so on. When Q is positive, the total mass decreases and the total kinetic energy increases. Such a reaction is called an *exoergic reaction*. When Q is negative, the mass increases and the kinetic energy decreases, and the reaction is called an *endoergic reaction*. The terms *exothermal* and *endothermal*, borrowed from chemistry, are also used. In an endoergic reaction the reaction cannot occur at all unless the initial kinetic energy in the center-of-mass reference frame is at least as great as $|Q|$. That is, there is a **threshold energy,** the minimum kinetic energy to make an endoergic reaction go.

Example 43.11   Exoergic and endoergic reactions

(a) When a lithium-7 nucleus is bombarded by a proton, two alpha particles (^{4}He) are produced. Find the reaction energy. (b) Calculate the reaction energy for the reaction $^4_2\text{He} + {}^{14}_7\text{N} \rightarrow {}^{17}_8\text{O} + {}^1_1\text{H}$.

SOLUTION

IDENTIFY and SET UP: The reaction energy Q for any nuclear reaction equals c^2 times the difference between the total initial mass and the total final mass, as in Eq. (43.23). Table 43.2 gives the required masses.

EXECUTE: (a) The reaction is $^1_1\text{H} + {}^7_3\text{Li} \rightarrow {}^4_2\text{He} + {}^4_2\text{He}$. The initial and final masses and their respective sums are

A:	$^1_1$H	1.007825 u	C:	$^4_2$He	4.002603 u
B:	$^7_3$Li	7.016004 u	D:	$^4_2$He	4.002603 u
		8.023829 u			8.005206 u

The mass decreases by 0.018623 u. From Eq. (43.23), the reaction energy is

$$Q = (0.018623\ \text{u})(931.5\ \text{MeV/u}) = +17.35\ \text{MeV}$$

(b) The initial and final masses are

A:	$^4_2$He	4.002603 u	C:	$^{17}_8$O	16.999132 u
B:	$^{14}_7$N	14.003074 u	D:	$^1_1$H	1.007825 u
		18.005677 u			18.006957 u

The mass increases by 0.001280 u, and the corresponding reaction energy is

$$Q = (-0.001280\ \text{u})(931.5\ \text{MeV/u}) = -1.192\ \text{MeV}$$

EVALUATE: The reaction in part (a) is *exoergic:* The final total kinetic energy of the two separating alpha particles is 17.35 MeV greater than the initial total kinetic energy of the proton and the lithium nucleus. The reaction in part (b) is *endoergic:* In the center-of-mass system—that is, in a head-on collision with zero total momentum—the minimum total initial kinetic energy required for this reaction to occur is 1.192 MeV.

Ordinarily, the endoergic reaction of part (b) of Example 43.11 would be produced by bombarding stationary ^{14}N nuclei with alpha particles from an accelerator. In this case an alpha's kinetic energy must be *greater than* 1.192 MeV. If all the alpha's kinetic energy went solely to increasing the rest energy, the final kinetic energy would be zero, and momentum would not be conserved. When a particle with mass m and kinetic energy K collides with a stationary particle with mass M, the total kinetic energy K_{cm} in the center-of-mass coordinate system (the energy available to cause reactions) is

$$K_{cm} = \frac{M}{M + m} K \qquad (43.24)$$

This expression assumes that the kinetic energies of the particles and nuclei are much less than their rest energies. We leave the derivation of Eq. (43.24) to you (see Problem 43.77). In the present example, $K_{cm} = (14.00/18.01)K$, so K must be at least $(18.01/14.00)(1.192\ \text{MeV}) = 1.533\ \text{MeV}$.

For a charged particle such as a proton or an α particle to penetrate the nucleus of another atom and cause a reaction, it must usually have enough initial kinetic energy to overcome the potential-energy barrier caused by the repulsive electrostatic forces. In the reaction of part (a) of Example 43.11, if we treat the proton and the ^{7}Li nucleus as spherically symmetric charges with radii given by Eq. (43.1), their centers will be 3.5×10^{-15} m apart when they touch. The repulsive potential energy of the proton (charge $+e$) and the ^{7}Li nucleus (charge $+3e$) at this separation r is

$$U = \frac{1}{4\pi\epsilon_0}\frac{(e)(3e)}{r} = (9.0 \times 10^9 \text{ N} \cdot \text{m}^2/\text{C}^2)\frac{(3)(1.6 \times 10^{-19} \text{ C})^2}{3.5 \times 10^{-15} \text{ m}}$$

$$= 2.0 \times 10^{-13} \text{ J} = 1.2 \text{ MeV}$$

Even though the reaction is exoergic, the proton must have a minimum kinetic energy of about 1.2 MeV for the reaction to occur, unless the proton *tunnels* through the barrier (see Section 40.4).

Neutron Absorption

Absorption of *neutrons* by nuclei forms an important class of nuclear reactions. Heavy nuclei bombarded by neutrons can undergo a series of neutron absorptions alternating with beta decays, in which the mass number A increases by as much as 25. Some of the *transuranic elements,* elements having Z larger than 92, are produced in this way. These elements have not been found in nature. Many transuranic elements, having Z possibly as high as 118, have been identified.

The analytical technique of *neutron activation analysis* uses similar reactions. When bombarded by neutrons, many stable nuclides absorb a neutron to become unstable and then undergo β^- decay. The energies of the β^- and associated γ emissions depend on the unstable nuclide and provide a means of identifying it and the original stable nuclide. Quantities of elements that are far too small for conventional chemical analysis can be detected in this way.

Test Your Understanding of Section 43.6 The reaction described in part (a) of Example 43.11 is exoergic. Can it happen naturally when a sample of solid lithium is placed in a flask of hydrogen gas? ❙

PhET: Nuclear Fission

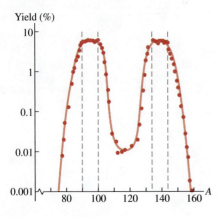

43.11 Mass distribution of fission fragments from the fission of ^{236}U* (an excited state of ^{236}U), which is produced when ^{235}U absorbs a neutron. The vertical scale is logarithmic.

43.7 Nuclear Fission

Nuclear fission is a decay process in which an unstable nucleus splits into two fragments of comparable mass. Fission was discovered in 1938 through the experiments of Otto Hahn and Fritz Strassman in Germany. Pursuing earlier work by Fermi, they bombarded uranium ($Z = 92$) with neutrons. The resulting radiation did not coincide with that of any known radioactive nuclide. Urged on by their colleague Lise Meitner, they used meticulous chemical analysis to reach the astonishing but inescapable conclusion that they had found a radioactive isotope of barium ($Z = 56$). Later, radioactive krypton ($Z = 36$) was also found. Meitner and Otto Frisch correctly interpreted these results as showing that uranium nuclei were splitting into two massive fragments called *fission fragments.* Two or three free neutrons usually appear along with the fission fragments and, very occasionally, a light nuclide such as ^{3}H.

Both the common isotope (99.3%) ^{238}U and the uncommon isotope (0.7%) ^{235}U (as well as several other nuclides) can be easily split by neutron bombardment: ^{235}U by slow neutrons (kinetic energy less than 1 eV) but ^{238}U only by fast neutrons with a minimum of about 1 MeV of kinetic energy. Fission resulting from neutron absorption is called *induced fission.* Some nuclides can also undergo *spontaneous fission* without initial neutron absorption, but this is quite rare. When ^{235}U absorbs a neutron, the resulting nuclide ^{236}U* is in a highly excited state and splits into two fragments almost instantaneously. Strictly speaking, it is ^{236}U*, not ^{235}U, that undergoes fission, but it's usual to speak of the fission of ^{235}U.

Over 100 different nuclides, representing more than 20 different elements, have been found among the fission products. Figure 43.11 shows the distribution of mass numbers for fission fragments from the fission of ^{235}U. Most of the fragments have mass numbers from 90 to 100 and from 135 to 145; fission into two fragments with nearly equal mass is unlikely.

Fission Reactions

You should check the following two typical fission reactions for conservation of nucleon number and charge:

$$^{235}_{92}\text{U} + ^{1}_{0}\text{n} \rightarrow ^{236}_{92}\text{U}^* \rightarrow ^{144}_{56}\text{Ba} + ^{89}_{36}\text{Kr} + 3^{1}_{0}\text{n}$$

$$^{235}_{92}\text{U} + ^{1}_{0}\text{n} \rightarrow ^{236}_{92}\text{U}^* \rightarrow ^{140}_{54}\text{Xe} + ^{94}_{38}\text{Sr} + 2^{1}_{0}\text{n}$$

The total kinetic energy of the fission fragments is enormous, about 200 MeV (compared to typical α and β energies of a few MeV). The reason for this is that nuclides at the high end of the mass spectrum (near $A = 240$) are less tightly bound than those nearer the middle ($A = 90$ to 145). Referring to Fig. 43.2, we see that the average binding energy per nucleon is about 7.6 MeV at $A = 240$ but about 8.5 MeV at $A = 120$. Therefore a rough estimate of the expected *increase* in binding energy during fission is about 8.5 MeV $-$ 7.6 MeV $= 0.9$ MeV per nucleon, or a total of $(235)(0.9 \text{ MeV}) \approx 200$ MeV.

CAUTION **Binding energy and rest energy** It may seem to be a violation of conservation of energy to have an increase in both the binding energy and the kinetic energy during a fission reaction. But relative to the total rest energy E_0 of the separated nucleons, the rest energy of the nucleus is E_0 *minus* E_B. Thus an *increase* in binding energy corresponds to a *decrease* in rest energy as rest energy is converted to the kinetic energy of the fission fragments. ▮

Fission fragments always have too many neutrons to be stable. We noted in Section 43.3 that the neutron–proton ratio (N/Z) for stable nuclides is about 1 for light nuclides but almost 1.6 for the heaviest nuclides because of the increasing influence of the electrical repulsion of the protons. The N/Z value for stable nuclides is about 1.3 at $A = 100$ and 1.4 at $A = 150$. The fragments have about the same N/Z as ^{235}U, about 1.55. They usually respond to this surplus of neutrons by undergoing a series of β^- decays (each of which increases Z by 1 and decreases N by 1) until a stable value of N/Z is reached. A typical example is

$$^{140}_{54}\text{Xe} \xrightarrow{\beta^-} ^{140}_{55}\text{Cs} \xrightarrow{\beta^-} ^{140}_{56}\text{Ba} \xrightarrow{\beta^-} ^{140}_{57}\text{La} \xrightarrow{\beta^-} ^{140}_{58}\text{Ce}$$

The nuclide ^{140}Ce is stable. This series of β^- decays produces, on average, about 15 MeV of additional kinetic energy. The neutron excess of fission fragments also explains why two or three free neutrons are released during the fission.

Fission appears to set an upper limit on the production of transuranic nuclei, mentioned in Section 43.6, that are relatively stable. There are theoretical reasons to expect that nuclei near $Z = 114$, $N = 184$ or 196, might be stable with respect to spontaneous fission. In the shell model (see Section 43.2), these numbers correspond to filled shells and subshells in the nuclear energy-level structure. Such *superheavy nuclei* would still be unstable with respect to alpha emission. In 2009 it was confirmed that there are at least four isotopes with $Z = 114$, the longest-lived of which has a half-life due to alpha decay of about 2.6 s.

Liquid-Drop Model

We can understand fission qualitatively on the basis of the liquid-drop model of the nucleus (see Section 43.2). The process is shown in Fig. 43.12 in terms of an electrically charged liquid drop. These sketches shouldn't be taken too literally, but they may help to develop your intuition about fission. A ^{235}U nucleus absorbs a neutron (Fig. 43.12a), becoming a $^{236}\text{U}^*$ nucleus with excess energy (Fig. 43.12b). This excess energy causes violent oscillations, during which a neck between two lobes develops (Fig. 43.12c). The electric repulsion of these two lobes stretches the neck farther (Fig. 43.12d), and finally two smaller fragments are formed (Fig. 43.12e) that move rapidly apart.

43.12 A liquid-drop model of fission.

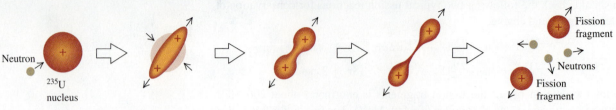

(a) A ^{235}U nucleus absorbs a neutron.

(b) The resulting ^{236}U* nucleus is in a highly excited state and oscillates strongly.

(c) A neck develops, and electric repulsion pushes the two lobes apart.

(d) The two lobes separate, forming fission fragments.

(e) The fragments emit neutrons at the time of fission (or occasionally a few seconds later).

43.13 Hypothetical potential-energy function for two fission fragments in a fissionable nucleus. At distances r beyond the range of the nuclear force, the potential energy varies approximately as $1/r$. Fission occurs if there is an excitation energy greater than U_B or an appreciable probability for tunneling through the potential-energy barrier.

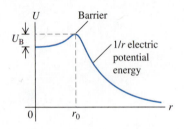

Application **Making Radioactive Isotopes for Medicine**
The fragments that result from nuclear fission are typically unstable, neutron-rich isotopes. A number of these are useful for medical diagnosis and cancer radiotherapy (see Section 43.5). This photograph shows a nuclear fission reactor used for producing such isotopes. The uranium fuel is kept in a large tank of water for cooling. Some of the neutron-rich fission fragments undergo beta decay and emit electrons that move faster than the speed of light in water (about $0.75c$). Like an airplane that produces an intense sonic boom when it flies faster than sound (see Section 16.9), these ultrafast electrons produce a "light boom" called Čerenkov radiation that has a characteristic blue color.

This qualitative picture has been developed into a more quantitative theory to explain why some nuclei undergo fission and others don't. Figure 43.13 shows a hypothetical potential-energy function for two possible fission fragments. If neutron absorption results in an excitation energy greater than the energy barrier height U_B, fission occurs immediately. Even when there isn't quite enough energy to surmount the barrier, fission can take place by quantum-mechanical *tunneling*, discussed in Section 40.4. In principle, many stable heavy nuclei can fission by tunneling. But the probability depends very critically on the height and width of the barrier. For most nuclei this process is so unlikely that it is never observed.

Chain Reactions

Fission of a uranium nucleus, triggered by neutron bombardment, releases other neutrons that can trigger more fissions, suggesting the possibility of a **chain reaction** (Fig. 43.14). The chain reaction may be made to proceed slowly and in a controlled manner in a nuclear reactor or explosively in a bomb. The energy release in a nuclear chain reaction is enormous, far greater than that in any chemical reaction. (In a sense, *fire* is a chemical chain reaction.) For example, when uranium is "burned" to uranium dioxide in the chemical reaction

$$U + O_2 \rightarrow UO_2$$

the heat of combustion is about 4500 J/g. Expressed as energy per atom, this is about 11 eV per atom. By contrast, fission liberates about 200 MeV per atom, nearly 20 million times as much energy.

Nuclear Reactors

A *nuclear reactor* is a system in which a controlled nuclear chain reaction is used to liberate energy. In a nuclear power plant, this energy is used to generate steam, which operates a turbine and turns an electrical generator.

On average, each fission of a ^{235}U nucleus produces about 2.5 free neutrons, so 40% of the neutrons are needed to sustain a chain reaction. A ^{235}U nucleus is much more likely to absorb a low-energy neutron (less than 1 eV) than one of the higher-energy neutrons (1 MeV or so) that are liberated during fission. In a nuclear reactor the higher-energy neutrons are slowed down by collisions with nuclei in the surrounding material, called the *moderator*, so they are much more likely to cause further fissions. In nuclear power plants, the moderator is often water, occasionally graphite. The *rate* of the reaction is controlled by inserting or withdrawing *control rods* made of elements (such as boron or cadmium) whose nuclei *absorb* neutrons without undergoing any additional reaction. The isotope ^{238}U can also absorb neutrons, leading to ^{239}U*, but not with high enough probability for it to sustain a chain reaction by itself. Thus uranium that is used in reactors is often "enriched" by increasing the proportion of ^{235}U above the natural value of 0.7%, typically to 3% or so, by isotope-separation processing.

43.14 Schematic diagram of a nuclear fission chain reaction.

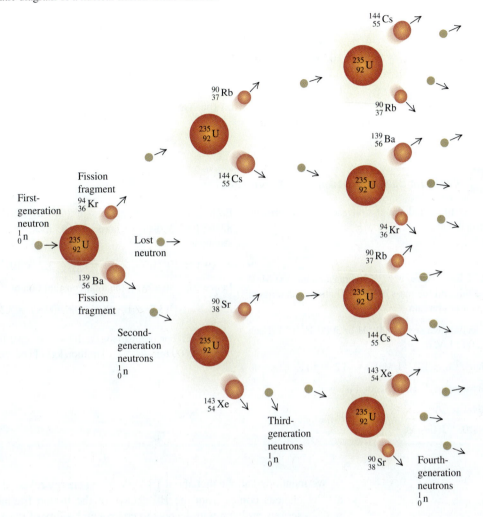

The most familiar application of nuclear reactors is for the generation of electric power. As was noted above, the fission energy appears as kinetic energy of the fission fragments, and its immediate result is to increase the internal energy of the fuel elements and the surrounding moderator. This increase in internal energy is transferred as heat to generate steam to drive turbines, which spin the electrical generators. Figure 43.15 is a schematic diagram of a nuclear power plant.

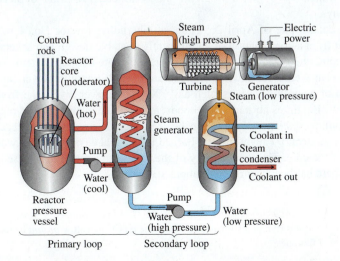

43.15 Schematic diagram of a nuclear power plant.

The energetic fission fragments heat the water surrounding the reactor core. The steam generator is a heat exchanger that takes heat from this highly radioactive water and generates nonradioactive steam to run the turbines.

A typical nuclear plant has an electric-generating capacity of 1000 MW (or 10^9 W). The turbines are heat engines and are subject to the efficiency limitations imposed by the second law of thermodynamics, discussed in Chapter 20. In modern nuclear plants the overall efficiency is about one-third, so 3000 MW of thermal power from the fission reaction is needed to generate 1000 MW of electrical power.

Example 43.12 Uranium consumption in a nuclear reactor

What mass of ^{235}U must undergo fission each day to provide 3000 MW of thermal power?

SOLUTION

IDENTIFY and SET UP: Fission of ^{235}U liberates about 200 MeV per atom. We use this and the mass of the ^{235}U atom to determine the required amount of uranium.

EXECUTE: Each second, we need 3000 MJ or 3000×10^6 J. Each fission provides 200 MeV, or

$$(200 \text{ MeV/fission})(1.6 \times 10^{-13} \text{ J/MeV}) = 3.2 \times 10^{-11} \text{ J/fission}$$

The number of fissions needed each second is

$$\frac{3000 \times 10^6 \text{ J}}{3.2 \times 10^{-11} \text{ J/fission}} = 9.4 \times 10^{19} \text{ fissions}$$

Each ^{235}U atom has a mass of $(235 \text{ u})(1.66 \times 10^{-27} \text{ kg/u}) = 3.9 \times 10^{-25}$ kg, so the mass of ^{235}U that undergoes fission each second is

$$(9.4 \times 10^{19})(3.9 \times 10^{-25} \text{ kg}) = 3.7 \times 10^{-5} \text{ kg} = 37 \ \mu\text{g}$$

In one day (86,400 s), the total consumption of ^{235}U is

$$(3.7 \times 10^{-5} \text{ kg/s})(86,400 \text{ s}) = 3.2 \text{ kg}$$

EVALUATE: For comparison, a 1000-MW coal-fired power plant burns 10,600 tons (about 10 million kg) of coal per day!

We mentioned above that about 15 MeV of the energy released after fission of a ^{235}U nucleus comes from the β^- decays of the fission fragments. This fact poses a serious problem with respect to control and safety of reactors. Even after the chain reaction has been completely stopped by insertion of control rods into the core, heat continues to be evolved by the β^- decays, which cannot be stopped. For a 3000-MW reactor this heat power is initially very large, about 200 MW. In the event of total loss of cooling water, this power is more than enough to cause a catastrophic meltdown of the reactor core and possible penetration of the containment vessel. The difficulty in achieving a "cold shutdown" following an accident at the Three Mile Island nuclear power plant in Pennsylvania in March 1979 was a result of the continued evolution of heat due to β^- decays.

The catastrophe of April 26, 1986, at Chernobyl reactor No. 4 in Ukraine resulted from a combination of an inherently unstable design and several human errors committed during a test of the emergency core cooling system. Too many control rods were withdrawn to compensate for a decrease in power caused by a buildup of neutron absorbers such as ^{135}Xe. The power level rose from 1% of normal to 100 times normal in 4 seconds; a steam explosion ruptured pipes in the core cooling system and blew the heavy concrete cover off the reactor. The graphite moderator caught fire and burned for several days, and there was a meltdown of the core. The total activity of the radioactive material released into the atmosphere has been estimated as about 10^8 Ci.

Test Your Understanding of Section 43.7 The fission of ^{235}U can be triggered by the absorption of a slow neutron by a nucleus. Can a slow *proton* be used to trigger ^{235}U fission?

43.16 The proton-proton chain.

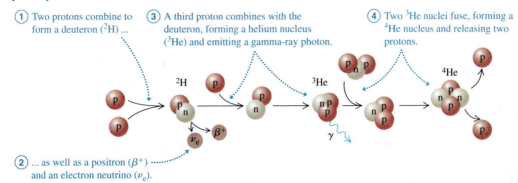

① Two protons combine to form a deuteron (^{2}H) ...

③ A third proton combines with the deuteron, forming a helium nucleus (^{3}He) and emitting a gamma-ray photon.

④ Two ^{3}He nuclei fuse, forming a ^{4}He nucleus and releasing two protons.

② ... as well as a positron (β^+) and an electron neutrino (ν_e).

43.8 Nuclear Fusion

ActivPhysics 19.3: Fusion

In a **nuclear fusion** reaction, two or more small light nuclei come together, or *fuse,* to form a larger nucleus. Fusion reactions release energy for the same reason as fission reactions: The binding energy per nucleon after the reaction is greater than before. Referring to Fig. 43.2, we see that the binding energy per nucleon increases with A up to about $A = 60$, so fusion of nearly any two light nuclei to make a nucleus with A less than 60 is likely to be an exoergic reaction. In comparison to fission, we are moving toward the peak of this curve from the opposite side. Another way to express the energy relationships is that the total mass of the products is less than that of the initial particles.

Here are three examples of energy-liberating fusion reactions, written in terms of the neutral atoms:

$$^1_1H + {}^1_1H \rightarrow {}^2_1H + \beta^+ + \nu_e$$

$$^2_1H + {}^1_1H \rightarrow {}^3_2He + \gamma$$

$$^3_2He + {}^3_2He \rightarrow {}^4_2He + {}^1_1H + {}^1_1H$$

In the first reaction, two protons combine to form a deuteron (^{2}H), with the emission of a positron (β^+) and an electron neutrino. In the second, a proton and a deuteron combine to form the nucleus of the light isotope of helium, ^{3}He, with the emission of a gamma ray. Now double the first two reactions to provide the two ^{3}He nuclei that fuse in the third reaction to form an alpha particle (^{4}He) and two protons. Together the reactions make up the process called the *proton-proton chain* (Fig. 43.16).

The net effect of the chain is the conversion of four protons into one α particle, two positrons, two electron neutrinos, and two γ's. We can calculate the energy release from this part of the process: The mass of an α particle plus two positrons is the mass of neutral ^{4}He, the neutrinos have zero (or negligible) mass, and the gammas have zero mass.

Mass of four protons	4.029106 u
Mass of ^{4}He	4.002603 u
Mass difference and energy release	0.026503 u and 24.69 MeV

The two positrons that are produced during the first step of the proton-proton chain collide with two electrons; mutual annihilation of the four particles takes place, and their rest energy is converted into $4(0.511 \text{ MeV}) = 2.044 \text{ MeV}$ of gamma radiation. Thus the total energy released is $(24.69 + 2.044) \text{ MeV} = 26.73 \text{ MeV}$. The proton-proton chain takes place in the interior of the sun and other stars (Fig. 43.17). Each gram of the sun's mass contains about 4.5×10^{23} protons. If all of these protons were fused into helium, the energy released would be about 130,000 kWh. If the sun were to continue to radiate at its present rate, it would take about 75×10^9 years to exhaust its supply of protons. As we will see below, fusion reactions can

43.17 The energy released as starlight comes from fusion reactions deep within a star's interior. When a star is first formed and for most of its life, it converts the hydrogen in its core into helium. As a star ages, the core temperature can become high enough for additional fusion reactions that convert helium into carbon, oxygen, and other elements.

occur only at extremely high temperatures; in the sun, these temperatures are found only deep within the interior. Hence the sun cannot fuse *all* of its protons, and can sustain fusion for a total of only about 10×10^9 years in total. The present age of the solar system (including the sun) is 4.54×10^9 years, so the sun is about halfway through its available store of protons.

Example 43.13   A fusion reaction

Two deuterons fuse to form a *triton* (a nucleus of tritium, or ^{3}H) and a proton. How much energy is liberated?

SOLUTION

IDENTIFY and SET UP: This is a nuclear reaction of the type discussed in Section 43.6. We find the energy released using Eq. (43.23).

EXECUTE: Adding one electron to each nucleus makes each a neutral atom; we find their masses in Table 43.2. Substituting into Eq. (43.23), we find

$$Q = [2(2.014102 \text{ u}) - 3.016049 \text{ u} - 1.007825 \text{ u}]$$
$$\times (931.5 \text{ MeV/u}) = 4.03 \text{ MeV}$$

EVALUATE: Thus 4.03 MeV is released in the reaction; the triton and proton together have 4.03 MeV more kinetic energy than the two deuterons had together.

Achieving Fusion

For two nuclei to undergo fusion, they must come together to within the range of the nuclear force, typically of the order of 2×10^{-15} m. To do this, they must overcome the electrical repulsion of their positive charges. For two protons at this distance, the corresponding potential energy is about 1.2×10^{-13} J or 0.7 MeV; this represents the total initial *kinetic* energy that the fusion nuclei must have—for example, 0.6×10^{-13} J each in a head-on collision.

Atoms have this much energy only at extremely high temperatures. The discussion of Section 18.3 showed that the average translational kinetic energy of a gas molecule at temperature T is $\frac{3}{2}kT$, where k is Boltzmann's constant. The temperature at which this is equal to $E = 0.6 \times 10^{-13}$ J is determined by the relationship

$$E = \frac{3}{2}kT$$

$$T = \frac{2E}{3k} = \frac{2(0.6 \times 10^{-13} \text{ J})}{3(1.38 \times 10^{-23} \text{ J/K})} = 3 \times 10^9 \text{ K}$$

43.18 This target chamber at the National Ignition Facility in California has apertures for 192 powerful laser beams. The lasers deliver 5×10^{14} W of power for a few nanoseconds to a millimeter-sized pellet of deuterium and tritium at the center of the chamber, thus triggering thermonuclear fusion.

Fusion reactions are possible at lower temperatures because the Maxwell–Boltzmann distribution function (see Section 18.5) gives a small fraction of protons with kinetic energies much higher than the average value. The proton-proton reaction occurs at "only" 1.5×10^7 K at the center of the sun, making it an extremely low-probability process; but that's why the sun is expected to last so long. At these temperatures the fusion reactions are called *thermonuclear* reactions.

Intensive efforts are under way to achieve controlled fusion reactions, which potentially represent an enormous new resource of energy (see Fig. 24.11). At the temperatures mentioned, light atoms are fully ionized, and the resulting state of matter is called a *plasma*. In one kind of experiment using *magnetic confinement,* a plasma is heated to extremely high temperature by an electrical discharge, while being contained by appropriately shaped magnetic fields. In another, using *inertial confinement,* pellets of the material to be fused are heated by a high-intensity laser beam (see Fig. 43.18). Some of the reactions being studied are

$$^2_1\text{H} + {}^2_1\text{H} \rightarrow {}^3_1\text{H} + {}^1_1\text{H} + 4.0 \text{ MeV} \qquad (1)$$

$$^3_1\text{H} + {}^2_1\text{H} \rightarrow {}^4_2\text{He} + {}^1_0\text{n} + 17.6 \text{ MeV} \qquad (2)$$

$$^2_1\text{H} + {}^2_1\text{H} \rightarrow {}^3_2\text{He} + {}^1_0\text{n} + 3.3 \text{ MeV} \qquad (3)$$

$$^3_2\text{He} + {}^2_1\text{H} \rightarrow {}^4_2\text{He} + {}^1_1\text{H} + 18.3 \text{ MeV} \qquad (4)$$

We considered reaction (1) in Example 43.13; two deuterons fuse to form a triton and a proton. In reaction (2) a triton combines with another deuteron to form an alpha particle and a neutron. The result of both of these reactions together is the conversion of three deuterons into an alpha particle, a proton, and a neutron, with the liberation of 21.6 MeV of energy. Reactions (3) and (4) together achieve the same conversion. In a plasma that contains deuterons, the two pairs of reactions occur with roughly equal probability. As yet, no one has succeeded in producing these reactions under controlled conditions in such a way as to yield a net surplus of usable energy.

Methods of achieving fusion that don't require high temperatures are also being studied; these are called *cold fusion*. One scheme that does work uses an unusual hydrogen molecule ion. The usual H_2^+ ion consists of two protons bound by one shared electron; the nuclear spacing is about 0.1 nm. If the protons are replaced by a deuteron (^2H) and a triton (^3H) and the electron by a *muon*, which is 208 times as massive as the electron, the spacing is made smaller by a factor of 208. The probability then becomes appreciable for the two nuclei to tunnel through the narrow repulsive potential-energy barrier and fuse in reaction (2) above. The prospect of making this process, called *muon-catalyzed fusion,* into a practical energy source is still distant.

Test Your Understanding of Section 43.8   Are *all* fusion reactions exoergic?   ❙

Nuclear properties: A nucleus is composed of A nucleons (Z protons and N neutrons). All nuclei have about the same density. The radius of a nucleus with mass number A is given approximately by Eq. (43.1). A single nuclear species of a given Z and N is called a nuclide. Isotopes are nuclides of the same element (same Z) that have different numbers of neutrons. Nuclear masses are measured in atomic mass units. Nucleons have angular momentum and a magnetic moment. (See Examples 43.1 and 43.2.)

$$R = R_0 A^{1/3}$$
$$(R_0 = 1.2 \times 10^{-15}\text{ m})$$
(43.1)

Nuclear binding and structure: The mass of a nucleus is always less than the mass of the protons and neutrons within it. The mass difference multiplied by c^2 gives the binding energy E_B. The binding energy for a given nuclide is determined by the nuclear force, which is short range and favors pairs of particles, and by the electric repulsion between protons. A nucleus is unstable if A or Z is too large or if the ratio N/Z is wrong. Two widely used models of the nucleus are the liquid-drop model and the shell model; the latter is analogous to the central-field approximation for atomic structure. (See Examples 43.3 and 43.4.)

$$E_B = (ZM_H + Nm_n - {}_Z^A M)c^2 \quad (43.10)$$

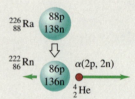

Radioactive decay: Unstable nuclides usually emit an alpha particle (a ${}_2^4$He nucleus) or a beta particle (an electron) in the process of changing to another nuclide, sometimes followed by a gamma-ray photon. The rate of decay of an unstable nucleus is described by the decay constant λ, the half-life $T_{1/2}$, or the lifetime T_{mean}. If the number of nuclei at time $t = 0$ is N_0 and no more are produced, the number at time t is given by Eq. (43.17). (See Examples 43.5–43.9.)

$$N(t) = N_0 e^{-\lambda t} \quad (43.17)$$
$$T_{\text{mean}} = \frac{1}{\lambda} = \frac{T_{1/2}}{\ln 2} = \frac{T_{1/2}}{0.693} \quad (43.19)$$

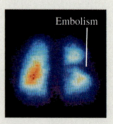

Biological effects of radiation: The biological effect of any radiation depends on the product of the energy absorbed per unit mass and the relative biological effectiveness (RBE), which is different for different radiations. (See Example 43.10.)

Embolism

Nuclear reactions: In a nuclear reaction, two nuclei or particles collide to produce two new nuclei or particles. Reactions can be exoergic or endoergic. Several conservation laws, including charge, energy, momentum, angular momentum, and nucleon number, are obeyed. Energy is released by the fission of a heavy nucleus into two lighter, always unstable, nuclei. Energy is also released by the fusion of two light nuclei into a heavier nucleus. (See Examples 43.11–43.13.)

BRIDGING PROBLEM   Saturation of ^{128}I Production

In an experiment, the iodine isotope ^{128}I is created by irradiating a sample of ^{127}I with a beam of neutrons, yielding 1.50×10^6 ^{128}I nuclei per second. Initially no ^{128}I nuclei are present. A ^{128}I nucleus decays by β^- emission with a half-life of 25.0 min. (a) To what nuclide does ^{128}I decay? (b) Could that nuclide decay back to ^{128}I by β^+ emission? Why or why not? (c) After the sample has been irradiated for a long time, what is the maximum number of ^{128}I atoms that can be present in the sample? What is the maximum activity that can be produced? (This steady-state situation is called *saturation*.) (d) Find an expression for the number of ^{128}I atoms present in the sample as a function of time.

SOLUTION GUIDE

See MasteringPhysics® study area for a Video Tutor solution.

IDENTIFY and SET UP

1. What happens to the values of Z, N, and A in β^- decay? What must be true for β^- decay to be possible? For β^+ decay to be possible?
2. You'll need to write an equation for the rate of change dN/dt of the number N of ^{128}I atoms in the sample, taking account of

both the creation of ^{128}I by the neutron irradiation and the decay of any ^{128}I present. In the steady state, how do the rates of these two processes compare?
3. List the unknown quantities for each part of the problem and identify your target variables.

EXECUTE

4. Find the values of Z and N of the nuclide produced by the decay of ^{128}I. What element is this?
5. Decide whether this nuclide can decay back to ^{128}I.
6. Inspect your equation for dN/dt. What is the value of dN/dt in the steady state? Use this to solve for the steady-state values of N and the activity.
7. Solve your dN/dt equation for the function $N(t)$. (*Hint:* See Section 26.4.)

EVALUATE

8. Your result from step 6 tells you the value of N after a long time (that is, for large values of t). Is this consistent with your result from step 7? What would constitute a "long time" under these conditions?

Problems

For instructor-assigned homework, go to www.masteringphysics.com

•, ••, •••: Problems of increasing difficulty. **CP**: Cumulative problems incorporating material from earlier chapters. **CALC**: Problems requiring calculus. **BIO**: Biosciences problems.

DISCUSSION QUESTIONS

Q43.1 BIO Neutrons have a magnetic dipole moment and can undergo spin flips by absorbing electromagnetic radiation. Why, then, are protons rather than neutrons used in MRI of body tissues? (See Fig. 43.1.)

Q43.2 In Eq. (43.11), as the total number of nucleons becomes larger, the importance of the second term in the equation decreases relative to that of the first term. Does this make physical sense? Explain.

Q43.3 Why aren't the masses of all nuclei integer multiples of the mass of a single nucleon?

Q43.4 Can you tell from the value of the mass number A whether to use a plus value, a minus value, or zero for the fifth term of Eq. (43.11)? Explain.

Q43.5 What are the six known elements for which Z is a magic number? Discuss what properties these elements have as a consequence of their special values of Z.

Q43.6 The binding energy per nucleon for most nuclides doesn't vary much (see Fig. 43.2). Is there similar consistency in the *atomic* energy of atoms, on an "energy per electron" basis? If so, why? If not, why not?

Q43.7 Heavy, unstable nuclei usually decay by emitting an α or β particle. Why don't they usually emit a single proton or neutron?

Q43.8 The only two stable nuclides with more protons than neutrons are $^1_1$H and $^3_2$He. Why is $Z > N$ so uncommon?

Q43.9 Since lead is a stable element, why doesn't the ^{238}U decay series shown in Fig. 43.7 stop at lead, ^{214}Pb?

Q43.10 In the ^{238}U decay series shown in Fig. 43.7, some nuclides in the series are found much more abundantly in nature than others, even though every ^{238}U nucleus goes through every step in the series before finally becoming ^{206}Pb. Why don't the intermediate nuclides all have the same abundance?

Q43.11 Compared to α particles with the same energy, β particles can much more easily penetrate through matter. Why is this?

Q43.12 If ^A_ZEl$_i$ represents the initial nuclide, what is the decay process or processes if the final nuclide is (a) $^A_{Z+1}$El$_f$; (b) $^{A-4}_{Z-2}$El$_f$; (c) $^A_{Z-1}$El$_f$?

Q43.13 In a nuclear decay equation, why can we represent an electron as $^{\ 0}_{-1}\beta^-$? What are the equivalent representations for a positron, a neutrino, and an antineutrino?

Q43.14 Why is the alpha, beta, or gamma decay of an unstable nucleus unaffected by the *chemical* situation of the atom, such as the nature of the molecule or solid in which it is bound? The chemical situation of the atom can, however, have an effect on the half-life in electron capture. Why is this?

Q43.15 In the process of *internal conversion*, a nucleus decays from an excited state to a ground state by giving the excitation energy directly to an atomic electron rather than emitting a gamma-ray photon. Why can this process also produce x-ray photons?

Q43.16 In Example 43.9 (Section 43.4), the activity of atmospheric carbon *before* 1900 was given. Discuss why this activity may have changed since 1900.

Q43.17 BIO One problem in radiocarbon dating of biological samples, especially very old ones, is that they can easily be contaminated with modern biological material during the measurement process. What effect would such contamination have on the estimated age? Why is such contamination a more serious problem for samples of older material than for samples of younger material?

Q43.18 The most common radium isotope found on earth, ^{226}Ra, has a half-life of about 1600 years. If the earth was formed well over 10^9 years ago, why is there any radium left now?

Q43.19 Fission reactions occur only for nuclei with large nucleon numbers, while exoergic fusion reactions occur only for nuclei with small nucleon numbers. Why is this?

Q43.20 When a large nucleus splits during nuclear fission, the daughter nuclei of the fission fly apart with enormous kinetic energy. Why does this happen?

Q43.21 As stars age, they use up their supply of hydrogen and eventually begin producing energy by a reaction that involves the fusion of three helium nuclei to form a carbon nucleus. Would you expect the interiors of these old stars to be hotter or cooler than the interiors of younger stars? Explain.

EXERCISES

Section 43.1 Properties of Nuclei

43.1 • How many protons and how many neutrons are there in a nucleus of the most common isotope of (a) silicon, $^{28}_{14}$Si; (b) rubidium, $^{85}_{37}$Rb; (c) thallium, $^{205}_{81}$Tl?

43.2 •• CP Hydrogen atoms are placed in an external 1.65-T magnetic field. (a) The *protons* can make transitions between states where the nuclear spin component is parallel and antiparallel to the field by absorbing or emitting a photon. Which state has lower energy: the state with the nuclear spin component parallel or antiparallel to the field? What are the frequency and wavelength of the photon? In which region of the electromagnetic spectrum does it lie? (b) The *electrons* can make transitions between states where the electron spin component is parallel and antiparallel to the field by absorbing or emitting a photon. Which state has lower energy: the state with the electron spin component parallel or antiparallel to the field? What are the frequency and wavelength of the photon? In which region of the electromagnetic spectrum does it lie?

43.3 • Hydrogen atoms are placed in an external magnetic field. The protons can make transitions between states in which the nuclear spin component is parallel and antiparallel to the field by absorbing or emitting a photon. What magnetic-field magnitude is required for this transition to be induced by photons with frequency 22.7 MHz?

43.4 •• Neutrons are placed in a magnetic field with magnitude 2.30 T. (a) What is the energy difference between the states with the nuclear spin angular momentum components parallel and antiparallel to the field? Which state is lower in energy: the one with its spin component parallel to the field or the one with its spin component antiparallel to the field? How do your results compare with the energy states for a proton in the same field (see Example 43.2)? (b) The neutrons can make transitions from one of these states to the other by emitting or absorbing a photon with energy equal to the energy difference of the two states. Find the frequency and wavelength of such a photon.

Section 43.2 Nuclear Binding and Nuclear Structure

43.5 • The most common isotope of boron is $^{11}_{5}$B. (a) Determine the total binding energy of $^{11}_{5}$B from Table 43.2 in Section 43.1. (b) Calculate this binding energy from Eq. (43.11). (Why is the fifth

term zero?) Compare to the result you obtained in part (a). What is the percent difference? Compare the accuracy of Eq. (43.11) for $^{11}_{5}$B to its accuracy for $^{62}_{28}$Ni (see Example 43.4).

43.6 • The most common isotope of uranium, $^{238}_{92}$U, has atomic mass 238.050783 u. Calculate (a) the mass defect; (b) the binding energy (in MeV); (c) the binding energy per nucleon.

43.7 • CP What is the maximum wavelength of a γ ray that could break a deuteron into a proton and a neutron? (This process is called photodisintegration.)

43.8 • Calculate (a) the total binding energy and (b) the binding energy per nucleon of ^{12}C. (c) What percent of the rest mass of this nucleus is its total binding energy?

43.9 • CP A photon with a wavelength of 3.50×10^{-13} m strikes a deuteron, splitting it into a proton and a neutron. (a) Calculate the kinetic energy released in this interaction. (b) Assuming the two particles share the energy equally, and taking their masses to be 1.00 u, calculate their speeds after the photodisintegration.

43.10 • Calculate the mass defect, the binding energy (in MeV), and the binding energy per nucleon of (a) the nitrogen nucleus, $^{14}_{7}$N, and (b) the helium nucleus, $^{4}_{2}$He. (c) How does the binding energy per nucleon compare for these two nuclei?

43.11 • Use Eq. (43.11) to calculate the binding energy per nucleon for the nuclei $^{86}_{36}$Kr and $^{180}_{73}$Ta. Do your results confirm what is shown in Fig. 43.2—that for A greater than 62 the binding energy per nucleon deceases as A increases?

Section 43.3 Nuclear Stability and Radioactivity

43.12 • (a) Is the decay $n \rightarrow p + \beta^- + \bar{\nu}_e$ energetically possible? If not, explain why not. If so, calculate the total energy released. (b) Is the decay $p \rightarrow n + \beta^+ + \nu_e$ energetically possible? If not, explain why not. If so, calculate the total energy released.

43.13 • What nuclide is produced in the following radioactive decays? (a) α decay of $^{239}_{94}$Pu; (b) β^- decay of $^{24}_{11}$Na; (c) β^+ decay of $^{15}_{8}$O.

43.14 •• CP ^{238}U decays spontaneously by α emission to ^{234}Th. Calculate (a) the total energy released by this process and (b) the recoil velocity of the ^{234}Th nucleus. The atomic masses are 238.050788 u for ^{238}U and 234.043601 u for ^{234}Th.

43.15 •• The atomic mass of ^{14}C is 14.003242 u. Show that the β^- decay of ^{14}C is energetically possible, and calculate the energy released in the decay.

43.16 • What particle (α particle, electron, or positron) is emitted in the following radioactive decays? (a) $^{27}_{14}$Si $\rightarrow$ $^{27}_{13}$Al; (b) $^{238}_{92}$U $\rightarrow$ $^{234}_{90}$Th; (c) $^{74}_{33}$As $\rightarrow$ $^{74}_{34}$Se.

43.17 •• (a) Calculate the energy released by the electron-capture decay of $^{57}_{27}$Co (see Example 43.7). (b) A negligible amount of this energy goes to the resulting $^{57}_{26}$Fe atom as kinetic energy. About 90% of the time, the $^{57}_{26}$Fe nucleus emits two successive gamma-ray photons after the electron-capture process, of energies 0.122 MeV and 0.014 MeV, respectively, in decaying to its ground state. What is the energy of the neutrino emitted in this case?

43.18 • Tritium ($^{3}_{1}$H) is an unstable isotope of hydrogen; its mass, including one electron, is 3.016049 u. (a) Show that tritium must be unstable with respect to beta decay because the decay products ($^{3}_{2}$He plus an emitted electron) have less total mass than the tritium. (b) Determine the total kinetic energy (in MeV) of the decay products, taking care to account for the electron masses correctly.

Section 43.4 Activities and Half-Lives

43.19 • If a 6.13-g sample of an isotope having a mass number of 124 decays at a rate of 0.350 Ci, what is its half-life?

43.20 • **BIO** Radioactive isotopes used in cancer therapy have a "shelf-life," like pharmaceuticals used in chemotherapy. Just after it has been manufactured in a nuclear reactor, the activity of a sample of ^{60}Co is 5000 Ci. When its activity falls below 3500 Ci, it is considered too weak a source to use in treatment. You work in the radiology department of a large hospital. One of these ^{60}Co sources in your inventory was manufactured on October 6, 2004. It is now April 6, 2007. Is the source still usable? The half-life of ^{60}Co is 5.271 years.

43.21 •• The common isotope of uranium, ^{238}U, has a half-life of 4.47×10^9 years, decaying to ^{234}Th by alpha emission. (a) What is the decay constant? (b) What mass of uranium is required for an activity of 1.00 curie? (c) How many alpha particles are emitted per second by 10.0 g of uranium?

43.22 •• **BIO Radiation Treatment of Prostate Cancer.** In many cases, prostate cancer is treated by implanting 60 to 100 small seeds of radioactive material into the tumor. The energy released from the decays kills the tumor. One isotope that is used (there are others) is palladium (^{103}Pd), with a half-life of 17 days. If a typical grain contains 0.250 g of ^{103}Pd, (a) what is its initial activity rate in Bq, and (b) what is the rate 68 days later?

43.23 •• A 12.0-g sample of carbon from living matter decays at the rate of 180.0 decays/min due to the radioactive ^{14}C in it. What will be the decay rate of this sample in (a) 1000 years and (b) 50,000 years?

43.24 •• **BIO Radioactive Tracers.** Radioactive isotopes are often introduced into the body through the bloodstream. Their spread through the body can then be monitored by detecting the appearance of radiation in different organs. ^{131}I, a β^- emitter with a half-life of 8.0 d, is one such tracer. Suppose a scientist introduces a sample with an activity of 375 Bq and watches it spread to the organs. (a) Assuming that the sample all went to the thyroid gland, what will be the decay rate in that gland 24 d (about $3\frac{1}{2}$ weeks) later? (b) If the decay rate in the thyroid 24 d later is actually measured to be 17.0 Bq, what percentage of the tracer went to that gland? (c) What isotope remains after the I-131 decays?

43.25 •• The unstable isotope ^{40}K is used for dating rock samples. Its half-life is 1.28×10^9 y. (a) How many decays occur per second in a sample containing 1.63×10^{-6} g of ^{40}K? (b) What is the activity of the sample in curies?

43.26 • As a health physicist, you are being consulted about a spill in a radiochemistry lab. The isotope spilled was 500 μCi of ^{131}Ba, which has a half-life of 12 days. (a) What mass of ^{131}Ba was spilled? (b) Your recommendation is to clear the lab until the radiation level has fallen 1.00 μCi. How long will the lab have to be closed?

43.27 • Measurements on a certain isotope tell you that the decay rate decreases from 8318 decays/min to 3091 decays/min in 4.00 days. What is the half-life of this isotope?

43.28 • The isotope ^{226}Ra undergoes α decay with a half-life of 1620 years. What is the activity of 1.00 g of ^{226}Ra? Express your answer in Bq and in Ci.

43.29 • The radioactive nuclide ^{199}Pt has a half-life of 30.8 minutes. A sample is prepared that has an initial activity of 7.56×10^{11} Bq. (a) How many ^{199}Pt nuclei are initially present in the sample? (b) How many are present after 30.8 minutes? What is the activity at this time? (c) Repeat part (b) for a time 92.4 minutes after the sample is first prepared.

43.30 •• **Radiocarbon Dating.** A sample from timbers at an archeological site containing 500 g of carbon provides 3070 decays/min. What is the age of the sample?

Section 43.5 Biological Effects of Radiation

43.31 •• **BIO** (a) If a chest x ray delivers 0.25 mSv to 5.0 kg of tissue, how many *total* joules of energy does this tissue receive? (b) Natural radiation and cosmic rays deliver about 0.10 mSv per year at sea level. Assuming an RBE of 1, how many rem and rads is this dose, and how many joules of energy does a 75-kg person receive in a year? (c) How many chest x rays like the one in part (a) would it take to deliver the same *total* amount of energy to a 75-kg person as she receives from natural radiation in a year at sea level, as described in part (b)?

43.32 •• **BIO** A person exposed to fast neutrons receives a radiation dose of 200 rem on part of his hand, affecting 25 g of tissue. The RBE of these neutrons is 10. (a) How many rad did he receive? (b) How many joules of energy did this person receive? (c) Suppose the person received the same rad dosage, but from beta rays with an RBE of 1.0 instead of neutrons. How many rem would he have received?

43.33 •• **BIO** A nuclear chemist receives an accidental radiation dose of 5.0 Gy from slow neutrons (RBE = 4.0). What does she receive in rad, rem, and J/kg?

43.34 • **BIO To Scan or Not to Scan?** It has become popular for some people to have yearly whole-body scans (CT scans, formerly called CAT scans) using x rays, just to see if they detect anything suspicious. A number of medical people have recently questioned the advisability of such scans, due in part to the radiation they impart. Typically, one such scan gives a dose of 12 mSv, applied to the *whole body*. By contrast, a chest x ray typically administers 0.20 mSv to only 5.0 kg of tissue. How many chest x rays would deliver the same *total* amount of energy to the body of a 75-kg person as one whole-body scan?

43.35 • **BIO Food Irradiation.** Food is often irradiated with either x rays or electron beams to help prevent spoilage. A low dose of 5–75 kilorads (krad) helps to reduce and kill inactive parasites, a medium dose of 100–400 krad kills microorganisms and pathogens such as salmonella, and a high dose of 2300–5700 krad sterilizes food so that it can be stored without refrigeration. (a) A dose of 175 krad kills spoilage microorganisms in fish. If x rays are used, what would be the dose in Gy, Sv, and rem, and how much energy would a 220-g portion of fish absorb? (See Table 43.3.) (b) Repeat part (a) if electrons of RBE 1.50 are used instead of x rays.

43.36 • **BIO** In an industrial accident a 65-kg person receives a lethal whole-body equivalent dose of 5.4 Sv from x rays. (a) What is the equivalent dose in rem? (b) What is the absorbed dose in rad? (c) What is the total energy absorbed by the person's body? How does this amount of energy compare to the amount of energy required to raise the temperature of 65 kg of water 0.010 C°?

43.37 •• **BIO** A 67-kg person accidentally ingests 0.35 Ci of tritium. (a) Assume that the tritium spreads uniformly throughout the body and that each decay leads on the average to the absorption of 5.0 keV of energy from the electrons emitted in the decay. The half-life of tritium is 12.3 y, and the RBE of the electrons is 1.0. Calculate the absorbed dose in rad and the equivalent dose in rem during one week. (b) The β^- decay of tritium releases more than 5.0 keV of energy. Why is the average energy absorbed less than the total energy released in the decay?

43.38 •• **CP BIO** In a diagnostic x-ray procedure, 5.00×10^{10} photons are absorbed by tissue with a mass of 0.600 kg. The x-ray wavelength is 0.0200 nm. (a) What is the total energy absorbed by the tissue? (b) What is the equivalent dose in rem?

Section 43.6 Nuclear Reactions, Section 43.7 Nuclear Fission, and Section 43.8 Nuclear Fusion

43.39 • Consider the nuclear reaction

$$^2_1H + ^{14}_7N \rightarrow X + ^{10}_5B$$

where X is a nuclide. (a) What are Z and A for the nuclide X? (b) Calculate the reaction energy Q (in MeV). (c) If the ^2_1H nucleus is incident on a stationary $^{14}_7N$ nucleus, what minimum kinetic energy must it have for the reaction to occur?

43.40 • **Energy from Nuclear Fusion.** Calculate the energy released in the fusion reaction

$$^3_2He + ^2_1H \rightarrow ^4_2He + ^1_1H$$

43.41 • Consider the nuclear reaction

$$^2_1H + ^9_4Be \rightarrow X + ^4_2He$$

where X is a nuclide. (a) What are the values of Z and A for the nuclide X? (b) How much energy is liberated? (c) Estimate the threshold energy for this reaction.

43.42 • The United States uses 1.0×10^{20} J of electrical energy per year. If all this energy came from the fission of ^{235}U, which releases 200 MeV per fission event, (a) how many kilograms of ^{235}U would be used per year and (b) how many kilograms of uranium would have to be mined per year to provide that much ^{235}U? (Recall that only 0.70% of naturally occurring uranium is ^{235}U.)

43.43 •• At the beginning of Section 43.7 the equation of a fission process is given in which ^{235}U is struck by a neutron and undergoes fission to produce ^{144}Ba, ^{89}Kr, and three neutrons. The measured masses of these isotopes are 235.043930 u (^{235}U), 143.922953 u (^{144}Ba), 88.917630 u (^{89}Kr), and 1.0086649 u (neutron). (a) Calculate the energy (in MeV) released by each fission reaction. (b) Calculate the energy released per gram of ^{235}U, in MeV/g.

43.44 •• Consider the nuclear reaction

$$^{28}_{14}Si + \gamma \rightarrow ^{24}_{12}Mg + X$$

where X is a nuclide. (a) What are Z and A for the nuclide X? (b) Ignoring the effects of recoil, what minimum energy must the photon have for this reaction to occur? The mass of a $^{28}_{14}Si$ atom is 27.976927 u, and the mass of a $^{24}_{12}Mg$ atom is 23.985042 u.

43.45 • The second reaction in the proton-proton chain (see Fig. 43.16) produces a ^3_2He nucleus. A ^3_2He nucleus produced in this way can combine with a ^4_2He nucleus:

$$^3_2He + ^4_2He \rightarrow ^7_4Be + \gamma$$

Calculate the energy liberated in this process. (This is shared between the energy of the photon and the recoil kinetic energy of the beryllium nucleus.) The mass of a ^7_4Be atom is 7.016929 u.

43.46 • Consider the nuclear reaction

$$^4_2He + ^7_3Li \rightarrow X + ^1_0n$$

where X is a nuclide. (a) What are Z and A for the nuclide X? (b) Is energy absorbed or liberated? How much?

43.47 •• **CP** In a 100.0-cm^3 sample of water, 0.015% of the molecules are D_2O. Compute the energy in joules that is liberated if all the deuterium nuclei in the sample undergo the fusion reaction of Example 43.13.

PROBLEMS

43.48 • **Comparison of Energy Released per Gram of Fuel.** (a) When gasoline is burned, it releases 1.3×10^8 J of energy per gallon (3.788 L). Given that the density of gasoline is 737 kg/m^3, express the quantity of energy released in J/g of fuel. (b) During fission, when a neutron is absorbed by a ^{235}U nucleus, about 200 MeV of energy is released for each nucleus that undergoes fission. Express this quantity in J/g of fuel. (c) In the proton-proton chain that takes place in stars like our sun, the overall fusion reaction can be summarized as six protons fusing to form one ^4He nucleus with two leftover protons and the liberation of 26.7 MeV of energy. The fuel is the six protons. Express the energy produced here in units of J/g of fuel. Notice the huge difference between the two forms of nuclear energy, on the one hand, and the chemical energy from gasoline, on the other. (d) Our sun produces energy at a measured rate of 3.86×10^{26} W. If its mass of 1.99×10^{30} kg were all gasoline, how long could it last before consuming all its fuel? (*Historical note:* Before the discovery of nuclear fusion and the vast amounts of energy it releases, scientists were confused. They knew that the earth was at least many millions of years old, but could not explain how the sun could survive that long if its energy came from chemical burning.)

43.49 •• Use conservation of mass-energy to show that the energy released in alpha decay is positive whenever the mass of the original neutral atom is greater than the sum of the masses of the final neutral atom and the neutral ^4He atom. (*Hint:* Let the parent nucleus have atomic number Z and nucleon number A. First write the reaction in terms of the nuclei and particles involved, and then add Z electron masses to both sides of the reaction and allot them as needed to arrive at neutral atoms.)

43.50 •• Use conservation of mass-energy to show that the energy released in β^- decay is positive whenever the neutral atomic mass of the original atom is greater than that of the final atom. (See the hint in Problem 43.49.)

43.51 •• Use conservation of mass-energy to show that the energy released in β^+ decay is positive whenever the neutral atomic mass of the original atom is at least two electron masses greater than that of the final atom. (See the hint in Problem 43.49.)

43.52 •• (a) Calculate the minimum energy required to remove one proton from the nucleus $^{12}_6C$. This is called the proton-removal energy. (*Hint:* Find the difference between the mass of a $^{12}_6C$ nucleus and the mass of a proton plus the mass of the nucleus formed when a proton is removed from $^{12}_6C$.) (b) How does the proton-removal energy for $^{12}_6C$ compare to the binding energy per nucleon for $^{12}_6C$, calculated using Eq. (43.10)?

43.53 •• (a) Calculate the minimum energy required to remove one neutron from the nucleus $^{17}_8O$. This is called the neutron-removal energy. (See Problem 43.52.) (b) How does the neutron-removal energy for $^{17}_8O$ compare to the binding energy per nucleon for $^{17}_8O$, calculated using Eq. (43.10)?

43.54 •• The neutral atomic mass of $^{14}_6C$ is 14.003242 u. Calculate the proton removal energy and the neutron removal energy for $^{15}_7N$. (See Problems 43.52 and 43.53.) What is the percentage difference between these two energies, and which is larger?

43.55 • **BIO Radioactive Fallout.** One of the problems of in-air testing of nuclear weapons (or, even worse, the *use* of such weapons!) is the danger of radioactive fallout. One of the most problematic nuclides in such fallout is strontium-90 (^{90}Sr), which breaks down by β^- decay with a half-life of 28 years. It is chemically similar to calcium and therefore can be incorporated into bones and teeth, where, due to its rather long half-life, it remains for years as an internal source of radiation. (a) What is the daughter nucleus of the ^{90}Sr decay? (b) What percentage of the original level of ^{90}Sr is left after 56 years? (c) How long would you have to wait for the original level to be reduced to 6.25% of its original value?

43.56 •• **CP** Thorium $^{230}_{90}$Th decays to radium $^{226}_{88}$Ra by α emission. The masses of the neutral atoms are 230.033127 u for $^{230}_{90}$Th and 226.025403 u for $^{226}_{88}$Ra. If the parent thorium nucleus is at rest, what is the kinetic energy of the emitted α particle? (Be sure to account for the recoil of the daughter nucleus.)

43.57 •• The atomic mass of $^{25}_{12}$Mg is 24.985837 u, and the atomic mass of $^{25}_{13}$Al is 24.990429 u. (a) Which of these nuclei will decay into the other? (b) What type of decay will occur? Explain how you determined this. (c) How much energy (in MeV) is released in the decay?

43.58 •• The polonium isotope $^{210}_{84}$Po has atomic mass 209.982857 u. Other atomic masses are $^{206}_{82}$Pb, 205.974449 u; $^{209}_{83}$Bi, 208.980383 u; $^{210}_{83}$Bi, 209.984105 u; $^{209}_{84}$Po, 208.982416 u; and $^{210}_{85}$At, 209.987131 u. (a) Show that the alpha decay of $^{210}_{84}$Po is energetically possible, and find the energy of the emitted α particle. (b) Is $^{210}_{84}$Po energetically stable with respect to emission of a proton? Why or why not? (c) Is $^{210}_{84}$Po energetically stable with respect to emission of a neutron? Why or why not? (d) Is $^{210}_{84}$Po energetically stable with respect to β^- decay? Why or why not? (e) Is $^{210}_{84}$Po energetically stable with respect to β^+ decay? Why or why not?

43.59 •• **BIO Irradiating Ourselves!** The radiocarbon in our bodies is one of the naturally occurring sources of radiation. Let's see how large a dose we receive. ^{14}C decays via β^- emission, and 18% of our body's mass is carbon. (a) Write out the decay scheme of carbon-14 and show the end product. (A neutrino is also produced.) (b) Neglecting the effects of the neutrino, how much kinetic energy (in MeV) is released per decay? The atomic mass of ^{14}C is 14.003242 u. (c) How many grams of carbon are there in a 75-kg person? How many decays per second does this carbon produce? (*Hint:* Use data from Example 43.9.) (d) Assuming that all the energy released in these decays is absorbed by the body, how many MeV/s and J/s does the ^{14}C release in this person's body? (e) Consult Table 43.3 and use the largest appropriate RBE for the particles involved. What radiation dose does the person give himself in a year, in Gy, rad, Sv, and rem?

43.60 •• **BIO Pion Radiation Therapy.** A neutral pion (π^0) has a mass of 264 times the electron mass and decays with a lifetime of 8.4×10^{-17} s to two photons. Such pions are used in the radiation treatment of some cancers. (a) Find the energy and wavelength of these photons. In which part of the electromagnetic spectrum do they lie? What is the RBE for these photons? (b) If you want to deliver a dose of 200 rem (which is typical) in a single treatment to 25 g of tumor tissue, how many π^0 mesons are needed?

43.61 • Gold, $^{198}_{79}$Au, undergoes β^- decay to an excited state of $^{198}_{80}$Hg. If the excited state decays by emission of a γ photon with energy 0.412 MeV, what is the maximum kinetic energy of the electron emitted in the decay? This maximum occurs when the antineutrino has negligible energy. (The recoil energy of the $^{198}_{80}$Hg nucleus can be ignored. The masses of the neutral atoms in their ground states are 197.968225 u for $^{198}_{79}$Au and 197.966752 u for $^{198}_{80}$Hg.)

43.62 •• Calculate the mass defect for the β^+ decay of $^{11}_{6}$C. Is this decay energetically possible? Why or why not? The atomic mass of $^{11}_{6}$C is 11.011434 u.

43.63 •• Calculate the mass defect for the β^+ decay of $^{13}_{7}$N. Is this decay energetically possible? Why or why not? The atomic mass of $^{13}_{7}$N is 13.005739 u.

43.64 •• The results of activity measurements on a radioactive sample are given in the table. (a) Find the half-life. (b) How many radioactive nuclei were present in the sample at $t = 0$? (c) How many were present after 7.0 h?

Time (h)	Decays/s
0	20,000
0.5	14,800
1.0	11,000
1.5	8,130
2.0	6,020
2.5	4,460
3.0	3,300
4.0	1,810
5.0	1,000
6.0	550
7.0	300

43.65 •• **BIO** A person ingests an amount of a radioactive source with a very long lifetime and activity 0.63 μCi. The radioactive material lodges in the lungs, where all of the 4.0-MeV α particles emitted are absorbed within a 0.50-kg mass of tissue. Calculate the absorbed dose and the equivalent dose for one year.

43.66 •• **Measuring Very Long Half-Lives.** Some radioisotopes such as samarium (^{149}Sm) and gadolinium (^{152}Gd) have half-lives that are much longer than the age of the universe, so we can't measure their half-lives by watching their decay rate decrease. Luckily, there is another way of calculating the half-life, using Eq. (43.16). Suppose a 12.0-g sample of ^{149}Sm is observed to decay at a rate of 2.65 Bq. Calculate the half-life of the sample in years. (*Hint:* How many nuclei are there in the 12.0-g sample?)

43.67 • **We Are Stardust.** In 1952 spectral lines of the element technetium-99 (^{99}Tc) were discovered in a red giant star. Red giants are very old stars, often around 10 billion years old, and near the end of their lives. Technetium has *no* stable isotopes, and the half-life of ^{99}Tc is 200,000 years. (a) For how many half-lives has the ^{99}Tc been in the red-giant star if its age is 10 billion years? (b) What fraction of the original ^{99}Tc would be left at the end of that time? This discovery was extremely important because it provided convincing evidence for the theory (now essentially known to be true) that most of the atoms heavier than hydrogen and helium were made inside of stars by thermonuclear fusion and other nuclear processes. If the ^{99}Tc had been part of the star since it was born, the amount remaining after 10 billion years would have been so minute that it would not have been detectable. This knowledge is what led the late astronomer Carl Sagan to proclaim that "we are stardust."

43.68 • **BIO** A 70.0-kg person experiences a whole-body exposure to α radiation with energy 4.77 MeV. A total of 6.25×10^{12} α particles are absorbed. (a) What is the absorbed dose in rad? (b) What is the equivalent dose in rem? (c) If the source is 0.0320 g of ^{226}Ra (half-life 1600 y) somewhere in the body, what is the activity of this source? (d) If all the alpha particles produced are absorbed, what time is required for this dose to be delivered?

43.69 •• Measurements indicate that 27.83% of all rubidium atoms currently on the earth are the radioactive ^{87}Rb isotope. The rest are the stable ^{85}Rb isotope. The half-life of ^{87}Rb is 4.75×10^{10} y. Assuming that no rubidium atoms have been formed since, what percentage of rubidium atoms were ^{87}Rb when our solar system was formed 4.6×10^9 y ago?

43.70 •• A $^{186}_{76}$Os nucleus at rest decays by the emission of a 2.76-MeV α particle. Calculate the atomic mass of the daughter nuclide produced by this decay, assuming that it is produced in its ground state. The atomic mass of $^{186}_{76}$Os is 185.953838 u.

43.71 •• **BIO** A ^{60}Co source with activity 2.6×10^{-4} Ci is embedded in a tumor that has mass 0.200 kg. The source emits γ photons with average energy 1.25 MeV. Half the photons are absorbed in the tumor, and half escape. (a) What energy is delivered to the tumor per second? (b) What absorbed dose (in rad) is delivered per second? (c) What equivalent dose (in rem) is delivered per second if the RBE for these γ rays is 0.70? (d) What exposure time is required for an equivalent dose of 200 rem?

43.72 • The nucleus $^{15}_{8}$O has a half-life of 122.2 s; $^{19}_{8}$O has a half-life of 26.9 s. If at some time a sample contains equal amounts of $^{15}_{8}$O and $^{19}_{8}$O, what is the ratio of $^{15}_{8}$O to $^{19}_{8}$O (a) after 4.0 minutes and (b) after 15.0 minutes?

43.73 • A bone fragment found in a cave believed to have been inhabited by early humans contains 0.29 times as much ^{14}C as an equal amount of carbon in the atmosphere when the organism containing the bone died. (See Example 43.9 in Section 43.4.) Find the approximate age of the fragment.

43.74 •• **An Oceanographic Tracer.** Nuclear weapons tests in the 1950s and 1960s released significant amounts of radioactive tritium ($^{3}_{1}$H, half-life 12.3 years) into the atmosphere. The tritium atoms were quickly bound into water molecules and rained out of the air, most of them ending up in the ocean. For any of this tritium-tagged water that sinks below the surface, the amount of time during which it has been isolated from the surface can be calculated by measuring the ratio of the decay product, $^{3}_{2}$He, to the remaining tritium in the water. For example, if the ratio of $^{3}_{2}$He to $^{3}_{1}$H in a sample of water is 1:1, the water has been below the surface for one half-life, or approximately 12 years. This method has provided oceanographers with a convenient way to trace the movements of subsurface currents in parts of the ocean. Suppose that in a particular sample of water, the ratio of $^{3}_{2}$He to $^{3}_{1}$H is 4.3 to 1.0. How many years ago did this water sink below the surface?

43.75 •• Consider the fusion reaction $^{2}_{1}$H + $^{2}_{1}$H → $^{3}_{2}$He + $^{1}_{0}$n. (a) Estimate the barrier energy by calculating the repulsive electrostatic potential energy of the two $^{2}_{1}$H nuclei when they touch. (b) Compute the energy liberated in this reaction in MeV and in joules. (c) Compute the energy liberated *per mole* of deuterium, remembering that the gas is diatomic, and compare with the heat of combustion of hydrogen, about 2.9×10^{5} J/mol.

43.76 •• **BIO** In the 1986 disaster at the Chernobyl reactor in the Soviet Union (now Ukraine), about $\frac{1}{8}$ of the ^{137}Cs present in the reactor was released. The isotope ^{137}Cs has a half-life for β decay of 30.07 y and decays with the emission of a total of 1.17 MeV of energy per decay. Of this, 0.51 MeV goes to the emitted electron and the remaining 0.66 MeV to a γ ray. The radioactive ^{137}Cs is absorbed by plants, which are eaten by livestock and humans. How many ^{137}Cs atoms would need to be present in each kilogram of body tissue if an equivalent dose for one week is 3.5 Sv? Assume

that all of the energy from the decay is deposited in that 1.0 kg of tissue and that the RBE of the electrons is 1.5.

43.77 •• **CP** (a) Prove that when a particle with mass m and kinetic energy K collides with a stationary particle with mass M, the total kinetic energy K_{cm} in the center-of-mass coordinate system (the energy available to cause reactions) is

$$K_{cm} = \frac{M}{M + m}K$$

Assume that the kinetic energies of the particles and nuclei are much lower than their rest energies. (b) If K_{th} is the minimum, or threshold, kinetic energy to cause an endoergic reaction to occur in the situation of part (a), show that

$$K_{th} = -\frac{M + m}{M}Q$$

43.78 • Calculate the energy released in the fission reaction $^{235}_{92}$U + $^{1}_{0}$n → $^{140}_{54}$Xe + $^{94}_{38}$Sr + 2^{1}_{0}n. You can ignore the initial kinetic energy of the absorbed neutron. The atomic masses are $^{235}_{92}$U, 235.043923 u; $^{140}_{54}$Xe, 139.921636 u; and $^{94}_{38}$Sr, 93.915360 u.

CHALLENGE PROBLEMS

43.79 ••• The results of activity measurements on a mixed sample of radioactive elements are given in the table. (a) How many different nuclides are present in the mixture? (b) What are their half-lives? (c) How many nuclei of each type are initially present in the sample? (d) How many of each type are present at $t = 5.0$ h?

Time (h)	Decays/s
0	7500
0.5	4120
1.0	2570
1.5	1790
2.0	1350
2.5	1070
3.0	872
4.0	596
5.0	414
6.0	288
7.0	201
8.0	140
9.0	98
10.0	68
12.0	33

43.80 ••• **Industrial Radioactivity.** Radioisotopes are used in a variety of manufacturing and testing techniques. Wear measurements can be made using the following method. An automobile engine is produced using piston rings with a total mass of 100 g, which includes 9.4 μCi of ^{59}Fe whose half-life is 45 days. The engine is test-run for 1000 hours, after which the oil is drained and its activity is measured. If the activity of the engine oil is 84 decays/s, how much mass was worn from the piston rings per hour of operation?

Answers

Chapter Opening Question

When an organism dies, it stops taking in carbon from atmospheric CO_2. Some of this carbon is radioactive ^{14}C, which decays with a half-life of 5730 years. By measuring the proportion of ^{14}C that remains in the specimen, scientists can determine how long ago the organism died. (See Section 43.4.)

Test Your Understanding Questions

43.1 Answers: (a) (iii), (b) (v) The radius R is proportional to the cube root of the mass number A, while the volume is proportional to R^3 and hence to $(A^{1/3})^3 = A$. Therefore, doubling the volume requires increasing the mass number by a factor of 2; doubling the radius implies increasing both the volume and the mass number by a factor of $2^3 = 8$.

43.2 Answer: (ii), (iii), (iv), (v), (i) You can find the answers by inspecting Fig. 43.2. The binding energy per nucleon is lowest for very light nuclei such as $^4_2$He, is greatest around $A = 60$, and then decreases with increasing A.

43.3 Answer: (v) Two protons and two neutrons are lost in an α decay, so Z and N each decrease by 2. A β^+ decay changes a proton to a neutron, so Z decreases by 1 and N increases by 1. The net result is that Z decreases by 3 and N decreases by 1.

43.4 Answer: (iii) The activity $-dN(t)/dt$ of a sample is the product of the number of nuclei in the sample $N(t)$ and the decay constant $\lambda = (\ln 2)/T_{1/2}$. Hence $N(t) = (-dN(t)/dt)T_{1/2}/(\ln 2)$. Taking the ratio of this expression for ^{240}Pu to this same expression for ^{243}Am, the factors of $\ln 2$ cancel and we get

$$\frac{N_{Pu}}{N_{Am}} = \frac{(-dN_{Pu}/dt)T_{1/2-Pu}}{(-dN_{Am}/dt)T_{1/2-Am}} = \frac{(5.00\ \mu Ci)(6560\ y)}{(4.45\ \mu Ci)(7370\ y)} = 1.00$$

The two samples contain *equal* numbers of nuclei. The ^{243}Am sample has a longer half-life and hence a slower decay rate, so it has a lower activity than the ^{240}Pu sample.

43.5 Answer: (ii) We saw in Section 43.3 that alpha particles can travel only a very short distance before they are stopped. By contrast, x-ray photons are very penetrating, so they can easily pass into the body.

43.6 Answer: no The reaction $^1_1$H $+ \, ^7_3$Li $\rightarrow \, ^4_2$He $+ \, ^4_2$He is a *nuclear* reaction, which can take place only if a proton (a hydrogen nucleus) comes into contact with a lithium nucleus. If the hydrogen is in atomic form, the interaction between its electron cloud and the electron cloud of a lithium atom keeps the two nuclei from getting close to each other. Even if isolated protons are used, they must be fired at the lithium atoms with enough kinetic energy to overcome the electric repulsion between the protons and the lithium nuclei. The statement that the reaction is exoergic means that more energy is released by the reaction than had to be put in to make the reaction occur.

43.7 Answer: no Because the neutron has no electric charge, it experiences no electric repulsion from a ^{235}U nucleus. Hence a slow-moving neutron can approach and enter a ^{235}U nucleus, thereby providing the excitation needed to trigger fission. By contrast, a slow-moving *proton* (charge $+e$) feels a strong electric repulsion from a ^{235}U nucleus (charge $+92e$). It never gets close to the nucleus, so it cannot trigger fission.

43.8 Answer: no Fusion reactions between sufficiently light nuclei are exoergic because the binding energy per nucleon E_B/A increases. If the nuclei are too massive, however, E_B/A decreases and fusion is *endoergic* (i.e., it takes in energy rather than releasing it). As an example, imagine fusing together two nuclei of $A = 100$ to make a single nucleus with $A = 200$. From Fig. 43.2, E_B/A is more than 8.5 MeV for the $A = 100$ nuclei but is less than 8 MeV for the $A = 200$ nucleus. Such a fusion reaction is possible, but requires a substantial input of energy.

Bridging Problem

Answers: **(a)** ^{128}Xe

   (b) no; β^+ emission would be endoergic

   (c) 3.25×10^9 atoms, 1.50×10^6 Bq

   (d) $N(t) = (3.25 \times 10^9 \text{ atoms})(1 - e^{-(4.62\times10^{-4}\text{ s}^{-1})t})$

44 PARTICLE PHYSICS AND COSMOLOGY

? Images made using infrared and x-ray wavelengths were combined to produce this view of the dynamic center of our Milky Way galaxy. The image shows atoms in various states: as isolated atoms in glowing, diffuse clouds of gas (shown in blue), as clumps of atoms and molecules in immense, cold dust clouds (shown in red), and in dense accumulations that we call stars. What fraction of the mass and energy in the universe is composed of "normal" matter—that is, atoms and their constituents?

What is the world made of? What are the most fundamental constituents of matter? Philosophers and scientists have been asking these questions for at least 2500 years. We still don't have the final answer, but as we'll see in this chapter, we've come a long way.

The chapter title, "Particle Physics and Cosmology," may seem strange. Fundamental particles are the *smallest* things in the universe, and cosmology deals with the *biggest* thing there is—the universe itself. Nonetheless, we'll see in this chapter that physics on the most microscopic scale plays an essential role in determining the nature of the universe on the largest scale.

Fundamental particles, we'll find, are not permanent entities; they can be created and destroyed. The development of high-energy particle accelerators and associated detectors has been crucial in our emerging understanding of particles. We can classify particles and their interactions in several ways in terms of conservation laws and symmetries, some of which are absolute and others of which are obeyed only in certain kinds of interactions. We'll conclude by discussing our present understanding of the nature and evolution of the universe as a whole.

44.1 Fundamental Particles—A History

The idea that the world is made of fundamental particles has a long history. In about 400 B.C. the Greek philosophers Democritus and Leucippus suggested that matter is made of indivisible particles that they called *atoms,* a word derived from *a-* (not) and *tomos* (cut or divided). This idea lay dormant until about 1804, when the English scientist John Dalton (1766–1844), often called the father of modern chemistry, discovered that many chemical phenomena could be explained if atoms of each element are the basic, indivisible building blocks of matter.

The Electron and the Proton

Toward the end of the 19th century it became clear that atoms are *not* indivisible. The existence of characteristic atomic spectra of elements suggested that atoms have internal structure, and J. J. Thomson's discovery of the negatively charged *electron* in 1897 showed that atoms could be taken apart into charged particles. Rutherford's experiments in 1910–11 (see Section 39.2) revealed that an atom's positive charge resides in a small, dense nucleus. In 1919 Rutherford made an additional discovery: When alpha particles are fired into nitrogen, one of the products is hydrogen gas. He reasoned that the hydrogen nucleus is a constituent of the nuclei of heavier atoms such as nitrogen, and that a collision with a fast-moving alpha particle can dislodge one of those hydrogen nuclei. Thus the hydrogen nucleus is an elementary particle, to which Rutherford gave the name *proton*. The following decade saw the blossoming of quantum mechanics, including the Schrödinger equation. Physicists were on their way to understanding the principles that underlie atomic structure.

The Photon

Einstein explained the photoelectric effect in 1905 by assuming that the energy of electromagnetic waves is quantized; that is, it comes in little bundles called *photons* with energy $E = hf$. Atoms and nuclei can emit (create) and absorb (destroy) photons (see Section 38.1). Considered as particles, photons have zero charge and zero rest mass. (Note that any discussions of a particle's mass in this chapter will refer to its rest mass.) In particle physics, a photon is denoted by the symbol γ (the Greek letter gamma).

The Neutron

In 1930 the German physicists Walther Bothe and Herbert Becker observed that when beryllium, boron, or lithium was bombarded by alpha particles, the target material emitted a radiation that had much greater penetrating power than the original alpha particles. Experiments by the English physicist James Chadwick in 1932 showed that the emitted particles were electrically neutral, with mass approximately equal to that of the proton. Chadwick christened these particles *neutrons* (symbol n or $^1_0$n). A typical reaction of the type studied by Bothe and Becker, using a beryllium target, is

$$^4_2\text{He} + {}^9_4\text{Be} \rightarrow {}^{12}_6\text{C} + {}^1_0\text{n} \tag{44.1}$$

Elementary particles are usually detected by their electromagnetic effects—for instance, by the ionization that they cause when they pass through matter. (This is the principle of the cloud chamber, described below.) Because neutrons have no charge, they interact hardly at all with electrons and produce little ionization when they pass through matter and so are difficult to detect directly. However, neutrons can be slowed down by scattering from nuclei, and they can penetrate a nucleus. Hence slow neutrons can be detected by means of a nuclear reaction in which a neutron is absorbed and an alpha particle is emitted. An example is

$$^1_0\text{n} + {}^{10}_5\text{B} \rightarrow {}^7_3\text{Li} + {}^4_2\text{He} \tag{44.2}$$

The ejected alpha particle is easy to detect because it is charged. Later experiments showed that neutrons, like protons and electrons, are spin-$\frac{1}{2}$ particles (see Section 43.1).

The discovery of the neutron cleared up a mystery about the composition of the nucleus. Before 1930 the mass of a nucleus was thought to be due only to protons, but no one understood why the charge-to-mass ratio was not the same for all nuclides. It soon became clear that all nuclides (except $^1_1$H) contain both protons

44.1 Photograph of the cloud-chamber track made by the first positron ever identified. The photograph was made by Carl D. Anderson in 1932.

The positron follows a curved path owing to the presence of a magnetic field.

The track is more strongly curved above the lead plate, showing that the positron was traveling upward and lost energy and speed as it passed through the plate.

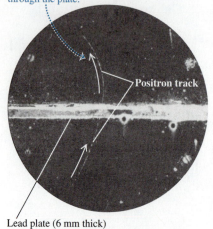

Positron track

Lead plate (6 mm thick)

and neutrons. Hence the proton, the neutron, and the electron are the building blocks of atoms. One might think that would be the end of the story. On the contrary, it is barely the beginning. These are not the only particles, and they can do more than build atoms.

The Positron

The positive electron, or positron, was discovered by the American physicist Carl D. Anderson in 1932, during an investigation of particles bombarding the earth from space. Figure 44.1 shows a historic photograph made with a *cloud chamber*, an instrument used to visualize the tracks of charged particles. The chamber contained a supercooled vapor; ions created by the passage of charged particles through the vapor served as nucleation centers, and liquid droplets formed around them, making a visible track.

The cloud chamber in Fig. 44.1 is in a magnetic field directed into the plane of the photograph. The particle has passed through a thin lead plate (which extends from left to right in the figure) that lies within the chamber. The track is more tightly curved above the plate than below it, showing that the speed was less above the plate than below it. Therefore the particle had to be moving upward; it could not have gained energy passing through the lead. The thickness and curvature of the track suggested that its mass and the magnitude of its charge equaled those of the electron. But the directions of the magnetic field and the velocity in the magnetic force equation $\vec{F} = q\vec{v} \times \vec{B}$ showed that the particle had *positive* charge. Anderson christened this particle the *positron*.

To theorists, the appearance of the positron was a welcome development. In 1928 the English physicist Paul Dirac had developed a relativistic generalization of the Schrödinger equation for the electron. In Section 41.5 we discussed how Dirac's ideas helped explain the spin magnetic moment of the electron.

One of the puzzling features of the Dirac equation was that for a *free* electron it predicted not only a continuum of energy states greater than its rest energy $m_e c^2$, as should be expected, but also a continuum of *negative* energy states *less than* $-m_e c^2$ (Fig. 44.2a). That posed a problem. What was to prevent an electron from emitting a photon with energy $2m_e c^2$ or greater and hopping from a positive state to a negative state? It wasn't clear what these negative-energy states meant, and there was no obvious way to get rid of them. Dirac's ingenious interpretation was that all the negative-energy states were filled with electrons, and that these electrons were for some reason unobservable. The exclusion principle (see Section 41.6) would then forbid a transition to a state that was already occupied.

A vacancy in a negative-energy state would act like a positive charge, just as a hole in the valence band of a semiconductor (see Section 42.6) acts like a positive charge. Initially, Dirac tried to argue that such vacancies were protons. But after

44.2 (a) Energy states for a free electron predicted by the Dirac equation. (b) Raising an electron from an $E < 0$ state to an $E > 0$ state corresponds to electron–positron pair production. (c) An electron dropping from an $E > 0$ state to a vacant $E < 0$ state corresponds to electron–positron pair annihilation.

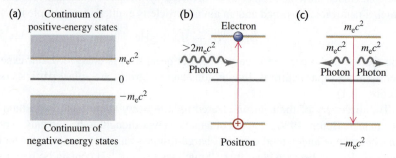

Anderson's discovery it became clear that the vacancies were observed physically as *positrons*. Furthermore, the Dirac energy-state picture provides a mechanism for the *creation* of positrons. When an electron in a negative-energy state absorbs a photon with energy greater than $2m_ec^2$, it goes to a positive state (Fig. 44.2b), in which it becomes observable. The vacancy that it leaves behind is observed as a positron; the result is the creation of an electron–positron pair. Similarly, when an electron in a positive-energy state falls into a vacancy, both the electron and the vacancy (that is, the positron) disappear, and photons are emitted (Fig. 44.2c). Thus the Dirac theory leads naturally to the conclusion that, like photons, *electrons can be created and destroyed*. While photons can be created and destroyed singly, electrons can be produced or destroyed only in electron–positron pairs or in association with other particles. (Creating or destroying an electron alone would mean creating or destroying an amount of charge –e, which would violate the conservation of electric charge.)

In 1949 the American physicist Richard Feynman showed that a positron could be described mathematically as an electron traveling backward in time. His reformulation of the Dirac theory eliminated difficult calculations involving the infinite sea of negative-energy states and put electrons and positrons on the same footing. But the creation and destruction of electron–positron pairs remain. The Dirac theory provides the beginning of a theoretical framework for creation and destruction of all fundamental particles.

Experiment and theory tell us that the masses of the positron and electron are identical, and that their charges are equal in magnitude but opposite in sign. The positron's spin angular momentum $\vec{S}$ and magnetic moment $\vec{\mu}$ are parallel; they are opposite for the electron. However, $\vec{S}$ and $\vec{\mu}$ have the same magnitude for both particles because they have the same spin. We use the term **antiparticle** for a particle that is related to another particle as the positron is to the electron. Each kind of particle has a corresponding antiparticle. For a few kinds of particles (necessarily all neutral) the particle and antiparticle are identical, and we can say that they are their own antiparticles. The photon is an example; there is no way to distinguish a photon from an antiphoton. We'll use the standard symbols e^- for the electron and e^+ for the positron, and the generic term "electron" will often include both electrons and positrons. Other antiparticles are often denoted by a bar over the particle's symbol; for example, an antiproton is $\bar{p}$. We'll see several other examples of antiparticles later.

Positrons do not occur in ordinary matter. Electron–positron pairs are produced during high-energy collisions of charged particles or γ rays with matter. This process is called e^+e^- *pair production* (Fig. 44.3). Enough energy E must be available to account for the rest energy $2m_ec^2$ of the two particles. The minimum energy for electron–positron pair production is

$$E_{min} = 2m_ec^2 = 2(9.109 \times 10^{-31} \text{ kg})(2.998 \times 10^8 \text{ m/s})^2$$

$$= 1.637 \times 10^{-13} \text{ J} = 1.022 \text{ MeV}$$

The inverse process, e^+e^- *pair annihilation,* occurs when a positron and an electron collide (see Example 38.6 in Section 38.3). Both particles disappear, and two (or occasionally three) photons can appear, with total energy of at least $2m_ec^2 = 1.022$ MeV. Decay into a *single* photon is impossible: Such a process could not conserve both energy and momentum.

Positrons also occur in the decay of some unstable nuclei, in which they are called beta-plus particles (β^+). We discussed β^+ decay in Section 43.3.

It's often convenient to represent particle masses in terms of the equivalent rest energy using $m = E/c^2$. Then typical mass units are MeV/c^2; for example, $m = 0.511$ MeV/c^2 for an electron or positron. We'll use these units frequently in this chapter.

44.3 (a) Photograph of bubble-chamber tracks of electron–positron pairs that are produced when 300-MeV photons strike a lead sheet. A magnetic field directed out of the photograph made the electrons and positrons curve in opposite directions. (b) Diagram showing the pair-production process for two of the photons.

(a)

Electron–positron pair

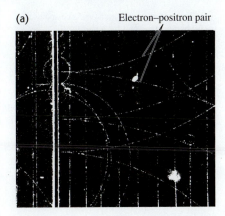

(b)

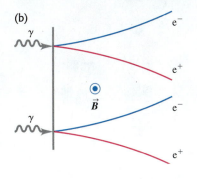

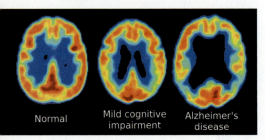

Normal   Mild cognitive impairment   Alzheimer's disease

44.4 An analogy for how particles act as force mediators.

(a) Two skaters exert repulsive forces on each other by tossing a ball back and forth.

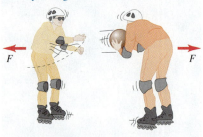

F F

(b) Two skaters exert attractive forces on each other when one tries to grab the ball out of the other's hands.

F F

Particles As Force Mediators

In classical physics we describe the interaction of charged particles in terms of electric and magnetic forces. In quantum mechanics we can describe this interaction in terms of emission and absorption of photons. Two electrons repel each other as one emits a photon and the other absorbs it, just as two skaters can push each other apart by tossing a heavy ball back and forth between them (Fig. 44.4a). For an electron and a proton, in which the charges are opposite and the force is attractive, we imagine the skaters trying to grab the ball away from each other (Fig. 44.4b). The electromagnetic interaction between two charged particles is *mediated* or transmitted by photons.

If charged-particle interactions are mediated by photons, where does the energy to create the photons come from? Recall from our discussion of the uncertainty principle (see Sections 38.4 and 39.6) that a state that exists for a short time Δt has an uncertainty ΔE in its energy such that

$$\Delta E\,\Delta t \geq \frac{\hbar}{2} \tag{44.3}$$

This uncertainty permits the creation of a photon with energy ΔE, provided that it lives no longer than the time Δt given by Eq. (44.3). A photon that can exist for a short time because of this energy uncertainty is called a *virtual photon*. It's as though there were an energy bank; you can borrow energy, provided that you pay it back within the time limit. According to Eq. (44.3), the more you borrow, the sooner you have to pay it back.

Mesons

Is there a particle that mediates the *nuclear* force? By the mid-1930s the nuclear force between two nucleons (neutrons or protons) appeared to be described by a potential energy $U(r)$ with the general form

$$U(r) = -f^2\left(\frac{e^{-r/r_0}}{r}\right) \qquad \text{(nuclear potential energy)} \tag{44.4}$$

The constant f characterizes the strength of the interaction, and r_0 describes its range. Figure 44.5 shows a graph of the absolute value of this function and compares it with the function f^2/r, which would be analogous to the *electric* interaction of two protons:

$$U(r) = \frac{1}{4\pi\epsilon_0}\frac{e^2}{r} \qquad \text{(electric potential energy)} \tag{44.5}$$

In 1935 the Japanese physicist Hideki Yukawa suggested that a hypothetical particle that he called a **meson** might mediate the nuclear force. He showed that the range of the force was related to the mass of the particle. Yukawa argued that the particle must live for a time Δt long enough to travel a distance comparable to the range r_0 of the nuclear force. This range was known from the sizes of nuclei and other information to be about 1.5×10^{-15} m = 1.5 fm. If we assume that an average particle's speed is comparable to c and travels about half the range, its lifetime Δt must be about

$$\Delta t = \frac{r_0}{2c} = \frac{1.5 \times 10^{-15}\ \text{m}}{2(3.0 \times 10^8\ \text{m/s})} = 2.5 \times 10^{-24}\ \text{s}$$

From Eq. (44.3), the minimum necessary uncertainty ΔE in energy is

$$\Delta E = \frac{\hbar}{2\Delta t} = \frac{1.05 \times 10^{-34}\ \text{J}\cdot\text{s}}{2(2.5 \times 10^{-24}\ \text{s})} = 2.1 \times 10^{-11}\ \text{J} = 130\ \text{MeV}$$

The mass equivalent Δm of this energy is

$$\Delta m = \frac{\Delta E}{c^2} = \frac{2.1 \times 10^{-11}\,\text{J}}{(3.00 \times 10^8\,\text{m/s})^2} = 2.3 \times 10^{-28}\,\text{kg} = 130\,\text{MeV}/c^2$$

This is about 250 times the electron mass, and Yukawa postulated that an as yet undiscovered particle with this mass serves as the messenger for the nuclear force.

A year later, Carl Anderson and his colleague Seth Neddermeyer discovered in cosmic radiation two new particles, now called **muons.** The μ^- has charge equal to that of the electron, and its antiparticle the μ^+ has a positive charge with equal magnitude. The two particles have equal mass, about 207 times the electron mass. But it soon became clear that muons were *not* Yukawa's particles because they interacted with nuclei only very weakly.

In 1947 a family of three particles, called π *mesons* or **pions,** were discovered. Their charges are $+e$, $-e$, and zero, and their masses are about 270 times the electron mass. The pions interact strongly with nuclei, and they *are* the particles predicted by Yukawa. Other, heavier mesons, the ω and ρ, evidently also act as shorter-range messengers of the nuclear force. The complexity of this explanation suggests that it has simpler underpinnings; these involve the quarks and gluons that we'll discuss in Section 44.4. Before discussing mesons further, we'll describe some particle accelerators and detectors to see how mesons and other particles are created in a controlled fashion and observed.

Test Your Understanding of Section 44.1   Each of the following particles can be exchanged between two protons, two neutrons, or a neutron and a proton as part of the nuclear force. Rank the particles in order of the range of the interaction that they mediate, from largest to smallest range. (i) the π^+ (pi-plus) meson of mass 140 MeV/c^2; (ii) the ρ^+ (rho-plus) meson of mass 776 MeV/c^2; (iii) the η^0 (eta-zero) meson of mass 548 MeV/c^2; (iv) the ω^0 (omega-zero) meson of mass 783 MeV/c^2.

44.2 Particle Accelerators and Detectors

Early nuclear physicists used alpha and beta particles from naturally occurring radioactive elements for their experiments, but they were restricted in energy to the few MeV that are available in such random decays. Present-day particle accelerators can produce precisely controlled beams of particles, from electrons and positrons up to heavy ions, with a wide range of energies. These beams have three main uses. First, high-energy particles can collide to produce new particles, just as a collision of an electron and a positron can produce photons. Second, a high-energy particle has a short de Broglie wavelength and so can probe the small-scale interior structure of other particles, just as electron microscopes (see Section 39.1) can give better resolution than optical microscopes. Third, they can be used to produce nuclear reactions of scientific or medical use.

Linear Accelerators

Particle accelerators use electric and magnetic fields to accelerate and guide beams of charged particles. A *linear accelerator* (linac) accelerates particles in a straight line. J. J. Thomson's cathode-ray tubes were early examples of linacs. Modern linacs use a series of electrodes with gaps to give the particles a series of boosts. Most present-day high-energy linear accelerators use a traveling electromagnetic wave; the charged particles "ride" the wave in more or less the way that a surfer rides an incoming ocean wave. In the highest-energy linac in the world today, at the SLAC National Accelerator Laboratory, electrons and positrons can be accelerated to 50 GeV in a tube 3 km long. At this energy their de Broglie wavelengths are 0.025 fm, much smaller than the size of a proton or a neutron.

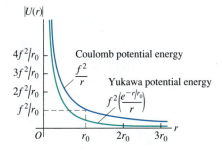

44.5 Graph of the magnitude of the Yukawa potential-energy function for nuclear forces, $|U(r)| = f^2 e^{-r/r_0}/r$. The function $U(r) = f^2/r$, proportional to the potential energy for Coulomb's law, is also shown. The two functions are similar at small r, but the Yukawa potential energy drops off much more quickly at large r.

44.6 Layout and operation of a cyclotron.

(a) Schematic diagram of a cyclotron

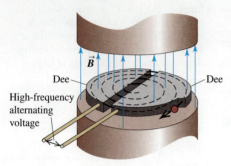

High-frequency alternating voltage

(b) As the positive particle reaches the gap, it is accelerated by the electric-field force ...

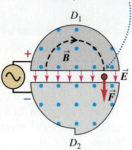

(c) ... and the next semicircular orbit has a larger radius.

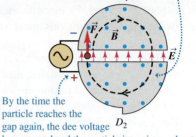

By the time the particle reaches the gap again, the dee voltage has reversed and the particle is again accelerated.

The Cyclotron

Many accelerators use magnets to deflect the charged particles into circular paths. The first of these was the *cyclotron,* invented in 1931 by E. O. Lawrence and M. Stanley Livingston at the University of California (Fig. 44.6a). Particles with mass m and charge q move inside a vacuum chamber in a uniform magnetic field $\vec{B}$ that is perpendicular to the plane of their paths. In Section 27.4 we showed that in such a field, a particle with speed v moves in a circular path with radius r given by

$$r = \frac{mv}{|q|B} \tag{44.6}$$

and with angular speed (angular frequency) ω given by

$$\omega = \frac{v}{r} = \frac{|q|B}{m} \tag{44.7}$$

An alternating potential difference is applied between the two hollow electrodes D_1 and D_2 (called *dees*), creating an electric field in the gap between them. The polarity of the potential difference and electric field is changed precisely twice each revolution (Figs. 44.6b and 44.6c), so that the particles get a push each time they cross the gap. The pushes increase their speed and kinetic energy, boosting them into paths of larger radius. The maximum speed v_{max} and kinetic energy K_{max} are determined by the radius R of the largest possible path. Solving Eq. (44.6) for v, we find $v = |q|Br/m$ and $v_{max} = |q|BR/m$. Assuming nonrelativistic speeds, we have

$$K_{max} = \tfrac{1}{2}mv_{max}^2 = \frac{q^2B^2R^2}{2m} \tag{44.8}$$

Example 44.1 **Frequency and energy in a proton cyclotron**

One cyclotron built during the 1930s has a path of maximum radius 0.500 m and a magnetic field of magnitude 1.50 T. If it is used to accelerate protons, find (a) the frequency of the alternating voltage applied to the dees and (b) the maximum particle energy.

SOLUTION

IDENTIFY and SET UP: The frequency f of the applied voltage must equal the frequency of the proton orbital motion. Equation (44.7) gives the *angular* frequency ω of the proton orbital motion; we find f using $f = \omega/2\pi$. The proton reaches its maximum energy

K_{max}, given by Eq. (44.8), when the radius of its orbit equals the radius of the dees.

EXECUTE: (a) For protons, $q = 1.60 \times 10^{-19}$ C and $m = 1.67 \times 10^{-27}$ kg. From Eq. (44.7),

$$f = \frac{\omega}{2\pi} = \frac{|q|B}{2\pi m} = \frac{(1.60 \times 10^{-19}\,\text{C})(1.50\,\text{T})}{2\pi(1.67 \times 10^{-27}\,\text{kg})}$$

$$= 2.3 \times 10^7\,\text{Hz} = 23\,\text{MHz}$$

(b) From Eq. (44.8) the maximum kinetic energy is

$$K_{max} = \frac{(1.60 \times 10^{-19}\,\text{C})^2(1.50\,\text{T})^2(0.50\,\text{m})^2}{2(1.67 \times 10^{-27}\,\text{kg})}$$

$$= 4.3 \times 10^{-12}\,\text{J} = 2.7 \times 10^{7}\,\text{eV} = 27\,\text{MeV}$$

This proton kinetic energy is much larger than that available from natural radioactive sources.

EVALUATE: From Eq. (44.6) or Eq. (44.7), the proton speed is $v = 7.2 \times 10^7$ m/s, which is about 25% of the speed of light. At such speeds, relativistic effects are beginning to become important. Since we ignored these effects in our calculation, the above results for f and K_{max} are in error by a few percent; this is why we kept only two significant figures.

The maximum energy that can be attained with a cyclotron is limited by relativistic effects. The relativistic version of Eq. (44.7) is

$$\omega = \frac{|q|B}{m}\sqrt{1 - v^2/c^2}$$

As the particles speed up, their angular frequency ω *decreases,* and their motion gets out of phase with the alternating dee voltage. In the *synchrocyclotron* the particles are accelerated in bursts. For each burst, the frequency of the alternating voltage is decreased as the particles speed up, maintaining the correct phase relationship with the particles' motion.

Another limitation of the cyclotron is the difficulty of building very large electromagnets. The largest synchrocyclotron ever built has a vacuum chamber that is about 8 m in diameter and accelerates protons to energies of about 600 MeV.

The Synchrotron

To attain higher energies, another type of machine, called the *synchrotron,* is more practical. Particles move in a vacuum chamber in the form of a thin doughnut, called the *accelerating ring.* The particle beam is bent to follow the ring by a series of electromagnets placed around the ring. As the particles speed up, the magnetic field is increased so that the particles retrace the same trajectory over and over. The Large Hadron Collider (LHC) near Geneva, Switzerland, is the highest-energy accelerator in the world (Fig. 44.7). It is designed to accelerate protons to a maximum energy of 7 TeV, or 7×10^{12} eV. (As we'll discuss in Section 44.3, *hadrons* are a class of elementary particles that includes protons and neutrons.)

As we pointed out in Section 32.1, accelerated charges radiate electromagnetic energy. In an accelerator in which the particles move in curved paths, this radiation is often called *synchrotron radiation.* High-energy accelerators are typically constructed underground to provide protection from this radiation. From the accelerator standpoint, synchrotron radiation is undesirable, since the energy given to an accelerated particle is radiated right back out. It can be minimized by making the accelerator radius r large so that the centripetal acceleration v^2/r is small. On the positive side, synchrotron radiation is used as a source of well-controlled high-frequency electromagnetic waves.

Available Energy

When a beam of high-energy particles collides with a stationary target, not all the kinetic energy of the incident particles is *available* to form new particle states. Because momentum must be conserved, the particles emerging from the collision must have some net motion and thus some kinetic energy. The discussion following Example 43.11 (Section 43.6) presented a nonrelativistic example of this principle. The maximum available energy is the kinetic energy in the frame of reference in which the total momentum is zero. We call this the *center-of-momentum system;* it is the relativistic generalization of the center-of-mass system that we discussed in Section 8.5. In this system the total kinetic energy after the collision can be zero, so that the maximum amount of the initial kinetic energy becomes available to cause the reaction being studied.

44.7 (a) The Large Hadron Collider at the European Organization for Nuclear Research (CERN). The underground accelerating ring (shown by the red circle) is 100 m underground and 8.5 km in diameter, so large that it spans the border between Switzerland and France. (Note the Alps in the background.) When accelerated to 7 TeV, protons travel around the ring more than 11,000 times per second. (b) An engineer working on one of the 9593 superconducting electromagnets around the LHC ring.

(a)

(b)

Consider the *laboratory system,* in which a target particle with mass M is initially at rest and is bombarded by a particle with mass m and total energy (including rest energy) E_m. The total available energy E_a in the center-of-momentum system (including rest energies of all the particles) can be shown to be given by

$$E_a^2 = 2Mc^2E_m + (Mc^2)^2 + (mc^2)^2 \quad \text{(available energy)} \quad (44.9)$$

When the masses of the target and projectile particles are equal, this can be simplified to

$$E_a^2 = 2mc^2(E_m + mc^2) \quad \text{(available energy, equal masses)} \quad (44.10)$$

If in addition E_m is much greater than mc^2, we can neglect the second term in the parentheses in Eq (44.10). Then E_a is

$$E_a = \sqrt{2mc^2E_m} \quad \text{(available energy, equal masses, } E_m \gg mc^2) \quad (44.11)$$

The square root in Eq. (44.11) is a disappointing result for an accelerator designer: Doubling the energy E_m of the bombarding particle increases the available energy E_a by only a factor of $\sqrt{2} = 1.414$. Examples 44.2 and 44.3 explore the limitations of having a stationary target particle.

Example 44.2 Threshold energy for pion production

A proton (rest energy 938 MeV) with kinetic energy K collides with a proton at rest. Both protons survive the collision, and a neutral pion (π^0, rest energy 135 MeV) is produced. What is the threshold energy (minimum value of K) for this process?

SOLUTION

IDENTIFY and SET UP: The final state includes the two original protons (mass m) and the pion (mass m_π). The threshold energy corresponds to the minimum-energy case in which all three particles are at rest in the center-of-momentum system. The total available energy E_a in that system must be at least the total rest energy, $2mc^2 + m_\pi c^2$. We use this to solve Eq. (44.10) for the total energy E_m of the bombarding proton; the kinetic energy K (our target variable) is then E_m minus the proton rest energy mc^2.

EXECUTE: We substitute $E_a = 2mc^2 + m_\pi c^2$ into Eq. (44.10), simplify, and solve for E_m:

$$4m^2c^4 + 4mm_\pi c^4 + m_\pi^2c^4 = 2mc^2E_m + 2(mc^2)^2$$

$$E_m = mc^2 + m_\pi c^2\left(2 + \frac{m_\pi}{2m}\right) = mc^2 + K$$

$$K = m_\pi c^2\left(2 + \frac{m_\pi}{2m}\right)$$

We see that the bombarding proton's kinetic energy K must be somewhat greater than twice the pion rest energy $m_\pi c^2$. With $mc^2 = 938$ MeV and $m_\pi c^2 = 135$ MeV, we have $m_\pi/2m = 0.072$ and

$$K = (135 \text{ MeV})(2 + 0.072) = 280 \text{ MeV}$$

EVALUATE: Compare this result with the result of Example 37.11 (Section 37.8), where we found that a pion can be produced in a head-on collision of two protons, each with only 67.5 MeV of kinetic energy. We discuss the energy advantage of such collisions in the next subsection.

Example 44.3 Increasing the available energy

The Fermilab accelerator in Illinois was designed to bombard stationary targets with 800-GeV protons. (a) What is the available energy E_a in a proton-proton collision? (b) What is E_a if the beam energy is increased to 980 GeV?

SOLUTION

IDENTIFY and SET UP: Our target variable is the available energy E_a in a stationary-target collision between identical particles. In both parts (a) and (b) the beam energy E_m is much larger than the proton rest energy $mc^2 = 938$ MeV $= 0.938$ GeV, so we can safely use the approximation of Eq. (44.11).

EXECUTE: (a) For $E_m = 800$ GeV, Eq. (44.11) gives

$$E_a = \sqrt{2(0.938 \text{ GeV})(800 \text{ GeV})} = 38.7 \text{ GeV}$$

(b) For $E_m = 980$ GeV,

$$E_a = \sqrt{2(0.938 \text{ GeV})(980 \text{ GeV})} = 42.9 \text{ GeV}$$

EVALUATE: With a stationary-proton target, increasing the proton beam energy by 180 GeV increases the available energy by only 4.2 GeV! This shows a major limitation of experiments in which one of the colliding particles is initially at rest. Below we describe how physicists can overcome this limitation.

Colliding Beams

The limitation illustrated by Example 44.3 is circumvented in *colliding-beam* experiments. In these experiments there is no stationary target; instead, beams of particles moving in opposite directions are tightly focused onto one another so that head-on collisions can occur. Usually the two colliding particles have momenta of equal magnitude and opposite direction, so the total momentum is zero. Hence the laboratory system is also the center-of-momentum system, and the available energy is maximized.

The highest-energy colliding beams available are those at the Large Hadron Collider (see Fig. 44.7). In operation, 2808 bunches of 7-TeV protons circulate around the ring, half in one direction and half in the opposite direction. Each bunch contains about 10^{11} protons. Magnets steer the oppositely moving bunches to collide at interaction points. The available energy E_a in the resulting head-on collisions is the *total* energy of the two colliding particles: $E_a = 2 \times 7 \text{ TeV} = 14 \text{ TeV}$. (Strictly, E_a is 14 TeV minus the rest energy of the two colliding protons. But this rest energy is only $2mc^2 = 2(938 \text{ MeV}) = 1.876 \times 10^{-3} \text{ TeV}$, which is so small compared to 14 TeV that it can be ignored.) Physicists expect that the very large available energy at the Large Hadron Collider will make it possible to produce particles that have never been seen before.

Detectors

Ordinarily, we can't see or feel individual subatomic particles or photons. How, then, do we measure their properties? A wide variety of devices have been designed. Many detectors use the ionization caused by charged particles as they move through a gas, liquid, or solid. The ions along the particle's path act as nucleation centers for droplets of liquid in the supersaturated vapor of a cloud chamber (Fig. 44.1) or cause small volumes of vapor in the superheated liquid of a bubble chamber (Fig. 44.3a). In a semiconducting solid the ionization can take the form of electron–hole pairs. We discussed their detection in Section 42.7. *Wire chambers* contain arrays of closely spaced wires that detect the ions. The charge collected and time information from each wire are processed using computers to reconstruct the particle trajectories. The detectors at the Large Hadron Collider use an array of devices to follow the tracks of particles produced by collisions between protons (Fig. 44.8). The giant solenoid in the photo that opens Chapter 28 is at the heart of one these detector arrays. The intense magnetic field of the solenoid helps identify newly produced particles, which curve in different directions and along paths of different radii depending on their charge and energy.

Cosmic-Ray Experiments

Large numbers of particles called *cosmic rays* continually bombard the earth from sources both within and beyond our galaxy. These particles consist mostly of neutrinos, protons, and heavier nuclei, with energies ranging from less than 1 MeV to more than 10^{20} eV. The earth's atmosphere and magnetic field protect us from much of this radiation. This means that cosmic-ray experimentation often must be carried out above all or most of the atmosphere by means of rockets or high-altitude balloons.

In contrast, neutrino detectors are buried below the earth's surface in tunnels or mines or submerged deep in the ocean. This is done to screen out all other types of particles so that only neutrinos, which interact only very weakly with matter, reach the detector. It would take a light-year's thickness of lead to absorb a sizable fraction of a beam of neutrinos. Thus neutrino detectors consist of huge amounts of matter: The Super-Kamiokande detector looks for flashes of light produced when a neutrino interacts in a tank containing 5×10^7 kg of water (see Section 44.5).

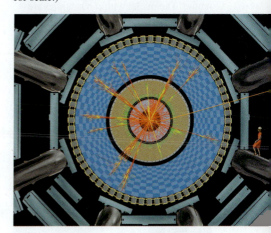

44.8 This computer-generated image shows the result of a simulated collision between two protons (not shown) in one of the interaction regions at the Large Hadron Collider. The view is along the beampipe. The different color tracks show different types of particles emerging from the collision. A variety of different detectors surround the collision region. (Note the drawing of woman in a red dress, shown for scale.)

Cosmic rays were important in early particle physics, and their study currently brings us important information about the rest of the universe. Although cosmic rays provide a source of high-energy particles that does not depend on expensive accelerators, most particle physicists use accelerators because the high-energy cosmic-ray particles they want are too few and too random.

Test Your Understanding of Section 44.2 In a colliding-beam experiment, a 90-GeV electron collides head-on with a 90-GeV positron. The electron and the positron annihilate each other, forming a single virtual photon that then transforms into other particles. Does the virtual photon obey the same relationship $E = pc$ as real photons do? ❙

44.3 Particles and Interactions

We have mentioned the array of subatomic particles that were known as of 1947: photons, electrons, positrons, protons, neutrons, muons, and pions. Since then, literally hundreds of additional particles have been discovered in accelerator experiments. The vast majority of known particles are *unstable* and decay spontaneously into other particles. Particles of all kinds, whether stable or unstable, can be created or destroyed in interactions between particles. Each such interaction involves the exchange of virtual particles, which exist on borrowed energy allowed by the uncertainty principle.

Although the world of subatomic particles and their interactions is complex, some key results bring order and simplicity to the seeming chaos. One key simplification is that there are only four fundamental types of interactions, each mediated or transmitted by the exchange of certain characteristic virtual particles. Furthermore, not all particles respond to all four kinds of interaction. In this section we will examine the fundamental interactions more closely and see how physicists classify particles in terms of the ways in which they interact.

Four Forces and Their Mediating Particles

In Section 5.5 we first described the four fundamental types of forces or interactions (Fig. 44.9). They are, in order of decreasing strength:

1. The strong interaction
2. The electromagnetic interaction
3. The weak interaction
4. The gravitational interaction

The *electromagnetic* and *gravitational* interactions are familiar from classical physics. Both are characterized by a $1/r^2$ dependence on distance. In this scheme, the mediating particles for both interactions have mass zero and are stable as ordinary particles. The mediating particle for the electromagnetic interaction is the familiar photon, which has spin 1. (That means its spin quantum number is $s = 1$, so the magnitude of its spin angular momentum is $S = \sqrt{s(s + 1)}\hbar = \sqrt{2}\hbar$.) That particle for the gravitational force is the spin-2 *graviton* ($s = 2$, $S = \sqrt{s(s + 1)}\hbar = \sqrt{6}\hbar$). The graviton has not yet been observed experimentally because the gravitational force is very much weaker than the electromagnetic force. For example, the gravitational attraction of two protons is smaller than their electrical repulsion by a factor of about 10^{36}. The gravitational force is of primary importance in the structure of stars and the large-scale behavior of the universe, but it is not believed to play a significant role in particle interactions at the energies that are currently attainable.

The other two forces are less familiar. One, usually called the *strong interaction,* is responsible for the nuclear force and also for the production of pions and several other particles in high-energy collisions. At the most fundamental level, the mediating particle for the strong interaction is called a *gluon.* However, the force between nucleons is more easily described in terms of mesons as the mediating particles. We'll discuss the spin-1, massless gluon in Section 44.4.

44.9 The ties that bind us together originate in the fundamental interactions of nature. The nuclei within our bodies are held together by the strong interaction. The electromagnetic interaction binds nuclei and electrons together to form atoms, binds atoms together to form molecules, and binds molecules together to form us.

Equation (44.4) is a possible potential-energy function for the nuclear force. The strength of the interaction is described by the constant f^2, which has units of energy times distance. A better basis for comparison with other forces is the dimensionless ratio $f^2/\hbar c$, called the *coupling constant* for the interaction. (We invite you to verify that this ratio is a pure number and so must have the same value in all systems of units.) The observed behavior of nuclear forces suggests that $f^2/\hbar c \approx 1$. The dimensionless coupling constant for *electromagnetic* interactions is

$$\frac{1}{4\pi\epsilon_0} \frac{e^2}{\hbar c} = 7.297 \times 10^{-3} = \frac{1}{137.0} \tag{44.12}$$

Thus the strong interaction is roughly 100 times as strong as the electromagnetic interaction; however, it drops off with distance more quickly than $1/r^2$.

The fourth interaction is called the *weak* interaction. It is responsible for beta decay, such as the conversion of a neutron into a proton, an electron, and an antineutrino. It is also responsible for the decay of many unstable particles (pions into muons, muons into electrons, and so on). Its mediating particles are the short-lived particles W^+, W^-, and Z^0. The existence of these particles was confirmed in 1983 in experiments at CERN, for which Carlo Rubbia and Simon van der Meer were awarded the Nobel Prize in 1984. The $W^\pm$ and Z^0 have spin 1 like the photon and the gluon, but they are *not* massless. In fact, they have enormous masses, 80.4 GeV/c^2 for the W's and 91.2 GeV/c^2 for the Z^0. With such massive mediating particles the weak interaction has a much shorter range than the strong interaction. It also lives up to its name by being weaker than the strong interaction by a factor of about 10^9.

Table 44.1 compares the main features of these four fundamental interactions.

More Particles

In Section 44.1 we mentioned the discoveries of muons in 1937 and of pions in 1947. The electric charges of the muons and the charged pions have the same magnitude e as the electron charge. The positive muon μ^+ is the antiparticle of the negative muon μ^-. Each has spin $\frac{1}{2}$, like the electron, and a mass of about $207m_e = 106 \text{ MeV}/c^2$. Muons are unstable; each decays with a lifetime of 2.2×10^{-6} s into an electron of the same sign, a neutrino, and an antineutrino.

There are three kinds of pions, all with spin 0; they have *no* spin angular momentum. The π^+ and π^- have masses of $273m_e = 140 \text{ MeV}/c^2$. They are unstable; each $\pi^\pm$ decays with a lifetime of 2.6×10^{-8} s into a muon of the same sign along with a neutrino for the π^+ and an antineutrino for the π^-. The π^0 is somewhat less massive, $264m_e = 135 \text{ MeV}/c^2$, and it decays with a lifetime of 8.4×10^{-17} s into two photons. The π^+ and π^- are antiparticles of one another, while the π^0 is its own antiparticle. (That is, there is no distinction between particle and antiparticle for the π^0.)

The existence of the *antiproton* $\bar{p}$ had been suspected ever since the discovery of the positron. The $\bar{p}$ was found in 1955, when proton–antiproton ($p\bar{p}$) pairs were created by use of a beam of 6-GeV protons from the Bevatron at the University of

ActivPhysics 19.5: Particle Physics

Table 44.1 Four Fundamental Interactions

Interaction	Relative Strength	Range	Mediating Particle			
			Name	Mass	Charge	Spin
Strong	1	Short ($\sim$1 fm)	Gluon	0	0	1
Electromagnetic	$\frac{1}{137}$	Long ($1/r^2$)	Photon	0	0	1
Weak	10^{-9}	Short ($\sim$0.001 fm)	$W^\pm, Z^0$	80.4, 91.2 GeV/c^2	$\pm e$, 0	1
Gravitational	10^{-38}	Long ($1/r^2$)	Graviton	0	0	2

California, Berkeley. The *antineutron* $\bar{n}$ was found soon afterward. After 1960, as higher-energy accelerators and more sophisticated detectors were developed, a veritable blizzard of new unstable particles were identified. To describe and classify them, we need a small blizzard of new terms.

Initially, particles were classified by mass into three categories: (1) leptons ("light ones" such as electrons); (2) mesons ("intermediate ones" such as pions); and (3) baryons ("heavy ones" such as nucleons and more massive particles). But this scheme has been superseded by a more useful one in which particles are classified in terms of their *interactions*. For instance, *hadrons* (which include mesons and baryons) have strong interactions, and *leptons* do not.

In the following discussion we will also distinguish between **fermions,** which have half-integer spins, and **bosons,** which have zero or integer spins. Fermions obey the exclusion principle, on which the Fermi-Dirac distribution function (see Section 42.5) is based. Bosons do not obey the exclusion principle and have a different distribution function, the Bose-Einstein distribution.

Leptons

The **leptons,** which do not have strong interactions, include six particles; the electron (e^-) and its neutrino (ν_e), the muon (μ^-) and its neutrino (ν_μ), and the tau particle (τ^-) and its neutrino (ν_τ). Each of the six particles has a distinct antiparticle. All leptons have spin $\frac{1}{2}$ and thus are fermions. The family of leptons is shown in Table 44.2. The taus have mass $3478m_e = 1777 \text{ MeV}/c^2$. Taus and muons are unstable; a τ^- decays into a μ^- plus a tau neutrino and a muon antineutrino, or an electron plus a tau neutrino and an electron antineutrino. A μ^- decays into a electron plus a muon neutrino and an electron antineutrino. They have relatively long lifetimes because their decays are mediated by the weak interaction. Despite their zero charge, a neutrino is distinct from an antineutrino; the spin angular momentum of a neutrino has a component that is opposite its linear momentum, while for an antineutrino that component is parallel to its linear momentum. Because neutrinos are so elusive, physicists have only been able to place upper limits on the rest masses of the ν_e, the ν_μ, and the ν_τ. Until recently, it was thought that the rest masses of the neutrinos were zero; compelling evidence now indicates that they have small but nonzero masses. We'll return to this point and its implications later.

Leptons obey a *conservation principle.* Corresponding to the three pairs of leptons are three lepton numbers L_e, L_μ, and L_τ. The electron e^- and the electron neutrino ν_e are assigned $L_e = 1$, and their antiparticles e^+ and $\bar{\nu}_e$ are given $L_e = -1$. Corresponding assignments of L_μ and L_τ are made for the μ and τ particles and their neutrinos. **In all interactions, each lepton number is separately conserved.** For example, in the decay of the μ^-, the lepton numbers are

$$\mu^- \quad \rightarrow \quad e^- \quad + \quad \bar{\nu}_e \quad + \quad \nu_\mu$$
$$L_\mu = 1 \qquad L_e = 1 \qquad L_e = -1 \qquad L_\mu = 1$$

These conservation principles have no counterpart in classical physics.

Table 44.2 The Six Leptons

Particle Name	Symbol	Anti-particle	Mass (MeV/c^2)	L_e	L_μ	L_τ	Lifetime (s)	Principal Decay Modes
Electron	e^-	e^+	0.511	+1	0	0	Stable	
Electron neutrino	ν_e	$\bar{\nu}_e$	$<3 \times 10^{-6}$	+1	0	0	Stable	
Muon	μ^-	μ^+	105.7	0	+1	0	2.20×10^{-6}	$e^- \bar{\nu}_e \nu_\mu$
Muon neutrino	ν_μ	$\bar{\nu}_\mu$	<0.19	0	+1	0	Stable	
Tau	τ^-	τ^+	1777	0	0	+1	2.9×10^{-13}	$\mu^- \bar{\nu}_\mu \nu_\tau$
Tau neutrino	ν_τ	$\bar{\nu}_\tau$	<18.2	0	0	+1	Stable	or $e^- \bar{\nu}_e \nu_\tau$

Example 44.4 | **Lepton number conservation**

Check conservation of lepton numbers for these decay schemes:

(a) $\mu^+ \rightarrow e^+ + \nu_e + \bar{\nu}_\mu$

(b) $\pi^- \rightarrow \mu^- + \bar{\nu}_\mu$

(c) $\pi^0 \rightarrow \mu^- + e^+ + \nu_e$

SOLUTION

IDENTIFY and SET UP: Lepton number conservation requires that L_e, L_μ, and L_τ (given in Table 44.2) separately have the same sums after the decay as before.

EXECUTE: We tabulate L_e and L_μ for each decay scheme. An antiparticle has the opposite lepton number from its corresponding particle listed in Table 44.2. No τ particles or τ neutrinos appear in any of the schemes, so $L_\tau = 0$ both before and after each decay and L_τ is conserved.

(a) $\mu^+ \rightarrow e^+ + \nu_e + \bar{\nu}_\mu$

$L_e: 0 = -1 + 1 + 0$

$L_\mu: -1 = 0 + 0 + (-1)$

(b) $\pi^- \rightarrow \mu^- + \bar{\nu}_\mu$

$L_e: 0 = 0 + 0$

$L_\mu: 0 = 1 + (-1)$

(c) $\pi^0 \rightarrow \mu^- + e^+ + \nu_e$

$L_e: 0 = 0 + (-1) + 1$

$L_\mu: 0 \neq 1 + 0 + 0$

EVALUATE: Decays (a) and (b) are consistent with lepton number conservation and are observed. Decay (c) violates the conservation of L_μ and has *never* been observed. Physicists used these and other experimental results to deduce the principle that all three lepton numbers must separately be conserved.

Hadrons

Hadrons, the strongly interacting particles, are a more complex family than leptons. Each hadron has an antiparticle, often denoted with an overbar, as with the antiproton $\bar{p}$. There are two subclasses of hadrons: *mesons* and *baryons*. Table 44.3 shows some of the many hadrons that are currently known. (We'll explain *strangeness* and *quark content* later in this section and in the next one.)

Mesons include the pions that have already been mentioned, K mesons or *kaons,* η mesons, and others that we will discuss later. Mesons have spin 0 or 1 and therefore are all bosons. There are no stable mesons; all can and do decay to less massive particles, obeying all the conservation laws for such decays.

Table 44.3 Some Hadrons and Their Properties

Particle	Mass (MeV/c^2)	Charge Ratio, Q/e	Spin	Baryon Number, B	Strangeness, S	Mean Lifetime (s)	Typical Decay Modes	Quark Content
Mesons								
π^0	135.0	0	0	0	0	8.4×10^{-17}	$\gamma\gamma$	$u\bar{u}, d\bar{d}$
π^+	139.6	+1	0	0	0	2.60×10^{-8}	$\mu^+\nu_\mu$	$u\bar{d}$
π^-	139.6	−1	0	0	0	2.60×10^{-8}	$\mu^-\bar{\nu}_\mu$	$\bar{u}d$
K^+	493.7	+1	0	0	+1	1.24×10^{-8}	$\mu^+\nu_\mu$	$u\bar{s}$
K^-	493.7	−1	0	0	−1	1.24×10^{-8}	$\mu^-\bar{\nu}_\mu$	$\bar{u}s$
η^0	547.3	0	0	0	0	$\approx 10^{-18}$	$\gamma\gamma$	$u\bar{u}, d\bar{d}, s\bar{s}$
Baryons								
p	938.3	+1	$\frac{1}{2}$	1	0	Stable	—	uud
n	939.6	0	$\frac{1}{2}$	1	0	886	$pe^-\bar{\nu}_e$	udd
Λ^0	1116	0	$\frac{1}{2}$	1	−1	2.63×10^{-10}	$p\pi^-$ or $n\pi^0$	uds
Σ^+	1189	+1	$\frac{1}{2}$	1	−1	8.02×10^{-11}	$p\pi^0$ or $n\pi^+$	uus
Σ^0	1193	0	$\frac{1}{2}$	1	−1	7.4×10^{-20}	$\Lambda^0\gamma$	uds
Σ^-	1197	−1	$\frac{1}{2}$	1	−1	1.48×10^{-10}	$n\pi^-$	dds
Ξ^0	1315	0	$\frac{1}{2}$	1	−2	2.90×10^{-10}	$\Lambda^0\pi^0$	uss
Ξ^-	1321	−1	$\frac{1}{2}$	1	−2	1.64×10^{-10}	$\Lambda^0\pi^-$	dss
Δ^{++}	1232	+2	$\frac{3}{2}$	1	0	$\approx 10^{-23}$	$p\pi^+$	uuu
Ω^-	1672	−1	$\frac{3}{2}$	1	−3	8.2×10^{-11}	$\Lambda^0 K^-$	sss
Λ_c^+	2285	+1	$\frac{1}{2}$	1	0	2.0×10^{-13}	$pK^-\pi^+$	udc

Baryons include the nucleons and several particles called *hyperons,* including the Λ, Σ, Ξ, and Ω. These resemble the nucleons but are more massive. Baryons have half-integer spin, and therefore all are fermions. The only stable baryon is the proton; a free neutron decays to a proton, and the hyperons decay to other hyperons or to nucleons by various processes. Baryons obey the principle of *conservation of baryon number,* analogous to conservation of lepton numbers, again with no counterpart in classical physics. We assign a baryon number $B = 1$ to each baryon (p, n, Λ, Σ, and so on) and $B = -1$ to each antibaryon ($\bar{\text{p}}$, $\bar{\text{n}}$, $\bar{\Lambda}$, $\bar{\Sigma}$, and so on).

In all interactions, the total baryon number is conserved.

This principle is the reason the mass number A was conserved in all of the nuclear reactions that we studied in Chapter 43.

Example 44.5   Baryon number conservation

Check conservation of baryon number for these reactions:

(a) $n + p \rightarrow n + p + p + \bar{\text{p}}$

(b) $n + p \rightarrow n + p + \bar{\text{n}}$

SOLUTION

IDENTIFY and SET UP: This example is similar to Example 44.4. We compare the total baryon number before and after each reaction, using data from Table 44.3.

EXECUTE: We tabulate the baryon numbers, noting that a baryon has $B = 1$ and an antibaryon has $B = -1$:

(a) $n + p \rightarrow n + p + p + \bar{\text{p}}$: $1 + 1 = 1 + 1 + 1 + (-1)$

(b) $n + p \rightarrow n + p + \bar{\text{n}}$: $1 + 1 \neq 1 + 1 + (-1)$

EVALUATE: Reaction (a) is consistent with baryon number conservation. It can occur if enough energy is available in the $n + p$ collision. Reaction (b) violates baryon number conservation and has never been observed.

Example 44.6   Antiproton creation

What is the minimum proton energy required to produce an antiproton in a collision with a stationary proton?

SOLUTION

IDENTIFY and SET UP: The reaction must conserve baryon number, charge, and energy. Since the target and bombarding protons are of equal mass and the target is at rest, we determine the minimum energy E_m of the bombarding proton using Eq. (44.10).

EXECUTE: Conservation of charge and conservation of baryon number forbid the creation of an antiproton by itself; it must be created as part of a proton–antiproton pair. The complete reaction is

$$p + p \rightarrow p + p + p + \bar{\text{p}}$$

For this reaction to occur, the minimum available energy E_a in Eq. (44.10) is the final rest energy $4mc^2$ of three protons and an antiproton. Equation (44.10) then gives

$$(4mc^2)^2 = 2mc^2(E_m + mc^2)$$
$$E_m = 7mc^2$$

EVALUATE: The energy E_m of the bombarding proton includes its rest energy mc^2, so its minimum *kinetic* energy must be $6mc^2 = 6(938 \text{ MeV}) = 5.63 \text{ GeV}$.

The search for the antiproton was a principal reason for the construction of the Bevatron at the University of California, Berkeley, with beam energy of 6 GeV. The search succeeded in 1955, and Emilio Segrè and Owen Chamberlain were later awarded the Nobel Prize for this discovery.

Strangeness

The K mesons and the Λ and Σ hyperons were discovered during the late 1950s. Because of their unusual behavior they were called *strange particles.* They were produced in high-energy collisions such as $\pi^- + p$, and a K meson and a hyperon were always produced *together.* The relatively high rate of production of these particles suggested that it was a *strong*-interaction process, but their relatively long lifetimes suggested that their decay was a *weak*-interaction process. The K^0 appeared to have *two* lifetimes, one about 9×10^{-11} s and another nearly 600 times longer. Were the K mesons strongly interacting hadrons or not?

The search for the answer to this question led physicists to introduce a new quantity called **strangeness.** The hyperons Λ^0 and $\Sigma^{\pm,0}$ were assigned a strangeness quantum number $S = -1$, and the associated K^0 and K^+ mesons were assigned $S = +1$. The corresponding antiparticles had opposite strangeness, $S = +1$ for $\overline{\Lambda}^0$ and $\overline{\Sigma}^{\pm,0}$ and $S = -1$ for $\overline{K}^0$ and K^-. Then strangeness was *conserved* in production processes such as

$$p + \pi^- \rightarrow \Sigma^- + K^+$$
$$p + \pi^- \rightarrow \Lambda^0 + K^0$$

The process

$$p + \pi^- \rightarrow p + K^-$$

does not conserve strangeness and it does not occur.

When strange particles decay individually, strangeness is usually *not* conserved. Typical processes include

$$\Sigma^+ \rightarrow n + \pi^+$$
$$\Lambda^0 \rightarrow p + \pi^-$$
$$K^- \rightarrow \pi^+ + \pi^- + \pi^-$$

In each of these decays, the initial strangeness is 1 or -1, and the final value is zero. All observations of these particles are consistent with the conclusion that *strangeness is conserved in strong interactions but it can change by zero or one unit in weak interactions.* There is no counterpart to the strangeness quantum number in classical physics.

> **CAUTION** **Strangeness vs. spin** Take care not to confuse the symbol S for strangeness with the identical symbol for the magnitude of the spin angular momentum. |

Conservation Laws

The decay of strange particles provides our first example of a *conditional conservation law*, one that is obeyed in some interactions and not in others. By contrast, several conservation laws are obeyed in *all* interactions. These include the familiar conservation laws; energy, momentum, angular momentum, and electric charge. These are called *absolute conservation laws.* Baryon number and the three lepton numbers are also conserved in all interactions. Strangeness is conserved in strong and electromagnetic interactions but *not* in all weak interactions.

Two other quantities, which are conserved in some but not all interactions, are useful in classifying particles and their interactions. One is *isospin,* a quantity that is used to describe the charge independence of the strong interactions. The other is *parity,* which describes the comparative behavior of two systems that are mirror images of each other. Isospin is conserved in strong interactions, which are charge independent, but not in electromagnetic or weak interactions. (The electromagnetic interaction is certainly *not* charge independent.) Parity is conserved in strong and electromagnetic interactions but not in weak ones. The Chinese-American physicists T. D. Lee and C. N. Yang received the Nobel Prize in 1957 for laying the theoretical foundations for nonconservation of parity in weak interactions.

This discussion shows that conservation laws provide another basis for classifying particles and their interactions. Each conservation law is also associated with a *symmetry* property of the system. A familiar example is angular momentum. If a system is in an environment that has spherical symmetry, there can be no torque acting on it because the direction of the torque would violate the symmetry. In such a system, total angular momentum is *conserved.* When a conservation law is violated, the interaction is often described as a *symmetry-breaking interaction.*

Test Your Understanding of Section 44.3 From conservation of energy, a particle of mass m and rest energy mc^2 can decay only if the decay products have a total mass less than m. (The remaining energy goes into the kinetic energy of the decay products.) Can a proton decay into less massive mesons?

44.4 Quarks and the Eightfold Way

The leptons form a fairly neat package: three particles and three neutrinos, each with its antiparticle, and a conservation law relating their numbers. Physicists believe that leptons are genuinely fundamental particles. The hadron family, by comparison, is a mess. Table 44.3 contains only a sample of well over 100 hadrons that have been discovered since 1960, and it has become clear that these particles *do not* represent the most fundamental level of the structure of matter.

Our present understanding of the structure of hadrons is based on a proposal made initially in 1964 by the American physicist Murray Gell-Mann and his collaborators. In this proposal, hadrons are not fundamental particles but are composite structures whose constituents are spin-$\frac{1}{2}$ fermions called **quarks.** (The name is found in the line "Three quarks for Muster Mark!" from *Finnegans Wake* by James Joyce.) Each baryon is composed of three quarks (qqq), each antibaryon of three antiquarks ($\overline{qqq}$), and each meson of a quark–antiquark pair ($q\overline{q}$). Table 44.3 of the preceding section gives the quark content of many hadrons. No other compositions seem to be necessary. This scheme requires that quarks have electric charges with magnitudes $\frac{1}{3}$ and $\frac{2}{3}$ of the electron charge e, which had previously been thought to be the smallest unit of charge. Each quark also has a fractional value $\frac{1}{3}$ for its baryon number B, and each antiquark has a baryon-number value $-\frac{1}{3}$. In a meson, a quark and antiquark combine with net baryon number 0 and can have their spin angular momentum components parallel to form a spin-1 meson or antiparallel to form a spin-0 meson. Similarly, the three quarks in a baryon combine with net baryon number 1 and can form a spin-$\frac{1}{2}$ baryon or a spin-$\frac{3}{2}$ baryon.

The Three Original Quarks

The first (1964) quark theory included three types (called *flavors*) of quarks, labeled *u* (up), *d* (down), and *s* (strange). Their principal properties are listed in Table 44.4. The corresponding antiquarks $\overline{u}$, $\overline{d}$, and $\overline{s}$ have opposite values of charge Q, B, and S. Protons, neutrons, π and K mesons, and several hyperons can be constructed from these three quarks. For example, the proton quark content is *uud*. Checking Table 44.4, we see that the values of Q/e add to 1 and that the values of the baryon number B also add to 1, as we should expect. The neutron is *udd*, with total $Q = 0$ and $B = 1$. The π^+ meson is $u\overline{d}$, with $Q/e = 1$ and $B = 0$, and the K^+ meson is $u\overline{s}$. Checking the values of S for the quark content, we see that the proton, neutron, and π^+ have strangeness 0 and that the K^+ has strangeness 1, in agreement with Table 44.3. The antiproton is $\overline{p} = \overline{uud}$, the negative pion is $\pi^- = \overline{u}d$, and so on. The quark content can also be used to explain hadron excited states and magnetic moments. Figure 44.10 shows the quark content of two baryons and two mesons.

44.10 Quark content of four different hadrons. The various color combinations that are needed for color neutrality are not shown.

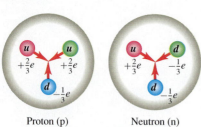

Proton (p)   Neutron (n)

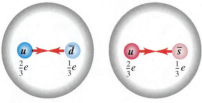

Positive pion (π^+)   Positive kaon (K^+)

Table 44.4 Properties of the Three Original Quarks

Symbol	Q/e	Spin	Baryon Number, B	Strangeness, S	Charm, C	Bottomness, B'	Topness, T
u	$\frac{2}{3}$	$\frac{1}{2}$	$\frac{1}{3}$	0	0	0	0
d	$-\frac{1}{3}$	$\frac{1}{2}$	$\frac{1}{3}$	0	0	0	0
s	$-\frac{1}{3}$	$\frac{1}{2}$	$\frac{1}{3}$	-1	0	0	0

Example 44.7   **Determining the quark content of baryons**

Given that they contain only u, d, s, $\bar{u}$, $\bar{d}$, and/or $\bar{s}$, find the quark content of (a) Σ^+ and (b) $\overline{\Lambda}{}^0$. The Σ^+ and Λ^0 (the antiparticle of the $\overline{\Lambda}{}^0$) are both baryons with strangeness $S = -1$.

SOLUTION

IDENTIFY and SET UP: We use the idea that the total charge of each baryon is the sum of the individual quark charges, and similarly for the baryon number and strangeness. We use the quark properties given in Table 44.4.

EXECUTE: Baryons contain three quarks. If $S = -1$, exactly *one* of the three must be an s quark, which has $S = -1$ and $Q/e = -\frac{1}{3}$.

(a) The Σ^+ has $Q/e = +1$, so the other two quarks must both be u quarks (each of which has $Q/e = +\frac{2}{3}$). Hence the quark content of Σ^+ is uus.

(b) First we find the quark content of the Λ^0. To yield zero total charge, the other two quarks must be u ($Q/e = +\frac{2}{3}$) and d ($Q/e = -\frac{1}{3}$), so the quark content of the Λ^0 is uds. The quark content of the $\overline{\Lambda}{}^0$ is therefore $\bar{u}\,\bar{d}\,\bar{s}$.

EVALUATE: Although the Λ^0 and $\overline{\Lambda}{}^0$ are both electrically neutral and have the same mass, they are different particles: Λ^0 has $B = 1$ and $S = -1$, while $\overline{\Lambda}{}^0$ has $B = -1$ and $S = 1$.

Motivating the Quark Model

What caused physicists to suspect that hadrons were made up of something smaller? The magnetic moment of the neutron (see Section 43.1) was one of the first reasons. In Section 27.7 we learned that a magnetic moment results from a circulating current (a motion of electric charge). But the neutron has *no* charge, or, to be more accurate, no *total* charge. It could be made up of smaller particles whose charges add to zero. The quantum motion of these particles within the neutron would then give its surprising nonzero magnetic moment. To verify this hypothesis by "seeing" inside a neutron, we need a probe with a wavelength that is much less than the neutron's size of about a femtometer. This probe should not be affected by the strong interaction, so that it won't interact with the neutron as a whole but will penetrate into it and interact electromagnetically with these supposed smaller charged particles. A probe with these properties is an electron with energy above 10 GeV. In experiments carried out at SLAC, such electrons were scattered from neutrons and protons to help show that nucleons are indeed made up of fractionally charged, spin-$\frac{1}{2}$ pointlike particles.

The Eightfold Way

Symmetry considerations play a very prominent role in particle theory. Here are two examples. Consider the eight spin-$\frac{1}{2}$ baryons we've mentioned: the familiar p and n; the strange Λ^0, Σ^+, Σ^0, and Σ^-; and the doubly strange Ξ^0 and Ξ^-. For each we plot the value of strangeness S versus the value of charge Q in Fig. 44.11. The result is a hexagonal pattern. A similar plot for the nine spin-0 mesons (six shown in Table 44.3 plus three others not included in that table) is shown in Fig. 44.12; the particles fall in exactly the same hexagonal pattern! In each plot, all the particles have masses that are within about ± 200 MeV/c^2 of the median mass value of that plot, with variations due to differences in quark masses and internal potential energies.

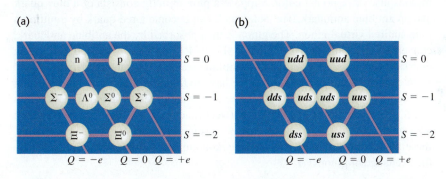

44.11 (a) Plot of S and Q values for spin-$\frac{1}{2}$ baryons, showing the symmetry pattern of the eightfold way. (b) Quark content of each spin-$\frac{1}{2}$ baryon. The quark contents of the Σ^0 and Λ^0 are the same; the Σ^0 is an excited state of the Λ^0 and can decay into it by photon emission.

44.12 (a) Plot of S and Q values for nine spin-0 mesons, showing the symmetry pattern of the eightfold way. Each particle is on the opposite side of the hexagon from its antiparticle; each of the three particles in the center is its own antiparticle. (b) Quark content of each spin-0 meson. The particles in the center are different mixtures of the three quark–antiquark pairs shown.

(a)

K^0	K^+	$S = +1$
η^0	π^0	
π^-	π^+	$S = 0$
	η'^0	
K^-	$\overline{K}^0$	$S = -1$

$Q = -e \quad Q = 0 \quad Q = +e$

(b)

$d\bar{s}$	$u\bar{s}$	$S = +1$
$d\bar{d}$	$u\bar{u}$	
$\bar{u}d$	$u\bar{d}$	$S = 0$
	$s\bar{s}$	
$\bar{u}s$	$\bar{d}s$	$S = -1$

$Q = -e \quad Q = 0 \quad Q = +e$

The symmetries that lead to these and similar patterns are collectively called the **eightfold way.** They were discovered in 1961 by Murray Gell-Mann and independently by Yu'val Ne'eman. (The name is a slightly irreverent reference to the Noble Eightfold Path, a set of principles for right living in Buddhism.) A similar pattern for the spin-$\frac{3}{2}$ baryons contains *ten* particles, arranged in a triangular pattern like pins in a bowling alley. When this pattern was first discovered, one of the particles was missing. But Gell-Mann gave it a name anyway (Ω^-), predicted the properties it should have, and told experimenters what they should look for. Three years later, the particle was found during an experiment at Brookhaven National Laboratory, a spectacular success for Gell-Mann's theory. The whole series of events is reminiscent of the way in which Mendeleev used gaps in the periodic table of the elements to predict properties of undiscovered elements and to guide chemists in their search for these elements.

What binds quarks to one another? The attractive interactions among quarks are mediated by massless spin-1 bosons called **gluons** in much the same way that photons mediate the electromagnetic interaction or that pions mediated the nucleon–nucleon force in the old Yukawa theory.

Color

Quarks, having spin $\frac{1}{2}$, are fermions and so are subject to the exclusion principle. This would seem to forbid a baryon having two or three quarks with the same flavor and same spin component. To avoid this difficulty, it is assumed that each quark comes in three varieties, which are whimsically called *colors*. Red, green, and blue are the usual choices. The exclusion principle applies separately to each color. A baryon always contains one red, one green, and one blue quark, so the baryon itself has no net color. Each gluon has a color–anticolor combination (for example, blue–antired) that allows it to transmit color when exchanged, and color is conserved during emission and absorption of a gluon by a quark. The gluon-exchange process changes the colors of the quarks in such a way that there is always one quark of each color in every baryon. The color of an individual quark changes continually as gluons are exchanged.

Similar processes occur in mesons such as pions. The quark–antiquark pairs of mesons have canceling color and anticolor (for example, blue and antiblue), so mesons also have no net color. Suppose a pion initially consists of a blue quark and an antiblue antiquark. The blue quark can become a red quark by emitting a blue–antired virtual gluon. The gluon is then absorbed by the antiblue antiquark, converting it to an antired antiquark (Fig. 44.13). Color is conserved in each emission and absorption, but a blue–antiblue pair has become a red–antired pair. Such changes occur continually, so we have to think of a pion as a superposition of three quantum states: blue–antiblue, green–antigreen, and red–antired. On a larger scale, the strong interaction between nucleons was described in Section 44.3 as due to the exchange of virtual mesons. In terms of quarks and gluons, these mediating virtual mesons are quark–antiquark systems bound together by the exchange of gluons.

44.13 (a) A pion containing a blue quark and an antiblue antiquark. (b) The blue quark emits a blue–antired gluon, changing to a red quark. (c) The gluon is absorbed by the antiblue antiquark, which becomes an antired antiquark. The pion now consists of a red–antired quark–antiquark pair. The actual quantum state of the pion is an equal superposition of red–antired, green–antigreen, and blue–antiblue pairs.

(a)

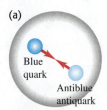

Blue quark

Antiblue antiquark

(b)

Blue–antired gluon

Red quark

Antiblue antiquark

(c)

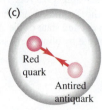

Red quark

Antired antiquark

The theory of strong interactions is known as *quantum chromodynamics* (QCD). No one has been able to isolate an individual quark, and indeed QCD predicts that quarks are bound in such a way that it is impossible to obtain a free quark. An impressive body of experimental evidence supports the correctness of the quark model and the idea that quantum chromodynamics is the key to understanding the strong interactions.

Three More Quarks

Before the tau particles were discovered, there were four known leptons. This fact, together with some puzzling decay rates, led to the speculation that there might be a fourth quark flavor. This quark is labeled c (the *charmed* quark); it has $Q/e = \frac{2}{3}$, $B = \frac{1}{3}$, $S = 0$, and a new quantum number **charm** $C = +1$. This was confirmed in 1974 by the observation at both SLAC and the Brookhaven National Laboratory of a meson, now named ψ, with mass 3097 MeV/c^2. This meson was found to have several decay modes, decaying into e^+e^-, $\mu^+\mu^-$, or hadrons. The mean lifetime was found to be about 10^{-20} s. These results are consistent with ψ being a spin-1 $c\bar{c}$ system. Almost immediately after this, similar mesons of greater mass were observed and identified as excited states of the $c\bar{c}$ system. A few years later, individual mesons with a nonzero net charm quantum number, D^0 ($c\bar{u}$) and D^+ ($c\bar{d}$), and a charmed baryon, Λ_c^+ (udc), were also observed.

In 1977 a meson with mass 9460 MeV/c^2, called upsilon (Υ), was discovered at Brookhaven. Because it had properties similar to ψ, it was conjectured that the meson was really the bound system of a new quark, b (the *bottom* quark), and its antiquark, $\bar{b}$. The bottom quark has the value 1 of a new quantum number B' (not to be confused with baryon number B) called *bottomness*. Excited states of the Υ were soon observed, as were the B^+ ($\bar{b}u$) and B^0 ($\bar{b}d$) mesons.

With the five flavors of quarks (u, d, s, c, and b) and the six flavors of leptons (e, μ, τ, ν_e, ν_μ, and ν_τ) it was an appealing conjecture that nature is symmetric in its building blocks and that therefore there should be a *sixth* quark. This quark, labeled t (top), would have $Q/e = \frac{2}{3}$, $B = \frac{1}{3}$, and a new quantum number, $T = 1$. In 1995, groups using two different detectors at Fermilab's Tevatron announced the discovery of the top quark. The groups collided 0.9-TeV protons with 0.9-TeV antiprotons, but even with 1.8 TeV of available energy, a top–antitop ($t\bar{t}$) pair was detected in fewer than two of every 10^{11} collisions! Table 44.5 lists some properties of the six quarks. Each has a corresponding antiquark with opposite values of Q, B, S, C, B', and T.

Table 44.5 Properties of the Six Quarks

Symbol	Q/e	Spin	Baryon Number, B	Strangeness, S	Charm, C	Bottomness, B'	Topness, T
u	$\frac{2}{3}$	$\frac{1}{2}$	$\frac{1}{3}$	0	0	0	0
d	$-\frac{1}{3}$	$\frac{1}{2}$	$\frac{1}{3}$	0	0	0	0
s	$-\frac{1}{3}$	$\frac{1}{2}$	$\frac{1}{3}$	-1	0	0	0
c	$\frac{2}{3}$	$\frac{1}{2}$	$\frac{1}{3}$	0	$+1$	0	0
b	$-\frac{1}{3}$	$\frac{1}{2}$	$\frac{1}{3}$	0	0	$+1$	0
t	$\frac{2}{3}$	$\frac{1}{2}$	$\frac{1}{3}$	0	0	0	$+1$

Test Your Understanding of Section 44.4   Is it possible to have a baryon with charge $Q = +e$ and strangeness $S = -2$?

44.5 The Standard Model and Beyond

The particles and interactions that we've discussed in this chapter provide a reasonably comprehensive picture of the fundamental building blocks of nature. There is enough confidence in the basic correctness of this picture that it is called the **standard model**.

The standard model includes three families of particles: (1) the six leptons, which have no strong interactions; (2) the six quarks, from which all hadrons are made; and (3) the particles that mediate the various interactions. These mediators are gluons for the strong interaction among quarks, photons for the electromagnetic interaction, the $W^{\pm}$ and Z^0 particles for the weak interaction, and the graviton for the gravitational interaction.

Electroweak Unification

Theoretical physicists have long dreamed of combining all the interactions of nature into a single unified theory. As a first step, Einstein spent much of his later life trying to develop a field theory that would unify gravitation and electromagnetism. He was only partly successful.

Between 1961 and 1967, Sheldon Glashow, Abdus Salam, and Steven Weinberg developed a theory that unifies the weak and electromagnetic forces. One outcome of their **electroweak theory** is a prediction of the weak-force mediator particles, the Z^0 and $W^{\pm}$ bosons, including their masses. The basic idea is that the mass difference between photons (zero mass) and the weak bosons ($\approx 100 \text{ GeV}/c^2$) makes the electromagnetic and weak interactions behave quite differently at low energies. At sufficiently high energies (well above 100 GeV), however, the distinction disappears, and the two merge into a single interaction. This prediction was verified in 1983 in experiments with proton-antiproton collisions at CERN. The weak bosons were found, again with the help provided by the theoretical description, and their observed masses agreed with the predictions of the electroweak theory, a wonderful convergence of theory and experiment. The electroweak theory and quantum chromodynamics form the backbone of the standard model. Glashow, Salam, and Weinberg received the Nobel Prize in 1979.

A remaining difficulty in the electroweak theory is that photons are massless but the weak bosons are very massive. To account for the broken symmetry among these interaction mediators, a particle called the Higgs boson has been proposed. Its mass is expected to be less than $1 \text{ TeV}/c^2$, but to produce it in the laboratory may require a much greater available energy. The search for the Higgs boson is an important mission of the Large Hadron Collider at CERN.

Grand Unified Theories

Perhaps at sufficiently high energies the strong interaction and the electroweak interaction have a convergence similar to that between the electromagnetic and weak interactions. If so, they can be unified to give a comprehensive theory of strong, weak, and electromagnetic interactions. Such schemes, called **grand unified theories** (GUTs), are still speculative.

One interesting feature of some grand unified theories is that they predict the decay of the proton (in violation of conservation of baryon number), with an estimated lifetime of more than 10^{28} years. (For comparison the age of the universe is known to be 1.37×10^{10} years.) With a lifetime of 10^{28} years, six metric tons of protons would be expected to have only one decay per day, so huge amounts of material must be examined. Some of the neutrino detectors that we mentioned in Section 44.2 originally looked for, and failed to find, evidence of proton decay. Nevertheless, experimental work continues, with current estimates setting the proton lifetime well over 10^{33} years. Some GUTs also predict the existence of magnetic monopoles, which we mentioned in Chapter 27. At present there is no confirmed experimental evidence that magnetic monopoles exist.

In the standard model, the neutrinos have zero mass. Nonzero values are controversial because experiments to determine neutrino masses are difficult both to perform and to analyze. In most GUTs the neutrinos *must* have nonzero masses. If neutrinos do have mass, transitions called *neutrino oscillations* can occur, in which one type of neutrino (ν_e, ν_μ, or ν_τ) changes into another type.

In 1998, scientists using the Super-Kamiokande neutrino detector in Japan (Fig. 44.14) reported the discovery of oscillations between muon neutrinos and tau neutrinos. Subsequent measurements at the Sudbury Neutrino Observatory in Canada have confirmed the existence of neutrino oscillations. This discovery is evidence for exciting physics beyond that predicted by the standard model.

The discovery of neutrino oscillations has cleared up a long-standing mystery about the sun. Since the 1960s, physicists have been using sensitive detectors to look for electron neutrinos produced by nuclear fusion reactions in the sun's core (see Section 43.8). However, the observed flux of solar electron neutrinos is only one-third of the predicted value. The explanation was provided in 2002 by the Sudbury Neutrino Observatory, which can detect neutrinos of all three flavors. The results showed that the combined flux of solar neutrinos of *all* flavors is equal to the theoretical prediction for the flux of *electron* neutrinos. The explanation is that the sun is indeed producing electron neutrinos at the rate predicted by theory, but that two-thirds of these electron neutrinos are transformed into muon or tau neutrinos during their flight from the sun's core to a detector on earth.

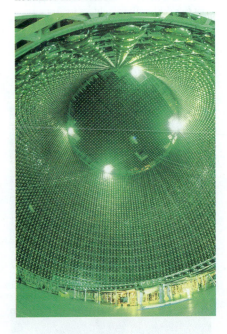

44.14 This photo shows the interior of the Super-Kamiokande neutrino detector in Japan. When in operation, the detector is filled with 5×10^7 kg of water. A neutrino passing through the detector can produce a faint flash of light, which is detected by the 13,000 photomultiplier tubes lining the detector walls. Data from this detector were the first to indicate that neutrinos have mass.

Supersymmetric Theories and TOEs

The ultimate dream of theorists is to unify all four fundamental interactions, adding gravitation to the strong and electroweak interactions that are included in GUTs. Such a unified theory is whimsically called a Theory of Everything (TOE). It turns out that an essential ingredient of such theories is a space-time continuum with more than four dimensions. The additional dimensions are "rolled up" into extremely tiny structures that we ordinarily do not notice. Depending on the scale of these structures, it may be possible for the next generation of particle accelerators to reveal the presence of extra dimensions.

Another ingredient of many theories is *supersymmetry,* which gives every boson and fermion a "superpartner" of the other spin type. For example, the proposed supersymmetric partner of the spin-$\frac{1}{2}$ electron is a spin-0 particle called the *selectron,* and that of the spin-1 photon is a spin-$\frac{1}{2}$ *photino.* As yet, no superpartner particles have been discovered, perhaps because they are too massive to be produced by the present generation of accelerators. Within a few years, new data from the Large Hadron Collider and other accelerators will help us decide whether these intriguing theories have merit.

Test Your Understanding of Section 44.5   One aspect of the standard model is that a *d* quark can transform into a *u* quark, an electron, and an antineutrino by means of the weak interaction. If this happens to a *d* quark inside a neutron, what kind of particle remains afterward in addition to the electron and antineutrino? (i) a proton; (ii) a Σ^-; (iii) a Σ^+; (iv) a Λ^0 or a Σ^0; (v) any of these.

44.6 The Expanding Universe

In the last two sections of this chapter we'll explore briefly the connections between the early history of the universe and the interactions of fundamental particles. It is remarkable that there are such close ties between physics on the smallest scale that we've explored experimentally (the range of the weak interaction, of the order of 10^{-18} m) and physics on the largest scale (the universe itself, of the order of at least 10^{26} m).

Gravitational interactions play an essential role in the large-scale behavior of the universe. One of the great achievements of Newtonian mechanics, including the law of gravitation, was the understanding it brought to the motion of planets in the solar system. Astronomical evidence shows that gravitational

44.15 (a) The galaxy M101 is a larger version of the Milky Way galaxy of which our solar system is a part. Like all galaxies, M101 is held together by the mutual gravitational attraction of its stars, gas, dust, and other matter, all of which orbit around the galaxy's center of mass. M101 is 25 million light-years away. (b) This image shows part of the Coma cluster, an immense grouping of over 1000 galaxies that lies 300 million light-years from us. The galaxies within the cluster are all in motion. Gravitational forces between the member galaxies of this cluster prevent them from escaping.

(a)

(b)

forces also dominate in larger systems such as galaxies and clusters of galaxies (Fig. 44.15).

Until early in the 20th century it was usually assumed that the universe was *static;* stars might move relative to each other, but there was not thought to be any overall expansion or contraction. But if everything is initially sitting still in the universe, why doesn't gravity just pull it all together into one big clump? Newton himself recognized the seriousness of this troubling question.

Measurements that were begun in 1912 by Vesto Slipher at Lowell Observatory in Arizona, and continued in the 1920s by Edwin Hubble with the help of Milton Humason at Mount Wilson in California, indicated that the universe is *not* static. The motions of galaxies relative to the earth can be measured by observing the shifts in the wavelengths of their spectra. For distant galaxies these shifts are always toward longer wavelength, so they appear to be receding from us and from each other. Astronomers first assumed that these were Doppler shifts and used a relationship between the wavelength λ_0 of light measured now from a source receding at speed v and the wavelength λ_S measured in the rest frame of the source when it was emitted. We can derive this relationship by inverting Eq. (37.25) for the Doppler effect, making subscript changes, and using $\lambda = c/f$; the result is

$$\lambda_0 = \lambda_S\sqrt{\frac{c + v}{c - v}} \qquad (44.13)$$

Wavelengths from receding sources are always shifted toward longer wavelengths; this increase in λ is called the **redshift.** We can solve Eq. (44.13) for v; the result is

$$v = \frac{(\lambda_0/\lambda_S)^2 - 1}{(\lambda_0/\lambda_S)^2 + 1}c \qquad (44.14)$$

CAUTION **Redshift, not Doppler shift** Equations (44.13) and (44.14) are from the *special* theory of relativity and refer to the Doppler effect. As we'll see, the redshift from *distant* galaxies is caused by an effect that is explained by the *general* theory of relativity and is *not* a Doppler shift. However, as the ratio v/c and the fractional wavelength change $(\lambda_0 - \lambda_S)/\lambda_S$ become small, the general theory's equations approach Eqs. (44.13) and (44.14), and those equations may be used.

Example 44.8 Recession speed of a galaxy

The spectral lines of various elements are detected in light from a galaxy in the constellation Ursa Major. An ultraviolet line from singly ionized calcium ($\lambda_S = 393$ nm) is observed at wavelength $\lambda_0 = 414$ nm, redshifted into the visible portion of the spectrum. At what speed is this galaxy receding from us?

SOLUTION

IDENTIFY and SET UP: This example uses the relationship between redshift and recession speed for a distant galaxy. We can use the wavelengths λ_S at which the light is emitted and λ_0 that we detect on earth in Eq. (44.14) to determine the galaxy's recession speed v if the fractional wavelength shift is not too great.

EXECUTE: The fractional redshift is $\lambda_0/\lambda_S = (414$ nm$)/(393$ nm$) = 1.053$. This is only a 5.3% increase, so we can use Eq. (44.14) with reasonable accuracy:

$$v = \frac{(1.053)^2 - 1}{(1.053)^2 + 1}c = 0.0516c = 1.55 \times 10^7 \text{ m/s}$$

EVALUATE: The galaxy is receding from the earth at 5.16% of the speed of light. Rather than going through this calculation, astronomers often just state the *redshift* $z = (\lambda_0 - \lambda_S)/\lambda_S = (\lambda_0/\lambda_S) - 1$. This galaxy has redshift $z = 0.053$.

The Hubble Law

Analysis of redshifts from many distant galaxies led Edwin Hubble to a remarkable conclusion: The speed of recession v of a galaxy is proportional to its distance r from us (Fig. 44.16). This relationship is now called the **Hubble law;** expressed as an equation,

$$v = H_0 r \qquad (44.15)$$

where H_0 is an experimental quantity commonly called the *Hubble constant,* since at any given time it is constant over space. Determining H_0 has been a key goal of the Hubble Space Telescope, which can measure distances to galaxies with unprecedented accuracy. The current best value is 2.3×10^{-18} s^{-1}, with an uncertainty of 5%.

Astronomical distances are often measured in *parsecs* (pc); one parsec is the distance at which there is a one-arcsecond $(1/3600°)$ angular separation between two objects 1.50×10^{11} m apart (the average distance from the earth to the sun). A distance of 1 pc is equal to 3.26 *light-years* (ly), where 1 ly $= 9.46 \times 10^{12}$ km is the distance that light travels in one year. The Hubble constant is then commonly expressed in the mixed units (km/s)/Mpc (kilometers per second per megaparsec), where 1 Mpc $= 10^6$ pc:

$$H_0 = (2.3 \times 10^{-18}\ \text{s}^{-1})\left(\frac{9.46 \times 10^{12}\ \text{km}}{1\ \text{ly}}\right)\left(\frac{3.26\ \text{ly}}{1\ \text{pc}}\right)\left(\frac{10^6\ \text{pc}}{1\ \text{Mpc}}\right) = 71\ \frac{\text{km/s}}{\text{Mpc}}$$

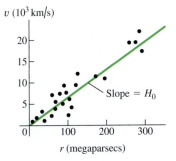

44.16 Graph of recession speed versus distance for several galaxies. The best-fit straight line illustrates Hubble's law. The slope of the line is the Hubble constant, H_0.

Example 44.9 | **Determining distance with the Hubble law**

Use the Hubble law to find the distance from earth to the galaxy in Ursa Major described in Example 44.8.

SOLUTION

IDENTIFY and SET UP: The Hubble law relates the redshift of a distant galaxy to its distance r from earth. We solve Eq. (44.15) for r and substitute the recession speed v from Example 44.8.

EXECUTE: Using $H_0 = 71\ (\text{km/s})/\text{Mpc} = 7.1 \times 10^4\ (\text{m/s})/\text{Mpc}$,

$$r = \frac{v}{H_0} = \frac{1.55 \times 10^7\ \text{m/s}}{7.1 \times 10^4\ (\text{m/s})/\text{Mpc}} = 220\ \text{Mpc}$$

$$= 2.2 \times 10^8\ \text{pc} = 7.1 \times 10^8\ \text{ly} = 6.7 \times 10^{24}\ \text{m}$$

EVALUATE: A distance of 220 million parsecs (710 million light-years) is truly stupendous, but many galaxies lie much farther away. To appreciate the immensity of this distance, consider that our farthest-ranging unmanned spacecraft have traveled only about 0.001 ly from our planet.

Another aspect of Hubble's observations was that, *in all directions,* distant galaxies appeared to be receding from us. There is no particular reason to think that our galaxy is at the very center of the universe; if we lived in some other galaxy, every distant galaxy would still seem to be moving away. That is, at any given time, *the universe looks more or less the same, no matter where in the universe we are.* This important idea is called the **cosmological principle.** There are local fluctuations in density, but on average, the universe looks the same from all locations. Thus the Hubble constant is constant in space although not necessarily constant in time, and the laws of physics are the same everywhere.

The Big Bang

The Hubble law suggests that at some time in the past, all the matter in the universe was far more concentrated than it is today. It was then blown apart in an immense explosion called the **Big Bang,** giving all observable matter more or less the velocities that we observe today. When did this happen? According to the

Hubble law, matter at a distance r away from us is traveling with speed $v = H_0 r$. The time t needed to travel a distance r is

$$t = \frac{r}{v} = \frac{r}{H_0 r} = \frac{1}{H_0} = 4.3 \times 10^{17} \text{ s} = 1.4 \times 10^{10} \text{ y}$$

By this hypothesis the Big Bang occurred about 14 billion years ago. It assumes that all speeds are *constant* after the Big Bang; that is, it neglects any change in the expansion rate due to gravitational attraction or other effects. We'll return to this point later. For now, however, notice that the age of the earth determined from radioactive dating (see Section 43.4) is 4.54 billion (4.54×10^9) years. It's encouraging that our hypothesis tells us that the universe is older than the earth!

Expanding Space

The general theory of relativity takes a radically different view of the expansion just described. According to this theory, the increased wavelength is *not* caused by a Doppler shift as the universe expands into a previously empty void. Rather, the increase comes from *the expansion of space itself* and everything in intergalactic space, including the wavelengths of light traveling to us from distant sources. This is not an easy concept to grasp, and if you haven't encountered it before, it may sound like doubletalk.

Here's an analogy that may help to develop some intuition on this point. Imagine we are all bugs crawling around on a horizontal surface. We can't leave the surface, and we can see in any direction along the surface, but not up or down. We are then living in a two-dimensional world; some writers have called such a world *flatland*. If the surface is a plane, we can locate our position with two Cartesian coordinates (x, y). If the plane extends indefinitely in both the x- and y-directions, we described our space as having *infinite* extent, or as being *unbounded*. No matter how far we go, we never reach an edge or a boundary.

An alternative habitat for us bugs would be the surface of a sphere with radius R. The space would still seem infinite in the sense that we could crawl forever and never reach an edge or a boundary. Yet in this case the space is *finite* or *bounded*. To describe the location of a point in this space, we could still use two coordinates: latitude and longitude, or the spherical coordinates θ and ϕ shown in Fig. 41.5.

Now suppose the spherical surface is that of a balloon (Fig. 44.17). As we inflate the balloon more and more, increasing the radius R, the coordinates of a point don't change, yet the distance between any two points gets larger and larger. Furthermore, as R increases, the *rate of change* of distance between two points (their recession speed) is proportional to their distance apart. *The recession speed is proportional to the distance,* just as with the Hubble law. For example, the distance from Pittsburgh to Miami is twice as great as the distance from Pittsburgh to Boston. If the earth were to begin to swell, Miami would recede from Pittsburgh twice as fast as Boston would.

We see that although the quantity R isn't one of the two coordinates giving the position of a point on the balloon's surface, it nevertheless plays an essential role in any discussion of distance. It is the radius of curvature of our two-dimensional space, and it is also a varying *scale factor* that changes as this two-dimensional universe expands.

Generalizing this picture to three dimensions isn't so easy. We have to think of our three-dimensional space as being embedded in a space with four or more dimensions, just as we visualized the two-dimensional spherical flatland as being embedded in a three-dimensional Cartesian space. Our real three-space is *not* Cartesian; to describe its characteristics in any small region requires at least one additional parameter, the curvature of space, which is analogous to the radius of the sphere. In a sense, this scale factor, which we'll continue to call R, describes the *size* of the universe, just as the radius of the sphere described the size of our

44.17 An inflating balloon as an analogy for an expanding universe.

(a) Points (representing galaxies) on the surface of a balloon are described by their latitude and longitude coordinates.

(b) The radius R of the balloon has increased. The coordinates of the points are the same, but the distance between them has increased.

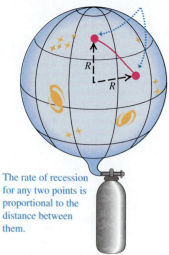

The rate of recession for any two points is proportional to the distance between them.

two-dimensional spherical universe. We'll return later to the question of whether the universe is bounded or unbounded.

Any length that is measured in intergalactic space is proportional to R, so the wavelength of light traveling to us from a distant galaxy increases along with every other dimension as the universe expands. That is,

$$\frac{\lambda_0}{\lambda} = \frac{R_0}{R} \qquad (44.16)$$

The zero subscripts refer to the values of the wavelength and scale factor *now,* just as H_0 is the current value of the Hubble constant. The quantities λ and R without subscripts are the values at *any* time—past, present, or future. In the situation described in Example 44.8, we have $\lambda_0 = 414$ nm and $\lambda = \lambda_S = 393$ nm, so Eq. (44.16) gives $R_0/R = 1.053$. That is, the scale factor *now* (R_0) is 5.3% larger than it was 710 million years ago when the light was emitted from that galaxy in Ursa Major. This increase of wavelength with time as the scale factor increases in our expanding universe is called the *cosmological redshift.* The farther away an object is, the longer its light takes to get to us and the greater the change in R and λ. The current largest measured wavelength ratio for galaxies is about 7, meaning that the volume of space itself is about $7^3 \approx 340$ times larger than it was when the light was emitted. Do *not* attempt to substitute $\lambda_0/\lambda_S = 7$ into Eq. (44.14) to find the recession speed; that equation is accurate only for small cosmological redshifts and $v \ll c$. The actual value of v depends on the density of the universe, the value of H_0, and the expansion history of the universe.

Here's a surprise for you: If the distance from us in the Hubble law is large enough, then the speed of recession will be greater than the speed of light! This does *not* violate the special theory of relativity because the recession speed is *not* caused by the motion of the astronomical object relative to some coordinates in its region of space. Rather, we can have $v > c$ when two sets of coordinates move apart fast enough as space itself expands. In other words, there are objects whose coordinates have been moving away from our coordinates so fast that light from them hasn't had enough time in the entire history of the universe to reach us. What we see is just the *observable* universe; we have no direct evidence about what lies beyond its horizon.

CAUTION **The universe isn't expanding into emptiness** The balloon shown in Fig. 44.17 is expanding into the empty space around it. It's a common misconception to picture the universe in the same way as a large but finite collection of galaxies that's expanding into unoccupied space. The reality is quite different! All the accumulated evidence shows that our universe is *infinite:* It has no edges, so there is nothing "outside" it and it isn't "expanding into" anything. The expansion of the universe simply means that the scale factor of the universe is increasing. A good two-dimensional analogy is to think of the universe as a flat, infinitely large rubber sheet that's stretching and expanding much like the surface of the balloon in Fig. 44.17. In a sense, the infinite universe is simply becoming more infinite! ▌

Critical Density

We've mentioned that the law of gravitation isn't consistent with a static universe. We need to look at the role of gravity in an *expanding* universe. Gravitational attractions should slow the initial expansion, but by how much? If these attractions are strong enough, the universe should expand more and more slowly, eventually stop, and then begin to contract, perhaps all the way down to what's been called a *Big Crunch.* On the other hand, if gravitational forces are much weaker, they slow the expansion only a little, and the universe should continue to expand forever.

The situation is analogous to the problem of escape speed of a projectile launched from the earth. We studied this problem in Example 13.5 (Section 13.3); now would be an excellent time to review that discussion. The total energy

$E = K + U$ when a projectile of mass m and speed v is at a distance r from the center of the earth (mass m_E) is

$$E = \tfrac{1}{2}mv^2 - \frac{Gmm_E}{r}$$

If E is positive, the projectile has enough kinetic energy to move infinitely far from the earth ($r \rightarrow \infty$) and have some kinetic energy left over. If E is negative, the kinetic energy $K = \tfrac{1}{2}mv^2$ becomes zero and the projectile stops when $r = -Gmm_E/E$. In that case, no greater value of r is possible, and the projectile can't escape the earth's gravity.

We can carry out a similar analysis for the universe. Whether the universe continues to expand indefinitely should depend on the average *density* of matter. If matter is relatively dense, there is a lot of gravitational attraction to slow and eventually stop the expansion and make the universe contract again. If not, the expansion should continue indefinitely. We can derive an expression for the *critical density* ρ_c needed to just barely stop the expansion.

Here's a calculation based on Newtonian mechanics; it isn't relativistically correct, but it illustrates the idea. Consider a large sphere with radius R, containing many galaxies (Fig. 44.18), with total mass M. Suppose our own galaxy has mass m and is located at the surface of this sphere. According to the cosmological principle, the average distribution of matter within the sphere is uniform. The total gravitational force on our galaxy is just the force due to the mass M inside the sphere. The force on our galaxy and potential energy U due to this spherically symmetric distribution are the same as though m and M were both points, so $U = -GmM/R$, just as in Section 13.3. The net force from all the uniform distribution of mass *outside* the sphere is zero, so we'll ignore it.

The total energy E (kinetic plus potential) for our galaxy is

$$E = \tfrac{1}{2}mv^2 - \frac{GmM}{R} \qquad (44.17)$$

If E is *positive,* our galaxy has enough energy to escape from the gravitational attraction of the mass M inside the sphere; in this case the universe should keep expanding forever. If E is negative, our galaxy cannot escape and the universe should eventually pull back together. The crossover between these two cases occurs when $E = 0$, so that

$$\tfrac{1}{2}mv^2 = \frac{GmM}{R} \qquad (44.18)$$

The total mass M inside the sphere is the volume $4\pi R^3/3$ times the density ρ_c:

$$M = \tfrac{4}{3}\pi R^3 \rho_c$$

We'll assume that the speed v of our galaxy relative to the center of the sphere is given by the Hubble law: $v = H_0 R$. Substituting these expressions for m and v into Eq. (44.18), we get

$$\tfrac{1}{2}m(H_0 R)^2 = \frac{Gm}{R}\left(\tfrac{4}{3}\pi R^3 \rho_c\right) \qquad \text{or}$$

$$\rho_c = \frac{3H_0^2}{8\pi G} \qquad \text{(critical density of the universe)} \qquad (44.19)$$

This is the *critical density.* If the average density is less than ρ_c, the universe should continue to expand indefinitely; if it is greater, the universe should eventually stop expanding and begin to contract.

Putting numbers into Eq. (44.19), we find

$$\rho_c = \frac{3(2.3 \times 10^{-18}\,\text{s}^{-1})^2}{8\pi(6.67 \times 10^{-11}\,\text{N}\cdot\text{m}^2/\text{kg}^2)} = 9.5 \times 10^{-27}\,\text{kg/m}^3$$

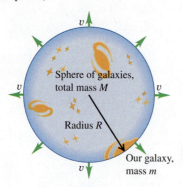

44.18 An imaginary sphere of galaxies. The net gravitational force exerted on our galaxy (at the surface of the sphere) by the other galaxies is the same as if all of their mass were concentrated at the center of the sphere. (Since the universe is infinite, there's also an infinity of galaxies outside this sphere.)

Sphere of galaxies, total mass M

Radius R

Our galaxy, mass m

The mass of a hydrogen atom is 1.67×10^{-27} kg, so this density is equivalent to about six hydrogen atoms per cubic meter.

Dark Matter, Dark Energy, and the Accelerating Universe

Astronomers have made extensive studies of the average density of matter in the universe. One way to do so is to count the number of galaxies in a patch of sky. Based on the mass of an average star and the number of stars in an average galaxy, this effort gives an estimate of the average density of *luminous* matter in the universe—that is, matter that emits electromagnetic radiation. (You are made of luminous matter because you emit infrared radiation as a consequence of your temperature; see Sections 17.7 and 39.5.) It's also necessary to take into account other luminous matter within a galaxy, including the tenuous gas and dust between the stars.

Another technique is to study the motions of galaxies within clusters of galaxies (Fig. 44.19; see also Fig. 44.15b). The motions are so slow that we can't actually see galaxies changing positions within a cluster. However, observations show that different galaxies within a cluster have somewhat different redshifts, which indicates that the galaxies are moving relative to the center of mass of the cluster. The speeds of these motions are related to the gravitational force exerted on each galaxy by the other members of the cluster, which in turn depends on the total mass of the cluster. By measuring these speeds, astronomers can determine the average density of *all* kinds of matter within the cluster, whether or not the matter emits electromagnetic radiation.

Observations using these and other techniques show that the average density of *all* matter in the universe is 27.4% of the critical density, but the average density of *luminous* matter is only 4.6% of the critical density. In other words, most of the matter in the universe is not luminous: It does not emit electromagnetic radiation of *any* kind. At present, the nature of this **dark matter** remains an outstanding mystery. Some proposed candidates for dark matter are WIMPs (weakly interacting massive particles, which are hypothetical subatomic particles far more massive than those produced in accelerator experiments) and MACHOs (massive compact halo objects, which include objects such as black holes that might form "halos" around galaxies). Whatever the true nature of dark matter, it is by far the dominant form of matter in the universe. For every kilogram of the ordinary matter that has been our subject for most of this book—including electrons, protons, atoms, molecules, blocks on inclined planes, planets, and stars—there are *five* kilograms of dark matter.

Since the average density of matter in the universe is less than the critical density, it might seem fair to conclude that the universe will continue to expand indefinitely, and that gravitational attraction between matter in different parts of the universe should slow the expansion down (albeit not enough to stop it). One way to test this prediction is to examine the redshifts of extremely distant objects. When astronomers look at a galaxy 10^9 light-years away, the light they receive has been in transit for 10^9 years, so they are seeing 10^9 years into the past. If the expansion of the universe has been slowing down, the expansion must have been more rapid in the distant past. Thus we would expect very distant galaxies to have *greater* redshifts than predicted by the Hubble law, Eq. (44.15).

Only since the 1990s has it become possible to accurately measure both the distances and the redshifts of extremely distant galaxies. The results have been totally surprising: Very distant galaxies actually have *smaller* redshifts than predicted by the Hubble law! The implication is that the expansion of the universe was slower in the past than it is now, so the expansion has been *speeding up* rather than slowing down.

If gravitational attraction should make the expansion slow down, why is it speeding up instead? The explanation generally accepted by astronomers and physicists is that space is suffused with a kind of energy that has no gravitational

44.19 The bright spots in this image are not stars, but entire galaxies. They are part of a cluster of galaxies about 10.2 billion ly (3.13 billion pc, or 3130 Mpc) away. (The blue glow is x-ray emission from hot gas within the cluster.) When the galaxies emitted the light used to make this image, the scale factor of the universe was only about 35% as large as it is now. By comparison, we see the relatively nearby Coma cluster (see Fig. 44.15b) as it was 300 million years ago, when the scale factor was 98% of the present-day value.

effect and emits no electromagnetic radiation, but rather acts as a kind of "anti-gravity" that produces a universal *repulsion*. This invisible, immaterial energy is called **dark energy.** As the name suggests, the nature of dark energy is poorly understood but is the subject of very active research.

Observations show that the *energy* density of dark energy (measured in, say, joules per cubic meter) is 72.6% of the critical density times c^2; that is, it is equal to $0.726\rho_c c^2$. As described above, the average density of matter of all kinds is 27.4% of the critical density. From the Einstein relationship $E = mc^2$, the average *energy* density of matter in the universe is therefore $0.274\rho_c c^2$. Because the energy density of dark energy is nearly three times greater than that of matter, the expansion of the universe will continue to accelerate. This expansion will never stop, and the universe will never contract.

If we account for energy of *all* kinds, the average energy density of the universe is equal to $0.726\rho_c c^2 + 0.274\rho_c c^2 = 1.00\rho_c c^2$. Of this, 72.6% is the mysterious dark energy, 22.8% is the no less mysterious dark matter, and a mere 4.6% is well-understood conventional matter. How little we know about the contents of our universe! When we take account of the density of matter in the universe (which tends to slow the expansion of space) and the density of dark energy (which tends to speed up the expansion), the age of the universe turns out to be 13.7 billion (1.37×10^{10}) years.

What is the significance of the result that within observational error, the average energy density of the universe is equal to $\rho_c c^2$? It tells us that the universe is infinite and unbounded, but just barely so. If the average energy density were even slightly larger than $\rho_c c^2$, the universe would be finite like the surface of the balloon depicted in Fig. 44.17. As of this writing, the observational error in the average energy density is still large enough (about 1%) that we can't be totally sure that the universe *is* unbounded. Improving these measurements will be an important task for physicists and astronomers in the years ahead.

Test Your Understanding of Section 44.6 Is it accurate to say that your body is made of "ordinary" matter?

44.7 The Beginning of Time

What an odd title for the very last section of a book! We will describe in general terms some of the current theories about the very early history of the universe and their relationship to fundamental particle interactions. We'll find that an astonishing amount happened in the very first second. A lot of loose ends will be left untied, and many questions will be left unanswered. This is, after all, one of the frontiers of physics.

Temperatures

The early universe was extremely dense and extremely hot, and the average particle energies were extremely large, all many orders of magnitude beyond anything that exists in the present universe. We can compare particle energy E and absolute temperature T using the equipartition principle (see Section 18.4):

$$E = \tfrac{3}{2}kT \tag{44.20}$$

In this equation k is Boltzmann's constant, which we'll often express in eV/K:

$$k = 8.617 \times 10^{-5} \text{ eV/K}$$

Thus we can replace Eq. (44.20) by $E \approx (10^{-4} \text{ eV/K})T = (10^{-13} \text{ GeV/K})T$ when we're discussing orders of magnitude.

Example 44.10 **Temperature and energy**

(a) What is the average kinetic energy E (in eV) of particles at room temperature $(T = 290 \text{ K})$ and at the surface of the sun $(T = 5800 \text{ K})$? (b) What approximate temperature corresponds to the ionization energy of the hydrogen atom and to the rest energies of the electron and the proton?

SOLUTION

IDENTIFY and SET UP: In this example we are to apply the equipartition principle. We use Eq. (44.20) to relate the target variables E and T.

EXECUTE: (a) At room temperature, from Eq. (44.20),

$$E = \tfrac{3}{2}kT = \tfrac{3}{2}(8.617 \times 10^{-5} \text{ eV/K})(290 \text{ K}) = 0.0375 \text{ eV}$$

The temperature at the sun's surface is higher than room temperature by a factor of $(5800 \text{ K})/(290 \text{ K}) = 20$, so the average kinetic energy there is $20(0.0375 \text{ eV}) = 0.75 \text{ eV}$.

(b) The ionization energy of hydrogen is 13.6 eV. Using the approximation $E \approx (10^{-4} \text{ eV/K})T$, we have

$$T \approx \frac{E}{10^{-4} \text{ eV/K}} = \frac{13.6 \text{ eV}}{10^{-4} \text{ eV/K}} \approx 10^{5} \text{ K}$$

Repeating this calculation for the rest energies of the electron $(E = 0.511 \text{ MeV})$ and proton $(E = 938 \text{ MeV})$ gives temperatures of 10^{10} K and 10^{13} K, respectively.

EVALUATE: Temperatures in excess of 10^5 K are found in the sun's interior, so most of the hydrogen there is ionized. Temperatures of 10^{10} K or 10^{13} K are not found anywhere in the solar system; as we will see, temperatures were this high in the very early universe.

Uncoupling of Interactions

We've characterized the expansion of the universe by a continual increase of the scale factor R, which we can think of very roughly as characterizing the *size* of the universe, and by a corresponding decrease in average density. As the total gravitational potential energy increased during expansion, there were corresponding *decreases* in temperature and average particle energy. As this happened, the basic interactions became progressively uncoupled.

To understand the uncouplings, recall that the unification of the electromagnetic and weak interactions occurs at energies that are large enough that the differences in mass among the various spin-1 bosons that mediate the interactions become insignificant by comparison. The electromagnetic interaction is mediated by the massless photon, and the weak interaction is mediated by the weak bosons $\text{W}^{\pm}$ and Z^0 with masses of the order of 100 GeV/c^2. At energies much *less* than 100 GeV the two interactions seem quite different, but at energies much *greater* than 100 GeV they become part of a single interaction.

The grand unified theories (GUTs) provide a similar behavior for the strong interaction. It becomes unified with the electroweak interaction at energies of the order of 10^{14} GeV, but at lower energies the two appear quite distinct. One of the reasons GUTs are still very speculative is that there is no way to do controlled experiments in this energy range, which is larger by a factor of 10^{11} than energies available with any current accelerator.

Finally, at sufficiently high energies and short distances, it is assumed that gravitation becomes unified with the other three interactions. The distance at which this happens is thought to be of the order of 10^{-35} m. This distance, called the *Planck length* l_{P}, is determined by the speed of light c and the fundamental constants of quantum mechanics and gravitation, h and G, respectively. The Planck length l_{P} is defined as

$$l_{\text{P}} = \sqrt{\frac{\hbar G}{c^3}} = 1.616 \times 10^{-35} \text{ m} \qquad (44.21)$$

You should verify that this combination of constants does indeed have units of length. The *Planck time* $t_{\text{P}} = l_{\text{P}}/c$ is the time required for light to travel a distance l_{P}:

$$t_{\text{P}} = \frac{l_{\text{P}}}{c} = \sqrt{\frac{\hbar G}{c^5}} = 0.539 \times 10^{-43} \text{ s} \qquad (44.22)$$

If we mentally go backward in time, we have to stop when we reach $t = 10^{-43}$ s because we have no adequate theory that unifies all four interactions. So as yet we have no way of knowing what happened or how the universe behaved at times earlier than the Planck time or when its size was less than the Planck length.

The Standard Model of the History of the Universe

The description that follows is called the *standard model* of the history of the universe. The title indicates that there are substantial areas of theory that rest on solid experimental foundations and are quite generally accepted. The figure on pages 1512–1513 is a graphical description of this history, with the characteristic sizes, particle energies, and temperatures at various times. Referring to this figure frequently will help you to understand the following discussion.

In this standard model, the temperature of the universe at time $t = 10^{-43}$ s (the Planck time) was about 10^{32} K, and the average energy per particle was approximately

$$E \approx (10^{-13} \text{ GeV/K})(10^{32} \text{ K}) = 10^{19} \text{ GeV}$$

In a totally unified theory this is about the energy below which gravity begins to behave as a separate interaction. This time therefore marked the transition from any proposed TOE to the GUT period.

During the GUT period, roughly $t = 10^{-43}$ to 10^{-35} s, the strong and electroweak forces were still unified, and the universe consisted of a soup of quarks and leptons transforming into each other so freely that there was no distinction between the two families of particles. Other, much more massive particles may also have been freely created and destroyed. One important characteristic of GUTs is that at sufficiently high energies, baryon number is not conserved. (We mentioned earlier the proposed decay of the proton, which has not yet been observed.) Thus by the end of the GUT period the numbers of quarks and antiquarks may have been unequal. This point has important implications; we'll return to it at the end of the section.

By $t = 10^{-35}$ s the temperature had decreased to about 10^{27} K and the average energy to about 10^{14} GeV. At this energy the strong force separated from the electroweak force (Fig. 44.20), and baryon number and lepton numbers began to be separately conserved. This separation of the strong force was analogous to a phase

44.20 Schematic diagram showing the times and energies at which the various interactions are thought to have uncoupled. The energy scale is backward because the average energy decreased as the age of the universe increased.

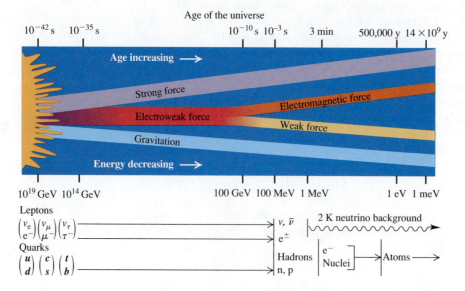

change such as boiling a liquid, with an associated heat of vaporization. Think of it as being similar to boiling a heavy nucleus, pulling the particles apart beyond the short range of the nuclear force. As a result, the universe underwent a dramatic expansion (far more rapid than the present-day expansion rate) called *cosmic inflation*. In one model, the scale factor R increased by a factor of 10^{50} in 10^{-32} s.

At $t = 10^{-32}$ s the universe was a mixture of quarks, leptons, and the mediating bosons (gluons, photons, and the weak bosons $W^{\pm}$ and Z^0). It continued to expand and cool from the inflationary period to $t = 10^{-6}$ s, when the temperature was about 10^{13} K and typical energies were about 1 GeV (comparable to the rest energy of a nucleon; see Example 44.11). At this time the quarks began to bind together to form nucleons and antinucleons. Also there were still enough photons of sufficient energy to produce nucleon–antinucleon pairs to balance the process of nucleon–antinucleon annihilation. However, by about $t = 10^{-2}$ s, most photon energies fell well below the threshold energy for such pair production. There was a slight excess of nucleons over antinucleons; as a result, virtually all of the antinucleons and most of the nucleons annihilated one another. A similar equilibrium occurred later between the production of electron–positron pairs from photons and the annihilation of such pairs. At about $t = 14$ s the average energy dropped to around 1 MeV, below the threshold for e^+e^- pair production. After pair production ceased, virtually all of the remaining positrons were annihilated, leaving the universe with many more protons and electrons than the antiparticles of each.

Up until about $t = 1$ s, neutrons and neutrinos could be produced in the endoergic reaction

$$e^- + p \rightarrow n + \nu_e$$

After this time, most electrons no longer had enough energy for this reaction. The average neutrino energy also decreased, and as the universe expanded, equilibrium reactions that involved *absorption* of neutrinos (which occurred with decreasing probability) became inoperative. At this time, in effect, the flux of neutrinos and antineutrinos throughout the universe uncoupled from the rest of the universe. Because of the extraordinarily low probability for neutrino absorption, most of this flux is still present today, although cooled greatly by expansion. The standard model of the universe predicts a present neutrino temperature of about 2 K, but no experiment has yet been able to test this prediction.

Nucleosynthesis

At about $t = 1$ s, the ratio of protons to neutrons was determined by the Boltzmann distribution factor $e^{-\Delta E/kT}$, where ΔE is the difference between the neutron and proton rest energies: $\Delta E = 1.294$ MeV. At a temperature of about 10^{10} K, this distribution factor gives about 4.5 times as many protons as neutrons. However, as we have discussed, free neutrons (with a half-life of 887 s) decay spontaneously to protons. This decay caused the proton–neutron ratio to increase until about $t = 225$ s. At this time, the temperature was about 10^9 K, and the average energy was well below 2 MeV.

This energy distribution was critical because the binding energy of the *deuteron* (a neutron and a proton bound together) is 2.22 MeV (see Section 43.2). A neutron bound in a deuteron does not decay spontaneously. As the average energy decreased, a proton and a neutron could combine to form a deuteron, and there were fewer and fewer photons with 2.22 MeV or more of energy to dissociate the deuterons again. Therefore the combining of protons and neutrons into deuterons halted the decay of free neutrons.

The formation of deuterons starting at about $t = 225$ s marked the beginning of the period of formation of nuclei, or *nucleosynthesis*. At this time, there were about seven protons for each neutron. The deuteron (^{2}H) can absorb a neutron and form a triton (^{3}H), or it can absorb a proton and form ^{3}He. Then ^{3}H can absorb a proton, and ^{3}He can absorb a neutron, each yielding ^{4}He (the alpha particle).

**AGE OF QUARKS AND
GLUONS (GUT Period)**
Dense concentration of matter and
antimatter; gravity a separate force;
more quarks than antiquarks.
Inflationary period (10^{-35} s): rapid expansion,
strong force separates from
electroweak force.

**AGE OF NUCLEONS AND
ANTINUCLEONS**
Quarks bind together to form
nucleons and antinucleons; energy
too low for nucleon–antinucleon
pair production at 10^{-2} s.

**AGE OF
NUCLEOSYNTHESIS**
Stable deuterons; matter
74% H, 25% He, 1%
heavier nuclei.

AGE OF LEPTONS
Leptons distinct from quarks;
$W^\pm$ and Z^0 bosons mediate
weak force (10^{-12} s).

**BIG
BANG** 10^{-43} s 10^{-32} s 10^{-6} s   225 s 10^3 s

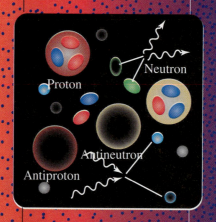

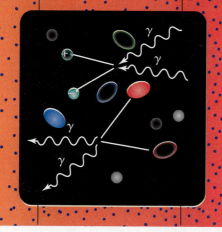

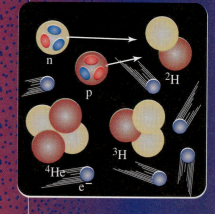

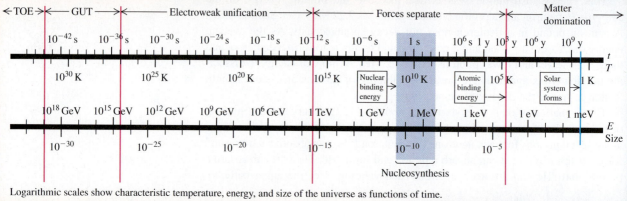

Logarithmic scales show characteristic temperature, energy, and size of the universe as functions of time.

A Brief History of the Universe

AGE OF IONS
Expanding, cooling gas of ionized H and He.

AGE OF ATOMS
Neutral atoms form; universe becomes transparent to most light.

AGE OF STARS AND GALAXIES
Thermonuclear fusion begins in stars, forming heavier nuclei.

10^{13} s

10^{15} s

NOW

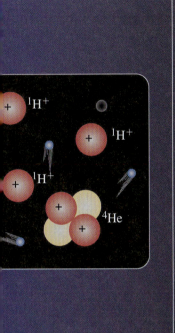

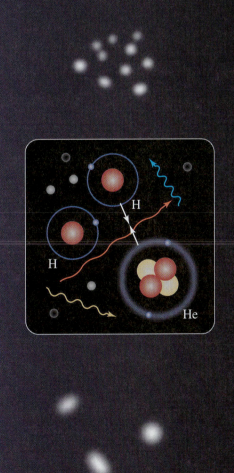

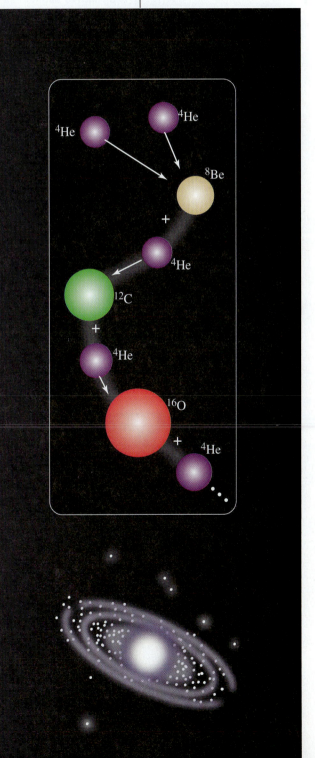

A few ^{7}Li nuclei may also have been formed by fusion of ^{3}H and ^{4}He nuclei. According to the theory, essentially all the ^{1}H and ^{4}He in the present universe was formed at this time. But then the building of nuclei almost ground to a halt. The reason is that *no* nuclide with mass number $A = 5$ has a half-life greater than 10^{-21} s. Alpha particles simply do not permanently absorb neutrons or protons. The nuclide ^{8}Be that is formed by fusion of two ^{4}He nuclei is unstable, with an extremely short half-life, about 7×10^{-17} s. Note also that at this time, the average energy was still much too large for electrons to be bound to nuclei; there were not yet any atoms.

Conceptual Example 44.11   **The relative abundance of hydrogen and helium in the universe**

Nearly all of the protons and neutrons in the seven-to-one ratio at $t = 225$ s either formed ^{4}He or remained as ^{1}H. After this time, what was the resulting relative abundance of ^{1}H and ^{4}He, by mass?

SOLUTION

The ^{4}He nucleus contains two protons and two neutrons. For every two neutrons present at $t = 225$ s there were 14 protons. The two neutrons and two of the 14 protons make up one ^{4}He nucleus, leaving 12 protons (^{1}H nuclei). So there were eventually 12 ^{1}H nuclei for every ^{4}He nucleus. The masses of ^{1}H and ^{4}He are about 1 u and 4 u, respectively, so there were 12 u of ^{1}H for every 4 u of ^{4}He. Therefore the relative abundance, by mass, was 75% ^{1}H and 25% ^{4}He. This result agrees very well with estimates of the present H–He ratio in the universe, an important confirmation of this part of the theory.

44.21 The Veil Nebula in the constellation Cygnus is a remnant of a supernova explosion that occurred more than 20,000 years ago. The gas ejected from the supernova is still moving very rapidly. Collisions between this fast-moving gas and the tenuous material of interstellar space excite the gas and cause it to glow. The portion of the nebula shown here is about 40 ly (12 pc) in length.

Further nucleosynthesis did not occur until very much later, well after $t = 10^{13}$ s (about 380,000 y). At that time, the temperature was about 3000 K, and the average energy was a few tenths of an electron volt. Because the ionization energies of hydrogen and helium atoms are 13.6 eV and 24.5 eV, respectively, almost all the hydrogen and helium was electrically neutral (not ionized). With the electrical repulsions of the nuclei canceled out, gravitational attraction could slowly pull the neutral atoms together to form clouds of gas and eventually stars. Thermonuclear reactions in stars then produced all of the more massive nuclei. In Section 43.8 we discussed one cycle of thermonuclear reactions in which ^{1}H becomes ^{4}He.

For stars whose mass is 40% of the sun's mass or greater, as the hydrogen is consumed the star's core begins to contract as the inward gravitational pressure exceeds the outward gas and radiation pressure. The gravitational potential energy decreases as the core contracts, so the kinetic energy of nuclei in the core increases. Eventually the core temperature becomes high enough to begin another process, *helium fusion*. First two ^{4}He nuclei fuse to form ^{8}Be, which is highly unstable. But because a star's core is so dense and collisions among nuclei are so frequent, there is a nonzero probability that a third ^{4}He nucleus will fuse with the ^{8}Be nucleus before it can decay. The result is the stable nuclide ^{12}C. This is called the *triple-alpha process,* since three ^{4}He nuclei (that is, alpha particles) fuse to form one carbon nucleus. Then successive fusions with ^{4}He give ^{16}O, ^{20}Ne, and ^{24}Mg. All these reactions are exoergic. They release energy to heat up the star, and ^{12}C and ^{16}O can fuse to form elements with higher and higher atomic number.

For nuclides that can be created in this manner, the binding energy per nucleon peaks at mass number $A = 56$ with the nuclide ^{56}Fe, so exoergic fusion reactions stop with Fe. But successive neutron captures followed by beta decays can continue the synthesis of more massive nuclei. If the star is massive enough, it may eventually explode as a *supernova,* sending out into space the heavy elements that were produced by the earlier processes (Fig. 44.21; see also Fig. 37.7). In space, the debris and other interstellar matter can gravitationally bunch together to form a new generation of stars and planets. Our own sun is one such "second-generation" star. This means that the sun's planets and everything on them (including you) contain matter that was long ago blasted into space by an exploding supernova.

Background Radiation

In 1965 Arno Penzias and Robert Wilson, working at Bell Telephone Laboratories in New Jersey on satellite communications, turned a microwave antenna skyward and found a background signal that had no apparent preferred direction. (This signal produces about 1% of the "hash" you see on a TV screen when you turn to an unused channel.) Further research has shown that the radiation that is received has a frequency spectrum that fits Planck's blackbody radiation law, Eq. (39.24) (Section 39.5). The wavelength of peak intensity is 1.063 mm (in the microwave region of the spectrum), with a corresponding absolute temperature $T = 2.725$ K. Penzias and Wilson contacted physicists at nearby Princeton University who had begun the design of an antenna to search for radiation that was a remnant from the early evolution of the universe. We mentioned above that neutral atoms began to form at about $t = 380,000$ years when the temperature was 3000 K. With far fewer charged particles present than previously, the universe became transparent at this time to electromagnetic radiation of long wavelength. The 3000-K blackbody radiation therefore survived, cooling to its present 2.725-K temperature as the universe expanded. The *cosmic background radiation* is among the most clear-cut experimental confirmations of the Big Bang theory. Figure 44.22 shows a modern map of the cosmic background radiation.

44.22 This false-color map shows microwave radiation from the entire sky mapped onto an oval. When this radiation was emitted 380,000 years after the Big Bang, the regions shown in blue were slightly cooler and denser than average. Within these cool, dense regions formed galaxies, including the Milky Way galaxy of which our solar system, our earth, and our selves are part.

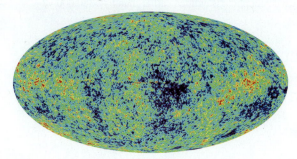

Example 44.12 — Expansion of the universe

By approximately what factor has the universe expanded since $t = 380,000$ y?

SOLUTION

IDENTIFY and SET UP: We use the idea that as the universe has expanded, all intergalactic wavelengths have expanded with it. The Wien displacement law, Eq. (39.21), relates the peak wavelength λ_m in blackbody radiation to the temperature T. Given the temperatures of the cosmic background radiation today (2.725 K) and at $t = 380,000$ y (3000 K) we can determine the factor by which wavelengths have changed and hence determine the factor by which the universe has expanded.

EXECUTE: We rewrite Eq. (39.21) as

$$\lambda_m = \frac{2.90 \times 10^{-3}\ \text{m} \cdot \text{K}}{T}$$

Hence the peak wavelength λ_m is inversely proportional to T. As the universe expands, all intergalactic wavelengths (including λ_m) increase in proportion to the scale factor R. The temperature has decreased by the factor $(3000\ \text{K})/(2.725\ \text{K}) \approx 1100$, so λ_m and the scale factor must both have *increased* by this factor. Thus, between $t = 380,000$ y and the present, the universe has expanded by a factor of about 1100.

EVALUATE: Our results show that since $t = 380,000$ y, any particular intergalactic *volume* has increased by a factor of about $(1100)^3 = 1.3 \times 10^9$. They also show that when the cosmic background radiation was emitted, its peak wavelength was $\frac{1}{1100}$ of the present-day value of 1.063 mm, or 967 nm. This is in the infrared region of the spectrum.

Matter and Antimatter

One of the most remarkable features of our universe is the asymmetry between matter and antimatter. One might think that the universe should have equal numbers of protons and antiprotons and of electrons and positrons, but this doesn't appear to be the case. Theories of the early universe must explain this imbalance.

We've mentioned that most GUTs include violation of conservation of baryon number at energies at which the strong and electroweak interactions have converged. If particle–antiparticle symmetry is also violated, we have a mechanism for making more quarks than antiquarks, more leptons than antileptons, and eventually more matter than antimatter. One serious problem is that any asymmetry that is created in this way during the GUT era might be wiped out by the electroweak interaction after the end of the GUT era. If so, there must be some mechanism that creates particle–antiparticle asymmetry at a much *later* time. The problem of the matter–antimatter asymmetry is still very much an open one.

There are still many unanswered questions at the intersection of particle physics and cosmology. Is the energy density of the universe precisely equal to $\rho_c c^2$, or are there small but important differences? What is dark energy? Has the density of dark energy remained constant over the history of the universe, or has the density changed? What is dark matter? What happened during the first 10^{-43} s after the Big Bang? Can we see evidence that the strong and electroweak interactions undergo a grand unification at high energies? The search for the answers to these and many other questions about our physical world continues to be one of the most exciting adventures of the human mind.

Test Your Understanding of Section 44.7 Given a sufficiently powerful telescope, could we detect photons emitted earlier than $t = 380{,}000$ y? ❚

Fundamental particles: Each particle has an antiparticle; some particles are their own antiparticles. Particles can be created and destroyed, some of them (including electrons and positrons) only in pairs or in conjunction with other particles and antiparticles.

Particles serve as mediators for the fundamental interactions. The photon is the mediator of the electromagnetic interaction. Yukawa proposed the existence of mesons to mediate the nuclear interaction. Mediating particles that can exist only because of the uncertainty principle for energy are called virtual particles.

Particle accelerators and detectors: Cyclotrons, synchrotrons, and linear accelerators are used to accelerate charged particles to high energies for experiments with particle interactions. Only part of the beam energy is available to cause reactions with targets at rest. This problem is avoided in colliding-beam experiments. (See Examples 44.1–44.3.)

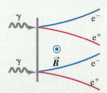

Particles and interactions: Four fundamental interactions are found in nature: the strong, electromagnetic, weak, and gravitational interactions. Particles can be described in terms of their interactions and of quantities that are conserved in all or some of the interactions.

Fermions have half-integer spins; bosons have integer spins. Leptons, which are fermions, have no strong interactions. Strongly interacting particles are called hadrons. They include mesons, which are always bosons, and baryons, which are always fermions. There are conservation laws for three different lepton numbers and for baryon number. Additional quantum numbers, including strangeness and charm, are conserved in some interactions and not in others. (See Examples 44.4–44.6.)

Quarks: Hadrons are composed of quarks. There are thought to be six types of quarks. The interaction between quarks is mediated by gluons. Quarks and gluons have an additional attribute called color. (See Example 44.7.)

Symmetry and the unification of interactions: Symmetry considerations play a central role in all fundamental-particle theories. The electromagnetic and weak interactions become unified at high energies into the electroweak interaction. In grand unified theories the strong interaction is also unified with these interactions, but at much higher energies.

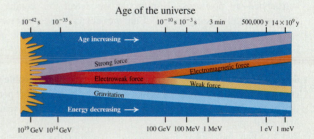

The expanding universe and its composition: The Hubble law shows that galaxies are receding from each other and that the universe is expanding. Observations show that the rate of expansion is accelerating due to the presence of dark energy, which makes up 72.6% of the energy in the universe. Only 4.6% of the energy in the universe is in the form of ordinary matter; the remaining 22.8% is dark matter, whose nature is poorly understood. (See Examples 44.8 and 44.9.)

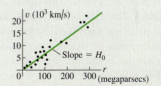

The history of the universe: In the standard model of the universe, a Big Bang gave rise to the first fundamental particles. They eventually formed into the lightest atoms as the universe expanded and cooled. The cosmic background radiation is a relic of the time when these atoms formed. The heavier elements were manufactured much later by fusion reactions inside stars. (See Examples 44.10–44.12.)

BRIDGING PROBLEM | Hyperons, Pions, and the Expanding Universe

A Λ^0 hyperon at rest decays into a neutron and a π^0. (a) Find the kinetic energies of the decay products. (b) What fraction of the total kinetic energy is carried off by each particle? (c) A physicist on earth detects one of the two photons that was emitted in the decay of a π^0. The π^0 was at rest in the cluster shown in Fig. 44.19 before it decayed. What is the energy of the photon that is detected on earth?

SOLUTION GUIDE

See MasteringPhysics® study area for a Video Tutor solution.

IDENTIFY and SET UP

1. Which quantities are conserved in the Λ^0 decay? In the π^0 decay?
2. The universe expanded during the time that the photon traveled from the cluster to earth. How does this affect the wavelength and energy of the photon that the physicist detects?
3. List the unknown quantities for each part of the problem and identify the target variables.
4. Select the equations that will allow you to solve for the target variables.

EXECUTE

5. Write the conservation equations for the decay of the Λ^0. (*Hint:* It's useful to write the energy E of a particle in terms of its momentum p and mass m using $E = (p^2c^2 + m^2c^4)^{1/2}$.)
6. Solve the conservation equations for the energy of one of the decay products. (*Hint:* Rearrange the energy conservation equation so that one of the $(p^2c^2 + m^2c^4)^{1/2}$ terms is on one side of the equation. Then square both sides.) Then use $K = E - mc^2$.
7. Find the fraction of the total kinetic energy that goes into the neutron and into the pion.
8. Write the conservation equations for the decay of the π^0 at rest and find the energy of each emitted photon. By what factor does the wavelength of this photon change as it travels from the galaxy cluster to earth? By what factor does the photon *energy* change? (*Hint:* See Fig. 44.19.)

EVALUATE

9. Which of the Λ^0 decay products should have the greater kinetic energy? Should the detected π^0 decay photon have more or less energy than when it was emitted?

Problems

For instructor-assigned homework, go to www.masteringphysics.com

•, ••, •••: Problems of increasing difficulty. **CP**: Cumulative problems incorporating material from earlier chapters. **CALC**: Problems requiring calculus. **BIO**: Biosciences problems.

DISCUSSION QUESTIONS

Q44.1 Is it possible that some parts of the universe contain antimatter whose atoms have nuclei made of antiprotons and antineutrons, surrounded by positrons? How could we detect this condition without actually going there? Can we detect these antiatoms by identifying the light they emit as composed of antiphotons? Explain. What problems might arise if we actually *did* go there?

Q44.2 Given the Heisenberg uncertainty principle, is it possible to create particle–antiparticle pairs that exist for extremely short periods of time before annihilating? Does this mean that empty space is really empty?

Q44.3 When they were first discovered during the 1930s and 1940s, there was confusion as to the identities of pions and muons. What are the similarities and most significant differences?

Q44.4 The gravitational force between two electrons is weaker than the electrical force by the order of 10^{-40}. Yet the gravitational interactions of matter were observed and analyzed long before electrical interactions were understood. Why?

Q44.5 When a π^0 decays to two photons, what happens to the quarks of which it was made?

Q44.6 Why can't an electron decay to two photons? To two neutrinos?

Q44.7 According to the standard model of the fundamental particles, what are the similarities between baryons and leptons? What are the most important differences?

Q44.8 According to the standard model of the fundamental particles, what are the similarities between quarks and leptons? What are the most important differences?

Q44.9 What are the main advantages of colliding-beam accelerators compared with those using stationary targets? What are the main disadvantages?

Q44.10 Does the universe have a center? Explain.

Q44.11 Does it make sense to ask, "If the universe is expanding, what is it expanding into?"

Q44.12 Assume that the universe has an edge. Placing yourself at that edge in a thought experiment, explain why this assumption violates the cosmological principle.

Q44.13 Explain why the cosmological principle requires that H_0 must have the same value everywhere in space, but does not require that it be constant in time.

EXERCISES

Section 44.1 Fundamental Particles—A History

44.1 • A neutral pion at rest decays into two photons. Find the energy, frequency, and wavelength of each photon. In which part of the electromagnetic spectrum does each photon lie? (Use the pion mass given in terms of the electron mass in Section 44.1.)

44.2 •• **CP** Two equal-energy photons collide head-on and annihilate each other, producing a $\mu^+\mu^-$ pair. The muon mass is given in terms of the electron mass in Section 44.1. (a) Calculate the maximum wavelength of the photons for this to occur. If the photons have this wavelength, describe the motion of the μ^+ and μ^- immediately after they are produced. (b) If the wavelength of each photon is half the value calculated in part (a), what is the speed of each muon after they have moved apart? Use correct relativistic expressions for momentum and energy.

44.3 •• A positive pion at rest decays into a positive muon and a neutrino. (a) Approximately how much energy is released in the decay? (Assume the neutrino has zero rest mass. Use the muon and pion masses given in terms of the electron mass in Section 44.1.) (b) Why can't a positive muon decay into a positive pion?

44.4 • A proton and an antiproton annihilate, producing two photons. Find the energy, frequency, and wavelength of each photon (a) if the p and $\bar{\text{p}}$ are initially at rest and (b) if the p and $\bar{\text{p}}$ collide head-on, each with an initial kinetic energy of 830 MeV.

44.5 •• **CP** For the nuclear reaction given in Eq. (44.2) assume that the initial kinetic energy and momentum of the reacting particles are negligible. Calculate the speed of the α particle immediately after it leaves the reaction region.

44.6 • Estimate the range of the force mediated by an ω^0 meson that has mass 783 MeV/c^2.

44.7 • The starship *Enterprise,* of television and movie fame, is powered by combining matter and antimatter. If the entire 400-kg antimatter fuel supply of the *Enterprise* combines with matter, how much energy is released? How does this compare to the U.S. yearly energy use, which is roughly 1.0×10^{20} J?

Section 44.2 Particle Accelerators and Detectors

44.8 • An electron with a total energy of 20.0 GeV collides with a stationary positron. (a) What is the available energy? (b) If the electron and positron are accelerated in a collider, what total energy corresponds to the same available energy as in part (a)?

44.9 • Deuterons in a cyclotron travel in a circle with radius 32.0 cm just before emerging from the dees. The frequency of the applied alternating voltage is 9.00 MHz. Find (a) the magnetic field and (b) the kinetic energy and speed of the deuterons upon emergence.

44.10 • The magnetic field in a cyclotron that accelerates protons is 1.30 T. (a) How many times per second should the potential across the dees reverse? (This is twice the frequency of the circulating protons.) (b) The maximum radius of the cyclotron is 0.250 m. What is the maximum speed of the proton? (c) Through what potential difference would the proton have to be accelerated from rest to give it the same speed as calculated in part (b)?

44.11 • (a) A high-energy beam of alpha particles collides with a stationary helium gas target. What must the total energy of a beam particle be if the available energy in the collision is 16.0 GeV? (b) If the alpha particles instead interact in a colliding-beam experiment, what must the energy of each beam be to produce the same available energy?

44.12 •• (a) What is the speed of a proton that has total energy 1000 GeV? (b) What is the angular frequency ω of a proton with the speed calculated in part (a) in a magnetic field of 4.00 T? Use both the nonrelativistic Eq. (44.7) and the correct relativistic expression, and compare the results.

44.13 • In Example 44.3 it was shown that a proton beam with an 800-GeV beam energy gives an available energy of 38.7 GeV for collisions with a stationary proton target. (a) You are asked to design an upgrade of the accelerator that will double the available energy in stationary-target collisions. What beam energy is required? (b) In a colliding-beam experiment, what total energy of each beam is needed to give an available energy of $2(38.7 \text{ GeV}) = 77.4$ GeV?

44.14 •• Calculate the minimum beam energy in a proton–proton collider to initiate the $\text{p} + \text{p} \rightarrow \text{p} + \text{p} + \eta^0$ reaction. The rest energy of the η^0 is 547.3 MeV (see Table 44.3).

Section 44.3 Particles and Interactions

44.15 • A K^+ meson at rest decays into two π mesons. (a) What are the allowed combinations of π^0, π^+, and π^- as decay products? (b) Find the total kinetic energy of the π mesons.

44.16 • How much energy is released when a μ^- muon at rest decays into an electron and two neutrinos? Neglect the small masses of the neutrinos.

44.17 • What is the mass (in kg) of the Z^0? What is the ratio of the mass of the Z^0 to the mass of the proton?

44.18 • Table 44.3 shows that a Σ^0 decays into a Λ^0 and a photon. (a) Calculate the energy of the photon emitted in this decay, if the Λ^0 is at rest. (b) What is the magnitude of the momentum of the photon? Is it reasonable to ignore the final momentum and kinetic energy of the Λ^0? Explain.

44.19 • If a Σ^+ at rest decays into a proton and a π^0, what is the total kinetic energy of the decay products?

44.20 • The discovery of the Ω^- particle helped confirm Gell-Mann's eightfold way. If an Ω^- decays into a Λ^0 and a K^-, what is the total kinetic energy of the decay products?

44.21 • In which of the following decays are the three lepton numbers conserved? In each case, explain your reasoning. (a) $\mu^- \rightarrow e^- + \nu_e + \bar{\nu}_\mu$; (b) $\tau^- \rightarrow e^- + \bar{\nu}_e + \nu_\tau$; (c) $\pi^+ \rightarrow e^+ + \gamma$; (d) $\text{n} \rightarrow \text{p} + e^- + \bar{\nu}_e$.

44.22 • Which of the following reactions obey the conservation of baryon number? (a) $\text{p} + \text{p} \rightarrow \text{p} + e^+$; (b) $\text{p} + \text{n} \rightarrow 2e^+ + e^-$; (c) $\text{p} \rightarrow \text{n} + e^- + \bar{\nu}_e$; (d) $\text{p} + \bar{\text{p}} \rightarrow 2\gamma$.

44.23 • In which of the following reactions or decays is strangeness conserved? In each case, explain your reasoning. (a) $K^+ \rightarrow \mu^+ + \nu_\mu$; (b) $\text{n} + K^+ \rightarrow \text{p} + \pi^0$; (c) $K^+ + K^- \rightarrow \pi^0 + \pi^0$; (d) $\text{p} + K^- \rightarrow \Lambda^0 + \pi^0$.

44.24 •• **CP** (a) Show that the coupling constant for the electromagnetic interaction, $e^2/4\pi\epsilon_0\hbar c$, is dimensionless and has the numerical value 1/137.0. (b) Show that in the Bohr model the orbital speed of an electron in the $n = 1$ orbit is equal to c times the coupling constant $e^2/4\pi\epsilon_0\hbar c$.

44.25 • Show that the nuclear force coupling constant $f^2/\hbar c$ is dimensionless.

Section 44.4 Quarks and the Eightfold Way

44.26 • Nine of the spin-$\frac{3}{2}$ baryons are four Δ particles, each with mass 1232 MeV/c^2, strangeness 0, and charges $+2e$, $+e$, 0, and $-e$; three Σ* particles, each with mass 1385 MeV/c^2, strangeness -1, and charges $+e$, 0, and $-e$; and two Ξ* particles, each with mass 1530 MeV/c^2, strangeness -2, and charges 0 and $-e$. (a) Place these particles on a plot of S versus Q. Deduce the Q and S values of the tenth spin-$\frac{3}{2}$ baryon, the Ω^- particle, and place it on your diagram. Also label the particles with their masses. The mass of the Ω^- is 1672 MeV/c^2; is this value consistent with your diagram? (b) Deduce the three-quark combinations (of u, d, and s) that make up each of these ten particles. Redraw the plot of S versus Q from part (a) with each particle labeled by its quark content. What regularities do you see?

44.27 • Determine the electric charge, baryon number, strangeness quantum number, and charm quantum number for the following quark combinations: (a) uds; (b) $c\bar{u}$; (c) ddd; and (d) $d\bar{c}$. Explain your reasoning.

44.28 •• Determine the electric charge, baryon number, strangeness quantum number, and charm quantum number for the following quark combinations: (a) uus, (b) $c\bar{s}$, (c) $\overline{ddu}$, and (d) $\bar{c}b$.

44.29 • The weak force may change quark flavor in an interaction. Explain how β^+ decay changes quark flavor. If a proton undergoes β^+ decay, determine the decay reaction.

44.30 • What is the total kinetic energy of the decay products when an upsilon particle at rest decays to $\tau^+ + \tau^-$?

44.31 • The quark content of the neutron is udd. (a) What is the quark content of the antineutron? Explain your reasoning. (b) Is the neutron its own antiparticle? Why or why not? (c) The quark content of the ψ is $c\bar{c}$. Is the ψ its own antiparticle? Explain your reasoning.

44.32 • Given that each particle contains only combinations of u, d, s, $\bar{u}$, $\bar{d}$, and $\bar{s}$, use the method of Example 44.7 to deduce the quark content of (a) a particle with charge $+e$, baryon number 0, and strangeness $+1$; (b) a particle with charge $+e$, baryon number -1, and strangeness $+1$; (c) a particle with charge 0, baryon number $+1$, and strangeness -2.

Section 44.6 The Expanding Universe

44.33 • The spectrum of the sodium atom is detected in the light from a distant galaxy. (a) If the 590.0-nm line is redshifted to 658.5 nm, at what speed is the galaxy receding from the earth? (b) Use the Hubble law to calculate the distance of the galaxy from the earth.

44.34 • **Redshift Parameter.** The definition of the redshift parameter z is given in Example 44.8. (a) Show that Eq. (44.13) may be written as $1 + z = ([1 + \beta]/[1 - \beta])^{1/2}$, where $\beta = v/c$. (b) The observed redshift parameter for a certain galaxy is $z = 0.500$. Find the speed of the galaxy relative to the earth, if the redshift is due to the Doppler shift. (c) Use the Hubble law to find the distance of this galaxy from the earth.

44.35 • A galaxy in the constellation Pisces is 5210 Mly from the earth. (a) Use the Hubble law to calculate the speed at which this galaxy is receding from earth. (b) What redshifted ratio λ_0/λ_S is expected for light from this galaxy?

44.36 •• (a) According to the Hubble law, what is the distance r from us for galaxies that are receding from us with a speed c? (b) Explain why the distance calculated in part (a) is the size of our observable universe (ignoring any change in the expansion rate of the universe due to gravitational attraction or dark energy).

44.37 •• The critical density of the universe is 9.5×10^{-27} kg/m³. (a) Assuming that the universe is all hydrogen, express the critical density in the number of H atoms per cubic meter. (b) If the density of the universe is equal to the critical density, how many atoms, on the average, would you expect to find in a room of dimensions 4 m × 7 m × 3 m? (c) Compare your answer in part (b) with the number of atoms you would find in the same room under normal conditions on the earth.

Section 44.7 The Beginning of Time

44.38 • (a) Show that the expression for the Planck length, $\sqrt{\hbar G/c^3}$, has dimensions of length. (b) Evaluate the numerical value of $\sqrt{\hbar G/c^3}$, and verify the value given in Eq. (44.21).

44.39 • Calculate the energy released in each reaction: (a) p + ^{2}H → ^{3}He; (b) n + ^{3}He → ^{4}He.

44.40 • Calculate the energy (in MeV) released in the triple-alpha process 3 ^{4}He → ^{12}C.

44.41 • Calculate the reaction energy Q (in MeV) for the reaction $e^- + p \rightarrow n + \nu_e$. Is this reaction endoergic or exoergic?

44.42 • Calculate the reaction energy Q (in MeV) for the nucleosynthesis reaction

$$^{12}_6C + {}^4_2He \rightarrow {}^{16}_8O$$

Is this reaction endoergic or exoergic?

44.43 • **CP** The 2.728-K blackbody radiation has its peak wavelength at 1.062 mm. What was the peak wavelength at $t =$ 700,000 y when the temperature was 3000 K?

PROBLEMS

44.44 • **CP** A positronium atom consists of an electron and a positron. In the Bohr model the two particles orbit around their common center of mass. In the Bohr model, what is the ionization energy for a positronium atom when it is in its ground state?

44.45 • In the LHC, each proton will be accelerated to a kinetic energy of 7.0 TeV. (a) In the colliding beams, what is the available energy E_a in a collision? (b) In a fixed-target experiment in which a beam of protons is incident on a stationary proton target, what must the total energy (in TeV) of the particles in the beam be to produce the same available energy as in part (a)?

44.46 •• A proton and an antiproton collide head-on with equal kinetic energies. Two γ rays with wavelengths of 0.780 fm are produced. Calculate the kinetic energy of the incident proton.

44.47 •• **CP BIO Radiation Therapy with π^- Mesons.** Beams of π^- mesons are used in radiation therapy for certain cancers. The energy comes from the complete decay of the π^- to *stable* particles. (a) Write out the complete decay of a π^- meson to stable particles. What are these particles? (b) How much energy is released from the complete decay of a single π^- meson to stable particles? (You can ignore the very small masses of the neutrinos.) (c) How many π^- mesons need to decay to give a dose of 50.0 Gy to 10.0 g of tissue? (d) What would be the equivalent dose in part (c) in Sv and in rem? Consult Table 43.3 and use the largest appropriate RBE for the particles involved in this decay.

44.48 •• Calculate the threshold kinetic energy for the reaction $\pi^- + p \rightarrow \Sigma^0 + K^0$ if a π^- beam is incident on a stationary proton target. The K^0 has a mass of 497.7 MeV/c^2.

44.49 •• Calculate the threshold kinetic energy for the reaction $p + p \rightarrow p + p + K^+ + K^-$ if a proton beam is incident on a stationary proton target.

44.50 •• An η^0 meson at rest decays into three π mesons. (a) What are the allowed combinations of π^0, π^+, and π^- as decay products? (b) Find the total kinetic energy of the π mesons.

44.51 • Each of the following reactions is missing a single particle. Calculate the baryon number, charge, strangeness, and the three lepton numbers (where appropriate) of the missing particle, and from this identify the particle. (a) $p + p \rightarrow p + \Lambda^0 + $?; (b) $K^- + n \rightarrow \Lambda^0 + $?; (c) $p + \bar{p} \rightarrow n + $?; (d) $\bar{\nu}_\mu + p \rightarrow n + $?

44.52 • Estimate the energy width (energy uncertainty) of the ψ if its mean lifetime is 7.6×10^{-21} s. What fraction is this of its rest energy?

44.53 • The ϕ meson has mass 1019.4 MeV/c^2 and a measured energy width of 4.4 MeV/c^2. Using the uncertainty principle, estimate the lifetime of the ϕ meson.

44.54 •• A ϕ meson (see Problem 44.53) at rest decays via $\phi \rightarrow K^+ + K^-$. It has strangeness 0. (a) Find the kinetic energy of the K^+ meson. (Assume that the two decay products share kinetic energy equally, since their masses are equal.) (b) Suggest a reason the decay $\phi \rightarrow K^+ + K^- + \pi^0$ has not been observed. (c) Suggest reasons the decays $\phi \rightarrow K^+ + \pi^-$ and $\phi \rightarrow K^+ + \mu^-$ have not been observed.

44.55 •• **CP BIO** One proposed proton decay is $p^+ \rightarrow e^+ + \pi^0$, which violates both baryon and lepton number conservation, so the proton lifetime is expected to be very long. Suppose the proton half-life were 1.0×10^{18} y. (a) Calculate the energy deposited per kilogram of body tissue (in rad) due to the decay of the protons in your body in one year. Model your body as consisting entirely of water. Only the two protons in the hydrogen atoms in each H_2O molecule would decay in the manner shown; do you see why? Assume that the π^0 decays to two γ rays, that the positron annihilates with an electron, and that all the energy produced in the primary decay and these secondary decays remains in your body. (b) Calculate the equivalent dose (in rem) assuming an RBE of 1.0 for all the radiation products, and compare with the 0.1 rem due to the natural background and the 5.0-rem guideline for industrial workers. Based on your calculation, can the proton lifetime be as short as 1.0×10^{18} y?

44.56 ••• **CP** A Ξ^- particle at rest decays to a Λ^0 and a π^-. (a) Find the total kinetic energy of the decay products. (b) What fraction of the energy is carried off by each particle? (Use relativistic expressions for momentum and energy.)

44.57 •• **CALC** Consider the spherical balloon model of a two-dimensional expanding universe (see Fig. 44.17 in Section 44.6). The shortest distance between two points on the surface, measured along the surface, is the arc length r, where $r = R\theta$. As the balloon expands, its radius R increases, but the angle θ between the two points remains constant. (a) Explain why, at any given time, $(dR/dt)/R$ is the same for all points on the balloon. (b) Show that $v = dr/dt$ is directly proportional to r at any instant. (c) From your answer to part (b), what is the expression for the Hubble constant H_0 in terms of R and dR/dt? (d) The expression for H_0 you found in part (c) is constant in space. How would R have to depend on time for H_0 to be constant in time? (e) Is your answer to part (d) consistent with the observed rate of expansion of the universe?

44.58 ••• **CALC** Suppose all the conditions are the same as in Problem 44.57, except that $v = dr/dt$ is constant for a given θ, rather than H_0 being constant in time. Show that the Hubble constant is $H_0 = 1/t$ and, hence, that the current value is $1/T$, where T is the age of the universe.

44.59 •• **Cosmic Jerk.** The densities of ordinary matter and dark matter have decreased as the universe has expanded, since the same amount of mass occupies an ever-increasing volume. Yet observations suggest that the density of dark energy has remained constant over the entire history of the universe. (a) Explain why the expansion of the universe actually slowed down in its early history but is speeding up today. "Jerk" is the term for a change in acceleration, so the change in cosmic expansion from slowing down to speeding up is called *cosmic jerk*. (b) Calculations show that the change in acceleration took place when the combined density of matter of all kinds was equal to twice the density of dark energy. Compared to today's value of the scale factor, what was the scale factor at that time? (c) We see the galaxy clusters in Figs. 44.15b and 44.19 as they were 300 million years ago and 10.2 billion years ago. Was the expansion of the universe slowing down or speeding up at these times? (*Hint:* See the caption for Fig. 44.19.)

44.60 ••• **CP** The K^0 meson has rest energy 497.7 MeV. A K^0 meson moving in the $+x$-direction with kinetic energy 225 MeV decays into a π^+ and a π^-, which move off at equal angles above and below the $+x$-axis. Calculate the kinetic energy of the π^+ and the angle it makes with the $+x$-axis. Use relativistic expressions for energy and momentum.

44.61 ••• **CP** A Σ^- particle moving in the $+x$-direction with kinetic energy 180 MeV decays into a π^- and a neutron. The π^- moves in the $+y$-direction. What is the kinetic energy of the neutron, and what is the direction of its velocity? Use relativistic expressions for energy and momentum.

CHALLENGE PROBLEM

44.62 ••• **CP** Consider a collision in which a stationary particle with mass M is bombarded by a particle with mass m, speed v_0, and total energy (including rest energy) E_m. (a) Use the Lorentz transformation to write the velocities v_m and v_M of particles m and M in terms of the speed v_{cm} of the center of momentum. (b) Use the fact that the total momentum in the center-of-momentum frame is zero to obtain an expression for v_{cm} in terms of m, M, and v_0. (c) Combine the results of parts (a) and (b) to obtain Eq. (44.9) for the total energy in the center-of-momentum frame.

Answers

Chapter Opening Question

Only 4.6% of the mass and energy of the universe is in the form of "normal" matter. Of the rest, 22.8% is poorly understood dark matter and 72.6% is even more mysterious dark energy.

Test Your Understanding Questions

44.1 Answer: (i), (iii), (ii), (iv) The more massive the virtual particle, the shorter its lifetime and the shorter the distance that it can travel during its lifetime.

44.2 Answer: no In a head-on collision between an electron and a positron of equal energy, the net momentum is zero. Since both momentum and energy are conserved in the collision, the virtual photon also has momentum $p = 0$ but has energy $E = 90$ GeV + 90 GeV = 180 GeV. Hence the relationship $E = pc$ is definitely *not* true for this virtual photon.

44.3 Answer: no Mesons all have baryon number $B = 0$, while a proton has $B = 1$. The decay of a proton into one or more mesons would require that baryon number *not* be conserved. No violation of this conservation principle has ever been observed, so the proposed decay is impossible.

44.4 Answer: no Only the s quark, with $S = -1$, has nonzero strangeness. For a baryon to have $S = -2$, it must have two s quarks and one quark of a different flavor. Since each s quark has

charge $-\frac{1}{3}e$, the nonstrange quark must have charge $+\frac{5}{3}e$ to make the net charge equal to $+e$. But *no* quark has charge $+\frac{5}{3}e$, so the proposed baryon is impossible.

44.5 Answer: (i) If a d quark in a neutron (quark content udd) undergoes the process $d \rightarrow u + e^- + \bar{\nu}_e$, the remaining baryon has quark content uud and hence is a proton (see Fig. 44.11). An electron is the same as a β^- particle, so the net result is beta-minus decay: $n \rightarrow p + \beta^- + \bar{\nu}_e$.

44.6 Answer: yes . . . and no The material of which your body is made is ordinary to us on earth. But from a cosmic perspective your material is quite *extraordinary*: Only 4.6% of the mass and energy in the universe is in the form of atoms.

44.7 Answer: no Prior to $t = 380,000$ y the temperature was so high that atoms could not form, so free electrons and protons were plentiful. These charged particles are very effective at scattering photons, so light could not propagate very far and the universe was opaque. The oldest photons that we can detect date from the time $t = 380,000$ y when atoms formed and the universe became transparent.

Bridging Problem

Answers: (a) 5.78 MeV for the neutron, 35.62 MeV for the pion
(b) 0.140 for the neutron, 0.860 for the pion
(c) 24 MeV

APPENDIX A

THE INTERNATIONAL SYSTEM OF UNITS

The Système International d'Unités, abbreviated SI, is the system developed by the General Conference on Weights and Measures and adopted by nearly all the industrial nations of the world. The following material is adapted from the National Institute of Standards and Technology (**http://physics.nist.gov/cuu).**

Quantity	Name of unit	Symbol	
SI base units			
length	meter	m	
mass	kilogram	kg	
time	second	s	
electric current	ampere	A	
thermodynamic temperature	kelvin	K	
amount of substance	mole	mol	
luminous intensity	candela	cd	
SI derived units			**Equivalent units**
area	square meter	m^2	
volume	cubic meter	m^3	
frequency	hertz	Hz	s^{-1}
mass density (density)	kilogram per cubic meter	kg/m^3	
speed, velocity	meter per second	m/s	
angular velocity	radian per second	rad/s	
acceleration	meter per second squared	m/s^2	
angular acceleration	radian per second squared	rad/s^2	
force	newton	N	$kg \cdot m/s^2$
pressure (mechanical stress)	pascal	Pa	N/m^2
kinematic viscosity	square meter per second	m^2/s	
dynamic viscosity	newton-second per square meter	$N \cdot s/m^2$	
work, energy, quantity of heat	joule	J	$N \cdot m$
power	watt	W	J/s
quantity of electricity	coulomb	C	$A \cdot s$
potential difference, electromotive force	volt	V	J/C, W/A
electric field strength	volt per meter	V/m	N/C
electric resistance	ohm	Ω	V/A
capacitance	farad	F	$A \cdot s/V$
magnetic flux	weber	Wb	$V \cdot s$
inductance	henry	H	$V \cdot s/A$
magnetic flux density	tesla	T	Wb/m^2
magnetic field strength	ampere per meter	A/m	
magnetomotive force	ampere	A	
luminous flux	lumen	lm	$cd \cdot sr$
luminance	candela per square meter	cd/m^2	
illuminance	lux	lx	lm/m^2
wave number	1 per meter	m^{-1}	
entropy	joule per kelvin	J/K	
specific heat capacity	joule per kilogram-kelvin	$J/kg \cdot K$	
thermal conductivity	watt per meter-kelvin	$W/m \cdot K$	

Quantity	Name of unit	Symbol	Equivalent units
radiant intensity	watt per steradian	W/sr	
activity (of a radioactive source)	becquerel	Bq	s^{-1}
radiation dose	gray	Gy	J/kg
radiation dose equivalent	sievert	Sv	J/kg
SI supplementary units			
plane angle	radian	rad	
solid angle	steradian	sr	

Definitions of SI Units

meter (m)   The *meter* is the length equal to the distance traveled by light, in vacuum, in a time of 1/299,792,458 second.

kilogram (kg)   The *kilogram* is the unit of mass; it is equal to the mass of the international prototype of the kilogram. (The international prototype of the kilogram is a particular cylinder of platinum-iridium alloy that is preserved in a vault at Sévres, France, by the International Bureau of Weights and Measures.)

second (s)   The *second* is the duration of 9,192,631,770 periods of the radiation corresponding to the transition between the two hyperfine levels of the ground state of the cesium-133 atom.

ampere (A)   The *ampere* is that constant current that, if maintained in two straight parallel conductors of infinite length, of negligible circular cross section, and placed 1 meter apart in vacuum, would produce between these conductors a force equal to 2×10^{-7} newton per meter of length.

kelvin (K)   The *kelvin*, unit of thermodynamic temperature, is the fraction 1/273.16 of the thermodynamic temperature of the triple point of water.

ohm (Ω)   The *ohm* is the electric resistance between two points of a conductor when a constant difference of potential of 1 volt, applied between these two points, produces in this conductor a current of 1 ampere, this conductor not being the source of any electromotive force.

coulomb (C)   The *coulomb* is the quantity of electricity transported in 1 second by a current of 1 ampere.

candela (cd)   The *candela* is the luminous intensity, in a given direction, of a source that emits monochromatic radiation of frequency 540×10^{12} hertz and that has a radiant intensity in that direction of 1/683 watt per steradian.

mole (mol)   The *mole* is the amount of substance of a system that contains as many elementary entities as there are carbon atoms in 0.012 kg of carbon 12. The elementary entities must be specified and may be atoms, molecules, ions, electrons, other particles, or specified groups of such particles.

newton (N)   The *newton* is that force that gives to a mass of 1 kilogram an acceleration of 1 meter per second per second.

joule (J)   The *joule* is the work done when the point of application of a constant force of 1 newton is displaced a distance of 1 meter in the direction of the force.

watt (W)   The *watt* is the power that gives rise to the production of energy at the rate of 1 joule per second.

volt (V)   The *volt* is the difference of electric potential between two points of a conducting wire carrying a constant current of 1 ampere, when the power dissipated between these points is equal to 1 watt.

weber (Wb)   The *weber* is the magnetic flux that, linking a circuit of one turn, produces in it an electromotive force of 1 volt as it is reduced to zero at a uniform rate in 1 second.

lumen (lm)   The *lumen* is the luminous flux emitted in a solid angle of 1 steradian by a uniform point source having an intensity of 1 candela.

farad (F)   The *farad* is the capacitance of a capacitor between the plates of which there appears a difference of potential of 1 volt when it is charged by a quantity of electricity equal to 1 coulomb.

henry (H)   The *henry* is the inductance of a closed circuit in which an electromotive force of 1 volt is produced when the electric current in the circuit varies uniformly at a rate of 1 ampere per second.

radian (rad)   The *radian* is the plane angle between two radii of a circle that cut off on the circumference an arc equal in length to the radius.

steradian (sr)   The *steradian* is the solid angle that, having its vertex in the center of a sphere, cuts off an area of the surface of the sphere equal to that of a square with sides of length equal to the radius of the sphere.

SI Prefixes   To form the names of multiples and submultiples of SI units, apply the prefixes listed in Appendix F.

APPENDIX B

USEFUL MATHEMATICAL RELATIONS

Algebra

$$a^{-x} = \frac{1}{a^x} \qquad a^{(x+y)} = a^x a^y \qquad a^{(x-y)} = \frac{a^x}{a^y}$$

Logarithms: If $\log a = x$, then $a = 10^x$. $\quad \log a + \log b = \log (ab) \quad \log a - \log b = \log (a/b) \quad \log (a^n) = n \log a$

If $\ln a = x$, then $a = e^x$. $\quad \ln a + \ln b = \ln (ab) \qquad \ln a - \ln b = \ln (a/b) \qquad \ln (a^n) = n \ln a$

Quadratic formula: If $ax^2 + bx + c = 0,$ $\quad x = \dfrac{-b \pm \sqrt{b^2 - 4ac}}{2a}.$

Binomial Theorem

$$(a + b)^n = a^n + na^{n-1}b + \frac{n(n-1)a^{n-2}b^2}{2!} + \frac{n(n-1)(n-2)a^{n-3}b^3}{3!} + \cdots$$

Trigonometry

In the right triangle ABC, $x^2 + y^2 = r^2$.

Definitions of the trigonometric functions:
$\sin \alpha = y/r \qquad \cos \alpha = x/r \qquad \tan \alpha = y/x$

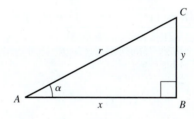

Identities: $\quad \sin^2 \alpha + \cos^2 \alpha = 1 \qquad\qquad \tan \alpha = \dfrac{\sin \alpha}{\cos \alpha}$

$$\sin 2\alpha = 2 \sin \alpha \cos \alpha \qquad\qquad \cos 2\alpha = \cos^2 \alpha - \sin^2 \alpha = 2\cos^2 \alpha - 1$$
$$= 1 - 2 \sin^2 \alpha$$

$$\sin \tfrac{1}{2}\alpha = \sqrt{\frac{1 - \cos \alpha}{2}} \qquad\qquad \cos \tfrac{1}{2}\alpha = \sqrt{\frac{1 + \cos \alpha}{2}}$$

$$\sin(-\alpha) = -\sin \alpha \qquad\qquad \sin(\alpha \pm \beta) = \sin \alpha \cos \beta \pm \cos \alpha \sin \beta$$
$$\cos(-\alpha) = \cos \alpha \qquad\qquad \cos(\alpha \pm \beta) = \cos \alpha \cos \beta \mp \sin \alpha \sin \beta$$
$$\sin(\alpha \pm \pi/2) = \pm\cos \alpha \qquad\qquad \sin \alpha + \sin \beta = 2 \sin \tfrac{1}{2}(\alpha + \beta) \cos \tfrac{1}{2}(\alpha - \beta)$$
$$\cos(\alpha \pm \pi/2) = \mp\sin \alpha \qquad\qquad \cos \alpha + \cos \beta = 2 \cos \tfrac{1}{2}(\alpha + \beta) \cos \tfrac{1}{2}(\alpha - \beta)$$

For *any* triangle $A'B'C'$ (not necessarily a right triangle) with sides a, b, and c and angles α, β, and γ:

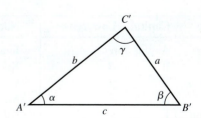

Law of sines: $\quad \dfrac{\sin \alpha}{a} = \dfrac{\sin \beta}{b} = \dfrac{\sin \gamma}{c}$

Law of cosines: $\quad c^2 = a^2 + b^2 - 2ab \cos \gamma$

Geometry

Circumference of circle of radius r:	$C = 2\pi r$	Surface area of sphere of radius r:	$A = 4\pi r^2$
Area of circle of radius r:	$A = \pi r^2$	Volume of cylinder of radius r and height h:	$V = \pi r^2 h$
Volume of sphere of radius r:	$V = 4\pi r^3/3$		

Calculus

Derivatives:

$$\frac{d}{dx}x^n = nx^{n-1}$$

$$\frac{d}{dx}\ln ax = \frac{1}{x}$$

$$\frac{d}{dx}e^{ax} = ae^{ax}$$

$$\frac{d}{dx}\sin ax = a\cos ax$$

$$\frac{d}{dx}\cos ax = -a\sin ax$$

Integrals:

$$\int x^n\, dx = \frac{x^{n+1}}{n+1} \quad (n \neq -1)$$

$$\int \frac{dx}{x} = \ln x$$

$$\int e^{ax}\, dx = \frac{1}{a}e^{ax}$$

$$\int \sin ax\, dx = -\frac{1}{a}\cos ax$$

$$\int \cos ax\, dx = \frac{1}{a}\sin ax$$

$$\int \frac{dx}{\sqrt{a^2 - x^2}} = \arcsin\frac{x}{a}$$

$$\int \frac{dx}{\sqrt{x^2 + a^2}} = \ln\left(x + \sqrt{x^2 + a^2}\right)$$

$$\int \frac{dx}{x^2 + a^2} = \frac{1}{a}\arctan\frac{x}{a}$$

$$\int \frac{dx}{(x^2 + a^2)^{3/2}} = \frac{1}{a^2}\frac{x}{\sqrt{x^2 + a^2}}$$

$$\int \frac{x\, dx}{(x^2 + a^2)^{3/2}} = -\frac{1}{\sqrt{x^2 + a^2}}$$

Power series (convergent for range of *x* shown):

$$(1 + x)^n = 1 + nx + \frac{n(n-1)x^2}{2!} + \frac{n(n-1)(n-2)}{3!}x^3 + \cdots \; (|x| < 1)$$

$$\sin x = x - \frac{x^3}{3!} + \frac{x^5}{5!} - \frac{x^7}{7!} + \cdots \; (\text{all } x)$$

$$\cos x = 1 - \frac{x^2}{2!} + \frac{x^4}{4!} - \frac{x^6}{6!} + \cdots \; (\text{all } x)$$

$$\tan x = x + \frac{x^3}{3} + \frac{2x^2}{15} + \frac{17x^7}{315} + \cdots \; (|x| < \pi/2)$$

$$e^x = 1 + x + \frac{x^2}{2!} + \frac{x^3}{3!} + \cdots \; (\text{all } x)$$

$$\ln(1 + x) = x - \frac{x^2}{2} + \frac{x^3}{3} - \frac{x^4}{4} + \cdots \; (|x| < 1)$$

APPENDIX C

THE GREEK ALPHABET

Name	Capital	Lowercase	Name	Capital	Lowercase	Name	Capital	Lowercase
Alpha	A	α	Iota	I	ι	Rho	P	ρ
Beta	B	β	Kappa	K	κ	Sigma	Σ	σ
Gamma	Γ	γ	Lambda	Λ	λ	Tau	T	τ
Delta	Δ	δ	Mu	M	μ	Upsilon	Y	υ
Epsilon	E	ϵ	Nu	N	ν	Phi	Φ	ϕ
Zeta	Z	ζ	Xi	Ξ	ξ	Chi	X	χ
Eta	H	η	Omicron	O	o	Psi	Ψ	ψ
Theta	Θ	θ	Pi	Π	π	Omega	Ω	ω

APPENDIX D

PERIODIC TABLE OF THE ELEMENTS

Group	1	2	3	4	5	6	7	8	9	10	11	12	13	14	15	16	17	18
Period																		
1	1 **H** 1.008																	2 **He** 4.003
2	3 **Li** 6.941	4 **Be** 9.012											5 **B** 10.811	6 **C** 12.011	7 **N** 14.007	8 **O** 15.999	9 **F** 18.998	10 **Ne** 20.180
3	11 **Na** 22.990	12 **Mg** 24.305											13 **Al** 26.982	14 **Si** 28.086	15 **P** 30.974	16 **S** 32.065	17 **Cl** 35.453	18 **Ar** 39.948
4	19 **K** 39.098	20 **Ca** 40.078	21 **Sc** 44.956	22 **Ti** 47.867	23 **V** 50.942	24 **Cr** 51.996	25 **Mn** 54.938	26 **Fe** 55.845	27 **Co** 58.933	28 **Ni** 58.693	29 **Cu** 63.546	30 **Zn** 65.409	31 **Ga** 69.723	32 **Ge** 72.64	33 **As** 74.922	34 **Se** 78.96	35 **Br** 79.904	36 **Kr** 83.798
5	37 **Rb** 85.468	38 **Sr** 87.62	39 **Y** 88.906	40 **Zr** 91.224	41 **Nb** 92.906	42 **Mo** 95.94	43 **Tc** (98)	44 **Ru** 101.07	45 **Rh** 102.906	46 **Pd** 106.42	47 **Ag** 107.868	48 **Cd** 112.411	49 **In** 114.818	50 **Sn** 118.710	51 **Sb** 121.760	52 **Te** 127.60	53 **I** 126.904	54 **Xe** 131.293
6	55 **Cs** 132.905	56 **Ba** 137.327	71 **Lu** 174.967	72 **Hf** 178.49	73 **Ta** 180.948	74 **W** 183.84	75 **Re** 186.207	76 **Os** 190.23	77 **Ir** 192.217	78 **Pt** 195.078	79 **Au** 196.967	80 **Hg** 200.59	81 **Tl** 204.383	82 **Pb** 207.2	83 **Bi** 208.980	84 **Po** (209)	85 **At** (210)	86 **Rn** (222)
7	87 **Fr** (223)	88 **Ra** (226)	103 **Lr** (262)	104 **Rf** (261)	105 **Db** (262)	106 **Sg** (266)	107 **Bh** (264)	108 **Hs** (269)	109 **Mt** (268)	110 **Ds** (271)	111 **Rg** (272)	112 **Uub** (285)	113 **Uut** (284)	114 **Uuq** (289)	115 **Uup** (288)	116 **Uuh** (292)	117 **Uus** (294)	118 **Uuo**

Lanthanoids

57 **La** 138.905	58 **Ce** 140.116	59 **Pr** 140.908	60 **Nd** 144.24	61 **Pm** (145)	62 **Sm** 150.36	63 **Eu** 151.964	64 **Gd** 157.25	65 **Tb** 158.925	66 **Dy** 162.500	67 **Ho** 164.930	68 **Er** 167.259	69 **Tm** 168.934	70 **Yb** 173.04

Actinoids

89 **Ac** (227)	90 **Th** (232)	91 **Pa** (231)	92 **U** (238)	93 **Np** (237)	94 **Pu** (244)	95 **Am** (243)	96 **Cm** (247)	97 **Bk** (247)	98 **Cf** (251)	99 **Es** (252)	100 **Fm** (257)	101 **Md** (258)	102 **No** (259)

For each element the average atomic mass of the mixture of isotopes occurring in nature is shown. For elements having no stable isotope, the approximate atomic mass of the longest-lived isotope is shown in parentheses. For elements that have been predicted but not yet confirmed, no atomic mass is given. All atomic masses are expressed in atomic mass units ($1\ u = 1.660538782(83) \times 10^{-27}\ kg$), equivalent to grams per mole (g/mol).

APPENDIX E

UNIT CONVERSION FACTORS

Length
1 m = 100 cm = 1000 mm = 10^6 μm = 10^9 nm
1 km = 1000 m = 0.6214 mi
1 m = 3.281 ft = 39.37 in.
1 cm = 0.3937 in.
1 in. = 2.540 cm
1 ft = 30.48 cm
1 yd = 91.44 cm
1 mi = 5280 ft = 1.609 km
1 Å = 10^{-10} m = 10^{-8} cm = 10^{-1} nm
1 nautical mile = 6080 ft
1 light year = 9.461 × 10^{15} m

Area
1 cm^2 = 0.155 $in.^2$
1 m^2 = 10^4 cm^2 = 10.76 ft^2
1 $in.^2$ = 6.452 cm^2
1 ft^2 = 144 $in.^2$ = 0.0929 m^2

Volume
1 liter = 1000 cm^3 = 10^{-3} m^3 = 0.03531 ft^3 = 61.02 $in.^3$
1 ft^3 = 0.02832 m^3 = 28.32 liters = 7.477 gallons
1 gallon = 3.788 liters

Time
1 min = 60 s
1 h = 3600 s
1 d = 86,400 s
1 y = 365.24 d = 3.156 × 10^7 s

Angle
1 rad = 57.30° = 180°/π
1° = 0.01745 rad = π/180 rad
1 revolution = 360° = 2π rad
1 rev/min (rpm) = 0.1047 rad/s

Speed
1 m/s = 3.281 ft/s
1 ft/s = 0.3048 m/s
1 mi/min = 60 mi/h = 88 ft/s
1 km/h = 0.2778 m/s = 0.6214 mi/h
1 mi/h = 1.466 ft/s = 0.4470 m/s = 1.609 km/h
1 furlong/fortnight = 1.662 × 10^{-4} m/s

Acceleration
1 m/s^2 = 100 cm/s^2 = 3.281 ft/s^2
1 cm/s^2 = 0.01 m/s^2 = 0.03281 ft/s^2
1 ft/s^2 = 0.3048 m/s^2 = 30.48 cm/s^2
1 mi/h·s = 1.467 ft/s^2

Mass
1 kg = 10^3 g = 0.0685 slug
1 g = 6.85 × 10^{-5} slug
1 slug = 14.59 kg
1 u = 1.661 × 10^{-27} kg
1 kg has a weight of 2.205 lb when g = 9.80 m/s^2

Force
1 N = 10^5 dyn = 0.2248 lb
1 lb = 4.448 N = 4.448 × 10^5 dyn

Pressure
1 Pa = 1 N/m^2 = 1.450 × 10^{-4} $lb/in.^2$ = 0.209 lb/ft^2
1 bar = 10^5 Pa
1 $lb/in.^2$ = 6895 Pa
1 lb/ft^2 = 47.88 Pa
1 atm = 1.013 × 10^5 Pa = 1.013 bar
         = 14.7 $lb/in.^2$ = 2117 lb/ft^2
1 mm Hg = 1 torr = 133.3 Pa

Energy
1 J = 10^7 ergs = 0.239 cal
1 cal = 4.186 J (based on 15° calorie)
1 ft·lb = 1.356 J
1 Btu = 1055 J = 252 cal = 778 ft·lb
1 eV = 1.602 × 10^{-19} J
1 kWh = 3.600 × 10^6 J

Mass–Energy Equivalence
1 kg ↔ 8.988 × 10^{16} J
1 u ↔ 931.5 MeV
1 eV ↔ 1.074 × 10^{-9} u

Power
1 W = 1 J/s
1 hp = 746 W = 550 ft·lb/s
1 Btu/h = 0.293 W

APPENDIX F

NUMERICAL CONSTANTS

Fundamental Physical Constants*

Name	Symbol	Value
Speed of light in vacuum	c	2.99792458×10^{8} m/s
Magnitude of charge of electron	e	$1.602176487(40) \times 10^{-19}$ C
Gravitational constant	G	$6.67428(67) \times 10^{-11}$ N$\cdot$m^2/kg^2
Planck's constant	h	$6.62606896(33) \times 10^{-34}$ J$\cdot$s
Boltzmann constant	k	$1.3806504(24) \times 10^{-23}$ J/K
Avogadro's number	N_A	$6.02214179(30) \times 10^{23}$ molecules/mol
Gas constant	R	$8.314472(15)$ J/mol$\cdot$K
Mass of electron	m_{e}	$9.10938215(45) \times 10^{-31}$ kg
Mass of proton	m_{p}	$1.672621637(83) \times 10^{-27}$ kg
Mass of neutron	m_{n}	$1.674927211(84) \times 10^{-27}$ kg
Permeability of free space	μ_0	$4\pi \times 10^{-7}$ Wb/A$\cdot$m
Permittivity of free space	$\epsilon_0 = 1/\mu_0 c^2$	$8.854187817\ldots \times 10^{-12}$ C^2/N$\cdot$m^2
	$1/4\pi\epsilon_0$	$8.987551787\ldots \times 10^{9}$ N$\cdot$m^2/C^2

Other Useful Constants*

Name	Symbol	Value
Mechanical equivalent of heat		4.186 J/cal (15° calorie)
Standard atmospheric pressure	1 atm	1.01325×10^{5} Pa
Absolute zero	0 K	-273.15°C
Electron volt	1 eV	$1.602176487(40) \times 10^{-19}$ J
Atomic mass unit	1 u	$1.660538782(83) \times 10^{-27}$ kg
Electron rest energy	$m_{\mathrm{e}}c^2$	$0.510998910(13)$ MeV
Volume of ideal gas (0°C and 1 atm)		$22.413996(39)$ liter/mol
Acceleration due to gravity (standard)	g	9.80665 m/s^2

*Source: National Institute of Standards and Technology (**http://physics.nist.gov/cuu**). Numbers in parentheses show the uncertainty in the final digits of the main number; for example, the number 1.6454(21) means 1.6454 ± 0.0021. Values shown without uncertainties are exact.

Astronomical Data[†]

Body	Mass (kg)	Radius (m)	Orbit radius (m)	Orbit period
Sun	1.99×10^{30}	6.96×10^{8}	—	—
Moon	7.35×10^{22}	1.74×10^{6}	3.84×10^{8}	27.3 d
Mercury	3.30×10^{23}	2.44×10^{6}	5.79×10^{10}	88.0 d
Venus	4.87×10^{24}	6.05×10^{6}	1.08×10^{11}	224.7 d
Earth	5.97×10^{24}	6.38×10^{6}	1.50×10^{11}	365.3 d
Mars	6.42×10^{23}	3.40×10^{6}	2.28×10^{11}	687.0 d
Jupiter	1.90×10^{27}	6.91×10^{7}	7.78×10^{11}	11.86 y
Saturn	5.68×10^{26}	6.03×10^{7}	1.43×10^{12}	29.45 y
Uranus	8.68×10^{25}	2.56×10^{7}	2.87×10^{12}	84.02 y
Neptune	1.02×10^{26}	2.48×10^{7}	4.50×10^{12}	164.8 y
Pluto[‡]	1.31×10^{22}	1.15×10^{6}	5.91×10^{12}	247.9 y

[†]Source: NASA Jet Propulsion Laboratory Solar System Dynamics Group (**http://ssd.jpl.nasa.gov**), and P. Kenneth Seidelmann, ed., ***Explanatory Supplement to the Astronomical Almanac*** (University Science Books, Mill Valley, CA, 1992), pp. 704–706. For each body, "radius" is its radius at its equator and "orbit radius" is its average distance from the sun or (for the moon) from the earth.

[‡]In August 2006, the International Astronomical Union reclassified Pluto and other small objects that orbit the sun as "dwarf planets."

Prefixes for Powers of 10

Power of ten	Prefix	Abbreviation	Pronunciation
10^{-24}	yocto-	y	*yoc*-toe
10^{-21}	zepto-	z	*zep*-toe
10^{-18}	atto-	a	*at*-toe
10^{-15}	femto-	f	*fem*-toe
10^{-12}	pico-	p	*pee*-koe
10^{-9}	nano-	n	*nan*-oe
10^{-6}	micro-	μ	*my*-crow
10^{-3}	milli-	m	*mil*-i
10^{-2}	centi-	c	*cen*-ti
10^{3}	kilo-	k	*kil*-oe
10^{6}	mega-	M	*meg*-a
10^{9}	giga-	G	*jig*-a or *gig*-a
10^{12}	tera-	T	*ter*-a
10^{15}	peta-	P	*pet*-a
10^{18}	exa-	E	*ex*-a
10^{21}	zetta-	Z	*zet*-a
10^{24}	yotta-	Y	*yot*-a

Examples:

1 femtometer = 1 fm = 10^{-15} m

1 picosecond = 1 ps = 10^{-12} s

1 nanocoulomb = 1 nC = 10^{-9} C

1 microkelvin = 1μK = 10^{-6} K

1 millivolt = 1 mV = 10^{-3} V

1 kilopascal = 1 kPa = 10^{3} Pa

1 megawatt = 1 MW = 10^{6} W

1 gigahertz = 1 GHz = 10^{9} Hz

ANSWERS TO ODD-NUMBERED PROBLEMS

Chapter 37

37.1 bolt A
37.3 $0.867c$, no
37.5 a) $0.998c$ b) 126 m
37.7 1.12 h, in the spacecraft
37.9 92.5 m
37.11 a) 0.66 km b) 49 μs, 15 km c) 0.45 km
37.13 a) 3570 m b) 90.0 μs c) 89.2 μs
37.15 a) $0.806c$ b) $0.974c$ c) $0.997c$
37.17 a) toward b) $0.385c$
37.19 $0.784c$
37.21 $0.611c$
37.23 $0.837c$, away from
37.25 a) $0.159c$ b) 172 million dollars
37.27 $3.06p_0$
37.29 a) $0.866c$ b) $0.608c$
37.31 a) 5.49×10^{15} m/s^2
b) 9.26×10^{14} m/s^2
37.33 a) $0.866c$ b) $0.986c$
37.35 a) 0.450 nJ b) 1.94×10^{-18} kg·m/s
c) $0.968c$
37.37 a) $3.3 \times 10^{-13}\%$ b) 4.0×10^{-13} g, increases, no
37.39 a) 1.11×10^3 kg b) 52.1 cm
37.41 a) 0.867 nJ b) 0.270 nJ c) 0.452
37.43 a) 5.34 pJ (nonrel), 5.65 pJ (rel), 1.06
b) 67.8 pJ (nonrel), 331 pJ (rel), 4.88
37.45 a) 2.06 MV
b) 0.330 pJ = 2.06 MeV
37.47 a) 4.2×10^9 kg = 4.6×10^6 tons
b) 1.5×10^{13} y
37.49 a) $\Delta = 8.42 \times 10^{-6}$ b) 34.0 GeV
37.51 $0.700c$
37.53 a) $0.995c$ b) 1%
37.55 a) $\Delta = 9 \times 10^{-9}$ b) $7000m$
37.57 0.168 MeV
37.59 a) $4c/5$ b) c c) (i) 145 MeV (ii) 625 MeV
d) (i) 117 MeV (ii) 469 MeV
37.65 b) $\Delta x' = \sqrt{(\Delta x)^2 - c^2(\Delta t)^2}$ c) 14.4 ns
37.67 $0.357c$, receding
37.69 a) 1000 y, 866 y, 1.4×10^{19} J, 14%
b) 505 y, 71 y, 5.5×10^{20} J, 550%
c) 501 y, 7.1 y, 6.3×10^{21} J, 6300%
37.71 2.04×10^{-13} N
37.75 a) toward us at 13.1 km/s, 39.4 km/s
b) 5.96×10^9 m, about 0.040 times the earth-sun distance; 5.55×10^{29} kg = $0.279\, m_{\mathrm{sun}}$
37.77 a) 2494 MeV b) 2.526 c) 987.4 MeV

Chapter 38

38.1 a) $K_2 = 4K_1$ b) $E_2 = 2E_1$
38.3 5.77×10^{14} Hz, 1.28×10^{-27} kg·m/s, 3.84×10^{-19} J = 2.40 eV
38.5 a) 5.00×10^{14} Hz b) 1.13×10^{19} photons/s
c) no
38.7 a) 4.8 eV b) 6.1×10^{-34} J·s
d) f_{th}, ϕ, and the horizontal-axis intercept are different, the slope is the same
38.9 249 km/s
38.11 2.14 eV
38.13 a) 264 nm b) 4.70 eV
38.15 0.311 nm, same
38.17 1.13 keV
38.19 0.0714 nm, 180°
38.21 a) 4.39×10^{-4} nm b) 0.04294 nm
c) 300 eV, loss d) 300 eV
38.23 51.0°
38.25 a) 1.27×10^{-14} J b) 9.46×10^{-14} J
c) 2.10 pm, less
38.27 1.19×10^{-27} kg·m/s, 1.96×10^{-29} kg·m/s
38.29 a) 1.04 eV b) 1200 nm c) 2.50×10^{14} Hz
d) 4.14×10^{-7} eV

38.31 a) 4.56×10^{14} Hz b) 658 nm
c) 1.89 eV d) 6.58×10^{-34} J·s
38.33 a) 5.07 mJ b) 11.3 W
c) 1.49×10^{16} photons/s
38.35 a) 6.99×10^{-24} kg·m/s b) 705 eV
38.37 6.28×10^{-24} kg·m/s, 59.4°
38.39 a) 5×10^{-33} m b) 4×10^{-9} deg
c) 0.1 mm
38.41 a) 319 eV, 1.06×10^7 m/s b) 3.89 nm
38.43 a) 4.85 pm b) 0.256 MeV

Chapter 39

39.1 a) 0.155 nm b) 8.46×10^{-14} m
39.3 a) 2.37×10^{-24} kg·m/s
b) 3.08×10^{-18} J = 19.3 eV
39.5 a) 0.332 nm, equals the circumference of the orbit b) 1.33 nm, $^1/_4$ the circumference of the orbit
39.7 a) 8.8×10^{-36} m b) no
39.9 a) 62.0 nm (photon), 0.274 nm (electron)
b) 4.96 eV (photon), 2.41×10^{-5} eV (electron) c) ≈ 250 nm, electron
39.11 3.90×10^{-34} m, no
39.13 a) 0.0607 V b) 248 eV c) 20.5 μm
39.15 a) 7.3×10^6 m/s b) 150 eV
c) 12 keV d) electron
39.17 0.432 eV
39.19 a) 2.07°, 4.14° b) 1.81 cm
39.21 a) 8260 b) electron
39.23 a) 3.63 MeV b) 3.63 MeV
c) 1.32×10^7 m/s
39.25 3.16×10^{-34} kg·m^2/s
39.27 a) -218 eV, 16 times b) 218 eV, 16 times
c) 7.60 nm d) $^1/_4$ hydrogen radius
39.29 a) 2.18×10^6 m/s, 1.09×10^6 m/s, 7.27×10^5 m/s
b) 1.53×10^{-16} s, 1.22×10^{-15} s, 4.13×10^{-15} s c) 8.2×10^6
39.31 a) -17.50 eV, -4.38 eV, -1.95 eV, -1.10 eV, -0.71 eV b) 378 nm
39.33 a) -5.08 eV b) -5.64 eV
39.35 5.32×10^{21} photons/s
39.37 4.00×10^{17} photons/s
39.39 a) 1.2×10^{-33} b) 3.5×10^{-17}
c) 5.9×10^{-9}
39.41 a) 2060 K b) 1410 nm
39.43 1.06 mm, microwave
39.45 a) $1.7T$ b) 0.58
39.47 a) 97 nm, no b) 8.2×10^9 m, $12R_{\mathrm{sun}}$ c) no
39.49 a) 4.40×10^{-32} m/s b) no
39.51 not valid
39.53 6.34×10^{-14} eV
39.55 a) 1.69×10^{-28} kg b) -2.53 keV
c) 0.655 nm
39.57 a) 12.1 eV b) 3; 103 nm, 122 nm, 657 nm
39.59 a) 0.90 eV
39.61 a) 5×10^{49} photons/s b) 30,000
39.63 29,800 K
39.65 a) $h/2mc$
b) 6.61×10^{-16} m, independent of n
39.67 a) $I(f) = \dfrac{2\pi h f^5}{c^3(e^{hf/kT} - 1)}$
c) $\sigma = 5.67 \times 10^{-8}$ W/m^2·K^4
39.69 a) 12 eV b) 0.15 mV, 7300 m/s
c) 8.2×10^{-8} V, 4.0 m/s
39.71 a) no b) 2.52 V
39.73 a) $E = c\sqrt{2mK}$ b) photon
39.75 1.66×10^{-17} m, no
39.77 b) $\Delta = \dfrac{m^2 c^2 \lambda^2}{2h^2}$ c) $\Delta = 8.50 \times 10^{-8}$
39.79 a) $\dfrac{h}{mc\sqrt{15}}$ b) (i) 1.53 MeV, 6.26×10^{-13} m

(ii) 2810 MeV, 3.41×10^{-16} m
39.81 a) 1.1×10^{-20} kg·m/s b) 19 MeV
c) $U_{\mathrm{coul}} = -0.29$ MeV, no
39.83 7.0×10^{-36} kg, 2.9×10^{-8}
39.85 a) 1.1×10^{-35} m/s b) 2.3×10^{27} y, no
39.87 a) no b) 1.51 V c) 1.51 eV, about $^1/_4$ the potential energy of the NaCl crystal
d) 1240 eV, yes
39.89 a) $2d\sin\theta = m\lambda$ b) 53.3° c) smaller
39.91 a) 248 eV b) 0.0603 eV
39.93 a) $F = -\dfrac{A|x|}{x}, x \neq 0$ b) $E = \dfrac{3}{2}\left(\dfrac{h^2 A^2}{m}\right)^{1/3}$

Chapter 40

40.1 $\Psi(x, t) = Ae^{-i[4.27 \times 10^{10}\ \mathrm{m}^{-1}]x}e^{-i[1.05 \times 10^{17}\ \mathrm{s}^{-1}]t}$
40.3 a) $8\pi/k$ b) $4\omega/k$, yes
40.5 a) $\lambda/4, 3\lambda/4, 5\lambda/4, \ldots$ b) 0, $\lambda/2$, λ, $3\lambda/2, \ldots$
40.7 no
40.11 a) 1.6×10^{-67} J b) 1.3×10^{-33} m/s, 1.0×10^{33} s c) 4.9×10^{-67} J d) no
40.13 1.66×10^{-10} m
40.15 0.61 nm
40.19 a) $0, L/2, L$ b) $L/4, 3L/4$ c) yes
40.23 a) 6.0×10^{-10} m (twice the width of the box), 1.1×10^{-24} kg·m/s b) 3.0×10^{-10} m (same as the width of the box), 2.2×10^{-24} kg·m/s
c) 2.0×10^{-10} m (2/3 the width of the box), 3.3×10^{-24} kg·m/s
40.25 b) yes
40.27 3.43×10^{-10} m
40.29 $-A\left(\dfrac{2mE}{\hbar^2}\right)\sin\left(\dfrac{\sqrt{2mE}}{\hbar}x\right) - B\left(\dfrac{2mE}{\hbar^2}\right)\cos\left(\dfrac{\sqrt{2mE}}{\hbar}x\right)$
40.31 22 fm
40.33 a) 4.4×10^{-8} b) 4.2×10^{-4}
40.35 $1/\sqrt{2}$
40.37 a) 0.0013 b) 10^{-143}
40.39 1.11×10^{-33} J = 6.93×10^{-15} eV, 2.22×10^{-33} J = 1.39×10^{-14} eV
40.41 a) 0.21 eV b) 5900 N/m
40.43 $(2n + 1)\dfrac{\hbar}{2}$, increases with n
40.45 a) 5.89×10^{-3} eV b) 106 μm c) 0.0118 eV
40.47 a) $|\Psi(x, t)|^2 = \dfrac{2}{L}\left[1 - \cos\left(\dfrac{4\pi^2\hbar t}{mL^2}\right)\right]$
b) $\dfrac{4\pi^2\hbar}{mL^2}$
40.49 a) $\psi(x) = \dfrac{\sin k_0 x}{k_0 x}$
b) $0, \pm \pi/k_0$; $w_x = 2\pi/k_0, L$ c) $2L$
d) h, which is greater than $\hbar/2$
40.51 $B = \left(\dfrac{k_1 - k_2}{k_1 + k_2}\right)A, C = \left(\dfrac{2k_2}{k_1 + k_2}\right)A$
40.53 a) 19.2 μm b) 11.5 μm
40.55 a) 0.818 b) 0.500 c) yes
40.57 a) $(2/L)\,dx$ b) 0 c) $(2/L)\,dx$
40.61 a) $A = C, B\sin kL + A\cos kL = De^{-\kappa L}$,
where $k = \dfrac{\sqrt{2mE}}{\hbar}$ b) $kB = \kappa C$,
$kB\cos kL - kA\sin kL = -\kappa De^{-\kappa L}$

40.65 6.63×10^{-34} J $= 4.14 \times 10^{-15}$ eV;
1.33×10^{-33} J $= 8.30 \times 10^{-15}$ eV

40.69 b) 134 eV

40.71 a) $E_n = \dfrac{(2n)^2 h^2}{8mL^2}$, $n = 1, 2, \ldots$

b) $E_n = \dfrac{(2n + 1)^2 h^2}{8mL^2}$, $n = 0, 1, 2, \ldots$

c) same

d) part (a)'s are odd, part (b)'s are even

40.73 a) $x = \pm \sqrt{2E/K'}$

c) underestimates

Chapter 41

41.1 a) 1   b) 3

41.3 3.51 nm

41.5 (2, 2, 1): $x = L/2$, $y = L/2$; (2, 1, 1): $x = L/2$;
(1, 1, 1): none

41.7 a) 0   b) 3.65×10^{-34} kg·m²/s
c) 3.16×10^{-34} kg·m²/s
d) 5.27×10^{-35} kg·m²/s   e) 1/6

41.9 4

41.11 $1.414\hbar$, $19.49\hbar$, $199.5\hbar$; as n increases, the
maximum L gets closer to $n\hbar$.

41.13 a) 18   b) $m_l = -4$, 153.4°
c) $m_l = +4$, 26.6°

41.17 a) 0.468 T   b) 3

41.19 a) 9   b) 3.47×10^{-5} eV
c) 2.78×10^{-4} eV

41.21 a) 2.5×10^{30} rad/s   b) 2.5×10^{13} m/s, not
valid since $v > c$

41.23 1.68×10^{-4} eV, $m_s = +\frac{1}{2}$

41.25 g

41.27 $n = 1, l = 0, m_l = 0, m_s = \pm\frac{1}{2}$: 2 states;
$n = 2, l = 0, m_l = 0, m_s = \pm\frac{1}{2}$: 2 states;
$n = 2, l = 1, m_l = \pm1, m_s = \pm\frac{1}{2}$: 6 states

41.29 a) $1s^2 2s^2$ b) $1s^2 2s^2 2p^6 3s^2$, magnesium
c) $1s^2 2s^2 2p^6 3s^2 3p^6 4s^2$, calcium

41.31 4.18 eV

41.33 a) $1s^2 2s^2 2p$ b) -30.6 eV   c) $1s^2 2s^2 2p^6 3s^2 3p$
d) -13.6 eV

41.35 a) -13.6 eV   b) -3.4 eV

41.37 a) 8.95×10^{17} Hz, 3.71 keV, 3.35×10^{-10} m
b) 1.68×10^{18} Hz, 6.96 keV, 1.79×10^{-10} m
c) 5.48×10^{18} Hz, 22.7 keV, 5.47×10^{-11} m

41.39 $3E_{1,1,1}$

41.41 a) $1/64 = 0.0156$ b) 7.50×10^{-4}
c) 2.06×10^{-3}

41.43 a) $4\pi A^2 r^2 e^{-2\alpha r^2} dr$ b) $1/\sqrt{2\alpha}$, no

41.45 a) $E = \hbar\left[(n_x + n_y + 1)\omega_1^2 + \left(n_z + \dfrac{1}{2}\right)\omega_2^2\right]$,
with n_x, n_y and n_z all nonnegative integers
b) $\hbar\left(\omega_1^2 + \dfrac{1}{2}\omega_2^2\right)$, $\hbar\left(\omega_1^2 + \dfrac{3}{2}\omega_2^2\right)$,   c) one

41.47 b) $n = 5$ shell

41.49 a) 0, $\sqrt{2}\hbar$, $\sqrt{6}\hbar$, $\sqrt{12}\hbar$, $\sqrt{20}\hbar$
b) 7470 nm, no

41.51 a) $1.51e$, 9.49 electrons
b) (i) 1.8   (ii) -2.75 eV

41.53 a) $2a$ b) 0.238

41.55 b) 0.176

41.57 b) $(\theta_L)_{\max} = \arccos(-\sqrt{1 - 1/n})$

41.59 $4a$, same

41.61 $2 \to 1, 1 \to 0, 0 \to -1, \dfrac{e\hbar B}{2m}$;
$1 \to 1, 0 \to 0, -1 \to -1, 0$;
$0 \to 1, -1 \to 0, -2 \to -1, -\dfrac{e\hbar B}{2m}$

41.63 3.00 T

41.65 a) $0.99999978 = 1 - 2.2 \times 10^{-7}$ b) 0.9978
c) 0.978

41.67 a) 4, 20   b) $1s^4 2s^4 2p^3$

41.69 a) 122 nm   b) 1.52 pm, increase

41.71 a) 0.188 nm, 0.250 nm
b) 0.0471 nm, 0.0624 nm

41.75 a) $2a/Z$ b) 0.238, independent of Z

Chapter 42

42.1 277 nm, ultraviolet

42.3 a) 6.1 K   b) 34,600 K   c) He₂ no, H₂ yes

42.5 40.8 μm

42.7 5.65×10^{-13} m

42.9 a) 0.0644 nm (carbon), 0.0484 nm (oxygen)
b) 1.45×10^{-46} kg·m², yes

42.11 a) 1.03×10^{12} rad/s
b) 66.3 m/s (carbon), 49.8 m/s (oxygen)
c) 6.10×10^{-12} s

42.13 a) 7.49×10^{-3} eV   b) 166 μm

42.15 b) $\dfrac{l\hbar}{2\pi I}$

42.17 2170 kg/m³

42.19 a) 1.12 eV

42.21 1.20×10^6

42.23 1.5×10^{22} states/eV

42.25 a) $0.0233R$ b) 0.767%
c) no, motion of the ions

42.27 31.2%

42.29 0.20 eV below the bottom of the conduction band

42.31 a) (i) 0.0204 mA   (ii) -0.0196 mA
(iii) 26.8 mA   (iv) -0.491 mA
b) good for V between ±1.0 mV, otherwise no

42.33 a) 5.56 mA   b) -5.18 mA, -3.77 mA

42.35 a) 977 N/m   b) 1.25×10^{14} Hz

42.37 a) 3.8×10^{-29} C·m   b) 1.3×10^{-19} C
c) 0.81   d) 0.058

42.39 a) 0.96 nm   b) 1.8 nm

42.41 a) 0.129 nm   b) 8, 7, 6, 5, 4   c) 485 μm
d) 118 μm, 134 μm, 156 μm, 188 μm, 234 μm

42.43 b) (i) 2.95   (ii) 4.73   (iii) 7.57   (iv) 0.838
(v) 5.69×10^{-9}

42.45 a) 1.146 cm, 2.291 cm
b) 1.171 cm, 2.341 cm; 0.025 cm $(2 \to 1)$,
0.050 cm $(1 \to 0)$

42.47 0.274 eV, much less

42.49 a) 4.24×10^{-47} kg·m²   b) (i) 4.30 μm
(ii) 4.28 μm   (iii) 4.40 μm

42.51 2.03 eV

42.53 a) 4.66×10^{28} atoms/m³   b) 4.7 eV

42.55 b) 3.81×10^{10} Pa $= 3.76 \times 10^5$ atm

42.57 a) 1.67×10^{33} m⁻³   b) yes
c) 6.66×10^{35} m⁻³   d) no

42.59 a) $\dfrac{-2p^2}{4\pi\epsilon_0 r^3}$, attractive   b) $\dfrac{+2p^2}{4\pi\epsilon_0 r^3}$, repulsive

Chapter 43

43.1 a) 14 proton, 14 neutrons
b) 37 protons, 48 neutrons
c) 81 protons, 124 neutrons

43.3 0.533 T

43.5 a) 76.21 MeV   b) 76.68 MeV, 0.6%

43.7 0.5575 pm

43.9 a) 1.32 MeV   b) 1.13×10^7 m/s

43.11 $^{86}_{36}$Kr: 8.73 MeV/nucleon,
$^{180}_{73}$Ta: 8.08 MeV/nucleon

43.13 a) $^{235}_{92}$U   b) $^{24}_{12}$Mg   c) $^{15}_{7}$N

43.15 156 keV

43.17 a) 0.836 MeV   b) 0.700 MeV

43.19 5.01×10^4 y

43.21 a) 4.92×10^{-18} s⁻¹   b) 2990 kg
c) 1.24×10^5 decays/s

43.23 a) 159 decays/min   b) 0.43 decays/min

43.25 a) 0.421 decays/s   b) 11.4 pCi

43.27 2.80 days

43.29 a) 2.02×10^{15}
b) 1.01×10^{15}, 3.78×10^{11} decays/s
c) 2.53×10^{14}, 9.45×10^{10} decays/s

43.31 a) 1.2 mJ   b) 10 mrad, 10 mrem, 7.5 mJ
c) 6.2

43.33 500 rad, 2000 rem, 5.0 J/kg

43.35 a) 1.75 kGy, 175 krem, 1.75 kSv, 385 J
b) 1.75 kGy, 2.625 kSv, 262.5 krem, 385 J

43.37 a) 9.32 rad, 9.32 rem

43.39 a) $Z = 3, A = 6$ b) -10.14 MeV
c) 11.59 MeV

43.41 a) $Z = 3, A = 7$ b) 7.152 MeV   c) 1.4 MeV

43.43 a) 173.3 MeV   b) 4.42×10^{23} MeV/g

43.45 1.586 MeV

43.47 324 MJ

43.53 a) 4.14 MeV   b) 7.75 MeV/nucleon

43.55 a) $^{90}_{39}$Y   b) 25%   c) 112 y

43.57 a) $^{25}_{13}$Al will decay into $^{25}_{12}$Mg   b) β^+ or
electron capture
c) 3.255 MeV, 4.277 MeV

43.59 a) $^{14}_{6}$C $\to$ e⁻ + $^{14}_{7}$N + $\bar{\nu}_e$ b) 0.156 MeV
c) 13.5 kg, 3400 decays/s
d) 530 MeV/s $= 8.5 \times 10^{-11}$ J/s
e) 36 μGy, 3.6 mrad, 36 μSv, 3.6 mrem

43.61 0.960 MeV

43.63 0.001286 u, yes

43.65 94.3 rad, 1900 rem

43.67 a) 5.0×10^4 b) $10^{-15,000}$

43.69 29.2%

43.71 a) 0.96 μJ/s   b) 0.48 mrad/s   c) 0.34 mrem
d) 6.9 days

43.73 1.0×10^4 y

43.75 a) 0.48 MeV
b) 3.270 MeV $= 5.239 \times 10^{-13}$ J
c) 3.155×10^{11} J/mol, more than a million
times larger

43.79 a) two   b) 0.400 h, 1.92 h
c) 1.04×10^7 (short-lived), 2.49×10^7
(long-lived)
d) 1800 (short-lived), 4.10×10^6 (long-lived)

Chapter 44

44.1 69 MeV, 1.7×10^{22} Hz, 18 fm, gamma ray

44.3 a) 32 MeV

44.5 9.26×10^6 m/s

44.7 7.2×10^{19} J, 70%

44.9 a) 1.18 T   b) 3.42 MeV, 1.81×10^7 m/s

44.11 a) 30.6 GeV   b) 8.0 GeV

44.13 a) 3200 GeV   b) 38.7 GeV

44.15 a) π^0, π^+ b) 219.1 MeV

44.17 1.63×10^{-25} kg, 97.2

44.19 116 MeV

44.21 (b) and (d)

44.23 (c) and (d)

44.27 a) 0, 1, -1, 0   b) 0, 0, 0, 1
c) $-e$, 1, 0, 0   d) $-e$, 0, 0, -1

44.29 p $\to$ e⁺ + n + ν_e

44.31 a) $\bar{u}\,\overline{dd}$ b) no   c) yes

44.33 a) 3.28×10^7 m/s   b) 1510 Mly

44.35 a) 1.1×10^5 km/s   b) 1.5

44.37 a) 3.8 atoms/m³   b) 320   c) 2.0×10^{27}

44.39 a) 5.494 MeV   b) 20.58 MeV

44.41 -0.783 MeV, endoergic

44.43 966 nm

44.45 a) 14.0 TeV   b) 1.0×10^5 TeV

44.47 a) $\pi^- \to \mu^-$ + neutrino $\to$ e⁻ + 3 neutrinos,
an electron and neutrinos   b) 139 MeV
c) 2.24×10^{10} d) 50 Sv, 5.0 krem

44.49 2.494 GeV

44.51 a) 0, $+e$, 1, all lepton numbers are 0, K⁺
b) 0, $-e$, 0, all lepton numbers are 0, π^-
c) -1, 0, 0, all lepton numbers are 0,
antineutron (n̄)
d) 0, $+e$, 0, muonic lepton number is -1, all
other lepton numbers are 0, μ^+

44.53 7.5×10^{-23} s

44.55 a) 0.70 rad   b) 0.70 rem, no

44.57 c) $H_0 = \dfrac{dR/dt}{R}$ d) $R(t) = R_0 e^{H_0 t}$ e) no

44.59 b) $R/R_0 = 0.574$
c) speeding up at 300 My, slowing down at
10.2 Gy

44.61 230 MeV, 12.5° below the $+x$-axis

PHOTO CREDITS

INDEX

For users of the three-volume edition: pages 1–686 are in Volume 1; pages 687–1260 are in Volume 2; and pages 1223–1522 are in Volume 3. Pages 1261–1522 are not in the Standard Edition.

Note: Page numbers followed by f indicate figures; those followed by t indicate tables.

A

Abdus, Salam, 1500
Absolute conservation laws, 1495
Absolute pressure, 377–378
Absolute temperature, 517
Absolute temperature scale, 556, 668–669
Absolute zero, 556, 669
Absorption lines, 1203
Absorption spectra
  line, 1293, 1297–1300, 1310–1314
  X-ray, 1396
AC source, 1022. *See also* Alternating-current circuits
Acceleration
  angular, 282–285, 284t, 311–314
  around curve, 74, 88
  average, 42–43. *See also* Average acceleration
  calculating by integration, 55–57
  centripetal, 86–87, 154
  centripetal component of, 286–287
  changing, 55–57
  circular motion and, 85–87
  constant, 46–52. *See also* Constant acceleration
  definition of, 43
  fluid resistance and, 152–154
  inertial frame of reference and, 110–112
  instantaneous, 43–44. *See also* Instantaneous acceleration
  linear, 282, 284t
  mass and, 113, 114, 118–120
  net force and, 112–118
  Newton's first law and, 108–112
  Newton's second law and, 112–117, 140–146
  of particle in wave, 480–482
  projectile motion and, 77–80, 87
  of rocket, 262–264
  of rolling sphere, 319–320
  signs for, 45, 46
  in simple harmonic motion, 444, 448
  tangential component of, 286
  units of, 117
  vs. velocity, 42
  on v_x-t graph, 44–46
  weight and, 118–119
  of yo-yo, 319
Acceleration due to gravity, 700
  apparent weight and, 143, 422
  definition of, 52
  at different latitudes and elevations, 422t
  in free fall, 52–55, 118, 143
  magnitude of, 405
  mass vs. weight and, 118–120
  variation with location and, 119
  vs. gravitation, 403
  weightlessness and, 143
Acceleration vectors, 35, 72–77, 283
  average, 73–75
  instantaneous, 73–75
  parallel and perpendicular components of, 75–77
Accelerators, 1485–1488. *See also* Particle accelerators
Acceptor level, 1425
Acceptors, 1425
Accretion disk, 425
Accuracy, 8
  vs. precision, 9
Acrobats, in unstable equilibrium, 228
Action-reaction pairs, 120–123
  gravitational forces as, 403
Activation energy, 610
Activity, in radioactive decay, 1456
Addition

significant figures in, 9
of vectors, 12–18
Adiabatic process, 634–635
  Carnot cycle and, 663
  for ideal gas, 640–642
Aging, relativity and, 1233
Air
  dielectric strength of, 768, 805
  as insulator, 571
  ionization of, 768–771
Air conditioners, 660–661
Air drag, 152–154
Air pressure, 375–376
Air resistance, projectile motion and, 77, 79–80
Airplanes
  banked curves and, 158
  noise control for, 531, 532
  sonic boom from, 538
  wing lift and, 388–389
  wing resonance in, 460
Airy disk, 1209
Airy, George, 1209
Alkali metals, 1390
  Bohr atomic model for, 1306
Alkaline earth elements, 1390
Alkaline earth metals, 1390
Alpha decay, 1450–1452
Alpha particles, 1294–1295
  emission of, 1450–1451
  tunneling and, 1349–1350
Alternating current, 822, 850, 1021
  applications of, 868
  dangers of, 1040
  lagging, 1036
  measurement of, 1022–1024
  rectified, 1022–1023
  rectified average, 1023
  root-mean-square value of, 1023–1024
Alternating-current circuits
  capacitors in, 1027–1028, 1029–1030, 1029t
  complex numbers in, 1049–1050
  impedance of, 1031–1033
  inductors in, 1025–1027, 1029, 1029t
  L-R-C series, 1030–1034
  phase angle and, 1026, 1031–1032
  phasors and, 1022
  power in, 1034–1037
  resistance and reactance in, 1024–1030
  resistors in, 1025, 1029, 1029t
  resonance in, 1037–1039
  tailoring, 1038–1039
  transformers and, 1040–1042
Alternators, 963–964, 1021–1022
Ammeters, 831, 860–861
  voltmeters and, 862–863, 1024
Amorphous solids, 1412
Ampere, 695, 820, 931–932
Ampère, André, 885
Ampere's law, 935–941. *See also* Maxwell's equations
  applications of, 938–941
  displacement current and, 975–977
  electromagnetic waves and, 1052, 1057, 1063
  general statement of, 937–938
  generalization of, 975–976
Amplitude
  displacement, 510, 518–519
  of electromagnetic waves, 1061
  of oscillation, 438
  of pendulum, 454–455
  pressure, 511–512, 519–521
  of sound waves, 510–513
Analyzers, 1095–1096
Anderson, Carl D., 1482, 1485
Angle(s)

notation for, 71
polarizing, 1097
radians and, 279, 287
Angle of deviation, 1111
Angle of incidence, critical, 1089
Angle of reflection, 1084
Angular acceleration, 282–285
  angular velocity and, 282
  calculation of, 283
  constant, 283–285, 284t
  torque and, 311–314
  as vector, 283
  vs. linear acceleration, 284t
Angular displacement, 279
  torque and, 320–322
Angular frequency, 438–439
  of electromagnetic waves, 1061
  natural, 459–460
  of particle waves, 1331
  period and, 438–439
  in simple harmonic motion, 441–442
  vs. frequency, 442
Angular magnification, vs. lateral magnification, 1147
  of microscope, 1147, 1148–1149
Angular momentum
  axis of symmetry and, 323–324
  of the body, 324
  conservation of, 325–328
  definition of, 322
  of electrons, 942
  of gyroscope, 328–330
  nuclear, 1442
  orbital, 1373–1374, 1384, 1387
  precession and, 328–330
  rate of change of, 323, 324
  rotation and, 322–328
  spin, 1384–1385, 1387, 1442
  torque and, 323, 324
  total, 1387, 1442
  as vector, 324, 328
Angular simple harmonic motion, 451
Angular size, 1146
Angular speed, 280
  instantaneous, 286
  precession, 329
  rate of change of, 286
Angular velocity, 279–282
  angular acceleration and, 282
  average, 279
  calculation of, 281
  instantaneous, 280
  rate of change of, 286
  as vector, 281–282
  vs. linear velocity, 280
Angular vs. linear kinematics, 285–288
Anomalous magnetic moment, 1443
Antimatter, 1516
Antineutrinos, 1492
Antineutrons, 1492
Antinodal curves, 1166
Antinodal planes, 1070
Antinodes, 492
  displacement, 523
  pressure, 523
Antiparallel vectors, 11, 12
Antiparticles, 1483
Antiprotons, 1491–1492
Antiquarks, 1496, 1499
Aphelion, 415
Apparent weight, 142
  acceleration due to gravity and, 143, 422
  Earth's rotation and, 421–423
  magnitude of, 422
Appliances, power distribution systems in, 868–872

Archimedes' principle, 380
Arecibo telescope, 1218
Aristotle, 52
Astigmatism, 1144
Aston, Francis, 897
Astronomical distances, units for, 1503
Astronomical telescopes, 1119, 1149–1151
Atmosphere, 355, 375
Atmospheric pressure, 355, 375–376
    elevation and, 594–595
    measurement of, 378–380
Atom(s)
    energy levels of. *See* Energy levels
    excited, 1298
    hydrogen. *See* Hydrogen atom
    interactions between, 451–453
    in magnetic field, 1379–1382
    many-electron, 1387–1393
    muonic, 1379
    nucleus of, 689, 1295
    Rydberg, 1306, 1402
    structure of. *See* Atomic structure
    Thomson's model of, 751–752
Atomic mass, 598, 690, 1305–1306
    measurement of, 897, 1441
Atomic models
    Bohr's, 1297–1306
    Rutherford's, 1294–1296
    Thomson's, 1293, 1294–1295
Atomic number, 690, 1306, 1379, 1387–1388, 1441
Atomic particles, 689–690. *See also* Electron(s);
        Neutron(s); Proton(s)
Atomic spectra, 1292, 1297–1300
    in Balmer series, 1304
    in Brackett series, 1304
    in Lyman series, 1304
    in Pfund series, 1304
Atomic structure, 689, 1364–1398
    central-field approximation and, 1388
    electron spin and, 1383–1387
    exclusion principle and, 1388–1393
    of hydrogen atom, 1372–1378
    of hydrogenlike atoms, 1378–1379
    of many-electron atoms, 1387–1393
    Moseley's law and, 1394–1396
    particle in three-dimensional box and, 1366–1371
    periodic table and, 1389, 1390–1393
    Schrödinger equation and, 1365–1366
    X-ray spectra and, 1393–1396
    Zeeman effect and, 1379–1382
Attenuator chains, 881–882
Audible range, 509
Automobiles
    gas compression in, 593–595
    ignition systems in, 1000
    Newton's second law and, 115
    power distribution systems in, 868, 870–871
    vertical simple harmonic motion in, 451
    weight distribution for, 349
Autoradiography, 1461
Available energy, 1487–1488
Avalanche breakdown, 1428
Average acceleration, 42–43
    definition of, 42
    units of, 42
    vs. average velocity, 42
    on v_x-t graph, 44–46
    x-component of, 42
Average acceleration vectors, 73–75
Average angular acceleration, 282
Average angular velocity, 279
Average density, 374
Average power, 193, 487
Average speed, 39
Average velocity, 36–40
    definition of, 36
    instantaneous velocity and, 38–41
    straight-line motion and, 36–38
    x-component of, 36–38
    on x-t graph, 37–38

Average velocity vectors, 70–72
Avogadro's number, 598
Axis
    elliptical, 415
    optic, 1118
    polarizing, 1094–1095
    semi-major, 415, 416
Axis of rotation, 281–282, 286–287
    change in direction of, 281–282, 286–287
    fixed, 278–279
    moment of inertia for, 312
    moving, 324–320
    parallel-axis theorem and, 293–294
    through center of mass, 294
Axis of symmetry, angular momentum and, 323–324
Axis system, right-handed, 24
Axons, 716, 781, 881–882
a_x-t graphs
    for changing acceleration, 55–57
    for constant acceleration, 46–49

B

Back emf, 908
Background radiation, 1515
Back-of-the-envelope calculations, 10
Bacteria, rotation in, 283f
Bainbridge's mass spectrometer, 897
Balance
    Cavendish (torsion), 404
    spring, 106
Ballistic pendulums, 253
Balmer series, 1304
Band spectra, 1411
Banked curves, 158
Bar, 375
Bar magnets, 883, 905–907
Bardeen, John, 1430
Barometers, mercury, 378–380
Baryons, 1492, 1493–1494, 1496
Baseball, curve ball in, 391
Batteries
    charging, 835–836
    as current source, 831
    energy in, 834
Beams, 1083
Beats, 531–532
Becker, Herbert, 1481
Becquerel, 1457
Bednorz, Johannes, 824
Bees, vision in, 1101
Bell, Alexander Graham, 521
Bernoulli's equation, 385–389
Beryllium, 1390
Beta decay, 1452–1453
Beta-minus particles, 1452–1453
Beta-plus particles, 1453
Bias conditions, 1426, 1427–1428
Big Bang, 1503–1504
Big Crunch, 1505
Bimetallic strip thermometer, 553
Binary star systems, 425–426, 1259
Binding energy, 1406, 1445–1446
Binoculars, 1150
Binomial theorem, 451, 452–453
Biological efficiency, 655
Biologically equivalent dose, 1460
Biot-Savart law, 927
Bipolar junction transistors, 1429
Bird song, 522
Bird vision, 1144
Bird wings
    flapping frequencies of, 438f
    moment of inertia of, 290f
Birefringence, 1100
Black holes, 423–426
Blackbody, 576
Blackbody radiation, 1310–1314
Blackett, Patrick, 1271
Bloch, Felix, 1416f
Blood flow, turbulence in, 390f

Blubber, as insulator, 572
Blue-ray discs, 1210, 1309
Body. *See* Human body
Bohr magneton, 941–942, 1380, 1442–1443
Bohr, Niels, 1273, 1389f
Bohr's atomic model, 1297–1306
    energy levels and, 1297–1300
    for hydrogen, 1300–1305, 1372
    limitations of, 1372
    photon emission and absorption and, 1297
    uncertainty principle and, 1317
    vs. Schrödinger analysis, 1373
Boiling, 566
Boltzmann constant, 601
Bonds, 1405–1408
    covalent, 1406–1407
    hydrogen, 1407–1408
    ionic, 1406, 1407
    metallic, 1408
    in solids, 1414–1415
    strong, 1407
    van der Waals, 1407
    weak, 1407
Bone cancer, radioisotope imaging for, 1391
Born, Max, 1333f
Bose-Einstein distribution, 1492
Bosons, 1492, 1500, 1501
Bothe, Walther, 1481
Bottomness, 1499
Bound charges, 806
Bound state, 1343–1344
Boundary conditions
    for harmonic oscillator, 1351–1351
    for waves, 489–490
Bourdon pressure gauge, 379f
Brackett series, 1304
Bragg condition, 1207
Bragg reflection, 1207
Brahe, Tycho, 414–415
Breaking stress, 358
Bremsstrahlung, 1267, 1268
Brewster's law, 1097–1098
Bridge circuits, 855–860, 880–881
Bright fringes, 1194, 1195
Brillouin, Léon, 1361
British system, 5–6, 117. *See also* Units of measure
British thermal unit (Btu), 562
Brittle material, 358
Bulk modulus, 355
Bulk strain, 354–356
Bulk stress, 352f, 354–356
Buoyancy, 380–382
Butterfly wings, interference and, 1179

C

Cables, winding/unwinding, 291–292, 313–314
Calcite, birefringence in, 1100
Calculations
    back-of-the-envelope, 10
    calorimetry, 568–570
    estimation and, 10
    units of measure in, 6. *See also* Units of measure
Calorie (cal), 562
Calorimetry calculations, 568–570
Calorimetry, phase changes and, 565–570
Cameras, 1139–1142
    flash unit of, 797
    focusing, 1137
    gamma, 1461
    resolving power and, 1210
Cancer
    imaging methods in, 1180, 1391
    magnetic nanoparticles for, 946
    radiation effects in, 1269
Capacitance
    calculation of, 789–793
    definition of, 788, 789
    equivalence, 794
    units for, 789
    vs. coulombs, 789

Capacitive reactance, 1028–1029
Capacitors, 788–800
  in ac circuits, 1027–1028, 1029–1030, 1029t,
    1035–1036
  applications of, 788, 797–798, 802–803
  capacitance of, 789. See also Capacitance
  capacitive reactance of, 1028–1029
  charge storage in, 797
  charging, 864–866, 975–976
  cylindrical, 792–793
  definition of, 788
  dielectrics in, 800–805
  discharging, 867–868
  electric-field energy and, 798, 799
  electrolytic double-layer, 802
  energy storage in, 788, 796–800
  in networks, 796
  in pacemakers, 866
  in parallel, 794–796, 852–853
  parallel-plate, 790, 791, 800–805
  in series, 793–794, 795–796, 852
  spherical, 792, 808
  symbols for, 789
  in touch screens, 794
  in vacuum, 789–791, 798
Carbon dating, 1458
Carbon dioxide, greenhouse effect and, 576–577
Carnot cycle, 663–669
  efficiency of, 667–668
  of heat engine, 663–666
  for ideal gas, 664–665
  Kelvin temperature scale and, 668–669
  of refrigerator, 666–667
  reversibility of, 667–668
  second law of thermodynamics and, 667–668
Cars. See Automobiles
Cathode ray tubes, for magnetic field measurement,
  887–888
Cavendish balance, 404
Celestial dynamics, 402
Cell imaging, 1180
Cell membrane
  dielectric, 805
  potential gradient across, 774
Celsius temperature scale, 553
Center of curvature, 1118
Center of gravity, 345–348
Center of mass, 258–262
  center of gravity and, 345–347
  combined rotational-translational motion and,
    315–316
  external forces and, 261–262
  motion of, 259–262
  planetary motion and, 417–418
  torque and, 312
Center-of-momentum system, 1272, 1487
Centigrade scale, 553
Central force, 416
Central-field approximation, 1388
Centrifugal force, 155
Centripetal acceleration, 86–87, 154
Centripetal component of acceleration, 286–287
Čerenkov radiation, 1257
Cgs metric system, 117
Chadwick, James, 1481
Chain reactions, 1466
Charge distribution, 703, 1390
  electric fields and, 725–728, 734–735, 746t. See also
    Gauss's law
  static, 759–760
Charged particles, motion in magnetic fields, 892–898
Charging by induction, 692
  polarization and, 693
Charm, 1499
Chemical reactions. See Reaction(s)
Chernobyl accident, 1468
Chokes, 995–998
Chromatic resolving power, 1203–1204, 1210
Circle
  circumference of, 279

reference, 440–441
Circuit breakers, 870
Circuit diagrams, 831–833
Circuits. See Electric circuits
Circular apertures, diffraction and, 1208–1211
Circular motion, 85–88, 440–442
  acceleration and, 85–87, 88
  dynamics of, 154–159
  nonuniform, 88, 159
  period of, 87
  uniform. See Uniform circular motion
  velocity and, 85–87
  vs. projectile motion, 87
Circular orbits, 412–413, 416–417
Circular polarization, 1099–1100
Circumference, 279
Classical mechanics, 104
Classical turning point, 1361–1362
Clausius statement, 662
Climate change, 576–577
Close packing, 1415
Closed orbits, 412
Closed surface, electric flux through, 725
Clotheslines, waves on, 480
Cloud chambers, 1482
Coefficient of kinetic friction, 147
Coefficient of linear expansion, 557–558, 559
Coefficient of performance, 659
Coefficient of resistivity, 824–825
Coefficient of static friction, 148
Coefficient of volume expansion, 558–560
Coherent waves, 1165
Coils
  Hemholtz, 954
  inductance of. See Inductance
  magnetic fields of, 932–935
  magnetic torque on, 904–905
  search, 983
  Tesla, 993–994
Cold fusion, 1471
Cold reservoirs, 655
Colliding-beam experiments, 1489
Collisions, 251–258
  atomic energy levels and, 1297–1298
  classification of, 254–255
  definition of, 251
  elastic, 251, 254, 255–258
  inelastic, 251–255
  kinetic energy in, 252
  momentum conservation and, 251–258
Combustion, 568
Comet Halley, orbit of, 417
Common-emitter circuits, 1429
Commutators, 907–908, 964
Compensated semiconductors, 1425
Complementarity principle, 1273
Complete circuits, 822, 828–831
Completely inelastic collisions, 251–255
Component vectors, 14–18, 106
Components, of vectors, 14–19, 21–22, 106–107
  vs. component vectors, 14, 106
Compound microscope, 1147–1149
Compressibility, fluid, 356, 382
Compression
  definition of, 354
  fluid density and, 476
Compression ratio, 657
Compressive strain, 354
Compressive stress, 354
Compton scattering, 1269–1271
Computed tomography, 1268–1269
Concave mirrors, 1118–1122
Concentration of particles, in current, 820
Condensation, 566
Condensed matter, 1412. See also Liquid(s); Solids
Condensor microphones, 790
Conditional conservation laws, 1495
Conduction, 570–574
Conduction bands, 1416–1417
Conduction current, 975

Conductivity
  electrical, 823
  intrinsic, 1424
  microscopic model of, 838–840
  thermal, 571, 823
Conductors, 691–692
  in capacitors, 789
  conductivity of, 823
  current density in, 821–822
  current flow in, 820–822
  diodes in, 827
  electric charge on, 736, 741–745
  electric fields at, 701, 744–745
  electron motion in, 819
  energy bands in, 1417
  equipotential surfaces and, 772–773
  holes in, 909
  interaction force between, 931–932
  magnetic fields of, 928–932
  magnetic force of, 931–932
  magnetic forces on, 898–901
  metallic, 838–839
  nonohmic (nonlinear), 823
  ohmic (linear), 823
  particle concentration in, 820–821
  resistance of, 825–828, 830, 833
  resistivity of, 822–825
  semiconductors, 823, 827, 909
  superconductors, 824, 968
  thermal, 552–553
Conservation laws
  absolute, 1495
  conditional, 1495
  universal, 690
Conservation of angular momentum, 325–328
  of planets, 415–416
Conservation of baryon number, 1494
Conservation of electric charge, 690
  Kirchoff's junction rule and, 856
Conservation of electrostatic force, 856
Conservation of energy, 176, 209, 224
  with electric force, 758–759
  in simple harmonic motion, 446–449
Conservation of lepton number, 1492
Conservation of mass and energy, 1247–1248
Conservation of mass in fluid, 383–384
Conservation of mechanical energy, 755
Conservation of momentum, 247, 1243
  collisions and, 251–258
Conservative forces, 221–229
  elastic collisions and, 255
  work done by, 755
Consonance, 532
Constant acceleration, 46–52
  due to gravity, 52–55
  equations of motion for, 49
  of freely falling bodies, 52–55
  Newton's second law and, 112–117
Constant angular acceleration, 284t
  rotation with, 283–285
Constant forces, 177
Constant linear acceleration, 284t
Constant torque, 321
Constant velocity, 51
  Newton's first law and, 108–112
Constant-pressure process, 635
Constant-temperature process, 635
Constant-volume process, 635
Constructive interference, 492, 530, 1164–1166,
  1168–1170
  in holography, 1211–1213
  in X-ray diffraction, 1206–1207
Contact force, 105, 146
Contact lenses, 1143–1145
Continuity equation, 383–384
Continuous lasers, 1309
Continuous spectra, 1310–1314
Convection, 570, 574
Conventional current, 820
Converging lenses, 1131–1133

Converging mirrors, 1119
Convex mirrors, 1122–1124
Cooling, evaporative, 568
Cooper pairs, 1430
Coordinate system, right-handed, 24
Coordinates
    spacetime, 1238
    spherical, 1366
Copernicus, Nicolaus, 414
Cornea, 1142, 1156, 1157, 1159
Corona discharge, 768–769
Correspondence principle, 1249
Cosmic background radiation, 1515
Cosmic inflation, 1511
Cosmic-ray experiments, 1489–1490
Cosmological principle, 1503
Cosmological redshift, 1505
Coulomb, 695–696
    vs. capacitance, 789
Coulomb's law, 597, 693–698
    Gauss's law and, 732
    proportionality constant in, 695
    statement of, 694
    superposition of forces and, 696
Coupling constant, 1491
Covalent bonds, 1406–1407
Covalent crystals, 1414–1415
Cowan, Clyde, 145
Critical angle, 1089
Critical damping, 458
Critical density, 1505–1507
Critical fields, 979
Critical point, 611
Critical temperature, 596, 979
Critically damped circuits, 1010
Cross (vector) product, 23–25
Crystal(s)
    covalent, 1414–1415
    ideal single, 1413
    imperfect, 1415
    ionic, 1414–1415
    liquid, 1412
    metallic, 1415
    perfect, 1413–1415
    structure of, 1412–1415
    types of, 1414–1415
Crystal lattice, 597, 1413–1414
Crystalline lens, 1142–1143
Crystalline solids, 1412–1415
Cube, electric flux through, 731
Curie, 1457
Curie constant, 944
Curie, Marie, 1454
Curie, Pierre, 944, 1454
Curie's law, 944
Current
    alternating, 822
    capacitor, 1027–1028, 1029, 1029t
    in circuits, 828–831
    concentration of particles in, 820
    conduction, 975
    conventional, 820
    definition of, 818, 819
    direct, 822, 850. See also Direct-current circuits
    direction of, 819–820, 825
    displacement, 975–977
    drift velocity and, 819, 820
    eddy, 974–975, 1042
    electric charge in, 819–820
    electric field and, 819–820
    electromotive force and, 991. See also Inductance
    electron motion in, 819
    full-wave rectifier, 1023
    generation, 1427
    heat, 571
    induced, 958, 967–968
    inductance and, 991. See also Inductance
    inductor, 1025–1026, 1029, 1029t
    Kirchoff's rules for, 855–860
    lagging, 1036

    measurement of, 860–861
    notation for, 865
    Ohm's law and, 822, 825–826
    recombination, 1427
    rectified average, 1023
    resistance and, 825–828
    resistor, 1025, 1029, 1029t
    root-mean-square, 1023–1024
    saturation, 1426
    sinusoidal, 1022–1024. See also Alternating
        current
    time-varying, 865
    units of, 695, 820
    "using up," 830
    voltage and, 825–828
    vs. current density, 821–822
Current amplitude, 1022
Current density, 821–822
    definition of, 821
    resistivity and, 823
    vector, 821
    vs. current, 821–822
Current flow, direction of, 819–820
Current loops. See also Magnetic dipoles
    force and torque on, 901–907
    magnetic fields of, 932–935
    magnetic moment of, 903, 934
    in magnetization, 941–942
Current-carrying conductor, magnetic forces on,
        898–901
Curve ball, 391
Curves
    acceleration around, 74, 88
    antinodal, 1166
    banked, 158
    gravitational potential energy and, 212–216
    magnetization, 945
    motion along, 191–193
    nodal, 1166
    resonance, 528, 1038–1039
    response, 1038–1039
    work-energy theorem for, 187–191
Cycles, 438
Cyclic process, in heat engines, 654
Cyclotron frequency, 893
Cyclotrons, 893, 918, 1486–1487
Cylinders, moment of inertia of, 295–296
Cystic fibrosis, sweat chloride test for, 695

D

Dalton, John, 1480–1481
Damped harmonic motion, 1009
Damped oscillations, 457–460
Damping, 457
    critical, 458
Dark energy, 1508
Dark fringes, 1193–1195
Dark matter, 1507
D'Arsonval galvanometer, 860, 863, 904, 1022
Daughter nucleus, 1454
Davisson, Clinton, 1287–1288
Davisson-Germer experiment, 1287–1288
DC circuits. See Direct-current circuits
De Broglie, Louis, 1286–1287
De Broglie wavelength, 1287, 1290
Decay. See Radioactive decay
Decay constant, 1456
Deceleration, 45
Decibel scale, 521
Deformation
    Hooke's law and, 352, 357–358
    plastic, 358
    reversible, 358
    stress and strain and, 352–357
Degeneracy, 1370–1371, 1374–1375
Degrees, 553, 555
Degrees of freedom, 606
Density, 373–375, 374t
    average, 374
    current, 821–822

    definition of, 373
    displacement current, 976
    of Earth, 408
    energy, 798, 1064–1065
    fluid, 380–381, 476
    linear charge, 704
    linear mass, 482
    magnetic energy, 999–1000
    magnetic flux, 892
    mass:volume ratio and, 373–374
    measurement of, 374
    nuclear, 1440
    probability, 1333
    of states, 1418–1419
    surface charge, 704
    volume charge, 704
    vs. pressure, 592
Depth, fluid pressure and, 376–377
Derivatives, 39
    partial, 226–227, 481
Destructive interference, 492, 529, 1164–1166,
        1168–1170
Detectors, 1489–1490
Deuterium, 1445
Deuterons, 1305–1306, 1445, 1511
Dewar flask, 576
Dewar, James, 576
Diamond structure, 1413–1414
Diamonds, 1090, 1092
Diatomic molecules, 605–606
Dichromism, 1094
Dielectric breakdown, 800, 804–805, 1417
Dielectric cell membrane, 805
Dielectric constant, 800–801
Dielectric function, 1064
Dielectric strength, 805
    of air, 768, 805
Dielectrics, 800–805
    electromagnetic waves in, 1063–1064
    Gauss's law and, 807–808
    permittivity of, 802
    polarization of, 801–803, 805–807
Diesel engines, 658–659
    adiabatic compression in, 642
Difference tone, 532
Differential principle, 245
Diffraction, 1190–1214
    bright fringes in, 1194, 1195
    with circular apertures, 1208–1211
    complementarity principle and, 1273–1274
    dark fringes in, 1193–1195, 1194
    definition of, 1191
    electron, 1287–1288
    Fraunhofer, 1192, 1193
    Fresnel, 1192, 1193
    holography and, 1211–1213
    Huygen's principle and, 1191–1192
    image formation and, 1209–1210
    intensity in, 1195–1199
    multiple-slit, 1199–1201
    photons and, 1273–1274
    resolving power of, 1209–1211
    single-slit, 1192–1199
    of sound, 1189
    vs. interference, 1192, 1200
    X-ray, 1205–1208
Diffraction gratings, 1201–1205
Diffuse reflection, 1083, 1115
Digital multimeters, 863
Dimagnetism, 943t, 944, 980
Dimensional consistency, 6
Dinosaurs, physical pendulum and, 456–457
Diode(s), 827
    light-emitting, 1428
    p-n junction, 1426
    tunnel, 1349
    Zener, 1428
Diode rectifier, 1426
Diopters, 1144–1145
Dipoles

electric, 709–713
magnetic, 903–904
Dirac distribution, 1419
Dirac equation, 1482–1483
Dirac, Paul, 1385, 1482–1483
Direct current, 822
dangers of, 1040
Direct-current circuits, 822, 850–873
in automobiles, 868
definition of, 850
Kirchoff's rules for, 855–860
measuring instruments for, 860–864
in power distribution systems, 868–872
R-C, 864–868
resistors in series and parallel in, 850–855
Direct-current generators, 964–965
Direct-current motors, 907–908
Direction, 10
of force, 10
of vectors, 11, 16
of waves, 479, 481–482
Discus throwing, 287
Disk, electric flux through, 730–731
Dislocations, in crystals, 1415
Dispersion, 1085, 1091–1093
Displacement, 11, 12
angular, 279
definition of, 11
multiplication of, 13
in oscillation, 438
in simple harmonic motion, 439–440, 443–445
straight-line motion and, 36–38
superposition principle and, 490–491
vector sum (resultant) of, 12
wave pulse and, 489–491
work and, 177–178, 181–183
Displacement amplitude, 510
sound intensity and, 518–519
Displacement current, 975–977
Displacement nodes/antinodes, 523
Dissipative forces, 222
Dissonance, 532
Distance
astronomical, 1503
image, 1116
object, 1116
relativity of, 1233–1237
Distribution function, 608
Diverging lenses, 1133
Division
significant figures in, 9
of vectors, 70
DNA
base pairing in, 717
measurement of, 1204
X-ray diffraction of, 1207
Dogs, panting by, 460
Donor, 1424
Donor level, 1424
Doping, 1424–1425
Doppler effect
for electromagnetic waves, 537–538, 1241–1243
for sound waves, 533–537, 1242
Doppler frequency shift, 1242
Doppler shift, vs. redshift, 1502
Dosimetry, radiation, 1459
Dot (scalar) product, 20–22
Down (quark), 1496
Drag, 152–154
Drift velocity, 819
current and, 820
Hall effect and, 909
Driven oscillation, 459–460
Driving force, 459–460
Ducks, swimming speed of, 486
Ductile material, 358
Dulong-Petit rule, 565, 608
DVD players, 1202, 1210, 1309
Dynamics. See also Force; Mass; Motion
celestial, 402

of circular motion, 154–159
definition of, 35, 104
fluid, 373, 389–390
Newton's second law and, 112–117, 140–146
of rotational motion, 308–331
Dyne, 117

E

Ear, sensitivity of, 528
Earth
density of, 407–408
magnetic fields of, 884, 887
rotation of, 421–423
surface temperature of, 576
Eccentricity, orbital, 415
Eddy currents, 974–975
in transformers, 1042
Edge dislocation, 1415
Edison, Thomas, 1021
Efflux, speed of, 387
Eggs, mechanical energy of, 209
Eightfold way, 1497–1498
Einstein, Albert, 91, 1223, 1263–1264, 1481
Elastic collisions, 251, 254, 255–258
relative velocity and, 256
Elastic deformations, Hooke's law and, 352, 357–358
Elastic hysteresis, 358
Elastic limit, 358
Elastic modulus, 352
bulk, 354–356
shear, 356–357
Young's, 353–354
Elastic potential energy, 216–221, 225
definition of, 217
gravitational potential energy and, 218
Elasticity, 344–359
Electric charge, 688–691
attraction and repulsion and, 688
bound, 806
in capacitors, 788. See also Capacitors
in closed surface, 726
on conductors, 736, 741–745
conservation of, 690, 856
definition of, 688
density of, 704
distribution of. See Charge distribution
electric dipole and, 709–713
electric field and, 725–728, 734–736. See also Gauss's law
flux and, 725–732. See also Electric flux
free, 806
induced, 692, 693, 805–807
magnetic force on, 886–887
magnitude of, 694
negative, 688
in nerve cells, 741
notation for, 865
point. See Point charges
positive, 688
quantitized, 691
structure of matter and, 689–690
superposition of forces and, 696
time-varying, 865
typical values for, 696
vs. magnetic poles, 885
Electric circuits
alternating-current, 822, 850
bridge, 855–860, 880–881
common-emitter, 1429
complete, 822, 828
critically damped, 1010
diagrams of, 831–833
direct-current, 822, 850
electromotive force and, 828–831
energy in, 834–836
incomplete, 828
inductors in, 994–998. See also Inductance
integrated, 1429–1430
junctions in, 855
Kirchoff's rules for, 855–860

L-C, 1005–1009
loops in, 855
L-R-C series, 1009–1011, 1030–1034
open, 870
oscillating, 1005–1009
overdamped, 1010
overloaded, 869–870
potential changes around, 833–834
power in, 834–838
R-C, 864–868
relaxation times of, 866–867
R-L, 1001–1005
self-inductance and, 995–998
short, 869–870, 870
time constants for, 866–867, 1003
underdamped, 1010–1011
Electric constants, fundamental, 695–696
Electric current. See Current
Electric dipole(s), 709–713, 805
definition of, 710
electric potential of, 765
field of, 712–713
force on, 710–711
potential energy of, 711
torque on, 710–711, 904
Electric dipole moment, 710–711
Electric energy, 194
units of, 194
Electric field(s), 699–700
calculation of, 703–708, 775
of capacitor, 790
charge distribution and, 725–728, 734–735, 746t. See also Gauss's law
of charged conducting sphere, 737–738
at conductors, 701, 744–745
current and, 819–820
definition of, 699
direction of, 701, 708
of Earth, 745
electric dipole and, 709–713
electric forces and, 698–703
electric potential and, 761, 763–764, 774
of electromagnetic waves, 1053, 1061, 1069–1070
energy storage in, 788
flux of, 725–732, 729. See also Electric flux
Gauss's law for, 732–746, 935, 1052, 1056
of hollow sphere, 741
impossible, 773f
induced, 971–974
line integral of, 937
magnetic fields and, 975–979
magnitude of, 701, 708
molecular orientation in, 805–807
nodal/antinodal planes of, 1070
nonelectrostatic, 959, 972–973
of parallel conducting plates, 739–740, 790
of plane sheet of charge, 739
resistivity and, 823
sharks and, 699
superposition of, 703–704
of symmetric charge distributions, 746t
test charge for, 699, 700
uniform, 709
of uniform line charge, 738
of uniformly charged sphere, 740
units for, 699, 764
in vacuum, 798
as vector quantity, 775
work done by, 755–761. See also Electric potential energy
Electric field lines, 708–709
electromagnetic waves and, 1053
equipotential surfaces and, 771–772
point charges and, 734, 1053
Electric flux
calculation of, 728–732
charge and, 725–732
enclosed charge and, 726–728
fluid-flow analogy for, 728–729
Gauss's law and, 732–741

Electric flux (*Continued*)
  of nonuniform electric field, 730–732
  of uniform electric field, 729–730
Electric force, 160
  conservation of energy with, 758–759
  Coulomb's law and, 693–698
  direction of, 694
  electric field and, 698–703
  electric potential and, 764–765
  potential energy and, 226–227
  on uncharged particles, 693
  units of, 695
  vector addition of, 697–698
  vs. electric potential energy, 758
  vs. gravitational force, 695
  work done by, 757–758, 761
Electric lines
  hot side, 869, 870
  household, 868–872
  neutral side of, 868–869, 870
Electric motors
  direct-current, 907–908
  magnetic force in, 898
Electric oscillation, 1005–1009
  in *L-C* circuits, 1005–1009
Electric potential, 761–771. *See also under* Potential
  calculation of, 762–771
  definition of, 761, 763
  electric circuits and, 834–836
  of electric dipole, 765
  electric field and, 761, 762–764, 774
  electric force and, 764–765
  electric potential energy and, 763, 766
  equipotential surfaces and, 771–773
  field lines and, 771–772
  maximum, 768–769
  of *a* with respect to *b*, 762
  as scalar quantity, 761, 775
  of two point charges, 765
  units of, 761, 764
  work done by, 762
Electric potential energy, 754–761
  alternative concepts of, 760
  in capacitors, 788. *See also* Capacitors
  definitions of, 760
  electric potential and, 763, 766
  electric-field energy and, 798
  of several point charges, 759–760
  of two point charges, 757–758
  in uniform field, 755
  vs. electric force, 758
Electric power, 194
Electric rays, 830
Electric stud finders, 802–803
Electric-field energy, 798, 799
Electricity, conductors of, 691–692
Electrocardiography, 762
Electrodynamics, quantum, 1081
Electrolytic double-layer capacitors, 802
Electromagnetic energy flow, 1065–1067
Electromagnetic induction, 957–981
  changing magnetic flux and, 958–959
  eddy currents and, 974–975
  experiments in, 958–959
  Faraday's law and, 957, 959–967
  induced electric fields and, 971–974
  Lenz's law and, 967–969
  Maxwell's equations and, 957, 977–979
  motional electromotive force and, 969–971
  superconductors and, 968, 979–980
Electromagnetic interaction, 159–160, 1490
Electromagnetic momentum flow, 1068–1069
Electromagnetic radiation, 574–577, 1053. *See also*
    Electromagnetic wave(s)
Electromagnetic spectrum, 1054–1055
Electromagnetic wave(s), 1051–1073
  amplitude of, 1061
  angular frequency of, 1061
  applications of, 1054
  definition of, 1052

in dielectrics, 1063–1064
direction of, 1058
Doppler effect for, 537–538, 1241–1243
electric fields of, 1053, 1061, 1069–1070
energy in, 1064–1067
frequency of, 1060
generation of, 1053
intensity of, 1066
magnetic fields of, 1061, 1069–1070
magnitude of, 1058
in matter, 1063–1064
Maxwell's equations and, 1052–1057
momentum of, 1068–1069
plane, 1055–1057
polarization of, 1058. *See also* Polarization
power in, 488
Poynting vector of, 1065–1066
properties of, 1058
radiation pressure and, 1068–1069
reflected, 1069–1071
right-hand rule for, 1058f
sinusoidal, 1060–1063
speed of, 1058, 1060, 1071
standing, 1053, 1069–1072
superposition of, 1069–1070
transverse, 1056, 1060
units for, 1053
wave functions for, 1061
wave number for, 1061
Electromagnetic wave equation, 1058–1060
Electromagnetism, 687, 885
Electrometers, 800
Electromotive force (emf), 828–831
  back, 908
  current and, 991. *See also* Inductance
  of electric motor, 908
  Hall, 909–910
  induced, 908, 958–959
  measurement of, 863–864
  motional, 969–971
  self-induced, 995, 998, 1026–1027
  sinusoidal alternating, 1021
  source of, 828, 830–831, 835–836
  theory of relativity and, 1224
  in transformers, 1040–1041
Electromyography, 861
Electron(s)
  angular momentum of, 942
  bonds and, 1405–1408
  charge of, 689–690, 695. *See also* Electric charge
  charge:mass ratio for, 896–897
  concentration of, 1420–1421
  creation and destruction of, 1483
  discovery of, 897, 1481
  excited-state, 1411
  exclusion principle and, 1388–1389
  ground-state configurations of, 1388–1390, 1391t
  in magnetic fields, 894
  magnetic moment of, 1379–1382
  mass of, 689, 897, 1440
  orbital angular momentum of, 1373–1374,
      1384, 1387
  orbital motion of, 819, 1386
  photoelectric effect and, 1261–1266
  probability distributions for, 1376–1378
  quantum states of, 1389t
  screening by, 1391–1392
  spin angular momentum of, 1384–1385, 1387
  spin of, 942
  spin-orbit coupling and, 1386
  valence, 1390, 1416
  Zeeman effect and, 1379–1382
Electron affinity, 1406
Electron capture, 1453
Electron diffraction, 1287–1288
Electron microscopes, 1290–1292
Electron shells, 1375, 1389, 1390–1391
  holes in, 1394–1395
Electron spin, 1383–1387
Electron volts, 764

Electron waves, 1286–1292
  atomic structure and, 1292–1296
  Bohr's hydrogen model and, 1300
Electron-gas model, 1415
Electron-positron pair annihilation, 1272
Electron-positron pair production, 1271–1272,
    1482–1483
Electrophoresis, 722
Electrostatic force
  conservation of, 856
  line integral for, 755, 937
Electrostatic painting, 683
Electrostatic precipitators, 784
Electrostatic shielding, 743–744
Electrostatics, 688
Electrostatics problems, 696
Electroweak interactions, 160, 1500
Electroweak theory, 1500
Electroweak unification, 1500
Elements
  ground state of, 1388–1390, 1391t
  isotopes of, 1441
  periodic table of, 1389, 1390–1391
  properties of, 1389
Elevation, atmospheric pressure and, 594–595
Elliptical orbits, 415–416
Elliptical polarization, 1099–1100
Emf. *See* Electromotive force (emf)
Emission line spectra, 1292–1293, 1297–1300
  continuous, 1310–1314
Emissivity, 575
Endoergic reactions, 1463
Endoscopes, 1090
Endothermal reactions, 1463
Energy
  activation, 610
  available, 1487–1488
  binding, 1406, 1445–1446
  conservation of, 176, 224
  conversion of, 224
  in damped oscillations, 458–459
  dark, 1508
  in electric circuits, 834–836
  electrical, 194
  electric-field, 798, 799
  in electromagnetic waves, 1064–1067
  equipartition of, 606
  Fermi, 1419–1421
  internal. *See* Internal energy
  ionization, 1304–1305, 1406
  kinetic. *See* Kinetic energy
  in *L-C* circuits, 1005–1009
  magnetic-field, 998–1001
  molecular, 597
  potential, 207–231. *See also* Potential energy
  power and, 871
  purchasing, 871
  quantized, 1261
  reaction, 1462–1464
  relativistic kinetic, 1246–1247
  rest, 1247–1249
  in simple harmonic motion, 446–449
  threshold, 1463
  total, 176, 1247
  uncertainty in, 1278
  units for, 764
  in wave motion, 486–489
  work and, 177–193. *See also* Work
Energy bands, 1416–1417
  in insulators, 1416–1417
Energy density, 798, 1064–1065
Energy diagrams, 228–229
Energy flow, electromagnetic, 1065–1067
Energy levels, 1297–1305
  degeneracy and, 1370–1371
  for harmonic oscillator, 1351–1352
  for hydrogen atom, 1302–1305, 1379–1382
  Moseley's law and, 1394–1396
  for particle in a box, 1340–1341
  quantization of, 1311–1312

rotational, 1408–1412
Schrödinger equation and, 1379–1382
selection rules and, 1382
vibrational, 1410
vs. states, 1307–1308
Zeeman effect and, 1379–1383
Energy storage, in capacitors, 788, 796–800. *See also*
    Capacitors
Energy transfer
    heat and, 562–565
    rates of, 570
Energy-flow diagrams
    for heat engines, 655–656
    for refrigerators, 659
Energy-mass conservation law, 1247–1248
Energy-time uncertainty principle, 1274–1275, 1278,
    1315–1316
Engine(s)
    Carnot, 663–669
    heat. *See* Heat engines
    internal combustion, 642, 657–659
Engine statement, of second law of thermodynamics,
    661
Enhancement-type MOSFETs, 1429
Entropy, 669
    calculation of, 676
    in cyclic processes, 672–673
    definition of, 670
    disorder and, 669–670
    internal energy and, 670
    in irreversible processes, 673
    in living organisms, 673
    microscopic interpretation of, 675–677
    Newton's second law and, 674, 677
    reversibility of, 670
Enzymes, electron tunneling in, 1349
Equation(s)
    Bernoulli's, 385–389
    continuity, 383–384
    dimensional consistency in, 6
    Dirac, 1482–1483
    electromagnetic wave, 1058–1060
    ideal-gas, 591–595
    lensmaker's, 1133–1135
    Maxwell's. *See* Maxwell's equations
    of motion, with constant acceleration, 49
    Schrödinger. *See* Schrödinger equation
    of simple harmonic motion, 440–442
    of state, 591–596, 612
    units of measure for, 6
    van der Waals, 595–596
    wave. *See* Wave equation
Equilibrium, 344–359
    center of gravity and, 345–348
    definition of, 32, 109
    extended-body, 345–352
    first condition of, 345
    for mechanical waves, 473
    net force and, 135
    Newton's first law and, 108–112, 134–139
    one-dimensional, 136–137
    phase, 566, 611
    potential energy and, 228–229
    problem-solving strategies for, 136–139
    rigid-body, 345, 348–352
    rotation and, 345
    second condition of, 345
    stable, 228
    static, 345
    tension and, 136–139
    thermal, 552
    torque and, 345
    two-dimensional, 137–138
    unstable, 228
    weight and, 345–347
    weight lifting and, 351
Equilibrium processes, 652
Equipartition, of energy, 606
Equipotential surfaces, 771–773
    conductors and, 772–773

definition of, 771
    vs. Gaussian surface, 773
Equipotential volume, 773
Equivalent capacitance, 794
Equivalent resistance, 851, 852
Erect image, 1117
Errors
    fractional (percent), 8
    in measured values, 8
Escape speed, 410–411, 413, 423, 1505–1506
Estimates, order-of-magnitude, 10
Ether, light travel through, 1224, 1881
Euler's formula, 1340
Evaporation, 568
Event horizons, 424
Event, in frame of reference, 1227
Excited levels, 1298
Excited states, 1454
Exclusion principle, 1388–1393
    bonds and, 1406, 1407
    periodic table and, 1389, 1390–1393
    quark colors and, 1498–1499
Exoergic reactions, 1463
Exothermal reactions, 1463
Expanding universe, 1501–1508
Experiments, 2
    thought, 1227
Extended objects
    definition of, 1115
    gravitational potential energy for, 293, 317
    image formation by lenses and, 1131–1133
    image formation by mirrors and, 1120–1122
Extended-body equilibrium, 345–352
External forces, 247
    center-of-mass motion and, 261–262
    torque and, 312
Extracorporeal shock wave lithotripsy, 539
Eye, 1142–1146
    index of refraction of, 1143
    laser surgery for, 1076, 1309
    resolution of, 1220
    structure of, 1142–1143
Eyeglasses, corrective, 1143–1146
Eyepiece, microscope, 1147–1148

F

Fahrenheit scale, 554
Far point, of eye, 1143
Farad, 789, 790
Faraday cage, 743–744
Faraday disk dynamo, 971
Faraday, Michael, 708, 789, 885
Faraday's icepail experiment, 742–744
Faraday's law of induction, 957, 959–967
    electromagnetic waves and, 1052, 1056, 1057, 1063.
        See also Maxwell's equations
Farsightedness, 1143–1146
Fermat's principle of least time, 1111
Fermi energy, 1419–1421
    electron concentration and, 1420–1421
Fermi-Dirac distribution, 1419, 1492
Fermions, 1492, 1494, 1501
Ferromagnetism, 943t, 944–946
Feynman, Richard, 1483
Field lines. *See* Electric field lines; Magnetic field
        lines
Field point, 700, 924
Field-effect transistors, 1429
Fields, 406
Filters
    high-pass, 1028
    low-pass, 1027
    polarizing, 1093, 1094–1097, 1098
Fine structure, 1387
Finite wells, 1343
Firecrackers, entropy and, 669f
First condition for equilibrium, 345
First law of thermodynamics, 624–643
    cyclic processes and, 631–634
    definition of, 630

internal energy and, 629–634
    isolated systems and, 631–634
Fish
    bulk stress on, 355f
    fluorescent, 1300f
Fission, nuclear, 785, 1247, 1464–1468
Fixed-axis rotation, 278–279, 283
Flash unit, of camera, 797
Flow line, 382–383. *See also* Fluid flow
Flow tubes, 383
Fluid(s)
    compressibility of, 356, 382
    ideal, 382
    motion of, 382–384
    speed of sound in, 514–515
    viscous, 389–390
Fluid density, 373–375
    buoyancy and, 380–382
    compression and, 476
    measurement of, 381
    rarefaction and, 476
Fluid dynamics, 373
    viscosity and, 389–390
Fluid flow, 382–384. *See also* Flow
    Bernoulli's equation and, 385–389
    continuity equation and, 383–384
    laminar, 383, 390
    measurement of, 388
    pressure and, 385–389
    rate of, 383–384
    speed of, 385–389
    steady, 382–383
    turbulent, 383, 390–391
Fluid mechanics, 373–392
    Bernoulli's equation and, 385–389
    buoyancy and, 380–382
    density and, 373–375, 380–381
    fluid flow and, 382–384
    pressure and, 375–380
    surface tension and, 382
Fluid pressure, 355, 375–380
    depth and, 376–377
    measurement of, 378–380
    Pascal's law and, 376–377
Fluid resistance, 151–154
Fluid statics, 373
Fluorescence, 1300
Fluorescent fish, 1300f
Fluorescent lights, 996–997, 1081, 1300
Fluorine, 1390
Flux. *See* Electric flux; Magnetic flux
f-number, 1140–1141
Focal length
    of camera lens, 1140
    of microscope lens, 1148–1149
    of mirror, 1119–1120
    of telescope lens, 1149–1150
    of thin lens, 1131, 1133–1135
Focal point
    of microscope lens, 1149
    of mirror, 1119–1120
    of thin lens, 1131
    virtual, 1123
Fog, 567
Food, energy value of, 568
Foot-pound, 177, 194
Force(s)
    acting at a distance, 406
    action-reaction, 120–123
    buoyant, 380
    central, 416
    centrifugal, 155
    components of, 226–227
    conservative, 221–224, 228–229
    constant, 177–178
    contact, 105, 146
    definition of, 105
    direction of, 10, 13. *See also* Vector(s)
    dissipative, 222
    driving, 459–460

Force(s) (Continued)
electric. *See* Electric force
electromotive, 828–831
electrostatic, 856
external, 247
fluid resistance, 151–154
free-body diagrams for, 124–125
friction, 105, 146–154. *See also* Friction
fundamental, 159–160, 1490–1492
gravitational. *See* Gravitation
interaction, 931–932
intermolecular, 597–598
internal, 247, 312
line of action of, 309
long-range, 105
magnetic, 159–160, 886–887
magnitudes of, 105
mass and, 113–114
measurement of, 106
motion and, 112–116. *See also* Newton's second law
of motion
net. *See* Net force
nonconservative, 222–224
normal, 105, 146
nuclear, 1446–1449, 1484, 1491
particle interactions and, 159–160
periodic driving, 459–460
potential energy and, 225–228
power and, 194
properties of, 105
restoring, 438
strong interactions, 160, 1446, 1490–1491
strong nuclear, 160, 689
superposition of, 106–108, 404, 405–406
tension, 105, 123. *See also* Tension
tidal, 425
torque of, 308–312
units of, 5–6, 105, 117
vs. pressure, 355, 376
weak interactions, 160, 1491
weight as, 105
Force constant, 188
Force diagrams, 106–107
Force fields, 406
Force mediators, 1484–1485
Force per unit area, 353
Force vectors, 105
components of, 106–107
Forced oscillations, 459–460, 527
Forensics, X-rays in, 1395
Forward bias, 1426, 1427–1428
Fosbury flop, 293f
Fossil fuels, climate change and, 577
Fossils, 1508
Fourier analysis, 513
Fourier series, 497
Fractional error (uncertainty), 8
Fracture, 358
Frame of reference, 89
event in, 1227
inertial, 110–112, 115–116, 1223–1227
simultaneity and, 1227–1228
Franck, James, 1300
Franck-Hertz experiment, 1300
Franklin, Benjamin, 688
Franklin, Rosalind, 1207f
Fraunhofer diffraction, 1192, 1193
Free charges, 806
Free expansion, 629
Free fall
acceleration due to gravity and, 52–55, 118–119
definition of, 52
fluid resistance and, 152–154
Free particle, 1330
Free-body diagrams, 124–125, 140, 144
Free-electron energy, average, 1421
Free-electron model, 1415, 1418–1422
Free-particle state, 1346
Frequency, 438
angular, 438–439

beat, 531–532
fundamental, 496
normal-mode, 496
period and, 438–439
of standing waves, 496
vs. angular frequency, 442
Fresnel diffraction, 1193
Friction, 146–154
coefficients of, 147, 151
definition of, 108, 147
fluid resistance and, 151–152
kinetic, 147, 149, 150–151, 222
magnitude of, 147
rolling, 151–154, 320
static, 147–149
stick-slip phenomenon and, 148, 149
Friction force, 105, 147
Full-wave rectifier current, 1023
Fundamental electric constants, 695–696
Fundamental forces, 159–160, 1490–1492
Fundamental frequency, 496
Fundamental particles, 1480–1501. *See also*
Particle(s)
historical perspective on, 1480–1485
Fur, as insulator, 572
Fuses, 870
Fusion. *See* Nuclear fusion

G
Galaxies, recession speed of, 1502–1503
Galilean coordinate transformation, 1225–1226
Galilean telescope, 1161
Galilean velocity transformation, 91, 1226
Galileo Galilei, 2, 52, 1080
Gallium, melting temperature of, 566
Galvanometer, d'Arsonval, 860, 863, 904, 1022
Gamma camera, 1461
Gamma decay, 1454
Gamma rays, 1454, 1462
pair production and, 1271–1272
Gas
bulk modulus of, 355
heat capacities of, 605–607
ideal. *See* Ideal gas
intermolecular forces in, 597
isotherms and, 596
kinetic energy of, 605–606
mass of, 591–592
molecules in, 597
noble, 1390
p-V diagrams for, 596
sound waves in, 517–518
volume of, 593
Gas constant, 517, 592
Gas pressure
molecular collisions and, 597–600
molecular kinetic energy and, 600–602
temperature and, 555
Gas thermometers, 554–556, 593, 669
Gaseous phase, 566
Gasoline engines, 546–548
Gauge pressure, 377–378
Gauges, pressure, 378–380
Gauss, 887
Gauss, Carl Friedrich, 732
Gaussian surface, 734
vs. equipotential surface, 773
Gauss's law, 732–746
applications of, 736–741
charge and electric flux and, 725–732
conductors with cavities and, 741–742, 773
dielectrics and, 807–808
for electric fields, 732–746, 935, 1052, 1056. *See
also* Maxwell's equations
experimental testing of, 742–744
general form of, 734–736
for gravitation, 752, 935
for magnetic fields, 891, 935, 1052, 1056. *See also*
Maxwell's equations
overview of, 732–736

point charge inside nonspherical surface and,
733–734
point charge inside spherical surface and, 732–733
qualitative statement of, 728
solid conductors and, 736–741
Geiger counters, 783
Geiger, Hans, 1294
Gell-Mann, Murray, 1496, 1498
Gemstones, 1090, 1092
General theory of relativity, 1249–1251, 1504. *See also*
Relativity
Generation currents, 1427
Generators
alternating-current, 1021
direct-current, 964–965
energy conversion in, 966–967
homopolar, 971
slidewire, 965–966, 967, 970
Geometric optics, 1082, 1114–1153
cameras and, 1139–1142
eye and, 1142–1146
magnifiers and, 1146–1147
microscopes and, 1147–1149
reflection at plane surface and, 1115–1118
reflection at spherical surface and, 1118–1126
refraction at plane surface and, 1115–1118
refraction at spherical surface and, 1126–1130
sign rules for, 1116
telescopes and, 1149–1151
thin lenses and, 1131–1139
Gerlach, Walter, 1383
Germanium semiconductors, 1422–1425
Germer, Lester, 1287–1288
Glashow, Sheldon, 1500
Glass, as amorphous solid, 1412
Global positioning systems, 1078, 1187, 1250–1251
Global warming, 576–577
Gluons, 1490, 1491, 1498–1499, 1500
GPS systems, 1078, 1187, 1250–1251
Gradient
definition of, 774
potential, 774–776
potential energy, 227
Grams, 5, 5t
Grand unified theories (GUTs), 160, 1500–1501,
1509, 1516
Graphical method, for image location, 1124–1126
Graphs
curvature of, 46, 47
parabolic, 48
a_x-t, 46–49, 55–57
of sound waves, 511
v_x-t, 44–46, 47
of wave function, 478–480
x-t, 37–38, 40–41. *See also* x-t graphs
Grating spectrographs, 1203–1204
Gravitation, 105, 159–160, 402–427
acceleration due to, 52–55, 700
action-reaction pairs and, 403
black holes and, 423–426
as conservative force, 409, 410
on cosmic scale, 406
escape speed and, 410–411, 413, 1505–1506
expanding universe and, 1501–1502, 1505–1507
as fundamental force, 159–160
Gauss's law for, 752, 934
general theory of relativity and, 1250
importance of, 406
measurement of, 404–405
Newton's law of, 402–406
satellite orbits and, 411–413
specific gravity and, 374
spherical mass distributions and, 418–421
spherically symmetric bodies and, 403–404
superposition of forces and, 405–406
weight and, 406–409
work done by, 409–410
Gravitational constant, 403
calculation of, 404–405
Gravitational field, 700

Gravitational force(s), 105, 159–160
  as action-reaction pairs, 403
  per unit mass, 700
  vs. electric force, 695
  vs. gravitational potential energy, 410
Gravitational interaction, 159–160, 1490
Gravitational potential energy, 208–216, 409–411
  definition of, 208, 409–410
  elastic potential energy and, 218
  for extended bodies, 293, 317
  motion along curves and, 212–216
  as negative value, 410
  nongravitational forces and, 211
  vs. gravitational force, 410
Gravitational red shift, 425, 1250
Gravitational torque, 346–347
Gravitons, 1490, 1500
Gravity. *See* Gravitation
Gray, 1459
Greenhouse effect, 576–577
Ground level, 1297
Ground state
  atomic, 1388–1390, 1391t
  nuclear, 1454
Ground-fault interrupters, 870
Grounding wires, 869, 870
Gyroscopes, 328–330

H

h vs. *h-bar,* 1276
Hadrons, 1492, 1493–1494, 1496, 1497
Hahn, Otto, 1464
Hale Telescope, 1219
Half-life, 1456–1457
Hall effect, 909–910
Halley's comet, 417
Halogens, 1390
Harmonic analysis, 497
Harmonic content, 497, 513
Harmonic motion, damped, 1009
Harmonic oscillators
  anisotropic, 1401
  Hermite functions for, 1350–1354
  isotropic, 1401–1402
  Newtonian, 439–440, 1352–1354
  quantum, 1350–1354
Harmonic series, 496
Harmonics, 496
Hearing loss, sound intensity and, 513, 522
Heat
  added in thermodynamic process, 628–629
  of combustion, 568
  definition of, 562, 563
  as energy in transit, 605
  energy transfer and, 562–565. *See also* Heat
    transfer
  of fusion, 566
  global warming and, 576–577
  mechanical energy and, 562
  melting and, 565–566
  phase changes and, 565–570
  quantity of, 562–565
  sign rules for, 625
  specific, 562–563
  steam, 567–568
  of sublimation, 567
  units of measure for, 562
  of vaporization, 566, 568
  vs. temperature, 562
Heat calculations, 568–570
Heat capacity, 605–608
  of gases, 605–607
  of ideal gas, 637–639
  molar, 564–565, 605–607
  point-molecule model of, 605
  ratio of, 639
  of solids, 607–608
  temperature variation of, 608
  vibration and, 606–608
Heat current, 571

Heat engines, 654–656
  Carnot, 663–666
  energy-flow diagrams and, 655
  hot and cold reservoirs and, 654–655
  internal combustion, 642, 657–659
  thermal efficiency of, 655
Heat flow, 562
Heat pumps, 661
Heat transfer, 562–565
  by conduction, 570–574
  by convection, 574
  mechanisms of, 570–577
  by radiation, 574–577
Heavy hydrogen, 1305–1306
Heisenberg uncertainty principle, 1276–1277,
    1314–1317
  Bohr model and, 1317
  energy-time, 1278, 1315–1316
  harmonic oscillator and, 1353–1354
  for matter, 1315–1316
  momentum-position, 1274–1275, 1278, 1315–1316
Helium, 1390
Helium atom, Bohr model of, 1306
Helium fusion, 1514
Hemholtz coils, 954
Henry, 993
Henry, Joseph, 885
Hermite functions, 1350–1354
Hertz, 438, 1053
Hertz, Gustav, 1300
Hertz, Heinrich, 1053, 1081
High-pass filters, 1028
Hole conduction, 909, 1423–1424
Holes, 820, 1423–1424
Holography, 1211–1213
Homopolar generators, 971
Hooke's law, 188–189, 352, 356
  elastic deformations and, 352, 357–358
  limits of, 357–358
  simple harmonic motion and, 439
Horse, acceleration around curve, 74
Horsepower, 194
Hot reservoirs, 654–655
Hot side of line, 869, 870
House wiring systems, 868–872, 1040–1041
Hubble constant, 1503
Hubble, Edwin, 1502
Hubble law, 1503
Hubble Space Telescope, 1119, 1218, 1375f,
    1503
Human body
  angular momentum of, 324
  fat measurement for, 1032
  magnetic fields of, 887
  radiation from, 575–576
  as thermodynamic system, 630
Humason, Milton, 1502
Huygen's principle, 1102–1104
  diffraction and, 1191–1192
Hybrid wave function, 1407
Hydrogen
  in fusion reactions, 1469–1471
  ground state of, 1390
  heavy, 1305–1306
Hydrogen atom, 1372–1379
  Bohr's model of, 1300–1305
  electron probability distributions for, 1376–1378
  energy levels in, 1302–1305, 1379–1382
  hydrogen-like atoms and, 1378–1379
  ionization energy of, 1304–1305
  in magnetic field, 1379–1382
  nuclear motion in, 1305–1306
  orbital angular momentum of, 1373–1374
  quantum states of, 1373–1374, 1375t
  reduced mass of, 1305–1306
  Schrödinger equation for, 1372–1373
Hydrogen bonds, 1407–1408
Hydrogenlike atoms
  Bohr model of, 1306
  Schrödinger analysis of, 1378–1379

Hydrometers, 381
Hyperfine structure, 1387, 1444
Hyperons, 1494–1495
Hyperopia, 1143–1146
Hysteresis, 945–946
Hysteresis loops, 945–946

I

I SEE acronym, 3
Ice, melting of, 565–566
Ideal fluid, 382
Ideal gas, 592
  adiabatic process for, 640–642
  Carnot cycle for, 664–665
  heat capacities of, 637–639
  internal energy of, 636
  isothermal expansion of, 627
  kinetic-molecular model of, 599–605
  volume of, 593
Ideal single crystals, 1413
Ideal-gas constant, 517, 592
Ideal-gas equation, 591–595
Idealized models, 3–4, 3f
Image
  erect, 1117
  inverted, 1117, 1141
  in optics, 1115
  real, 1115
  virtual, 1115, 1137
Image distance, 1116
Image formation
  by cameras, 1139–1142
  by diffraction, 1209–1210
  by lenses, 1131–1133
  by reflection, 1115–1126. *See also* Mirrors
  by refraction, 1126–1130
Image point, 1114, 1115
Imaging studies. *See* Medical imaging
Impedance, 1031–1033
Impulse, 243
Impulse-momentum theorem, 242–244, 483–484
Incident waves, 492–493
Incubators, 576
Index of refraction, 1063, 1083–1088
  birefringence and, 1100
  definition of, 1083
  dispersion and, 1091–1093
  of eye, 1143
  of gemstones, 1090
  laws of reflection and refraction and, 1085
  of lens, 1133–1134
  of reflective/nonreflective coatings, 1178–1179
  total internal reflection and, 1088–1090
  transparency and, 1085
  wave aspects of light and, 1086–1088
Induced charges, 692
  molecular model of, 805–807
  polarization and, 693, 801–803
Induced current, 958
  direction of, 967–968, 970
  magnitude of, 968
Induced electric fields, 971–974
Induced emf, 908, 958–959. *See also* Electromagnetic
    induction
  applications of, 959, 961
  direction of, 961
  magnetic flux and, 959, 962
Inductance, 991–1012
  definition of, 995
  magnetic-field energy and, 998–1001
  mutual, 991–994
  R-L circuits and, 1001–1005
  self-inductance, 994–998
Inductive reactance, 1026–1027
Inductors, 994–998
  in ac circuits, 1025–1027, 1029, 1029t, 1035
  energy stored in, 998–1000
  inductive reactance of, 1026
  vs. resistors, 999
Inelastic collisions, 251–255

Inertia, 109
  definition of, 109
  mass and, 118
  moment of. *See* Moment of inertia
  rotational, 289
Inertial confinement, 1470
Inertial frame of reference, 110–112, 1223–1227
  Newton's first law and, 110–112
  Newton's second law and, 115–116
  simultaneity and, 1227–1228
  theory of relativity and, 1223
Inertial mass, 113. *See also* Mass
Inertial navigation systems, 56f
Infrasonic sound, 510
Inkjet printers, 722
Instantaneous acceleration, 43–44. *See also* Acceleration
  angular, 282
  definition of, 43
  on v_x-t graph, 44–46
  x-component of, 43
Instantaneous acceleration vectors, 73–75. *See also* Acceleration vectors
Instantaneous angular acceleration, 282
Instantaneous angular speed, 286
Instantaneous angular velocity, 280
Instantaneous power, 193, 194
  in waves, 487–488
Instantaneous speed, 39. *See also* Speed
  angular, 286
Instantaneous velocity, 38–42, 39f–41f. *See also* Velocity
  definition of, 38, 70
  straight-line motion and, 38–41
  vs. instantaneous speed, 39–40
  x-component of, 39
  on x-t graph, 40–41
Instantaneous velocity vectors, 70–72
Insulators, 552, 691
  energy bands in, 1416–1417
Integral(s)
  line, 192, 755
  moment of inertia, 295
  surface, 730
Integral principles, 244
Integrated circuits, 1429–1430
Integration, velocity and position by, 55–57
Intensity
  of electromagnetic radiation, 1066
  inverse-square law for, 488–489, 520
  pressure amplitude and, 519–521
  in single-slit diffraction, 1195–1199
  sound, 518–522
  vs. spectral emittance, 1310–1311
  wave, 488–489
Intensity maxima, 1197
Interactions. *See* Particle interactions
Interference, 489–492, 529–531, 1163–1183
  amplitude in, 1170–1171
  butterfly wings and, 1179
  coherent sources and, 1164–1166
  complementarity principle and, 1273–1274
  constructive, 492, 530, 1164–1166, 1168–1170, 1206–1207
  definition of, 489, 1164
  destructive, 492, 529, 1164–1166, 1168–1170
  in holography, 1211–1213
  Michelson interferometer and, 1179–1181
  Michelson-Morley experiment and, 1180–1181
  Newton's rings and, 1178
  nodal/antinodal curves and, 1166
  in noise control, 531
  path difference and, 1171–1173
  phase difference and, 1171–1173
  phase shifts and, 1174–1175
  photons and, 1273–1274
  during reflection, 1174–1175
  reflective/nonreflective coatings and, 1178–1179
  sinusoidal waves and, 1164
  in sound waves, 529–531

sound waves and, 529–531
standing waves and, 492, 1164, 1166
superposition and, 1164
in thick films, 1176
in thin films, 1173–1179
in three dimensions, 1164
in two dimensions, 1164
two-source/slit, 1166–1173, 1315–1316
vs. diffraction, 1192, 1200
in water waves, 1166–1167
water waves and, 1166–1167
Young's experiment for, 1167–1169, 1179
Interference fringes, 1168, 1174, 1178
Newton's, 1178
Interference maxima, 1199
Interference patterns, 1166
intensity in, 1170–1173
Interferometer, 1179–1181
Intermolecular forces, 597–598
Internal combustion engines, 657–659
Internal energy, 224, 624, 670
  change in, 630–631, 639
  of cyclic processes, 631–634
  definition of, 629, 631
  entropy and, 670
  first law of thermodynamics and, 629–634
  of ideal gas, 636
  of isolated systems, 631–634
  notation for, 629
  temperature and, 636
Internal forces, 247
  torque and, 312
Internal resistance, 830, 833
International System, 4, 5t
Interplanetary travel, biological hazards of, 416f
Interstellar gas clouds, 1293
Intrinsic semiconductors, 1423, 1424
Inverse-square law, for intensity, 488–489, 520
Inverted image, 1117
  in camera lens, 1141
Iodine-127, 1461
Ionic bonds, 1406, 1407
Ionic crystals, 1414–1415
Ionization, 690
  corona discharge and, 768–771
Ionization energy, 1406
  of hydrogen atom, 1304–1305
Ions, 690
Irreversible process, 652–653
Isobaric process, 635
Isochoric process, 635
Isolated systems, 247
  internal energy of, 631–634
Isospin, 1495
Isothermal expansion, of ideal gas, 627
Isothermal process, 635–636
  Carnot cycle and, 663
Isotherms, 596
Isotopes, 897, 1441

J

Jet propulsion, in squids, 262
Josephson junctions, 1349
Joule, 177, 183, 562
Joule per coulomb, 761
Junctions, in circuits, 855

K

K mesons, 1493
Kaons, 1259, 1493
Keck telescopes, 1151
Kelvin, 555
Kelvin scale, 555–556, 665, 668–669
Kelvin-Planck statement, 661
Kepler, Johannes, 414–415
Kepler's first law, 414–415
Kepler's second law, 415–416
Kepler's third law, 416–417, 426
Killowat-hour, 194
Kilocalorie (kcal), 562

Kilograms, 5, 113
Kilohm, 826
Kilowatt, 193
Kilowatt-hour, 871
Kinematics. *See also* Motion
  definition of, 35
  linear vs. angular, 285–288
Kinetic energy
  in collisions, 252
  in composite systems, 186–187
  conservative forces and, 221–222
  with constant forces, 177–178
  definition of, 182
  equipartition of, 606
  gas pressure and, 600–602
  heat capacities and, 605–608
  molecular, 597, 600–602, 605–606, 636
  moment of inertia and, 288–291
  of photons, 1263–1264
  potential energy and, 207, 208, 221–222
  relativistic, 1246–1247
  rotational, 288–293, 315–316
  as scalar quantity, 182
  in simple harmonic motion, 446–449
  stopping potential and, 1262–1263
  torque and, 321
  units of, 183
  with varying forces, 187–191
  vs. momentum, 242–246, 2420
  work-energy theorem and, 181–187
Kinetic friction, 147, 149, 150–151
  coefficient of, 147
  as nonconservative force, 222
Kinetic-molecular model, of ideal gas, 599–605
Kirchoff's rules, 855–860, 976–977
Kramers, Hendrik, 1361
Kundt's tube, 523

L

Ladder, stability of, 350
Lagging current, 1036
Laminar flow, 383, 390
Land, Edwin H., 1094
Large Hadron Collider, 1257, 1487, 1489, 1501
Laser(s), 1307–1309
  continuous, 1309
  definition of, 1307
  metastable states and, 1308
  population inversions and, 1308–1309
  production of, 1308–1309
  pulsed, 1309
  semiconductor, 1309
  spontaneous emission and, 1307
  stimulated emission and, 1307–1309
Laser eye surgery, 1076, 1309
Laser light, 1055, 1081, 1164, 1166
Laser printers, 689, 769, 1309
Latent heat of fusion, 566
Lateral magnification, 1117, 1120–1121
  of camera, 1140
  of microscope, 1148–1149
  vs. angular magnification, 1147
Laue pattern, 1205f
Law of Biot and Savart, 927
Law of conservation of energy, 176, 209, 224
Law of reflection, 1084–1086
Law of refraction, 1084–1086
Lawrence, E.O., 1486
Laws, physical, 2
L-C circuits, 1005–1009
Leaning Tower of Pisa, 2, 53, 118
Length
  Planck, 1509
  proper, 1235
  relativity of, 1233–1237
  units of, 4–5, 6f
Length contraction, 1235, 1236
Lens(es)
  of camera, 1139–1142
  corrective, 1143–1146

definition of, 1131
of eye, 1142–1143
magnifying, 1146–1147
of microscopes, 1147–1149
nonreflective coatings for, 1178
parabolic, 1151
properties of, 1131
reflective coatings for, 1178, 1179
of telescopes, 1149–1151
thin, 1131–1139. *See also* Thin lenses
Lensmaker's equation, 1133–1135
Lenz's law, 967–969
Leptons, 1492–1493, 1496, 1499, 1500
Lever arm, 309
Light, 1080–1105
absorption of, 1261–1266
beams of, 1083
coherent, 1165
diffraction of, 1190–1214
dispersion of, 1085, 1091–1093
Doppler effect for, 537–538, 1241–1243
early studies of, 1080–1081
fluorescent, 996, 1081
Huygen's principle and, 1102–1104
intensities of, 1085
interference and, 1166–1170. *See also* Interference
laser, 1055, 1081, 1164, 1166
monochromatic, 1054–1055, 1164
natural, 1094
photoelectric effect and, 1261–1266
as photons, 1261–1280. *See also* Photons
polarized, 1093–1100
rays of, 1082
reflection of, 1082–1091
refraction of, 1082–1088
scattering of, 1100–1101, 1269–1273
speed of, 4–5, 1054, 1063f, 1081, 1224–1226
total internal reflection of, 1088–1091
unpolarized, 1094
visible, 1054
as wave and particle. *See* Wave-particle duality
wave fronts of, 1081–1082
wavelengths of, 1054–1055
Light pipes, 1090
Light-emitting diodes, 1428
Lightning rods, 769
Lightning strikes, inductors and, 995
Light-years, 1503
Limit of resolution, 1209
Linacs, 1485–1486
Line integral, 192, 755
of electric fields, 937
of electrostatic force, 755, 937
of magnetic fields, 937
Line of action, 309
Line spectra, 1292, 1297–1300, 1304
continuous, 1310
molecular, 1300
Zeeman effect and, 1379
Linear acceleration, 282
constant, 284t
in rigid-body rotation, 286–288
vs. angular acceleration, 284t
Linear accelerators, 1485–1486
Linear charge density, 704
Linear conductors, 823
Linear expansion, 557–558, 558t, 559
Linear mass density, 482
Linear momentum, 242, 322
Linear polarization, 1058, 1093, 1095
of electromagnetic wave, 1058
Linear speed, in rigid-body rotation, 285–286
Linear superposition, 491
Linear velocity, 282
vs. angular velocity, 280
Linear vs. angular kinematics, 285–288
Liquid(s)
compressibility of, 356
as condensed matter, 1412
molecular speed in, 610

molecules in, 597
phases of, 610–613
properties of, 1412
Liquid crystals, 1412
Liquid phase, 566
Liquid-drop model
of nuclear fission, 1465–1466
of nucleus, 1447–1448
Lithium, 1390
Bohr model of, 1306
Livingston, M. Stanley, 1486
Longitudinal waves, 473. *See also* Mechanical waves; Wave(s)
periodic, 475–476
sound, 476
wave function for, 482
Long-range forces, 105
Loops, in circuits, 855
Lorentz transformations, 1237–1241
coordinate, 1237–1238
velocity, 1238–1239
Loudness, 513
Loudspeakers, 1029
magnetic forces in, 899–900
Low-pass filters, 1027
L-R-C parallel circuits, resonance in, 1039, 1048
L-R-C series circuits, 1009–1011
with ac source, 1030–1034
impedance in, 1031–1032, 1038
phase angle and, 1031–1032
power in, 1034–1037
resonance in, 1037–1039
Luminous matter, 1507
Lyman series, 1304

M
Mach number, 539
Macroscopic properties
theories of matter and, 599
vs. microscopic properties, 590
Macroscopic state, 675
Magic numbers, 1449
Magnet(s)
attracting unmagnetized objects, 906–907
bar, 883, 905–907
magnetic dipoles of, 906–907
magnetic moment of, 906
permanent, 883, 941
Magnetic ballast, 997
Magnetic bottles, 893
Magnetic confinement, 1470
Magnetic declination, 884
Magnetic dipole moment. *See* Magnetic moment
Magnetic dipoles, 903–904, 906–907
definition of, 903
force and torque on, 901–903
of magnets, 905–907
in nonuniform magnetic fields, 905–907
potential energy for, 903–904
Magnetic domains, 944–945
Magnetic energy density, 999–1000
Magnetic field(s), 884, 885–889
on axis of coil, 933–934
calculation of, 927
of circular current loops, 932–935
critical, 979
of current element, 926–927
definition of, 885
direction of, 887–888
of Earth, 884, 887
of electromagnetic waves, 1055–1057, 1061–1062, 1069–1070
Gauss's law for, 891, 935, 1052, 1056
Hall effect and, 909
of human body, 887
hydrogen atom in, 1379–1382
line integral of, 937
of long cylindrical conductor, 939, 948t
of long straight conductor, 935–937, 938, 948t

magnitude of, 887–888, 892
measurement of, 887–889
motion in, 892–898
of motors, 898
of moving charge, 886, 923–926
nodal/antinodal planes of, 1070
notation for, 886
of solenoid, 939–941, 948t
sources of, 923–946, 975–979, 1444
of straight current-carrying conductor, 928–931
superposition of, 926, 931
test charges for, 887–889
vector, 886, 924
Zeeman effect and, 1379–1382
Magnetic field lines, 884–885, 889–890
for current element, 927
direction of, 924
end points of, 891
magnetic flux and, 890–891
for moving charge, 924–925
vs. magnetic lines of force, 889
Magnetic flux, 890–892
calculation of, 959–960
definition of, 890–891
Faraday's law and, 959–967
Gauss's law of magnetism and, 891
induced electric fields and, 972–974
induced emf and, 958–959
Lenz's law and, 967–969
Meissner effect and, 980
as scalar quantity, 891
superconductivity and, 980
in transformers, 1040
units for, 891
Magnetic flux density, 892
Magnetic force(s)
on current loops, 901–907
on current-carrying conductors, 898–901
direction of, 886–887
in electric motors, 898
as fundamental force, 159–160
Hall effect and, 909
in loudspeakers, 899–900
magnitude of, 887
between parallel conductors, 931–932
units for, 887
Magnetic inclination, 884
Magnetic lines of force, 889–892
Magnetic materials, 907, 941–946
Bohr magneton, 941–942
diamagnetic, 943t, 944
ferromagnetic, 943t, 944–946
paramagnetic, 943–944, 943t
relative permeability of, 943
Magnetic moment, 903, 906, 1379
alignment of, 941–945
anomalous, 1443
of current loop, 903, 934
definition of, 903
direction of, 903
magnitude of, 942–945
of neutron, 1442–1443, 1497
nuclear, 1442–1443
of orbiting electron, 1379–1382
of proton, 1442–1443
spin, 1443
vector, 903
Zeeman effect and, 1379–1382
Magnetic monopoles, 885, 1500
Magnetic nanoparticles, for cancer, 946
Magnetic poles, 884
vs. electric charge, 885
Magnetic quantum number, 1374
Magnetic resonance imaging (MRI), 904–905, 934, 1444
Magnetic susceptibility, 943–944
Magnetic torque, 901–905
Magnetic variation, 884
Magnetic-field energy, 998–1001

Magnetism, 883–885
  electron motion and, 885
  Gauss's law for, 891, 935
Magnetization, 906–907, 927, 941–946
  saturation, 945
Magnetization curve, 945
Magnetons
  Bohr, 941–942, 1380
  nuclear, 1443
Magnetrons, 893
Magnification
  angular, 1147–1148
  lateral, 1117, 1120–1121, 1147, 1148–1149
Magnifiers, 1146–1147
Magnitude, of vector, 11, 12, 16
Malus' law, 1096
Manometers, 378–380
Maple seed, motion of, 316
Mars, gravitation on, 408–409
Marsden, Ernest, 1294
Mass
  acceleration and, 113, 114, 118–120
  of atom, 598, 690, 897, 1305–1306, 1441
  center of, 258–262
  definition of, 113
  density and, 373–374
  of electron, 689, 896–897, 1440
  force and, 113–114
  of gas, 591–592
  inertia and, 118
  measurement of, 114, 119
  molar, 517, 564, 591, 598
  of molecule, 598, 897
  of neutrino, 1500–1501
  of neutron, 689, 1440
  of nucleus, 1440
  of proton, 689, 1440
  rest, 1243–1246
  of star, 1259
  terminal speed and, 153–154
  units of, 5, 113–114, 117, 119
  weight and, 114, 117–120
Mass number, 1440
Mass per unit length, 482
Mass spectrograph, 918
Mass spectrometers, 897
Mass-energy conservation law, 1247–1248
Mass:volume ratio, density and, 373–374
Matter
  antimatter and, 1516
  condensed, 1412. See also Liquid(s); Solids
  luminous, 1507
  molecular properties of, 596–598
  phases of, 610–613
Maxwell, James Clerk, 976, 1052f, 1081
Maxwell-Boltzmann distribution, 608–609, 1307,
    1419–1420
Maxwell's equations, 885, 957, 977–979
  electromagnetic waves and, 1052–1057
  Huygen's principle and, 1102–1104
  in optics, 1085
Maxwell's wave theory, 1052–1057, 1262–1263, 1267
Mean free path, 604
Mean free time, 838
Measurement
  accuracy of, 8
  errors in, 8
  significant figures in, 8–9
  uncertainty in, 8
  units of, 4–6. See also Units of measure
Mechanical energy
  conservation of, 209, 446–448, 755
  conservative vs. nonconservative forces and,
      221–223
  heat and, 562
  in simple harmonic motion, 446–449
  total, 209
Mechanical waves, 472–499. See also under Wave
  boundary conditions for, 489–490
  definition of, 473

direction of, 479, 481–482
  energy of, 486–489
  energy transport by, 474
  equilibrium state for, 473
  incident, 492–493
  intensity of, 488–489
  interference in, 489
  longitudinal. See Longitudinal waves
  mathematical description of, 477–482
  normal-mode patterns of, 496
  periodic, 474–477. See also Periodic waves
  power of, 487–488
  propagation of, 474
  sinusoidal, 475–482. See also Sinusoidal waves
  sound, 476
  speed of. See Wave speed
  standing, 491–498
  superposition of, 490–491, 497
  transverse. See Transverse waves
  traveling, 492, 494
  types of, 473–474
  wave equation for, 481, 485
  wavelength of, 475
Mechanics
  classical (Newtonian), 104
  definition of, 35
Medical imaging
  radioactive isotopes in, 1391, 1461, 1466
  X rays in, 1268–1269
Medicine
  nuclear, 1391, 1461, 1466
  pair annihilation in, 1484
Medium, 473
Megohm, 826
Meissner effect, 979–980
Meitner, Lise, 1464
Melting, 565–566
Melting points, of solids, 1412
Membrane, dielectric, 805
Membrane potential, 774
Mercury barometers, 378–380
Mesons, 1484–1485, 1492, 1493–1495, 1499
  colors of, 1489
  quarks in, 1496
Metallic bonds, 1408
Metallic conduction, 838–840
Metallic crystals, 1415
Metals
  alkali, 1306, 1390
  alkaline earth, 1390
  average free-electron energy of, 1421
  as conductors, 571
  electron configurations of, 1390–1391
  electron-gas model of, 1415
  free-electron model of, 1415, 1418–1422
  rare earth, 1390
Metastable states, 1308
Meters, 4–5, 438
Methane, structure of, 1407
Michelson, Albert, 1180–1181
Michelson interferometer, 1179–1181
Michelson-Morley experiment, 1180–1181, 1224
Microcoulomb, 696
Microfarad, 790–791
Micrographs, 1149
Microphones, condensor, 790
Microscopes, 1147–1149
  electron, 1290–1292
  resolving power of, 1210
  scanning tunneling, 1349
Microscopic state, 675–677
Microscopic vs. macroscopic properties, 590
Microwave ovens, 1071f
Milliampere, 820
Millibar, 375
Millikan, Robert, 897, 1264, 1293
Millikan's oil-drop experiment, 786
Mirages, 1103
Mirrors. See also Reflection
  concave, 1118–1122

converging, 1119
  convex, 1122–1124
  graphical methods for, 1124–1126
  image formation by, 1115–1126
  parabolic, 1120
  plane, 1115–1118
  spherical, 1118–1126
Mitchell, John, 423
Models
  definition of, 3
  idealized, 3–4
Molar heat capacity, 564–565, 605–607, 637–639
Molar mass, 517, 564, 591, 598
Molar specific heat, 564–565
Molecular bonds. See Bonds
Molecular clouds, 1435–1436
Molecular collisions, 603–605
  gas pressure and, 597–600
Molecular kinetic energy, 597, 605–606
  gas pressure and, 600–602
  temperature and, 636
Molecular mass, 598
  measurement of, 897
Molecular rotation, vibration and, 1410–1412
Molecular spectra, 1300, 1408–1412
Molecular speed, 602–603, 608–610
  Maxwell-Boltzmann distribution and, 608–609
Molecular vibration, 451–453, 597, 606–608
  rotation and, 1410–1412
Molecular weight, 564, 591
Molecular zippers, 1408
Molecules, 1405–1431
  gas, 597, 599
  intermolecular forces and, 597–598
  liquid, 597
  polar, 805–806, 1407
  polyatomic, 605–606
  solid, 597
Moles, 564, 598
Moment arm, 309
Moment of inertia, 288–291
  of bird's wing, 290f
  calculation of, 290, 294–296, 456
  of cylinder, 295–296
  definition of, 289
  parallel-axis theorem and, 293–294
  in simple harmonic motion, 451
  of sphere, 296
  torque and, 312
Moment, vs. torque, 309
Momentum, 241–266
  angular. See Angular momentum
  collisions and, 251–258
  components of, 242
  conservation of, 247, 255, 1243
  definition of, 242
  electromagnetic, 1068–1069
  impulse and, 241–246
  impulse-momentum theorem and, 242–244, 483–484
  linear, 242, 322
  magnitude of, 242
  net force and, 242
  Newton's second law and, 242
  Newton's third law and, 247
  of photons, 1277
  rate of change of, 242
  relativistic, 1243–1246
  rocket propulsion and, 262–265
  in simple harmonic motion, 449
  total, 247, 260
  transverse, 484–485
  units of, 242
  as vector, 242, 248
  vs. kinetic energy, 240, 242–246
  wave speed and, 484–485
Momentum-position uncertainty principle, 1274–1275,
    1278, 1315–1316
Monochromatic light, 1054–1055, 1164
Monopoles, magnetic, 885, 1500
Moon walking, 407f

Morley, Edward, 1180–1181
Moseley's law, 1394–1396
Motion
    along curve. *See* Curves
    of center of mass, 259–262
    circular, 85–88, 154–159, 440–442. *See also* Circular
        motion
    forces and, 112–116
    Kepler's laws of, 414–416
    Newton's laws of, 104–133. *See also* Newton's laws
        of motion
    orbital, 411–413
    period of, 87
    periodic, 437–462. *See also* Oscillation
    planetary, 414–418
    projectile, 77–85
    rotational. *See* Rotation/rotational motion
    of satellites, 411–413
    simple harmonic, 439–453
    straight-line, 35–68. *See also* Straight-line motion
    translational, 308, 314–320, 606
    in two/three dimensions, 69–103
Motion diagrams, 41, 41f, 46f
Motional electromotive force, 969–971
Motors, electric, 898, 907–908
Moving-axis rotation, 314–320
MRI (magnetic resonance imaging), 904–905, 934,
    1444
Muller, Karl, 824
Multimeters, digital, 863
Multiplets, 1383
Multiplication
    of displacement, 13
    significant figures in, 8f, 9
    of vectors, 13, 16, 20–22, 70
Muon-catalyzed fusion, 1471
Muonic atoms, 1379
Muons, 1254, 1485, 1491
Muscle fibers, work done by, 177
Music, sound waves in, 513–514
Musical instruments
    pipe organs, 524–527
    standing waves and, 497–498
    string, 497–498
    wind organs, 524–527
Mutual inductance, 991–994
Myopia, 1143–1146

N

Natural angular frequency, 459–460
Natural light, 1094
Near point, of eye, 1143
Nearsightedness, 1143–1146
Neddermeyer, Seth, 1485
Ne'eman, Yu'val, 1498
Negative ions, 690
Negative work, 179–180, 183
Neon, 1390
Nerve cells, electric charge in, 741
Nerve conduction, resistivity in, 824
Net force, 107, 247–248
    acceleration and, 112–118
    center-of-mass motion and, 261–262
    definition of, 107
    equilibrium and, 135
    momentum and, 242–244
    Newton's first law and, 109
    Newton's second law and, 112–118
    torque and, 311–312, 323, 324
Net torque, 321
Neurons, 716
    electric charge in, 741
Neutral side of line, 868–869, 870
Neutrino detectors, 1489–1490
Neutrino oscillations, 1500–1501
Neutrinos, 1452, 1492
    mass of, 1500–1501
Neutron(s)
    absorption of, 1464
    discovery of, 1481–1482

magnetic moment of, 1442–1443, 1497
    mass of, 689, 1440
    spin angular momentum of, 1442
Neutron number, 1441
Neutron-proton pair binding, 1446–1447
Newton, 6, 105, 113–114, 119
Newton, Isaac, 1080
Newtonian mechanics, 104
Newtonian synthesis, 418
Newton-meter, 177, 309
Newton's first law of motion, 108–112
    application of, 124–125, 134–139
    equilibrium and, 109
    inertia and, 109
    inertial frame of reference and, 110–112
    net force and, 109
    particles in equilibrium and, 134–139
    statement of, 108–109
Newton's law of gravitation, 402–406
Newton's laws of motion, 104–133
    application of, 124–125
    first law, 108–112. *See also* Newton's first law of
        motion
    free-body diagrams and, 124–125
    Kepler's laws and, 414–416, 418
    overview of, 104–108
    particle model and, 1274
    relativity and, 1244–1245, 1249–1251
    second law, 112–117. *See also* Newton's second law
        of motion
    statement of, 104
    third law, 120–125. *See also* Newton's third law of
        motion
    uncertainty and, 1274–1275
Newton's rings, 1178
Newton's second law of motion, 112–117
    application of, 115, 124–125, 140–146
    component equations for, 115
    constant mass and, 115
    entropy and, 674, 677
    external forces and, 115
    fluid resistance and, 152–154
    inertial frame of reference and, 115–116, 118
    momentum and, 242
    relativity and, 1244–1245, 1249–1251
    rotational analog of, 312, 318–320
    statement of, 114–115
Newton's third law of motion, 120–125
    action-reaction pairs and, 120–123
    application of, 124–125
    fluid resistance and, 151–154
    momentum and, 247
    statement of, 120
    tension and, 123
Noble gases, 1390
Nodal curves, 1166
Nodal planes, 1070
Nodes, 492
    displacement, 523
    pressure, 523
    of Ranvier, 882
Noise, 514
Noise control
    beat synchronization in, 532
    wave interference in, 531
Nonconservative forces, 222–224
Nonelectrostatic fields, 959, 972–973
Nonlinear conductors, 823
Nonreflective coatings, 1178–1179
Nonuniform circular motion, 159
Normal force, 105, 146
Normal mode, 496
Normalization condition, 1342, 1365
North (N) pole, 884
Notation
    for angles, 71
    scientific (powers-of-ten), 9
    spectroscopic, 1375
    for units of measure, 5
    for vectors, 11, 19

n-type semiconductors, 1424
Nuclear accidents, 1468
Nuclear angular momentum, 1442
Nuclear binding, 1444–1449
Nuclear fission, 785, 1247, 1464–1468
    chain reactions in, 1466
    liquid-drop model of, 1465–1466
    reaction dynamics in, 1465
    in reactors, 1466–1468
Nuclear force, 1446–1447
    mesons and, 1484–1485
    potential-energy function for, 1448–1449, 1491
Nuclear fusion, 785, 1284, 1469–1471
    heat of, 566
    helium, 1514
    solar, 1501
    tunneling in, 1349–1350
Nuclear magnetic moment, 1442–1443
Nuclear magnetic resonance, 1444
Nuclear magneton, 1443
Nuclear medicine, 1391, 1461, 1466
Nuclear physics, 1439–1471
Nuclear power plants, 1247, 1466–1468
Nuclear reactions, 1462–1471
    chain, 1466
    endoergic, 1463
    endothermic, 1463
    exoergic, 1463
    exothermic, 1463
    fission, 785, 1247, 1464–1468
    fusion, 1469–1471
    neutron absorption in, 1464
    reaction energy for, 1462–1464
    thermonuclear, 1469–1471, 1470
    threshold energy for, 1463
Nuclear reactors, 1247, 1466–1468
Nuclear spin, 1442–1443
Nuclear stability, 1449–1456
Nucleon number, 1440
Nucleons, 1439–1440
Nucleosynthesis, 1511–1515
Nucleus, 1439–1471
    atomic, 689, 1295
    daughter, 1454
    density of, 1440
    in excited states, 1454
    formation of, 1511–1514
    in ground state, 1454
    half-life of, 1456–1457
    lifetime of, 1457
    liquid-drop model of, 1447–1448
    mass of, 1440
    parent, 1454
    properties of, 1439–1444
    radius of, 1440
    shell model of, 1448–1449
    structure of, 1444
Nuclides, 1441
    decay of, 1450–1458. *See also* Radioactive decay
    odd-odd, 1449
    radioactive, 1450–1454. *See also* Radioactivity
    stable, 1449–1450
    synthesis of, 1511–1515

O

Object distance, 1116
Object, in optics, 1114
Object point, 1115
Objective, microscope, 1147–1148
Occhialini, Guiseppe, 1271
Oculars, 1148
Odd-odd nuclides, 1449
Oersted, Hans Christian, 885
Ohm, 826
Ohmic conductors, 823
Ohmmeters, 863
Ohm's law, 822, 825–826
1 newton per coulomb, 699
Onnes, Heike Kamerlingh, 824
Open circuits, 870

Open orbits, 412
Operational definition, 4
Optic axis, of mirror, 1118
Optical fibers, 1090
Optics, 1080. *See also* Light
  geometric. *See* Geometric optics
  image in, 1115
  object in, 1114
  physical, 1082, 1163
Orbit(s)
  center of mass and, 417–418
  circular, 412–413, 416–417
  closed, 412
  of Comet Halley, 417
  elliptical, 415–416
  open, 412
  satellite, 411–413
  sector velocity and, 415
  semi-major axis of, 415, 416
Orbital angular momentum
  quantization of, 1373–1374, 1384
  spin angular momentum and, 1387
Orbital eccentricity, 415
Orbital magnetic quantum number, 1374
Orbital period, 416–417
Orbital quantum number, 1373
Orbital speed, 416–417
Order-of-magnitude estimates, 10
Organ pipes, 524–527
Oscillation, 437–462
  amplitude of, 438
  damped, 457–460
  definition of, 437
  displacement in, 438
  driven, 459–460
  electrical, 1005–1009
  forced, 459–460, 527
  frequency of, 438
  in harmonic oscillators, 439–440
  molecular vibration and, 451–453
  neutrino, 1500–1501
  overview of, 437–438
  of pendulum, 453–457
  of periodic waves, 474–476
  resonance and, 460, 527
  simple harmonic motion and, 439–453. *See also*
    Simple harmonic motion
  in spring, 437–438
Oscillation cycle, 438
Oscillation period, 438
Otto cycle, 657–658
Overdamped circuits, 1010
Overdamping, 458
Overloaded circuits, 869–870
Overtones, 496

P

Pacemakers, 866
Painting, electrostatic, 683
Pair annihilation, 1483, 1484
Pair production, 1482–1483
  photons in, 1271–1272
  positrons in, 1271–1272, 1482–1483
Parabolic graphs, 48
Parabolic lenses, 1151
Parabolic mirrors, 1120
Parabolic trajectories, 79, 79f
Parallel connection, 794–796
Parallel, resistors in, 851, 852–855
Parallel vectors, 11, 12
Parallel-axis theorem, 293–294
Parallel-plate capacitors, 790, 791
  dielectrics in, 800–805
Paramagnetism, 942–944, 943t
Paraxial approximation, 1119
Paraxial rays, 1119
Parent nucleus, 1454
Parity, 1495
Parsec, 1503
Partial derivatives, 226–227, 481

Particle(s), 36
  alpha, 1294–1295, 1349–1350
  antiparticles and, 1483
  in bound state, 1343–1344
  definition of, 3
  distinguishable, 1419
  as force mediators, 1484
  free, 1330
  fundamental, 1480–1501
  light waves as, 1261–1280
  in Newtonian mechanics, 1274
  photons as, 1263. *See also* Photons; Wave-particle
    duality
  in standard model, 1499–1500
  strange, 1494–1495
  wave function for, 1328–1335
Particle accelerators, 1485–1488
  cyclotrons, 893, 918, 1486–1487
  linear, 1485–1486
  synchrotrons, 1487
Particle collisions
  in accelerators, 1485–1488
  available energy and, 1487–1488
  in colliding-beam experiments, 1489
Particle detectors, 1489
  neutrino, 1489–1490
Particle in a box
  in one dimension, 1338–1343, 1371
  in three dimensions, 1366–1371
Particle interactions, 1490–1495
  conservation laws for, 1495
  electromagnetic, 159–160, 1490
  fundamental types of, 159–160, 1490–1492
  gravitational, 159–160, 1490
  isospin and, 1495
  parity in, 1495
  strangeness in, 1494–1495
  strong, 160, 1490–1491
  symmetry-breaking, 1495
  weak, 160, 1491
Particle motion, vs. wave motion, 475
Particle physics, historical perspective on, 1480–1485
Particle speed, vs. wave speed, 519
Particle velocity, vs. wave velocity, 519
Particle waves
  angular frequency of, 1331
  one-dimensional, 1329–1333
  vs. mechanical waves, 1329
  wave equation for, 1330–1333
  wave number for, 1331
Pascal, 353, 375
Pascal, Blaise, 353, 354, 377
Pascal's law, 376–377
Paths, in thermodynamic system, 628–629
Pauli, Wolfgang, 1389
Pendulum
  ballistic, 253
  periodic motion of, 453–457
  physical, 455–457
  simple, 453–455, 456
Pendulum bob, 454
Penzias, Arno, 1515
Percent error (uncertainty), 8
Perfect crystals, 1413–1415
Perihelion, 415
  precession of, 1250
Period, 87
  frequency and, 438–439
  orbital, 416–417
  oscillation, 438
  in simple harmonic motion, 443
Periodic driving force, damped oscillation and,
  459–460
Periodic motion, 437–462. *See also* Oscillation
  amplitude of, 438
  definition of, 437
  displacement in, 438
  frequency of, 438
  in harmonic oscillators, 439–440
  molecular vibration and, 451–453

overview of, 437–438
  of pendulum, 453–457
  resonance and, 460
  simple harmonic motion and, 439–453. *See also*
    Simple harmonic motion
  of spring, 437–438
  of waves, 474–476
Periodic table, 1389, 1390–1391
Periodic waves, 474–477. *See also* Mechanical waves
  longitudinal, 475–476. *See also* Longitudinal waves
  mathematical description of, 477–482
  sinusoidal, 475, 477–482. *See also* Sinusoidal waves
  transverse, 474–475. *See also* Transverse waves
Permanent magnets, 883, 941
Permanent set, 358
Permeability, 943
Permittivity, of dielectric, 802
PET (positron emission tomography), 1484
Pfund series, 1304
Phase, 565
  wave, 479
Phase angle, 444, 1026, 1031–1032
Phase change, 565–570
Phase diagrams, 611
Phase equilibrium, 566, 611
Phase shifts, interference and, 1174–1175
Phase transitions, 611–612
Phase velocity, 479
Phased-array radar, 1220
Phases of matter, 610–613
  critical point and, 611
  molecular interactions and, 610
  sublimation and, 611
  triple point and, 611
Phases of state
  p-V diagrams and, 596
  pVT-surfaces and, 612–613
Phasor diagrams, 1022
Phasors, 441, 1022
Photinos, 1501
Photocells, 1425
Photocopying machines, 769
Photoelasticity, 1100
Photoelectric effect, 1081, 1261
Photoelectrons, 1262
Photography. *See* Cameras
Photomicrographs, 1149
Photomultipliers, 1273
Photons, 1081, 1248, 1261–1280
  absorption of, 1261–1266, 1484
  in Bohr's atomic model, 1297–1306
  in charged-particle interactions, 1484
  Compton scattering and, 1269–1271
  definition of, 1261
  diffraction and, 1273–1274
  discovery of, 1481
  Einstein's explanation for, 1263–1264
  electroweak interactions and, 1500
  emission of, 1266–1269, 1484
  as force mediators, 1484
  gamma ray, 1454
  interference and, 1273–1274
  light emitted as, 1266–1269, 1484
  momentum of, 1264–1265, 1274–1275, 1277
  pair production and, 1271–1272
  as particles, 1263
  photoelectric effect and, 1261–1266
  position of, 1274–1275, 1277
  probability and, 1274–1275
  spontaneous emission of, 1307
  in standard model, 1500
  stimulated emission of, 1307–1309
  stopping potential and, 1262–1263
  threshold frequency and, 1263
  uncertainty and, 1274–1278
  virtual, 1484
  wave-particle duality and, 1273–1279. *See also*
    Wave-particle duality
  X-ray, 1266–1269
Photovoltaic effect, 1428

Physical laws (principles), 2
Physical optics, 1082, 1163
Physical pendulum, 455–457
  vs. simple pendulum, 456
Physical quantities
  definition of, 4
  units of, 4–6. *See also* Units of measure
Physical theories, 2
Physics
  as experimental science, 2
  nuclear, 1439–1471
  overview of, 2
  particle, 1480–1485
  as process, 2
  quantum, 1328–1356
Pi, value of, 8, 8f
Pianos, string vibration in, 528
Picoampere, 820
Picofarad, 791
Pileated woodpecker impulse, 243
Pions, 1485, 1489, 1491, 1493–1494
Pipe organs, 524–527
Pitch, 513
Planck length, 1509
Planck, Max, 1311–1312
Planck radiation law, 1311–1314
Planck time, 1509
Planck's constant, 942, 1263
Plane mirrors
  graphical methods for, 1124–1126
  image formation by, 1115–1118
Plane of incidence, 1084
Plane surface
  reflection at, 1115–1118
  refraction at, 1129
Plane waves, electromagnetic, 1055–1057, 1060
Planet Imager, 1221
Planetary motion, 414–418. *See also* Orbit(s)
  center of mass and, 417–418
  Kepler's laws of, 414–417
Plant growth, deuterium and, 1445
Plastic deformation, 358
Plastic flow, 358
Plasticity, 357–358
p-n junctions, 1425–1428
Point charges, 693–694
  electric dipole and, 709–713
  electric fields of, 725–726. *See also* Electric charge
  electric potential energy of, 757–760
  electric potential of, 765
  electromagnetic waves from, 1053
  force between, 697
  inside closed surface, 725–726
  inside nonspherical surface, 732
  inside spherical surface, 732
  magnetic field lines for, 924–925
  superposition of, 696
Point objects
  definition of, 1115
  image formation for, 1118–1119, 1127–1130, 1209–1210
  resolution of, 1209–1210
Point-molecule model, of gas heat capacity, 605
Polar molecules, 805–806, 1407
Polarization, 693, 805–806, 1093–1100
  bee vision and, 1101
  charged bodies and, 693, 807
  circular, 1099–1100
  definition of, 788, 1093
  of dielectrics, 801–803, 805–807
  electric field lines and, 709
  of electromagnetic waves, 1058, 1093–1100
  elliptical, 1099–1100
  induced charges and, 693, 807
  of light waves, 1093–1100
  linear, 1058, 1093, 1095
  partial, 1097
  photoelasticity and, 1100
  by reflection, 1097–1098
Polarizers, 1093

Polarizing angle, 1097
Polarizing axis, 1094–1095
Polarizing filters, 1093, 1094–1097, 1098
Polaroid filters, 1094–1095
Pollen, fluid resistance and, 152
Polyatomic molecules, 605–606
Population inversions, 1308–1309
Porro prism, 1089–1090
Position
  by integration, 55–57, 55f, 56f
  potential energy and, 208
  x-t graphs and, 37–38, 38f, 40–42
Position vectors, 70–72, 70f
Position-momentum uncertainty principle, 1274–1275, 1278, 1315–1316
Positive ions, 690
Positive work, 179, 183
Positron emission tomography (PET), 1484
Positroniums, 1379
Positrons, 1453, 1482–1483
  motion in magnetic fields, 894
  in pair annihilation, 1483
  in pair production, 1271–1272, 1482–1483
Potential. *See* Electric potential
Potential barriers, 1347–1350
Potential difference, 763–764. *See also* Voltage
  capacitance and, 789
  measurement of, 861–862
  notation for, 865
  resistance and, 852
  time-varying, 865
Potential energy, 207–231, 755
  around circuits, 833–834
  of capacitor, 796–800
  conservative forces and, 221–229
  definition of, 208
  elastic, 216–221, 225
  electric, 754–761. *See also* Electric potential energy
  of electric dipole, 711
  electric forces and, 226–227
  energy diagrams and, 228–229
  equilibrium and, 228–229
  force and, 225–228
  gradient of, 227
  gravitational, 208–216, 293, 317, 409–411
  intermolecular forces and, 597
  kinetic energy and, 207, 208, 221–222
  for magnetic dipoles, 903–904
  of molecules, 597
  of particle in a box, 1339
  position and, 208
  potential barriers and, 1347–1350
  potential wells and, 597, 1343–1347
  in simple harmonic motion, 446–449
  work and, 755
Potential gradient, 774–776
Potential wells, 597, 1343–1347
Potential-energy function
  for harmonic oscillator, 1350, 1352, 1353–1354
  for nuclear force, 1448–1449, 1491
  for particle in a box, 1339
Potentiometers, 863–864
Pound, 117
Pounds per square inch, 353
Pounds per square inch absolute (psia), 378
Pounds per square inch gauge (psig), 378
Power, 193–195
  in ac circuits, 1034–1037
  average, 193, 487
  of corrective lens, 1144
  definition of, 193
  in electric circuits, 834–838
  for electric motors, 907
  electrical, 194
  energy and, 871
  force and, 194
  instantaneous, 193, 194
  measurement of, 862–863
  rotational motion and, 321–322
  of sound waves, 519

  velocity and, 194
  of waves, 487–488
  work and, 487
Power distribution systems, 868–872
Power factor, 1036
Power plants, nuclear, 1247, 1466–1468
Power transmission systems, lightning strikes
  on, 995
Powers-of-10 notation, 9
Poynting vector, 1065–1067
Precession, 328–330
  of perihelion, 1250
Precession angular speed, 329
Precipitators, electrostatic, 784
Precision, vs. accuracy, 9
Prefixes, for units of measure, 5
Presbyopia, 1143
Pressure
  absolute, 377–378
  atmospheric, 355, 375–376
  bulk stress/strain and, 355–356
  definition of, 355, 375
  fluid flow and, 385–389
  in fluids, 355, 375–380
  gauge, 377–378
  measurement of, 378–380
  radiation, 1068–1069
  reciprocal, 356
  residential water, 387
  as scalar quantity, 355
  speed and, 385–389
  units of, 353, 375
  vs. density, 592
  vs. force, 355, 376
Pressure amplitude, 511–512
  sound intensity and, 519–521
Pressure gauges, 378–380
Pressure nodes/antinodes, 523
Primary windings, 1040, 1042f
Principal maxima, 1201
Principal quantum number, 1301, 1373
Principal rays
  for lenses, 1135–1136
  for mirrors, 1124–1125
Principles
  differential, 245
  integral, 244
  physical, 2
Printers
  inkjet, 722
  laser, 689, 769, 1309
Prism
  dispersion by, 1091
  Porro, 1089–1090
Prism binoculars, 1150
Probability density, 1333
Probability distribution, 1376
Probability distribution function, 1333
  for harmonic oscillator, 1353
  one-dimensional, 1333
  radial, 1376
  three-dimensional, 1365–1366
Probability, wave-particle duality and, 1274–1278
Problem-solving strategies, 2–3
Product
  scalar, 20–22
  vector, 23–25
Projectile, 77
Projectile motion, 77–85
  acceleration and, 77–80, 87
  air resistance and, 77, 79–80
  components of, 77–78
  trajectory and, 77
  velocity and, 77–80
  vs. circular motion, 87
Projectors, 1141
Propagation speed, 474, 475
Propeller design, 288
Proper length, 1235
Proper time, 1230–1231, 1232

Proton(s)
  charge of, 695
  electron screening of, 1391–1392
  lifetime of, 1500
  magnetic moment of, 1442–1443
  mass of, 689, 1440
  spin angular momentum of, 1442
Proton decay, 1500
Proton-antiproton pairs, 1491–1492
Proton-neutron pair binding, 1446–1447
Proton-proton chains, 1469
Psia (pounds per square inch absolute), 378
Psig (pounds per square inch gauge), 378
p-type semiconductors, 1425
Pulsed lasers, 1309
Purcell, Edward, 1416f
Pure semiconductors, 1423, 1424
p-V diagrams, 596
p-V isotherms, 596
pVT-surfaces, 612–613

Q
Quality factor, 1459
Quanta, 1081. *See also* Photons
Quantitized energy, 1261
Quantity of heat, 562–565
Quantum dots, 1363
Quantum electrodynamics, 1081
Quantum hypothesis, 1310–1314
Quantum mechanics, 1328–1356
  atomic structure and, 1364–1398
  bound states and, 1343–1345
  definition of, 1329
  harmonic oscillator in, 1350–1354
  one-dimensional waves in, 1329–1333
  particle in a box and, 1338–1343, 1366–1371
  potential barriers and, 1347–1350
  potential wells and, 1343–1347
  probability distribution function and, 1333, 1353, 1365–1366
  Schrödinger equation and, 1332–1333, 1336–1337, 1365–1366
  stationary states and, 1337–1338, 1366
  tunneling and, 1347–1350
  wave functions and, 1328–1335
  wave packets and, 1335–1336
Quantum number
  notation for, 1374–1375
  orbital magnetic, 1374
  principal, 1373
  spin, 1385
  spin magnetic, 1385
Quarks, 689, 920, 1443, 1496–1499
  antiquarks and, 1496, 1499
  colors of, 1498–1499
  down, 1496
  eightfold way and, 1497–1498
  flavors of, 1496
  in standard model, 1500
  strange, 1496
  types of, 1496
  up, 1496
Quarter-wave plates, 1100
Quasars, 1220

R
Rad, 1459
Radar
  Doppler, 537
  phased-array, 1220
Radar guns, 1242f
Radial probability distribution function, 1376, 1392
Radians, 279, 287
Radiation, 570, 574–577
  absorption of, 575
  applications of, 576
  background, 1515
  beneficial uses of, 1461–1462
  biological effects of, 1459–1462
  blackbody, 576, 1310–1314

cancer and, 1269
Čerenkov, 1257
definition of, 574
electromagnetic, 574–577, 1053–1054. *See also* Electromagnetic wave(s)
global warming and, 576–577
from human body, 575–576
quality factor for, 1459
solar, 576, 1262
Stefan-Boltzmann law/constant and, 575
synchrotron, 1487
thermal, 1081
X. *See* X-ray(s)
Radiation doses, 1459–1460
Radiation exposure
  hazards of, 1460–1461, 1476
  limits on, 1459–1460
  sources of, 1458–1459, 1461
Radiation pressure, 1068–1069
Radiator, ideal, 576
Radioactive dating, 1458
Radioactive decay, 1450–1458
  activity in, 1456–1457
  alpha, 1450–1452
  beta, 1452–1453
  gamma, 1454
  half-life and, 1456–1457
  natural radioactivity and, 1454–1455
  rate of, 1456–1457
Radioactive decay series, 1454–1455
Radioactive fallout, 1476–1477
Radioactive isotopes, in medicine, 1391, 1461, 1466
Radioactive nuclides, decay of, 1450–1454
Radioactive tracers, 1461–1462
Radioactivity
  definition of, 1449
  natural, 1454–1455
  units for, 1457
Radioisotope imaging, 1391
Radiology, 1268–1269
Radios
  transmitters and receivers for, 1054
  tuning, 1038, 1039
Radium, alpha decay of, 1451–1452
Radius
  of nucleus, 1440
  Schwarzschild, 424
Radius of curvature
  for lens, 1133–1134
  for spherical surface, 1116, 1128–1129
Radon, 1458–1459
Rainbows, 1092–1093
Randomness, in thermodynamic processes, 653
Range of validity, 2
Rare earth metals, 1390
Rarefaction, 476
Ratio of heat capacities, 517, 639
Rayleigh, Lord, 1209, 1311
Rayleigh's criterion, 1209, 1210
Rays, 1082, 1135
  paraxial, 1119
  principal, 1124–1125, 1135–1136
R-C circuits, 864–868
Reaction(s)
  activation energy for, 610
  chain, 1466
  chemical, 610
  nuclear, 1462–1471. *See also* Nuclear reactions
Reaction energy, 1462–1464
Real image, 1115
Recession speed, 1502–1503, 1504, 1505
Reciprocal pressure, 356
Recombination currents, 1427
Rectified alternating current, 1022–1023
Rectified average current, 1023
Redshifts, 1502, 1507
  cosmological, 1505
Reduced mass, 1305–1306, 1408–1409
Reference circle, 440–441
Reference point, 440–441

Reference standards, 4
Reflected waves, 489–490
  sinusoidal, 491–495
Reflecting telescopes, 1150–1151
Reflection, 1082–1091
  Bragg, 1207
  definition of, 1082
  diffuse, 1083, 1115
  of electromagnetic waves, 1071
  Huygen's principle and, 1102
  image formation and, 1115–1118
  interference during, 1174–1175
  law of, 1084–1086
  of light waves, 1082–1088
  phase shifts during, 1174–1175
  at plane surface, 1115–1118
  polarization by, 1097–1098
  specular, 1083, 1115
  at spherical surface, 1118–1126
  total internal, 1088–1091
  in X-ray diffraction, 1206
Reflective coatings, 1178, 1179
  interference and, 1178–1179
Reflector, ideal, 576
Refracting telescopes, 1149
Refraction, 1082–1088
  definition of, 1082
  in eye, 1143
  Huygen's principle and, 1102–1104
  index of. *See* Index of refraction
  law of, 1084–1086
  at plane surface, 1115–1118
  at spherical surface, 1126–1130
Refractive index. *See* Index of refraction
Refractors, 1154
Refrigerator(s), 659–661
  Carnot, 666–667
  practical, 660–661
  workless, 661
Refrigerator statement, of second law of thermodynamics, 662
Reines, Frederick, 145
Relative biological effectiveness, 1459–1460
Relative permeability, 943
Relative velocity, 88–93. *See also* Velocity
  definition of, 88
  elastic collisions and, 256–258
  frame of reference and, 89
  Galilean velocity transformation and, 91
  in one dimension, 88–90, 89f
  in two or three dimensions, 90–93
Relativistic momentum, 1243–1246
Relativistic work and energy, 1246–1249
Relativity, 1223–1252
  aging and, 1233
  Doppler effect and, 537–538, 1241–1243
  Einstein's postulates for, 1224–1225
  Galilean coordinate transformation and, 1225–1226
  general theory of, 1249–1251, 1504
  inertial frame of reference and, 1223, 1224, 1226
  invariance of physical laws and, 1223–1226
  of length, 1233–1237
  Lorentz transformations and, 1237–1241
  Newtonian mechanics and, 1244–1245, 1249
  principle of, 1224
  of simultaneity, 1226, 1227–1228
  special theory of, 1223–1249
  speed of light and, 1224–1225
  of time intervals, 1228–1233
  twin paradox and, 1232–1233
Relativity principle, 1224
Relaxation time, 866–867
Rem, 1460
Resistance, 825–828
  equivalent, 851, 852
  internal, 830, 833
  measurement of, 860–861, 862–864
Resistance thermometer, 553–554
Resistivity, 822–825
  of metal, 838–839

in nerve conduction, 824
temperature and, 824–825
Resistors, 826–827
in ac circuits, 1025, 1029, 1034–1035
in dc circuits, 850–855
energy dissipation in, 835
equivalent resistance and, 851–855
in parallel, 851, 852–855
power in, 1034–1035
power input to, 834–835
power rating of, 835
in series, 851–852, 853–855
shunt, 861
vs. inductors, 999
Resolving power (resolution), 1209–1211
chromatic, 1203–1204, 1210
in diffraction, 1209–1210
of grating spectrograph, 1203–1205
limit of, 1209
of microscope, 1290–1291
Rayleigh's criterion and, 1209, 1210
Resonance, 460, 527–529
in ac circuits, 1037–1039
definition of, 460, 1038
in mechanical systems, 460
Resonance angular frequency, 1038
Resonance curves, 1038–1039
Resonance frequency, 1038
Resonance width, 1049
Response curves, 1038–1039
Rest energy, 1247–1249
Rest mass, 1243–1246
Restoring force, 438
in pendulum, 454
in simple harmonic motion, 439–440
Resultant, of displacements, 12
Reverse bias, 1426
Reversed, 1117
Reversed image, 1117
Reversible processes, 653
Right-hand rule, 23
Right-handed system, 24
Rigid body, 278
Rigid-body equilibrium, 345, 348–352
Rigid-body rotation, 278–297. *See also*
    Rotation/rotational motion
    about moving axis, 314–320
    angular acceleration in, 282–285
    angular velocity in, 279–282
    around fixed axis, 278–279
    dynamics of, 308–331
    kinetic energy in, 288–293
    linear acceleration in, 286–288
    linear speed in, 285–286
    moment of inertia and, 288–291
    with translational motion, 314–320
R-L circuits, 1001–1005
    current decay in, 1004–1005
Rms speed, 602
Rocket propulsion, 262–265
Roller coasters, 74, 88
Rolling friction, 151–154, 320
Rolling without slipping, 316–318
Röntgen, Wilhelm, 1205, 1267, 1454
Root-mean-square current, 1023–1024
Root-mean-square speed, 602
Root-mean-square values, 602, 1023–1024
Root-mean-square voltage, 869
Rotational energy levels, 1408–1412
Rotational inertia, 289
Rotational kinetic energy, 288–293, 315–316
Rotation/rotational motion
    about axis of symmetry, 323–324
    angular acceleration and, 282–285, 311–314
    angular momentum and, 322–328
    angular velocity in, 279–282
    around fixed axis, 278–279
    in bacteria, 283f
    with constant angular acceleration, 283–285
    coordinates for, 279

direction of, 279, 309
dynamics of, 308–331
of Earth, 421–423
energy in, 288–293
equilibrium and, 345
fixed-axis, 278–279, 283
of gyroscope, 328–330
kinetic energy and, 288–293
linear acceleration in, 286–288
linear speed in, 285–286
molecular, 1410–1412
moving-axis, 314–320
Newton's second law of motion and, 312, 318–320
power and, 321–322
precession and, 328–330
rigid-body, 278–297. *See also* Rigid-body rotation
in rolling without slipping, 316–318
torque and, 308–314
with translational motion, 314–320
units of, 279
work and, 320–322
Rotors, 907–908
Rubbia, Carlo, 1491
Rule of Dulong and Petit, 565, 608
Running on moon, 407f
Rutherford, Ernest, 1294–1296, 1297, 1349–1350,
    1439–1440, 1454, 1462, 1481
Rutherford's atomic model, 1294–1296
Rutherford's scattering experiments, 1294–1296
Rydberg atom, 1306, 1402
Rydberg constant, 1303

S

Satellite orbits, 411–413
Saturation, 1447
Saturation current, 1426
Saturation magnetization, 945
Scalar (dot) product, 20–22
Scalar quantities, 11
    in vector multiplication, 13
Scale factor, 1504, 1505
Scanning electron microscope, 1291–1292
Scanning tunneling microscope, 1349
Scattering of light, 1100–1101
Schrieffer, Robert, 1430
Schrödinger equation, 1332–1333
    for hydrogen atom, 1372–1373
    for hydrogenlike atoms, 1378–1379
    one-dimensional, 1332–1333, 1336–1337
    with potential energy, 1336–1337
    three-dimensional, 1365–1371
    time-independent, 1338, 1366
    X-ray spectra and, 1393–1396
Schrödinger, Erwin, 1332
Schwarzschild radius, 424
Scientific notation, 9
Scintigram, 1461
Scintillation, 1294
Screening, 1391–1392
Scuba tank, air mass in, 594
Search coils, 983
Second condition for equilibrium, 345
Second law of thermodynamics, 652, 661–677
    Carnot cycle and, 667–668
    Clausius statement of, 662–663
    engine statement of, 661
    Kevin-Planck statement of, 661
    refrigerator statement of, 662–663
Secondary windings, 1040, 1042f
Seconds, 4, 5t, 438
Seconds per cycle, 438
Sector velocity, 415
Segré chart, 1449–1450, 1454
Segré Emilio, 1449
Selection rules, 1382
Selectrons, 1501
Self-induced emf, inductive reactance and, 1026–1027
Self-inductance, 991, 994–998. *See also* Inductance
Semiconductor(s), 909, 1422–1425
    bias conditions and, 1426, 1427–1428

compensated, 1425
conduction in, 819–820
diodes of, 827
doping and, 1424–1425
energy bands in, 1417
holes in, 909, 1423–1424
impurities in, 1415, 1424–1425
intrinsic, 1423, 1424
moving charges in, 819–820
n-type, 1424
p-type, 1425
resistivity of, 823
silicon, 1422–1425
Semiconductor devices, 1425–1430
    integrated circuits, 1429–1430
    light-emitting diodes, 1428
    photocells, 1425
    p-n junctions in, 1425–1428
    solar cells, 1428
    transistors, 1429
Semiconductor lasers, 1309
Semiempirical mass formula, 1448
Semi-major axis, 415, 416
Separation of variables, 1367
Series connection, 793–794, 795–796
Series motors, 908
Series, resistors in, 851–852, 853–855
Sharks
    electric field detection by, 699f
    flux through mouth of, 729f
Shear modulus, 357
Shear strain, 356–357
Shear stress, 352f, 356–357
Shell model, 1448–1449
Shells, electron, 1375, 1389, 1390–1391, 1394–1395
Shock waves, 538
Short circuits, 869–870, 870
Shunt motors, 908
Shunt resistors, 861
SI units, 4, 5t. *See also* Units of measure
Sievert, 1460
Significant figures, 8–9
Silicon semiconductors, 1422–1425
Simple harmonic motion, 439–453. *See also*
    Oscillation
    acceleration in, 444, 448
    amplitude in, 442–443
    angular, 451
    applications of, 450–453
    circular motion and, 440–442
    definition of, 440
    displacement in, 443–444
    energy in, 446–449
    equations of, 440–442
    as model of periodic motion, 440
    momentum in, 449
    period in, 442–443
    velocity in, 444, 448
    vertical, 450–451
    vs. electric oscillation, 1008
Simple pendulum, 453–455
    vs. physical pendulum, 456
Simultaneity, relativity of, 1226, 1227–1228
Sinusoidal alternating emf, 1021
Sinusoidal current, 1022–1024. *See also* Alternating
    current
Sinusoidal electromagnetic waves, 1060–1063
Sinusoidal waves, 475, 477–482. *See also* Mechanical
    waves; Wave(s)
    electromagnetic, 1022–1024
    energy of, 486–489
    interference and, 1164
    particle velocity/acceleration in, 480
    reflected, 491–495
    standing, 491–498. *See also* Standing waves
    traveling, 492, 494
    wave function for, 477–479
Sledding, Newton's first law and, 109
Slidewire generators, 965–966, 967, 970
Slipher, Vesto, 1502

Slug, 117
Snakes, wave motion of, 473f
Snell's law, 1084
Sodium doublet, 1204
Solar cells, 1428
Solar neutrinos, 1501
Solar radiation, 576, 1264
Solenoids, 904, 906, 939–941, 948t
Solids
  amorphous, 1412
  bonds in, 1414–1415
  as condensed matter, 1412
  crystalline, 1412–1415
  energy bands in, 1416–1417
  heat capacities of, 607–608
  melting points of, 1412
  molecules in, 597
  phases of, 610–613
  sound waves in, 515–517
  structure of, 1412–1415
Sonar waves, 516–517
Sound
  definition of, 509
  infrasonic, 510
  loudness of, 513
  pitch of, 513
  resonance and, 527–529
  timbre of, 513–514
  ultrasonic, 510
Sound intensity, 518–522
  decibel scale for, 521
  hearing loss and, 513, 522
  representative values for, 521t
Sound intensity level, 521
Sound waves, 476, 509–542. See also Mechanical
      waves; Wave(s)
  audible range of, 509
  beats and, 531–532
  diffraction of, 1198
  direction of, 510
  displacement amplitude of, 510, 518–519
  Doppler effect and, 533–538, 1242
  in fluid, 514–515
  frequency of, 513, 520
  in gas, 517–518
  graphing of, 511
  harmonic content of, 497
  interference and, 529–531
  musical, 513–514
  perception of, 513–514
  pipe organs and, 524–527
  power of, 519
  pressure amplitude of, 511–512, 519–521
  as pressure fluctuations, 510–513
  shock, 538
  in solid, 515–517
  speed of, 476, 485, 514–518
  standing, 522–527
  superposition of, 491
  wind instruments and, 527
Source of emf, 828
  internal resistance of, 830–831
  power input to, 835–836
  power output of, 835–836
Source point, 700, 924
South (S) pole, 884
Space. See also Universe
  curvature of, 1250f
  dimensions of, 1504–1505
  expansion of, 1503–1508
Space travel, aging and, 1233
Spacecraft, in interplanetary travel, 416f
Spacetime, 1238
Spacetime coordinates, 1238
Spark plugs, 1000
Special theory of relativity, 91, 1223–1249. See also
      Relativity
Specific gravity, 374
Specific heat, 562–563
  molar, 564–565

Spectra, 1091
  absorption line, 1293
  atomic, 1292, 1297–1300
  band, 1411
  continuous, 1310–1314
  emission line, 1292–1293
  molecular, 1300, 1408–1412
  X-ray, 1393–1396
Spectral emittance, 1310–1314
  definition of, 1310
  quantum hypothesis and, 1311–1313
  vs. intensity, 1310–1311
Spectral lines, 1292, 1297
  Zeeman effect and, 1379–1382
Spectrographs, grating, 1203–1204
Spectroscopic notation, 1375
Specular reflection, 1083, 1115
Speed, 39
  air drag and, 152–154
  angular, 280, 286, 329
  average, 39
  of efflux, 387
  of electromagnetic waves, 1058, 1071
  escape, 410–411, 413, 423, 1505–1506
  instantaneous, 39
  molecular, 602–603, 608–610
  orbital, 416–417
  recession, 1502–1503, 1504, 1505
  of rocket, 265
  root-mean-square (rms), 602
  of sound waves, 514–518
  supersonic, 539
  terminal, 152–154
  units of, 6f
  vs. velocity, 39–40, 286
  wave, 474, 475, 479, 483–486
  work and, 181–183
  of yo-yo, 317
Speed of light, 1054, 1063f, 1081, 1224–1226
  measurement of, 4–5
  relativity and, 1224
Spheres
  electric field of, 737–738, 740–741
  electric flux through, 731–732
  gravitation and, 403–404
  mass distributions and, 418–421
  moment of inertia of, 296
  point charge inside, 732–733
  rolling, acceleration of, 319–320
Spherical aberration, 1119, 1134
Spherical coordinates, 1366
Spherical mirrors
  concave, 1118–1122
  convex, 1122–1124
  extended objects in, 1120–1122
  focal point/length of, 1119–1120
  graphical methods for, 1124–1126
  image formation by, 1118–1126
Spherical surface
  radius of curvature for, 1116, 1128–1129
  reflection at, 1118–1126
  refraction at, 1126–1130
Spherical symmetry, 1366, 1372, 1388
  gravitation and, 403–404
Spin
  electron, 942
  nuclear, 1442–1443
Spin angular momentum, 1384–1385,
      1442
  orbital angular momentum and, 1387
Spin magnetic moment, 1443
Spin magnetic quantum number, 1385
Spin quantum number, 1385
Spin-2 graviton, 1490
Spin-orbit coupling, 1386
Spiny lobsters, magnetic compasses in, 886
Spring(s)
  elastic potential energy of, 216–221
  ideal, 439
  oscillation in, 437–438. See also Oscillation

  simple harmonic motion in, 439–453. See also
      Simple harmonic motion
  tendons ad, 190
  work done on/by, 188–189
Spring balance, 106
Spring constant, 188
Square wells, finite vs. infinite, 1345–1347
Square-well potential, 1343–1347
  bound states of, 1343–1345
Squids, jet propulsion in, 262f
Stable equilibrium, 228
Stable isotope ratio analysis (SIRA), 919
Stable nuclides, 1449–1450
Standard deviation, 1275
Standard model, 1499–1500, 1510–1511
Standards, reference, 4
Standing waves, 491–498
  complex, 497
  electromagnetic, 1053, 1069–1072
  on fixed string, 495–498
  frequencies of, 496
  harmonics and, 496
  interference and, 492, 1164, 1166
  nodes and antinodes and, 492
  sound, 522–527
  string instruments and, 497–498
Stars
  binary, 425–426, 1259
  helium fusion in, 1514
  mass of, 1259
  second-generation, 1514
  supernova, 160, 1230f, 1514
  systems of, 405–406
  white dwarf, 1437
State(s)
  bound, 1343–1344
  degenerate, 1370–1371, 1374–1375
  density of, 1418–1419
  free-particle, 1346
  of matter, 565
  metastable, 1308
  stationary, 1337–1338, 1366–1369
  vs. energy levels, 1307–1308
State variables, 591
Static charge distribution, 759–760
Static equilibrium, 345
Static friction, 147–149
Stationary state
  one-dimensional, 1337–1338
  three-dimensional, 1366–1369
Steady flow, 382–383
Steam heat, 567–568
Stefan-Boltzmann constant, 575, 1310
Stefan-Boltzmann law, 575, 1310, 1313
Stern-Gerlach experiment, 1383–1384
Stick-slip phenomenon, 148, 149
Stimulated emission, 1307–1309
Stopping potential, 1262–1263
Straight-line motion, 35–68
  with average acceleration, 42–46
  average velocity and, 36–38
  with constant acceleration, 46–52
  with constant force, 141
  displacement and, 36–38
  of freely falling bodies, 52–55
  with friction, 141
  with instantaneous acceleration, 42–46
  instantaneous velocity and, 38–42
  relative velocity and, 88–90
  time and, 37–38
  work-energy theorem for, 187–191
Strain
  bulk, 354–356
  compressive, 354
  definition of, 352
  deformation and, 352–357
  elastic modulus and, 352
  elasticity and, 357–358
  shear, 356–357
  stress, 356–357

tensile, 352–354
volume, 355–356
Strange (quark), 1496
Strange particles, 1494–1495
Strangeness, 1495, 1499t
Strassman, Fritz, 1464
Streamline, 383, 708
Strength
  tensile, 358
  ultimate, 358
Stress
  breaking, 358
  bulk, 352f, 354–356
  compressive, 354
  definition of, 352
  deformation and, 352–357
  elastic modulus and, 352
  elasticity and, 357–358
  shear, 352f, 356–357
  tensile, 352–354, 560–561
  thermal, 560–561
  units of, 353
  volume, 355–356
Stress-strain diagram, 358
String instruments, standing waves and, 497–498
String, standing waves on. See Standing waves
Strong bonds, 1407
Strong interactions, 160, 1446, 1490–1491
Strong nuclear force, 160, 689
Stud finders, 802–803
Sublimation, 567, 611
Substitutional impurities, 1415, 1424–1425
Subtraction
  significant figures in, 8f, 9
  of vectors, 13
Sudbury Neutrino Observatory, 1501
Sum, vector, 12
Sun. See also under Solar
  magnetic eruption on, 1000
Sunglasses, polarized, 1094, 1096f, 1098
Sunlight, radiation pressure of, 1068–1069
Sunsets, 1100–1101
Suntans, 1262
Superconductivity, 1430
Superconductors, 824, 968, 979–980
Supercooling, 567
Superheating, 567
Super-Kamiokande detector, 1489, 1501, 1501f
Supermassive black holes, 426
Supernovas, 160, 1230f, 1514
Superposition
  of electric fields, 703–704
  of forces, 106–108, 404, 405–406, 696
  of magnetic fields, 926, 931
  principle of, 490–491, 696, 1164
  of waves, 490–491, 497
Supersonic speed, 539
Supersymmetric theories, 1501
Surface charge density, 704
Surface integral, 730
Surface tension, 382
Sweat chloride test, 695
Symmetry
  conservation laws and, 1495
  in particle theory, 1497–1498
  spherical, 1366, 1372
  supersymmetry, 1501
Symmetry properties, of systems, 725
Symmetry-breaking interactions, 1495
Synchrocyclotrons, 1487
Synchrotrons, 1487
Systems
  isolated, 247
  symmetry properties of, 725

T

Tangential component of acceleration, 286
Target variables, 3
Taus, 1492
Technetium-99, 1461

Telephoto lens, 1140
Telescopes, 1119, 1149–1151, 1161
  Hubble Space Telescope, 1119, 1218, 1375f, 1503
  infrared, 1221
  resolving power of, 1210
Temperature, 552–553
  absolute, 517
  boiling, 566
  critical, 596, 979
  of early universe, 1508
  gas pressure and, 555
  internal energy and, 636
  macroscopic definition of, 552
  melting, 566
  molecular kinetic energy and, 636
  resistivity and, 824–825
  units of measure for, 553, 555
  vs. heat, 562
  vs. temperature interval, 554
Temperature coefficient, of resistivity, 824–825
Temperature gradient, 571
Temperature interval, 554
Temperature scale(s), 552
  absolute, 556, 668–669
  Celsius, 553
  conversion between, 554
  Fahrenheit, 554
  Kelvin, 555–556, 665, 668–669
Temporal artery thermometer, 554
Tendons
  as nonideal springs, 190
  Young's modulus of, 353f
Tensile strain, 352–354
  elasticity and, 357–358
  plasticity and, 357–358
Tensile strength, 358
Tensile stress, 352–354
  elasticity and, 357–358
  plasticity and, 357–358
  thermal stress and, 560–561
Tension, 105, 123, 353, 354
  definition of, 105
  Newton's first law and, 136–139
  Newton's second law and, 142
  static friction and, 148
  surface, 382
Terminal speed, 152–154
Terminal voltage, 830–831
Tesla, 887
Tesla coils, 993–994
Test charge, 699, 700
  for magnetic fields, 887–889
Test mass, 700
Theory, definition of, 2
Theory of Everything (TOE), 160, 1501
Theory of relativity. See Relativity
Thermal conductivity, 571, 823
Thermal conductors, 552–553
Thermal efficiency, of heat engine, 655–656, 658
Thermal equilibrium, 552
Thermal expansion, 557–561
  linear, 557–558, 559
  in object with hole, 558
  volume, 558–560
  of water, 560
Thermal properties of matter, 590–614
Thermal radiation, 1081
Thermal resistance, 571
Thermal stress, 560–561
Thermionic emission, 1267
Thermistors, 824
Thermodynamic processes, 624, 625
  adiabatic, 634–635, 640–642, 663
  direction of, 652–653
  disorder in, 653–654
  equilibrium in, 653
  heat added in, 628–629
  in heat engines, 654–656
  infinitesimal changes of state in, 634
  intermediate states (paths) in, 628–629

  isobaric, 635
  isochoric, 635
  isothermal, 635, 663
  reversibility of, 652–653, 663
  types of, 634–636
  work done in, 628
Thermodynamic systems
  human body as, 630
  internal energy of. See Internal energy
  paths in, 628–629
  work done in, 625–626, 628
Thermodynamics
  applications of, 625
  definition of, 551
  first law of, 624–643. See also First law of thermodynamics
  second law of, 652–679. See also Second law of thermodynamics
  sign rules for, 625
  third law of, 669
  zeroth law of, 552–553
Thermometers, 552
  bimetallic strip, 553
  gas, 554–556, 593, 669
  resistance, 553–554
  temporal artery, 554
Thermonuclear reactions, 1470
Thermos bottles, 576
Thick-film interference, 1176
Thin lenses, 1131–1139
  converging, 1131–1133
  diverging, 1133
  focal length of, 1131, 1133–1135
  focal point of, 1131
  graphical methods for, 1135–1137
  image formation by, 1135–1139
  index of refraction of, 1133–1134
  positive, 1131
  properties of, 1131
  radius of curvature for, 1133–1134
Thin-film interference, 1173–1179
Third law of thermodynamics, 669
Thomson, G.P., 1288–1289
Thomson, J.J., 751, 896, 1289, 1293
Thomson's atomic model, 751–752, 1293, 1294–1295
Thomson's e/m experiment, 896–897
Thought experiments, 1227
Three Mile Island accident, 1468
Threshold energy, 1463
Threshold frequency, 1263
Throwing, discus, 287
Tidal forces, 425
Timbre, 513–514
Time
  history of, 1508–1516
  mean free, 838
  Planck, 1509
  power and, 193
  proper, 1230–1231, 1232
  spacetime and, 1238
  straight-line motion and, 36–38
  units of, 4, 5t
  x-t graphs and, 37–38, 40–42
Time constant, for circuit, 866–867, 1003
Time dilation, 425, 1229–1230, 1231–1232
Time intervals
  measurement of, 1230
  relativity of, 1228–1233
Time-energy uncertainty principle, 1274–1275, 1278, 1315–1316
Time-independent Schrödinger equation, 1338, 1366
Tolman-Stewart experiment, 849
Topnes, 1499
Topography, potential energy gradient and, 227
Toroidal solenoid, magnetic field of, 940–941, 948t
Torque, 308–314
  angular acceleration and, 311–314
  angular displacement and, 320–322
  angular momentum and, 323, 324
  application of, 310–311

Torque (*Continued*)
  calculation of, 309–310, 349
  center of mass and, 312
  constant, 321
  on current loops, 901–905
  definition of, 309
  direction of, 309
  on electric dipole, 904
  equilibrium and, 345
  friction and, 320
  gravitational, 346–347
  of internal vs. external forces, 312
  kinetic energy and, 321
  magnetic, 901–905
  magnitude of, 310
  measurement of, 309
  moment of inertia and, 312
  net, 321
  net force and, 311–312, 323
  positive vs. negative, 309
  unit of, 309
  as vector, 310–311
  vs. moment, 309
  weight and, 312
  work done by, 320–322
Torr, 379
Torsion balance, 404
Torsion constant, 451
Total angular momentum, 1387, 1442
Total energy, 176, 1247
Total internal reflection, 1088–1091
Total mechanical energy, 209
Total momentum, 247, 260
Total work, 180, 244
Touch screens, 794
Tracers, radioactive, 1461–1462
Tractive resistance, 151
Traffic light sensors, 997
Trajectory, 77–79
Trampolines, 218–219
Transcranial magnetic stimulation, 961
Transformers, 1040–1042
Transients, 1032
Transistors, 1429
Translational motion
  definition of, 308
  molecular kinetic energy and, 606
  with rotational motion, 314–320
  vibrational, 606
Transmission electron microscope, 1291
Transmission grating, 1201
Transparency, index of refraction and, 1085
Transverse waves, 473. *See also* Mechanical waves;
    Wave(s)
  electromagnetic, 1056, 1060
  periodic, 474–475
  speed of, 482–486
  wave function for, 480–482
Traveling waves, 492, 494, 529
Trilobite fossils, 1508
Triple point, 611
Triple-alpha process, 1514
Tritium, 1462
Tritons, 1471
True weight, 421
Tsunamis, 1216
Tuning forks, 442–443
Tunnel diodes, 1349
Tunneling, 1347–1350
Tunneling probability, 1348
Turbulent flow, 383, 390–391
Tweeters, 2039
Twin paradox, 1232–1233
Tyrannosaurus rex, physical pendulum and, 456–457

U

Ultimate strength, 358
Ultrasonic sound, 510
Ultrasound, 516–517
Ultraviolet catastrophe, 1311

Ultraviolet vision, 1055
Uncertainty, 1274–1279
  fractional (percent), 8
  in measurement, 8
  wave-particle duality and, 1274–1278, 1314–1317
Uncertainty principle, 1275–1277, 1314–1317
  Bohr model and, 1317
  energy-time, 1278, 1315–1316
  harmonic oscillator and, 1353–1354
  for matter, 1315–1316
  momentum-position, 1274–1275, 1278, 1315–1316
Underdamped circuits, 1010–1011
Underdamping, 458
Uniform circular motion, 85–87, 88, 154–159
  definition of, 85
  dynamics of, 154–159
  in vertical circle, 158–159
  vs. nonuniform circular motion, 88
  vs. projectile motion, 87
Unit multipliers, 7
Unit vectors, 19–20
Units of measure, 4–6. *See also* Measurement
  for acceleration, 42
  for amplitude, 438
  for angular frequency, 438
  for astronomical distances, 1503
  in British system, 5–6, 117
  in calculations, 6
  for capacitance, 789
  in cgs metric system, 117
  consistency for, 6
  conversion of, 6–7
  for electric current, 695, 820
  for electric field, 699, 764
  for electric force, 695
  for electric potential, 761, 764
  for electromagnetic waves, 1053
  in equations, 6
  for force, 5–6, 105, 117
  for frequency, 438
  for heat, 562
  for kinetic energy, 183
  for length, 4–5
  for magnetic flux, 891
  for magnetic force, 887
  for mass, 5, 119
  for momentum, 242
  for mutual inductance, 993
  for period, 438
  prefixes for, 5
  for pressure, 353, 375
  for radiation dose, 1460
  for radioactivity, 1457
  for resistance, 826
  for rotation, 279
  in SI system, 4, 5
  significant figures and, 8–9
  for speed, 6f
  for temperature, 553, 555
  for time, 4, 5t, 7
  for torque, 309
  uncertainty and, 8
  for velocity, 38
  for volume, 7
  for weight, 119
  for work, 177–178
Universal conservation law, 690
Universe. *See also* Space
  critical density of, 1505–1507
  expansion of, 1501–1508. *See also* Expanding
    universe
  history of, 1508–1516
  scale factor for, 1504, 1505
  size of, 1504–1505
  standard model of, 1510–1511
  temperature of, 1508
  timeline for, 1512–1513
  uncoupling of interactions and, 1509–1510
Unpolarized light, 1094
Unstable equilibrium, 228

Up (quark), 1496
Upsilon, 1499
Uranium
  decay series for, 1454–1455
  in nuclear fission, 1464–1468

V

Vacancies, 820
Vacuum
  capacitors in, 789–791, 798
  electric fields in, 798
  electric-field energy in, 798
  permittivity of, 802
Vacuum bottles, 576
Valence bands, 1416–1417
Valence electrons, 1390, 1416
Validity, range of, 2
Van Allen radiation belts, 893
Van de Graaff electrostatic generator, 743–744, 768
Van der Meer, Simon, 1491
Van der Waals bonds, 1407
Van der Waals equation, 595–596
Van der Waals interactions, 452–453, 595
Vaporization, 566
  heat of, 566, 568
Variables
  separation of, 1367
  state, 591
  target, 3
Vector(s), 10–25
  acceleration, 35, 72–77, 283. *See also* Acceleration
    vectors
  addition of, 12–18
  angular momentum, 324, 328
  angular velocity, 281–282
  antiparallel, 11, 12
  component, 14, 106
  components of, 14–19, 21–22, 106–107
  direction of, 11, 16
  displacement and, 11, 12, 36–38
  division of, 70
  drawing diagrams with, 11–12
  force, 105–107
  heads (tips) of, 12
  magnitude of, 11, 16
  momentum, 242, 248
  multiplication of, 13, 16, 20–22, 70
  negative of, 11
  notation for, 11, 19
  parallel, 11, 12
  position, 70–72
  Poynting, 1065–1067
  products of, 20–25
  right-hand rule for, 23
  subtraction of, 13
  tails of, 12
  torque, 310–311
  unit, 19–20
  velocity, 35, 70–72, 281–282
Vector current density, 821
Vector field, 701
Vector magnetic field
  for current element, 926
  for moving charge, 924
Vector magnetic moment, 903
Vector (cross) product, 23–25
Vector quantities, 11
Vector sum, of displacements, 12
Velocity, 10
  angular, 279–282, 286
  average, 36–40. *See also* Average velocity
  circular motion and, 85–87
  constant, 51
  definition of, 39
  drift, 819, 820
  instantaneous, 38–42. *See also* Instantaneous
    velocity
  by integration, 55–57
  linear, 280, 282
  Lorentz transformation for, 1238–1239

magnitude of, 38t
motion diagram for, 41
Newton's first law of motion and, 108–112
of particle in wave, 480–482
phase, 479
power and, 194
projectile motion and, 77–80
relative, 88–93. *See also* Relative velocity
sector, 415
signs for, 37
in simple harmonic motion, 444, 448
units of, 38
vs. acceleration, 42
vs. speed, 39–40, 286
on *x-t* graph, 37–38, 40–41
Velocity selectors, 896
Velocity vectors, 35, 70–72, 281–282
Venturi meter, 388
Verne, Jules, 410
Vertex, of mirror, 1118
Vertical circle, uniform circular motion in, 158–159
Vibration, 438
heat capacities and, 606–608
molecular, 451–453, 597, 606, 1410–1412
Vibrational energy levels, 1410
Virtual focal point, 1123
Virtual image, 1115, 1137
Virtual object, 1137
Virtual photons, 1484
Viscosity, 382, 389–390
Visible light, 1054
Vision
in animals, 1144
defects in, 1143–1145
laser surgery for, 1076, 1309
normal, 1143
ultraviolet, 1055
Volcanoes, on Io, 975
Volt, 761
electron, 764
Volt per meter, 764
Voltage. *See also* Potential difference
capacitor, 1027–1028, 1029–1030
current and, 826–828
definition of, 762
Hall, 909–910
household, 869
inductor, 1025–1026
measurement of, 861–862
resistor, 1025, 1029
root-mean-square, 869
sinusoidal, 1022
terminal, 830–831
transformers and, 1040–1042
Voltage amplitude, 1022, 1025
Voltmeters, 762, 831, 861–863
ammeters and, 862–863, 1024
digital, 863
Volume
density and, 373–374
equipotential, 773
of gas, 593
units of, 7
Volume change, work done during, 596
Volume charge density, 704
Volume expansion, 558–560
Volume strain, 355–356
Volume stress, 355–356
v_x-*t* graphs, 44–46, 45f, 46f
acceleration on, 44–46, 47

W

Wa$^+$, 1491, 1500
Wa$^-$, 1491, 1500
Walking on moon, 407f
Water
supercooled, 567
thermal expansion of, 560
Water pressure, in home, 387
Water waves, interference and, 1166–1167

Watt, 193–194
Wave(s)
coherent, 1165
de Broglie, 1290
electromagnetic, 1051–1073
electron, 1286–1296, 1300
light as. *See* Wave-particle duality
mechanical, 472–499. *See also* Mechanical waves
medium for, 473
particle, 1328–1338. *See also* Particle waves
polarization of. *See* Polarization
reflected, 489–490
shock, 538
in snake movement, 473
sonar, 516–517
sound, 476, 509–542
transverse, 1056, 1060
uncertainty and, 1276–1277
Wave equation, 481, 485
for electromagnetic waves, 1058–1060
for mechanical waves, 481–482, 485, 1329–1330
for particle waves, 1330–1333
potential wells and, 1343–1347
statement of, 1329
Wave fronts, 1081–1082
Wave function
additive property of, 491
definition of, 477
for electromagnetic waves, 1061
graphing of, 478–480
for harmonic oscillator, 1350–1354
Hermite, 1352
hybrid, 1407
for longitudinal waves, 482
for mechanical waves, 477–479
normalized, 1342, 1365
notation for, 1329
one-dimensional Schrödinger equation and, 1332–1333, 1336–1337
for particle in a box, 1339–1343
for particle waves, 1328–1335
probability interpretation of, 1333–1334
for sinusoidal waves, 477–479
stationary-state, 1337–1338, 1366–1369
superposition principle and, 491
three-dimensional Schrödinger equation and, 1365–1371
time dependence of, 1337–1338, 1343
for transverse waves, 480–482
wave packets and, 1335–1336
Wave intensity, 488–489
Wave interference. *See* Interference
Wave motion, vs. particle motion, 475
Wave number, 478, 1061, 1330
Wave packets, 1335–1336
Wave phase, 479
Wave pulse, 474
Wave speed, 474, 475, 479
calculation of, 483–486
impulse-momentum theory and, 483–484
on a string, 482–485
for transverse waves, 482–486
vs. particle speed, 519
Wave velocity, vs. particle velocity, 519
Wavelengths, 475
in Balmer series, 1304
in Brackett series, 1304
de Broglie, 1287, 1290
frequency and, 1060
of light, 1054–1055
in Lyman series, 1304
measurement of, 1179–1181
in Pfund series, 1304
Wave-particle duality, 1073, 1081, 1261, 1286
atomic spectra and, 1292–1296
complementarity principle and, 1273–1274
electron waves and, 1286–1292
index of refraction and, 1086–1088
light and, 1261–1263, 1273. *See also* Photons

Maxwell's wave theory and, 1052–1057, 1262–1263, 1267
probability and uncertainty and, 1274–1278, 1314–1317
Weak bonds, 1407
Weak interactions, 160, 1491
Weber, 891
Weight
acceleration and, 118–119
apparent, 142–143, 421–423
definition of, 105, 117
equilibrium and, 345–347
as force, 118
gravitation and, 406–409
mass and, 114, 117–120
measurement of, 119
molecular, 564, 591
Newton's second law and, 142–143
torque and, 312
true, 421
units of, 119
Weight lifting, equilibrium and, 351
Weightlessness
apparent, 142, 413
true, 413
Weinberg, Steven, 1500
Wentzel, Gregor, 1361
Westinghouse, George, 1021
Wheatstone Bridge, 880–881
White dwarf stars, 1437
Wide-angle lens, 1140
Wien displacement law, 1311
Wilson, Robert, 1515
Wind instruments, 527
Windings, 1040, 1042f
Windshield wipers, 148
Wings. *See* Airplanes; Bird wings; Butterfly wings
Wire(s)
Ampere's law for, 935–937, 938
interaction force between, 931–932
magnetic field of, 928–931, 935–937, 938
Wire chambers, 1489
Wiring systems
automobile, 868, 870–871
household, 868–872, 1040–1041
WKB approximation, 1361–1362
Woodpecker impulse, 243
Woofers, 1029
Work, 177–193
change in speed and, 181–183
in composite systems, 186–187
definition of, 177
displacement and, 177–178, 181–183
done by conservative force, 755
done by constant forces, 177–178
done by electric fields, 755–761. *See also* Electric potential energy
done by electric force, 757–758, 761
done by electric potential, 762
done by electromotive force, 829
done by fluid flow, 385–389
done by gravitation, 409–410
done by muscle fibers, 177
done by torque, 320–322
done by varying forces, 187–191
done by waves, 487–488
done by working substance, 655
done during volume change, 596, 652
done in thermodynamic system, 625–626, 628
done on/by springs, 188–189
done to charge capacitor, 796–797
kinetic energy and, 181–187
negative, 179–180, 183
positive, 179, 183
potential energy and, 755
power and, 193–195, 487
rate of, 193–195
relativistic kinetic energy and, 1246–1247

Work (*Continued*)
  as scalar quantity, 178
  sign rules for, 625
  total, 180, 244
  units of, 177–178
  zero, 179
Work-energy theorem, 181–187,
      755
  for composite systems, 186–187
  for constant forces, 177–178
  for motion along curve, 191–193
  for straight-line motion, 187–191
  for varying forces, 187–191
Working substance, 654, 655
Workless refrigerators, 661

X

X-ray(s), 1266–1269. *See also* Radiation
  applications of, 1268–1269, 1395
X-ray diffraction, 1205–1208
X-ray spectra, 1393–1396
  absorption, 1396
x-t graphs, 37–38, 40–41
  definition of, 37
  velocity on, 37–38, 40–41

Y

Yeager, Chuck, 539
Young's interference experiment, 1167–1169,
      1179
Young's modulus, 353–354

Yo-yo
  acceleration of, 319
  speed of, 317
Yukawa, Hideki, 1484

Z

Z machine, 797–798
Z^0, 1491, 1500
Zeeman effect, 1379–1382
Zener breakdown, 1428
Zener diodes, 1428
Zero, absolute, 556, 669
Zero work, 179
Zeroth law of thermodynamics, 552–553
Zipper, molecular, 1408
Zoom lenses, 1141